Distributed Sensor Networks

Second Edition

Image and Sensor Signal Processing

Distributed Sensor Networks, Second Edition

Distributed Sensor Networks, Second Edition: Image and Sensor Signal Processing

Distributed Sensor Networks, Second Edition: Sensor Networking and Applications

CHAPMAN & HALL/CRC
COMPUTER and INFORMATION SCIENCE SERIES

Series Editor: Sartaj Sahni

PUBLISHED TITLES

ADVERSARIAL REASONING: COMPUTATIONAL
APPROACHES TO READING THE OPPONENT'S MIND
Alexander Kott and William M. McEneaney

DISTRIBUTED SENSOR NETWORKS, SECOND EDITION
S. Sitharama Iyengar and Richard R. Brooks

DISTRIBUTED SYSTEMS: AN ALGORITHMIC APPROACH
Sukumar Ghosh

ENERGY-AWARE MEMORY MANAGEMENT FOR EMBEDDED
MULTIMEDIA SYSTEMS: A COMPUTER-AIDED DESIGN
APPROACH
Florin Balasa and Dhiraj K. Pradhan

ENERGY EFFICIENT HARDWARE-SOFTWARE
CO-SYNTHESIS USING RECONFIGURABLE HARDWARE
Jingzhao Ou and Viktor K. Prasanna

FUNDAMENTALS OF NATURAL COMPUTING: BASIC CONCEPTS,
ALGORITHMS, AND APPLICATIONS
Leandro Nunes de Castro

HANDBOOK OF ALGORITHMS FOR WIRELESS NETWORKING
AND MOBILE COMPUTING
Azzedine Boukerche

HANDBOOK OF APPROXIMATION ALGORITHMS
AND METAHEURISTICS
Teofilo F. Gonzalez

HANDBOOK OF BIOINSPIRED ALGORITHMS
AND APPLICATIONS
Stephan Olariu and Albert Y. Zomaya

HANDBOOK OF COMPUTATIONAL MOLECULAR BIOLOGY
Srinivas Aluru

HANDBOOK OF DATA STRUCTURES AND APPLICATIONS
Dinesh P. Mehta and Sartaj Sahni

HANDBOOK OF DYNAMIC SYSTEM MODELING
Paul A. Fishwick

HANDBOOK OF ENERGY-AWARE AND GREEN COMPUTING
Ishfaq Ahmad and Sanjay Ranka

HANDBOOK OF PARALLEL COMPUTING: MODELS,
ALGORITHMS AND APPLICATIONS
Sanguthevar Rajasekaran and John Reif

HANDBOOK OF REAL-TIME AND EMBEDDED SYSTEMS
Insup Lee, Joseph Y-T. Leung, and Sang H. Son

HANDBOOK OF SCHEDULING: ALGORITHMS, MODELS, AND
PERFORMANCE ANALYSIS
Joseph Y.-T. Leung

HIGH PERFORMANCE COMPUTING IN REMOTE SENSING
Antonio J. Plaza and Chein-I Chang

INTRODUCTION TO NETWORK SECURITY
Douglas Jacobson

LOCATION-BASED INFORMATION SYSTEMS:
DEVELOPING REAL-TIME TRACKING APPLICATIONS
Miguel A. Labrador, Alfredo J. Pérez, and
Pedro M. Wightman

METHODS IN ALGORITHMIC ANALYSIS
Vladimir A. Dobrushkin

PERFORMANCE ANALYSIS OF QUEUING AND COMPUTER
NETWORKS
G. R. Dattatreya

THE PRACTICAL HANDBOOK OF INTERNET COMPUTING
Munindar P. Singh

SCALABLE AND SECURE INTERNET SERVICES AND
ARCHITECTURE
Cheng-Zhong Xu

SOFTWARE APPLICATION DEVELOPMENT: A VISUAL C++®,
MFC, AND STL TUTORIAL
Bud Fox, Zhang Wenzu, and Tan May Ling

SPECULATIVE EXECUTION IN HIGH PERFORMANCE
COMPUTER ARCHITECTURES
David Kaeli and Pen-Chung Yew

VEHICULAR NETWORKS: FROM THEORY TO PRACTICE
Stephan Olariu and Michele C. Weigle

Distributed Sensor Networks

Second Edition

Image and Sensor Signal Processing

Edited by

S. Sitharama Iyengar

Richard R. Brooks

CRC Press
Taylor & Francis Group
Boca Raton London New York

CRC Press is an imprint of the
Taylor & Francis Group, an **informa** business

A CHAPMAN & HALL BOOK

CRC Press
Taylor & Francis Group
6000 Broken Sound Parkway NW, Suite 300
Boca Raton, FL 33487-2742

© 2013 by Taylor & Francis Group, LLC
CRC Press is an imprint of Taylor & Francis Group, an Informa business

No claim to original U.S. Government works

Printed in the United States of America on acid-free paper
Version Date: 20120709

International Standard Book Number: 978-1-4398-6282-7 (Hardback)

Library of Congress Cataloging-in-Publication Data

Distributed sensor networks. Image and sensor signal processing / edited by S. Sitharama Iyengar and Richard R. Brooks. -- 2nd ed.
    p. cm. -- (Chapman & Hall/CRC computer & information science series)
    Summary: "The six years since the first edition appeared has seen the field of sensor networks ripen. The number of researchers working in the field and the number of technical (journal and conference) publications related to sensor networks have exploded. Those of us working in the field since its infancy have been gratified by the large number of practical sensor network applications that are being used. Not surprisingly, a number of lessons have been learned. The current state of affairs could hardly have been predicted when the first edition went to press. Partly because of this, we were extremely gratified when Chapman & Hall contacted us and suggested printing a revised second edition of Distributed Sensor Networks"-- Provided by publisher.
    Includes bibliographical references and index.
    ISBN 978-1-4398-6282-7 (hardback)
    1. Sensor networks. 2. Multisensor data fusion. 3. Intelligent agents (Computer software) I. Iyengar, S. S. (Sundararaja S.) II. Brooks, R. R. (Richard R.)

TK7872.D48D5726 2013
681'.25--dc23
    2012017622

Visit the Taylor & Francis Web site at
http://www.taylorandfrancis.com

and the CRC Press Web site at
http://www.crcpress.com

I would like to dedicate this book to Professor C.N.R. Rao for his outstanding and sustainable contributions to science in India and around the world. Professor Rao has been able to create and establish opportunities in science education in India and the world.

Last but not the least, I am grateful to my graduate students who make everything possible in my career.

S. Sitharama Iyengar

Dedicated to my graduate students, who make the research possible.

Richard R. Brooks

Contents

Preface to the Second Edition xiii

Preface to the First Edition xv

Editors...xvii

Contributors xix

PART I Overview

1 Microsensor Applications 3
   David Shepherd and Sri Kumar

2 Interfacing Distributed Sensor Actuator Networks with the Physical World 21
   Shivakumar Sastry and S. Sitharama Iyengar

3 Contrast with Traditional Systems 33
   Richard R. Brooks

PART II Distributed Sensing and Signal Processing

4 Digital Signal Processing Backgrounds... 41
   Yu Hen Hu

5 Image Processing Background.. 59
   Lynne L. Grewe and Ben Shahshahani

6 Object Detection and Classification... 93
   Akbar M. Sayeed

7 Parameter Estimation 111
   David S. Friedlander

8 Target Tracking with Self-Organizing Distributed Sensors 131
   Richard R. Brooks, Christopher Griffin, David S. Friedlander, J.D. Koch, and İlker Özçelik

9 Collaborative Signal and Information Processing 191
   Feng Zhao, Jie Liu, Juan Liu, Leonidas Guibas, and James Reich

10 Environmental Effects 207
    David C. Swanson

11 Detecting and Counteracting Atmospheric Effects 219
    Lynne L. Grewe

12 Signal Processing and Propagation for Aeroacoustic Sensor Networks 245
    Richard J. Kozick, Brian M. Sadler, and D. Keith Wilson

13 Distributed Multi-Target Detection in Sensor Networks 291
    Xiaoling Wang, Hairong Qi, and Steve Beck

14 Symbolic Dynamic Filtering for Pattern Recognition in Distributed
    Sensor Networks 307
    Xin Jin, Shalabh Gupta, Kushal Mukherjee, and Asok Ray

15 Beamforming 335
    J.C. Chen and Kung Yao

PART III Information Fusion

16 Foundations of Data Fusion for Automation 377
    S. Sitharama Iyengar, Shivakumar Sastry, and N. Balakrishnan

17 Measurement-Based Statistical Fusion Methods for Distributed Sensor
    Networks 387
    Nageswara S.V. Rao

18 Soft Computing Techniques 415
    Richard R. Brooks

19 Estimation and Kalman Filters 429
    David L. Hall

20 Data Registration 455
    Richard R. Brooks, Jacob Lamb, Lynne L. Grewe, and Juan Deng

21 Signal Calibration, Estimation for Real-Time Monitoring and Control 495
    Asok Ray and Shashi Phoha

22 Semantic Information Extraction 513
    David S. Friedlander

23 Fusion in the Context of Information Theory 523
    Mohiuddin Ahmed and Gregory Pottie

24 Multispectral Sensing 541
    N.K. Bose

25 Chebyshev's Inequality-Based Multisensor Data Fusion
    in Self-Organizing Sensor Networks 553
    *Mengxia Zhu, Richard R. Brooks, Song Ding, Qishi Wu, Nageswara S.V. Rao,
    and S. Sitharama Iyengar*

26 Markov Model Inferencing in Distributed Systems 569
Chen Lu, Jason M. Schwier, Richard R. Brooks, Christopher Griffin,
and Satish Bukkapatnam

27 Emergence of Human-Centric Information Fusion 581
David L. Hall

PART IV Power Management

28 Designing Energy-Aware Sensor Systems 595
N. Vijaykrishnan, M.J. Irwin, M. Kandemir, L. Li, G. Chen, and B. Kang

29 Operating System Power Management 609
Vishnu Swaminathan and Krishnendu Chakrabarty

30 An Energy-Aware Approach for Sensor Data Communication 639
H. Saputra, N. Vijaykrishnan, M. Kandemir, Richard R. Brooks, and M.J. Irwin

31 Compiler-Directed Communication Energy Optimizations for
Microsensor Networks 655
I. Kadayif, M. Kandemir, A. Choudhary, M. Karakoy, N. Vijaykrishnan, and M.J. Irwin

32 Jamming Games in Wireless and Sensor Networks 679
Rajgopal Kannan, Costas Busch, and Shuangqing Wei

33 Reasoning about Sensor Networks 693
Manuel Peralta, Supratik Mukhopadhyay, and Ramesh Bharadwaj

34 Testing and Debugging Sensor Network Applications 711
Sally K. Wahba, Jason O. Hallstrom, and Nigamanth Sridhar

Index 729

Preface to the Second Edition

The six years since the first edition appeared have seen the field of sensor networks ripen. The number of researchers working in the field and technical (journal and conference) publications related to sensor networks has exploded. Those of us working in the field since its infancy have been gratified by the large number of practical sensor network applications that are being used.

Not surprisingly, many lessons have been learned. The current state of affairs could hardly have been predicted when the first edition went to press. Partly because of this, we were extremely gratified when Chapman & Hall contacted us and suggested printing a revised second edition of *Distributed Sensor Networks*.

It was a daunting task to bring together, once again, the distinguished set of researchers whom we relied on for the first edition. We are proud to have been able to expand and revise the original tome of over 1100 pages. The size of the new book forced us to deliver it in two books:

1. *Distributed Sensor Networks, Second Edition: Image and Sensor Signal Processing* comprises the following parts:
   I. Overview
   II. Distributed Sensing and Signal Processing
   III. Information Fusion
   IV. Power Management

2. *Distributed Sensor Networks, Second Edition: Sensor Networking and Applications* comprises the following parts:
   I. Sensor Deployment
   II. Adaptive Tasking
   III. Self-Configuration
   IV. System Control
   V. Engineering Examples

Although half of the chapters remain the same as in the first edition, 13 chapters have been revised, and there are 22 new chapters. Readers of the first edition will recognize many of the same authors. Several of them were veterans of the DARPA ISO SENSIT program. We were privileged to get inputs from a number of distinguished new contributors. We were also saddened by the demise of Dr. Bose from Penn State who wrote the chapter on multispectral sensing.

We believe that the new books have managed to maintain the feel of our earlier text. The chapters serve both as tutorials and archive research material. Our target audience is students, researchers, and engineers who want a book with practical insights providing current information on the field.

MATLAB® is a registered trademark of The MathWorks, Inc. For product information, please contact:

The MathWorks, Inc.
3 Apple Hill Drive
Natick, MA 01760-2098 USA
Tel: 508 647 7000
Fax: 508-647-7001
E-mail: info@mathworks.com
Web: www.mathworks.com

Preface to the First Edition

In many ways, this book started ten years ago, when the editors started their collaboration at Louisiana State University in Baton Rouge. At that time, sensor networks were a somewhat arcane topic. Since then, many new technologies have ripened and prototype devices have emerged in the market. We were lucky enough to be able to continue our collaboration under the aegis of the DARPA IXO Sensor Information Technology Program and the Emergent Surveillance Plexus Multi-disciplinary University Research Initiative.

What was clear ten years ago, and has become more obvious, is that the only way to adequately monitor the real world is to use a network of devices. Many reasons for this will be given in this book. These reasons range from financial considerations to statistical inference constraints. Once you start using a network situated in the real world, the need for adaptation and self-configuration also becomes obvious.

What was probably not known ten years ago was the breadth and depth of research needed to adequately design these systems. The book in front of you contains chapters from acknowledged leaders in sensor network design. The contributors work at leading research institutions and have expertise in a broad range of technical fields.

The field of sensor networks has matured greatly within the last few years. The editors are grateful to have participated in this process. We are especially pleased to have been able to interact with the research groups whose work is presented here. This growth has only been possible with support from many government agencies, especially within the Department of Defense. Visionary program managers at DARPA, ONR, AFRL, and ARL have made a significant impact on these technologies.

It is the editors' sincere hope that the field continues to mature. We also hope that the cross fertilization of ideas between technical fields that has enabled these advances deepens.

The editors thank all the authors who contributed to this book. Jamie Sigal and Bob Stern from CRC Press were invaluable. We also recognize the efforts of Rose Sweeney in helping to organize and maintain this enterprise.

Editors

S. Sitharama Iyengar is the director and Ryder Professor of the School of Computing and Information Sciences at Florida International University, Miami, and is also the chaired professor at various institutions around the world. His publications include 6 textbooks, 5 edited books, and over 400 research papers. His research interests include high-performance algorithms, data structures, sensor fusion, data mining, and intelligent systems. Dr. Iyengar is a world class expert in computational aspects of sensor networks, data structures, and algorithms for various distributed applications. His techniques have been used by various federal agencies (Naval Research Laboratory [NRL], ORNL, National Aeronautics and the Space Administration [NASA]) for their projects. His work has been cited very extensively by researchers and scientists around the world.

Dr. Iyengar is an SIAM distinguished lecturer, ACM national lecturer, and IEEE distinguished scientist. He is a fellow of IEEE, ACM, AAAS, and SDPS. Dr. Iyengar is a recipient of IEEE awards, best research paper awards, the Distinguished Alumnus award of the Indian Institute of Science, Bangalore, and other awards. He has served as the editor of several IEEE journals and is the founding editor in chief of the *International Journal of Distributed Sensor Networks*.

Dr. Iyengar's research has been funded by the National Science Foundation (NSF), Defense Advanced Research Projects Agency (DARPA), Multi-University Research Initiative (MURI Program), Office of Naval Research (ONR), Department of Energy/Oak Ridge National Laboratory (DOE/ORNL), NRL, NASA, U.S. Army Research Office (URO), and various state agencies and companies. He has served on U.S. National Science Foundation and National Institutes of Health panels to review proposals in various aspects of computational science and has been involved as an external evaluator (ABET-accreditation) for several computer science and engineering departments.

Dr. Iyengar has had 40 doctoral students under his supervision, and the legacy of these students can be seen in prestigious laboratories (JPL, Oak Ridge National Lab, Los Alamos National Lab, Naval Research Lab) and universities round the world. He has been the program chairman of various international conferences and has given more than 50 keynote talks at these conferences.

Richard R. Brooks is an associate professor of electrical and computer engineering at Clemson University in Clemson, South Carolina. He received his PhD in computer science from Louisiana State University and his BA in mathematical sciences from The Johns Hopkins University. He has also studied operations research at the Conservatoire National des Arts et Metiers in Paris, France.

Dr. Brooks is a senior member of the IEEE. He has written the book *Disruptive Security Technologies with Mobile Code and Peer-to-Peer Networks* and has cowritten *Multi-Sensor Fusion*. He has coedited both versions of *Distributed Sensor Networks* in collaboration with S. S. Iyengar.

Dr. Brooks was principal investigator (PI) of the Reactive Sensor Networks Project sponsored by the Defence Advanced Research Projects Agency (DARPA) ITO Sensor Information Technology initiative, which has explored collaborative signal processing to aggregate information moving through the network and the use of mobile code for coordination among intelligent sensor nodes. He was co-PI of a DARPA IXO JFACC program that has used distributed discrete event controllers for air combat C2 planning. He has coordinated a DARPA MURI program that uses cooperating automata in a cellular space to coordinate sensor network planning and execution. He is PI of an ONR URI on cybersecurity issues relating to mobile code and the construction of secure information infrastructures.

Dr. Brooks' current research concentrates on adaptation in distributed systems. His research interests include network security, sensor networks, and self-organizing systems. His research has been sponsored by ONR, DARPA, ARO, AFOSR, NIST, U.S. Department of State, NSF, and BMW Manufacturing Corporation.

Dr. Brooks' PhD dissertation has received an exemplary achievement certificate from the Louisiana State University graduate school. He is associate managing editor of the *International Journal of Distributed Sensor Networks*. He has had a broad professional background with computer systems and networks, and was head of the Pennsylvania State University Applied Research Laboratory Distributed Systems Department for over six years. Dr. Brooks was technical director of Radio Free Europe's computer network for many years, and is a consultant to the French stock exchange authority and the World Bank.

Contributors

Mohiuddin Ahmed
Department of Electrical Engineering
University of California
Los Angeles, California

N. Balakrishnan
Carnegie Mellon University
Pittsburgh, Pennsylvania

and

Indian Institute of Science
Bangalore, India

Steve Beck
BAE Systems
Austin, Texas

Ramesh Bharadwaj
Naval Research Laboratory
Washington, District of Columbia

N.K. Bose
Department of Electrical Engineering
The Pennsylvania State University
University Park, Pennsylvania

Richard R. Brooks
Holcombe Department of Electrical and
   Computer Engineering
Clemson University
Clemson, South Carolina

Satish Bukkapatnam
Oklahoma State University
Stillwater, Oklahoma

Costas Busch
Department of Computer Science
Louisiana State University
Baton Rouge, Louisiana

Krishnendu Chakrabarty
Department of Electrical and Computer
   Engineering
Duke University
Durham, North Carolina

G. Chen
Microsystems Design Laboratory
The Pennsylvania State University
University Park, Pennsylvania

J.C. Chen
Microsystems Design Laboratory
The Pennsylvania State University
University Park, Pennsylvania

A. Choudhary
Department of Electrical and Computer
   Engineering
Northwestern University
Evanston, Illinois

Juan Deng
Department of Electrical and Computer
   Engineering
Clemson University
Clemson, South Carolina

Song Ding
Department of Computer Science
Louisiana State University
Baton Rouge, Louisiana

David S. Friedlander
Defense Advanced Research Projects Agency
Arlington, Virginia

Lynne L. Grewe
Department of Computer Science
California State University
Hayward, California

Christopher Griffin
Applied Research Laboratory
The Pennsylvania State University
University Park, Pennsylvania

Leonidas Guibas
Department of Computer Science
Stanford University
Stanford, California

Shalabh Gupta
Department of Electrical and Computer
    Engineering
University of Connecticut
Storrs, Connecticut

David L. Hall
The Pennsylvania State University
University Park, Pennsylvania

Jason O. Hallstrom
School of Computing
Clemson University
Clemson, South Carolina

Yu Hen Hu
Department of Electrical and Computer
    Engineering
University of Wisconsin-Madison
Madison, Wisconsin

M.J. Irwin
Microsystems Design Laboratory
The Pennsylvania State University
University Park, Pennsylvania

S. Sitharama Iyengar
Department of Computer Science
Florida International University
Baton Rouge, Louisiana

Xin Jin
Department of Mechanical and Nuclear
    Engineering
The Pennsylvania State University
University Park, Pennsylvania

I. Kadayif
Microsystems Design Laboratory
The Pennsylvania State University
University Park, Pennsylvania

M. Kandemir
Microsystems Design Laboratory
The Pennsylvania State University
University Park, Pennsylvania

B. Kang
Microsystems Design Laboratory
The Pennsylvania State University
University Park, Pennsylvania

Rajgopal Kannan
Department of Computer Science
Louisiana State University
Baton Rouge, Louisiana

M. Karakoy
Department of Computing
Imperial College London
London, United Kingdom

J.D. Koch
Applied Research Laboratory
The Pennsylvania State University
University Park, Pennsylvania

Richard J. Kozick
Department of Electrical Engineering
Bucknell University
Lewisburg, Pennsylvania

Sri Kumar
BAE Systems
London, United Kingdom

Jacob Lamb
Applied Research Laboratory
The Pennsylvania State University
University Park, Pennsylvania

L. Li
Department of Computer Science and
  Engineering
The Pennsylvania State University
University Park, Pennsylvania

Jie Liu
Microsoft Research
Redmond, Washington

Juan Liu
Palo Alto Research Center
Palo Alto, California

Chen Lu
Department of Electrical and Computer
  Engineering
Clemson University
Clemson, South Carolina

Kushal Mukherjee
Department of Mechanical and Nuclear
  Engineering
The Pennsylvania State University
University Park, Pennsylvania

Supratik Mukhopadhyay
Department of Computer Science
Louisiana State University
Baton Rouge, Louisiana

İlker Özçelik
Clemson University
Clemson, South Carolina

Manuel Peralta
Department of Computer Science
Louisiana State University
Baton Rouge, Louisiana

Shashi Phoha
Applied Research Laboratory
The Pennsylvania State University
University Park, Pennsylvania

Gregory Pottie
Department of Electrical Engineering
University of California
Los Angeles, California

Hairong Qi
Department of Electrical and Computer
  Engineering
The University of Tennessee
Knoxville, Tennessee

Nageswara S.V. Rao
Department of Computer Science and
  Mathematics Division
Oak Ridge National Laboratory
Oak Ridge, Tennessee

Asok Ray
Department of Mechanical and Nuclear
  Engineering
The Pennsylvania State University
University Park, Pennsylvania

James Reich
Streetline Networks
Foster City, California

Brian M. Sadler
U.S. Army Research Laboratory
Adelphi, Maryland

H. Saputra
Department of Computer Science and
  Engineering
The Pennsylvania State University
University Park, Pennsylvania

Shivakumar Sastry
Department of Electrical and Computer
  Engineering
The University of Akron
Akron, Ohio

Akbar M. Sayeed
Department of Electrical and Computer
  Engineering
University of Wisconsin-Madison
Madison, Wisconsin

Jason M. Schwier
Department of Electrical and Computer
    Engineering
Clemson University
Clemson, South Carolina

Ben Shahshahani
Google
Mountain View, California

David Shepherd
Defense Advanced Research Projects Agency
Arlington, Virginia

Nigamanth Sridhar
Department of Electrical and Computer
    Engineering
Cleveland State University
Cleveland, Ohio

Vishnu Swaminathan
Department of Electrical and Computer
    Engineering
Duke University
Durham, North Carolina

David C. Swanson
The Pennsylvania State University
University Park, Pennsylvania

N. Vijaykrishnan
Microsystems Design Laboratory
The Pennsylvania State University
University Park, Pennsylvania

Sally K. Wahba
School of Computing
Clemson University
Clemson, South Carolina

Xiaoling Wang
Department of Electrical and Computer
    Engineering
The University of Tennessee
Knoxville, Tennessee

Shuangqing Wei
Department of Computer Science
Louisiana State University
Baton Rouge, Louisiana

D. Keith Wilson
U.S. Army Cold Regions Research and
    Engineering Laboratory
Hanover, New Hampshire

Qishi Wu
Department of Computer Science
University of Memphis
Memphis, Tennessee

Kung Yao
Department of Electrical Engineering
University of California, Los Angeles
Los Angeles, California

Feng Zhao
Microsoft Research Asia
Beijing, People's Republic of China

Mengxia Zhu
Department of Computer Science
Southern Illinois University
Carbondale, Illinois

I

Overview

1 Microsensor Applications *David Shepherd and Sri Kumar* 3
Introduction • Sensor Networks: Description • Sensor Network Applications,
Part 1: Military Applications • Sensor Network Applications, Part 2: Civilian
Applications • Conclusion • References

**2 Interfacing Distributed Sensor Actuator Networks with the Physical
World** *Shivakumar Sastry and S. Sitharama Iyengar* 21
Introduction • Benefits and Limitations of DSAN System • Taxonomy of DSAN System
Architectures • Conclusions • Further Reading

3 Contrast with Traditional Systems *Richard R. Brooks* 33
Problem Statement • Acknowledgments and Disclaimer • References

This part provides a brief overview of sensor networks. It introduces the topic by discussing what they are, their applications, and how they differ from traditional systems.

Iyengar et al. provide a definition for distributed sensor networks. They introduce many applications that will be dealt with in more detail later. A discussion is also provided of the technical challenges these systems present.

Sri Kumar of Defence Advanced Research Projects Agency (DARPA) and Shepherd provide an overview of sensor networks from the military perspective in Chapter 1. Of particular interest in this chapter is a summary of military applications since the 1960s. This chapter then proceeds to recent research advances. Many of these advances come from research groups whose works are presented in later parts of this book.

Sastry and Iyengar provide a taxonomy of distributed sensor and actuator networks (DSANs) in Chapter 2. The taxonomy should help readers in structuring their understanding of the field. It is built on laws describing the evolution of technology. These laws can help readers anticipate the future developments that are likely to appear in this domain. We feel that this chapter helps provide a perspective for the rest of the book.

Brooks describes briefly how distributed sensor networks (DSNs) differ from traditional systems in Chapter 3. The global system is composed of distributed elements that are failure prone and have a limited lifetime. Creating a reliable system from these components requires a new type of flexible system design.

The purpose of this part is to provide a brief overview of DSNs. The chapters presented concentrate on the applications of this technology and why the new technologies presented in this book are necessary.

1

Microsensor Applications

1.1 Introduction 3
1.2 Sensor Networks: Description 3
1.3 Sensor Network Applications, Part 1: Military Applications 4
     Target Detection and Tracking • Application 1: Artillery and
     Gunfire Localization • Application 2: Unmanned Aerial Vehicle
     Sensor Deployment • Target Classification • Application 3: Vehicle
     Classification • Application 4: Imaging-Based Classification and
     Identification
1.4 Sensor Network Applications, Part 2: Civilian Applications 15
     Application 5: Monitoring Rare and Endangered Species •
     Application 6: Personnel Heartbeat Detection
1.5 Conclusion 18
References...18

David Shepherd
*Defense Advanced
Research Projects Agency*

Sri Kumar
BAE Systems

1.1 Introduction

In recent years, the reliability of microsensors and the robustness of sensor networks have improved to the point that networks of microsensors have been deployed in large numbers for a variety of applications. The prevalence of localized networks and the ubiquity of the Internet have enabled automated and human-monitored sensing to be performed with an ease and expense acceptable to many commercial and government users. Some operational and technical issues remain unsolved, such as how to balance energy consumption against frequency of observation and node lifetime, the level of collaboration among the sensors, and distance from repeater units or reachback stations. However, the trend in sensor size is to become smaller, and at the same time the networks of sensors are becoming increasingly powerful. As a result, a wide range of applications use distributed microsensor networks for a variety of tasks, from battlefield surveillance and reconnaissance to environment monitoring and industrial controls.

1.2 Sensor Networks: Description

Sensor networks consist of multiple sensors, often multi-phenomenological and deployed in forward regions, containing or connected to processors and databases, with alerts exfiltrated for observation by human operators in rear areas or on the front lines. Sensor networks' configurations range from very flat, with few command nodes or exfiltration nodes, to hierarchical nets consisting of multiple networks layered according to operational or technical requirements. The topology of current sensor networks deployed in forward areas generally includes several or tens of nodes reporting to a single command node in a star topology, with multiple command nodes (which vary in number according to area coverage and operational needs) reporting to smaller numbers of data exfiltration points. Technological advances in recent years have enabled networks to "flatten": individual sensor nodes share information with each other and collaborate to improve detection probabilities while reducing the likelihood of false alarms [1,2]. Aside from the operational goal of increasing the likelihood of detection, another reason to flatten sensor networks is to reduce the likelihood of overloading command nodes or human operators, with numerous, spurious and even accurate detection signals. In addition, although a point of diminishing returns can be reached, increasing the amount of collaboration and processing performed among the nodes reduces the time to detect and classify targets, and saves power—the costs of communicating data, particularly for long distances, far outweighs the costs of processing it locally. Processing sophistication has enabled sensor networks in these configurations to scale to hundreds and thousands of nodes, with the expectation that sensor networks can easily scale to tens of thousands of nodes or more as required.

Sensor networks can be configured to sense a variety of target types. The networks themselves are mode-agnostic, enabling multiple types of sensor to be employed, depending on operational requirements. Phenomenologies of interest range over many parts of the electromagnetic spectrum, including infrared, ultraviolet, radar and visible-range radiations, and also include acoustic, seismic, and magnetic ranges. Organic materials can be sensed using biological sensors constructed of organic or inorganic materials. Infrared and ultraviolet sensors are generally used to sense heat. When used at night, they can provide startlingly clear depictions of the environment that are readily understandable by human operators. Radar can be used to detect motion, including motion as slight as heartbeats or the expansion and contraction of the chest due to breathing. More traditional motion detectors sense in the seismic mode, since the Earth is a remarkably good transmitter of shock waves. Acoustic and visible-range modes are widely used and readily understood—listening for people talking or motors running, using cameras to spot trespassers, etc. Traditional magnetic sensors can be used to detect large metal objects such vehicles or weapons, whereas they are unlikely to detect small objects that deflect the Earth's magnetic field only slightly. More recent sensors, such as those used in the past few decades, are able to detect small metal objects. They induce a current in nonmagnetic metal, with the response giving field sensitivity and direction.

Early generations of sensors functioned primarily as tripwires, alerting users to the presence of any and all interlopers or targets without defining the target in any way. Recent advances have enabled far more than mere presence detection, though. Target tracking can be performed effectively with sensors deployed as a three-dimensional field and covering a large geographic area. Sufficient geographic coverage can be accomplished by connecting several (or many) smaller sensor fields, but this raises problems of target handoff and network integration, as well as difficulties in the deployment of the sensors themselves. Deployment mechanisms currently in use include hand emplacement, air-dropping, unmanned air or ground vehicles, and cannon-firing. Effective tracking is further enhanced by target classification schemes. Single targets can be more fully identified through classification, either by checking local or distributed databases, or by utilizing reachback to access more powerful databases in rear locations. Furthermore, by permitting disambiguation, classification systems can enable multiple targets to be tracked as they simultaneously move through a single sensor field [3].

1.3 Sensor Network Applications, Part 1: Military Applications

Prior to the 1990s, warfighters planning for and engaging in battle focused their attentions on maneuvering and massing weapons *platforms*: ships at sea; tanks and infantry divisions and artillery on land; aircraft in the air. The goal was to bring large quantities of weapons platforms to bear on the enemy, to provide overwhelming momentum and firepower to guarantee success regardless of the quantity and quality of information about the enemy. Warfighting had been conducted this way for at least as long as technology has enabled the construction of weapons platforms, and probably before then also. In the 1990s, however, the United States Department of Defense (DoD) began to reorient fundamental thinking about warfighting. As a result of technological advances and changes in society as a whole, thinkers in the military began advocating a greater role for *networked* operations. While planners still emphasize bringing overwhelming force to bear, that force would no longer be employed by independent actors who controlled separate platforms and had only a vague understanding of overall battlefield situations or commanders' intentions. Instead, everyone involved in a conflict would be connected, physically via electronic communications, and cognitively by a comprehensive awareness and understanding of the many dimensions of a battle. Thanks to a mesh of sensors at the point of awareness and to computers at all levels of engagement and analysis, information about friendly and enemy firepower levels and situation awareness can be shared among appropriate parties. Information assumed a new place of prominence in warfighters' thinking as a result of this connectedness and information available to commanders. Even more comprehensive understandings of network-centric warfare also considered the ramifications of a completely information-based society, including how a fully networked society would enable operations to be fought using financial, logistical, and social relationships [4].

Critical to the success of network-centric warfare is gathering, analyzing, and sharing information, about the enemy and about friendly forces. Because sensors exist at the point of engagement and provide valuable information, they have begun to figure larger in the planning and conduct of warfare. Especially with the ease of packaging and deployment afforded by miniaturization of electronic components, sensors can provide physical or low-level data about the environment and opponents. They can be the eyes and ears of warfighters and simultaneously enable humans to remain out of harm's way. Sensors can provide information about force structures, equipment and personnel movement; they can provide replenishment and logistics data; they can be used for chemical or biological specimen detection; sensors can provide granular data points or smoothed information about opposition and friendly forces. In this way, microsensor data can be an important part of what defense department theorists term "intelligence preparation of the battlespace." Sensor data can be folded into an overall picture of the enemy and inform speculations of enemy intent or action.

1.3.1 Target Detection and Tracking

A primary application of networked microsensors is the detection and tracking of targets. The first modern sensor application and system was intended for this purpose. During the Viet Nam war in the 1960s, Secretary of Defense Robert McNamara wanted knowledge of North Vietnamese troop activity, and ordered the construction of an electronic anti-infiltration barrier below the Demilitarized Zone (DMZ), the line of demarcation between North and South Vietnam. The principal purpose of this "McNamara Line" would be to sound the alarm when the enemy crossed the barrier. The mission of the program was later changed to use sensors along the Ho Chi Minh Trail to detect North Vietnamese vehicle and personnel movements southward. Named Igloo White, the program deployed seismic and acoustic sensors by hand and by air-drop. Seismic intrusion detectors (SIDs) consisted of geophones to translate ground movements induced by footsteps (at ranges of up to 30 m) or by explosions into electrical signals. SIDs could be hand emplaced or attached to ground-penetrating spikes and airdropped (air-delivered SIDs [ADSIDs; Figure 1.1]). Up to eight SIDs could transmit to one receiver unit over single frequency channels. All SID receiver units transmitted to U.S. Air Force patrol planes orbiting 24 h per day, with data displayed using self-contained receive and display (i.e. lamp) units or the aircraft's transceiver. Hand-emplaced versions of the SIDs weighed 7 lb each and were contained in a 4.5 in. × 5 in. × 9 in. metal box. Smaller versions, called patrol SIDs (PSIDs), were intended to be carried by individual soldiers and came in sets of four sensors and one receiver. Each sensor weighed 1 lb, could be fitted into clothes pockets, and operated continuously for 8 h. Sensor alarms sounded by transmitting beeps, with one beep per number of the sensor [1–4]. Target tracking was performed by human operators listening for the alarms and gauging target speed and direction from the numbers and directions of alarm hits [5].

The Igloo White sensing systems required human operators to listen for detections, disambiguate noise from true detections, correlate acoustic and seismic signals, and transmit alarms to rear areas. It also employed a hub-and-spoke topology, with many sensors reporting to a single user or exfiltration point. More recent research into target detection and tracking has reduced or removed the requirement for human intervention in the detection and tracking loop. It has also proved the superiority of

FIGURE 1.1   An ADSID sensor from the Vietnam-era Igloo White program.

a mesh topology, where all sensors are peers, signal processing is performed collaboratively, and data are routed according to user need and location. Routing in a distributed sensor network is performed optimally using *diffusion* methods, with nodes signifying interests in certain types of data (i.e. about prospective targets) and supplying data if their data match interests published by other nodes. This data-centric routing eliminates the dependency on IP addresses or pre-set routes. The redundancy provided by multiple sensor nodes eliminates single points of failure in the network and enables sensors to use multiple inputs for classification and to disambiguate targets. Reaching decisions at the lowest level possible also conserves bandwidth and power by minimizing longer range transmissions. If necessary (for operational reasons, power conservation, or other reasons), exfiltration is still often performed by higher power, longer range nodes flying overhead or contained in vehicles [6].

Collaborative target tracking using distributed nodes begins as nodes local to a target and these nodes cluster dynamically to share data about targets; the virtual cluster follows the target as it moves, drawing readings from whichever nodes are close to the target at any instant. One method of obtaining readings is the closest point of approach (CPA) method. Points of maximum signal amplitude are considered to correspond to the point of the target's closest physical proximity to the sensor. Spurious features, such as ambient or network noise, are eliminated by considering amplitudes within a space–time window and resolving the energy received within that window to a single maximum amplitude. The size of the window can be adjusted dynamically to ensure that signal strength within the window remains approximately constant to prevent unneeded processing. This can be achieved because it is unlikely that a ground target moves significantly enough to change signal strength appreciably. As the target moves through the sensor field, maximum readings from multiple sensors can be analyzed to determine target heading. As many sensors as possible (limited by radio range) should be used to compute the CPA, because too few sensors (fewer than four or five) provides an insufficient number of data points for accurate computations.

Improved tracker performance can be achieved by implementing an extended Kalman filter (EKF) or with the use of techniques such as least-squares linear regression and lateral inhibition. Kalman filters compute covariance matrixes that vary according to the size of the sensor field, even if the size varies dynamically to follow a moving target. Least squares is a straightforward method of approximating linear relations between observed data, with noise considered to be independent, white Gaussian noise [7]. This is an appropriate approximation for sensor data owing to the large sample size and large proportion of noise achieved with the use of a field of sensors. In lateral inhibition, nodes broadcast intentions to continue tracks to candidate nodes further along the target's current track, and then wait a period of time, the duration of which is proportional to how accurate they consider their track to be (based on the number and strength of past readings, for example). During this waiting period, they listen for messages from other nodes that state they are tracking the target more accurately. If other nodes broadcast superior tracks, then the first node ceases tracking the target; if no better track is identified, then the node continues the track. The performance of these tracking schemes has varied. This variability has shown that certain trackers and certain modalities are better suited to certain applications. For example, in tracking tests performed on sensor data obtained by researchers in the Defense Advanced Research Projects Agency (DARPA) Sensor Information Technology (SensIT) program at the U.S. Marine Corps base in Twentynine Palms, CA, trackers using EKFS produced slightly more accurate tracks, but lateral inhibition can more ably track targets that do not have a linear target trajectory (such as when traveling on a road) [3,8,9].

Another target-tracking algorithm using intelligent processing among distributed sensors is the information directed sensor querying (IDSQ) algorithm. IDSQ forms belief states about objects by combining existing sensor readings with new inputs and with estimates of target position. To estimate belief states (posterior distributions) derived from the current belief state and sensor positions, sensors use entropy and predefined dynamics models. IDSQ's goal is to update the belief states as efficiently as possible, by selecting the sensor that provides the greatest improvement to the belief state for the lowest cost (power, latency, processing cycles, etc.). The tradeoff between information utility and cost defines the objective function used to determine which nodes should be used in routing target information, as well as to select clusterhead nodes. The information utility metric is determined using the information theoretic measure of entropy, the Mahalanobis distance (the distance to the average normalized by the variation in each dimension measured) and expected posterior distribution. The use of expected posterior distribution is considered particularly applicable to targets not expected to maintain a certain heading, such as when following a road. The belief states are passed from leader node (clusterhead) to leader node, with leaders selected by nearby sensors in a predefined region. This enables other sensors to become the leader when the original leader fails, increasing network robustness [10].

1.3.2 Application 1: Artillery and Gunfire Localization

The utility of distributed microsensors for tracking and classification can be shown in applications such as gunfire localization and caliber classification. Artillery localization illuminates several issues involving sensing. Artillery shell impact or muzzle blasts can be located using seismic and acoustic sensors; however, the physics behind locating seismic or acoustic disturbances goes beyond mere triangulation based on one-time field disturbances. When artillery shells are fired from their launchers, the shells themselves do not undergo detonation. Instead, a process called "deflagration" occurs. Deflagration is the chemical reaction of a substance in which the reaction front advances into the unreacted substance (the warhead) at less than sonic velocity. This slow burn permits several shock waves to emanate from the muzzle, complicating the task of finding the muzzle blast. This difficulty is magnified by the multiple energy waves that result from acoustic reflections and energy absorption rates of nearby materials, and seismic variations caused by different degrees of material hardness. Furthermore, differences in atmospheric pressures and humidity levels can affect blast dispersion rates and directions. Causing additional difficulty is the presence of ambient acoustical events, such as car backfires and door slams in urban areas. Each of these factors hampers consistent impulse localization [11].

Whereas the sonic frequencies of artillery blasts are low enough to prevent much dispersion of the sounds of artillery fire, the sounds of handgun and rifle fire are greatly affected by local conditions. One distributed sensor system that tracks the source of gunfire is produced by Planning Systems,

Incorporated (Reston, VA). The company has tested systems consisting of tens of acoustic sensors networked to a single base station. The system employs acoustic sensors contained in 6 in. × 6 in. × 4 in. boxes that can be mounted on telephone poles and building walls, although sensors in recent generations of production are the size of a hearing aid [12]. When elevated above ground by 10 m or more using telephone poles approximately 50 m apart, sensors can locate gunshots to within 1–2 m. The system uses triangulation of gunshot reports to locate the source of the firings; at least five sensor readings are needed to locate targets accurately. Primary challenges in implementing this system include acoustic signal dispersion, due to shifting and strong winds, and elimination of transient acoustic events. To counter the effects of transient noise, engineers are designing adaptive algorithms to take into account local conditions. In the past these algorithms have been implemented on large arrays, but recent implementations have been performed on a fully distributed network. For use on one such distributed net, researchers at BBN Technologies have designed a parametric model of gunshot shock waves and muzzle blast space-time waveforms. When at least six distributed omnidirectional microphones are used, the gunshot model and waveforms can be inverted to determine bullet trajectory, speed and caliber. When a semipermanent distributed system can be constructed, two four-element tetrahedral microphone arrays can be used. A three-coordinate location of the shooter can also be estimated if the muzzle blast can be measured.

The researchers have developed a wearable version of the gunshot localization system for the implementation with a fully distributed, ad hoc sensor network. Each user wears a helmet, mounted with 12 omnidirectional microphones, and a backpack with communications hardware and a global positioning system location device. The use of omnidirectional sensors eliminates the need for orientation sensors to determine the attitude of an array of directional sensors. The system detects low frequency (<10 kHz) gunshot sounds, and so is relatively impervious to waveform scattering and dispersion typical of higher frequency transmissions. It works primarily by detecting the supersonic shock wave made by the bullet as it travels through the air, thus avoiding a reliance on sensing muzzle blasts or other effects local to the shooter which can be masked or minimized. Its performance is enhanced if microphones can be located laterally compared with the bullet trajectory, to ease comparisons between the Mach cone angle (the cone produced by the edges of the expanding sound waves) and trajectory angles. Of course, this enhancement cannot be relied upon in a distributed system with microphones located on mobile users. If a spatially distributed system can be employed, then adequate muzzle and shock arrival time estimates can be obtained with less than 8 kHz bandwidth (20 kHz sampling). The 8 kHz bandwidth is also sufficient to classify bullet caliber. System tests in 1997 showed that the helmet-mounted system detected all 20 rounds fired, using 22, 30, and 50 caliber weapons. Some 90% of shots were determined to within 5° of azimuth, and 100% to within 20°. All shots were detected to 5° of elevation. Some 50% of shots were detected to be within 5% range accuracy, and 100% to within 20% of the actual range. Designers blamed the poor range-estimation performance on muzzle detection performance, attributable to algorithmic difficulties [13].

1.3.3 Application 2: Unmanned Aerial Vehicle Sensor Deployment

Distributed ground sensors without organic long-range exfiltration avenues do not have standoff capabilities. Their advantage lies in providing in situ sensing, or sensing done at or very near the location of the target. Sensing very near the target minimizes signal attenuation due to distance, atmospheric changes, radiation interference, or similar factors. On the other hand, in situ sensing suffers from challenges that standoff sensing avoids, such as an absence of robust power sources, narrowed or inflexible field of regard, and reliance on extended communications. While enabling many of its advantages, many difficulties of in situ sensing can be avoided or overcome through effective deployment of sensors, particularly through the use of unmanned aerial vehicles (UAVs). Present generations of smart sensors can be deployed via UAV quite effectively, because the sensors are self-aware and self-configuring. This eliminates the need for pre-established links, clear lines of sight, or defined routes around topological and connectivity disturbances, such as hills or buildings. Distributed sensors employ collaborative signal processing and mesh-style networking methods, such as diffusion routing, to guarantee message delivery despite uncertain connectivity, node failure, or variations in the number of nodes needed to determine target location or type. With this sort of intelligence built into its sensors, a UAV flown from a nearby ground station or with a preplanned route and instructions to drop sensors at waypoints can deploy a network of sensors with sufficient sensing power and longevity to perform numerous tracking tasks and to exfiltrate data. As UAV and sensor network capabilities improve, the need for preplanned routes, human operators, preset communications relays, and portage of redundant batteries will also fade. One result of all these improvements is that networked battlefields of the future will use UAVs heavily to scout enemy positions, track targets, and relay messages—as well as deliver sensors and exfiltrated sensor reports.

Through the use of multiple UAVs, tracking can be performed in a decentralized manner. This eliminates single points of failure, lessens power requirements on central nodes, and eliminates reliance on single, high-latency links. By using fusion algorithms to combine readings from multiple UAVs covering overlapping regions, all targets within the covered area can be mapped, with their headings and directions determined. In one multiple-UAV fusion algorithm, when multiple observations are reported, Bayes's theorem is used to provide position estimates through linear combinations of observed information states. One advantage of this system is its maintenance of full information states, because each node maintains its complete information state estimate and transmits only incremental information updates. This system suffers, however, from a lack of processing power at the nodes. In one test, this resulted in nodes being unable to merge inputs from other nodes consistently. As long as few targets exist, the multiple nodes do not have trouble maintaining a consistent understanding of the targets' locations. When more targets appear than can be processed by the system, nodes unable to clear their buffers show a small number of targets not common to the fused picture. Another difficulty was seen in inaccuracies in UAV location maintenance. Whereas a large number of UAV passes over the target range will smooth this error to zero, location errors can be introduced for circumstances with few UAV passes [14].

UAVs are so new to the military that operational units do not currently use multiple, networked UAVs to sense in single regions. In fact, the current field-tested and operational UAVs organic to forward-deployed military units have limited flight and communications ranges and limited networking power. One such small UAV currently in predeployment testing and use by the U.S. Marine Corps is the Dragon Eye. Developed by the Marine Corps Systems Command in Quantico, VA, the Dragon Eye is a "backpackable" UAV. It has a 4 ft wingspan and weighs 5.5 lb itself; the ground station, required to fly the UAV and to communicate (radio frequency) with the UAV, adds an additional 7 lb and includes goggles for video viewing. The UAVs wings can be detached from the fuselage for transport. Its sensor package currently includes off-the-shelf color video and low-light black-and-white video cameras. Uncooled infrared cameras are also planned [15]. A UAV that is slightly larger than the Dragon Eye and is in use for maritime applications is produced by Insitu Group. The Seascan UAV is intended specifically for shipboard imaging, and thus can be used for littoral operations. Like the Dragon Eye, Seascan UAVs must be launched using a catapult or elastic to shoot the UAV into the air; recovery is done by landing the UAV in an area of flat ground nearby (in the case of the Dragon Eye) or into a shipboard recovery net (in the case of the Seascan). Current sensor packages (summer 2003) include a color video camera installed in the nose of the craft. The daylight camera has a 45° field of view, pan-tilt capability, and the capacity to remain locked on a target while the UAV maneuvers. Other sensors being built have included infrared cameras, LIDAR systems, air particle sensors, and hydrometers [16].

Like the sensors deployed in the Dragon Eye, sensors in the Seascan use a hub-and-spoke topology rather than the mesh that a fully networked system would employ. Likewise, the nodes do not collaborate to reach decisions about a target; instead, a single video camera reports to the base station user what it shows. Both the Seascan and the Dragon Eye employ a single UAV deployed by a single operator; no plans exist to use a network of Dragon Eyes. However, research is under way in swarming UAVs; these would be fully networked and capable of acting collaboratively. Future UAV-mounted sensor systems

FIGURE 1.2   MAV in flight, with video sensor feed transmitted to ground computer.

will report multimodal information, with multiple nodes collaborating to lower false alarm rates, verify target types, and perform other more sophisticated tasks. One direction research into UAVs is headed is into micro-air vehicles (MAVs). MAVs are only several inches in wingspan, with further reductions to insect-sized vehicles expected in the future. One MAV built by a group from the University of Florida at Gainesville and the NASA Langley Research Center relayed live video streams to ground users (Figure 1.2) [17]. However, tremendous challenges exist in enabling MAVs to sense over long periods of time and extended geographic areas, such as power requirements, communications distances to users and databases, and aerodynamic obstacles (e.g. flying in the presence of winds).

1.3.4 Target Classification

A significant advancement over previous detection and tracking applications is the capacity to classify targets rather than merely to detect them. Early sensing applications such as Igloo White, and even later applications such as REMBASS, were notorious for triggering alarms as a result of animal intrusions. Accurate classification of targets would reduce the probability of false alarm in each sensor and network, as well as provide the user with knowledge of target type and ideally even identify the precise target itself within the target type. This enables superior knowledge of the target, which in turn facilitates tracking over long distances and tracking in a multitarget, multisensor situation. In the latter case, a target such as a vehicle is identified by characteristics specific to that particular target, through distinctive engine noises (such as piston knocks), suspension irregularities (such as a squeaky spring), etc. This sort of identification is very helpful in solving the computation- and communications-intensive multiple-target tracking situations, when targets are not constrained to roads or other predetermined pathways.

Classification occurs through the comparison of unknown target signatures with so-called training data obtained from known target types. Data on targets is often reduced to feature vectors, or sets of data with the lowest dimensionality that can identify the target distinctly. Detection in the acoustic, seismic, infrared or magnetic modes occurs when feature vector energies exceed a predetermined threshold; classification occurs when the specifics of the feature vector match the signatures in the training data. Classification in the visual mode operates not through energy exceeding a threshold, but (in the case of unmanned classifiers) by matching signature data to the training data. This makes selection of training data and accuracy of test data obtained from the field critical. Yet field data are always obtained under less than ideal conditions. Factors such as variability in sensor location, instrumentation quality, and latency due to network configurations amplify other, less avoidable variations, such as Doppler effects due to target motion and small differences in target locations, stealthiness, and mechanical functioning [3]. However, despite these opportunities for uncertainty, small differences can be eliminated in the process of reducing data to feature vectors to reduce target dimensionality. To obtain feature vectors, target spectra of various modalities can be obtained efficiently through Fourier analysis of the data to yield power spectral densities, grouped with data points corresponding to frequencies of interest [8]. Given that classification involves more than the comparison of target energy levels to preset thresholds, consistent determination of the feature vectors from the full data set is important to ensure accurate classification. Each detection event yields a single feature vector; multiple detection events can be collapsed into a single, representative vector to save bandwidth.

Correlations and relationships between feature vectors can be determined using any of several signal-processing algorithms. Common algorithms include k-nearest neighbor, maximum likelihood, support vector machine, singular value decomposition (SVD), and principal component analysis. The k-nearest neighbor algorithms test fitness between training data and test vectors, and then combine classifications from the nearest k neighbors using algorithms such as majority vote to decide on a single vector. Maximum-likelihood classifiers model feature vectors from similar targets based on Gaussian density functions and classify the models through algorithms such as expectation-maximization. Support-vector-machine classifiers are learning machines (like neural networks) that can perform binary classification for pattern recognition and real-valued function approximation for regression estimation. SVD can infer relationships between data through the use of matrix decomposition of principal input values. In principal component analysis, the principal values are analyzed from feature vectors, broken down according to their most basic components, and grouped into regions where deviations from maxima and minima are most extreme (to retain the variation in the original data set as much as possible), and uncorrelated [18].

In distributed sensor networks, targets are classified using multiple measurements performed (sometimes multiple times) at multiple nodes. Fusion of data obtained in multiple modalities is also a beneficial method to classify targets, because most modalities are largely orthogonal, thereby increasing the likelihood that additional data will contribute significantly. This process is facilitated by nodes that have sensors from several modes colocated; commonly used modes include infrared, acoustic, seismic and magnetic. The multiple measurements, whether from a single node using multiple modes or from multiple nodes, are integrated by the network to arrive at a single result for each classification event. To arrive at this single "answer," all classifiers, regardless of application, use data fusion, decision fusion, or a combination of the two. When conducting data fusion, classifiers fuse feature vectors of all measurements prior to arriving at a final classification decision. In decision fusion, each classifier individually classifies targets and then sends this information to other nodes for final determination of target type [19]. Decision fusion is superior when target information is uncorrelated and statistically independent, enabling each node to arrive at a separate, independently valid decision. When data are correlated, data fusion is needed because separate nodes are likely to decide similar outcomes, making the fused decisions less than ideal. Fusion decisions must be based on observations that are as distinct from each other as possible to avoid skewed results. Decision fusion offers the advantage of minimizing communications burdens, because the messages passed among nodes are smaller (scalar yes/no classification results per event rather than multiple, high-dimensional vectors specifying target characteristics). Decision fusion also offers the benefit of minimizing computational burdens at the nodes and requiring less data to train nodes accurately, due to the reduced dimensionality of the transmitted data [20].

1.3.5 Application 3: Vehicle Classification

Performance statistics of a distributed classifier were gained from data taken during several experiments performed using a network of distributed microsensors deployed at the Marine Corps Air–Ground Combat Center (MCAGGC) in Twentynine Palms, CA. In each of these experiments, conducted from

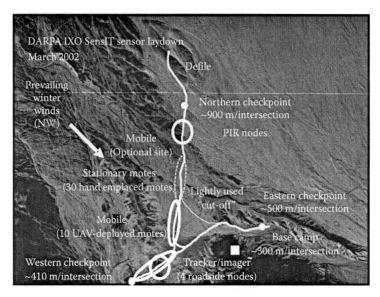

FIGURE 1.3   Deployment of sensors for the March 2002 SITEX experiments at Twentynine Palms MCAGCC, CA.

2000 to 2002 by researchers from the DARPA SensIT program, microsensors deployed along roadways and near an intersection located and tracked vehicles using multimodality sensing, including at various times via acoustic, seismic and magnetic modes (Figure 1.3). The sensors were connected in an ad hoc network and reported the positions of the vehicles, their speed and direction of travel, and displayed the results to a base station located some kilometers away. In some cases, classification algorithms analyzed the seismic and acoustic data to determine vehicle type. In other cases, passive infrared sensor nodes detected the presence and motion of ground vehicles and triggered an imager, located in the middle of the sensor field, at the appropriate time. During a third type of experiment, 15 magnetometers were deployed on a road, by hand emplacement and UAV drop. The UAV returned to query the ground network from the air for vehicle detection, timestamp and speed, and then exfiltrated the responses to a remote base camp. Targets employed during the experiments included military vehicles (amphibious assault vehicles [AAVs], Dragon Wagons [DWs], HMMVW [HVs], and M1 tanks) and civilian vehicles (SUVs).

Numerical classification results are available from several of the SensIT experiments. One series of experiments, employing acoustic data taken from runs of AAVs, DWs, and HVs, illustrates the superiority of a classifier employing decision fusion and data fusion simultaneously. Local exchange of high-bandwidth feature vectors can improve measurement signal-to-noise ratios, with the results combined using a global exchange of low-bandwidth decisions across regions to stabilize signal fluctuations. In fact, reliable, high-level classification decisions can be reached through the use of fairly unreliable local decisions from multiple regions where signals are strongly coherent. Furthermore, the superiority of decision fusion utilizing local data fusion holds for networks regardless of whether their decisions are hard, soft or noisy (where hard fusion transmits ones or zeroes only, depending on the measured values, soft fusion employs real-valued measurements, and noisy fusion uses hard decisions transmitted over noisy and unreliable links). Research has shown that, under mild conditions for soft, hard and noisy fusion, the probability of misclassification approaches zero as the number of sensor reports from distinct but coherent regions increases. These results show that a network configuration currently preferred by planners of a distributed microsensor network, in which local sensors share high-bandwidth feature vectors but transmit to regional sensor clusters using low-bandwidth decisions, offers superior sensing performance: a relatively small number (under 50) of local decisions with a high probability of error (0.2 or 0.3) can be fused to yield a very reliable final decision (probability of error near 0.01) [21].

TABLE 1.1 Classifier Results

Actual Vehicle	Sensed Vehicle			Classified Correctly (%)
	AAV	DW	HV	
AAV	117	4	7	94
DW	0	106	2	98
HV	0	7	117	94

Source: Data from SensIT experiment, 2000.

SensIT researchers also used data gathered during experiments at Twentynine Palms to analyze the efficiency of multimodality fusion based on SVD. Employing time series data from acoustic and seismic sensor readings, CPA data were determined to construct feature vectors for each target. Data within a time window of 4–5 s around the peak of each sample were used, to ensure accuracy of the CPA and eliminate superfluous noise. These data were divided into consecutive frames of 512 data points sampled at 5 kHz (0.5 s each), with a 0.07 s overlap with each of their neighbors. Power spectral densities of each frame were stored as a column vector of 513 data points corresponding to frequencies from 0 to 512 Hz. Using an eigenvalue analysis, the algorithm computed an unknown sample vector's distance from the feature space of a database of targets. The distance between the sample vector and the target database is considered the likelihood that the sampled frame corresponds to the frame of the target in the database, where the databases are grouped by attribute. The algorithm was employed to classify the three target vehicles (i.e. AAVs, DWs, and HVs), with CPA peaks detected by hand rather than by the algorithm. Only one vehicle was driven through the sensor field at a time, under environmentally noisy conditions (particularly wind). As shown in Table 1.1, the algorithm classified vehicles correctly roughly 95% of the time, even under suboptimal conditions [9].

1.3.6 Application 4: Imaging-Based Classification and Identification

Target classification based on target images (still or moving) offers significant advantages over classification based on other sensing modalities. Whereas sensor fields operating in seismic, acoustic and magnetic modes can detect and track targets readily, the ability to discriminate between targets using these modes is not as simple. For instance, many vehicles sound similar or have similar magnetic signatures, and distinguishing personnel by the sounds or impacts of their footfalls is very difficult. Using video to classify targets, though, can provide more guaranteed results due to the rich feature set available with imagery. There is a large variability between target classes apparent in the visual mode, and a large number of target-specific features per class. For personnel identification, visually distinct characteristics, such as facial features and gait peculiarities, provide distinct signatures that are difficult to mimic. In the case of inorganic targets such as vehicles, identifying marks such as tire types, rust-spot locations, canvas versus hardtop distinctions and the like make targets readily identifiable. Imagery also makes a particularly compelling classification application because people rely so much on what their eyes provide—in the case of target classification, "seeing is believing." When connected to a large database of target identities, microsensor networks with video imaging systems can permit a field commander to ascertain his target's identity and be confident of the results of his mission. This capability also permits tracking systems to track multiple, similar targets. The primary drawback of an image-based classification system, though, is that the transmission of images occupies significant bandwidth. In a low-power, small-bandwidth system such as a microsensor network, this has prevented imagers from being employed on a scale similar that of the other modalities. However, as algorithms for feature extraction and image compression improve, and as camera sizes continue to diminish, microsensor networks that classify and identify targets using images will increase in utility and prominence.

Visual-mode and infrared-mode images are useful to imaging-classification systems. Uncooled microbolometer thermal imaging sensors can classify vehicles, for instance, according to temperature profiles based on normalized histograms or statistical properties of target pixel luminances compared with background information. This permits the classification of targets in daytime or nighttime, and through obscurants such as smoke, fog, or changes in lighting. Both infrared- and visual-mode classifiers have historically worked by comparing unknown target signatures with templates stored in a database. A disadvantage of this system is that templates must be available for multiple target orientations, as well as for each target. Shape-based classifiers compare features of targets with training data, using categories such as area and perimeter (both of which depend on range to target), orientation, spread (how unevenly an object's mass is distributed about the centroid of the target), elongation (the degree to which an object's mass is concentrated along the target's major axis), and compactness (which measures the complexity of the target shape) according to the formula:

$$\text{Compactness} = \frac{\text{Area}}{(\text{Perimeter})^2}$$

Comparisons between template and targets may be made using any of a number of decision-making algorithms, such as decision trees, neural networks, and Bayesian networks [22].

An alternative classifier compares target features rather than shapes, and offers the advantage of operating when the target is partially occluded. In one example of an image classifier, moving targets are illuminated first by eliminating the background using subtraction. Because background images are not static, pixels from the background are modeled as a mixture of Gaussians, or in the nonparametric case the actual probability density functions (pdfs) of the pixel intensities are used. The Gaussians or pdfs are compared with pixels of the target image, with only pixel intensities above a threshold considered to belong to the target. Next, depending on the proximity of the cameras detecting the target, a correlation of pixel neighborhood intensities is performed. For systems with cameras located far apart, a scale-invariant feature comparison is done. This is done using a scale-space approach, in which feature points that can be detected consistently in images from different viewpoints, i.e. points that vary in only two dimensions, are compared. Invariance is achieved further by obtaining rotation-invariant feature descriptors, consisting of vectors of 15 elements corresponding to higher order derivatives of the image. These descriptors can be chosen to isolate only certain, readily identifiable elements of the target. They can then be compared by obtaining the minimum Mahalanobis distance between each pair of descriptors, to enable target classification and identification [23].

Another imaging classifier uses distributed imaging sensors that have 360° cameras with software to piece together views that do not have the distorted, fish-eye look typically associated with panoramic cameras. This "normal view" can be achieved by projecting a viewplane at the angle and distance desired from the viewer to display an appropriate image; pixels from the panoramic image that show up on the viewplane behind the image are recorded and used to construct the usual perspective view. As a result, any number of views can be constructed from the same panoramic source. Because they operate independent of camera orientation, these sensing systems lend themselves to microsensor applications in which the sensor nodes are remotely deployed, such as by UAV or cannon-launch. One prototype vision node contains a 180° fish-eye lens and a 12 bit, 1 megapixel, 30 frames-per-second CCD color camera. The visual and infrared cameras are housed in a 2.5 cm × 2.5 cm × 1 cm unit. An alternate arrangement uses eight "regular" cameras, each with a 55° field of view, arranged in an octagon such that their fields of view overlap and provide a 360° view. One commercial camera suitable for this arrangement occupies less than 1/3 cm^3 and has a built-in lens and active pixel focal plane. Software merges the views and permits pan, tilt, and zoom capabilities.

Image and video transfer consumes considerable energy, whereas sensing in acoustic, seismic or magnetic modes does not. A power-efficient system, therefore, uses low-bandwidth sensors to cue an imager to power up only when a target comes into view. In one scenario envisioned by the designers, the energy requirements of detecting and classifying tanks and heavy trucks are determined. Hand-emplaced acoustic and seismic sensors detect targets, and cue an imager. Designers assumed a traffic usage of

TABLE 1.2 Power Requirements of a Sensor System with Satellite Image Exfiltration

	Average Daytime Power	Energy/Day (J)
Data acquisition	350 μW	20
Image processing	5 mW	300
Comms (target report)	360 μW	23
Comms (image chips)	3 J/target	n/a

FIGURE 1.4 A UAV magnetometer drop: the racetrack pattern flown by a UAV; image from the UAV nose of foam-wrapped magnetometer nodes being dropped. (Image of a DW taken from the UAV.)

200 vehicles (20 of them trucks) per hour during the day, and 30 vehicles/5 heavy trucks per hour at night. A target detection report consumes 200 bits, and an image chip consumes 10 kb (images sent on request). The camera takes approximately 20 still frames of an intersection and exfiltrates the images via satellite. A breakdown of system power requirements is provided in Table 1.2. Employing this usage model and these performance statistics, a 9 V battery would last 65 days; and it would last 34 days after taking 100 images per day and exfiltrating them via Iridium to a classification algorithm in the rear, which compares the target images with a database of templates and permits viewing by analysts [24].

Experiments and real deployments in recent years have shown the practicality and utility of image-based tracking and classification. During one system test and demonstration in March, 2001, performed at MCAGCC in Twentynine Palms, CA, researchers from the University of California at Berkeley hand-launched a UAV to deploy sensors, record target images, and transmit the images to a base camp. A UAV carrying eight nodes with magnetometers flew a preset pattern over a roadway intersection. According to drop points loaded into the UAV software, the craft deployed half the magnetometers on a road; seven other the nodes were hand emplaced along the road. Targets included DWs, HVs, and vehicles of opportunity (civilian and military vehicles). After deploying the nodes and waiting for targets to traverse the field, the UAV returned to fly over the deployed nodes and query the ground network for vehicle detection, time of occurrence, and vehicle speed. It exfiltrated the responses to a remote base camp and relayed images taken from cameras mounted in its nose and body. Although the system did not classify targets, it demonstrated the practicality of a full remote-deployment sensor–imager system. Classification software could be installed on the UAV or on the nodes with little difficulty to make this a truly distributed classification system (Figure 1.4).

1.4 Sensor Network Applications, Part 2: Civilian Applications

Networked microsensors can be used for a host of applications other than detection, tracking, and classification of military targets. Sensors' small size, the flexibility and robustness provided by ad hoc networks, and the variety of ways sensor data can be reported to users combine to ensure that sensor nets can be adapted to a variety of applications. Although privacy concerns have been raised due to the impending ubiquity of interconnected, distributed microsensor networks (especially considering visual sensing), the possibilities of distributed and remote sensing seem endless [25]. Some "civilian" applications of sensor networks include home security illegal-entry sensing, industrial machinery wear sensing, traffic control sensing, climate monitoring, long-term in situ medical monitoring of people and animals, agricultural crop monitoring (both pre- and post-harvest in situ and remote sensing), aircraft control-surface embedded sensing, and so on. Two such applications that have been tested outside of the laboratory include aiding the recovery of endangered plant species and personnel/heartbeat sensing.

1.4.1 Application 5: Monitoring Rare and Endangered Species

Owing to their capacity to provide data inexpensively over long periods of time and large geographic areas, sensor networks can be used to monitor biological and environmental conditions that would tax the resources of human monitors, such as in the case of temperature or rainfall extremes.

One such application is providing longitudinal environmental data at sites with endangered plants and animals, to enable analysis of environmental effects on the species. Data helpful in determining habitat conditions include temperature, humidity, rainfall, wind and solar radiation. The same information recorded for areas that do not contain the endangered species enables scientists to infer the role of climate in the species' distributions. The presence of endangered species adds an additional dimension to this use of sensor networks, though: the sensors themselves must not disturb the environment. In particular, the sensors must require little maintenance to ensure that technicians and researchers disturb the species' habitats as little as possible. This effort not to disturb the environment is amplified in deployments of distributed and remote sensing where the environment itself attracts visitors. The sensors themselves must attract as little attention as possible to minimize degradation of visitors' experiences.

In one such instance of these demanding requirements, researchers installed a network of sensors at the Hawaii Volcanoes National Park, to measure climate data for studies of environmental effects on the development of the endangered plant species *Silene hawaiiensis*. Some locations of Hawaii have steep environmental gradients due to the presence of extremely varied terrain and weather conditions. This enables researchers to measure the impacts of a large variety of climates on a relatively small region. In 2000 and 2001, to monitor environmental effects on phenological events, such as periods of flowering and seed-set over long periods of time, environmentalists and park officials installed a distributed network of wireless microsensors. Modalities sensed included temperature, light, wind and relative humidity. About 100 sensors, radiating in the 900 MHz band, were deployed, to connect to an Internet link approximately 2 km from the rare plants. Ranges between sensors varied according to the sensors' placement above the ground: 10 cm of height afforded approximately 30 m of range, and a range of 100 m could be attained when sensors were placed 2 m above the ground (in trees). Power was provided by four "C-size" batteries per sensor, enabling the network to operate for about 6 weeks when sensors turned on briefly only every 10 min to perform sensing and communication functions, including transmission of images from the base station to an off-island data repository. (The team is investigating using alternate energy sources, such as solar and wind power.) Researchers developed a power-efficient routing scheme, i.e. multi-path on-demand routing, based on node reachability rather than node positions, and also used a routing scheme similar to directed diffusion. For data flow and visualization, the team designed a data-reduction scheme such that nodes could store the data locally. This includes using an "exception reporting" scheme in which values are reported only when they fall outside of a normal range determined at the beginning of the sensing (a model for the environmental results was constructed from early data). Data visualization was aided by reporting the data using color-coded icons of the individual sensors on a map, with red representing values outside of the norm and green representing values inside it.

The most innovative aspect of the sensing application was in the packaging and deployment of the sensors themselves. The desire not to impact visitors' experiences at the parked researchers to disguise the sensors by deploying them in "fake rocks" and short, hollow, tree-like tubes (Figure 1.5). Micro-fiber filler (or "Bondo") and plaster of Paris molds were used to create a variety of rock-like enclosures. For the tree branches, researchers used PVC pipe, with both the "branches" and the "rocks" painted to blend

FIGURE 1.5 Fake "rocks" and PVC "tree limb" sensor enclosures.

into the environment. The sensors themselves were packaged in plastic bags inside the enclosures, to prevent weather damage. These sensor housings were transparent to the radio emissions of the network, and went largely undetected by park visitors [26].

1.4.2 Application 6: Personnel Heartbeat Detection

Advances in technology have enabled sensors to be sensitive enough to detect tiny changes in the environment. One application that capitalizes on such sensitivity is the use of radar to detect motionless people. This is possible because very sensitive radars detect infinitesimal environmental changes, including the shock waves sent through the human thorax produced by the beating heat or the expansion and contraction of the chest due to breathing. The motion of a person's chest due to breathing is easier for a radar sensor to detect—radar returns are typically 10 times higher for people breathing than for the motion due to a beating heart. However, one radar sensor measured changes as small as three heartbeats per minute; the sensor was used in the late 1990s to measure the heart rates of Olympic archers and riflemen to determine whether their training permitted them to avoid the approximately 5 mrad movement of the bow or rifle due to their heartbeat at the time of firing. A "flashlight" version of the radar has been mounted on a tripod for long-term remote sensing; the beam can be narrowed to 16° to allow directed sensing [27]. Other radars that can sense through foliage and nonmetal walls have also been produced recently. In one preproduction unit, manufactured by Advantaca, the beamwidths can be set to 90°, 120°, or 360°. They have a maximum range of 20 m and a minimum range of 0.01 m, making them ideal to sense the presence of humans in closed rooms. A visual liquid-crystal display is provided on a separate unit connected to the radar emitter via wire. The sensors can detect people through 4 ft of rubble (2 ft of which is considered void), and through reinforced concrete walls 1.6 ft thick. In the latter case, the subject was 10 ft from the radar, lying prone. The Advantaca personnel detection radars were used immediately following the attacks on September 11, 2001, to detect people in the rubble of the World Trade Center [28].

The short-range radars have been packaged into "micropower impulse radars" and networked for surveillance of larger areas. They can be dropped from UAVs or hand emplaced to cover areas of interest, with signals exfiltrated via satellite. The radars are packaged into nodes each weighing 21 oz and about the size of a tennis ball can, with a similarly shaped but smaller unit inside it. The internal unit rotates freely along its long axis to ensure that the antenna always points upright upon landing. A detection radius of 20–30 m per node can be achieved; in tests, a single UAV pass can drop enough nodes to cover a 16 km² area. The primary technology enabler for a network of microradars is ultrawideband communications. Ultrawideband has emerged in recent years as a technology important to many communications arenas, such as personal area communications, tagging, ranging, localization, and motion sensing.

It uses very short pulses of energy spread over very wide bandwidths. It operates at very low duty cycle and short pulse length and so requires little power. It can operate at 100 Mbps or more, with each pulse covering multi-gigahertz of spectrum [29]. For sensing applications, ultrawideband microradars can sense the presence of people, vehicles, animals, or any human-sized target that moves. Advantaca personnel envision the radars being useful for many commercial applications beyond heartbeat sensing. Other applications considered include vehicle/traffic sensing, security and personnel sensing, and anti-collision sensing, as well as imaging systems such as baggage detection, roadway and runway inspection, and piping and infrastructure detection.

1.5 Conclusion

Recent years have witnessed tremendous growth in the capabilities of networked sensors. As a result, people are finding more and more applications for them. Applications cover a spectrum of uses: short-term to long-term, perimeter security to surveillance in depth, trip-wire alerts to target identification using database access. Early funding for sensor networks was provided primarily by the U.S. Department of Defense, so the applications in which the most development has been performed are military ones. Area surveillance, target detection, target tracking, and remote deployment mechanisms have all been under development for at least several years to decades. Progress has been made to the point that miniaturized, long-lived, distributed sensors can be deployed to sense in multiple modalities and report results to rear areas fairly consistently. These sensor networks can perform a variety of functions that are ultimately useful to both the military and civilian worlds. As a result, more and more applications useful to the home and to industry are beginning to appear. For instance, weather and environmental data are being sought by scientists in regions heretofore not measured [30], and industry advocates see uses for sensors in both the production of goods and their consumption. Distributed sensors are a link between the phenomenological world all around us and the increasingly computerized world of databases, processors, and analysis; hence, the data and information provided by sensors will only become more useful as time goes on, and the number of applications will continue to grow.

References

1. Kumar, S. et al. eds., Collaborative information processing special issue, *IEEE Signal Processing Magazine*, 19(2), 2002.
2. Gharavi, H. and Kumar, S.P. eds., Special issue on sensor networks and applications, *Proceedings of the IEEE*, 19(8), 2003.
3. Li, D. et al., Detection, classification, and tracking of targets, *IEEE Signal Processing Magazine*, 19, 17–29, 2002.
4. Cebrowski, A. and Garstka, J., Network-centric warfare: Its origins and future, *Naval Institute Proceedings*, 124(1), 28–35, January 1998.
5. Jeppeson, C., Acoubuoy, SpikeBuoy, Muscle Shoals and Igloo White, home.att.net/~c.jeppeson/igloo_white.html, 1999.
6. Estrin, D. and Pottie, G., Directed diffusion: A scalable and robust communication paradigm for sensor networks, In *Proceedings of the Sixth Annual International Conference on Mobile Computing and Networking (MobiCOM '00)*, Boston, MA, August 2000.
7. Brooks, R. and Iyengar, S.S., *Mutli-Sensor Fusion: Fundamentals and Applications with Software*, Prentice Hall, Upper Saddle River, NJ, 1998.
8. Brooks, R. et al., Distributed target classification and tracking in sensor networks, *Proceedings of the IEEE*, 91, 1163–1171, 2003.
9. Freidlander, D. et al., Dynamic agent classification and tracking using an ad hoc mobile acoustic sensor network, *EURASIP Journal on Applied Signal Processing*, 4, 371–377, 2003.

10. Zhao, F. et al., Information-driven dynamic sensor collaboration, *IEEE Signal Processing Magazine*, 19, 61–72, 2002.

11. Swanson, D., Artillery localization using networked wireless ground sensors. Unattended ground sensor technologies and applications, *Proceedings of the SPIE*, 4743, 73–79, 2002.

12. Lewis, G., Planning Systems Incorporated. www.PlanningSystemsInc.com

13. Duckworth, G. et al., Fixed and wearable acoustic counter-sniper systems for law enforcement. Sensors, C3I, Information, and Training Technologies for Law Enforcement, *Proceedings of the SPIE*, 2577, 1998.

14. Ridley, M. et al., Decentralized ground target tracking with heterogeneous sensing nodes on multiple UAVs, In *Information Processing in Sensor Networks*, Zhao, F. and Guibas, L. eds., Springer-Verlag, Berlin, Germany, 2003, pp. 545–565.

15. Adams, C., Minidrones: Near term …, *Avionics Magazine*, November 2002. www.aviationtoday.com

16. www.insitugroup.com

17. Ettinger, S.M. et al., Towards mission-capable micro air vehicles: Vision-guided flight stability and control, *Advance Robotics*, in press.

18. Jolliffe, I.T., *Principal Component Analysis*, Springer-Verlag, New York, 1986.

19. Kokar, M. et al., Data vs. decision fusion in the category theory framework, In *Proceedings of Fusion 2001 Fourth International Conference on Information Fusion*, Vol. 1, Montréal, Quebec, Canada, 2001.

20. D'Costa, A. and Sayeed, A., Collaborative signal processing for distributed classification in sensor networks, In *Information Processing in Sensor Networks*, Zhao, F. and Guibas, L. eds., Springer-Verlag, Berlin, Germany, 2003, pp. 193–208.

21. D'Costa, A. et al., Distributed classification of Gaussian space-time sources in wireless sensor networks, *IEEE Journal on Selected Areas of Communications*, submitted.

22. Thomas, R. and Porter, R., Omnisense© visually enhanced tracking system. Unattended ground sensor technologies and applications, *Proceedings of the SPIE*, 4743, 129–140, 2002.

23. Pahalawatta, P. et al., Detection, classification, and collaborative tracking of multiple targets using video sensors, In *Information Processing in Sensor Networks*, Zhao, F. and Guibas, L. eds., Springer-Verlag, Berlin, Germany, 2003, pp. 529–544.

24. Boettcher, P. and Shaw, G., Energy-constrained collaborative processing for target detection, tracking and geolocation, In *Information Processing in Sensor Networks*, Zhao, F. and Guibas, L. eds., Springer-Verlag, Berlin, Germany, 2003, pp. 254–268.

25. Shepherd, D., Networked microsensors and the end of the world as we know IT, *IEEE Technology and Society Magazine*, 22, 16–22, 2003.

26. Biagioni, E. and Bridges, K., The application of remote sensor technology to assist the recovery of rare and endangered species, *The International Journal of High Performance Computing Applications*, 16, 112–121, 2002.

27. Greneker, E., Radar flashlight for through-the-wall detection of humans, In *Aerosense97*, Orlando, FL, 1997.

28. http://www.advantaca.com

29. Jones, E., Ultrawideband squeezes, *MIT Technology Review*, September, 71–79, 2002.

30. Lundquist, J. et al., Meteorology and hydrology in Yosemite National Park: A sensor network application, In *Information Processing in Sensor Networks*, Zhao, F. and Guibas, L. eds., Springer-Verlag, Berlin, Germany, 2003, pp. 518–528.

2

Interfacing Distributed Sensor Actuator Networks with the Physical World

Shivakumar Sastry
The University of Akron

S. Sitharama Iyengar
Florida International University

2.1 Introduction 21
2.2 Benefits and Limitations of DSAN System 21
2.3 Taxonomy of DSAN System Architectures 23
    Input/Output • Processing • Communication • System
    Attributes • System Integration • Security
2.4 Conclusions...30
Further Reading...30

2.1 Introduction

The interface between a distributed sensor-actuator network (DSAN) system and the physical world is the critical foundation that enables the nodes to acquire sensor data and effect actions in the environment. Rapid advances in a variety of disciplines offer us tiny devices—namely, microcontrollers, radio transceivers, sensors, and miniature actuators—that can be deeply embedded in applications. When coupled with advances in modeling, software, control theory, and real-time systems, such a collection of networked devices can sense and actuate the physical world to achieve novel system-level objectives.

In a DSAN system, data from the sensors must be integrated to synthesize new information in a reliable manner within fixed timing constraints. In applications such as automation or critical infrastructure monitoring systems, the sensing tasks must be performed periodically while satisfying additional performance constraints. The efficient synthesis of information from noisy and possibly faulty data from sensors makes it necessary to better understand the constraints imposed by the architecture of the system and the manner in which individual devices are connected locally—with each other and with the environment. Architectural considerations also impact the reliable operation of one or a group of spatially distributed actuators. Once deployed, a DSAN system must organize itself, adapt to changes in the environment and nodes, and continue to function reliably over an extended duration of time. Since the available technology enables several architectures for a DSAN system, we propose a taxonomy that is useful for designing such systems and planning future research.

2.2 Benefits and Limitations of DSAN System

The interface between a DSAN system and the physical world represents a fundamental departure from the traditional networked systems in which sensors and actuators were connected to a centralized computing system. Figure 2.1a depicts the traditional approach. Data from sensors are gathered by interface boards that are physically wired to the sensors; these data are presented to applications in formats

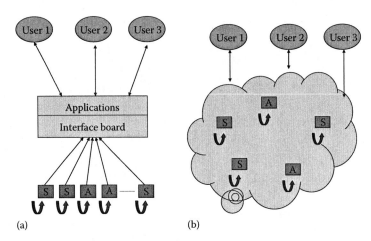

FIGURE 2.1 DSAN system represents a fundamental departure from the traditional approach to interfacing computers with sensors (S) and actuators (A). (a) Traditional centralized approach. (b) Decentralized approach.

specified by the applications that use the data. Data that are sent to actuators were also sent from the computer to the devices via similar interface boards. Power and resource constraints of the interface boards were imposed on the devices connected to the boards, and hence, on the data. The physical wiring, which was required to attach sensors and actuators to the interface boards and the interface boards to the computers, made it necessary for the entire system to be highly engineered. Figure 2.1b depicts the approach that is enabled by a DSAN system. Individual or groups of sensors and actuators are connected to a local microcontroller. The microcontrollers interact with each other over wired or wireless media to coordinate their local tasks. There is no central computer that coordinates activities at any of the microcontrollers. The network abstracts the spatial distribution of the nodes, and users interact with the network in batch, interactive, or proactive paradigms.

The spatial and temporal distribution of nodes in a DSAN system is designed to match the geometry of applications and simplify design through specialization. Such an approach offers important advantages such as

1. Lower cost and reliable implementation
2. Quick response times for demanding control loops
3. Incremental design and deployment
4. Heterogeneous sensors and actuators
5. Redundancy to improve reliability
6. Graceful degradation in fault or overload conditions
7. Increase concurrent sensing and actuation operations

Despite these advantages, significant challenges must be overcome to successfully design, deploy, and operate DSAN systems that operate over large time scales. Some questions that arise are, *how to*

1. Incorporate data from multiple modality sensors to improve confidence and reliability of the system?
2. Collect correlated data from heterogeneous sensors that operate over different time scales?
3. Effectively improve the quality of service (QoS) of the system (usually measured in terms of throughput, latency, and jitter) by exploiting concurrent and redundant devices?
4. Configure, program, and/or query DSAN systems?
5. Match the communication capacity with the data volume and rate limitations of individual devices on the one side and application demands on the other?
6. Process sensor data to reduce communication load?
7. Achieve scalable systems?

The taxonomy we present offer some solutions to these questions and will stimulate the readers to ask and answer other similar questions.

2.3 Taxonomy of DSAN System Architectures

Figure 2.2 depicts four aspects that must be addressed to realize any DSAN system. For each aspect, current device-level technologies offer several options that affect the structure and performance of the DSAN system. We capture many of the available options (and a few desirable options) in the proposed taxonomy.

It is important to distinguish between *function* and *implementation*. Function is the primary designed feature of an aspect of the DSAN system. It represents a basic operation or capability that the aspect supports. Implementation refers to the methods that are used to deploy the designed feature; one important consideration in the deployment is the location of the feature in the system. The choice of location is important because it is closely related to how the devices are packaged. The manner in which devices are packaged drives the cost of the system throughout its lifecycle. For each device, there is a certain minimum cost necessary for installing, operating, and maintaining the device throughout the system lifecycle. By colocating groups of devices, some of these costs can be amortized over the colocated devices. While it is necessary to distribute the devices from an architectural perspective, colocating devices is essential to control the costs. As a result, finding a balance between these considerations is the principal focus for packaging design.

The primary purpose of a DSAN system is to gather data from a physical environment and effect actions in the environment within a predictable, bounded time. Depending on costs associated with missed deadlines, we can view a DSAN system as being a hard real-time system or a soft real-time system. For example, if the data gathered are used to regulate the operation of a critical actuator (e.g., a coolant in a power plant), we need a hard real-time system; in contrast, if the data gathered are used to locate a nearby restaurant in an automobile, we need a soft real-time system. Some features of the implementation affect predictability of a DSAN system.

A DSAN system may also be viewed as being deterministic, quasi-deterministic, or nondeterministic. In a *deterministic* DSAN system, it is possible to accurately predict the output and performance of the system given a fixed set of inputs. In a *quasi-deterministic* DSAN system, although the output or performance cannot be predicted as accurately, we can establish worst-case bounds. In a *nondeterministic* DSAN system, it is not always possible to guarantee the output or the performance without relying on assumptions about various operating parameters at different levels of the system. Under "normal" operating conditions, the performance of nondeterministic systems can be significantly better than the other systems; however, the behavior under "abnormal" operating conditions is difficult to characterize.

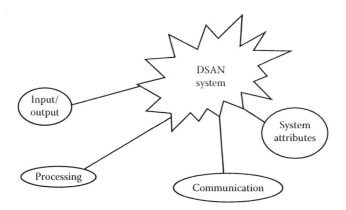

FIGURE 2.2   Major aspects of a DSAN system.

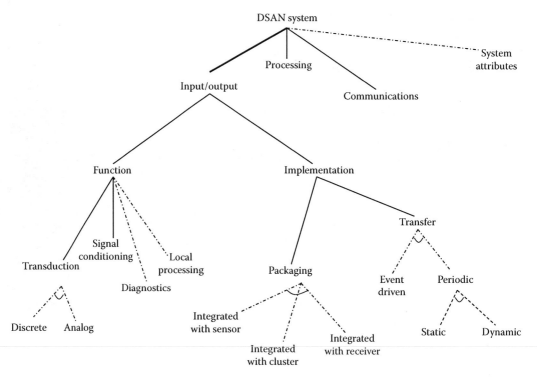

FIGURE 2.3 Taxonomy of the input/output aspect of a DSAN system.

The relationship between the options, selected in the four aspects of a DSAN system, and predictability will be discussed as a part of the taxonomy.

The taxonomy we present can be used as follows: to construct a DSAN system, one must select at least one of the options in the input/output, processing, and communication aspects. While it may be critical to do so in certain domains, none of the system attributes need to be selected. In the following taxonomy diagrams, solid lines represent mandatory items for a successful system realization; dotted lines indicate options (e.g., see Function in Figure 2.3). A set of dotted lines connected by a solid arc (e.g., see Transduction in Figure 2.3) indicate that at least one option must be selected.

2.3.1 Input/Output

The input aspect focuses on how signals are captured, converted, and processed at the device level. The output aspect affects how decisions are realized and/or presented to the actuator devices. As shown in Figure 2.3, there are four primary functions in the input/output aspect: *transduction* and *signal conditioning* are mandatory functions, while *diagnostics* and *local processing* are optional functions. Many sensor and actuator devices incorporate hardware support for signal transduction and conditioning.

Transduction is either of analog or discrete type. Discrete devices present digital data from the underlying sensors and receive digital data for their actuators. Analog devices usually have hardware support for converting between the digital and analog representations. Applications tend to be predominantly one or the other, although mixed systems are becoming more prevalent. Unlike the traditional approach, heterogeneous devices can be used in a DSAN system with minimal overhead.

Signal conditioning function includes activities such as amplifying, digitizing, filtering, forcing, or other basic signal-processing operations that improve the fidelity of the data. During input, the result of signal conditioning (irrespective of transduction type) is a digital representation of sensed values. Digital representations are desirable because they

1. Are robust with respect to noise
2. Can support error detection and correction schemes
3. Are easy to store and manipulate
4. Can be easily secured

Diagnostics for a sensor refers to methods for determining whether the sensor is functioning properly or has failed. Additional circuitry is required for performing self-test at the level of an individual sensor. It is sometimes feasible to use the values of other sensors or history tables to ascertain whether or not a sensor has failed. For an actuator, diagnostics ascertain whether the actuation signal that was sent to the device has actually been transferred to the environment and whether the intended effect in the environment has been achieved.

Local processing refers to a list of tasks that may be included at the level of a sensor or an actuator. Examples of such tasks include historical trending, data logging, alarming functions, and support for configuration and security management.

The *implementation* of the input functions described earlier significantly affects the overall architecture and performance of the DSAN system. The transduction function can be collocated with the sensor or actuator. The remaining input functions, that is, signal conditioning, diagnostics, and local processing, may be located

- By integrating these functions with the sensor or actuator device
- In special modules and provide these services to a cluster of devices
- In devices that use the data from the sensors or issue commands to the actuators

These options are depicted in Figure 2.3 as *packaging* options. A particular choice affects the response time under normal and fault conditions, wiring costs, and the processing requirements of the DSAN system. By performing the functions locally at each device, we can optimize the implementation with respect to specific devices and reduce the communication traffic, thus resulting in faster sampling rates. By locating these functions in special nodes, we can reduce the cost of design and maintenance, use more resource-constrained sensor or actuator devices, and apply the functions at the level of a cluster. Such a choice is likely to increase security risks because of the cost of securing the links between individual links and clusters. The third alternative is to transmit the raw data between the devices and the nodes that require or originate the data; typically, this choice tends to increase communication demands and precludes options for early recognition of critical status information.

As an example, let us consider the diagnostics function. This function can be implemented at the level of a sensor or actuator device, a cluster, that is, at a special node, or at the node that uses or originates the data. By locating the diagnostics function at the device, we can make local decisions within required time limits. The propagation of erroneous data is prevented. However, we need additional redundant circuitry and system resources such as memory and timers at the device level. By performing the diagnostics function at the cluster level, we reduce design, implementation, and maintenance costs. It is feasible to use values of other devices in the cluster to diagnose a device. The resources of sensors can be constrained while the resources at special nodes are less constrained and better utilized. On the other hand, if we locate diagnostics at the receivers or originators of the data, we will require redundant implementations of the function that transfers data between the device and the node, increase the resource requirements for receivers, increase communication traffic, and increase the risk of propagating erroneous data in the DSAN system. Despite the high overhead, this choice may be important when the device is deployed in a hazardous environment (e.g., chemical presence) that is difficult to access and maintain. The specific choice depends on the application and must be selected to balance system-wide cost and performance issues.

The *transfer* function refers to how sensed data are delivered to the DSAN system from a sensor and how the data from the DSAN system are presented to an actuator. The options selected here must be coordinated with corresponding options in the communication aspect of the DSAN system. From the input aspect's perspective, the implementation of a transfer method involves the specification of the method, that is, either periodic or event driven. This choice affects the manner in which the operating system and communication protocols at the level of sensors are designed.

Periodic input can either be static or dynamic. Depending on the packaging, such synchronization can be initiated either by the DSAN system that uses a master clock, by devices that use local clocks, or by special nodes that serve as data concentrators. Periodic transfer is said to be *static* if the data are gathered or presented deterministically within a fixed time period, called the scan time. The time period for each device in a DSAN may be different. Static-periodic systems have significant, unintentional variations in the scan time. For example, if the strategy is to scan *as fast as possible*, the scan time is affected if the time to process certain pieces of data is different from others. The scan time also varies when the DSAN system initiates special processing and transfers in response to certain abnormal events such as security breaches or multiple device faults. Periodic transfer is said to be *dynamic* if successive scan times are not equal. When using dynamic transfer mechanisms, it is important to track both the value and the time at which the data are acquired or generated before synthesizing information.

Event-driven input is fundamentally different from periodic input and is based on detecting either of the following:

- The change-of-state (COS) of one or more predefined variables
- Predefined events that are described as a sequence or an expression involving COS variables

The variables involved must have clearly defined states. For example, in the case of temperature sensor, the range of temperatures that the sensor can detect can be divided into a set of quantized states. The system detects changes that cause the value to change from one quantized state to another. The advantages of an event-driven system over a periodic system are (1) the event-driven approach is, on average, more responsive to changes in the environment and (2) the communication traffic can be reduced by not sending repetitive information. However, the disadvantages are (1) additional measures are necessary to guarantee the delivery of data, (2) methods to detect failures in a sensor–receiver path are required (since it is difficult to distinguish between a failure and a long period of no COS), and (3) mechanisms are necessary to prevent an avalanche of data from overwhelming the communication system under exceptional situations. Unlike periodic input, event-driven input is nondeterministic unless explicit techniques for bounding the performance are also designed and implemented (e.g., priority scheduling).

2.3.2 Processing

The processing aspect of the DSAN system is essential to effect actions in the environment based on information that is synthesized from data acquired via one or more sensors and prior state of the system. As shown in Figure 2.4, the primary functions for this aspect are *algorithm execution, exception management, data management,* and *system interfaces.*

Algorithm execution refers to the computational tasks in the DSAN system. Such tasks include information synthesis, data encryption and decryption, decision making, updating system state at multiple levels, determining values for actuators, clock synchronization, and detecting system-wide state. The operating environment of a node is responsible for ensuring that these algorithms are executed fairly and effectively.

Exception management tasks are additional computations that are performed at the level of a device or in special nodes that augment the local processing function of the device. Techniques that automatically embed code for diagnostics monitoring or distributed services are also a part of this function. As an example, code that is embedded in this manner can provide status information and alarm data to operator monitoring stations. Some diagnostics strategies require temporal information in addition to

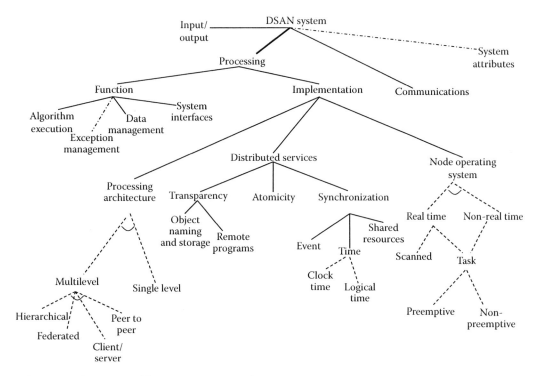

FIGURE 2.4   Taxonomy of the computing aspect of a DSAN system.

the input data. Since the exception management tasks can be reused across multiple applications, they are not considered to be a part of the algorithm execution.

Data management is a function that is becoming increasingly important for DSAN systems in which assured operations are critical such as monitoring of critical infrastructure. Because of the size of contemporary systems, the data in a DSAN are immense. Typically, it is not feasible to associate mass storage devices at the level of individual devices. Data management considerations for periodic systems are more critical because of issues of data freshness. Emerging opportunities that can integrate elastic computing and storage resources with a DSAN are therefore attractive.

System interfaces are another important component of the processing aspect. These interfaces and services are necessary to effectively integrate the DSAN system with other systems in the environment. For example, it is useful for a DSAN system to present its data in browser-ready formats that are based on XML. System interfaces to elastic computing and storage resources and tools for analytics and visualization are important in emerging systems.

The implementation of the aforementioned functions of the processing aspect is discussed in the following under the categories of *processing architecture*, *distributed services*, and *node operating system*.

There are two choices for the *processing architecture* of a DSAN system. Recall, a DSAN system is a networked collection of devices. In a flat architecture, all the devices are treated uniformly and there is no hierarchy. When the system is organized in this manner, we must capture and reason about contextual information to properly manage system evolution over time. Because of the immense scale of a DSAN system, multilevel architectures are easier to manage and operate. There are four options for multilevel architectures: (1) *hierarchical* in which there are tiers of authority in which devices in higher tiers are masters of devices (slaves) in lower tiers of the system; (2) *federated* in which certain responsibilities are granted to devices in a higher tier, but many functions are performed autonomously by devices in lower tiers; (3) *client–server* in which the devices are delineated into roles so that clients request services or data from the servers; and (4) *peer to peer* in which the devices can be clients, servers, or both.

These architectures are not always clearly separable. We expect most DSAN systems in the future to be federated, with many subsections organized as peer to peer or client–server.

Distributed services enable the coordinated operation of the multiple devices that comprise the DSAN system. *Transparency* refers to the ability to regard the distributed system as a single computer. Specifically, transparency concerns *object naming and storage* service that provides the ability to access system objects without regard to their physical location. *Remote task services* provide the ability to create, place, execute, or delete a task without regard to the physical location of the device. Typically, servers are necessary to perform the registration and lookup functions to provide these services.

The *atomicity* service is used to improve the reliability of the system by a sequence of operations, called transactions, occurring either in their entirety or not at all. Various forms of recovery mechanisms must be implemented to checkpoint and restore the component state to cope with the failure of one or more atomic operations. Typically, atomicity is more important at the level of information-based transactions and actuation decisions; it is less important at the level of periodic data gathering.

The order in which data from various devices are gathered and the nature of interactions among the multiple devices depends on the *synchronization* method. The *event* service allows a device to register an interest in particular events and to be notified when they occur. The *time* service is used to provide a system-wide notion of time. An important application of system time is in the diagnostics function where it is used to establish event causality. The management of *shared resources* across the network is supported through mechanisms that implement mutual exclusion schemes for concurrent access to resources.

All tasks in a sensor execute in an environment provided by the *node operating system*. This operating system (OS) provides services to manage resources, handle interrupts, and schedule tasks for execution. The operating system can either be a real-time OS that is designed to meet the deadline of every task or a non-real-time OS that aims to reduce the average response time of all the tasks. The OS can either support preemptive or non-preemptive tasks. Depending on the way in which the scheduler operates, the methods used to code computing, and the interaction with the communication interfaces, the execution in a sensor can be deterministic, quasi-deterministic, or nondeterministic. Designing deterministic or quasi-deterministic DSAN systems remains a significant research challenge.

2.3.3 Communication

The communication aspect is the primary infrastructure that enables the devices in a DSAN system to interact with each other. As depicted in Figure 2.5, *data transport* and *bridging* are the two key functions of this aspect. For *data transport*, it is useful to distinguish between three types of data. *Device data* are the data that are gathered via sensors or the data that are sent to actuators. Typically, such data are limited to a few bytes and must be delivered in a predictable manner to maintain system integrity. Data across multiple devices must be used to synchronize operations across the DSAN, diagnose system state, and recover from failures. Consequently, *interdevice* data are likely to be sporadic and bursty. Such data are typically aggregated over temporal or spatial regions and will hence contain more information. Thus, guaranteed delivery is an important attribute for these data. Such data are more suitable to quasi-deterministic or nondeterministic delivery mechanisms. *System data* refer to all the other data that are necessary for the proper operation of the DSAN including synchronization, startup, safety shutdown, monitoring, and status alarms. Some of these data may be critical and real time, while other data may not have a hard deadline constraint. Each of these types of data can include both discrete and continuous data.

The *bridging* function moves data between multiple networks. This function is important in contemporary DSAN systems that are likely to be integrated into existing engineering systems. The protocol used on the two networks on either end of the bridge device may not necessarily be the same. These intelligent devices also provide services such as data filtering, data fusion, routing, and broadcasting and serve to logically partition the system.

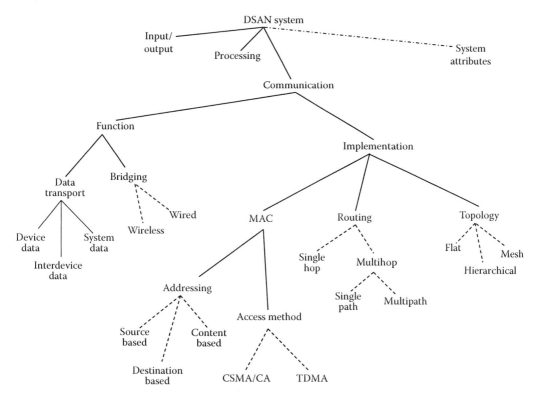

FIGURE 2.5　Taxonomy of the communication aspect of a DSAN system.

Instead of the traditional stack proposed in the open systems interface (OSI), devices in a DSAN system typically support a three-level stack comprising the physical, media access (MAC), and network layers. MAC, routing, and topology are the three dominant implementation considerations. For MAC, the addressing scheme and the access mechanism are important. The *addressing scheme* can be *source based* in which only the producing device's address is used in messages versus using the destination address to route the message. Source-based schemes can be extended to use *content-based* addressing in which codes are used to identify the type of data within the message. *Destination-based* schemes are used when there is usually one destination or when the routing is dynamically constructed.

The capability to provide deterministic service is strongly affected by the *access method* that establishes the rules for sharing the common communication medium. In the CSMA/CA method, the nodes avoid collisions by randomly delaying the time at which they attempt to send a packet based on the measured noise status of the environment. In contrast, the time division multiple access (TDMA) scheme that is based on a fixed time slot allocation is deterministic. Emerging systems are likely to be based on better modulation and media access schemes that can support higher data rates reliability.

The physical *topology* of the communication system is another important implementation choice. The system may be organized as a *flat* system in which all the devices are connected to each other as peers. Alternatively, the system topology can be hierarchical. In this arrangement, the devices are divided into a fixed number of layers and each device is connected with a device in a layer above it and another device in a layer below it. There is a single device at the highest layer of the hierarchy and the devices in the lower layer are only connected to devices in a higher layer. When the devices are interconnected using a mesh topology, every node is connected to a fixed set of its immediate neighbors. Such an arrangement provides interesting avenues for exploiting multiple paths in the system to improve the QoS.

The topology of the DSAN system affects how the messages in the system can be routed. In a system that has a flat topology, all messages are routed from the source node to the destination node in a *single hop*.

In contrast, when the messages must travel over *multiple hops*, the messages can be routed along a *single path* (usually the shortest path) or all the available paths (*multipath*). In all these cases, the routing strategy can either be source directed or receiver directed.

2.3.4 System Attributes

In addition to the aspects discussed in the preceding sections, a DSAN can be distinguished based on other attributes. Some of these important attributes are discussed here.

2.3.5 System Integration

DSAN systems are unlikely to operate as stand-alone applications; it is more likely that these systems would be deployed to complement existing infrastructures and hence it is critical for the DSAN system to integrate and interoperate with other systems. The systems integration capability refers to the capabilities supported under the bridging function of the communication aspect, the data management, and system interfaces of the processing aspect. In addition, the DSAN system must support efficient tools to enable the design, operation, and maintenance of the integrated system.

2.3.6 Security

DSAN security is challenging because these systems inherit security issues from distributed systems, wireless networks, sensor networks, and wired networks such as Ethernet-based factory systems. The unique characteristics of the DSAN system make it necessary to design new solutions to address the challenges of these security issues.

2.4 Conclusions

The landscape of architectures of DSAN systems is vast. This chapter presented a taxonomy of the architectural options. The major aspects of a DSAN system are input, computing, communication, and system attributes. The taxonomy proposed here provides a systematic approach to traverse this vast landscape. The taxonomy is a useful tool for research planning and system development.

Further Reading

Agre, J., L. Clare, and S. Sastry, A taxonomy for distributed real-time control systems, *Advances in Computers*, 49, 303–352, 1999.

Brooks, R.R. and S.S. Iyengar, *Multi-Sensor Fusion: Fundamentals and Applications with Software*, Prentice Hall, Upper Saddle River, NJ, 1997.

Coleman, B., Using sensor inputs to affect virtual and real environments, *IEEE Pervasive Computing*, 8(3), 16–23, 2009.

Computer Science Telecommunication Board, Embedded everywhere: A research agenda for networked systems of embedded computers, National Research Council, Washington, DC, 2001.

Estrin, D., D. Culler, K. Pister, and G. Sukhatme, Connecting the physical world with pervasive networks, *IEEE Pervasive Computing*, 1(1), 59–69, 2002.

Freris, N., S. Graham, and P.R. Kumar, Fundamental limits on synchronizing clocks over networks, *IEEE Transactions on Automatic Control*, 56(6), 1352–1364, 2011.

Hu, T. and Y. Fei, QELAR: A machine-learning-based adaptive routing protocol for energy-efficient and lifetime-extended underwater sensor networks, *IEEE Transactions on Mobile Computing*, 9(6), 796–809, 2011.

Ilic, M., L. Xie, U. Khan, and J. Moura, Modeling of future cyber–physical energy systems for distributed sensing and control, *IEEE Transactions on Systems, Man and Cybernetics, Part A: Systems and Humans*, 40(4), 825–838, 2010.

Iyengar, S.S., K. Chakrabarty, and H. Qi, Distributed sensor networks for real-time systems with adaptive configuration, *Journal of the Franklin Institute*, 338, 571–582, 2001.

Kim, K., Challenges and future directions of cyber-physical system software, *Proceedings of the IEEE Annual Computer Software and Applications Conference*, Seoul, Korea, July 19–23, pp. 10–13, 2010.

Li, M., D. Ganesan, and P. Shenoy, PRESTO: Feedback-driven data management in sensor networks, *IEEE/ACM Transactions on Networking*, 17(4), 1256–1269, 2009.

Mamidisetty, K., M. Duan, S. Sastry, and P.S. Sastry, Multipath dissemination in regular mesh topologies, *IEEE Transactions on Parallel and Distributed Systems*, 20(8), 1188–1201, 2009.

Poovendran, R., Cyber–physical systems: Close encounters between two parallel worlds, *Proceedings of the IEEE*, 98(8), 1363–1366, 2010.

Rasheed, A. and R. Mahapatra, The three-tier security scheme in wireless sensor networks with mobile sinks, *IEEE Transactions on Parallel and Distributed Systems*, 23(5), 958–965, 2010.

Weiser, M., The computer for the 21st century, *Scientific American*, 265(3), 94–104, 1991.

3

Contrast with Traditional Systems

Richard R. Brooks

Clemson University

3.1   Problem Statement..33
Acknowledgments and Disclaimer 35
References...35

3.1 Problem Statement

Sensor networks are a fundamentally new type of system. Computing systems up to now have been primarily user-centric. Desktop devices interact directly with a human operator following their instructions. Networked systems served primarily as communications devices between human users. These communications were often augmented by the existence of databases for transaction processing. Information was retained and processed for later use.

Sensor networks now provide distributed systems that interact primarily with their environment. The information extracted is processed and retained in databases as before, but the human is removed from many parts of the processing loop. Devices are designed to act more autonomously than was previously the case. They are also designed to work as a team.

Embedded systems are not new, but sensor networks greatly extend the capabilities of these systems. Up to now, embedded systems were resource-constrained devices providing limited intelligence in a strictly defined workspace. Sensor network research is aiming towards developing self-configuring systems working in either unknown or inherently chaotic environments.

Network communications needs to be handled in a new manner. Data are important because of their contents and not because of the machine from where it originates. Wireless transmission is expensive, making it attractive to process data close to the source. The problems of multi-path fading, interference, and limited node lifetimes combine to make data paths, even under good conditions, transient and unreliable. The implications this has for network design should not be underestimated.

Similarly, signal-processing technologies need to be aware of transmission delays and the volume of data needed. The resources consumed by an implementation, and the latencies incurred, are an integral part of any sensing solution. In short, all levels of the networking protocols need to be considered as a single gestalt. Beyond this, system design requires new power-aware computing hardware and operating systems.

Embedded sensors are now part of large, loosely coupled networks with mobile code and data. Data supply and demand are stochastic and unpredictable. Processing occurs concurrently on multiple processors. Applications are in hostile environments, with noise-corrupted communication and failure-prone components. Under these conditions, efficient operation cannot rely on static plans. Van Creveld [1] has defined five characteristics that hierarchical systems need to adapt to this type of environment [2]:

- Decision thresholds fixed far down in the hierarchy.
- Self-contained units exist at a low level.
- Information circulates from the bottom up and the top down.
- Commanders actively seek data to supplement routine reports.
- Informal communications are necessary.

Organizations based on this approach have been successful in market economies, war, and law enforcement [3].

Data requests from a declarative programming language *pull* data by making requests. Sensors *push* information by providing sensor data. Requests and data follow paths of least resistance through intermediate nodes. Mobile data and code allow signal processing and automatic target recognition to be done during data transmission. Processing includes decomposition, compression, and fusion of sensor data. This is done robustly and efficiently by giving each node the ability to make limited local optimizations based on locally available information.

The sensor network consists of nodes integrating these abilities. Requests for information form flexible ad hoc virtual enterprises of nodes, allowing the network to adapt to and compensate for failures and congestion. Complex adaptive behavior for the whole network emerges from straightforward choices made by individual nodes.

The final system will be a virtual enterprise. Groups of components form flexible ad hoc confederations to deliver data in response to changing needs and resources. This is applicable to any sensing modality and can be implemented on any testbed containing networked processors, sensors, and embedded processors.

Factors needing to be considered for system optimization include

- Data format for transmission
- Paths taken through the network for requests and data
- Points where fusion should occur during transmission
- Processing to be performed by mobile code during data transit

The final issue to consider is adaptation to system state. Information can be exchanged in a number of formats. It is reasonable to compress data for transmission over slow channels. Transmission over noisy channels requires redundant data. As noise (lack of noise) is detected in a channel, error checking increases (decreases). The meta-protocol starts with pessimistic assumptions of channel quality and modifies the protocol dynamically. Modifications are based on information from normal operations. Extra traffic for monitoring status is to be avoided.

Sensor networks are fundamentally different from their predecessors. Some of these differences are a matter of degree:

- Number of nodes required
- Power constraint severity
- Proximity to hostile environment
- Timeliness requirements

In the final analysis, the most important differences are the fundamental differences in the way information technology is used. Computer networks are no longer sensitive devices being coddled in air-conditioned clean rooms. They are working in hostile conditions. Sensor networks respond to the needs of human users. How they respond, and their internal configurations, will be decided independent of human intervention.

Acknowledgments and Disclaimer

This research is sponsored by the Defense Advance Research Projects Agency (DARPA), and administered by the Army Research Office under Emergent Surveillance Plexus MURI Award No. DAAD19-01-1-0504. Any opinions, findings, and conclusions or recommendations expressed in this publication are those of the author and do not necessarily reflect the views of the sponsoring agencies.

References

1. van Creveld, M.L., *Command in War*, Harvard University Press, Cambridge, MA, 1985.
2. Czerwinski, T., *Coping with the Bounds: Speculations on Nonlinearity in Military Affairs*, National Defense University, Washington, DC, 1998.
3. Cebrowski, A.K. and Garsta, J.J., Network-centric warfare: Its origin and future, *Proceedings of the Naval Institute*, 124(1), 28–35, January 1998. http://www.usni.org/Proceedings/Articles98/PROcebrowski.htm

II

Distributed
Sensing and
Signal Processing

4 Digital Signal Processing Backgrounds *Yu Hen Hu* 41
Introduction • Discrete-Time System Theory • Frequency Representation and
the DFT • Digital Filters • Sampling, Decimation, Interpolation • Conclusion •
Appendix • References

5 Image Processing Background *Lynne L. Grewe and Ben Shahshahani* 59
Introduction • Motivation • Image Creation • Image Domains: Spatial,
Frequency • Point-Based Operations • Area-Based Operations • Noise
Removal • Feature Extraction • Registration, Calibration, and Fusion
Issues • Compression and Transmission: Impacts on a Distributed Sensor
Network • Image Resampling • Image Processing for Output • More Imaging in Sensor
Network Applications • References

6 Object Detection and Classification *Akbar M. Sayeed* 93
Introduction • A Signal Model for Sensor Measurements • Object Detection • Object
Classification • Conclusions • Acknowledgment • References

7 Parameter Estimation *David S. Friedlander* 111
Introduction • Self-Organization of the Network • Velocity and Position
Estimation • Moving Target Resolution • Target Classification Using
Semantic Information Fusion • Stationary Targets • Peaks for Different Sensor
Types • Acknowledgments • References

8 Target Tracking with Self-Organizing Distributed Sensors *Richard R. Brooks,
Christopher Griffin, David S. Friedlander, J.D. Koch, and İlker Özçelik* 131
Introduction • Related Work • Computation Environment • Intercluster Tracking
Framework • Local Parameter Estimation • Track Maintenance Alternatives • Tracking
Examples • Cellular Automata Model • Cellular Automata Results • Collaborative
Tracking Network • Dependability Analysis • Resource Parsimony • Multiple Target
Tracking • Conclusion • Acknowledgments • References

   9 **Collaborative Signal and Information Processing: An Information-Directed
      Approach** *Feng Zhao, Jie Liu, Juan Liu, Leonidas Guibas, and James Reich*.................. 191
      Sensor Network Applications, Constraints, and Challenges • Tracking
      as a Canonical Problem for CSIP • Information-Driven Sensor Query:
      A CSIP Approach to Target Tracking • Combinatorial Tracking
      Problems • Discussion • Conclusion • Acknowledgments • References

  10 **Environmental Effects** *David C. Swanson*..207
      Introduction • Sensor Statistical Confidence Metrics • Atmospheric
      Dynamics • Propagation of Sound Waves • References

  11 **Detecting and Counteracting Atmospheric Effects** *Lynne L. Grewe* 219
      Motivation: The Problem • Sensor-Specific Issues • Physics-Based Solutions • Heuristics
      and Non-Physics-Based Solutions • Conclusion • References

  12 **Signal Processing and Propagation for Aeroacoustic Sensor
      Networks** *Richard J. Kozick, Brian M. Sadler, and D. Keith Wilson*...............................245
      Introduction • Models for Source Signals and Propagation • Signal Processing •
      Concluding Remarks • Acknowledgments • References

  13 **Distributed Multi-Target Detection in Sensor Networks** *Xiaoling Wang,
      Hairong Qi, and Steve Beck*... 291
      Blind Source Separation • Source Number Estimation • Distributed Source Number
      Estimation • Performance Evaluation • Conclusions • References

  14 **Symbolic Dynamic Filtering for Pattern Recognition in Distributed Sensor
      Networks** *Xin Jin, Shalabh Gupta, Kushal Mukherjee, and Asok Ray*............................307
      Introduction • Symbolic Dynamics and Encoding • Construction of Probabilistic
      Finite-State Automata for Feature Extraction • Pattern Classification Using SDF-
      Based Features • Validation I: Behavior Recognition of Mobile Robots in a Laboratory
      Environment • Validation II: Target Detection and Classification Using Seismic and PIR
      Sensors • Summary, Conclusions, and Future Work • Acknowledgments • References

  15 **Beamforming** *J.C. Chen and Kung Yao* 335
      Introduction • DOA Estimation and Source Localization • Array System Performance
      Analysis and Robust Design • Implementations of Two Wideband Beamforming
      Systems • References

This part discusses signal processing and sensor data interpretation issues of distributed sensor networks (DSNs). The chapters presented are tutorial in nature. Some chapters are overviews and surveys of important issues. Other chapters delve into recent advances in the field. Every effort has been made to make the material accessible to technically literate readers.

Hu provides a survey and review of digital signal processing in Chapter 4. This chapter describes what signals are and the most widely used transformations for time-series data. The discussion contains a number of practical examples.

Grewe and Shahshahani explain the basics of image processing in Chapter 5. This chapter was thoroughly revised for the second edition. While signal processing deals primarily with one-dimensional time series, image processing concentrates on two-dimensional images. These images are mainly in the visible wavelengths. They describe how imaging devices work and provide transformations that aid in extracting information from the images. Of particular interest is the discussion of image calibration, registration, and data association.

Sayeed discusses methods for object detection and classification in Chapter 6. Detection occurs when a signal of interest is located in a set of readings. Classification refers to differentiating between signals emanating from different classes of targets. For example, a tripwire sensor can usually detect the presence of a vehicle but is unable to differentiate between vehicle types. On the other hand, an acoustic sensor may be able to differentiate between tracked and wheeled vehicles. Sayeed develops a model describing sensor readings. He then provides statistical techniques for differentiating between signals belonging to classes of interest and background noise. In doing so, he concentrates on issues related to interpreting multiple sensor inputs.

Friedlander describes techniques that use distributed sensor inputs to estimate parameters that describe targets in Chapter 7. He provides a novel technique for estimating target heading and velocity. Results from a field test are given showing that this approach is very robust. Friedlander also explains how sensor networks can be used to count the number of targets present in the sensor field. This problem is deceptively difficult, and he derives limitations on the network's ability to perform this task.

Brooks et al. describe a distributed tracking approach in Chapter 8 that uses the parameter estimation technique described in Chapter 7. A clump head is chosen locally to estimate the heading and velocity of a target based on the local information. These data are propagated through the network to those nodes likely to see the target in the future. The chapter describes several techniques that have been tested for associating target detections to tracks. This chapter has been revised to include recent developments in distributed target tracking.

An alternative tracking approach is discussed by Zhao et al. in Chapter 9. Sensors are considered as providers of information to the system. Nodes evaluate their current beliefs and determine what information is needed to disambiguate their beliefs. For target counting, this approach leads to electing leaders for equivalence classes of nodes that all detect the same target.

Swanson discusses the problems faced by sensor nodes in the real world in Chapter 10. Environmental effects on sensor inputs are very difficult to account for. The effects of wind and weather on acoustic sensors make it very difficult to design robust algorithms that work dependably. The problem of calibrating seismic sensors to accommodate differences in soil and bedrock have made them very difficult to use in rapidly deployed systems. Swanson's discussion of the atmospheric issues that make the detection of chemical and biological agents challenging will be of interest to many readers.

Grewe's revised chapter (Chapter 11) for the second edition discusses how to counteract some of these environmental effects in imaging systems. She explains how atmospheric effects degrade the performance of imaging sensors. Results are given showing how the presence of fog in images can be detected and mitigated.

Kozick et al. discuss the use of narrow and wide band data sources in sensor networks in Chapter 12. They provide models of signals and their propagation. These models are used to derive methods for estimating the angle of arrival of a target and localizing targets.

Wang et al. discuss the problem of detecting multiple targets in a distributed network in Chapter 13. This chapter initially discusses methods for detecting the number of signal sources in readings. It then delves into the more complicated issue of combining these estimates in a hierarchical network. A Bayesian approach is proposed and field tested.

Jin et al. describe innovative techniques for translating continuous signal processing problems into symbolic reasoning applications in Chapter 14. This is a very innovative area of research. Symbolic abstractions are attractive for many reasoning tasks. They are particularly suited to detect asynchronous behaviors and tolerant to large volumes of noise. This is a very important and innovative contribution that is new in the second edition.

Chapter 15 discusses beamforming technology. As a sign of the importance of this technology, we have separated it from the other signal processing chapters. Beamforming uses signal processing techniques to infer information from multiple time signals. The individual time signals are collected from sensors located at different positions. The members of Yao's group at UCLA describe applications of beamforming, limitations of the approach, and how it can be used.

This part provides essential background on signal processing issues. Readers will find that the discussion progresses from tutorial information to in-depth discussion of research issues. A number of topics, such as signal detection, classification, and target tracking, are viewed from many different perspectives. These issues are one of the core aspects of sensor networks. Other aspects arise from these approaches being embedded in an unreliable distributed system.

4

Digital Signal Processing Backgrounds

4.1 Introduction 41
4.2 Discrete-Time System Theory...42
    Discrete Time Signal • Discrete-Time System • Impulse
    Response Characterization of a Linear and Shift (Time) Invariant
    Discrete-Time System
4.3 Frequency Representation and the DFT 44
    The z-Transform • Discrete-Time Fourier Transform • Frequency
    Response • The Discrete Fourier Transform • The Fast Fourier
    Transform
4.4 Digital Filters..47
    Frequency Response of Digital Filters • Structures of Digital
    Filters • Example: Baseline Wander Removal
4.5 Sampling, Decimation, Interpolation...50
    Sampling Continuous Analog Signal • Sampling Rate Conversion
4.6 Conclusion 54
4.A Appendix A..54
4.B Appendix B 55
4.C Appendix C..56
4.D Appendix D..57
References...58

Yu Hen Hu
*University of
Wisconsin-Madison*

4.1 Introduction

The purpose of this chapter is to review the fundamentals of deterministic digital signal processing techniques, with specific attention to wireless distributed sensor network applications. Signal processing concerns the acquisition, filtering, transformation, estimation, detection, compression, and recognition of signals represented in multiple media, multiple modalities, including sound, speech, image, video, and others. In *digital* signal processing, a natural or synthetic signal is first sampled and quantized using an analog-to-digital (A/D) converter. The result is a stream of finite-precision numbers that can be processed using a digital computer. The results may then be converted back to a continuous-time analog form using a digital-to-analog (D/A) converter.

Examples of natural signals in sensor network applications include temperature, humidity, wind speed, density of chemical agents, gas, solvent, sound, voice, image and video of targets, gesture, facial expressions, and many others. Essentially, every type of sensor will produce a stream of signals in the form of time-varying electrical voltage or current waveforms. Many sensor nodes also have built-in A/D converters sampling the signal at a prespecified sampling rate.

After sampling, the sampled signals need to be preprocessed before applying additional collaborative signal-processing algorithms. The preprocessing step may involve digital filtering, sampling rate conversion, discrete Fourier transform (DFT) and other deterministic digital signal processing operations.

The preprocessed digital signal may then be subject to further on-board signal processing so that the amount of data that needs to be transmitted through a wireless channel can be reduced. This will not only reduce the network congestion and hence improve communication efficiency, but also conserves the rate of energy consumption at individual sensor nodes. Deterministic signal processing tasks performed at this stage may involve discrete cosine transform, discrete wavelet transform (for data compression), and digital filtering.

In the rest of this chapter, the following topics will be introduced. First, the basic discrete-time system theory will be reviewed including basic definitions and properties. Next, the frequency domain representation of discrete time signals, the DFT, and fast Fourier transform (FFT) implementation will be reviewed. The topic of digital filters will then be discussed. This is followed by the introduction of a sampling theorem, decimation and interpolation techniques.

4.2 Discrete-Time System Theory

4.2.1 Discrete Time Signal

A discrete time signal, denoted $x[n]$, is a sequence of real-valued or complex-valued finite-precision numbers that are sampled from a continuous-time signal $x(t)$ at a regular interval T such that $x[n] = x(nT)$. For convenience, the time indices n may be shifted during processing so that the sequence may contain negative indices. The indices of a sequence n may range from $-\infty$ to $+\infty$. Hence, mathematically, a sequence can have infinite length. In reality, only sequences with finite length can be processed.

Some examples of basic sequences are listed in Table 4.1, and several properties of a sequence are introduced in Table 4.2 without extensive discussion. For more details, see Mitra [1] and Oppenheim and Schafer [2].

TABLE 4.1 Examples of Discrete-Time Sequences

Type	Definition
Unit sample (impulse) sequence	$\delta[n] = \begin{cases} 1, & n = 0 \\ 0, & n \neq 0 \end{cases}$
Unit step sequence	$u[n] = \begin{cases} 1, & n \geq 0 \\ 0, & n < 0 \end{cases}$
Sinusoidal sequence	$x[n] = A \cos(\omega_0 n + \phi)$
Exponential sequence	$x[n] = A\alpha^n$

TABLE 4.2 Properties of Discrete-Time Sequences

Property	Formula (For All n)
Conjugate-symmetric, even	$x[n] = x^*[-n]$
Conjugate-antisymmetric, odd	$x[n] = -x^*[-n]$
Periodic with period N	$x[n] = x[n + kN]$; k: integer
Energy of a sequence $x[n]$	$E = \displaystyle\sum_{n=-\infty}^{\infty} |x[n]|^2$
Bounded sequence	$|x[n]| \leq B_x < \infty$
Square-summable sequence	$\displaystyle\sum_{n=-\infty}^{\infty} |x[n]|^2 < \infty$

TABLE 4.3   Properties of Discrete-Time System

Type	Definition
Linear	Let a and b be two constants. The system is a linear system if and only if $x[n] = ax_1[n] + bx_2$ implies $y[n] = ay_1[n] + by_2[n]$
Shift (time) invariant	Let n_0 be a fixed integer. The system is shift invariant if and only if $x[n] = x_1[n - n_0]$ implies $y[n] = y_1[n - n_0]$
Causal	Let n_0 be a fixed integer. The system is casual if and only if $x_1[n] = x_2[n]$ for $n < n_0$ implies $y_1[n] = y_2[n]$ for $n < n_0$
Stability	Let A and B be two appropriately chosen constants. A system is bounded-input, bounded-output (BIBO) stable if and only if $x[n] < A \ \forall n$ implies $y[n] < B \ \forall n$

4.2.2 Discrete-Time System

A *system* maps sequences to sequences. A single-input–single-output (SISO) discrete-time system is a *mapping* from an input sequence $x[n]$, to an output sequence, also known as a *response*, $y[n]$. Therefore, it is possible that a particular output value $y[n_1]$ depends not only on corresponding input value $x[n_1]$, but also on other input values $x[n]$, $n \neq n_1$. Some examples of discrete-time systems are

$$y[n] = \sum_{k=0}^{\infty} x[n-k] = \sum_{m=-\infty}^{n} x[m]$$

$$y[n] = 3x[n-1] + 2x[n+1]$$

Some important properties of discrete-time systems are summarized in Table 4.3. For convenience, assume that $y_1[n]$ and $y_2[n]$ are the responses corresponding to the input sequences $x_1[n]$ and $x_2[n]$ respectively.

4.2.3 Impulse Response Characterization of a Linear and Shift (Time) Invariant Discrete-Time System

If a system is both linear and shift (time) invariant, it is called an LTI system. An LTI system can be uniquely characterized by its *impulse response* $h[n]$, defined as the response of the system when the input is an impulse sequence. In other words, given the impulse response $h[n]$ and an input sequence $x[n]$, the corresponding response of an LTI system, denoted by $y[n]$, can be found through the following convolution operation:

$$y[n] = \sum_{k=-\infty}^{\infty} x[k]h[n-k] = \sum_{k=-\infty}^{\infty} h[k]x[n-k]$$

If the impulse response sequence has finite length, i.e.

$$h[n] = 0 \quad \text{for } n < 0 \text{ and } n > N$$

then it is called *a finite* impulse response (FIR). Otherwise, it is an *infinite* impulse response (IIR).

4.3 Frequency Representation and the DFT

4.3.1 The z-Transform

The z-transform is a complex polynomial representation of a discrete time sequence. Given $x[n]$, its z-transform is defined as

$$X(z) = \sum_{n=-\infty}^{\infty} x[n]z^{-n} \tag{4.1}$$

where z is a complex variable over the complex z-plane. The region of convergence (ROC) of $X(z)$ is the region in the z-plane where $|X(z)|$ is finite. The z-transform pairs of some popular sequences are listed in Table 4.4.

Among the many interesting properties of the z-transform, perhaps the most useful is the convolution property that states if $y[n]$ is the results of the convolution of two sequences $x[n]$ and $h[n]$, and $Y(z), X(z), H(z)$, respectively are their corresponding z-transforms, then

$$Y(z) = X(z)H(z) \tag{4.2}$$

If the input sequence is a unit-sample (impulse) sequence, then $X(z) = 1$ according to Table 4.4. Hence, $Y(z) = H(z)$. Since $H(z)$ is the z-transform of the impulse response $h[n]$, it is called the *system function* or the *transfer function* of the underlying system.

The z-transform representation is useful in practical signal-processing applications in several ways. Firstly, for finite length sequence, its z-transform is a polynomial with a finite number of terms. As shown in Equation 4.2, the convolution of the two sequences can be easily obtained by multiplying the corresponding polynomials. Secondly, for a broad class of LTI systems, their transfer functions can be represented well with a quotient of two polynomials, such as those shown in Table 4.4. When an LTI system is represented in such an expression, it is possible to solve its time response analytically, and to analyze its behavior in great detail. Let us assume $A(z)$ and $B(z)$ are two finite polynomials such that

$$H(z) = \frac{A(z)}{B(z)} = K\frac{\prod_{k=1}^{P}(z - z_k)}{\prod_{l=1}^{Q}(z - p_l)} \tag{4.3}$$

where

$\{z_k; 1 \geq k \leq P\}$ are called *zeros*
$\{p_i; \leq l \leq Q\}$ are called *poles* of the transfer function $H(z)$

An LTI system is *stable* if all the poles of its transfer function locate within the unit circle $\{z; |z| = 1\}$ in the z-plane.

TABLE 4.4 The z-Transform Pairs Some Common Sequences

Sequence	z-Transform	ROC
$\delta[n]$	1	Entire z-plane
$u[n]$	$1/(1 - z^{-1})$	$|z| > 1$
$a^n u[n]$	$1/(1 - az^{-1})$	$|z| > |a|$
$r^n \cos(\omega_0 n)u[n]$	$\dfrac{1 - (r\cos\omega_0)z^{-1}}{1 - (2r\cos\omega_0)z^{-1} + r^2 z^{-2}}$	$|z| > |r|$
$r^n \sin(\omega_0 n)u[n]$	$\dfrac{1 - (r\sin\omega_0)z^{-1}}{1 - (2r\cos\omega_0)z^{-1} + r^2 z^{-2}}$	$|z| > |r|$

4.3.2 Discrete-Time Fourier Transform

The discrete-time Fourier transform (DTFT) pair of a sequence $x[n]$, denoted by $X(e^{j\omega})$, is defined as

$$X(e^{j\omega}) = \sum_{n=-\infty}^{\infty} x[n]e^{-j\omega n}$$

$$x[n] = \frac{1}{2\pi} \int_{-\pi}^{\pi} X(e^{j\omega})e^{j\omega n}d\omega$$

(4.4)

Note that $X(e^{j\omega})$ is a periodic function of ω with period equals to 2π. The DTFT and the z-transform are related such that

$$X(e^{j\omega}) = X(z)\big|_{z=e^{j\omega}}$$

4.3.3 Frequency Response

The DTFT of the impulse response $h[n]$ of an LTI system is called the *frequency response* of that system, and is defined as

$$H(e^{j\omega}) = \mathrm{DTFT}\{h[n]\} = H(z)\big|_{z=e^{j\omega}} \left|H(e^{j\omega})\right|e^{j\theta(\omega)}$$

where
$\left|H(e^{j\omega})\right|$ is called the *magnitude response*
$\theta(\omega) = \arg\{H(e^{j\omega})\}$ is called the *phase response*

If $\{h[n]\}$ is a real-valued impulse response sequence, then its magnitude response is an even function of ω and its phase response is an odd function of ω. The derivative of the phase response with respect to frequency ω is called the *group delay*. If the group delay is a constant at almost all ω, then the system is said to have *linear phase*. If an LTI system has unity magnitude response and is linear phase, then its output sequence will be a delayed version of its input without distortion.

4.3.4 The Discrete Fourier Transform

For real-world applications, only finite-length sequences are involved. In these cases, the DFT is often used in lieu of the DTFT. Given a finite-length sequence $\{x[n]; 0 \leq n \leq N-1\}$, the DFT and inverse DFT (IDFT) are defined thus:

$$\mathrm{DFT:}\quad X[k] = \sum_{n=0}^{N-1} x[n]\exp\left(-\frac{j2\pi kn}{N}\right) = \sum_{n=0}^{N-1} x[n]W_N^{kn} \quad 0 \leq k \leq N-1$$

$$\mathrm{IDFT:}\quad x[n] = \frac{1}{N}\sum_{k=0}^{N-1} X[k]\exp\left(\frac{j2\pi kn}{N}\right) = \frac{1}{N}\sum_{k=0}^{N-1} X[k]W_N^{-kn} \quad 0 \leq n \leq N-1$$

(4.5)

where $W_N = e^{-j2\pi/N}$.

TABLE 4.5   Properties of DFT

Property	Length-N Sequence	N-Point DFT
	$x[n], y[n]$	$X[k], Y[k]$
Linearity	$ax[n] + by[n]$	$aX[K] + bY[k]$
Circular shift	$x[\langle n - n_0 \rangle_N]$	$W_N^{kn_0} X[k]$
Modulation	$W_N^{-k_0 n} x[n]$	$X[\langle k - k_0 \rangle_N]$
Convolution	$\sum_{m=0}^{N-1} x[m] y[\langle n - m \rangle_N]$	$X[k] Y[k]$
Multiplication	$X[n] y[n]$	$\dfrac{1}{N} \sum_{m=0}^{N-1} X[m] Y[\langle k - m \rangle_N]$

Note that $\{X[k]\}$ is a periodic sequence in that $X[k + mN] = X[k]$ for any integer m. Similarly, the $x[n]$ sequence obtained in Equation 4.5 is also periodic, in that $x[n + mN] = x[n]$. Some practically useful properties of the DFT are listed in Table 4.5, where the *circular shift operation* is defined as

$$x[\langle n - n_0 \rangle_N] = \begin{cases} x[n - n_0] & n_0 \leq n \leq N - 1 \\ x[n - n_0 + N] & 0 \leq n < n_0 \end{cases} \tag{4.6}$$

The convolution property is very important, in that the response of an LTI system can be conveniently computed using the DFT if both the input sequence $x[n]$ and the impulse response sequence $h[n]$ are finite-length sequences. This can be accomplished using the following algorithm:

Algorithm.   Compute the output of an FIR LTI system.
Given: $\{x[n]; 0 \leq n \leq M - 1\}$ and $\{h[n]; 0 \leq n \leq L - 1\}$:

1. Let $N = M + L - 1$. Pad zeros to both $x[n]$ and $h[n]$ so that they both have length N.
2. Compute the respective DFTs of these two zero-padded sequences, $X[k]$ and $H[k]$.
3. Compute $Y[k] = X[k]H[k]$ for $0 \leq k \leq N - 1$.
4. Compute $y[n] = \text{IDFT}\{Y[k]\}$.

There are some useful symmetry properties of the DFT that can be explored in practical applications. We focus on the case when $x[n]$ is a real-valued sequence. In this case, the following symmetric relations hold:

$$X[k] = X^*[N - 1 - k]$$

Therefore, one may deduce

$$\text{Re}\, X[k] = \text{Re}\, X[N - 1 - k]$$

$$\text{Im}\, X[k] = -\text{Im}\, X[N - 1 - k]$$

$$\left| X[k] \right| = \left| X[N - 1 - k] \right|$$

$$\arg X[k] = \arg X[N - 1 - k]$$

where

$$\arg X[k] = \tan^{-1} \frac{\text{Im}\, X[k]}{\text{Re}\, X[k]} \in [-\pi \quad \pi]$$

4.3.5 The Fast Fourier Transform

The FFT is a computation algorithm that computes the DFT efficiently. The detailed derivation of the FFT is beyond the scope of this chapter. Readers who are interested in knowing more details about the FFT are referred to several excellent textbooks, e.g. [1–3].

4.4 Digital Filters

Digital filters are LTI systems designed to modify the frequency content of the input digital signal. These systems can be SISO systems or multiple-inputs–multiple-outputs (MIMO) systems.

4.4.1 Frequency Response of Digital Filters

Depending on the application, the frequency response of a digital filter can be characterized as all pass, band-pass, band-stop, high-pass, and low-pass. They describe which frequency band of the input sequence is allowed to pass through the filter while the remaining input signals are filtered out. All pass filters are often implemented as a MIMO system such that the input sequence is decomposed into complementary frequency bands. These components can be combined to perfectly reconstruct the original signal with a fixed delay. Hence, the overall system passes all the frequency bands and, therefore, the term *all pass*. With a MIMO system, the digital filter has a *filter bank* structure that can be exploited to implement various linear transformations, including DFT and discrete wavelet transform (DWT). Digital filter banks have found wide acceptance for applications such as data compression, multi-resolution signal processing, and orthogonal frequency division multiplexing. Low-pass filters are perhaps the most commonly encountered digital filter. They have found applications in removing high-frequency noise, extracting the low-frequency trend, and preventing alias before decimation of a digital sequence. High-pass filters are used for exposing the high-frequency content of a potential signal. They can be used for event detection. A band-stop filter will filter out unwanted interference from a frequency band that does not significantly overlap with the desired signal. For example, in a special band-stop filter, known as the notch-filter, it is possible to reject 60 Hz power-line noise without affecting the broadband signal. Band-pass digital filters are designed to pass a narrow-band signal while rejecting broadband background noise. Table 4.6 illustrates the magnitudes of that frequency responses of four types of digital filter. The corresponding transfer functions are also listed with $a = 0.8$ and $b = 0.5$. The constants are to ensure that the maximum magnitudes of the frequency responses are equal to unity. The MATLAB® program that generates these plots is given in Appendix 4.A.

4.4.2 Structures of Digital Filters

Based on whether the impulse response sequence is of finite length, digital filter structures can be categorized into FIR filters and IIR filters. An FIR filter has several desirable characteristics:

1. It can be designed to have exact an linear phase.
2. It can easily be implemented to ensure the BIBO stability of the system.
3. Its structure is less sensitive to quantization noise than an IIR filter.

In addition, numerous computer-aided design tools are available to design an arbitrarily specified FIR filter with relative ease.

Compared with the FIR filter, an IIR filter often requires fewer computation operations per input sample, and hence would potentially consumes less power on computing. However, the stability of an IIR filter is prone to accumulation of quantization noise, and linear phase usually cannot be guaranteed.

TABLE 4.6 Examples of Digital Filter Frequency Responses

Filter Type	Transfer Function	Magnitude Frequency Response Plot
Low pass	$\dfrac{1-a}{2}\dfrac{1+z^{-1}}{1-az^{-1}},\quad a=0.8$	(a)
High pass	$\dfrac{1+a}{2}\dfrac{1-z^{-1}}{1-az^{-1}},\quad a=0.8$	(b)
Band pass	$\dfrac{1-a}{2}\dfrac{1+z^{-2}}{1-b(1+a)z^{-1}+az^{-2}},\quad a=0.8, b=0.5$	(c)
Band stop	$\dfrac{1-a}{2}\dfrac{1-z^{-2}}{1-b(1+a)z^{-1}+az^{-2}},\quad a=0.8, b=0.5$	(d)

4.4.3 Example: Baseline Wander Removal

Consider a digital signal sequence $\{x[n]; 0 \leq n \leq 511\}$ as plotted in Figure 4.1a by the dotted line. A low-pass FIR filter is designed to have the impulse response shown in Figure 4.1b. This FIR filter is designed to use the Hamming window with $2L + 1$ non-zero impulse response components. In this example, $L = 20$, and $2L + 1 = 41$. It has a default normalized cut-off frequency $10/m$, where m is the length of the sequence (512 here). At the cut-off frequency, the magnitude response of the FIR filter is half of that at the zero frequency. The magnitude frequency response of the FIR filter, represented in decibel (dB) format, is shown in Figure 4.2a. In general, it contains a main lobe and several side lobes. The output of the FIR filter is the baseline signal shown by the solid line in Figure 4.1a. The longer the filter length (i.e. larger value of L), the narrower the main lobe is and the smoother the filtered output (baseline) will be. The differences between these two sequences are depicted in Figure 4.1c. The frequency response of the original sequence is shown in Figure 4.2b. The frequency response of the baseline, i.e. the output of this low-pass filter, is shown in Figure 4.2c. Figure 4.2b and c, uses log-magnitudes. The MATLAB program that generates Figures 4.1 and 4.2 is listed in Appendix 4.B.

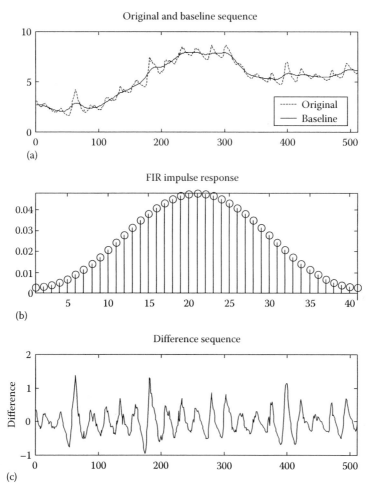

FIGURE 4.1 Baseline wander removal; (a) original time sequence (dotted line) and baseline wander (solid line), (b) low-pass filter impulse response, and (c) digital sequence with baseline wander removed.

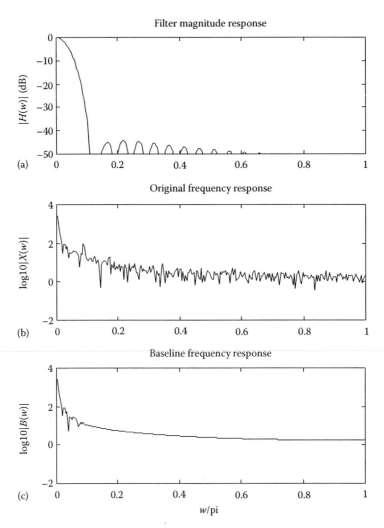

FIGURE 4.2 (a) Filter frequency response, (b) frequency representation of original sequence, (c) frequency representation of filtered baseline sequence.

4.5 Sampling, Decimation, Interpolation

4.5.1 Sampling Continuous Analog Signal

An important question in sensor application is how to set the sampling rate. Denote $x(t)$ to be a continuous time signal. When an A/D converter performs sampling, every T seconds, the quantized value of $x(nT) = x(n)$ is obtained. The question then is how much information is lost during this sampling process, and how important is the lost information in terms of reconstructing $x(t)$ from $x(nT)$. Suppose that the Fourier transform of $x(t)$

$$X(f) = \int_{-\infty}^{\infty} x(t) e^{-j\omega t} dt \tag{4.7}$$

is such that $|X(f)| = 0$ for $|f| < f_0$, and that the sampling frequency $f_s = 1/T \geq 2f_0$; then, according to the classical Shannon sampling theorem, $x(t)$ can be recovered completely through the interpolation formula

$$\hat{x}(t) = \sum_{n=-\infty}^{\infty} x(n)\frac{\sin[\pi(t-nT)/T]}{\pi(t-nT)/T} \sum_{n=-\infty}^{\infty} x(n)\text{sinc}\left(\frac{t-nT}{T}\right) \qquad (4.8)$$

where

$$\text{sinc}(t) = \frac{\sin \pi t}{\pi t}$$

In other words, if $x(t)$ is *band limited* with bandwidth f_0, then it can be recovered exactly using Equation 4.8 provided that the sampling frequency is at least twice that of f_0. $f_s = 2f_0$ is known as the Nyquist sampling rate. If the sampling rate is lower than the Nyquist rate, then a phenomenon known as *aliasing* will occur.

Two examples are depicted in Figures 4.3 and 4.4. The first example, in Figure 4.3, shows a continuous time signal

$$x(t) = \cos(2\pi f_0 t) \quad x \in [0 \quad 1]$$

with $f_0 = 2\,\text{Hz}$ to be sampled at a sampling rate of $f_s = 7\,\text{Hz}$. The waveform of $x(t)$ is shown as the solid line in the figure. Sampled values are shown in circles. Then, Equation 4.8 is applied to estimate $x(t)$, and the estimated waveform is shown as the dotted line in the same figure. Note that the solid line and the dotted line do not completely match. This is because $x(t)$ is truncated to the time interval [0 1]. In the second example, in to Figure 4.4, the same $x(t)$ is sampled at a rate that is lower than the Nyquist rate. As a result, the reconstructed curve (dotted line) exhibits a frequency that is lower than the original signal (solid line). Also note that both lines pass through every sampling point.

In practical applications, the bandwidth f_0 of the signal $x(t)$ can sometimes be estimated roughly based on the underlying physical process that generates $x(t)$. It may also depend on which physical phenomenon is to be monitored by the sensor, as well as on the sensor capability and power consumptions. Experiments may be employed to help determine the minimum sampling rate required. One may

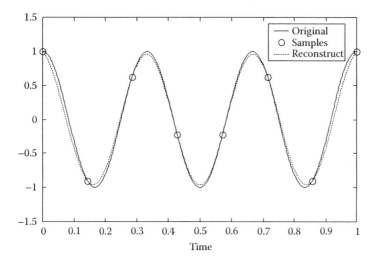

FIGURE 4.3 Sampling theory demonstration: $f_0 = 2\,\text{Hz}$, $f_s = 7\,\text{Hz}$. Solid line is the original signal $x(t)$, circles are sampled data, dotted line is reconstructed signal using Equation 4.8. The mismatch is due to the truncation of $x(t)$ to the interval [0 1].

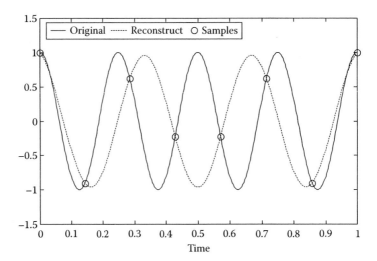

FIGURE 4.4 Illustration of aliasing effect. $f_0 = 4\,Hz$, $f_s = 7\,Hz$. Clearly, the reconstructed signal (dotted line) has a lower frequency than the original signal (solid line). Also note that both lines pass through every sampling point. The MATLAB® program that generates Figures 4.3 and 4.4 is listed in Appendix 4.C.

initially use the highest sampling rate available; then, by analyzing the frequency spectrum of the time series, it is possible to determine the best sampling frequency.

4.5.2 Sampling Rate Conversion

After a digital signal is sampled, the sampling rate may need to be changed to meet the needs of subsequent processing. For example, sensors of different modalities often require different sampling rates. However, later, during processing, one may want to compare signals of different modality at the same time scale. This would require a change of sampling rate. When the original sampling rate is an integer multiple of the new sampling rate the process is called *down-sampling*. Conversely, when the new sampling rate is an integer multiple of the original sampling rate, the process is called *up-sampling*.

4.5.2.1 Down-Sampling (Decimation)

For a factor of M down-sampling, the DTFT of the resulting signal, $Y(e^{j\omega})$, is related to the DTFT of the original digital signal $X(e^{j\omega})$ by the following expression:

$$Y(e^{j\omega}) = \frac{1}{M}\sum_{k=0}^{M-1}X(e^{j(\omega-2\pi k)/M}) \tag{4.9}$$

If the bandwidth of $X(e^{j\omega})$ is more than $2\pi/M$, then aliasing will occur.

Consider the example depicted in Figure 4.5a: a signal $x(t)$ is sampled at an interval of $T = 1\,min$ per sample and 100 samples are obtained over a period of 100 min. The sampled signal is denoted by $x(n)$. The magnitude of the DFT of $x(n)$, denoted by $|X(k)|$, is plotted in Figure 4.5b. Since $|X(k)| = |X(N-k)|$ ($N = 100$), only the first $N/2 = 50$ elements are plotted. Note that $|X(k)|$ is a periodic sequence with period f_s, which is equivalent to a normalized frequency of 2π. Hence, the frequency increment between $|X(k)|$ and $|X(k+1)|$ is f_s/N, and the x-axis range is $[0\ f_s/2]$. Note the two peaks at $k = 9$ and 10, representing two harmonic components of period $N/(9f_s) = 11.1\,min$ per sample and $N/(10f_s) = 10\,min$ per sample. This roughly coincides with the waveform of $x(n)$ shown in Figure 4.5a, where a 10 min cycle is clearly visible.

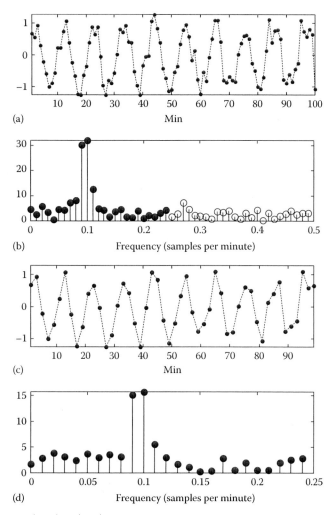

FIGURE 4.5 (a) $x(n)$, (b) $|X(k)|$, (c) $|Y(\ell)|$, and (d) $y(m)$.

Next, we consider a new sequence by sub-sampling $x(n)$ using a 2:1 ratio. Let us denote this new sequence $y(m) = x(2m + 1)$, $0 \le m \le 49$. This is depicted in Figure 4.5c. Note that the sampling period of $y(m)$ is 2 min per sample. Hence, the time duration of $y(m)$ is still 100 min. Also, note that the sampling frequency for $y(m)$ is now $1/2 = 0.5$ samples per minute. Since $y(m)$ has only 50 samples, there are only 50 harmonic components in its DFT magnitudes $|Y(\ell)|$. These harmonic components spread over the range of $[0\ 2\pi]$, which represents a frequency range of $[0\ 0.5]$ samples per minute. Since we plot only the first 25 of these harmonic components in Figure 4.5d, the frequency range of the x-axis is $[0\ 0.25]$ samples per minute. Comparing Figure 4.5d and b, the $|Y(\ell)|$ has a shape that is similar to the first 25 harmonics of $|X(k)|$. In reality, they are related as

$$Y(\ell) = \frac{1}{2}\left[X(\ell) + X(\langle \ell - N/2\rangle_N)\right] = \begin{cases} \dfrac{[X(\ell) + X(\ell - N/2)]}{2} & \ell \ge N/2 \\ \dfrac{[X(\ell) + X(\ell + N/2)]}{2} & 0 \le \ell < N/2 \end{cases} \tag{4.10}$$

In this example, the two major harmonic components in $|X(k)|$ have changed very little since they are much larger than other harmonics $\{|X(k)|; N/2 \le k \le N - 1\}$ shown in hollow circles in Figure 4.5b.

As such, if the sampling rate of this sensor is reduced to half of its original sampling rate, then it will have little effect of identifying the feature of the underlying signal, namely the two major harmonic components. The MATLAB program that generates Figure 4.5 is listed in Appendix 4.D.

4.5.2.2 Up-Sampling (Interpolation)

With an *L*-fold up-sampling, a new sequence $x_u(n)$ is constructed from the original digital signal $x(n)$ such that

$$x_u(m) = \begin{cases} x(n) & m = Ln \\ 0 & \text{otherwise} \end{cases}$$

It is easy to verify that

$$X_u(z) = X(z^L)$$

Hence

$$X_u(e^{j\omega}) = X(e^{jL\omega})$$

and

$$X_u(\ell) = X(\ell \bmod N)$$

However, the zeros in the $x_u(n)$ sequence must be interpolated with more appropriate value in real applications. This can be accomplished by low-pass filtering the $x_u(n)$ sequence so that only one copy of the frequency response $X(k)$ remains.

4.6 Conclusion

In this chapter, we briefly introduced the basic techniques for the processing of deterministic digital sensor signals. Specifically, the methods of frequency spectrum representation, digital filtering, and sampling are discussed in some detail. However, owing to space limitation, mathematical derivations are omitted. Readers interested in further reading on these topics should check out the many text books concerning digital signal processing, e.g. the three referred to in preparing this chapter.

4.A Appendix A

MATLAB® M-file that generates the plots in Table 4.6.

```
% examples of different frequency responses of digital filters
% (C) 2003 by Yu Hen Hu
% examples are taken from Digital Signal Processing, A computer based
% approach, 2nd ed. by S. K. Mitra, McGraw-Hill Irwin, 2001
%
clear all
w = [0:pi/255:pi]; % frequency domain axis
a =0.8;
b =0.5;
% Low pass IIR example
```

```
% H1(z)=0.5*(1-a)(1+ẑ-1)/(1-a*ẑ-1)
H1 = freqz(0.5*(1-a)*[1 1],[1 -a],w);
figure(1),clf
plot(w/pi,abs(H1)), ylabel('|H(w)|'), xlabel('w/pi')
title('(a) Low pass filter')
axis([0 10 1])
% High pass IIR
%H2(z)=0.5*(1+a)*(1-ẑ-1)/(1-a*ẑ-1)
H2=freqz(0.5*(1+a)*[1-1],[1-a],w);
figure(2),clf
plot(w/pi,abs(H2)), ylabel('|H(w)|'),xlabel('w/pi')
title('(b) High pass filter')
axis([0 10 1])
% Band pass IIR
% H3(z)=0.5*(1-a)*(1-ẑ-2)/(1-b*(1+a)*ẑ-1+a*ẑ-2)
H3 = freqz(0.5*(1-a)*[1 0-1],[1-b*(1+a) a],w);
figure(3), clf
plot(w/pi,abs(H3)), ylabel('|H(w)|'), xlabel('w/pi')
title('(c) Band pass filter')
axis([0 1 0 1])
% Band stop IIR
%H4(z)=0.5*(1+a)*(1-2*b*ẑ-1+ẑ-2)/(1-b*(1+a)*ẑ-1+a*ẑ-2)
H3 = freqz(0.5*(1+a)*[1-2*b1],[1-b*(1+a)a],w);
figure (4),clf
plot(w/pi,abs(H3)), ylabel('|H(w)|'),xlabel('w/pi')
title('(d) Band stop filter')
axis([0 1 0 1])
```

4.B Appendix B

Baseline Wander Removal Example Driver and Subroutine that produces Figures 4.1 and 4.2. Save trendrmv.m into a separate file.

```
% baseline wander removal example
% (C) 2003 by Yu Hen Hu
% call trendrmv.m
% which require signal processing toolbox routines fir1.m, filter.m
% input sequence gt is stored in file try.mat
clear all
load try; % load input sequence variable name gt m x 1 vector [m,m1] =
  size(gt);
wcutoff = 10/m; % 3dB cut off frequency is set to about 10/512 here
  L=input('filter length=2L+1, L=');
[y,ylow,b] = trendrmv(gt,L,wcutoff);
figure(1),clf
subplot(311),plot([1:m],gt, ':',[1:m],ylow, '-'),
legend('original', 'baseline')
title('(a) original and baseline sequence')
axis([0m 0 10])
subplot(312),stem([1:2*L+1],b),title('FIR impulse response')
axis([1 2*L+1 floor(min(b)) max(b)])
subplot(313),plot([1:m],y),ylabel('difference')
title('(c) difference sequence')
axis([0m floor(min(y)) ceil(max(y))])
w = [0:pi/255:pi];
```

```
Hz = freqz(b,1,w);
figure(2),clf
subplot(311), plot(w/pi, 20*log10(abs(Hz))),
ylabel('|H(w)| (db)')
axis([0 1 -50 0]),title('(a) filter magnitude response')
fgt = abs(fft(gt));
m2 = m/2;
subplot(312),plot([1:m2]/m2,log10(fgt(1:m2))),
ylabel('log10|X(w)|')
axis([0 1 -2 4]),title('(b) original frequency response')
fylow=abs(fft(ylow));
subplot(313),plot([1:m2]/m2,log10(fylow(1:m2))),
ylabel('log10|B(w)|')
axis([0 1 -2 4]),xlabel('w/pi'),
title('(c) baseline frequency response')
function [y,ylow,b]=trendrmv(x,L,wcutoff)
% Usage: [y,ylow,b]=trendrmv(x,L,wcutoff)
% trend removal using a low pass, symmetric FIR filter
% x is nrecord x N matrix each column is to be low-pass filtered
% L: length of the filter (odd integer>1)
% wcutoff: normalized frequency, a postive fraction
% in terms of normalized frquency
% y: high pass results
% ylow: baseline wander
% b: low pass filter of length 2*L+1
% (C) 2003 by Yu Hen Hu
nrecord=size(x,1);
Npt = 2*L; % length of FIR filter
b = fir1(Npt,wcutoff); % low pass filter
% since we want to apply a 2L+1 filter (L = 25 here)
% to a sequence of length nrecord
% we need to perform symmetric extension on both ends with L point each
% since matlab filter.m will return an output of same length nrecord + 2L
% the output we want are 2L+1:nrecord+2L of the results
temp0=[flipud(x(1:L,:));x;flipud(x(nrecord-L+1:nrecord,:))];
% temp0 is nrecord + 2L by nsensor
temp1=filter(b,1,temp0); % temp1 is nrecord + 2L by nsensor
ylow=temp1(2*L+1:nrecord+2*L,:);
y=x-ylow;
```

4.C Appendix C

MATLAB® M-files demonstrating sampling of continuous time signal. Save sinc.m into a separate file.

```
% demonstration of sampling and aliasing
% (c) 2003 by Yu Hen Hu
% call sinc.m
clear all
np=300;
% 1. generate a sinusoid signal
f0 = input('Enter freuency in Hertz (cycles/second):');
tx = [0:np-1]/np; % 512 sampling point within a second, time axis
x = cos(2*pi*f0*tx); % cos(2pi f0t), original continuous function
% 2. Enter sampling frequency
fs=input('Enter sampling frequency in Hertz:');
```

```
T=1/fs; % sampling period
ts=[0:T:1]; % sampling points
xs=cos(2*pi*f0*ts); % x(n)
nts=length(ts);
% 3 computer reconstructed signal.
xhat=zeros(size(tx)); for i=1:nts,
xhat=xhat + xs(i)*sinc(pi*fs*tx,pi*(i-1));
end
% plot
figure (1),clf
plot(tx,x, 'b-',ts,xs, 'bo',tx,xhat, 'r:');axis([0 1-1.5 1.5])
legend('original', 'samples''reconstruct')
title(['f_0=' int2str(f0)' hz, f_s=' int2str(fs) 'hz.'])
function y=sinc(x,a)
% Usage: y=sinc(x,a)
% (C) 2003 by Yu Hen Hu
% y=sin(x-a)/(x-a)
% x: a vector
% a: a constant
% if x=a, y=1;
%
if nargin==1, a=0; end % default, no shift
n=length(x); % length of vector x
y=zeros(size(x));
idx=find(x==a); % sinc(0)=1 needs to be computed separately
if ~isempty(idx),
y(idx)=1;
sidx=setdiff([1:n],idx);
y(sidx)=sin(x (sidx) - a)./(x(sidx) - a); else
y=sin(x-a)./(x - a); end
```

4.D Appendix D

MATLAB® program to produce Figure 4.5.

```
% demonstration on reading frequency spectrum
% (C) 2003 by Yu Hen Hu
%
clear all
n=100;
f0 = 0.095; % samples/min
%load onem.mat; % variable y(1440,2)
tx=[1:n]'; % sampled at 1min/sample period
tmp=0.2*randn(n,1);
x=sin(2*pi*f0*tx + rand(size(tx))) +tmp;
fs=1; % 1 sample per minute
% spectrum of various length of the sequence
% (a) n point
n2=floor(n/2);
xs=abs(fft(x(:))); %
% dc component not plotted
figure(1), clf
subplot(411),plot([1:n],x(:),'g:',[1:n],x(:),'b.'),
axis([1n min(x(:)) max(x(:))])
xlabel('min')
```

```
ylabe l('(a) x(n)')
% (b) subsample 2:1
xc0=x(1:2:n); nc=length(xc0); tc=tx(1:2:n);
xsc=abs(fft(xc0)); nsc=floor(nc/2);
subplot(413),plot(tc,xc0, 'g:',tc,xc0, 'b.')
axis([1 max(tc) min(x(:)) max(x(:))])
xlabel('min')
ylabel('(c) Y(l)')
tt0=[0:nc-1]/(nc)*(fs/2);
% frequency axis 0 to 2pi, half of sampling frequency
tc=tt0(1:nsc); % plot the first half due to symmetry of mag(DFT)
ltt=length(tc);
subplot(414),stem(tc,xsc(1:nsc), 'filled')
xlabel('frequency (samples/min.)')
axis([0 0.25 0 max(xsc)])
ylabel('(d) Y(m)')
t2=[0:n2-1]/n2*fs/2;
subplot(412),stem(t2(1:nsc),xs(1:nsc), 'filled'),hold on
stem(t2(nsc + 1:n2),xs(nsc + 1:n2)),hold off,xlabel('hz')
axis([0 0.5 0 max(xs)])
xlabel('frequency (samples/min.)')
ylabel('(b) |X(k)|')
```

References

1. Mitra, S.K., *Digital Signal Processing: A Computer-Based Approach*, McGraw Hill, New York, 2001.
2. Oppenheim, A.V. and Schafer, R.W., *Digital Signal Processing*, Prentice-Hall, Englewood Cliffs, NJ, 1975.
3. Mitra, S.K. and Kaiser, J.F., *Handbook for Digital Signal Processing*, John Wiley & Sons, New York, 1993.

5

Image Processing Background

5.1 Introduction 59
5.2 Motivation..60
5.3 Image Creation..61
 Image Spectrum • Image Dimensionality • Image Sensor
 Components • Analog-to-Digital Images
5.4 Image Domains: Spatial, Frequency 67
5.5 Point-Based Operations 69
 Thresholding • Conversion • Contrast Stretching and
 Histogram Equalization • Inversion • Level Slicing • Bitplane
 Slicing • Image Subtraction • Image Averaging
5.6 Area-Based Operations...72
 Low-Pass Filter • High-Pass Filter • Median Filter • Edge
 Detection • Morphological Operators
5.7 Noise Removal...76
5.8 Feature Extraction 77
 Edges • Hough Transform: Detecting Shapes • Segmentation:
 Surfaces • Examples
5.9 Registration, Calibration, and Fusion Issues..81
 Registration • Geometric Transformations • Calibration •
 Fusion Issues
5.10 Compression and Transmission: Impacts on a Distributed
 Sensor Network..85
5.11 Image Resampling 86
5.12 Image Processing for Output 86
5.13 More Imaging in Sensor Network Applications 87
References...89

Lynne L. Grewe
California State University

Ben Shahshahani
Google

5.1 Introduction

Images whether from visible spectrum, photometric cameras, or other sensors are often a key and primary source of data in distributed sensor networks. As such it is important to understand images and to effectively manipulate them. The common use of images in distributed sensor networks may be because of our own heavy reliance on images as human beings. As the old adage says: "an image is worth a thousand words." Prominence of images in sensor networks may also be due to the fact that there are many kinds of images not just those formed in the visual spectrum but also, infrared, multispectral, and sonar images. This chapter will give the reader an understanding of images from creation through manipulation and present applications of images and image processing in sensor networks.

Image processing is simply defined as the manipulation or processing of images. The goal of processing the image is first dependent on how it is used in a sensor network as well as the objectives of the network. For example, images may be used as informational data sources in the network or could be used to calibrate or monitor network progress. Image processing can take place at many stages. These stages are commonly called preprocessing, feature or information processing, and postprocessing. Another important image processing issue is that of compression and transmission. In distributed sensor networks this is particularly important, as images tend to be large in size causing heavy storage and transmission burdens.

We begin by providing motivation through a few examples of how images are used in sensor network applications. Our technical discussions begin with the topic of image creation; this is followed by a discussion of preprocessing and noise removal. The next sections concentrate on mid-level processing routines and a discussion of the spatial and frequency domain for images. Feature extraction is discussed and then the image processing related issues of registration and calibration are presented. Finally, compression and transmission of images are discussed. Our discussion comes full circle when we discuss some further examples of networks in which images play an integral part.

5.2 Motivation

Before beginning our discussion of image processing, let us look at a few examples of sensor networks where images take a prominent role. This is meant as motivational and we will discuss appropriate details of each system in the remaining sections of this chapter.

In [Ros 09], a person detection system that uses both visual and infrared (IR) images is discussed. In this system, the potential person is detected separately in each image first using image processing techniques starting with the simple stage of background subtraction. Results for person detection for this system are shown in Figure 5.1. After this separate person detection, a data fusion stage takes place using Bayesian occupancy filters.

In [Foresti 02], a distributed sensor network is described that uses images, both visible spectrum and IR to do outdoor surveillance. These images are the only source of informational data to find moving targets in the outdoor environment the sensors are monitoring. The use of both kinds of images allows the system to track heat-providing objects, such as humans and other animals in both day and night conditions. See Figure 5.2 for some images from the system.

A vehicle-mounted sensor network is discussed in [Bhatia 00] that uses various kinds of images for the task of mine detection in outdoor environments. Here, images play a key role in providing the information to detect the mines in the scene. In particular, IR, ground-penetrating radar (GPR), and metal detection images are used.

In [Verma 01], multiple cameras and laser-range finders are distributed in the work environment used by a robotic system to help with goal-oriented tasks that involve object detection, obstacle avoidance, and navigation. Figure 5.3 shows an example environment for this system.

Another of the many robotic systems that use multiple sensors, in this case a photometric camera and laser-range finder, is [Luo 09] system. The goal for this robot is human tracking and the human is detected separately using the image and laser-range finder data and then results are fused.

[Marcenaro 02] describes another surveillance type application that uses images as the only data source. Here, static and mobile cameras are used for gathering data as shown in Figure 5.4.

These are just a few of the many sensor networks that use images. We will revisit these and others in the remaining chapters as we learn about images and how to effectively manipulate images to achieve the goals of a network system.

FIGURE 5.1 System that performs person detection using visual and IR images: (a) visual image; (b) background subtraction mask for visual image; (c) IR image; and (d) background subtraction mask for IR image. (From Ros, J. and Mekhnacha, K., Multi-sensor human tracking with the Bayesian occupancy filter, *International Conference on Digital Signal Processing*, Santorini, Greece, 2009.)

5.3 Image Creation

Effective use of images in a sensor network requires understanding of how images are created. There are many "kinds" of images employed in networks. They differ in that different sensors can be used, which record fundamentally different information. Sensors can be classified by the spectrum of energy (light) they operate in, dimensionality of data produced, and whether they are active or passive.

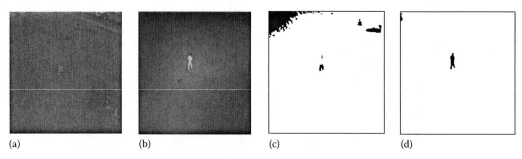

(a)   (b)   (c)   (d)

FIGURE 5.2 Describes a system for outdoor surveillance: (a) original image, (b) original IR image, (c) image (a) after processing to detect blobs, and (d) image (b) after processing to detect blobs. (From Foresti, G. and Snidaro, L., A distributed sensor network for video surveillance of for video of outdoor environments, *International Conference on Image Processing*, pp. 525–528, New York, 2002.)

FIGURE 5.3 A distributed sensor environment in which a robot system navigates to attempt to push a box from one location to another. Both cameras and laser-range finders are used. (From Verma, A. et al. Computational intelligence in robotics and automation. *Proceedings 2001 IEEE International Symposium on Computational Intelligence in Robotics and Automation*, pp. 212–217, Alberta, Canada, 2001.)

5.3.1 Image Spectrum

The images that we capture with "photometric" cameras measure information in the visible light spectrum. Figure 5.5a shows the spectrum of light and where our human visible spectrum falls. The other images in Figure 5.5 illustrate images created from sensors that sample different parts of this light spectrum.

The most commonly used images come from sensors measuring the visible light spectrum. The name of this band of frequencies comes from the fact that it is the range in which human see. Our common

(a)

(b)

FIGURE 5.4 System developed using static and mobile cameras to track outdoors events. This figure shows a pedestrian trespassing a gate: (a) image from static camera and (b) image from mobile camera. (From Marcenaro, L. et al., A multi-resolution outdoor dual camera system for robust video-event metadata extraction, *FUSION 2002: Proceedings of the Fifth International Conference on Information Fusion*, pp. 1184–1189, Annapolis, MD, 2002.)

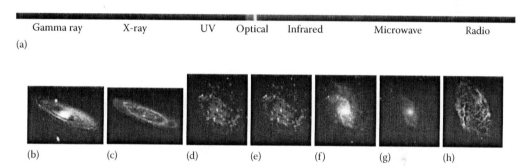

| Gamma ray | X-ray | UV | Optical | Infrared | Microwave | Radio |

(a)

(b)   (c)   (d)   (e)   (f)   (g)   (h)

FIGURE 5.5 Some of the images created via pseudocoloring image pixel values: (a) spectrum of light, (b) visible image of the Andromeda galaxy, (c) IR version of (b), (d) x-ray image, (e) UV image, (f) visible light image, (g) near-IR image, and (h) radio image. (Images courtesy of NASA, Multiwavelength milky way: Across the spectrum, http://adc.gsfc.nasa.gov/mw/mmw_across.html, Version 2, September 2001.)

understanding of the images we see is why this is probably the most commonly used imaging sensor in sensor networks. These sensors are commonly called photometric or optical cameras.

A sensor also used in network systems is the IR sensor. This sensor has elements that are sensitive to the thermal (IR) region of the spectrum. Forward looking infrared (FLIR) sensors are often used and involve a passive sensing scheme, which means it measures infrared energy emitted from the objects not reflected by some source. Near-IR is the portion of the IR spectrum that is closest to the visible spectrum. Viewing devices that are sensitive to this range are called night vision devices. This range of the spectrum is important to use when the sensor network needs to operate in no- or low-light conditions or when information from this band is important (e.g., sensing heat-emitting objects like animals).

The nearest high-energy neighbor to visible light is the ultraviolet (UV) region. The sun is a strong emitter of UV radiation, but the earth's atmosphere shields much of this radiation. UV radiation is used in photosynthesis and hence this band is used for vegetation detection in images as seen in many remote-sensing applications.

Some of the other frequency bands are less often employed in sensor network applications but, for completeness, they are worth mentioning. The highest energy electromagnetic waves (or photons) are the gamma rays. Many nuclear reactions and interactions result in the emission of gamma ray and they are used in medical applications, such as cancer treatments, where focused gamma rays can be used to eliminate malignant cells. Also, other galaxies produce gamma rays that are thought to be caused by very hot matter falling into a black hole. The earth's atmosphere shelters us from most of these gamma rays. X-rays, the next band of wavelengths, was discovered by Wilhelm Röentgen, a German physicist, who, in 1895, accidentally found these "light" rays when he put a radioactive source in a drawer with some unexposed photographic negatives and found the next day that the film had been exposed. The radioactive source had emitted x-rays and produced bright spots on the film. X-rays like gamma rays are used in medical applications, in particular, to see inside the body. Finally, microwaves like radio waves are used for communications, specifically the transmissions of signals. Microwaves are also a source of heat, as in microwave ovens.

5.3.2 Image Dimensionality

Images can also differ by their dimensionality. All of the images shown so far have two dimensions. However, there are also three-dimensional (3D) images like the one visualized in Figure 5.6. Here, the information at every point in our image represents the depth from the sensor or some other calibration point in our scene. The term 3D is used because at each point in the image we have information corresponding to the (x, y, z) coordinates of that point, meaning its position in a 3D space. Sensors producing this kind of information are called range sensors. There are a myriad of range sensors including various forms of radar (active sensors) like sonar- and laser-range finders, as well as triangulation-based sensors like stereo and structure light scanners.

Another kind of multidimensional image is the multispectral or hyperspectral image. Here, the image is composed of multiple bands (N-dimensions), each band has its own two-dimensional (2D) image that measures light in that specified frequency band. Multispectral images are typically used in remote sensing applications.

As the most commonly used imaging sensor in network systems is the photometric camera, in Section 5.2.4, we will discuss the 2D image structure. However, it is important to note that the image processing techniques described in this chapter can usually be extended to work with a N-dimensional image.

5.3.3 Image Sensor Components

Besides the spectrum or dimensionality, the configuration of the sensor equipment itself will greatly alter the information captured. In the case of photometric cameras, it is made up of two basic components,

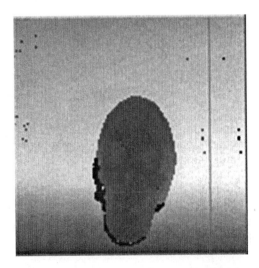

FIGURE 5.6 Visualization of 3D image of human skull. (From Grewe, L. and Brooks, R.R., Efficient registration in the compressed domain, *Wavelet Applications VI, SPIE Proceedings*, Vol. 3723, H. Szu (ed.), Aerosense, Orlando, FL, 1999.)

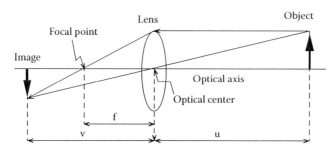

FIGURE 5.7   Components of a typical camera: sensor plane and lens.

a lens and a sensor array. The lens is used to focus the light onto the sensor array. As shown in Figure 5.7, what is produced is an upside-down image of the scene.

Photometric cameras are considered passive, meaning they only register incoming emissions. Other sensors such as the range sensors are active, meaning they actively alter the environment by sending out a signal and afterward measuring the response in the environment. Thus, active sensors will have the additional component of the signal generator.

One example of an active sensor is that of GPR. GPR is used to detect object buried underground where traditional photometric cameras cannot see. Here, a signal at a particular frequency range (i.e., 1–6 GHz) is transmitted in a frequency-stepped fashion. Then, an antenna is positioned to receive any signals reflected from objects underground. By scanning the antenna, an image of a 2D spatial area can be created.

Another sensor with different components is that of an electromagnetic induction (EMI) sensor. This sensor uses coils to detect magnetic fields present in its path. This can be used to detect objects, albeit metallic ones obscured by ground or other objects. An image can be comprised by mapping out signals obtained through scanning a 2D spatial area.

There are many ways in which you can alter your environment or configure your sensors to achieve better images for your sensor network application. Some of the performance influencing factors of a sensor include the dynamic range of the sensor, optical distortions introduced by the sensor, sensor blooming (overly larger response to higher intensity signals), and sensor shading (nonuniform response at outer edges of sensor array). Lighting, temperature, placement of sensors, focus, and lens settings are some of the factors that can be altered. Whether and how this is done should be in direct relation to the network's objectives. As the focus of this chapter is on image processing, implying we already have the image, we will not discuss this further, but it is important to stress how critical these factors are in determining the success of a sensor network system that uses images.

Figure 5.8 shows a sequence of images taken of the same scene from a camera mounted on a car that are taken under different exposure conditions [Goermer 10]. The image is darker in the shorter exposure situation, but this yields more detail in the naturally brighter area of the scene. Contrastingly, the image is overall brighter in the longer exposure situation that yields more detail in the darker tunnel area of the scene. Utilizing both exposure images can yield more optimal processing results. See [Razlighi 07] for a similar system applied to images from cell phones.

5.3.4 Analog-to-Digital Images

Figure 5.9 shows the two-step process of creating a digital image from the analog light information hitting a sensor plane. The first step is that of sampling. This involves taking measurements at specific location in the sensor plane, represented by the location of a sensor array element. These elements are usually distributed in a grid or near-grid pattern. Hence, when we think of a digital image, we often visualize it as shown in Figure 5.9 by a 2D grid of boxes. These boxes are referred to as pixels (picture elements). At this point, we can still have a continuous value at each pixel, and as we wish to store the image inside of a computer, we need to convert it to a discrete value. This process is referred to as quantization.

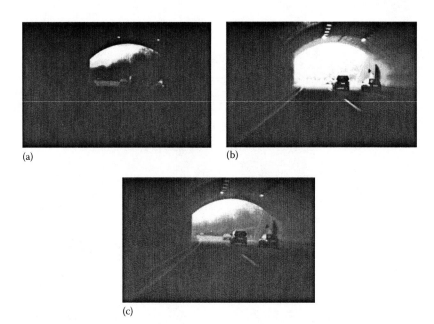

(a)   (b)

(c)

FIGURE 5.8   Results from Goermer et al.: (a) short exposure image, (b) longer exposure image, and (c) combined results. (From Goermer, S. et al., Multi-exposure image acquisition for automotive high dynamic range imaging, *Intelligent Transportation Systems, IEEE Conference on Digital Object Identifier*, pp. 1881–1886, Funchal, Portugal, 2010.)

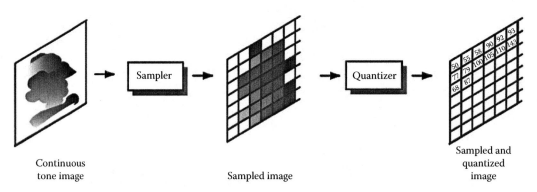

Continuous tone image   Sampled image   Sampled and quantized image

FIGURE 5.9   Creation of digital image: sampling and quantization. (From Baxes, G.A. *Digital Image Processing: Principles and Applications*, John Wiley & Sons, New York, 1994, ISBN-10: 0471009490, ISBN-13: 978-0471009498.)

You lose information in the process of quantization, meaning you cannot invert the process to obtain the original. However, sampling does not have to be a lossy procedure. If you sample at least two times the highest frequency in the analog image, you will not lose any information.

What results from sampling and quantization is a 2D array of pixels, illustrated in Figure 5.10. Any pixel in an image is referenced by its row and column location in the 2D array. The upper left-hand corner of the image is usually considered the origin as shown in the figure.

Through quantization, the range of the values stored in the pixel array can be selected to help achieve the system objectives. However, for the application of display and for the case of most photometric (visible light) images, we represent the information stored at each pixel as either a gray-scale value or a color value.

In the case of a gray-scale image, each pixel has a single value associated with it, which falls in the range of 0–255 (thus taking 8 bits). Zero represents black or the absence of any energy at this pixel location and 255 represents white, meaning the highest energy the senor can measure.

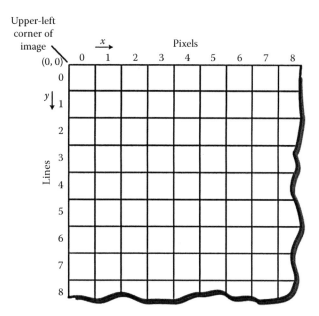

FIGURE 5.10 Discrete pixel numbering convention. (From Baxes, G.A. *Digital Image Processing: Principles and Applications*, John Wiley & Sons, New York, 1994, ISBN-10: 0471009490, ISBN-13: 978-0471009498.)

Color images by contrast typically have three values associated with each pixel representing red, green, and blue. In today's computers and monitors, this is the most common representation of color. Each color field (red, green, and blue) has a range of 0–255 (thus taking 8 bits). This kind of color is called 24 bit color or full color and allows us to store approximately 16.7 million different colors. While this may be sufficient for most display applications and many image processing and understanding applications, it is important to note that there is an entire area of imaging dealing with color science that is actively pursued.

As a note of interest, the difference between a digital and analog sensor is that for digital sensors the sensor array has its values directly read out into storage. However, analog sensors go through an inefficient digital-to-analog conversion (this is the output of the sensor) and then another analog-to-digital conversion (this time by an external digitizer) sequence before having the information placed in storage.

5.4 Image Domains: Spatial, Frequency

Domains are alternative spaces in which to express the information contained in an image. The process of going from one domain to another is referred to as a transformation. If you can go from one domain to another and return again, the transformation is termed invertible. If you do not lose any information in this transformation process, the transformation is considered lossless and is called one to one. We discuss the following commonly used image domains: the spatial domain, the frequency domain, and the wavelet domain.

We have already been exposed to the spatial domain; it is the original domain of the image data. The spatial domain is given its name from the fact that neighboring pixels represent spatially adjacent areas in the projected scene. Most image processing routines operate in the spatial domain. This is a consequence of our intuitive understanding of our physical, spatial world.

The frequency domain is an alternative to expressing the underlying information in the spatial domain in terms of the frequency components in the image data. Frequency measures the spatial variations of the image data. Rapid changes in the pixel values in the spatial domain indicate high-frequency components. Almost uniform data values mean there are lower frequency components. The frequency domain is used for many image processing applications like noise removal, compression, feature extraction, and even convolution-based pattern matching.

There are many transformations that yield different versions of the frequency domain. The most famous and frequently used is the Fourier frequency transformation. The following illustrates the forward and reverse transformation equations, where f(x, y) is the spatial domain array and F(u,v) is the frequency domain array:

$$F(u, v) = \left(\frac{1}{MN} \right) \sum_{x=0}^{M-1} \sum_{y=0}^{N-1} f(x, y) \exp\left(\frac{-2\pi jux}{M} \right) \exp\left(\frac{-2\pi jvy}{N} \right) \tag{5.1}$$

$$f(x, y) = \sum_{u=0}^{M-1} \sum_{v=0}^{N-1} F(u, v) \exp\left(\frac{2\pi jux}{M} \right) \exp\left(\frac{2\pi jvy}{N} \right) \tag{5.2}$$

A fast algorithm fast Fourier transform (FFT) is available for computing this transform, providing that N and M are powers of 2. In fact, a 2D FFT transform can be separated into a series of one-dimensional (1D) transforms. In other words, we transform each horizontal line of the image individually to yield an intermediate form in which the horizontal axis is frequency (u) and the vertical axis is space (y).

Figures 5.11 and 5.12 show some sample images and their corresponding frequency domain images. In Figure 5.12a, we see a very simple image that consists of one frequency component that is of repetitive lines horizontally spaced at equal distances, a sinusoidal brightness pattern. This image has a very simple frequency domain representation as shown in Figure 5.12b. In fact, only 3 pixels in this frequency domain image have nonzero values. The pixel at the center of the frequency domain represents the DC component, meaning the "average" brightness or color in the image. For Figure 5.12a, this is some mid-gray value. The other two nonzero pixel values straddling the DC component shown in Figure 5.12b are the positive and negative components of a single frequency value (represented by a complex number). We will not go into a discussion of complex variables but note their presence in Equations 5.1 and 5.2.

The wavelet domain is a more recently developed domain used by some image processing algorithms. For example, the JPEG standard went from using a Discrete Cosine transform, another frequency transform similar to the Fourier transform, to using the wavelet transform. Wavelet basis functions are localized in space and in frequency. This contrasts with a 2D gray-scale image, whose pixels show values at a given location in space, that is, localized in the spatial domain. It also contrasts with the sine and cosine basis functions of the Fourier transform, which represent a single frequency not localized in space, that is, localized in the frequency domain. Wavelets describe a limited range of frequencies found in a limited region of space; this gives the wavelet domain many of the positive attributes of both the spatial and frequency domains.

(a)

(b)

FIGURE 5.11 (a) Image and (b) its FFT image. (From Baxes, G.A. *Digital Image Processing: Principles and Applications*, John Wiley & Sons, New York, 1994, ISBN-10: 0471009490, ISBN-13: 978-0471009498.)

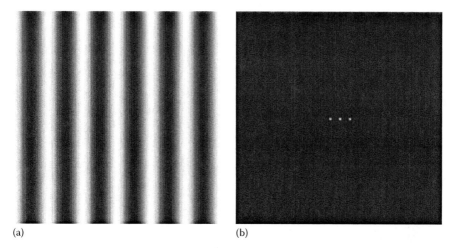

(a)   (b)

FIGURE 5.12 (a) Image and (b) its FFT image. (From Baxes, G.A. *Digital Image Processing: Principles and Applications*, John Wiley & Sons, New York, 1994, ISBN-10: 0471009490, ISBN-13: 978-0471009498.)

FIGURE 5.13 Image and its corresponding wavelet domain (visualization of data).

There exist a number of different wavelet basis functions in use. What is common to all of the variations of the wavelet transformation is that the wavelet domain is a hierarchical space where different levels of the transform represented repeated transformations at different scales of the original space. The reader is referred to [Brooks 01] for more details about the wavelet domain. Figure 5.13 shows an image and the visualization of the corresponding wavelet domain. [Wang 06] is one of many publications involving the use of wavelets in sensor fusion, and this work concentrates on how the selection of the wavelet basis can affect this process.

5.5 Point-Based Operations

In this section, we discuss some of the simplest image processing algorithms. These include thresholding, conversion, contrast stretching, threshold equalization, inversion, subtraction, averaging, gray-level slicing, and bitplane slicing. What all of these algorithms have in common is that they can be thought of as "point processes," meaning that they operate on one point or pixel at a time. Consider the case of producing a binary image from a gray-scale image using thresholding. This is accomplished by comparing each pixel value with the threshold value and consequently setting the pixel value to 0 or 1 (or 255) making it binary.

One way that we can write "point processes" is in terms of their transformation function, T( ), as follows: Pnew[r,c] = T(P[r,c]), where r represents row, c represents column, and P[r,c] is the original image's pixel value at r,c. Pnew[r,c] is the new pixel value.

5.5.1 Thresholding

Thresholding is typically used to create a binary image from a gray-scale image. This can be used to highlight areas of potential interest leading to simple feature extraction and object detection based on brightness information.

This technique can also be used to produce a gray-scale image with a reduced range of values or a color image with a reduced range of colors, etc. Notice that in the following algorithm we visit the pixels in a raster scan fashion, meaning one row at a time. This method of visiting all of the pixels in an image is prevalent in many image processing routines. Figure 5.14b shows the results of thresholding a gray-level image.

```
for(r=0; r<M; r++)
for(c=0; c<N; c++)
{
    if(P[r,c] < Threshold)
      Pnew[r,c] = 0;
    else
      Pnew[r,c] = 255;
}
```

5.5.2 Conversion

Conversion is simply converting from one "type" of image to another. Specifically, type here means the modality or dimensionality of the pixel values such as color or gray scale. For example, consider the conversion of a color image into a gray-scale image as performed by the following algorithm. Other types of conversions include converting a full color image (24 bit/pixel) to an 8 bit color image (using a Color LUT). The main application of conversion is to reduce the amount of information associated with an image. If it is possible to perform your task using a reduced set of information, your network system will run faster and require less energy and storage. Figure 5.14c and d shows the results of converting a color to gray-scale image.

```
for(r=0; r<M; r++)
for(c=0; c<N; c++)
   Pnew[r,c] = (Red[r,c] + Green[r,c] + Blue[r,c])/3;
          p1 = (orig 10, final 50) and p2 = (orig 40, final 170)
```

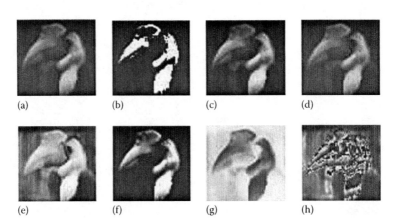

(a)   (b)   (c)   (d)

(e)   (f)   (g)   (h)

FIGURE 5.14 Various point-based operations: (a) original, (b) thresholding image (a), (c) color original, (d) conversion of (c) to gray scale, (e) contrast stretched version of image (a), (f) histogram equalized version of image (a), (g) inversion of image (a), and (h) bit-plane sliced version of image (a).

5.5.3 Contrast Stretching and Histogram Equalization

Contrast stretching and histogram equalization are examples of image enhancement algorithms that attempt to increase the contrast or range of pixel values present in the image. Low-contrast images can result from poor illumination, lack of dynamic range in the image sensor, or incorrect setting of the lens aperture. The result of a low-contrast image is that not all of the possible pixel value range is used with possibly only a relatively small part of the range being used. By too small, it is meant that the range of values used in the image is not large enough to clearly capture the detail and variation in the original scene. Hence, increasing the contrast of an image can make it easier for both humans and computer programs to extract information for interpretation. Caution should be applied in the use of these algorithms as their blind use can produce extreme results, which can introduce artifacts making it more difficult to interpret the images.

Contrast stretching is a simple algorithm that shifts each pixel value by a transformation value that is only a function of the pixel value itself. Commonly, although not required, the transformation can be represented by one or more linear functions. In general, we can define a set of such linear transformations by specifying sets of corresponding pixel values between the original and transformed image that define monotonically increasing lines. Consider Figure 5.15, where we consider a gray-level image and designate two correspondences, p1 and p2. The three linear equations spanning {0,0 to p1}, {p1 to p2}, and {p2 to 256,256} define the transformation. Figure 5.14e shows the results of contrast stretch on the image in Figure 5.14a.

Histogram equalization is a common technique for enhancing the appearance of images. Suppose we have a gray-scale image, which is predominantly dark. Then, its histogram would be skewed toward the lower end of the gray scale and all the image detail is compressed into the dark end of the histogram. If we could "stretch out" the gray levels at the dark end to produce a more uniformly distributed histogram, then the image would become much clearer. Note a histogram is simply a count of the number of pixels at each possible pixel value (gray level or color or whatever).

Unlike in contrast stretching, there are no parameters (i.e., p1, p2) to select. In histogram equalization, we are transforming the gray levels such that the resultant image will have a uniform density of gray levels. To achieve this, we use the following transformation:

$$T(x) = (L-1) \sum_{w=0}^{x} \frac{h(w)}{(\# \text{ pixels in image})}$$

where
  L is the number of gray levels in the image
  h( ) is the histogram of the original image

Note: h(w)/(# pixels in image) = Probability of gray-level w occurring in the image

Figure 5.14f shows the results of applying histogram equalization on an image that already has a lot of contrast. This illustrates the fact you should apply histogram equalization carefully and in general only to low-contrast images.

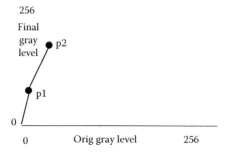

FIGURE 5.15 Illustration of contrast stretching using set of three linear functions.

5.5.4 Inversion

This algorithm produces what you can think of as the negative of the image as shown in Figure 5.14g. It is produced by simply inverting the image's pixel range. The application of this algorithm is mostly for data display.

5.5.5 Level Slicing

This is an effect that can be used to highlight a portion of an image's pixel value range. Consider a gray-scale image of some coins. There are coins made of silver and pennies made of copper. If you wanted to sort out the pennies, they would correspond to the darker gray circular objects in the image. Hence, you could highlight them in the image by mapping the mid-level gray values to white and the rest to black. Applications of this can be found in various object detection tasks and in multispectral remote sensing applications.

5.5.6 Bitplane Slicing

Bitplane slicing is similar to level slicing, but here we examine the bits used to represent each pixel and set certain bits to 0 and leave the others untouched. Figure 5.14h shows the results of Bitplane slicing on a gray-scale image. Some of the higher frequency components can be removed in an image by Bitplane slicing the lower-order bits. However, we recommend doing filtering in the frequency domain for this task.

5.5.7 Image Subtraction

Image subtraction is the pixel-by-pixel subtraction of one image from another. One use of image subtraction is the removal of the background. Consider the application of detecting people that come up to an ATM machine's camera. If you had a picture of the stationary background and subtracted it from a current image, then only the new items in the scene, for example, a person, would be visible.

5.5.8 Image Averaging

Image averaging is the pixel-by-pixel averaging of two or more images. Before this process takes place, registration of the images should occur so that averaging only happens between corresponding pixels. One use of image averaging is the reduction of noise in the scene.

5.6 Area-Based Operations

The next class of image processing algorithms involves looking at a neighborhood of a pixel and using the neighboring pixel values to alter its value. This kind of algorithm can be thought of as an area processing algorithm. Another name for it is spatial filtering. There are many examples of filters or area-based algorithms in existence and each can be used for different purposes. In addition to how you combine or use the neighboring pixel values, the size of the neighborhood is another variable. In this section, we discuss a few of these algorithms.

Let us begin by defining what a neighborhood is. For the pixel labeled (i, j), the neighboring pixels that share borders are highlighted in black in Figure 5.16a. These pixels are referred to as the 4-neighbors of pixel (i, j). All of the pixels surrounding the pixel (i, j) are called 8-neighbors and comprise the smallest neighborhood surrounding this pixel, which is called the 3 × 3 neighborhood shown in Figure 5.16b. Increasing the neighborhood size by one row and one column in all directions would yield a 5 × 5 neighborhood. Typically, most area-based processing algorithms surround a center pixel evenly, which yields an odd number of rows and columns to the processing neighborhood.

A basic distinction between area-based algorithms is whether they are linear or nonlinear. Linear algorithms replace the center pixel value as a linear function of their neighboring pixels.

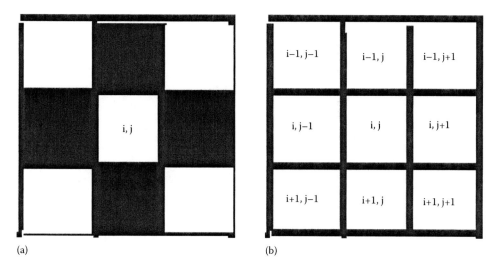

(a)   (b)

FIGURE 5.16   (a) 4-neighbors of pixel (i, j) and (b) 3 × 3 neighborhood of pixel (i, j).

5.6.1 Low-Pass Filter

Low-pass filtering is an example of a linear filter and another name for it is a smoothing filter. This filter is a weighted averaging of the neighboring pixels. This can be represented as convolution with an N × N (neighborhood size) positively valued convolution mask. A common implementation of a low-pass filter is to set all of the weights to 1 and divide the response by a normalization constant if N * N to keep the image pixel range the same. The following is the algorithm to implement such a smoothing operation. Figure 5.17b shows the low-pass filtered image resulting from Figure 5.17a.

```
for(r=0; r<M; r++)
for(c=0; c<N; c++)
{
    Pnew[r,c] = (P[r-1,c-1] + P[r-1,c] + P[r-1,c+1] +
                P[r,c-1] + P[r,c] + P[r,c+1] +
                P[r+1,c-1] + P[r+1,c] + P[r+1,c+1] )/ 9;
}
```

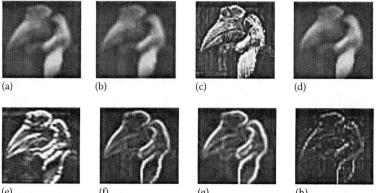

(a)   (b)   (c)   (d)

(e)   (f)   (g)   (h)

FIGURE 5.17   Area-based algorithms: (a) original image, (b) low-pass filtered version of image (a), (c) high-pass filtered version of image (a), (d) median filtered version of image (a), (e) detection of horizontal edges, (f) Prewitt edge detection, (g) Sobel edge detection, and (h) inversion of magnitude of LOG edge image.

Low-pass filtering performs smoothing on the image and removes high-frequency details from the image. This is one form of noise removal and is also used in imaging system to remove extraneous details not necessary to achieve the system's objectives. Low-pass filtering can be done more efficiently in the frequency domain by simply setting all of the high-frequency components to zero and then taking the inverse transform to obtain the smoothed spatial image.

5.6.2 High-Pass Filter

This is a filter that performs the opposite of the low-pass filter; it tries to accentuate the higher frequency information in the image and hence sharpen rather than smooth it. Instead of positive values at neighboring locations in the mask, they are negative. There are numerous kinds of high-pass filter convolution masks although setting all to −1 is common. High-pass filtering can be done using the frequency domain by adding to the original image a new image formed by setting all but the highest frequencies in the frequency domain to zero and taking the inverse back to the spatial domain. Figure 5.17c shows the results of applying a high-pass filter on our bird image.

5.6.3 Median Filter

Median filtering is an example of nonlinear filtering. If the objective is to remove noise and to not blur the image, this approach may be successful. It is particularly useful for removing shot noise as discussed in Section 5.7. The algorithm replaces a pixel's value with the median of the pixel values in its neighborhood. Figure 5.17d shows the results of applying median filtering to our bird image.

5.6.4 Edge Detection

Edges represent some of the important features of an image. They can describe the boundary of a physical object in the scene or a transition in material or lighting and other environmental effects. The extraction of edge-based features has been used in everything from compression algorithms to more commonly image interpretation. Edges can be extracted from many kinds of images including visible spectrum, IR, sonar, multispectral, and more.

If you were to look at pixel value transitions of a 1D signal (or you can think of this as looking for vertical kinds of edges in a single image row of a 2D image), the following would be "kinds" of edges that you may observe.

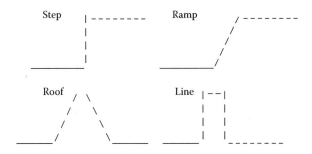

Of course, such perfect profiles rarely happen, but they illustrate the kind of variation that leads to the observance of an edge point.

There are many ways to detect edges. In the spatial domain, this can be done by convolution with a filter mask, as we describe in the following. In the frequency domain, one can use high-frequency filters to extract edge information. In either case, what results from the "edge detection" process is simply a new image with the nonzero values representing edge points of varying strength. Extracting features such as boundaries, lines, or other shapes from an edge-detected image is part of the feature extraction phase and is discussed later in this chapter.

There are many kinds of edge detectors and we will mention a few of the commonly used ones. One of the simplest edge detectors called the Roberts detector involves the following 2 × 2 convolution masks. The edge image is |Gx| + |Gy|. Figure 5.17f shows the resulting edge image of our bird picture. One unusual thing about this detector is that its masks are not "symmetric," meaning they do not surround the pixel in question with the same number of pixels in every direction.

$$
Gx = \begin{array}{|c|c|} \hline 1 & 0 \\ \hline 0 & -1 \\ \hline \end{array}
$$

$$
Gy = \begin{array}{|c|c|} \hline 0 & 1 \\ \hline -1 & 0 \\ \hline \end{array}
$$

An example of a "symmetric" edge detector is the Sobel, which uses the following masks to produce an edge image equal to the sqrt (Sx*Sx + Sy*Sy). Figure 5.17g show the Sobel edge-detected image for our bird picture. Comparing the results for Roberts and Sobel, you can see that edge detectors give slightly different results. Selecting the right edge detector for your system is often a function of empirical testing.

$$
Sx = \begin{array}{|c|c|c|} \hline -1 & 0 & 1 \\ \hline -2 & 0 & 2 \\ \hline -3 & 0 & 1 \\ \hline \end{array}
$$

$$
Sy = \begin{array}{|c|c|c|} \hline 1 & 2 & 1 \\ \hline 0 & 0 & 0 \\ \hline -1 & -2 & -1 \\ \hline \end{array}
$$

The last edge detector we will mention is the LOG or the "Laplacian of the Gaussian" operator. This is fundamentally different than the previous two operators, in that they took only a first-order difference, whereas the LOG takes the second-order difference (derivative) to measure the presence of an edge. Hence, instead of taking edge points as the maximums of the first-order derivative filters, we use the zero-crossings of the second-order derivative. The Laplacian function is the implementation of the second-order derivative. However, first, a Gaussian function is applied to do some smoothing. This two-step procedure modeled after how our human eye detects edges. The edge-detected image is produced using the following single convolution mask. Often the magnitude of the result values is used with zero values indicating the edge points. Here, we show a 5 × 5 mask, but larger masks can be employed. Figure 5.17h shows the LOG image of our bird picture.

Log =

0	0	−1	0	0
0	−1	−2	−1	0
−1	−2	16	−2	−1
0	−1	−2	−1	0
0	0	−1	0	0

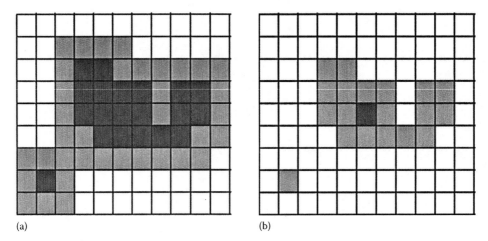

(a)   (b)

FIGURE 5.18   Morphological operations: (a) dilation by 1 pixel, where gray pixels are newly added and (b) erosion by 1 pixel, where gray pixels are the removed pixels.

5.6.5 Morphological Operators

Morphological operators get their name from the fact that they alter the form of the image. Typically, they are applied to images that have been made binary through thresholding or that have been taken into a "feature-like" space such as an edge-detected image. We will discuss dilation and erosion, two commonly used morphological operators.

Dilation is used to expand the boundary of a blob in the image. A blob is any contiguous set of non-zero pixels. In its simplest form, it is used to expand the boundary of blobs by 1 pixel in each direction. An example of this is shown in Figure 5.18a. Dilation is useful in joining nearby blobs that were separated erroneously in processing.

Erosion by contrast removes pixels at the boundary of a blob. Figure 5.18b illustrates this process with an erosion pattern of 1 pixel. Erosion can be used to reduce a blob to its center location or to trim it down.

Note that any pattern of pixels, rather than simply 1 pixel, can be used in the erosion or dilation process.

5.7 Noise Removal

An image can have noise introduced at many stages including creation and transmission. In distributed sensor networks, this can be a serious issue and hence the topic is discussed in a number of chapters of this book. Note that we differentiate between noise and artifacts or distortions introduced through manipulating the data. Depending on whether noise is signal dependent or independent, it is handled differently. Signal independent means that the noise is statistically independent of the image pixel values and hence is treated like an additive component: $I' = I + N$, where I is the perfect image and N is the noise added resulting in the noisy image I'. Noise that is added during transmission of an image is often signal independent.

Regardless of origin, it is possible to create a measure of the amount of signal-independent noise present at that time. For images, this is commonly done by taking an image, which has a known continuously valued area, and measuring how far the actual image diverges from this known value. This divergence is usually measured by the mean difference as well as the standard deviation. Another measurement of the significance of the noise is called the signal-to-noise ratio and is the ratio of these two magnitudes. This metric is commonly used is describing the performance of a sensor [Mirzu 98].

Signal-dependent noise means that the noise introduced is a function of the original image signal itself. This kind of noise is much more difficult to deal with because discovering this functional relationship is often not possible with great certainty.

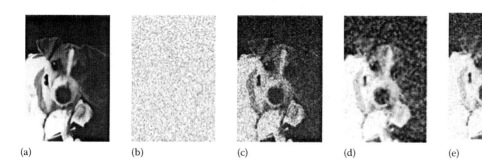

(a)   (b)   (c)   (d)   (e)

FIGURE 5.19 (a) Original image, (b) noise image, (c) noisy image, (d) median filter applied to image (c), and (e) low-pass filter applied to image (c).

Noise is usually described by its probabilistic characteristics. For example, white noise is given its name because it has a constant power spectrum (its intensity does not decrease with increasing frequency). Gaussian noise has a Gaussian function for its power spectrum. When nothing is known about the origin of a noise signal, it is often assumed to be either Gaussian or white.

There are a number of techniques that have been proposed to reduce the effects of noise. We have already mentioned use of median and low-pass filters for this task. Both can be used for noise reduction. The results on our noisy dog image are shown in Figure 5.19d and e. These are great techniques if not much is known about the noise. Median filtering works best when the noise is "shot" noise. This kind of noise occurs sometimes in transmission due to weather disturbances. As you can see, our random noise pattern (which is not shot) is improved the most through the low-pass filter, which simply eliminates the highest frequency patterns. Unfortunately, with any kind of noise reduction algorithm, we may also eliminate or subjugate non-noise information. As we can see, in Figure 5.19d and e, we remove some of the non-noise finest edge, high-frequency information.

Another technique used when we have more than one image of the scene, which is possible in sensor networks, is to average the registered images. As with most stochastic processes, if we sample enough images, the ensemble mean approaches the noise-free original signal. However, it is not always feasible to have multiple images and the registration has to be very good not to induce new artifacts.

[Zhang 01] discuss a system for image fusion that uses median filters and neural network-based fusion for the purpose of image noise reduction. First, the images, of the same scene and registered, are preprocessed by a weighted median filters for the purpose of removing noise. Next an image clustering/segmentation routine is applied using the neural networks. Finally, fusion takes place with the clustered information. Noise removal can also take place in the frequency domain. In this case, it is the higher frequency information that is assumed in part to originate from noise. In [Othman 06], noise reduction of multispectral imagery is performed using both spatial and spectral derivatives in the wavelet domain.

Specific kinds of noise including environmental and atmospheric produced noise are described in subsequent chapters of this book.

5.8 Feature Extraction

In distributed sensor networks that use images, some of the common tasks involve object recognition, navigation, and scene understanding. These high-level tasks must go beyond simply processing images at their pixel level to the extraction of higher-order information. This higher-order information is commonly called features and the process to obtain them called feature extraction.

There are two basic kinds of features, statistical and structural (there can be hybrids). Statistical features are those that are measured using statistics. An example might include the average pixel value of an area, the standard deviation, texture, etc. Structural features typically take on a physical or structural makeup. Some examples are edges or boundaries of objects and surfaces. Of course, you may gather statistics on structural elements, like average number of edges in an area, etc.

We discuss here a few of the many features that are used in sensor networks. It must be stressed that the features that will yield good results are entirely a function of network's objectives and the environment. The fact that there is not a definite language to our visual world is what makes image processing and understanding more challenging than speech recognition.

5.8.1 Edges

Edge following or linking is the process of "chaining" together edge pixels by starting at some edge pixel location, traveling to its neighbor, and so on, until the edge "chain" stops or comes to a junction. This process assumes that there is a definite chain. We discussed previously how to obtain an edge image. What is observed in an edge image is a continuum of edge values. Usually, there are prominent edge pixels, which are surrounded by edge points that to varying degrees are edge points themselves (see Figure 5.20). What we must do before we can run our "edge-following" algorithm on the image is to reduce down the number of edge pixels to the "essential" ones, meaning in Figure 5.20 to eliminate the gray-colored pixels.

This reduction can be achieved through the use of "closing," which involves the sequence of dilation followed by erosion (see Section 5.6.5). Dilation helps close any open boundary areas and erosion reduces the boundary. "Closing" can cause some problems including reducing the length of a boundary, which could impair recognition.

An alternative is to perform "thinning." Thinning is an algorithm that unlike "Closing" will never lead to removed blobs, reduced boundary lengths, or disconnection of boundaries. Thinning performs erosion but will not eliminate a pixel if it is by itself or is the only pixel connecting other edge pixels. Thinning would result in removal of the gray-colored pixels in Figure 5.20. After thinning we can easily apply our edge-following algorithm with consistent results.

There are many data structures that are used to store an edge as it is being traced. The most commonly used is the linked list. Once an edge is traced, various attributes can be calculated like length, curvature, end points, orientation, center, shape, etc.

5.8.2 Hough Transform: Detecting Shapes

The Hough transform is a technique to search for parameterized shapes like lines or circles in an image. While you could try to trace out these shapes using an algorithm like "edge following," if the

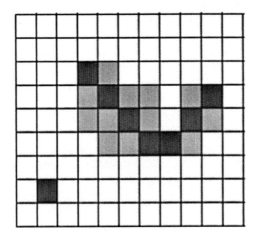

FIGURE 5.20 Small edge image where the strongest edge points are shown in black and lesser edge points in gray. These are exactly the pixels eliminated with "thinning."

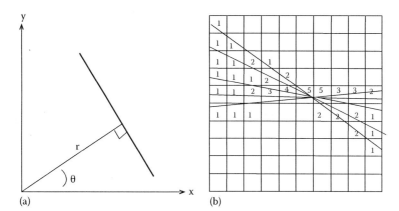

FIGURE 5.21 (a) A straight line can be described by its parameters r and θ; (b) Hough transform space for a straight line.

shapes you are looking for are parameterized, then the Hough transform is a more efficient way of detecting them.

Like edge following, the Hough transform is applied to an image that has been processed so that the nonzero pixels represent the presence of potential shape points. Often this may be an edge image, which has been thresholded such that its nonzero pixels represent edges of scene objects like their boundaries. Consider the simple idea of taking an image of a road scene. You could find the roads by looking for long straight lines in the image. To do this you would first create a thinned edge image and then you could apply the Hough transform to detect all of the long straight lines in the image.

For every parameterized shape you are searching for, you will need to apply its Hough transform to the image. The basic idea of a Hough transform is that every nonzero pixel (e.g., edge point) votes for all examples of the shape that pass through it. Votes are collected in the "parameter space" of that shape and the algorithm looks for peaks in this "parameter space." These peaks indicate the strongest instances of the shape in the image.

Let us look at the Hough transform for lines in more detail. A straight line can be described by its perpendicular distance from the origin and its angle from the horizontal axis as shown in Figure 5.21a. The parameters r and theta uniquely describe a straight line in an image. The Hough transform parametric space is the 2D space described by the values of r and theta. This space is discretely sampled and each bin, which acts as a counter, is set to 0. The Hough transform algorithm visits each nonzero pixel in the image (x', y') and determines the curve equation with unknowns (r, theta) and increments all of the (r, theta) cells in the discrete Hough transform space the curve intersects. Figure 5.21b shows the bins being increments in the Hough space. As you can see, peaks in the Hough space will be created when a number of pixels coincide on the same line, thus voting for that line.

5.8.3 Segmentation: Surfaces

Segmentation is the process of breaking up an image into regions that are similar in terms of some property. The goal is usually to divide an image into parts that have a strong correlation with parts of scene objects. Totally correct and complete segmentation of complex scenes usually cannot be achieved. Segmentation works best when you have contrasted objects on a different and ideally uniform background. Common properties used to segment the image include pixel value, statistical measures, and texture.

The simplest way to segment an image is through the process of thresholding. Single or multiple threshold values can be used. Often times and as a function of the system's environment, various image processing algorithms will be applied first to remove noise and emphasize the homogeneity of properties within a segment. Figure 5.22 illustrates this process.

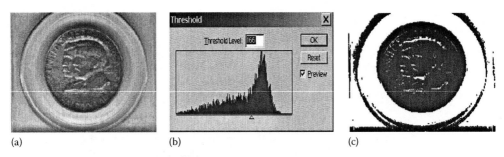

FIGURE 5.22 (a) Original image, (b) histogram of brightness values showing thresholding point, and (c) resulting binary image.

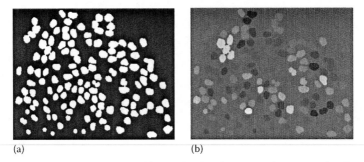

(a)   (b)

FIGURE 5.23 (a) Original binary image and (b) pseudocolored segmented image, each color shows a different segment detected.

More elaborate and better segmentation routines exist including run-length encoding, split-and-merge, and border tracing. The split-and-merge algorithm is a two-step process. First, the algorithm recursively splits an image region (starting with the entire image) by dividing it into quarters. Feature vectors are computed for each block based on some property. If the four blocks are judged to be similar, they are grouped back together and the process ends. If not, each of the blocks is recursively divided and analyzed in the same fashion. When we are finished with this stage, there will be many different sized blocks, each of which has a homogeneous property value. However, the arbitrary division of the image into rectangles (called a "quad-tree" decomposition) might have accidentally split up regions of homogeneous property. The second step, merging, then, tries to fix this by looking at adjacent regions that were not compared to each other during the split phase, and merging them if they are similar. Figure 5.23b shows the results of this algorithm on the binary image in Figure 5.23a.

5.8.4 Examples

In [Marcello 02], a system for feature extraction from multisensorial oceanographic imagery is discussed. After an initially preprocessing stage of noise removal is performed, the image is divided up into smaller images called regions of interest (ROI). The histogram of an ROI is used to threshold it and the segmentation takes place to find contiguous areas of "upwelling" and "filaments" in the water images.

Feature extraction can take place in either the spatial or the frequency domain. For example, in [Sveinsson 01], feature extraction takes place in the wavelet domain using cluster-based (segmentation) techniques.

In [Waxman 07], a system for target tracking is proposed that utilizes a special multisensor device that captures multiple kinds of images from which feature extraction is performed. Figure 5.24 shows the resulting feature stack, which in addition to the original visual, SWIR, MWIR, and LWIR images contains a fused image and the contrast, contours, texture, and 3D image maps that result from processing the original and fused images.

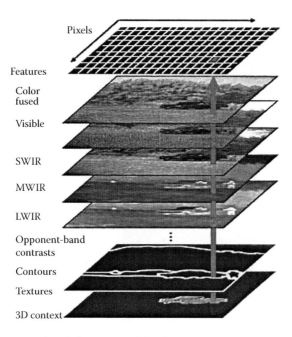

Pixels

Features

Color
fused

Visible

SWIR

MWIR

LWIR

Opponent-band
contrasts

Contours

Textures

3D context

FIGURE 5.24 Shows feature stack including raw pixel data from multiple images; the fused (combination of other images), visible, SWIR, MWIR, and LWIR images; and the opponent-band contrasts, contours, textures, and 3D content. (From Waxman, A. et al., Active tracking of surface targets in fused video, *International Conference on Information Fusion*, Québec City, Québec, Canada, 2007.)

5.9 Registration, Calibration, and Fusion Issues

Registration and calibration are steps dealing with understanding the data samples from a sensor source either in relation to other sensor data (registration) or in terms of some absolute scale (calibration). This understanding is critical if sensor networks are to be able to reason over the multiple sensor data acquired.

5.9.1 Registration

When we have multiple sources of data, as is often the case for sensor networks, it is common to want to understand the mapping between one point in a sensor's data to the corresponding point in another sensor's data. This is often critical for networks to be able to collaboratively use the different sensor data to understand the environment. Registration describes this mapping. As discussed earlier, registration is an important topic in distributed network systems and as such has an entire chapter devoted to it. Hence, in this section, we will not go over the detail of the registration process but instead highlight the kinds of image processing steps that registration systems utilize via discussion of some sample system.

In [Zamora 98], a system is created to perform registration for multisensor images. Here synthetic aperture radar (SAR) and electro-optic (EO) images are registered for the purpose of fusing them. As these are different kinds of images, registration can be challenging. This system performs preprocessing and feature extraction to assist in this task. In particular, edges are used as features, which are extracted following a preprocessing stage, which includes adaptive clustering segmentation. In the case of the SAR images, first the image is histogram equalized to increase contrast, and then a 3 × 3 median filter for smoothing is applied. After this, the SAR image is thresholded into a binary image. In the case of the EO image, the median filter is applied first, and then histogram equalization is applied followed by binarization. The EO image is first median filtered as the authors feel there is already a lot of contrast in EO images as compared to SAR images and they want to eliminate little unwanted variations. Figure 5.25 shows results on a scene.

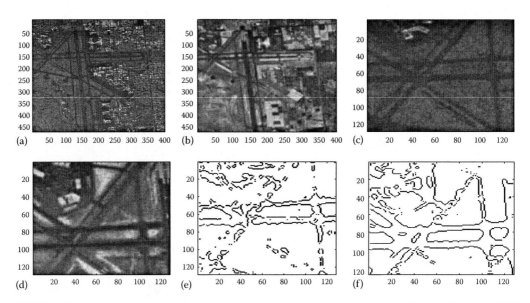

FIGURE 5.25 System developed for registration of multisensor images for the purpose of fusion. (a) original SAR image, (b) original EO image, (c) clustered SAR image, (d) clustered EO image, (e) edges in processed SAR image, and (f) edges in processed EO image. (From Zamora, G. et al., A robust registration technique for multi-sensor images, *IEEE Southwest Symposium on Image Analysis and Interpretation*, pp. 87–90, Tucson, AZ, 1998.)

In [Grewe 99], 3D and 2D photometric images are presented to the system for registration. In this case, a 3D wavelet domain is used to extract features. As described earlier, the wavelet domain is a variation of the frequency domain that captures both frequency and spatial information. In this work, 3D points of inflection are detected in the wavelet domain and used for features to calculate the registration information. Figure 5.26 shows the detected feature points.

See Chapter 20 on registration for more details of this and other systems. The reader is also referred to [Brown 92] for a survey on image registration techniques.

FIGURE 5.26 Detected feature points from wavelet domain in the 3D image. (From Grewe, L. and Brooks, R.R., Efficient registration in the compressed domain, *Wavelet Applications VI, SPIE Proceedings*, Vol. 3723, H. Szu (ed.), Aerosense, Orlando, FL, 1999.)

5.9.2 Geometric Transformations

The product of registration is a transformation matrix, which describes the scale, rotation, and translation necessary to transform from one to another image. We discuss each kind of transformation in this section.

Geometric transformations alter the position of pixels. In other words, the locations of a particular pixel value are altered. However, the pixel values also change when the new location falls between pixels requiring an interpolation process to calculate the new pixel values.

Scaling involves the shrinking or magnification of an image. Shrinking is useful if we need to reduce the number of pixels to be used by our system out of runtime, energy, or storage restrictions. Magnification can be useful when you want to view scenes at larger scale or even do subpixel calculations. Figure 5.25b shows a shrunk version of the original bird image in Figure 5.25a. Suppose we want to shrink our image from M × N to M/2 × N/2. A simple algorithm to do this would simply average 2 × 2 blocks of pixels to get a value for each pixel in the new smaller image. Instead, a faster algorithm, but arguably not as good in quality, would be to simply choose one of the pixels in the 2 × 2 block of the original image as the pixel value for the corresponding pixel in the new smaller image. The first algorithm averages or interpolates the values in a 2 × 2 block to get a single value. This kind of interpolation is called linear interpolation. There are many kinds of interpolation functions that are used but linear is the most common.

Rotation of an image means the laws of trigonometry are applied to rotate the image around its center point by the angle desired. The following equations govern the relationship between the old position (r,c) and the new position (r′, c′). Figure 5.27c shows a rotated version of our bird image.

```
r′ = r*cos(angle) + c*sin(angle)
c′ = c*cos(angle) - r*sin(angle)
```

When performing translation, usually you are moving parts of images to different locations within the same image. An example is shown in Figure 5.27d. To specify a translation, you specify the portion of the image and the destination in (row, column) values of this portion. This destination is usually signified as the new location of the upper left corner of the rectangular area you are translating. You may choose to set the original portion of the image not overlapping with the copied version to black as is done in the image of Figure 5.27d.

The last geometric transformation we discuss is mirroring. This is the "flipping" of the image around a specified axis or line. An example of this is shown in Figure 5.27e where we have mirrored our bird image around the vertical axis.

5.9.3 Calibration

The meaning of sensor calibration depends entirely on the sensor. In the case of visible spectrum images, camera calibration takes on multiple meanings. The first definition, the most common, is the determination of the parameters that map a position of a point in the scene to its position in the image. The second meaning is that of pixel value calibration where the dynamic range of the sensor is configured

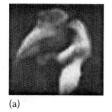 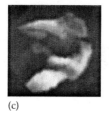 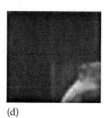 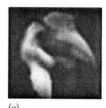

(a) (b) (c) (d) (e)

FIGURE 5.27 (a) Original image, (b) scaled (shrunk) version, (c) rotated version, (d) translated version, and (e) mirrored version around the vertical axis.

to produce expected values. Interestingly enough, the later meaning of adjusting the sensor to produce expected values is what is typically meant by sensor calibration for non-visible-spectrum sensors.

Currently, camera calibration is a cumbersome process of estimating the *intrinsic* and *extrinsic* parameters of a camera. There are four intrinsic camera parameters: two are for the position of the origin of the image coordinate frame, and two are for the scale factors of the axes of this frame. There are six extrinsic camera parameters: three are for the position of the center of projection, and three are for the orientation of the image plane coordinate frame.

Modern CCD cameras are usually capable of a spatial accuracy greater than 1/50 of the pixel size. However, such accuracy is not easily attained due to various errors that can affect the image formation process. Current calibration methods typically assume that the observations are unbiased, the only error is zero-mean independent and uniformly distributed random noise in the observed image coordinates, and the camera model completely explains the mapping between the 3D coordinates and the image coordinates. In general, these conditions are not met, causing the calibration results to be less accurate than expected [Zhang 03].

There are two basic techniques that are commonly employed in camera calibration. The first is called photogrammetric calibration and is done by observing an object whose 3D geometry is known with good precision. The best known such method was performed by TSAI, a system that automatically calibrates a camera, given two planes with a particular shape drawn on them (16 rectangles). The other technique is referred to as self-calibration and is done by moving the camera in a static scene. The rigidity of the scene can be used to produce two constraints on the intrinsic and extrinsic parameters of the camera. Therefore, by obtaining pictures of the camera at different places, we can estimate the intrinsic and extrinsic parameters of the camera.

Calibration can also mean pixel value calibration. In [Bear 03], they discuss color calibration for color cameras. Similar calibration schemes can be developed for each kind of imaging sensor.

5.9.4 Fusion Issues

A number of chapters of this book concentrate on fusion, as this is commonly an important stage of distributed sensor networks. Regarding images and distributed networks, fusion takes place at the pixel level

FIGURE 5.28 Smart camera that allows for on-node processing of an image in a distributed multi-smart camera network. (WiCa 1.1, wireless smart camera; IC3D, SIMD processor; ZigBee, transmitter; also an embedded processor 8051). (From Wu, C. et al., Real-time human posture reconstruction in wireless smart camera networks, *International Conference on Information Processing in Sensor Networks*, pp. 321–331, St. Louis, MO, 2008.)

FIGURE 5.29 Results of processing at a "smart" camera node in a multi-smart card network. White dots indicate body parts features to be transmitted and used for 3D human posture reconstruction. (From Wu, C. et al., Real-time human posture reconstruction in wireless smart camera networks, *International Conference on Information Processing in Sensor Networks*, pp. 321–331, St. Louis, MO, 2008.)

or feature level [Gunatilaka 01, Brooks 98, Wu 08]. When done at the pixel level, the step of registration is especially critical. Both of these types of fusion will be discussed in great detail in later chapters of the book.

In [Wu 08], a system to reconstruct real-time human posture in a wireless "smart" camera network is discussed. In this work, the system does much image processing up through the feature extraction stage at each "smart" camera node in the network. The goal in doing this is to reduce the amount of information that needs to be transmitted across the network. The feature information that represents important body parts like head, shoulders, and hands are then fused at a central location to calculate the 3D human posture. Figure 5.28 shows the hardware of the "smart" camera that allows for on-node processing. Figure 5.29 shows the results of processing in the "smart" camera yielding body part features.

5.10 Compression and Transmission: Impacts on a Distributed Sensor Network

Compression of data is important not only for storage but also for reducing transmission times. This is especially important when the data are large, as typically is the case for images. In distributed sensor networks, it is possible to compress the signal at the sensor before transmission. But compression along with any other processing you might do at the sensor is a trade-off with energy conservation [Maniezzo 02]. There have been a number of proposed systems that discuss the development of distributed compression schemes or sensor-based compression schemes for sensor networks and we will mention a few here.

In [Martina 02], a system is described that uses the wavelet transform for compression of images in a wireless sensor network system. Here to save transmission time, the authors have developed an integrated circuit to implement their algorithm in hardware. While the addition of hardware consumes more power and could be a burden on a distributed sensor network, the savings in transmission time and hence power far outweigh this.

In [Kusuma 01], a very different approach to the problem is taken. They have developed a distributed compression scheme that does not require internode communications in the distributed sensor network to exploit the correlation. While the authors show this scheme can reduce the transmission time and hence energy consumption at each sensor node in the network, it is not clear if this scheme or a hardware-based nondistributed compression scheme like that in [Martina 02] will yield superior results.

Beyond the typical algorithmic compression of images and video [Ahmad 09, Magli 03], there is a body of work dealing with optical compression. In this case, optical elements are utilized to perform calculations like FFT that are done by many other systems in software. See [Cottour 08] for a system that performs optical video compression using a multiplexing method based on spectral fusion.

Transmission of information and the related networking issues for a distributed sensor network is a major topic in this book and discussed in a series of chapters. We will only mention that there is a body of developing work related specifically to images and video and transmission for distributed sensor networks [Kostrzewski 01], [Wagner 03], [Nasri 10].

5.11 Image Resampling

Sometimes the system is presented with an image at a resolution that is not as high as desired. In this case, an image processing algorithm called image resampling can be applied. There are numerous techniques to do image resampling; many that focus on the use of only the image itself look at subpixel calculations often fitting splines between surrounding pixels to create new pixel samples for the higher resolution image.

There is another body of work that looks at fusing multiple images to create new higher resolution images. In [Ghassemian 09], a system is described that merges multispectral image with higher resolution SPOT PAN images resulting in a higher resolution multispectral image that the authors claim maintains more accurate spectral information than other techniques.

5.12 Image Processing for Output

Occasionally, the output of a fusion system will be an image. An example of this is where humans are in the loop such as in the inspection of multisensor data in security operations like detection of weapons in airport security. In these cases, it can be valuable to use image processing techniques to enhance the output image for better understanding of results. A simple image processing technique to perform image enhancement is called pseudocoloring. The idea is to alter the color of pixels in a meaningful way, but one that usually is a pseudo (not direct) conveyance of information through color.

This technique is often used in multispectral image outputs for the easier consumption by human viewers. In [Abidi 06], pseudocoloring is applied to x-ray images for improved weapons detection. Figure 5.30 shows one of the pseudocolor scheme employed in [Abidi 06] and these results prove how valuable a tool that enhances visualization of output images (fused or not) can be.

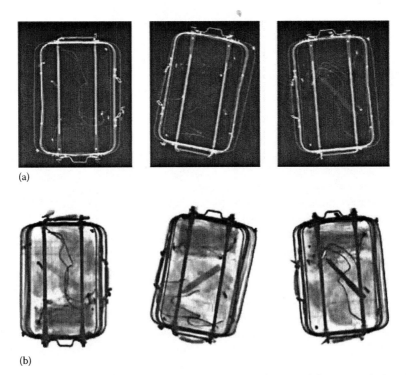

(a)

(b)

FIGURE 5.30 (a) Original images and (b) pseudocolored images using one of the proposed algorithms. (From Abidi, B. et al., *IEEE Trans. Syst. Man Cybern.*, 36(6), 784, 2006.)

5.13 More Imaging in Sensor Network Applications

As a closure for our discussion on image processing, we present a few more examples for sensor networks that use images and image processing. This is only a small sampling of the many that exist but it will give the reader some idea how the techniques discussed in this chapter are applied.

[Gunatilaka 01] describes a sensor fusion system for detection of buried land mines. Here, the system uses GPR, an IR camera, and an EMI sensor. The sensors were used independently and fusion took place via a feature-level fusion scheme that uses an integration of feature information to handle the fact that the sensor data were not necessarily taken at the same locations.

Of interest to us with regards to image processing is the fact that each kind of image was processed differently prior to fusion. Specifically, the GPR images were processed using an iterative technique to try to eliminate ground reflection, a common problem with GPR. Specifically, the onset time and duration of the ground reflection are estimated from low-pass filtered down-range (depth) profiles. The time-domain impulse response of the system is estimated and then iteratively subtracted from the data at points within the ground reflection window taking care not to remove near-surface targets. The features measured for the GPR image include the cumulative energy. The EMI sensor at each measurement point outputs a time-domain waveform. Here, the mean and standard deviation of each trace waveform is used as a feature. Finally, feature extraction for the IR image is performed via the application of a pattern matching subimage filter. This filter is circular in structure with a concentric negative outer ring. The result is a binary image containing "blobs," which indicate the potential presence of a mine. A segmentation algorithm is applied and each feature region is described by its area, pixel count, rectangle dimensions, center, mean temperature (IR measures temperature), and variance.

In [Yamamoto 98], a system using images taken from sensors in the air and on ground is discussed for assistance in detecting and avoiding obstacles for helicopter pilots. Both IR and color video

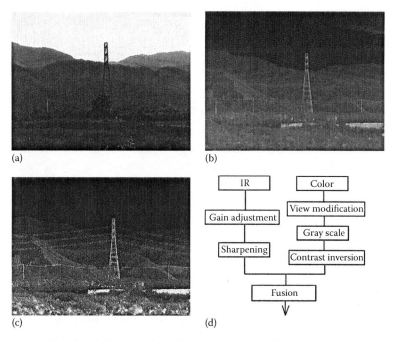

FIGURE 5.31 System for aiding helicopter pilots: (a) color video image, (b) IR image, (c) fused image, and (d) one processing sequence. (From Yamamoto, K. and Yamada, K., Image processing and fusion to detect navigation obstacles, *SPIE Conference on Signal Processing, Sensor Fusion, and Target Recognition VII*, pp. 337–346, Orlando, FL, 1998.)

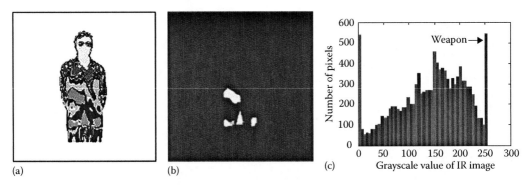

(a)   (b)   (c)

FIGURE 5.32 (a) Results of fuzzy k-means clustering algorithm on IR image, (b) resulting binary mask containing potential areas containing weapons, and (c) histogram of IR image. (From Blum, R. et al., Multisensor concealed weapon detection by using a multiresolution Mosaic approach, *IEEE Vehicular Technology Conference*, pp. 4597–4601, Los Angeles, CA, 2004.)

camera sensors were used. Each type of images is processed differently before a fusion process takes place. The fused information is used for obstacle detection. In the case of the IR image, a contrast inversion process takes place, followed by sharpening and finally histogram equalization. Some of the other filters applied on a case-by-case basis for the IR images included median filtering to reduce noise and some thresholding. In the case of the color video images, a geometric transformation takes place to handle registration followed by conversion to a gray-scale image. Figure 5.31 shows some results produced by the system.

In [Blum 04], a system for hidden weapons detection is used that first detects a weapon in either the IR or millimeter-wave image using a fuzzy k-means clustering algorithm. Figure 5.32 shows the results of the fuzzy k-means clustering algorithm that yields a kind of segmented image on a sample IR image. Next using a mosaic technique of combining only the subimages containing a potential weapon in the IR or millimeter-wave images is performed and fused with the visual camera image. The multiresolution image mosaic technique involves applying weighted averaging operations in the predefined area at different resolutions.

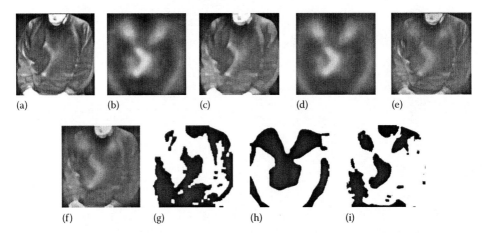

(a)   (b)   (c)   (d)   (e)

(f)   (g)   (h)   (i)

FIGURE 5.33 Image processing and fusion to detect concealed weapons: (a) original IR image, (b) original MMW image, (c) image (a) after 5 × 5 morphological filter applied, (d) image (b) after 5 × 5 morphological filter applied, (e) fused original images, (f) fused filtered images, (g) thresholded version image (c), (h) thresholded version image (d), and (i) fused thresholded images. (From Ramac, L. et al., *Proc. SPIE*, 3375, 110, 1998.)

A wavelet domain–based image fusion system for the detection of concealed weapons is discussed in [Ramac 98]. This system uses IR and millimeter-wave sensors. Unlike the previous system for concealed weapons detection, these two images are completely fused, but, first, the images from both sensors are morphologically filtered to remove image artifacts. These images are then converted to the wavelet domain where fusion takes place. Figure 5.33 shows some images from this work.

An IR sensor array is used in [Feller 02] for the purpose of tracking humans for law enforcement applications. They explore the distribution of these sensors geometrically and how it influences real-time recognition and response. The goal is to have the minimal sensors and yet be able to effectively view the entire scene area. Registration and geometric transformations is key to understanding this problem.

References

[Abidi 06] B. Abidi, Y. Zheng, A. Gribok, M. Abidi, Improving weapon detection in single energy x-ray images through pseudocoloring, *IEEE Transactions on Systems, Man and Cybernetics*, 36(6): 784, 2006.

[Ahmad 09] J. Ahmad, H. Khan, S. Khayam, Energy efficient video compression for wireless sensor networks, *43rd Annual Conference on Information Sciences and Systems*, pp. 629–634, 2009, Baltimore, MD.

[Baxes 94] G. A. Baxes, *Digital Image Processing: Principles and Applications*, John Wiley & Sons, New York, 1994, ISBN-10: 0471009490, ISBN-13: 978-0471009498.

[Bear 03] J. Bear, Picture perfection: Digital camera calibration http://desktoppub.about.com/library/weekly/aa072102a.htm, 2003.

[Bhatia 00] I. Bhatia, V. Diehl, T. Moore, J. Marble, K. Tran, S. Bishop, Sensor data fusion for mine detection from a vehicle-mounted system, *Detection and Remediation Technologies for Mines and Minelike Targets V*, A. Dubey, J. Harvey, J. Broach, R. Dugan (eds.), Vol. 4038, pp. 824–834, SPIE, Orlando, FL, 2000.

[Blum 04] R. Blum, X. Zhiyun, Z. Liu, D. Forsyth, Multisensor concealed weapon detection by using a multiresolution Mosaic approach, *IEEE Vehicular Technology Conference*, pp. 4597–4601, 2004, Los Angeles, CA.

[Brooks 01] R. Brooks, L. Grewe, S. Iyengar, Recognition in the wavelet domain, *Journal of Electronic Imaging*, 10(3): 757–784, July 2001.

[Brooks 98] R. R. Brooks, S. S. Iyengar, *Multi-Sensor Fusion: Fundamentals and Applications with Software*, Prentice Hall PTR, Saddle River, NJ, 1998.

[Brown 92] L. Brown, A survey of image registration techniques, *ACM Computing Surveys*, 24(4): 325–376, 1992.

[Cottour 08] A. Cottour, A. Alfalous, H. Hamam, Optical video image compression: A multiplexing method based on the spectral fusion of information, *Third International Conference on Information and Communication Technologies: From Theory to Applications*, 2008, Damascus, Syria.

[Feller 02] S. Feller, Y. Zheng, E. Cuull, D. Brady, Tracking and imaging humans on heterogeneous infrared sensor arrays for law enforcement applications, *Sensors, and Command, Control, Communications and Intelligence (C3I) Technologies for the Homeland Defense and Law Enforcement*, Vol. 4708, pp. 212–221, SPIE, Orlando, FL, 2002.

[Foresti 02] G. Foresti, L. Snidaro, A distributed sensor network for video surveillance of for video of outdoor environments, *International Conference on Image Processing*, pp. 525–528, 2002, New York.

[Ghassemian 09] H. Ghassemian, Multi-sensor remote sensing image fusion based on retina-inspired model, *IEEE Industrial Electronics and Applications*, 2009, Kuala Lumpur, Malaysia.

[Goermer 10] S. Goermer, S. Hold, A. Kummert, U. Iurgel, M. Meuter, Multi-exposure image acquisition for automotive high dynamic range imaging, *Intelligent Transportation Systems, IEEE Conference on Digital Object Identifier*, pp. 1881–1886, 2010, Funchal, Portugal.

[Grewe 99] L. Grewe, R. R. Brooks, Efficient registration in the compressed domain, *Wavelet Applications VI, SPIE Proceedings*, H. Szu (ed.), Vol. 3723, Aerosense, Orlando, FL, 1999.

[Gunatilaka 01] A. H. Gunatilaka, B. A. Baertlein, Feature-level and decision fusion of non coincidently sampled sensor for land mine detection pattern analysis and machine intelligence, *IEEE Transactions*, 23: 577–589, 2001.

[Kostrzewski 01] A. Kostrzewski, S. Ro, T. Jannson, Visual sensor network based on video streaming and IP-transparency, *Battlespace Digitization and Network-Centric Warfare*, Raja Suresh, Williams E. Roper (Eds.), SPIE, 4741: 4–81, Orlando, FL, 2001.

[Kusuma 01] J. Kusuma, L. Doherty, K. Ramchandran, Distributed compression for sensor networks, *International Conference on Image Processing*, pp. 82–85, 2001, Thessaloniki, Greece.

[Luo 09] R. Luo, C. Nai-Wen, L. Shih-Chi, W. Shish-Chiang, Human tracking and following using sensor fusion approach for mobile assistive companion robot, *35th Annual Conference of IEEE Industrial Electronics*, pp. 2234–2240, 2009, Porto, Portugal.

[Magli 03] E. Magli, M. Mancin, L. Merello, Low-complexity video compression for wireless sensor networks, *International Conference on Multimedia and Expo*, pp. 585–588, 2003, Baltimore, MD.

[Maniezzo 02] D. Maniezzo, K. Yao, G. Mazzini, Energetic trade-off between computing and communication resource in multimedia surveillance sensor network, *Fourth International Workshop on Mobile and Wireless Communications Network*, pp. 373–376, 2002, Stockholm, Sweden.

[Marcello 02] J. Marcello, F. Maques, F. Eugenio, Automatic feature extraction from multisensorial oceanographic imagery, *IEEE International Geoscience and Remote Sensing Symposium*, pp. 2483–2485, 2002, Honolulu, HI.

[Marcenaro 02] L. Marcenaro, L. Marchesotti, C. Regazzoni, A multi-resolution outdoor dual camera system for robust video-event metadata extraction, *FUSION 2002: Proceedings of the Fifth International Conference on Information Fusion*, pp. 1184–1189, 2002, Annapolis, MD.

[Martina 02] M. Martina, G. Masera, G. Piccinini, F. Vacca, M. Zamboni, Embedded IWT evaluation in reconfigurable wireless sensor network, *IEEE International Conference on Electronics Circuits and Systems*, pp. 855–858, 2002, Croatia, Europe.

[Mirzu 98] M. Mirzu, L. Cosereau, M. Jurba, G. Copot, D. Ralea, R. Marginean, About the real performance of image intensifier systems for night vision, *Proceedings of SPIE*, 3405: 926–929, 1998.

[Nasri 10] M. Nasri, A. Helali, H. Sghaier, H. Maaref, Adaptive image transfer for wireless sensor networks (WSNs), *Fifth International Conference on Design and Technology of Integrated Systems in Nanoscale*, pp. 1–7, 2010, Hammamet, Tunisia.

[Nasa 01] NASA, Multiwavelength milky way: Across the spectrum, http://adc.gsfc.nasa.gov/mw/mmw_across.html, Version 2, September 2001.

[Othman 06] H. Othman, S. Quian, Noise reduction of hyperspectral imagery using hybrid spatial-spectral derivative-domain wavelet shrinkage, *IEEE Transactions on Geoscience and Remote Sensing*, 44(2): 397–408, 2006.

[Razlighi 07] Q. Razlighi, N. Kehtarnavaz, Correction of over-exposed images captured by cell-phone cameras, *IEEE Symposium on Consumer Electronics*, pp. 1–6, 2007, Irving, TX.

[Ramac 98] L. Ramac, M. Uner, P. Varshneyu, Morphological filters and wavelet based image fusion for concealed weapons detection, *Proceedings of SPIE*, 3375: 110–119, 1998.

[Ros 09] J. Ros, K. Mekhnacha, Multi-sensor human tracking with the Bayesian occupancy filter, *International Conference on Digital Signal Processing*, 2009, Santorini, Greece.

[Sveinsson 01] J. Sveinsson, M. Ulfarsson, J. Benediktsson, Cluster-based feature extraction and data fusion in the wavelet domain, *IEEE International Geoscience and Remote Sensing Symposium*, pp. 867–869, 2001, Sydney, New South Wales, Australia.

[Verma 01] A. Verma, Bogoon Jung, G. S. Sukhatme, Computational Intelligence in Robotics and Automation. *IEEE International Symposium on computational Intelligence in Robotic and Automation*. pp. 212–217, July 2001, Alberta, Canada.

[Wang 06] Q. Wang, Y. Shen, Effects of wavelets selection on performances of hyperspectral image fusion, *Instrumental and Measurement Technology Conference*, pp. 812–815, 2006, Sorrento, Italy.

[Wagner 03] R. Wagner, R. Nowak, R. Baraniuk, Distributed image compression for sensor networks using correspondence analysis and super-resolution, *International Conference on Image Processing*, pp. 597–600, 2003, Barcelona, Spain.

[Waxman 07] A. Waxman, D. Fay, P. Ilardi, P. Arambel, J. Silver, Active tracking of surface targets in fused video, *International Conference on Information Fusion*, 2007, Québec City, Québec, Canada.

[Wu 08] C. Wu, H. Aghajan, R. Kleihorst, Real-time human posture reconstruction in wireless smart camera networks, *International Conference on Information Processing in Sensor Networks*, pp. 321–331, 2008, St. Louis, MO.

[Yamamoto 98] K. Yamamoto, K. Yamada, Image processing and fusion to detect navigation obstacles, *SPIE Conference on Signal Processing, Sensor Fusion, and Target Recognition VII*, pp. 337–346, 1998, Orlando, FL.

[Zamora 98] G. Zamora, M. Dickens, S. Mitra, A robust registration technique for multi-sensor images, *IEEE Southwest Symposium on Image Analysis and Interpretation*, pp. 87–90, 1998, Tucson, AZ.

[Zhang 01] Z. Zhang, S. Sun, F. Zheng, Image fusion based on median filters and SOM neural networks: A three-step scheme, *Signal Processing*, 81(6): 1325–1330, 2001.

[Zhang 03] Z. Zhang, A flexible new technique for camera calibration, http://research.microsoft.com/~zhang/Calib, 2003.

6

Object Detection and Classification

6.1 Introduction 93
6.2 A Signal Model for Sensor Measurements.....................................95
    Example: Temporal Point Sources
6.3 Object Detection 98
    Soft Decision Fusion • Hard Decision Fusion
6.4 Object Classification..102
    Soft Decision Fusion • Hard Decision Fusion • Numerical Results
6.5 Conclusions..108
    Realistic Modeling of Communication Links • Multi-Object
    Classification • Nonideal Practical Settings
Acknowledgment...109
References..109

Akbar M. Sayeed
*University of
Wisconsin-Madison*

6.1 Introduction

Wireless sensor networks are an emerging technology for monitoring the physical world with a densely distributed network of wireless nodes [1,2]. Each node has limited communication and computation ability and can sense the environment in a variety of modalities, such as acoustic, seismic, and infrared. In principle, sensor networks can be deployed anywhere: on the ground, in the air, or in the water. Once deployed, the nodes have the ability to communicate with each other and configure themselves into a well-connected network. A wide variety of applications are being envisioned for sensor networks, including disaster relief, border monitoring, condition-based machine monitoring, and surveillance in battlefield scenarios. Detection and classification of objects moving through the sensor field is an important task in many applications. Exchange of sensor information between different nodes in the vicinity of the object is necessary for reliable execution of such tasks for a variety of reasons, including limited (local) information gathered by each node, variability in operating conditions, and node failure. Consequently, development of theory and methods for collaborative signal processing (CSP) of the data collected by different nodes is an important research area for realizing the promise of sensor networks.

The exchange of information between nodes for CSP comes at the expense of network resources. The two most critical network resources are: (1) the bandwidth available for communication between nodes, and (2) the power available at each node for communication and computation. In the case of battery-powered nodes, both constraints are critical. Thus, a key goal in the design of CSP algorithms is to exchange the least amount of data between nodes to attain a desired level of performance. In this chapter, with the above goal in mind, we discuss CSP algorithms for detection and classification of objects using multiple measurements at different nodes.

Some form of region-based signal processing is needed in sensor networks in order to facilitate CSP and also for efficient routing of information through the network, e.g., see Ref. [3]. A region-based

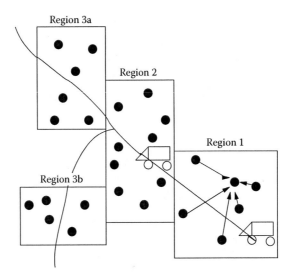

FIGURE 6.1 A region-based approach for object tracking in sensor networks.

approach for object tracking is illustrated in Figure 6.1. Typically, the nodes in the network are partitioned into a number of regions and a manager node is designated within each region to facilitate CSP between the nodes in the region and for communication of information from one region to another. Object detection, classification, and tracking generally involves the following steps [3]:

1. *Object detection and data collection*: An object is detected in a particular region which becomes the active region (e.g., region 1 in Figure 6.1). The detection of an object itself may involve CSP between nodes. For example, signal energy measurements at different nodes may be communicated to the manager node to make the final decision about object detection. The nodes within the active region also collect time series data in different modalities that may be used for more sophisticated tasks, such as object classification.
2. *Object localization*: Information related to object detection collected at different nodes (such as the time of closest point of approach and energy measurements) is used by the manager node to estimate the location of the target.
3. *Object location prediction*: Location estimates over a period of time are used by the manager node to predict object location at future time instants.
4. *Creation of new regions*: When the object gets close to exiting the current active region, the estimates of predicted object location are used to put new regions on alert for object detection (e.g., regions 3a and 3b in Figure 6.1).
5. *Determination of new active region*: Once the object is detected in a new region, it becomes the new active region. The above four steps are repeated in the new active region for object tracking through the sensor field.

The CSP techniques discussed in this chapter apply to data collected by different nodes in a particular active region. As we will see, CSP of data collected in different active regions can be done independently. While we specifically discuss CSP methods for detection and classification of a single object, the basic principles apply to distributed decision making in sensor networks in general.

There are two main forms of information exchange between nodes dictated by the statistics of measured signals. If two nodes yield correlated measurements, *data fusion* is needed for optimal performance—exchange of (low-dimensional) feature vectors that yield sufficient information for the desired task. For example, estimates of signal energy at different frequencies (Fourier/spectral feature vectors) may be used for classification. On the other hand, if two nodes yield statistically independent measurements, *decision fusion* is sufficient—exchange of soft decisions (real-valued scalars) or hard decisions (discrete-valued scalars) computed at the two nodes. In general, the measurements at different nodes would exhibit a mixture of correlated and independent components and would require a combination of data and decision fusion between nodes. In the context of sensor networks, decision fusion is clearly the more attractive choice. First, it imposes a significantly lower communication burden on the network, compared with data fusion, since only scalars are communicated to the manager node. Second, it also imposes a lower computational burden compared with data fusion, since lower dimensional data have to be jointly processed at the manager node. Third, a classifier based on decision fusion requires a much smaller amount of data for *training,* since fewer parameters characterize the classifier compared with data fusion.

An object in a region covered by a sensor network generates a signal field in space and time that can be sensed by the nodes in different modalities. In Section 6.2 we present a basic but general model for the signal field generated by an object. The model provides a simple characterization of the signal statistics in space and time associated with an object of interest. In particular, the model imposes a universal structure on all CSP algorithms for decision making in which costly* data fusion is confined to local sub-regions of an active region, and only cheaper decision fusion is needed across different sub-regions. This model forms the basis of the CSP algorithms presented in the remainder of the chapter. In any network query involving an object (such as a vehicle), the first task is to detect the presence of the object in a region of interest. Section 6.3 discusses CSP algorithms for object detection. Once an object has been detected, the next logical task is to classify the object as belonging to one of a finite number of classes. Section 6.4 discusses CSP algorithms for object classification. In both detection and classification, we discuss algorithms for soft and hard decision fusion and illustrate the performance gains due to multiple node measurements with numerical results. Section 6.5 concludes the chapter with an overview of areas of current and future research.

6.2 A Signal Model for Sensor Measurements

In this section we present a simple model for characterizing the statistics of signals associated with objects in the sensor field. This model will then be used to develop CSP algorithms for object detection and classification. Consider a region of interest $R = D_x \times D_y$, illustrated in Figure 6.2, associated with a network query involving a set of objects. It represents a rectangular region of area $D_x D_y$ (m^2). The network senses the object via the signals collected by the nodes, possibly in multiple modalities. Consider a single sensing modality (e.g., acoustic signals collected by microphones) and a single object (e.g., a vehicle).

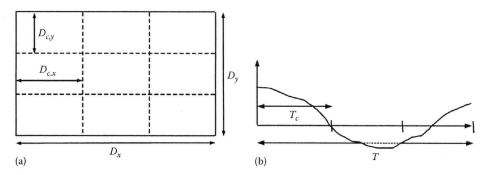

FIGURE 6.2 A schematic illustrating the notion of spatial coherence regions and coherence time over which the space–time signal of interest $s(x,y,t)$ remains approximately constant (strongly correlated). (a) Coherence regions in space. (b) Coherence intervals in time.

---

* This relates to network cost in terms of bandwidth and power expenditure.

Let $s(x,y,t)$ denote the signal due to the vehicle at spatial coordinates (x,y) and at time t. The signal $s(x,y,t)$ is best modeled as a random process due to a variety of sources of uncertainty. We assume that $s(x,y,t)$ is zero-mean Gaussian stationary process (or field) as a function of (x,y,t) [4]. As illustrated in Figure 6.2, we divide the region R into spatial coherence (sub-) regions (SCRs) of size $D_{c,x} \times D_{c,y}$ over which $s(x,y,t)$ is assumed constant as function of (x,y), and the constant value varies statistically independently from one SCR to another. The size of an SCR depends on the statistical signal characteristics, as explained next.

To appreciate the notion of SCRs, let us consider a random process $s(t)$ as a function of one variable (time). Since $s(t)$ is a zero-mean Gaussian stationary random process, its statistics are completely determined by the correlation function

$$r_s(\Delta t) = E[s(t)s(t - \Delta t)] = \int_{-B/2}^{B/2} \Phi_s(f)e^{j2\pi f \Delta t} df \qquad (6.1)$$

which is related to the power spectral density (PSD) $\Phi_s(f)$ of the process via a Fourier transform. B denotes the bandwidth of the process and determines how fast the process changes over time.

In particular, the process remains strongly correlated over any time interval of duration $T_c = 1/B$, which is called the coherence time and is illustrated in Figure 6.2b [5].* Thus, all samples of $s(t)$ taken within a duration T_c will be strongly correlated, whereas samples in disjoint time intervals of duration T_c will be approximately statistically independent. The same general principle applies to the space-time signal $s(x,y,t)$. Specifically, let B_x and B_y denote the bandwidths associated with spatial dimensions x and y respectively, analogous to the bandwidth B in time. Then, $D_{c,x} = 1/B_x$ and $D_{c,y} = 1/B_y$ denote the coherence distances in x and y dimensions over which the signal remains strongly correlated. The two coherence distances define SCRs of size $D_{c,x} \times D_{c,y}$, as illustrated in Figure 6.2a.

6.2.1 Example: Temporal Point Sources

In general, the spatial and temporal signal characteristics can be arbitrary. However, for this important class of signal sources they are coupled. Acoustic signals emitted by vehicles, as well as seismic (vibration) signals produced by moving vehicles, can be modeled in this fashion. Such space-time signals are completely characterized by an underlying temporal signal $s_o(t)$ via signal propagation in space. For isotropic spatial propagation, $s(x,y,t) = s(r,t) = s_o(t - r/v)$ where $r = \sqrt{x^2 + y^2}$ and v is the speed of propagation. Thus, the signal is stationary along radial lines. It can be shown that $B_r = B/v$, where B_r is the spatial bandwidth in the radial dimension and B is the temporal bandwidth of $s_o(t)$. The SCRs are concentric bands around the source and the radial coherence distance $D_{c,r}$ is given by $D_{c,r} = 1/B_r = v/B = vT_c$. For example, for $B = 500\,\text{Hz}$, $D_r = 0.66\,\text{m}$, whereas for $B = 20\,\text{Hz}$, $D_r = 17\,\text{m}$.

Based on the above discussion, we make two slightly idealized assumptions about signal variation as a function of (x,y):

1. $s(x,y,t)$ is perfectly correlated in each SCR; i.e., at any given time the signal in the (i,j)th SCR is constant as a function of (x,y), $s(x,y,t) = s_{i,j}(t)$, $(x,y) \in \text{SCR}_{i,j}$.
2. The signal values $s_{i,j}(t)$ in different SCRs are statistically independent.

In the region $R = D_x \times D_y$, there are $G = N_x N_y$ independent SCRs, where $N_x = D_x/D_{c,x} = D_x B_x$ and $N_y = D_y/D_{c,y} = D_y B_y$, and we label the SCRs as $\text{SCR}_{i,j}$, $i = 1, \ldots, N_x$, $j = 1, \ldots, N_y$. In some cases, for simplicity, we will label the SCRs by a single index: SCR_k, $k = 1, \ldots, G = N_x N_y$. We assume that there are n_G nodes in each SCR, resulting in a total of $K = G n_G$ nodes in the query region R from which the measurements are collected. Note that under our model, all n_G nodes in a particular SCR, say $\text{SCR}_{i,j}$ (or SCR_k), observe the same time signal $s_{i,j}(t)$ (or $s_k(t)$).

---

* Note that B is also the required Nyquist rate for sampling the process without any loss of information.

In practice, the sensor node measurements will be corrupted by noise. Mathematically, the time signal sensed by the kth node is given by

$$z_k(t) = s_k(t) + n_k(t), \quad k = 1,\dots,K = Gn_G \tag{6.2}$$

where

$s_k(t)$ denotes the stationary Gaussian signal due to the object of interest (as discussed above)
$n_k(t)$ denotes additive noise

We assume that $n_k(t)$ is a zero-mean Gaussian white-noise process and that the noise processes at different nodes (whether within an SCR or not) are statistically independent. However, all signal measurements $s_k(t)$ in any particular SCR are identical, and they vary statistically independently from one SCR to the other. At each node, the signal is sampled as

$$z_k[i] = z_k\left(\frac{i}{W}\right) = s_k\left(\frac{i}{W}\right) + n_k\left(\frac{i}{W}\right) = s_k[i] + n_k[i] \tag{6.3}$$

where W denotes the sensor bandwidth (in hertz). We assume that the sampled signal is processed in disjoint blocks of N time samples corresponding to a block duration of $T_o = N/W$ seconds. We denote the nth block of time samples at the kth node as an N-dimensional vector: $[z_k[n] = [z_k[nN], z_k[nN + 1], \dots, z_k[n(N + 1) - 1)]^T$. Thus, every T_o seconds, we collect $K = Gn_G$ sampled measurement vectors, $z_k, k = 1, \dots, K$, each of dimension N, and there are n_G vectors in each of the G SCRs.

The above signal model has important implications for distributed detection and classification algorithms, both from a fundamental decision theoretic viewpoint and from the viewpoint of information exchange between nodes. On a fundamental level, there are two sources of error in decision making: (1) the additive noise and (2) the inherent statistical variability in the source signal. The notion of SCRs illustrated in Figure 6.2 imposes a natural structure on optimal detectors and classifiers that enables us to mitigate both sources of error. Under the assumptions of our signal model, all CSP algorithms for optimal decision making share the following structure:

1. Since the source signal is nearly constant in each SCR, the n_G measurements in each SCR are averaged to mitigate the effect of noise and increase the effective signal-to-noise ratio (SNR) by a factor of n_G.
2. Since the G averaged measurements in different SCRs are statistically independent, they are combined appropriately to reduce the inherent statistical variability in the source signal.

Both of these aspects improve the performance of detection and classification algorithms; but, as we will see, the second effect is more critical in the context of random source signals. For the remainder of the chapter we assume that the n_G sampled vectors z_k in each SCR are averaged to yield a single N-dimensional vector z_k for each SCR:

$$z_k = \frac{1}{n_G} \sum_{i \in SCR_k} z_i = s_k + n_k, \quad k = 1,\dots,G \tag{6.4}$$

Note that this averaging corresponds to data fusion, since the N-dimensional feature vectors are exchanged between the nodes in each SCR. This data fusion in each SCR could be coordinated by the manager for the entire region R or separate manager nodes could be designated for each SCR. The net result of this averaging is that the signal component remains unchanged, since it is constant in each SCR, whereas the averaging of the noise reduces its variance by a factor of n_G. Thus, if the original noise variance is σ_n^2, then the variance of the averaged noise becomes σ_n^2/n_G. Thus, we work with a total of $G = N_x N_y$ statistically independent averaged measurement vectors $\{z_k\}$ from different SCRs as in Equation 6.4, where $s_k \sim \mathcal{N}(0, \Sigma)$ and $n_k \sim \mathcal{N}(0, \sigma_n^2 I/n_G)$. The notation $s \sim \mathcal{N}(\mu, \Sigma)$ means that the vector s is a vector of Gaussian (normal) random variables with mean μ and covariance matrix Σ, i.e.,

$$\mu[i] = E[s[i]] \quad \text{and} \quad \sum[i,j] = E[(s[i] - \mu[i])(s[j] - \mu[j])], \quad i, j = 1, \dots, N \tag{6.5}$$

I denotes an identity matrix and $n_k \sim \mathcal{N}(0, \sigma_n^2 I/n_G)$ means that the noise vector has zero-mean independent components with variance σ_n^2/n_G.

6.3 Object Detection

Consider a network query of the form: "Is there a vehicle in the region R"? This corresponds to object detection, which is a natural precursor to classification of the object involved. Mathematically, this corresponds to a binary hypothesis test

$$H_0: z_k = n_k, \quad k = 1, \dots, G \tag{6.6}$$

$$H_1: z_k = s_k + n_k, \quad k = 1, \dots, G \tag{6.7}$$

where z_k denotes the averaged N-dimensional sampled vector from the kth SCR in a given time block (we ignore the block index for simplicity). H_0 corresponds to the hypothesis that no object is present, i.e., $z_k = n_k \sim \mathcal{N}(0, \sigma_n^2 I/n_G)$. On the other hand, H_1 represents the presence of an object, i.e., $z_k = s_k + n_k \sim \mathcal{N}(0, (\sigma_s^2 + \sigma_n^2/n_G)I)$. For simplicity we have assumed that the different components of the signal vector s_k are independently identically distributed (i.i.d.) with variance σ_s^2. However, this assumption does not alter the nature of the detector. We consider two types of detector. First, we consider a detector that combines real-valued *soft decisions* from different SCRs and serves as an idealized detector. Second, we consider a practical detector that combines binary-valued *hard decisions* from the G SCRs to make the final decision.

6.3.1 Soft Decision Fusion

Under the above assumptions, the optimal detection statistic is given by

$$l(z_1, \dots, z_G) = \frac{1}{NG} \sum_{k=1}^{G} \|z_k\|^2 = \frac{1}{NG} \sum_{k=1}^{G} \sum_{n=1}^{N} z_k^2[n] \tag{6.8}$$

which is the average energy in the measurements. Note that in this case each node communicates the energy in its local measurement, $\|z_k\|^2$, to the manager node. The final detector implemented at the manager node is called an *energy detector* and it makes the decision d by comparing l with a threshold γ

$$d(z_1, \dots, z_k) = \begin{cases} 1 & l > \gamma \\ 0 & l \leq \gamma \end{cases} \tag{6.9}$$

The decision $d = 1$ corresponds to H_1 (object present) and $d = 0$ corresponds to H_0 (no object present). The threshold γ has to be chosen carefully to control detector performance. Two important performance criteria are: (1) the probability of false alarm (PFA)—the probability of declaring an event detection under H_0 (when only noise is present); and (2) probability of detection (PD)—the probability of declaring an event detection under H_1 (when the signal is actually present). Mathematically:

$$\text{PFA} = P(D=1|H_0) = P(L>\gamma|H_0), \quad \text{PD} = P(D=1|H_1) = P(L>\gamma|H_1) \tag{6.10}$$

Note that uppercase L denotes the random variable corresponding to the detection statistic l (lower case) in Equation 6.8. Similarly, uppercase D denotes the random variable corresponding to the decision d (lower case) in Equation 6.9. This customary notation for representing random variables with uppercase letters, and their particular values (realizations) with lowercase letters is used throughout the chapter.

Ideally, we would like PFA to be as small as possible and PD to be as large as possible. In practice, the threshold y is chosen to keep PFA below a prescribed value (e.g., less than 5%). Let $\sigma_0^2 = E_0[Z_k^2[n]] = E[N_k^2[n]] = \sigma_n^2/n_G$ denote the average power in the measurements under H_0, where $E_0[.]$ denotes expectation under H_0. Similarly, let $\sigma_1^2 = E_1[Z_k^2[n]] = E[S_k^2[n]] + E[N_k^2[n]] = \sigma_s^2 + \sigma_0^2 = \sigma_s^2 + \sigma_n^2/n_G$ denote the average measurement power under H_1, where σ_s^2 denotes the power in each signal vector component. Then, for any even value of NG ($NG = 2m$), the PFA and PD can be computed as a function of $\gamma > 0$ as [5]

$$\text{PFA}(\gamma) = e^{-NG\gamma/2\sigma_0^2} \sum_{k=0}^{m-1} \frac{1}{k!}\left(\frac{NG\gamma}{2\sigma_0^2}\right)^k \tag{6.11}$$

$$\text{PD}(\gamma) = e^{-NG\gamma/2\sigma_1^2} \sum_{k=0}^{m-1} \frac{1}{k!}\left(\frac{NG\gamma}{2\sigma_1^2}\right)^k \tag{6.12}$$

Note that PD and PFA depend on N and G as well as the SNR

$$\text{SNR} = \frac{E_1[Z_k^2[n]]}{E_0[Z_k^2[n]]} = \frac{\sigma_1^2}{\sigma_0^2} = 1 + n_G\frac{\sigma_s^2}{\sigma_n^2} = 1 + n_G\text{SNR}_0 \tag{6.13}$$

which clearly shows the improvement in SNR as a function of the number n_G of sensor measurements in each SCR (SNR$_0$ denotes the SNR at each sensor).

A typical way of characterizing detector performance is to plot PD(γ) as a function of PFA(γ) for different values of γ. The resulting plot is called the receiver operating characteristic (ROC). Figure 6.3

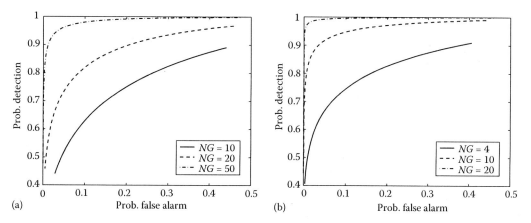

FIGURE 6.3 ROC curves for the soft decision fusion energy detector for different values of independent measurements NG. The curve for each value of NG is generated by varying γ between σ_0^2 and σ_1^2. (a) $\sigma_0^2 = 1$ and $\sigma_1^2 = 2$ (b) $\sigma_0^2 = 1$ and $\sigma_1^2 = 4$.

shows ROC curves for different values of *NG*. Figure 6.3a corresponds to SNR = 2 and Figure 6.3b corresponds to SNR = 4. Comparing the two plots, it is clear that the performance improves with SNR. Furthermore, for a given SNR, the performance improves with increasing *NG*, which could be increased by increasing *G*—number of independent measurements from different SCRs.

Detector design thus boils down to choosing the value of γ that corresponds to the desired operating point on the ROC. A value of γ between σ_0^2 and σ_1^2 usually suffices. To see this, note that the detection statistic random variable L in Equation 6.8 is the average of *NG* i.i.d. random variables $\{Z_k^2[n]\}$. Furthermore, since $\{Z_k[n]\}$ are zero-mean Gaussian, by the moment theorem for Gaussian random variables [5] we have

$$E_i[Z_k^2[n]] = \sigma_i^2, \quad \text{var}_i[Z_k^2[n]] = 2\sigma_i^4, \quad i = 0, 1 \tag{6.14}$$

where var$[Z] = E[Z^2] - (E[Z])^2$ denotes the variance of Z. Consequently, by the law of large numbers [5] we have

$$E_i[L] = \sigma_i^2, \quad \text{var}_i[L] = \frac{2\sigma_i^4}{NG}, \quad i = 0, 1 \tag{6.15}$$

The mean of L is σ_i^2 under H_i, and thus a value of γ between the two means usually suffices. Furthermore, since the variance of L goes to zero as *NG* increases (independent measurements reduce the statistical variability in the decision statistic), we expect PFA → 0 and PD → 1 as $G \to \infty$. To see this, let $\gamma = \sigma_0^2 + \delta$, where $0 < \delta < \sigma_1^2 - \sigma_0^2 = \sigma_s^2$. Then, using Tchebyshev's inequality [5], it can be shown that

$$\text{PFA}(\gamma) \le P\left(|L - \sigma_0^2|^2 \ge \delta^2 \mid H_0\right) \le \frac{\text{var}_0[L]}{\delta^2} = \frac{\sigma_0^4}{NG\delta^2} \tag{6.16}$$

$$\text{PD}(\gamma) \ge P\left(|L - \sigma_1^2|^2 \le \left(\sigma_s^2 - \delta\right)^2 \mid H_1\right) \ge 1 - \frac{\text{var}_1[L]}{\left(\sigma_s^2 - \delta\right)^2} = 1 - \frac{2\sigma_1^4}{NG\left(\sigma_s^2 - \delta\right)^2} \tag{6.17}$$

Thus, for any value of γ satisfying $\sigma_0^2 < \gamma < \sigma_1^2 = \sigma_0^2 + \sigma_s^2$ we get closer to perfect performance (PD = 1 and PFA = 0) as *NG* increases.

6.3.2 Hard Decision Fusion

Recall that the measurement energy at the kth SCR, $\|z_k\|^2$, is communicated to the manager node that implements the soft decision fusion detector based on the G energy values. However, energy is a nonnegative real number and, in general, each SCR has to encode it with a finite number of bits (quantized) to communicate it digitally to the manager node. In this section we consider a particular form of quantization in which the kth SCR makes a hard decision based on its local measurement vector z_k

$$u_k(z_k) = \begin{cases} 1 & \text{if } \|z_k\|^2 / N > \gamma \\ 0 & \text{if } \|z_k\|^2 / N \le \gamma \end{cases} \tag{6.18}$$

and then communicates its binary decision u_k to the manager node. The final detector at the manager node (for the entire region) computes the following averaged hard decision statistic

$$l'(u_1, \dots, u_G) = \frac{1}{G} \sum_{k=1}^{G} uk \tag{6.19}$$

and compares it to a threshold γ' to make the final decision

$$d_{hard}(u_1,\ldots,u_k) = \begin{cases} 1 & \text{if } l' > \gamma' \\ 0 & \text{if } l' \leq \gamma' \end{cases} \tag{6.20}$$

Since the $\{z_k\}$'s are i.i.d. so are $\{u_k\}$. Each u_k is a binary-valued Bernoulli random variable characterized by the following two probabilities under the two hypotheses:

$$p_1[1] = P(U_k = 1 \mid H_1), \quad p_1[0] = P(U_k = 0 \mid H_1) = 1 - p_1[1] \tag{6.21}$$

$$p_0[1] = P(U_k = 1 \mid H_0), \quad p_0[0] = P(U_k = 0 \mid H_0) = 1 - p_0[1] \tag{6.22}$$

We note from Equations 6.8 and 6.18 that $p_1[1]$ is the PD and $p_0[1]$ is the PFA for the soft decision fusion detector when $G = 1$. It follows that the hard decision statistic in Equation 6.19 is a (scaled) binomial random variable under both hypotheses, and thus the PD and PFA corresponding to d_{hard} can be computed as a function of γ' as [5]

$$PD(\gamma') = P(D_{hard} > \gamma' \mid H_1) = 1 - \sum_{k=0}^{\lfloor \gamma'G \rfloor} \binom{G}{k} p_1[1]^k (1 - p_1[1])^{G-k} \tag{6.23}$$

$$PFA(\gamma') = P(D_{hard} > \gamma' \mid H_0) = 1 - \sum_{k=0}^{\lfloor \gamma'G \rfloor} \binom{G}{k} p_0[1]^k (1 - p_0[1])^{G-k} \tag{6.24}$$

Thus, we see that the design of the hard decision fusion detector boils down to the choice of two thresholds: γ in Equation 6.18 that controls $p_1[1]$ and $p_0[1]$; and γ' in Equation 6.20 that, along with γ, controls the PFA and PD of the final detector. Since $E_i[\|Z_k\|^2/N] = \sigma_i^2$ under H_i, the threshold γ can, in general, be chosen between σ_0^2 and $\sigma_1^2 = \sigma_0^2 + \sigma_s^2$ to yield a sufficiently low $p_0[1]$ (local PFA) and a corresponding $p_1[1] > p_0[1]$ (local PD). The threshold γ' can then be chosen between $p_0[1]$ and $p_1[1]$. To see this, note that the mean and variance of each U_k are: $E_i[U_k] = p_i[1]$ and $var_i[U_k] = p_i[1](1 - pi[1])$, $i = 0,1$. Again, by the law of large numbers, $E_i[L'] = p_i[1]$ and $var_i[L'] = p_i[1](1 - p_i[1])/G$, $i = 0,1$. Thus, as long as $p_1[1] > p_0[1]$, which can be ensured via a proper choice of γ, the mean of l' is distinct under the two hypotheses and its variance goes to zero under both hypotheses as G increases. Let $\gamma' = p_0[1] + \delta$, where $0 < \delta < p_1[1] - p_0[1]$. Using Tchebyshev's inequality, as in soft decision fusion, it can be shown that for d_{hard}

$$PD(\gamma') = P(L' > \gamma' \mid H_1) \geq 1 - \frac{E_i[(L' - p_1[1])^2]}{(p_1[1] - \gamma')^2} = 1 - \frac{p_1[1](1 - p_1[1])}{G(p_1[1] - p_0[0] - \delta)^2} \tag{6.25}$$

$$PFA(\gamma') = P(L' > \gamma' \mid H_0) \leq \frac{E[(L' - p_0[1])^2]}{\delta^2} = \frac{p_0[1](1 - p_0[1])}{G\delta^2} \tag{6.26}$$

Thus, as long as γ' is chosen to satisfy $p_0[1] < \gamma' < p_1[1]$, we attain perfect detector performance as $G \to \infty$.

Figure 6.4 plots the ROC curves for the energy detector based on hard decision fusion. The two chosen sets of values for $(p_0[1], p_1[1])$ are based on two different operating points on the $NG = 10$ ROC curve in Figure 6.3a for the soft decision fusion detector. Thus, the local hard decisions (u_k) corresponding to Figure 6.4 can be thought of as being based on $N = 10$-dimensional vectors. Then, the $G = 5$ curves

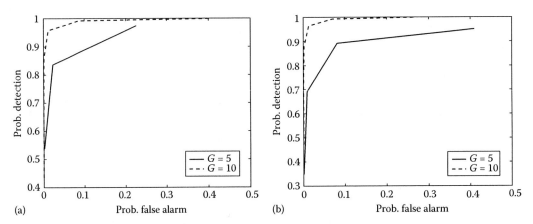

FIGURE 6.4 ROC curves for the hard decision fusion detector for different values of independent measurements G. The curve for each value of G is generated by varying γ' between $p_0[1]$ and $p_1[1]$. (a) $p_0[1] = 0.05$ and $p_1[1] = 0.52$; (b) $p_0[1] = 0.1$ and $p_1[1] = 0.63$.

for hard decision fusion in Figure 6.4 can be compared with the $NG = 50$ curve in Figure 6.3a for soft decision fusion. In soft decision fusion, the energies of the $N = 10$-dimensional vectors at the $G = 5$ independent nodes are combined to yield the $NG = 50$ curve in Figure 6.3a. On the other hand, hard decisions based on $N = 10$-dimensional vectors at the $G = 5$ independent nodes are combined in hard decision fusion to yield the $G = 5$ curve in Figure 6.4a. It is clear that the difference in performance between hard and soft decision fusion is significant. However, the $G = 10$ curve for hard decision fusion in Figure 6.4a yields better performance than the $NG = 50$ curve for soft decision fusion in Figure 6.3a. Thus, hard decision fusion from ten SCRs performs better than soft decision fusion from five SCRs. Thus, we conclude that hard decision fusion from a sufficient number of SCRs may be more attractive (lower communication cost) than soft decision fusion. However, a more complete comparison requires carefully accounting for the communication cost of the two schemes.

6.4 Object Classification

In Section 6.3 we discussed object detection, deciding whether there is a vehicle present in a region R or not. The optimal detector compares the energy in the measurements with a threshold. Suppose the answer to the detection query is positive, i.e., there is a vehicle present in the region of interest. The next logical network query is to classify the vehicle. An example query is: Does the vehicle belong to class A, B, or C? Such a classification query is the focus of this section. We assume that the vehicle can be from one of M possible classes. Mathematically, this corresponds to choosing one out of M possible hypotheses, as opposed to two hypotheses in object detection.

In event (vehicle) detection, the performance was solely determined by signal energy in the two hypotheses. In single vehicle classification, we have to decide between M hypotheses. Thus, we need to exploit more detailed statistical characteristics (rather than energy) of the N-dimensional source signal vector s_k at each node. An important issue is what kind of N-dimensional measurements z_k should be collected at each node? This is the called *feature selection* [6]. Essentially, the raw time series data collected over the block time interval T_o at each node is processed to extract a relevant feature vector that best facilitates discrimination between different classes. Feature selection is a big research area in its own right, and we will not discuss it here; we refer the reader to Duda et al. [6] for a detailed discussion.

We will assume a particular type of feature vector—spectral feature vector—that can be obtained by computing a Fourier transform of the raw data [3]. This is a natural consequence of our signal model, in which the signals emitted by objects of interest are modeled as a stationary process [7]. Thus, we assume that the N-dimensional feature vector z_k is obtained by Fourier transformation of each block of raw time series data (whose length can be longer than N). An important consequence of Fourier features is that the different components of s_k correspond to different frequencies and are approximately statistically independent with the power in each component, $E[S_k^2[n]]$, proportional to a sample of the PSD associated with the vehicle class as defined in Equation 6.1; that is, $E[S_k^2[n]]\alpha\,\Phi_s((n-1)/N), n=1,\ldots,N$. Furthermore, the statistics of n_k remain unchanged, since Fourier transformation does not change the statistics of white noise.

Mathematically, we can state the classification problem as an M-ary hypothesis testing problem

$$H_j:\ z_k = s_k + n_k,\quad k=1,\ldots,G,\quad j=1,\ldots,M \tag{6.27}$$

where $\{n_k\}$ are i.i.d. $N(0,\sigma_n^2 I/n_G)$ as before but $\{s_k\}$ are i.i.d. $\mathcal{N}(0,\Lambda_j)$ under H_j, where Λ_j is a diagonal matrix (since the different Fourier components of s_k are uncorrelated) with diagonal entries given by $\{\lambda_j[1],\ldots,\lambda_j[N]\}$, which are nonnegative and are proportional to samples of the PSD associated with class j as discussed above. Thus, under H_j, $\{z_k\}$ are i.i.d. $N(0,\tilde\Lambda_j)$, where $=\tilde\Lambda_j=\Lambda_j+\sigma_n^2 I/nG$. Based on the measurement vectors $\{z_k\}$ from the G SCRs, the manager node has to decide which one of the M classes the detected vehicle belongs to. We discuss CSP algorithms for classification based on fusion of both soft and hard decisions from each node.

6.4.1 Soft Decision Fusion

Assuming that different classes are equally likely, the optimal classifier chooses the class with the largest likelihood [5–7]:

$$C(z_1,\ldots,z_G)=\arg\max_{j=1,\ldots,M} p_j(z_1,\ldots,z_G) \tag{6.28}$$

where $p_j(z_1,\ldots,z_G)$ is the probability density function (pdf) of the measurements under H_j. Since the different measurements are i.i.d. zero-mean Gaussian, the joint pdf factors into marginal pdfs

$$p_j(z_1,\ldots,z_G)=\prod_{k=1}^{G} p_j(z_k) \tag{6.29}$$

$$p_j(z_k)=\frac{1}{(2\pi)^{N/2}\,|\tilde\Lambda_j|^{1/2}}e^{-1/2 z_k^T \tilde\Lambda_j^{-1} z_k} \tag{6.30}$$

where

$|\tilde\Lambda_j|=\prod_{n=1}^{N}(\lambda_j[n]+\sigma_n^2/n_G)$ denotes the determinant of $\tilde\Lambda_j$

$z_k^T\tilde\Lambda_j^{-1}z_k=\sum_{n=1}^{N}z_k^2[n]/\lambda_j[n]+\sigma_n^2/n_G$ is a weighted energy measure, where the weights depend on the vehicle class

It is often convenient to work with the negative log-likelihood functions

$$C(z_1,\ldots,z_G)=\arg\min_{j=1,\ldots,M} l_j(z_1,\ldots,z_G) \tag{6.31}$$

$$l_j(z_1,\ldots,z_G)=-\frac{\log p_j(z_1,\ldots,z_G)}{G}=\frac{1}{G}\sum_{k=1}^{G}\log p_j(z_k) \tag{6.32}$$

Note that the kth SCR has to communicate the log-likelihood functions for all classes, log $p_j(z_k)$, $j = 1, \ldots, M$, based on the local measurement z_k, to the manager node. Ignoring constants that do not depend on the class, the negative log-likelihood function for H_j takes the form

$$l_j(z_1, \ldots, z_G) = \log |\tilde{\mathbf{\Lambda}}_j| + \frac{1}{G} \sum_{k=1}^{G} z_k^T \tilde{\mathbf{\Lambda}}_j^{-1} z_k \qquad (6.33)$$

Thus, for each set of measurements $\{z_k\}$ for a detected object, the classifier at the manager node computes l_j for $j = 1, \ldots, M$ and declares that the object (vehicle) belongs to the class with the smallest l_j.

A usual way of characterizing the classifier performance is to compute the average probability of error P_e, which is given by

$$P_e = \frac{1}{M} \sum_{m=1}^{M} P_{e,m} \qquad (6.34)$$

$$P_{e,m} = P(l_j < l_m \text{ for some } j \neq m \mid H_m) \qquad (6.35)$$

where $P_{e,m}$ is the conditional error probability when the true class of the vehicle is m. Computing $P_{e,m}$ is complicated, in general, but we can bound it using the union bound [5]

$$P_{e,m} \leq \sum_{j=1, j \neq m}^{M} P(l_j < l_m \mid H_m) \qquad (6.36)$$

Note that $P_e = 1 - PD$, where PD denotes the average probability of correct classification

$$PD = \frac{1}{M} \sum_{m=1}^{M} PDm \qquad (6.37)$$

$$PD_m = P(l_m < l_j \text{ for all } j \neq m \mid H_m) \qquad (6.38)$$

and PD_m denotes the probability of correct classification conditioned on H_m. The pairwise error probabilities on the right-hand side of Equation 6.36 can be computed analytically, but they take on complicated expressions [5,7]. However, it is relatively easy to show that, as the number of independent measurements G increases, P_e decreases and approaches zero (perfect classification) in the limit. To see this, note from Equation 6.32 that, by the law of large numbers [5], under H_m we have

$$\lim_{G \to \infty} l_j(z_1, \ldots, z_G) = -E_m[\log p_j(Z)] = D(p_m \| p_j) + h_m(Z) \qquad (6.39)$$

where

$D(p_m \| p_j)$ is the Kullback–Leibler (K–L) distance between the pdfs p_j and p_m
$h_m(Z)$ is the differential entropy of Z under H_m [8]:

$$D(p_m \| p_j) = E_m\left[\log\left(\frac{p_m(Z)}{p_j(Z)}\right)\right] = \log\left(\frac{|\tilde{\mathbf{\Lambda}}_j|}{|\tilde{\mathbf{\Lambda}}_m|}\right) + \text{tr}\left(\tilde{\mathbf{\Lambda}}_j^{-1} \tilde{\mathbf{\Lambda}}_m - I\right) \qquad (6.40)$$

$$h_m(Z) = -E_m[\log(p_m(Z))] = \frac{1}{2}\log\left((2\pi e)^N |\tilde{\mathbf{\Lambda}}_m|\right) \qquad (6.41)$$

Note that tr($\cdot$) denotes the trace of a matrix (sum of the diagonal entries). An important property of the K–L distance is that $D(p_m\|p_j) > 0$ unless $p_m = p_j$, i.e., the densities for class j and m are identical (in which case there is no way to distinguish between the two classes). Thus, from Equation 6.39 we conclude that, under H_m, l_m will always give the smallest value and thus lead to the correct decision as $G \to \infty$ as long as $D(p_j\|p_m) > 0$ for all $j \neq m$. For more discussion on performance analysis of soft decision fusion, we refer the reader to D'Costa and Sayeed [7].

6.4.2 Hard Decision Fusion

In soft decision fusion, the kth SCR sends M log-likelihood values $\{z_k^T \tilde{\Lambda}_j^{-1} z_k : j = 1, \ldots, M\}$ for the M classes, computed from its local measurement vector z_k, to the manager node. All these local likelihood values are real-valued and thus require many bits for accurate and reliable digital communication. The number of bits required for accurate communication can be estimated from the differential entropy of the likelihoods [7,8]. While exchange of real-valued likelihoods puts much less communication burden on the network compared with data fusion in which the feature vectors $\{z_k\}$ are communicated from each SCR to the manager node, it is attractive to reduce the communication burden even further. One way is to quantize the M likelihood values from different SCRs with a sufficient number of bits. Another natural quantization strategy is to compute local hard decisions in each SCR based on the local measurement vector z_k, analogous to the approach in object detection. In this section we discuss this hard decision fusion approach.

We assume that in the kth SCR a hard decision is made about the object class based on the local measurement vector z_k:

$$u_k(z_k) = \arg\max_{j=1,\ldots,M} p_j(z_k), \quad k = 1,\ldots,G \tag{6.42}$$

Equivalently, the decision could be made based on the negative log-likelihood function. Note that u_k maps z_k to an element in the set of classes $\{1, \ldots, M\}$ and is thus a discrete random variable with M possible values. Furthermore, since all $\{z_k\}$ are i.i.d., so are $\{u_k\}$. Thus, the hard decision random variable U (we ignore the subscript k) is characterized by a probability mass function (pmf) under each hypothesis. Let $\{p_m[j]: j = 1, \ldots, M\}$ denote the M values of the pmf under H_m. The pmfs for all hypotheses are described by the following probabilities:

$$p_m[j] = P(U(z_k) = j \mid H_m) = P(p_j(z_k) > p_l(z_k) \text{ for all } l \neq j \mid H_m), \quad j, m = 1, \ldots, M \tag{6.43}$$

The hard decisions $\{u_k\}$ from all SCRs are communicated to the manager node, which makes the final decision as

$$C_{hard}(u_1, \ldots, u_G) = \arg\max_{j=1,\ldots,M} p_j[u_1, \ldots, u_G] \tag{6.44}$$

where

$$p_j[u_1, \ldots, u_G] = \prod_{k=1}^{G} p_j[u_k] \tag{6.45}$$

since the $\{u_k\}$'s are i.i.d. Again, we can write the classifier in terms of negative log-likelihoods:

$$C_{hard}(u_1, \ldots, u_G) = \arg\max_{j=1,\ldots,M} l'_j[u_1, \ldots, u_G] \tag{6.46}$$

$$l'_j[u_1,\ldots,u_G] = \frac{1}{G}\log p_j[u_1,\ldots,u_G] = -\frac{1}{G}\sum_{k=1}^{G}\log p_j[u_k] \tag{6.47}$$

and while the exact calculation of the probability of error is complicated, it can be bounded via pairwise error probabilities analogous to the soft decision classifier. Similarly, we can say something about the asymptotic performance of the hard decision classifier as $G \to \infty$. Note from Equation 6.47 that, because of the law of large numbers, under H_m we have

$$\lim_{G\to\infty} l'_j[u_1,\ldots,u_G] = -E_m[\log p_j[U]] = D(p_m \,\|\, p_j) + H_m(U) \tag{6.48}$$

where
  $D(p_m\|p_j)$ is the K–L distance between the pmfs p_m and p_j
  $H_m(U)$ is the entropy of the hard decision under H_m [8]:

$$D(p_m\|p_j) = \sum_{i=1}^{M} p_m[i]\log\left(\frac{p_m[i]}{p_j[i]}\right) \tag{6.49}$$

$$H_m(U) = -\sum_{i=1}^{M} p_m[i]\log p_m[i] \tag{6.50}$$

Thus, we see from Equation 6.48 that, in the limit of large G, we will attain perfect classification performance as long as $D(p_m\|p_j) > 0$ for all $j \neq m$.

6.4.3 Numerical Results

We present some numerical results to illustrate soft decision classification.* We consider classification of a single vehicle from $M = 2$ possible classes: Amphibious Assault Vehicle (AAV; tracked vehicle) and Dragon Wagon (DW; wheeled vehicle). We simulated $N = 25$-dimensional (averaged) acoustic Fourier feature vectors for $K = Gn_G = 10$ nodes in G SCRs (n_G nodes in each SCR) for different values of G and n_G. The diagonal matrices Λ_1 (AAV) and Λ_2 (DW) corresponding to PSD samples were estimated from measured experimental data. The PSD estimates are plotted in Figure 6.5 for the two vehicles. In addition to the optimal soft decision fusion classifier C, two sub-optimal classifiers were also simulated: (1) a decision-fusion classifier C_{df} that assumes that all measurements are independent (optimal for $K = G$); (2) a data-averaging classifier C_{da} that treats all measurements as perfectly correlated (optimal for $K = n_G$). For each H_j, the G statistically independent source signal vectors s_k were generated using Λ_j as

$$s_k = \Lambda_j^{1/2}\mathbf{v}_k, \quad k = 1,\ldots,G \tag{6.51}$$

where $\mathbf{v}_k \sim \mathcal{N}(0,I)$. Then, the n_G noisy measurements for the kth SCR were generated as

$$z_{k,i} = s_k + \boldsymbol{n}_{k,i}, \quad i = 1,\ldots,n_G, \quad k = 1,\ldots,G \tag{6.52}$$

where $\boldsymbol{n}_{k,i} \sim \mathcal{N}(0,\sigma_n^2 I)$. The average probability of correct classification, PD $= 1 - P_e$, for the three classifiers was estimated using Monte Carlo simulation over 5000 independent trials.

Figure 6.6 plots PD as a function of the SNR for the three classifiers for $K = 10$ and different combinations of G and n_G. As expected, C and C_{da} perform identically for $K = n_G$ (perfectly correlated

---

* The results are based on real data collected as part of the DARPA SensIT program.

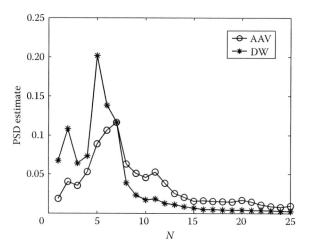

FIGURE 6.5 Covariance matrix eigenvalues (PSD estimates) for AAV and DW.

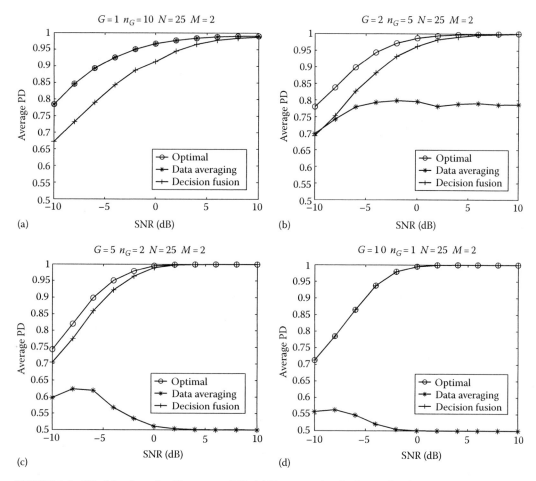

FIGURE 6.6 PD of the three classifiers versus SNR. (a) $K = n_G = 10$ (perfectly correlated measurements). (b) $G = 2$ and $n_G = 5$. (c) $G = 5$ and $n_G = 2$. (d) $K = G = 10$ (independent measurements).

measurements); see Figure 6.6a. On the other hand, C and C_{df} perform identically for $K = G$ (perfectly independent measurements); see Figure 6.6d). Note that C_{df} incurs a small loss in performance compared with C in the perfectly correlated (worst) case, which diminishes at high SNRs. The performance loss in C_{da} in the independent (worst) case is very significant and does not improve with SNR.* Thus, we conclude that the sub-optimal decision-fusion classifier C_{df} that ignores correlation in the measurements (and thus avoids the high-bandwidth data fusion of feature vectors in each SCR for signal averaging) closely approximates the optimal classifier, except for an SNR loss. It can be shown that the SNR loss is proportional to n_G, since C_{df} does not perform signal averaging in each SCR for noise reduction [7]. Furthermore, it can also be shown that C_{df} yields perfect classification performance (just as the optimal classifier) as $G \to \infty$ under mild conditions on the signal statistics, analogous to those for the optimal classifier [7]. Thus, the sub-optimal decision-fusion classifier (with either hard or soft decisions) is a very attractive choice in sensor networks because it puts the least communication burden on the network (avoids data fusion in each SCR).

6.5 Conclusions

Virtually all applications of sensor networks are built upon two primary operations: (1) distributed processing of data collected by the nodes; (2) communication and routing of processed data from one part of the network to another. Furthermore, the second operation is intimately tied to the first operation, since the information flow in a sensor network depends directly on the data collected by the nodes. Thus, distributed signal processing techniques need to be developed in the context of communication and routing algorithms and vice versa.

In this chapter we have discussed distributed decision making in a simple context—detection and classification of a single object—to illustrate some basic principles that govern the interaction between information processing and information routing in sensor networks. Our approach was based on modeling the object signal as a band-limited random field in space and time. This simple model partitions the network into disjoint SCRs whose size is inversely proportional to the spatial signal bandwidths. This partitioning of network nodes into SCRs suggests a structure on information exchange between nodes that is naturally suited to the communication constraints in the network: high-bandwidth feature-level data fusion is limited to spatially local nodes within each SCR, whereas global fusion of low-bandwidth local SCR decisions is sufficient at the manager node. We showed that data averaging within each SCR improves the effective measurement SNR, whereas decision-fusion across SCRs combats the inherent statistical variability in the signal. Furthermore, we achieve perfect classification in the limit of large number of SCRs (large number of independent measurements). This simple structure on the nature of information exchange between nodes applies to virtually all CSP algorithms, including distributed estimation and compression.

Our investigation based on the simple model suggests several interesting directions for future studies.

6.5.1 Realistic Modeling of Communication Links

We assumed an ideal noise-free communication link between nodes. In practice, the communication link will introduce some errors which must be taken into account to obtain more accurate performance estimates. In the context of detection, there is considerable work that can be made to bear on this problem [9]. Furthermore, the object signal strength sensed by a node will depend on the distance between the node and the object. This effect should also be included in a more detailed analysis. Essentially, this will limit the size of the region over which node measurements can be combined—the nodes beyond a certain range will exhibit very poor measurement SNR.

---

* It can be shown that, at high SNR, all events are classified as DW by C_{da}, since $\log |A_{DW}| < \log |\Lambda_{AAV}|$ due to the peakier eigenvalue distribution for DW [7], as evident from Figure 6.5.

6.5.2 Multi-Object Classification

Simultaneous classification of multiple objects is a much more challenging problem. For example, the number of possible hypotheses increases exponentially with the number of objects. Thus, simpler distributed classification techniques are needed. Several forms of sub-optimal algorithms, including tree-structured classifiers [6] and subspace-based approaches [10,11], could be exploited in this context. Furthermore, we have only discussed some particular forms of soft and hard decision fusion in this chapter. There are many (sub-optimal) possibilities in general [12] which could be explored to best suit the needs of a particular application.

6.5.3 Nonideal Practical Settings

We have investigated distributed decision making under idealized assumptions to underscore some basic underlying principles. The assumptions are often violated in practice and must be taken into account to develop robust algorithms [3]. Examples of nonideality include nonstationary signal statistics (which may arise due to motion or gear-shifts in a vehicle), variability in operating conditions compared with those encountered during training, and faulty sensors. Training of classifiers, which essentially amounts to estimating object statistics, is also a challenging problem [6]. Finally, Gaussian modeling of object statistics may not be adequate; non-Gaussian models may be necessary.

Acknowledgment

This work was supported in part by DARPA SensIT program under grant F30602-00-2-0555.

References

1. Estrin, D. et al., Instrumenting the world with wireless sensor networks, in *Proceedings of the IEEE International Conference on Acoustics, Speech, and Signal Processing 2001*, Vol. 4, pp. 2033, Salt Lake City, UT, 2001.
2. Kumar, S. et al. (eds), Special issue on collaborative signal and information processing in microsensor networks, *IEEE Signal Processing Magazine* (March), 2002.
3. Li, D. et al., Detection, classification, tracking of targets in microsensor networks, *IEEE Signal Processing Magazine* (March), pp. 17, 2002.
4. Stark, H. and Woods, J.W., *Probability, Random Processes, and Estimation Theory for Engineers*, Prentice Hall, Upper Saddle River, NJ, 1986.
5. Proakis, J.G., *Digital Communications*, 3rd edn., McGraw Hill, New York, 1995.
6. Duda, R. et al., *Pattern Classification*, 2nd edn., Wiley, New York, 2001.
7. D'Costa, A. and Sayeed, A.M., Collaborative signal processing for distributed classification in sensor networks, in *Lecture Notes in Computer Science* (*Proceedings of IPSN'03*), Zhao, F. and Guibas, L. (eds.), Springer-Verlag, Berlin, Germany, p. 193, 2003.
8. Cover, T.M. and Thomas, J.A., *Elements of Information Theory*, Wiley, Hoboken, NJ, 1991.
9. Varshney, P.K., *Distributed Detection and Data Fusion*, Springer, New York, 1996.
10. Fukunaga, K. and Koontz, W.L.G., Application of the Karhunen–Loeve expansion to feature selection and ordering, *IEEE Transactions on Computers*, C-19, 311, 1970.
11. Watanabe, S. and Pakvasa, N., Subspace method to pattern recognition, in *Proceedings of the 1st International Conference on Pattern Recognition*, pp. 25–32, Washington, DC, February 1973.
12. Kittler, J. et al., Advances in statistical feature selection, in *Advances in Pattern Recognition—ICAPR 2001: Second International Conference*, Vol. 2013, p. 425, Rio de Janeiro, Brazil, March 2001. ICAPR 2001—Electronic Edition (Springer LINK).

7

Parameter Estimation

7.1 Introduction 111
7.2 Self-Organization of the Network.. 112
7.3 Velocity and Position Estimation... 113
     Dynamic Space–Time Clustering • Experimental Results
     for Targets Velocities
7.4 Moving Target Resolution 116
7.5 Target Classification Using Semantic Information Fusion 116
     Experimental Results for SIF Classifier
7.6 Stationary Targets... 120
     Localization Using Signal Strengths • Localization Using Time
     Delays • Experimental Results for Localization Using Signal
     Strengths
7.7 Peaks for Different Sensor Types.. 123
Acknowledgments 128
References.. 128

David S. Friedlander
Defense Advanced
Research Projects Agency

7.1 Introduction

Most of the work presented in this chapter was done under two research projects: *Semantic Information Fusion* and *Reactive Sensor Networks*. These projects were performed at the Penn State University Applied Research Laboratory and funded under DARPA's Sensor Information Technology program (see Section 8.8). Experimental results were obtained from field tests performed jointly by the program participants.

Parameters measured by sensor networks usually fall into three categories: environmental parameters such as wind speed, temperature, or the presence of some chemical agent; target features used for classification; and estimates of position and velocity along target trajectories. The environmental parameters are generally local. Each measurement is associated with a point in space (the location of the sensor) and time (when the measurement was taken). These can be handled straightforwardly by sending them to the data sink via whatever networking protocol is being used. Other parameters may need to be determined by multiple sensor measurements integrated over a region of the network. Techniques for doing this are presented in this chapter.

These parameters are estimated by combining observations from sensor platforms distributed over the network. It is important for the network to be robust in the sense that the loss of any given platform or small number of platforms should not destroy its ability to function. We may not know ahead of time exactly where each sensor will be deployed, although its location can be determined by a global positioning system after deployment. For these reasons, it is necessary for the network to self-organize [1]. In order to reduce power consumption and delays associated with transmitting large amounts of information over long distances, we have designed an algorithm for dynamically organizing platforms into clusters along target trajectories.

This algorithm is based on the concept of space–time neighborhoods. The platforms in each neighborhood exchange information to determine target parameters, allowing multiple targets to be processed in parallel and distributed power requirements over multiple platforms. A neighborhood N is a set of space–time points defined by

$$N \equiv \left\{ (\overline{x}', t') : \left| \overline{x} - \overline{x}' \right| < \Delta x \text{ and } \left| t - t' \right| < \Delta t \right\}$$

where Δx and Δt define the size of the neighborhood in space and time. A dynamic space–time window $w(t)$ around a moving target with trajectory $\overline{g}(t)$ is defined by

$$w(t) \equiv \left\{ (\overline{x}', t) : \left| \overline{g}(t) - \overline{x}' \right| < \Delta x \text{ and } \left| t' - t \right| < \Delta t \right\}$$

We want to solve for $\overline{g}(t)$ based on sensor readings in the dynamic window $w(t)$. Most sensor readings will reach a peak at the closest point of approach (CPA) of the target to the sensor platform. We call these occurrences CPA events. In order to filter out noise and reflections, we count only peaks above a set threshold and do not allow more than one CPA event from a given platform within a given dynamic window.

Assuming we know the locations of each platform and that each platform has a reasonably accurate clock, we can assign a space–time point to each CPA event. We define the platforms with CPA events within a given dynamic window as a *cluster*. Platforms within a given cluster exchange information to define target parameters within the associated space and time boundaries. This technique can be easily extended to include moving platforms, as long as the platform trajectories are known and their velocities are small compared with the propagation speed of the energy field measured by the sensors. Typically, this would be the speed of light or the speed of mechanical vibrations, such as sound.

7.2 Self-Organization of the Network

We now show how to determine platform clusters along target trajectories [2]. The clusters are defined by dynamic space–time windows of size $\Delta x \times \Delta t$. Ideally, the window boundaries would also be dynamic. For example, we want Δx to be large compared with the platform density and small compared with the target density, and we want $\Delta x \approx v_t \Delta t$ where v_t is a rough estimate of the target velocity, possibly using the previously calculated value. In practice, we have obtained good results with constant values for Δx and Δt in experiments where the target density was low and the range of target velocities was not too large.

The algorithm for determining clusters is shown in Figure 7.1. Each platform contains two buffers, one for the CPA events it has detected and another for the events detected by its neighbors. The CPA Detector looks for CPA events. When it finds one, it stores the amplitude of the peak, time of the peak, and position of the platform in a buffer and broadcasts the same information to its neighbors. When it receives neighboring CPA events, it stores them in another buffer. The *Form Clusters* routine looks at each CPA event in the local buffer. A space–time window is determined around each local event. All of the neighboring events within the window are compared with the local event. If the peak amplitude of the local event is greater than that of its neighbors within the window, then the local platform elects itself as the *cluster head*. The cluster head processes its own and its neighbor's relevant information to determine target parameters. If a platform has determined that it is not the cluster head for a given local event, then the event is not processed by that platform. If the size of the window is reasonable, then

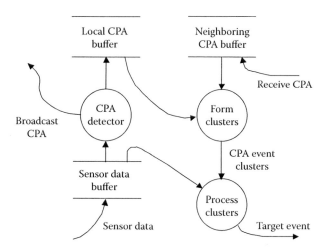

FIGURE 7.1   Cluster formation process.

For each local CPA event $k_{ij} = k(x_i, t_j)$

For each neighboring CPA event $n_{kl} = n(x_l, t_k)$

If n_{kl} is in the neighborhood $N_{ij} = N(\bar{x}_i, t_j)$

Add n_{kl} to the event set M

If the local peak amplitude $a(k_{ij}) \geq a(n_{kl}) \; \forall \, n_{kl} \in M$

Process CPA event cluster $F \equiv k_{ij} \cup M$

FIGURE 7.2   *Form Clusters* pseudo-code.

the method results in efficient use of the platforms and good coverage of the target track. Pseudo-code for this process is shown in Figure 7.2. The *Process Clusters* routine then determines the target position, velocity, and attributes as described below.

7.3 Velocity and Position Estimation

7.3.1 Dynamic Space–Time Clustering

We have extended techniques found in Hellebrant et al. [3] for velocity and position estimation [2]. We call the method *dynamic space–time clustering* [4]. The example shown below is for time and two spatial dimensions, $\bar{x} = (x, y)$; however, its extension to three spatial dimensions is straightforward. The technique is a parameterized linear regression. The node selected as the cluster head, n_0, located at position (x_0, y_0), estimates velocity and position. We estimate the target location and velocity at time t_0, the time of CPA for node n_0. This node has information and observations from a set of other nodes in a cluster around it. Denote the cluster around n_0 as $F \equiv \{n_i \mid |\bar{x}' - \bar{x}_0| < \Delta x$ and $|t_0 - t_i| < \Delta t\}$, where Δx and Δt are bounds in space and time. We defined the spatial extent of the neighborhoods so that vehicle velocities are approximately linear [3].

The position of a given node n_i in the cluster is $\bar{x}_i$ and the time of its CPA is t_i. This forms a space–time sphere around the position (x_0, y_0, t_0). The data are divided into one set for each special dimension; in our case: $(t_0, x_0), (t_1, x_1), \ldots, (t_n, x_n)$ and $(t_0, y_0), (t_1, y_1), \ldots, (t_n, y_n)$. We then weighted the observations based on the CPA peak amplitudes, on the assumption that CPA times are more accurate when the target passes closer to the sensor, to give $(x_0, t_0, w_0), (x_1, t_1, w_1), \ldots, (x_n, t_n, w_n)$ and $(y_0, t_0, w_0), (y_1, t_1, w_1), \ldots, (y_n, t_n, w_n)$, where w_i is the weight of the ith event in the cluster. This greatly improved the quality of the

Input: Time-sorted event cluster F of CPA values.
Output: Estimated velocity components v_x and v_y.

```
While |F|≥5 {
        Compute vx and vy using event cluster F;
        Compute rx and ry; the vx and vy velocity
                        ; correlation coefficients for F
        If rx > Rx || ry > Ry
        {
                Rx = rx;
                Ry = ry
                vx_store = vx;
                vy_store = vy;
        }
        PopBack(F);
};
```

FIGURE 7.3   Velocity calculation algorithm.

predicted velocities. Under these assumptions, we can apply least-squares linear regression to obtain the equations $x(t) = v_x t + c_1$ and $y(t) = v_y t + c_2$, where

$$
v_x = \frac{\sum_i t_i \sum_i x_i - \sum_i w_i \sum_i x_i t_i}{\left(\sum_i t_i\right)^2 - \sum_i w_i \sum_i t_i^2}, \quad v_y = \frac{\sum_i t_i \sum_i y_i - \sum_i w_i \sum_i y_i t_i}{\left(\sum_i t_i\right)^2 - \sum_i w_i \sum_i t_i^2}
$$

and the position $\bar{x}(t_0) = (c_1, c_2)$: The space–time coordinates of the target for this event are $(\bar{x}(t_0), t_0)$.

This simple technique can be augmented to ensure that changes in the vehicle trajectory do not degrade the quality of the estimated track. The correlation coefficients for the velocities in each spatial dimension (r_x, r_y) can be used to identify large changes in vehicle direction and thus limit the CPA event cluster to include only those nodes that will best estimate local velocity. Assume that the observations are sorted as follows: $o_i < o_j \Rightarrow |t_i - t_0| < |t_j - t_0|$, where o_t is an observation containing a time, location, and weight. The velocity elements are computed once with the entire event set. After this, the final elements of the list are removed and the velocity is recomputed. This process is repeated while at least five CPAs are present in the set; subsequently, the event subset with the highest velocity correlation is used to determine velocity. Estimates using fewer than five CPA points can bias the computed velocity and reduce the accuracy of our approximation. Figure 7.3 summarizes our technique.

Once a set of position and velocity estimates has been obtained, they are integrated into a track. The tracking algorithms improve the results by considering multiple estimates [5–7].

Beamforming is another method for determining target velocities [8]. Beamforming tends to be somewhat more accurate than dynamic space–time clustering, but it uses much greater resources. A comparison of the two methods is given in Phoha et al. [4].

7.3.2 Experimental Results for Targets Velocities

We have analyzed our velocity estimation algorithm using the field data these results appear in Table 7.1. Figures 7.4 and 7.5 show plots displaying the velocity estimations.

TABLE 7.1   Quality of Estimation

Computed vs. True Velocity	Percent
Within 1 m/s	81
Within 2 m/s	91
Within 5°	64
Within 11°	80
Within 17°	86

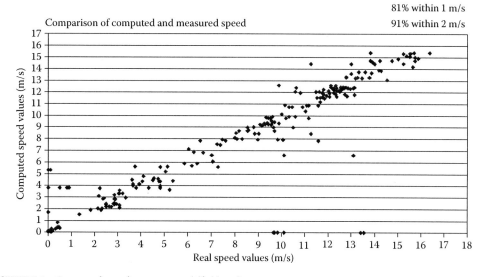

FIGURE 7.4   Computed speed vs. true speed (field test).

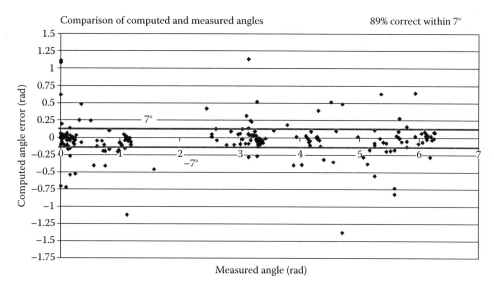

FIGURE 7.5   Computed angle vs. true angle (field test).

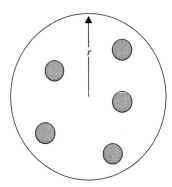

FIGURE 7.6   Sensor network area needed to determine target velocity.

7.4 Moving Target Resolution

We developed new results on estimating the capacity of a sensor network to handle moving targets. Theoretically, target velocity can be determined from three platforms. Our analysis of the data shows that five are necessary for good accuracy and stability; see Figure 7.6.

As shown in Figure 7.6, the radius of the spatial window for resolving a target's velocity is $r \approx \sqrt{5/\pi \rho_p}$ where r is the radius and ρ_p is the platform density. This gives us approximately five nodes in a space–time window, as required in Section 7.3. The amount of time needed to collect these data is determined by the time it takes the target to cross the spatial window Δt and the network latency $\Delta t \approx (2r/v) + \delta$, where v is the target velocity; i.e., platforms in the window can be separated by a distance of up to $2r$. Two given target trajectories can be resolved unless $\exists t, t' : |\bar{x}_1(t) - \bar{x}_2(t)| \leq 2r$ and $|t - t'| \leq \Delta t$, where $\bar{x}_i(t)$ is the trajectory of target i.

We can define the target density ρ_t as the density of targets in a reference frame moving with the target velocity v_t or, equivalently, the density of targets in a "snapshot" of the moving targets. We can have only one target at a time in the area shown in Figure 7.6, so $\rho_t \leq \rho_p/5$. The maximum capacity of the network in targets per second per meter of perimeter is given by $J_{max} \approx \rho_p v_t/5$. This is based on the assumption that the acoustic signals from two targets spaces approximately $2r = 2\sqrt{5/\pi \rho_p}$ meters apart will not interfere to the point where their peaks cannot be distinguished.

7.5 Target Classification Using Semantic Information Fusion

The semantic information fusion (SIF) technique described in this section was developed and applied to acoustic data [1]. It should be applicable to other scalar data, such as seismic sensors, but may not apply to higher dimensional data, such as radar. The method identifies the presence or absence of target features (attributes) detectable by one or more sensor types. Its innovation is to create a separate database for each attribute–value pair under investigation. Since it uses principal component analysis (PCA) [9], data from different types of sensor can be integrated in a natural way. The features can be transmitted directly or used to classify the target. PCA uses singular value decomposition (SVD), a matrix decomposition technique that can be used to reduce the dimension of time series data and improve pattern-matching results.

SIF processing consists of offline and online stages. The offline processing is computationally intensive and includes SVD of vectors whose components are derived from the time series of multiple channels. The attributes are expressed as mutually exclusive alternatives such as *wheeled* (or *tracked*), *heavy* (or *light*), *diesel engine* (or *piston engine*), etc.

Typically, the time series are transformed by a functional decomposition technique such as Fourier analysis. More recently developed methods, such as those by Goodwin and Vaidyanathan are promising. The spectral vectors are merged with the attribute data to from the pattern-matching database. The database vectors are then merged into a single matrix M, where each column of the matrix is one of the vectors. The order of the columns does not matter. The matrix is then transformed using SVD. The results include a set of reduced-dimension pattern-matching exemplars that are preloaded into the sensor platforms.

SVD produces a square matrix Σ and rectangular, unitary matrices U and V such that $M = U\Sigma V^T$. The dimension of M is $m \times n$, where n is the number of vectors and m is the dimension of each vector (the number of spectral components plus attribute dimensions). The dimension of Σ is $k \times k$, where k is the rank of M, where Σ is a diagonal matrix containing the eigenvalues of M in decreasing order. The dimension of U is $m \times k$ and the dimension of V^T is $k \times n$.

The number of significant eigenvalues r is then determined. The matrix Σ is truncated to a square matrix containing only the rth largest eigenvalues. The matrix U is truncated to be $m \times r$ and V^T to be $r \times n$. This results in a modified decomposition: $M \approx \hat{U}\hat{\Sigma}\hat{V}^T$ where $\hat{U}$ is an $m \times r$ matrix containing the first k columns of U, $\hat{\Sigma}$ is an $r \times r$ matrix containing the first r rows and columns of Σ, and $\hat{V}^T$ is $r \times n$ matrix containing the first r rows of V^T.

The online, real-time processing is relatively light. It consists of taking the power spectrum of the unknown time series data; forming the unknown sample vector; a matrix multiplication to convert the unknown sample vector into the reduced dimensional space of the pattern database; and vector dot products to determine the closest matches in the pattern database. Since the pattern matching is done on all of the peaks in the event neighborhood, an estimate of the uncertainty in the target attributes can also be calculated.

The columns of $\hat{V}^T$ comprise the database for matching against the unknown reduced-dimensional target vector. If we define the ith column of $\hat{V}^T$ as $\hat{p}^i$, the corresponding column of M as $\bar{p}^i$, the unknown full-dimensional target vector as $\bar{q}$, and $\hat{q} = (\bar{q})^T\hat{U}\hat{\Sigma}^{-1}$, as the reduced-dimensional target vector, then the value of the match between $\hat{p}^i$ and the target is $(\hat{p}^i \cdot \hat{q})$. We can define the closest vector to $\hat{q}$ as $\hat{p}^m$ where $m = $ index $\max_i(\hat{p}^i \cdot \hat{q})$. We then assign the attribute values of $\bar{p}^m$, the corresponding full dimensional vector, to the unknown target. The results might be improved using a weighted sum, say $w_i \sim 1/|\hat{q} - \hat{p}^m|$, of the attribute values (zero or one) of the k closest matches as the target attributes' values, i.e., $\bar{q}' = \Sigma_{i=1}^k w_i \bar{p}^i$. This would result in attributes with a value between zero and one instead of zero or one, which could be interpreted as the probability or certainty that the target has a given attribute.

Two operators, which are trivial to implement algorithmically, are defined for the SIF algorithms. If M is an $(r \times c)$ matrix, and $\bar{x}$ is a vector of dimension r, then

$$M \otimes \bar{x} \equiv \begin{bmatrix} m_{11} & \cdots & m_{1c} & x_1 \\ \vdots & \ddots & \vdots & \vdots \\ m_{r1} & \cdots & m_{rc} & x_r \end{bmatrix}$$

If $\bar{x}$ is a vector of dimension n and $\bar{y}$ is a vector of dimension m, then $\bar{x} \,||\, \bar{y} \equiv (x_1, x_2, \ldots, x_n, y_1, y_2, \ldots y_m)$. The offline algorithm for SIF is shown in Figure 7.7.

The matrices $\hat{U}$, $\hat{\Sigma}$, and $\hat{V}^T$ are provided to the individual platforms. When a target passes near the platform, the time series data is process and matched against the reduced-dimensional vector database as described above and shown in the online SIF algorithm, Figure 7.8.

In practice, we extend the results of Bhatnagar with those of Wu et al. [10], which contains processing techniques designed to improve results for acoustic data. CPA event data are divided into training and test sets. The training data are used with the data-processing algorithm and the test data are used with the data-classification algorithm to evaluate the accuracy of the method. The training set is further divided into databases for each possible value of each target attribute being used in the classification. Target attribute-values can be used to construct feature vectors for use in pattern classification. Alternatively, we can define "vehicle type" as a single attribute and identify the target directly.

A 4–5 s window is selected around the peak of each sample. All data outside of the window are discarded. This ensures that noise bias is reduced. The two long vertical lines in Figure 7.9 show what the

Let m = the number of target events, n = the number of channels per platform, p = the number of target attributes, and $\overline{T}^{ij}$ = the time series for event i, channel j.

For each target event i, and each sensor channel j

$\overline{s}^{ij} \leftarrow PowerSpectrum(\overline{T}^{ij})$; see [10,11]

For each target event i

    $\overline{v}^i \leftarrow \overline{a}^i \| \overline{s}^{i1} \| ... \| \overline{s}^{in}$

$M \leftarrow \overline{v}^1 \oplus \overline{v}^2 \oplus \cdots \oplus \overline{v}^m$

$(U, \Sigma, V^T) \leftarrow SVD(M)$; see [10,12]

$k \leftarrow NumSignificantEignvalues(\Sigma)$

$\hat{U} \leftarrow TruncateColumns(\hat{U}, k)$

$\hat{V}^T \leftarrow TrunscateRows(V^T, k)$

$\hat{\Sigma} \leftarrow TruncateRows(TruncateColumns(\Sigma, k), k)$

FIGURE 7.7   SIF offline algorithm.

Let m = the number of target events, n = the number of channels per platform, p = the number of target attributes, k = the reduced dimension size, and $\overline{T}^i$ = the time series from channel i.

For each sensor channel i

    $\overline{s}^i \leftarrow PowerSpectrum(\overline{T}^i)$

$\overline{q} \leftarrow (0, \cdots, 0) \| \overline{s}^1 \| \cdots \| \overline{s}^n$; the m components of the attribute vector
                                               ; are initially set to zero

$\hat{q} \leftarrow (\overline{q})^T \hat{U} \hat{\Sigma}^{-1}$

$r = index \, \underset{i}{max}(p^i . \hat{q}^j)$

$\hat{w} = MatrixColumn(V^T, r)$

$\overline{W} \leftarrow (\hat{w})^T \hat{\Sigma} \hat{U}$

$\overline{a}^q \leftarrow TrancateRows(\overline{w}, m)$; estimated attribute vector for target

FIGURE 7.8   Online SIF algorithm.

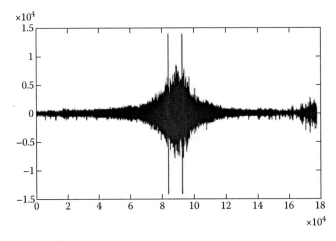

FIGURE 7.9   Time series window.

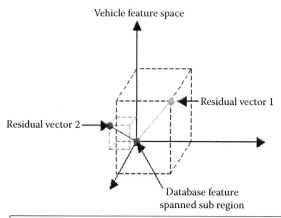

Vehicle feature space

Residual vector 1

Residual vector 2

Database feature
spanned sub region

Vehicle 2 is closer to the feature spanned sub region of the
database than vehicle 1, and is more likely to have the feature

FIGURE 7.10   Isolating qualities in the feature space.

boundaries of the window would be on a typical sample. The window corresponds to the period of time when a vehicle is closest to the platform. The data are divided into consecutive frames. A frame is 512 data points sampled at 5 kHz (0.5 s in length), and has a 12.5% (0.07 s) overlap with each of its neighbors. The power spectral density of each frame is found and stored as a column vector of 513 data points (grouped by originating sample), with data points corresponding to frequencies from 0 to 512 Hz.

Target identification combines techniques from Wu et al. [10] and makes use of an eigenvalue analysis to give an indication of the distance that an unknown sample vector is from the feature space of each database. This indication is called a residual. These residuals "can be interpreted as a measurement of the likelihood" that the frame being tested belongs to the class of vehicles represented by the database [10]. The databases are grouped by attribute and the residuals of each frame within each group are compared. The attribute value corresponding to the smallest total of the residuals within each group is assigned to the frame. Figure 7.10 illustrates this process.

7.5.1 Experimental Results for SIF Classifier

Penn state applied research laboratory (ARL) evaluated its classification algorithms against the data collected during field tests. Data are shown for three types of military vehicle, labeled armored attack vehicle (AAV), Dragon wagon (DW), and hummvee (HV). The CPA peaks were selected by hand rather than automatically detected by the software, and there was only a single vehicle present in the network at a time. Environmental noise due to wind was significant. The data in Table 7.2 show that classification of military vehicles in the field can be accurate under noisy conditions.

TABLE 7.2   Classification

Actual Vehicle	Classified Numbers			Correctly Classified (%)
	AAV	DW	HV	
AAV	117	4	7	94
DW	0	106	2	98
HV	0	7	117	94

7.6 Stationary Targets

7.6.1 Localization Using Signal Strengths

Another problem of interest for sensor networks is the counting and locating of stationary sources, such as vehicles with their engines running. Both theory and experiment suggest that acoustic energy for a single source is determined by $E = aJ/r^2$, where E is the energy measured by the sensor, r is the distance from the source to the sensor, J is the intensity of the source and a is approximately constant for a given set of synoptic measurements over the sensor network. Therefore $E_i(\overline{x}_s - \overline{x}_i)^2 = E_j(\overline{x}_s - \overline{x}_j)^2 \ \forall i, j \leq N$, where E_k is the energy measured by platform k, $\overline{x}_k$ is the location of platform k, N is the number of platforms in the network, and $\overline{x}_s$ is the location of the source. The location of the source is unknown, but it can be found iteratively by minimizing Var $(E_i(\overline{x}_s - \overline{x}_i)^2)$ as a function of $\overline{x}_s$.

If node i is at (u, v) and the source is at (x, y), $r_i^2 = (x - u_i)^2 + (y - v_i)^2$ the equation for all the sensors can be represented in matrix form as

$$\frac{1}{aJ} \begin{bmatrix} E_1 & -2u_1 & -2v_1 & E\left(u_1^2 + v_1^2\right) \\ \vdots & \vdots & \vdots & \vdots \\ E_n & -2u_n & -2v_n & E\left(u_n^2 + v_n^2\right) \end{bmatrix} \begin{bmatrix} x^2 + y^2 \\ x \\ y \\ 1 \end{bmatrix} = I \tag{7.1}$$

where
> n is the number of sensors
> I is the identity matrix

This over-determined set of equations can be solved for x, y, and J.

7.6.2 Localization Using Time Delays

We have also surveyed the literature to see whether signal time delays could be used in stationary-vehicle counting to enhance the results described above. A lot of work has been done in this area. We conclude that methods, as written, are not too promising when the vehicles are within the network, as opposed to the far field. The results and a suggestion for further research are summarized below.

The speed of sound in air is relatively constant. Therefore, given the delay between arrival times at multiple sensors of an acoustic signal from a single source, the distance from those sensors to the source can easily be calculated. Estimation of the location of that source then becomes a problem of triangulation, which, given enough microphones, can be considered an over-determined least-squares estimation problem. Thus, the problem of localization turns into one of time delay of arrival estimation.

When sources are far from the sensor network, the acoustic data can be thought of as arriving in an acoustic plane. The source is said to be in the far field. Finding the incidence of arrival of this plane and either enhancing or reducing its reception is the idea behind beam forming. When sources are close to the sensor array, the curvature of the surface of propagation through the array is pronounced; and now, more than an estimation of the direction of incidence of the wave front is required. The source is said to be in the near field, and is modeled as a single signal delayed by different amounts arriving at each sensor. Noise cancellation can be performed by removing any signals that arrive from the far field. Both of these topics are explored by Naidu [11].

Noise cancellation can be thought of as enhancing the signal of interest by finding its source in the near field. This is what we would like to use when trying to find the location of multiple vehicles in a sensor network. We want to find and count the various noise sources in the array. The task is difficult, and Emile et al. [12] claim that blind identification, in the presence of more than one source, when the signals are unknown, has only been solved well in the case of narrow-band spectra. There are many methods of performing time delay of arrival estimation, e.g., Ref. [13], and the main problem in the application of it to the problem of vehicle counting is selecting one that makes use only of the limited amount of information that we have of the array and environment. Even the size of the sample used to do the time delay of arrival may affect the performance significantly, as pointed out by Zou and Zhiping [14]. However, we may be able to make use of the idea of a local area of sensors, as we have in other algorithms such as the velocity estimator, to put bounds on the delay arrival times and the time the vehicle may be in the area. Therefore, we have information that may allow us to use one of the algorithms already in existence and remove enough of the errors associated with them to allow us to perform vehicle localization and/or counting in future work.

7.6.3 Experimental Results for Localization Using Signal Strengths

Acoustic data were recorded from 20 sensors placed in a sensor mesh. Vehicles were driven into the mesh and kept stationary during the recording. Four tests were run, containing one, two, three, and three cars (in a different configuration than the third test). For example, test two had the layout shown in Figure 7.11. Figures 7.12 and 7.13 show how the data look for accurate and inaccurate estimates of $\bar{x}_s$. Figure 7.14 shows the dependence between estimates of $\bar{x}_s$ and $\mathrm{Var}(E_i(\bar{x}_s - \bar{x}_i)^2)$.

As shown in the Figure 7.14, we could resolve single vehicles to within the grid spacing of 20 ft. We could not, however resolve the multiple vehicle tests. For the two-vehicle test, the vehicles in Figure 7.11 were positioned at (50, 30) and (15, 15), approximately 38 ft apart. Figure 7.15 shows the resulting acoustic energy field. The peaks due to each vehicle cannot be resolved. Figure 7.16 shows the theoretical energy field derived from Equation 7.1 using the actual source locations and fitting the constant K to the experimental data. The two surfaces are similar, so we would not expect to resolve the two vehicles with a 20 ft grid. Figure 7.17 shows the theoretical energy field for a fine sensor grid. It suggests that a sensor grid of 10–15 ft would be needed to resolve the two vehicles, depending on where they were placed.

We conclude that single vehicles in a sensor grid can be detected and located to within the sensor separation distance and that multiple vehicles can be resolved to within three to four times the sensor separation distance.

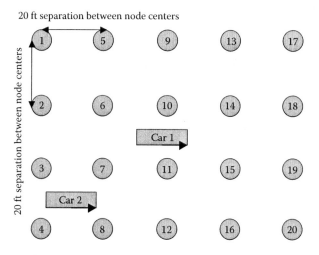

FIGURE 7.11 Experimental grid with two vehicles.

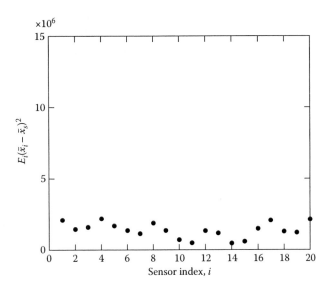

FIGURE 7.12 Data for an accurate location source estimate.

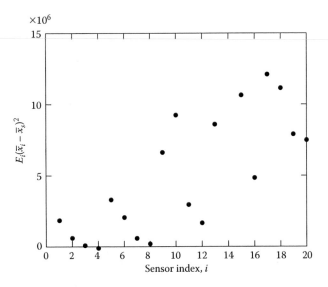

FIGURE 7.13 Data for an inaccurate source location estimate.

In the first experiment, with a single car revving its engine, the location of the car can be found consistently in the correct location, (58, –30). It is marked in Figure 7.18, which also contains the acoustic sensor values when the engine is revving.

Figure 7.19 shows the acoustic data when the engine is not being revved. The location estimate is inaccurate. This follows from a geometric interpretation of the estimator. Effectively, a $1/r^2$ surface is fit to the data in the best way available in a least-squares error sense. Therefore, if noise or the propagation time of the acoustic energy warps the data far from the desired surface, then the estimate will be seemingly random.

When tracking vehicles we are only concerned when the CPA has been detected; therefore, the vehicle sound should have a large intensity, creating a large signal-to-noise ratio. Furthermore, only the sensors in the local area of the CPA need to be consulted, removing the noise of sensors far from the source. This work shows, however, that we may have trouble finding stationary idling vehicles.

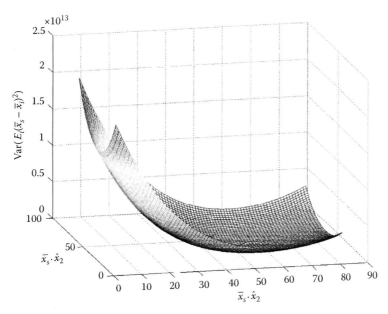

FIGURE 7.14   Minimization surface for a target located at (51, 31).

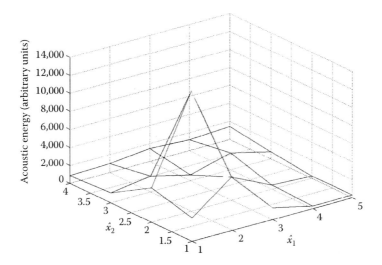

FIGURE 7.15   Acoustic energy field, two-vehicle test.

7.7 Peaks for Different Sensor Types

The quality of the local velocity determination algorithms is based in part on accurate determination of when a test vehicle is at its CPA to any given sensor node. We have begun analysis of how consistently this can be determined using three sensor modalities: acoustic, seismic and a two-pixel infrared camera. To facilitate this, we created a database of peak signal times for each node and sensor type. Three different military vehicles, labeled AAV, DW, and HV, are represented. There are database tables containing node information, global positioning system (GPS; ground truth), and CPA times. An overview of each table follows.

Node table. This table contains a listing of each node's UTM x and y locations (the unit is meters), indexed by node number. Figure 7.20 contains a plot of the node locations with the road overlaid.

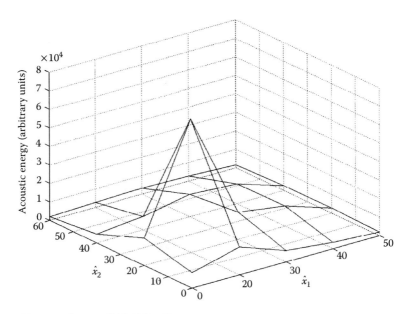

FIGURE 7.16   Theoretical energy field, 20 ft grid.

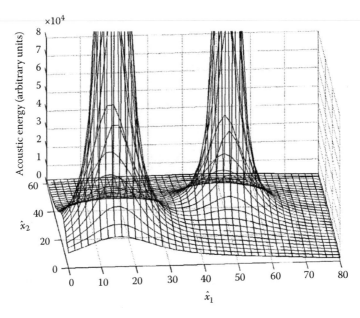

FIGURE 7.17   Theoretical energy field, 2 ft grid.

GPS tables. These three tables contain the "ground truth" locations vs. time. There is one table for each vehicle type. The GPS data were recorded every 2 s, and the second, minute, hour, day, month, and year recorded. The location of the vehicle at each time interval is recorded in two formats: UTM (the same as the node table) and latitude/longitude.

CPA table. We went through every acoustic, seismic, and passive infrared sensor file previously mentioned, and manually selected the peaks in each. Peak time is used to estimate the vehicle's CPA to the node that is recording the data. The CPA table contains records of the sensor, the vehicle causing the peak, the node associated with the peak, the peak time, and, for acoustic and seismic sensors, the maximum energy and amplitude of the signal.

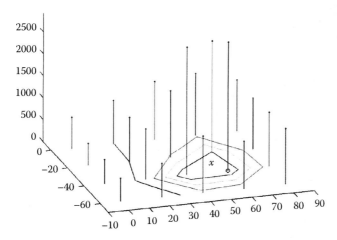

FIGURE 7.18 Acoustic sensor values and car location when engine is revving.

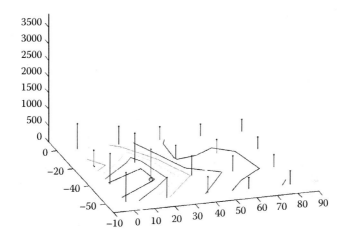

FIGURE 7.19 Acoustic data when the engine is not being revved.

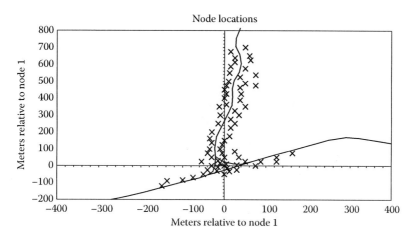

FIGURE 7.20 Node locations.

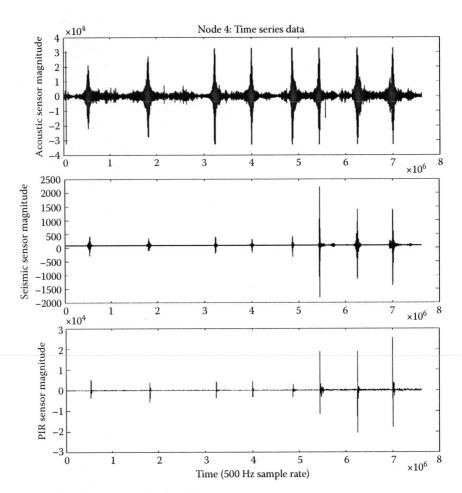

FIGURE 7.21   Infrared, seismic, and acoustic data.

Peaks for each sensor type were selected independently of one another. The CPA times for each sensor type tend to be similar; however, there are cases where one channel has a peak and the others do not. The data for node 4 are shown in Figure 7.21.

We have created a visualization of the combined CPA time and GPS data. The plot is three-dimensional, where the x and y axes are UTM coordinates and the z-axis is time in seconds since the first vehicle started. The path of the GPS device is the continuous line, and the dots are peak detections at the locations of the corresponding nodes. The "+" indicates a peak while the AAV was running, a "o" indicates a DW was running, and a "." indicates the HV runs. We have inclined the graph to get a good view of the situation, as shown in Figure 7.22. (The perspective results in a small display angle between the x and y axes.) A close-up of one of the vehicle traversals of the test region is shown in Figure 7.23.

All three sensors types do not always have a noticeable peak as a vehicle drives by. The peak database was used to examine the locality in time of the peaks from the different sensor types as a vehicle drives past a given node. We selected a time-window size (e.g., 10 s) and clustered all peaks at each specific node that occurred within the same window. This provided clusters of peaks for different sensor types that contained: just a single peak, two peaks, or three or more peaks. Figure 7.24 is a plot of the number of single, double and triple (or more) clusters vs. the size of time window selected.

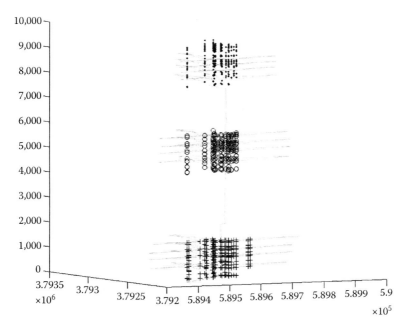

FIGURE 7.22 Three-dimensional plot of sensor peak and ground truth data.

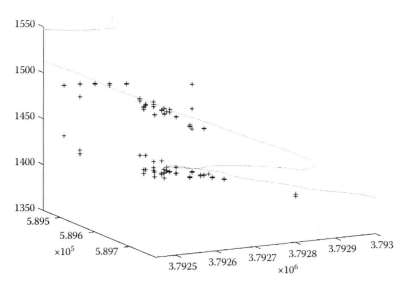

FIGURE 7.23 Close-up of AAV data.

It usually takes 20–30 s for a vehicle to traverse the network area, and ideally there should only be one peak of each type at each node during each traversal, i.e., a single traversal results in one CPA event for each node. Thus, the statistics at this window size and larger should be relatively stable, because all peaks that are going to occur due to a given traversal should happen sometime during that 20–30 s period. The graph support this. However, a reduction of the window size down to 10–15 s yields little change. Hence, the data suggest that the peaks that are going to occur at a given node, from a single event, for the three sensor modalities usually occur within a 10–15 s window.

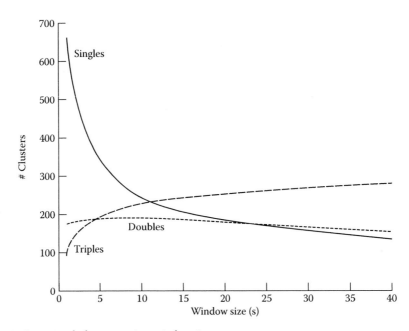

FIGURE 7.24 Sensor peak clusters vs. time window size.

Acknowledgments

This effort is sponsored by the Defense Advanced Research Projects Agency (DARPA) and the Space and Naval Warfare Systems Center, San Diego, CA (SSC-SD), under grant number N66001-00-C-8947 (Semantic Information Fusion in Scalable, Fixed and Mobile Node Wireless Networks), by the Defense Advance Research Projects Agency (DARPA) Air Force Research Laboratory, Air Force Materiel Command, USAF, under agreement number F30602-99-2-0520 (Reactive Sensor Network), and by the U.S. Army Robert Morris Acquisition under Award No. DAAD19-01-1-0504.

The U.S. Government is authorized to reproduce and distribute reprints for Governmental purposes notwithstanding any copyright annotation thereon. The views and conclusions contained herein are those of the author's and should not be interpreted as necessarily representing the official policies or endorsements, either expressed or implied, of the Defense Advanced Research Projects Agency (DARPA), the Space and Naval Warfare Systems Center, the Army Research Office, or the U.S. Government.

References

1. Friedlander, D.S. and Phoha, S., Semantic information fusion of coordinated signal processing in mobile sensor networks, *Special Issue on Sensor Networks of the International Journal of High Performance Computing Applications*, 16(3): 235, 2002.
2. Friedlander, D. et al., Dynamic agent classification and tracking using an ad hoc mobile acoustic sensor network, *Eurasip Journal on Applied Signal Processing*, 2003(4): 371, 2003.
3. Hellebrant, M. et al., Estimating position and velocity of mobiles in a cellular radio network, *IEEE Transactions Vehicular Technology*, 46(1): 65, 1997.
4. Phoha, S. et al., Sensor network based localization and target tracking through hybridization in the operational domains of beamforming and dynamic space–time clustering, in *Proceedings of IEEE Global Communications Conference*, San Francisco, CA, December 1–5, 2003.

5. Brooks, R.R. et al., Tracking targets with self-organizing distributed ground sensors, in *Proceedings to the IEEE Aerospace Conference Invited Session "Recent Advances in Unattended Ground Sensors,"* Big Sky, MT, March 10–15, 2003.
6. Brooks, R. et al., Distributed tracking and classification of land vehicles by acoustic sensor networks, *Journal of Underwater Acoustics*, in review, 2002.
7. Brooks, R. et al., Self-organized distributed sensor network entity tracking, *International Journal of High Performance Computer Applications*, 16(3): 207, 2002.
8. Yao, K. et al., Blind beamforming on a randomly distributed sensor array system, *IEEE Journal on Selected Areas in Communications*, 16: 1555, 1998.
9. Jolliffe, I.T., *Principal Component Analysis,* Springer-Verlag, New York, 1986.
10. Wu, H. et al., Vehicle sound signature recognition by frequency vector principal component analysis, *IEEE Transactions on Instrumentation and Measurement*, 48(5): 1005, 1999.
11. Naidu, P.S., *Sensor Array Signal Processing*, CRC Press LLC, New York, 2001.
12. Emile, B. et al., Estimation of time delays with fewer sensors than sources, *IEEE Transactions Signal Processing*, 46(7): 2012, 1998.
13. Krolik, J. et al., Time delay estimation of signals with uncertain spectra, *IEEE Transactions on Acoustics, Speech, and Signal Processing*, 36(12): 1801, 1988.
14. Zou, Q. and Zhiping, L., Measurement time requirement for generalized cross-correlation based time-delay estimation, in *Proceedings of IEEE International Symposium on Circuits and Systems (ISCAS 2002)*, Vol. 3, Phoenix, AZ, May 2002, p. 492.

8

Target Tracking with Self-Organizing Distributed Sensors

8.1 Introduction 131
8.2 Related Work.. 132
8.3 Computation Environment... 134
8.4 Intercluster Tracking Framework 136
8.5 Local Parameter Estimation 139
8.6 Track Maintenance Alternatives... 142
      Pheromone Routing • Extended Kalman Filter • Bayesian Entity
      Tracking
8.7 Tracking Examples 148
      Pheromone Routing • Extended Kalman Filter • Bayesian
      Belief Net
8.8 Cellular Automata Model... 151
8.9 Cellular Automata Results 154
      Linear Tracks • Crossing Tracks • Nonlinear Crossing
      Tracks • Intersecting Tracks • Track Formation Effect on Network
      Traffic • Effects of Network Pathologies
8.10 Collaborative Tracking Network.. 168
8.11 Dependability Analysis.. 176
8.12 Resource Parsimony 178
8.13 Multiple Target Tracking.. 179
8.14 Conclusion 186
Acknowledgments... 187
References... 188

Richard R. Brooks
Clemson University

Christopher Griffin
*The Pennsylvania
State University*

David S. Friedlander
*Defense Advanced
Research Projects Agency*

J.D. Koch
*The Pennsylvania
State University*

İlker Özçelik
Clemson University

8.1 Introduction

As computational devices have shrunk in size and cost, the Internet and wireless networking have become ubiquitous. Both trends are enabling technologies for the implementation of large-scale, distributed, embedded systems. Multiple applications exist for these systems in control and instrumentation. One area of particular importance is distributed sensing. Distributed sensing is necessary for applications of importance to government and industry such as defense, transportation, border patrol, arms control verification, contraband interdiction, and agriculture.

Sensors return information gained from physical interaction with their environment. To aid physical interaction, it is useful for sensors to be in close physical proximity with the objects observed. To adequately observe actions occurring over a large region, multiple sensors may be necessary. Battery-powered devices with wireless communication are advisable. It is also prudent to provide the devices with local intelligence.

It is rare for a sensor to directly measure information at the semantic level desired (e.g., how many cars have passed by this intersection in the last hour?). Sensors discern semantic information indirectly by interpreting entity interactions with the environment (e.g., cars are detected through acoustic vibrations emitted or ground vibrations caused by wheels moving over terrain.). Semantic information is inferred by interpreting one or more cues (often called features) detected by the sensor. Cues can be inexact and prone to misinterpretation. They contain noise and are sensitive to changes in the environment. Since the sensor has physical interaction with the environment, it is prone to failure, drift, and loss of calibration.

Creating a system from a large network of inexpensive intelligent sensors is an attractive means of overcoming these limitations. The use of multiple devices helps counter drift, component failure, and loss of calibration. It also allows for statistical analysis of data sources to better filter noise. Similarly, the use of multiple sensing modalities can make decisions more robust to environmental factors by increasing the number of cues available for interpretation. A central issue that needs to be overcome is the complexity inherent in creating a distributed system from multiple-failure-prone components. Batteries run out. Components placed in the field are prone to destruction. Wireless communication are prone to disruption. Manual installation and configuration of nontrivial networks would be onerous. Similarly, programming networks and interpreting readings are time consuming and expensive.

In this chapter, we present an example application that shows how to compensate for these issues. We discuss a flexible self-organizing approach to sensor network design, implementation, and tasking. The application described is entity tracking. Our approach decomposes the application into subtasks. Self-organizing implementations of each subtask are derived. Example problems are discussed.

The rest of the chapter is organized as follows. In Section 8.2, short survey of recent studies is presented. Section 8.3 describes the computational environment used. Network interactions and self-organization use diffusion routing as implemented at USC/ISI [1] and a mobile code API implemented at the Penn State Applied Research Laboratory [2]. The distributed multi-target tracking problem is decomposed into subproblems in Section 8.4. In Section 8.5, we discuss the first subproblem: how to use a cluster of nodes for collaborative parameter estimation. Once parameters have been estimated, they must be associated with tracks as described in Section 8.6. Simulations comparing alternative approaches are described in Section 8.7.

Section 8.8 describes the use of cellular automata (CA) tools to evaluate and contrast network-embedded tracking approaches. Section 8.9 presents statistical and anecdotal results gained from using the CA models from Section 8.8 to study entity-tracking algorithms. Section 8.10 provides a brief description of the collaborative tracking network (ColTraNe) and some results of field tests. Sections 8.11 and 8.12 provide a comparative analysis of dependability and power requirements for our distributed tracking approach versus a more typical centralized approach. We describe the results of simulated multiple target tracking in Section 8.13. Section 8.14 concludes the chapter by presenting conclusions based on our work.

8.2 Related Work

Many recent studies have been conducted on target tracking with distributed sensor networks. A detailed survey is presented in Duncan and Sameer [3]. This survey focused on multisensor data fusion in target tracking and explained data fusion using the well-known JDL data fusion model. This model divides the data fusion process into four levels:

- JDL Level 1: Object Refinement
- JDL Level 2: Situation Assessment
- JDL Level 3: Threat Assessment
- JDL Level 4: Process Assessment

JDL Level 1 "Object Refinement" covers data registration, data association, position attribute estimation, and identification. This level can be seen as a foundation for other JDL levels. Therefore, most studies focus on this level.

In the data registration step, observations are converted into known data forms, which enable them to be used in the next steps. In the data association step, measurements are compared in order to distinguish the ones that belong to the same target. These measurements are used to create a new route or to associate with an existing route. *Nearest neighbor, joint probabilistic data association, and fuzzy logic* algorithms can count as examples of data association algorithms. Position attribute estimation is calculating the target's state by using associated measurements. One of the most commonly used position attribute estimation techniques is Kalman filter (KF). In KF design, it is assumed that system is Gaussian. For a Gaussian system, KF can estimate the state of the system from noisy sensor measurements. Although originally KF and extended KF were used on data from single sensor, it is possible to improve the performance of multisensor systems by using KF at a central fusion node to fuse the results of local KFs. An important point to consider while making position attribute estimation is the system dynamics modes. A filter can represent only one mode. The simplest form of hybrid system interacting multiple model (IMM) solves this problem. IMM consists of multiple KFs, each representing a different system mode.

Because of KF's difficulty adjusting to dynamically changing system modes, particle filters (PF) have become an alternative. It is shown that "bootstrap filter" or sequential importance resampling PFs outperform the extended Kalman filter (EKF) for tracking in a system with nonlinear measurements. Kabaoglu proposed support vector regression (SVR)-based PF to further improve the performance of PF [4]. He showed that SVR-based particle-producing scheme provides more accurate approximation of the importance density function.

Another alternative to KF is Fuzzy logic adaptive filter. Sasiadek and Hartana proposed three extension of fuzzy logic adaptive KF (FL-AKF); fuzzy-logic-based C adaptive KF (FL-ACKF), fuzzy-logic-based D adaptive KF (FL-ADKF), and fuzzy-logic-based F adaptive KF (FL-AFKF). It is shown that these techniques are effective while observing same parameter in different dynamics and noise statistic with heterogenous sensors.

The identification step is used to classify the target. Bayesian interface, Dempster–Shafer (D–S) rule of combination, and voting and summing approaches are example techniques to identify the target. It is also possible to classify the target with a distributed manner. Many algorithms are proposed for distributed classification, but the main problem is deciding whether collecting the data from all sensors and fusing data to classify (data fusion) or classifying locally and fusing the decision results (decision fusion). Brooks et al. [5] suppose that data fusion is a better choice if measured data are correlated and decision fusion is a better choice if data are uncorrelated.

JDL Level 2 "Situation Assessment" and Level 3 "Threat Assessment" are generally referred as "information fusion." These two levels are not investigated as much as Level 1. Situation assessment fuses the data coming from Level 1 to come up with an idea about target's actions and infer the situation. At the Level 3, importance level of the threat is investigated. Final level of the JDL "Process Assessment" is a continuous assessment of the other fusion stages to ensure data acquisition and fusion give optimal results. This can be done by adjusting the fusion parameters, assigning target priorities, and/or selecting sensors to track the target. Especially, in distributed sensing literature, process assessment has been covered as selecting sensors to gain most information in an effective way. Aeron et al. propose an adaptive dynamic strategy for sensor and fusion center location selection by using certainty equivalence approach [6]. In their paper, they tried to optimize trade-off between tracking error and communication cost. They showed that optimal sensor selection and fusion center switching strategy is a hybrid switching strategy in which active sensors and fusion center location change decision is made based on a threshold radius.

Duncan and Sameer [3] also mentioned the challenges of the multisensor tracking. They pointed out two major problems: out of sequence measurements and data correlation. Several proposed algorithms to solve these problems are presented in their survey paper.

Allegretti et al. showed that CA algorithms can be used to detect and track moving targets in low signal to noise ratio (SNR) environment by communicating only neighbor sensors. In their paper, they present two methods for tracking targets in noisy environments. First one is called "mass algorithm," which calculates the average of the x- and y-coordinates of detections at a time step. On the other hand, second algorithm "motion algorithm" generates wave-like pulses of positive cell values. When these "waves" propagate at the speed of moving object, they interfere constructively around the moving object [7]. They tested the system both for sensors distributed in a perfect grid and randomly disabling certain amount of sensors in the same grid. Results showed that CA show outstanding performance in noisy environments. Therefore, they claim that it is convenient to use small, cheap, low-powered sensors to track moving targets in a distributed sensor network (DSN).

Unlike most of the studies on tracking, Crespi et al. formalized the trackability on sensor networks and they proposed a quantitative theory of trackability of weak models [8]. They explained the weak model using the analogy with hidden Markov models (HMMs) as the model, and only described the possible transitions and possible associations of transitions to states, without assigning any probabilities to them. In their paper, they showed that for a weak model, if the worst-case number of hypothesis consistent with a given set of observations grows polynomially, model is trackable; but if it grows exponentially, the model is untrackable. They also showed that trackability can be determined in polynomial time.

8.3 Computation Environment

This chapter describes three levels of work:

1. Theoretical derivations justify the approach taken.
2. Simulations provide proof of concept.
3. Prototype implementations give final verification.

We perform simulations at two levels. The first level uses in-house CA models for initial analysis. This analysis has been performed and is presented. Promising approaches will then be ported to the virtual Internet test bed (VINT) [9], which maintains data for statistical analysis. VINT also contains a network animation tool, nam, for visualization of network interactions. This supports replay and anecdotal study of pathologies when they occur. VINT has been modified to support sensor modeling.

In this chapter, we present mainly results based on simulations. Portions of the approach have been implemented and tested in the field. We differentiate clearly between the two. The hardware configuration of the prototype network nodes defines many factors that affect simulations as well as the final implementation. In this section, we describe the prototype hardware and software environment, which strongly influences much of the rest of the chapter. For experimentation purposes, we use prototype sensor nodes that are battery-powered nodes with wireless communication. Each node has limited local storage and CPU. For localization and clock synchronization, all nodes have GPS receivers. The sensor suite includes acoustic microphones, ground vibration detectors, and infrared motion detectors. All nodes have the same hardware configuration, with two exceptions: (1) the sensor suite can vary from node to node, and (2) some nodes have more powerful radios. The nodes with more powerful radios work as gateways between the sensor network and the Internet. Development has been done using both Linux and Windows CE operating systems.

Wireless communication range is limited for several reasons. Short-range communication require less power. Since nodes will be deployed at or near ground level, multipath fading significantly limits the effective range. The effective sensing range is significantly larger than the effective communication range of the standard radios. These facts have distinct consequences for the resulting network topology. Short-range wireless communication make multi-hop information transmission necessary. Any two nodes with direct radio communication will have overlapping sensor ranges. The sensor field is dense. In most cases, more than one node will detect an event. Finite battery lifetimes translate into finite node lifetimes, so that static network configurations cannot be maintained.

Manual organization and configuration of a network of this type of any size would be a Sisyphean task. The system needs to be capable of self-configuration and automatic reconfiguration. In fact, the underlying structure of the network seems chaotic enough to require an ad hoc routing infrastructure. Static routing tables are eschewed in this approach. Routing decisions are made at run time. It is also advisable to minimize the amount of housekeeping information transmitted between nodes, since this consumes power. Each bit transmitted by a node shortens its remaining useful lifetime.

To support this network infrastructure, a publish–subscribe paradigm has been used [1]. Nodes that are sources of information announce information availability to the network via a publish method provided by the networking substrate. When the information becomes available, a send method is used to transmit the information. Nodes that consume information inform the networking substrate of their needs by invoking a subscribe method. The subscribe method requires a parameter containing the address of a callback routine that is invoked when data arrive.

A set of user-defined attributes is associated with each publish and subscribe call. The parameters determine matches between the two. For example, a publish call can have attributes whose values correspond to the universal transverse mercator (UTM) coordinates of its position. The corresponding subscribe would use the same attributes and define a range of values that include the values given in the publish call. Alternatively, it is possible to publish to a region and subscribe calls provide values corresponding to the UTM coordinates. It is the application programmer's responsibility to define the attributes in an appropriate manner. The ad hoc routing software establishes correspondences and routes data appropriately. Proper application of this publish–subscribe paradigm to the entity-tracking problem is an important aspect of this chapter. We use it to support network self-organization.

In addition to supporting ad hoc network routing, the system contains a mobile code infrastructure for flexible tasking. Currently, embedded systems have real constraints on memory and storage. This severely limits the volume of software that can be used by an embedded node and directly limits the number of behaviors available. By allowing a node to download and execute code as required, the number of possible behaviors available can be virtually limitless. It also allows fielded nodes to be reprogrammed as required. Our approach manages code in a manner similar to the way a cache manages data. This encourages a coding style where mobile code is available in small packages.

In our approach, both code and data are mobile. They can be transferred as required. The only exceptions to this rule are sensors, which are data sources but tied to a physical location. We have implemented exec calls, which cause a mobile code package to execute on a remote node. Blocking and nonblocking versions of exec exist. Another important call is pipe. The semantics of this call is similar to a distributed form of pipes as used by most Unix shell programs. The call associates a program on a node with a vector of input files and a vector of output files. When one of the input files changes, the program runs. After the program terminates, the output files are transmitted to other nodes as needed. This allows the network to be reprogrammed dynamically, using what is effectively an extensible distributed data-flow scripting language.

Introspective calls provide programs with information about mobile code modules resident on the network and on a specific node. Programs can prefetch modules and lock them onto a node. Locking a module makes it unavailable for garbage collection.

The rest of this chapter uses entity tracking as an example application for this sensor network computational environment. The environment differs from traditional approaches to embedded systems in many ways:

- It is highly distributed.
- It is assumed that individual components are prone to failure and have finite lifetimes.
- Network routing is ad hoc.
- The roles played by nodes change dynamically.
- Sensing is done by collaboration between nodes.
- A node's software configuration is dynamic.

These aspects of the system require a new programming approach. They also provide the system with the ability to adapt and modify itself when needed.

8.4 Intercluster Tracking Framework

Sensor data interpretation takes place at multiple levels of abstraction. One example of this is the process from Ref. [10] shown in Figure 8.1. Sensor information enters the system. Objects are detected using signal-processing filters. Association algorithms determine which readings refer to the same object. Sequences of readings form tracks. Track information is used to estimate the future course of the entities and allocate sensors. Sensor allocation is done with human in the loop guidance. We consider entity tracking as the following sequence of problems:

1. *Object detection*: Signal processing extracts features that indicate the presence of entities of interest.
2. *Object classification*: Once a detection event occurs, signal-processing algorithms assign the entity to one of a set of known classes. This includes estimating parameters regarding position, speed, heading, and entity attributes. Attributes can be combined into discrete disjoint sets that are often referred to as codebook values.
3. *Data association*: After classification, the entity is associated with a track. Tracks are initiated for newly detected entities. If a track already exists, the new detection is associated with it. If it is determined that two separate tracks refer to the same entity, they are merged.
4. *Entity identification*: Given track information, it may be possible to infer details about the identity and intent of an entity.
5. *Track prediction*: Based on current information, the system needs to predict likely future trajectories and cue sensor nodes to continue tracking the entity.

Object detection, classification, and parameter estimation are discussed in this section and Section 8.5. Information is exchanged locally between clusters of nodes to perform this task. Data association, entity identification, and track prediction are all discussed in Section 8.6.

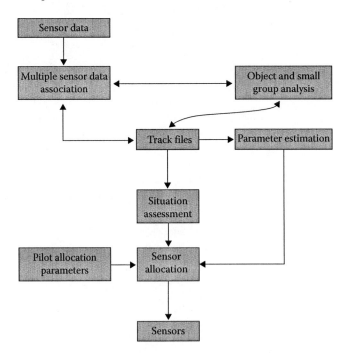

FIGURE 8.1 Concept for multiple sensor entity tracking. (From Blackman, S.S. and Broida, T.J., *J. Rob. Syst.*, 7(3), 445, 1990.)

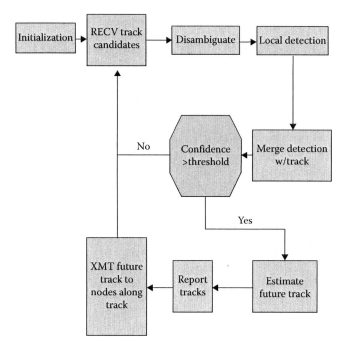

FIGURE 8.2 Flowchart of the processing performed at any given node to allow network-embedded entity tracking.

This chapter concentrates on embedding this sequence of problems defining entity tracking into the self-organizing network technologies described in Section 8.3. The approach given in this section supports multiple approaches to the individual problems. Figure 8.2 gives a flowchart of the logic performed at each node. The flowchart is not strictly correct, since multiple threads execute concurrently. It does, however, show the general flow of data through the system. In the current concept, each node is treated equally and all execute the same logic. This could be changed in the future.

The flowchart in Figure 8.2 starts with an initialization process. Initialization involves invoking appropriate publish and subscribe methods. Subscribe is invoked three times. One invocation has an associated parameter associating it with "near" tracks. Second is associated with "mid-tracks." The third uses the parameter to associate it with "far" tracks. Figure 8.3 illustrates near-, mid-, and far regions. All three subscribe calls have two parameters that contain the node's x and y UTM coordinates. The subscribe invocations announce to the network routing substrate the node's intent to receive candidate tracks of entities that may pass through its sensing range. Near, mid, and far differ in the distance between the node receiving the candidate track and the node broadcasting candidate track information. The flowchart in Figure 8.2 shows nodes receiving candidate tracks after initialization. For example, if the node in Figure 8.3 detects an entity passing by with a north-easterly heading, it will estimate the velocity and heading of the target; calculate and invoke publish to the relevant near-, mid-, and far regions; and invoke the network routing send primitive to transmit track information to nodes within those regions. Nodes in those regions can receive multiple candidate tracks from multiple nodes. The disambiguation process (see Figure 8.2) finds candidate tracks that are inconsistent and retains the tracks that are most likely. Section 8.6 describes example methods for performing this task. This step is important, since many parameter estimation methods do not provide unique answers. They provide a family of parallel solutions. It is also possible for more than one cluster to detect an entity. In the human retina and many neural network approaches, lateral inhibition is

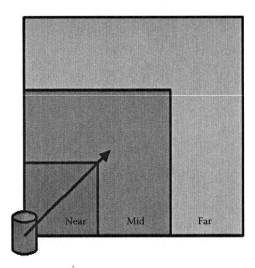

FIGURE 8.3 Example of dynamic regions used to publish candidate tracks. Solid arrow is the estimated target velocity and heading.

performed so that a strong response weakens other responses in its vicinity. We perform this function in the disambiguation task.

It is worth noting that the reason to publish (and send) to all three near-, mid-, and far regions shown in Figure 8.3 may be less than obvious. They increase system robustness. If no node is present in the near region or if nodes in the near region fail to detect an entity, the track is not necessarily lost. Nodes in the mid- and far regions have candidate track information and may continue the track when an appropriate entity is detected. The existence of three levels (near, mid, and far) is somewhat arbitrary. Future research may indicate the need for more or fewer levels.

Track candidate reception and disambiguation run in one thread that produces an up-to-date list of tracks of entities that may be entering the local node's sensing range. Local detections refer to detection events within a geographic cluster of nodes. In Section 8.5, we explain how local information can be exchanged to accurately estimate detection event parameters, including position, heading, closest point of approach (CPA), and detection time.

When local detections occur, as detailed in Section 8.5, they are merged with candidates. Each track has an associated certainty factor. To reduce the number of tracks propagated by the system, a threshold is imposed. Only those tracks with a confidence above the threshold value are considered for further processing.

Fused tracks are processed to predict their future trajectory. This is similar to predicting the state and error covariance at the next time step given current information using the Kalman filter algorithm described in Ref. [11]. The predicted future track determines the regions probably to detect the entity in the future. The algorithm invokes the send method to propagate this information to nodes in those regions, thus completing the processing loop. Figure 8.4 shows how track information can be propagated through a distributed network of sensor nodes.

This approach integrates established entity-tracking techniques with the self-organization abilities of the architecture described in Section 8.3. Association of entities with tracks is almost trivial in this approach, as long as sampling rates are high enough and entity distribution is low enough to avoid ambiguity. When that is not the case, local disambiguation is possible using established data association techniques [9]. A fully decentralized approach, like the one proposed here, should be more robust and efficient than current centralized methods. The main question is whether or not this approach consumes significantly more resources than a centralized entity-tracking approach. This chapter is a first step in considering this problem.

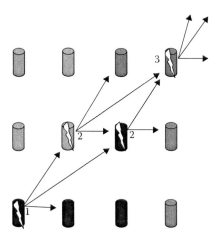

FIGURE 8.4 Example of how network-embedded entity tracking propagates track information in a network. Each cluster (a) receives candidate tracks, (b) merges local estimates, (c) publishes current track information, and (d) propagates new candidate track to nodes along trajectory.

8.5 Local Parameter Estimation

A number of methods exist for local parameter estimation. In August 2000, we tested a method of collaborative parameter estimation based on distributed computing fault tolerance algorithms given in Ref. [11]. Tests were run at the Marine Corp's Twentynine Palms test facility using the computational environment described in Section 8.3. Target detection was performed by signal-processing algorithms derived and implemented by BAE Systems, Austin, Texas. Network routing used the ISI data diffusion methodology [1]. Penn State ARL implemented the collaboration and local parameter estimation approach described in this section. For this test, only acoustic sensor data were used.

Each node executed the publish method with three attributes: the node's UTM x-coordinate, the node's UTM y-coordinate, and a nominal value associating the publish with local collaboration. The same nodes execute the subscribe method with three attributes: a UTM range in the x-dimension that approximates the coverage of the nodes acoustic sensor, a UTM range in the y-dimension that approximates the coverage of the node's acoustic sensor, and the nominal value used by the publish method. When an entity is detected by a node, the send method associated with the publish transmits a data structure describing the detection event. The data structure contains the target class, node location, and detection time. All active nodes whose sensor readings overlap with the detection receive the data structure. The callback routine identified by their subscribe method is activated at this point. The callback routine invokes the local send method to transmit the current state of the local node. Temporal limits stop nodes from responding more than once.

In this way, all nodes exchange local information about the detection. One node is chosen arbitrarily to combine the individual detection events into a single collaborative detection. The test at Twentynine Palms combined readings at the node that registered the first detection. It has also been suggested that the node with the most certain detection would be appropriate. Since the same set of data would be merged using the same algorithm, the point is moot.

At Twentynine Palms, the data were merged using the distributed agreement algorithm described in Ref. [11]. This algorithm is based on a solution to the "Byzantine Generals Problem." Arbitrary faults are tolerated as long as at least two-thirds of the participating nodes are correct and some connectivity restrictions are satisfied. The ad hoc network routing approach satisfies the connectivity requirements.

The algorithm uses computational geometry primitives to compute a region where enough sensors agree to guarantee the correctness of the reading in spite of a given number of possible false negatives or

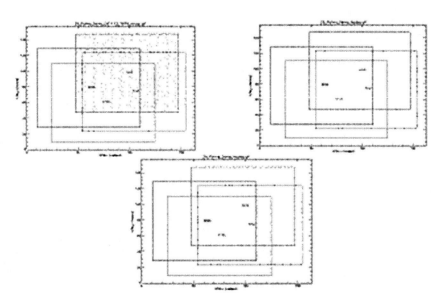

FIGURE 8.5 Entity enters the sensor field from the north–east. It is within sensing range of only one sensor (1619). Since only one sensor covers the target, no faults can be tolerated. Clockwise from upper left, we show results when agreement is such that no fault, one fault, and two faults can be tolerated.

false positives. Figures 8.5 through 8.8 show the results of this experiment. A sequence of four readings is shown. The entity was identified by its location during a given time window.

By modifying the number of sensors that had to agree, we were able to significantly increase the accuracy of our parameters. For each time slice, three decisions are shown proceeding clockwise from the top left: (1) no faults tolerated, (2) one fault tolerated, and (3) two faults tolerated. Note that in addition

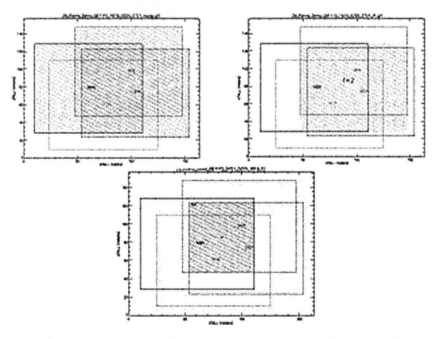

FIGURE 8.6 Same scenario as Figure 8.5. The entity is within sensing range of three sensors (1619, 5255, 5721). Up to two faults can be tolerated.

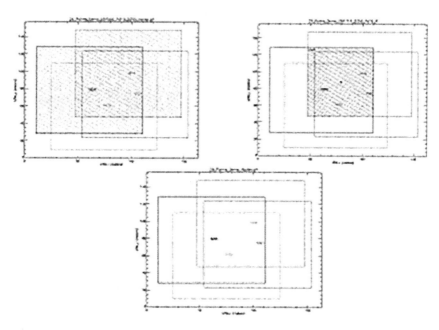

FIGURE 8.7 Same scenario as Figure 8.5. The entity is within sensing range of two sensors (1619, 5255). One fault can be tolerated.

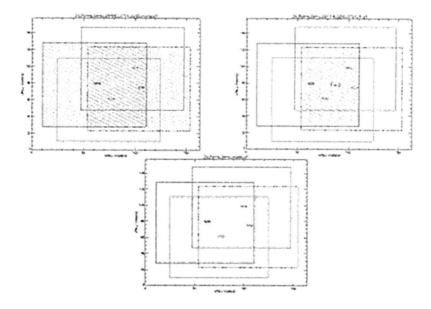

FIGURE 8.8 Same scenario as Figure 8.5. The entity is within sensing range of two sensors (5255, 5721). One fault can be tolerated.

to increasing system dependability, the ability of the system to localize entities has improved greatly. It appears that localization improves as the number of faults tolerated increases. For a dense sensor network like the one used, this type of fault tolerance is useful and it is likely that, excluding boundary regions, entities will always be covered by a significant number of sensors. For many sensing modalities, sensor range can be very sensitive to environmental influences. For that reason, it may be worthwhile to use alternative statistics such as the CPA. CPA can generally be detected as the point where the signal received from the entity is at its maximum. An alternative approach is being implemented, which uses the networking framework described in this section and CPA data. The CPA information is used to construct a trigonometric representation of the entities trajectory. Solution of the problem returns information as to entity heading and velocity. This approach is described in detail in Ref. [12]. Simulations indicate that this approach is promising. We tested it in the field in Fall 2001. The tracking methods given in Section 8.6 are designed using information from the CPA tracker but could also function using localization information from local collaboration.

8.6 Track Maintenance Alternatives

Given these methods for local collaboration to estimate entity-tracking parameters, we derive methods for propagating and maintaining tracks. Three separate methods will be derived: (1) pheromone tracking, (2) EKF, and (3) Bayesian. All three methods are encapsulated in the "disambiguate," "merge detection with track," and "estimate future track" boxes in Figure 8.3. All three require three inputs: (1) current track information, (2) current track confidence levels, and (3) current parameter estimates. They produce three outputs: (1) current best estimate, (2) confidence of current best estimate, and (3) estimated future trajectory. For each method, in turn, we derive methods to

- Disambiguate candidate tracks
- Merge the local detection with the track information
- Initiate a new track
- Extrapolate the continuation of the track

We consider one method at a time.

8.6.1 Pheromone Routing

For our initial approach, we will adapt pheromone routing to entity tracking. Pheromone routing is loosely based on the natural mechanisms used by insect colonies for distributed coordination [13].

When they forage, ants use two pheromones to collaboratively find efficient routes between the nest and food source. One pheromone is deposited by each ant, as it searches for food and moves away from the nest. This pheromone usually has its strongest concentration at the nest. Ants carrying food moving toward the nest deposit another pheromone. Its strongest concentration tends to be at the food source. Detailed explanations of exactly how these and similar pheromone mechanisms work can be found in Ref. [13]. Of interest to us is how pheromones can be an abstraction to aggregate information and allow information relevance to deteriorate over time as shown in Figure 8.9. Pheromones are scent hormones that trigger specific behaviors. After they are deposited, they evaporate slowly and are dissipated by the wind. This means that they become less strong and more diffuse over time. This is useful for track formation. Entities move and their exact position becomes less definite over time. The relevance of sightings also tends to abate over time. This is illustrated in the top portion of Figure 8.9. If two insects deposit the same pheromone in a region, the concentration of the pheromone increases. The sensory stimulation for other insects increases additively, as shown in the middle portion of Figure 8.9. Finally, multiple pheromones can exist. It is even possible for pheromones to trigger conflicting behaviors. In which case, as shown at the bottom of Figure 8.9, the stimuli can cancel each other, providing less or no net effect. These primitives provide a simple but robust method for distributed track formation. We consider each sensor node as a member of the insect society. When a node detects an entity of a specific type, it deposits pheromone for that entity type at its current location. We handle the pheromone as a random variable following a Gaussian distribution. The mean is at the current location. The height of the curve at the mean is determined by the certainty of the detection. The variance of the random variable increases as a function of time. Multiple detections are aggregated by summing individual detections. Note that the sum of normal distributions is a normal distribution.

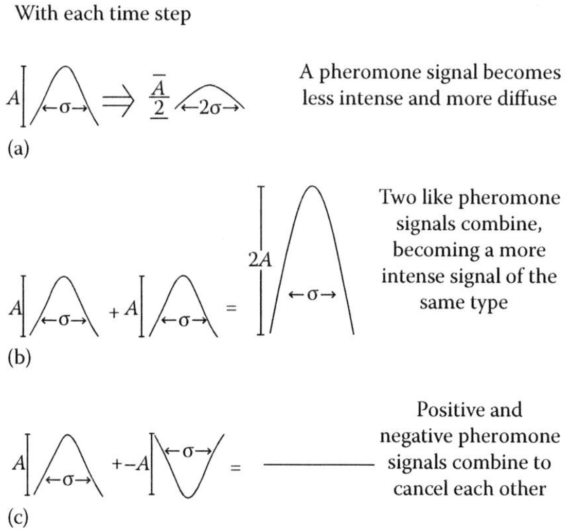

With each time step

(a)

A pheromone signal becomes
less intense and more diffuse

(b)

Two like pheromone
signals combine,
becoming a more
intense signal of the
same type

(c)

Positive and
negative pheromone
signals combine to
cancel each other

FIGURE 8.9 Diagram explains how pheromones can be used to diffuse (a), aggregate (b), and cancel (c) information from multiple sources.

Our method is only loosely based on nature. We try instead to adapt these principles to a different application domain. Track information is kept as a vector of random variables. Each random variable represents position information within a particular time window. Using this information, we derive our entity-tracking methods:

Track disambiguation performs the following steps for entity type and each time window:

1. Update the random variables of pheromones attached to current track information by changing variances using the current time.
2. For each time window, sum the pheromones associated with that time window. Create a new random variable representing the pheromone concentration. Since the weighted sum of a normal distribution is also normal, this maintains the same form.

This provides a temporal sequence of Gaussian distributions providing likely positions of each entity during increasing time windows.

Merging local detection with entity tracks is done by creating a probability density function for the current reading on the current node. The track is a list of probability density functions expressing the merged detections ordered by time windows.

Track initiation is done simply by creating a normal distribution that represents the current reading. This distribution is transmitted to nodes in the rectangular regions that are indicated by the heading parameter.

Track extrapolation is done by finding extreme points of the distribution error ellipses. For each node, four reference points are defined that define four possible lines. The region enclosed laterally by any two lines defines the side boundaries of where the entity is to be expected in the near future. The front and back boundaries are defined by the position of the local node (far and near back), the time step used times the estimated velocity plus a safety factor (near front), and twice the time step times the velocity estimate with safety (far front). The pheromone information is transmitted to nodes in these two regions.

8.6.2 Extended Kalman Filter

This section uses the same approach as in Ref. [11]. We will not derive the EKF algorithm here; numerous control theory textbooks cover this subject in detail. We will derive the matrices for an EKF application that embeds entity tracking in the distributed sensor network. In deriving the equations, we make some simple assumptions.

We derive a filter that uses the three most recent parameter estimates to produce a more reliable estimate. This filtering is required for several reasons. Differentiation amplifies noise. Filtering will smooth out the measurement noise. Local parameter estimates are derived from sensors whose proximity to the entity is unknown. Multiple answers are possible to the algorithms. Combining independently derived parameter estimates lowers the uncertainty caused by this.

When we collaboratively estimate parameters using local clusters of nodes, we assume that the path of the entity is roughly linear as it passes through the cluster. As described in Section 8.5, our velocity estimation algorithm is based on the supposition that sensor fields will contain a large number of sensor nodes densely scattered over a relatively large area. Hence, we presume the existence of a sensor web, through which a vehicle is free to move. The algorithm utilizes a simple weighted least squares regression approach in which a parameterized velocity estimate is constructed.

As a vehicle moves through the grid, sensors are activated. Because the sensor grid is assumed to be densely positioned, we may consider the position of an activated sensor node to be a good approximation of the position of the vehicle in question. Sensors are organized into families by spatial distance. That is, nodes within a certain radius form a family. The familial radii are generally small (6 m), so we may assume that the vehicle is moving in a relatively straight trajectory, free of sharp turns. If each sensor is able to estimate the certainty of its detection, then as a vehicle moves through a clique of sensors, a set of four-tuples is collected:

$$D = \{(x_1, y_1, t_1, w_1), (x_2, y_2, t_2, w_2), \ldots, (x_n, y_n, t_n, w_n)\} \tag{8.1}$$

Each four-tuple consists of the UTM coordinates, (x_i, y_i), of the detecting sensor, the time of detection, t_i, and the certainty of the detection $w_i \in [0,1]$. Applying our assumption that the vehicle is traveling in linear fashion, we hope to fit the points to the equations:

$$x(t) = v_x t + x_0 \tag{8.2}$$

$$y(t) = v_y t + y_0 \tag{8.3}$$

It may be assumed that nodes with a higher certainty were closer to the moving target than those with lower certainty. Therefore, we wish to filter out those nodes whose estimation of the target's position may be inaccurate. We do so by applying a weighted, linear regression [14] to the data mentioned earlier. The following equations show the result:

$$v_x = \frac{\sum_i w_i x_i \sum_i w_i t_i - \sum_i w_i \sum_i w_i x_i t_i}{\left(\sum_i w_i t_i\right)^2 - \sum_i w_i \sum_i w_i t_i^2} \tag{8.4}$$

$$v_y = \frac{\sum_i w_i y_i \sum_i w_i t_i - \sum_i w_i \sum_i w_i y_i t_i}{\left(\sum_i w_i t_i\right)^2 - \sum_i w_i \sum_i w_i t_i^2} \tag{8.5}$$

The primary use of the resulting velocity information is position estimation and track propagation for use in the tracking system. Position estimation is accomplished using an EKF [11]. For the remainder of this chapter, we use

$$x_{k+1} = \Phi_k x_k + w_{k+1} \tag{8.6}$$

$$y_k = M_k x_k + v_k \tag{8.7}$$

as our filter equations. For our Kalman filter, we have set

$$\vec{x}_k = \begin{pmatrix} x_k & y_k & v_k^x & v_k^y \end{pmatrix} \tag{8.8}$$

$$\Phi = \begin{pmatrix} 1 & 0 & \Delta t_k & 0 \\ 0 & 1 & 0 & \Delta t_k \\ 0 & 0 & 1 & 0 \\ 0 & 0 & 0 & 1 \end{pmatrix} \tag{8.9}$$

$$\Phi = \begin{pmatrix} 1 & 0 & 0 & 0 \\ 0 & 1 & 0 & 0 \\ 0 & 0 & 1 & 0 \\ 0 & 0 & 0 & 1 \\ 1 & 0 & \Delta t_k & 0 \\ 0 & 1 & 0 & \Delta t_k \\ 0 & 0 & 1 & 0 \\ 0 & 0 & 0 & 1 \\ 1 & 0 & \Delta t_{k-1} & 0 \\ 0 & 1 & 0 & \Delta t_{k-1} \\ 0 & 0 & 1 & 0 \\ 0 & 0 & 0 & 1 \end{pmatrix} \tag{8.10}$$

where Δt_k is the time differential between the previous detection and the current detection. We are considering the last three CPA readings as the measurements. The covariance matrix of the error in the estimator is given by

$$P_{k+1} = \Phi_k P_k \Phi_k^T + Q_{k+1} \tag{8.11}$$

where Q is the system noise covariance matrix. It is difficult to measure Q in a target-tracking application because there are no real control conditions. We have devised a method for estimating its actual value. The estimate, though certainly not perfect, has provided good experimental results in the laboratory. Acceleration bounding and breaking surfaces for an arbitrary vehicle are shown in Figure 8.10. We may construct an ellipse about these bounding surfaces as shown in Figure 8.11. Depending upon the area of the ellipse and our confidence in our understanding of the vehicle's motion, we may vary the probability p that the target is within the ellipse, given that we have readings from within the bounding surfaces. We can use the radii of the ellipse and our confidence that our target is somewhere within this ellipse to construct an approximation for Q. Assuming that the values are independent and identically distributed (IID), we have

$$p = \int_{-r_x}^{r_x} \exp\left(-\left(\frac{x}{\sigma_x \sqrt{2}} \right)^2 \right) \tag{8.12}$$

Thus, we may conclude that

$$\sigma_x = \frac{r_x \sqrt{2}}{2_{P/2}} \tag{8.13}$$

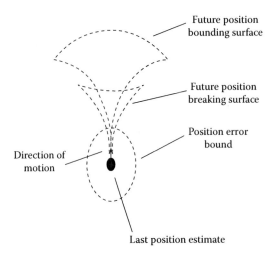

FIGURE 8.10 Target position uncertainty.

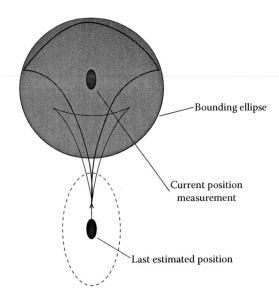

FIGURE 8.11 Target position bounding ellipse.

and likewise that

$$\sigma_y = \frac{r_y \sqrt{2}}{2_{P/2}}$$ (8.14)

We can thus approximate a value for Q as

$$Q = \begin{pmatrix} \sigma_x & 0 & 0 & 0 \\ 0 & \sigma_y & 0 & 0 \\ 0 & 0 & \sigma_{v_x} & 0 \\ 0 & 0 & 0 & \sigma_{v_y} \end{pmatrix}$$ (8.15)

In the aforementioned matrix, we do not mean to imply that these variances are actually independent, but this matrix provides us with a rough estimate for the noise covariance matrix, against which we can tune the Kalman filter. We have tested our velocity estimation and Kalman filter algorithms in the laboratory and at Twentynine Palms. The results of these tests are presented in the following sections.

8.6.3 Bayesian Entity Tracking

This section extends the Bayesian net concepts introduced in Ref. [15]. A belief network is constructed that connects beliefs that influence each other in the problem space. In this section, we derive the portion of this belief network, which is embedded in single nodes in the sensor network. This portion receives inputs from other nodes in the sensor network. In this manner, the global structure of the extended Bayesian net extends across the sensor network and is not known in advance. The structure evolves in response to detections of objects in the environment. Although this evolution, and the computation of how beliefs are quantified, differs from the Bayesian net framework provided in Ref. [15], the basic precepts are the same.

Figure 8.12 shows the belief network that is local to a node in the sensor network. Candidate near and far tracks are received from other nodes in the network. The current net is set up to consider one near and one far candidate track at a time. This is described later in more detail. Detections refer to the information inferred by combining data from the local cluster of nodes. All three entities have a probabilistic certainty factor. They also have associated values for speed and heading. We use the same state vectors as in Section 8.6. We describe the functionality of the belief network from the bottom up:

- No track represents the probability that the current system has neither a new track nor a continuation of an existing one. It has the value $1 - (P_n + P_c - P_n * P_c)$ where P_n is the probability that there is a new track and P_c is the probability that the detection is a continuation of an existing track. The sum of the probabilities assumes their independence.
- New track is the probability that a new track has been established. Its value is calculated by subtracting the likelihood that the current detection matches the near and far tracks under consideration from the certainty of the current detection.

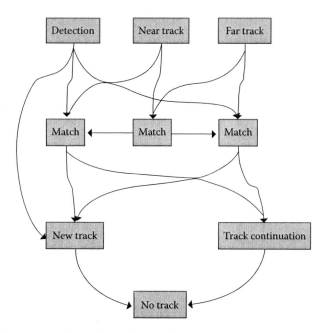

FIGURE 8.12   Belief network used in entity track evaluation.

- Track continuation expresses the probability that the current reading is a continuation of the near and far tracks under consideration. It is computed as $P_c = L_n + L_f - L_n * L_f$, where P_c is the probability that the track is a continuation of the near track under consideration or the far track under consideration, and $L_n (L_f)$ is the likelihood that the near (far) track matches the current detection.
- Matching detections to tracks (or tracks to tracks) is done by comparing the value in Equation 8.8 with the variance of the track readings. This provides a probabilistic likelihood value for the current detection belonging to the current track. This value is weighted by the certainty values attached to the current detection (L_d) and track (L_t). The weight we use is $1 - ((1 - L_t) * (1 - L_d))$, which is the likelihood that either the track or the detection are correct.
- Matches where a near and a far track match are favored by adding one-eighth of the matching value of the near and far track, as defined in the previous bullet, to the value calculated earlier. The addition is done as adding two probabilities with the assumption of independence.

Given this belief network, we can now describe how the specific entity-tracking methods.

Track disambiguation is performed by evaluating the belief network for every combination of near and far tracks. We retain the combination of tracks where the value of track continuation is a maximum. The decision is then made between continuing the tracks, starting a new track, or saying there is no current track by taking the decision with the highest probability.

Merging entity detection with local track is done by combining the current detection with the track(s) picked in the disambiguation step. If the match between the near and far tracks is significant, then the values of the near and far tracks are both merged with the current detection. If not, then the current detection is merged with the track it best matches. Merged parameters are their expected values. Parameter variance is calculated by assuming that all discrepancies follow a normal distribution.

Track initiation is performed when the decision is taken to start a new track during the disambiguation phase. Track parameters are the current estimate and no variance is available.

Track extrapolation is identical to the method given in Section 8.6.

8.7 Tracking Examples

In this section, we present a simple example illustrating how each proposed method functions.

8.7.1 Pheromone Routing

Figure 8.13 represents a field of sensor nodes. An entity moves through the field and sensor clusters form spontaneously in response to detection events. Each cluster produces a pheromone abstraction signaling the presence of an entity. In this example, we consider a single class of entity and a single pheromone. Multiple pheromones could be used to reflect multiple entity classes or entity orientation. Figure 8.14 shows the situation at time t_2. Each cluster in row 5 exudes its own pheromone. They also receive pheromone information from the nodes in row 0. The three detections in row 0 were each represented as a random variable with a Gaussian distribution. The nodes performing the tracking locally independently combine the pheromone distributions from time t_1. This creates a single pheromone random variable from step one. This random variable encloses a volume equivalent to the sum of the volumes of the three individual pheromone variables at time t_1. The mean of the distribution is the sum of the means of the individual pheromone variables at time t_1. The new variance is formed by summing the individual variances from t_1 and increasing it by a constant factor to account for the diffusion of the pheromone. The situation at time t_3 is shown in Figure 8.15. The pheromone cloud from time t_1 is more diffuse and smaller. This mimics the biological system where pheromone chemicals evaporate and diffuse over time. Evaporation is represented by reducing the volume enclosed by the distribution. Diffusion is represented by increasing the variance.

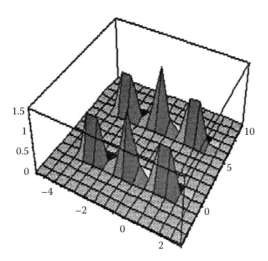

FIGURE 8.13 Entity detections from six sensor node clusters. The three detections at row 0 occur at time t_1. The three at row 5 occur at time t_2. Each detection results from a local collaboration. The pheromone is represented by a normal distribution.

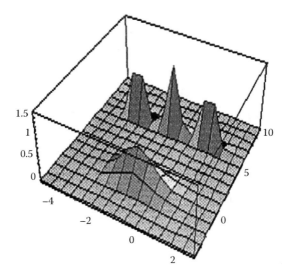

FIGURE 8.14 At time t_2 the pheromones exuded at time t_1 are combined into a single distribution.

8.7.2 Extended Kalman Filter

We present a numerical example of the EKF Network Track formation. Coordinates will be given in UTM instead of Lat/Long to simplify the example. Without loss of generality, assume that our node is positioned at UTM coordinates $(0, 0)$. We will not consider the trivial case when no detections have been made but instead assume that our node has been vested with some state estimate $\hat{x}$ and some covariance matrix P from a neighboring node. Finally, assume that a target has been detected and has provided an observation y_0. Let

$$y = \left[\langle 0.2, 3.1, 0.1, 3.3 \rangle, \quad \langle -0.5, 2.8, -0.6, 2.7 \rangle, \quad \langle -0.8, 3, 0, 2.5 \rangle \right] \tag{8.16}$$

$$\hat{x}(k \mid k) = [-0.225, 3.0, 0.17, 2.83] \tag{8.17}$$

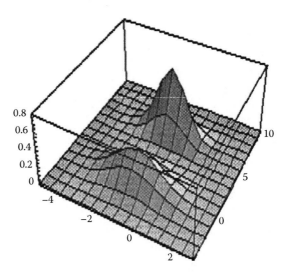

FIGURE 8.15 Situation at time t_3.

and assume $y_0 = [0, 3.1, 0, 2.9]$. Finally, let

$$Q = \begin{pmatrix} 0.4 & 0 & 0 & 0 \\ 0 & 1.1 & 0 & 0 \\ 0 & 0 & 0.3 & 0 \\ 0 & 0 & 0 & 1.2 \end{pmatrix} \tag{8.18}$$

be the covariance matrix last computed. Since there is only one track, it is clear that the minimum of $\sqrt{(y0 - \hat{x})^2}$ will match the state estimate given earlier. As soon as detection occurs, y becomes

$$y = \left[\langle 0, 3.1, 0, 2.9 \rangle, \langle 0.2, 3.1, 0.1, 3.3 \rangle, \langle 0.5, 2.8, -0.6, 2.7 \rangle \right] \tag{8.19}$$

We can now compute $P(k+1|k+1)$ and $\hat{x}(k+1|k+1)$; from Section 8.6.2 (Extended Kalman Filter), we have

$$\hat{x}(k+1|k+1) = [-0.2869, 3.1553, -0.293, 2.8789] \tag{8.20}$$

and

$$P(k+1|k+1) = \begin{bmatrix} 0.15 & 0.21 & 0.05 & 0.02 \\ 0.16 & -0.07 & 0.2 & -0.07 \\ -0.08 & 0.11 & 0.17 & 0.18 \\ 0.21 & 0.05 & 0.05 & 0.22 \end{bmatrix} \tag{8.21}$$

This information, along with the last observation, can now be sent to nodes in the direction of the vehicle's motion, namely, in a north-eastwardly direction heading away from $(0,0)$ toward $(1,1)$ in UTM.

8.7.3 Bayesian Belief Net

We will construct a numerical example of Bayesian network track formation from the node point of view. The values given will be used in Section 8.8 for comparing the tracking approaches derived here. Further research is needed to empirically determine appropriate likelihood functions. These values are used for initial testing. Without loss of generality assume that

$$L_n = L_f = \begin{cases} 0.5, & \text{if a detection exist} \\ 0, & \text{otherwise} \end{cases} \tag{8.22}$$

where L_n (L_f) is the likelihood that a near (far) track matches a current detection. Also, let

$$L_t = 1 - L_d = \begin{cases} 0.5, & \text{if a detection exist} \\ 0, & \text{otherwise} \end{cases} \tag{8.23}$$

Assume that detection has occurred in a node with no near or far track information; then the following conclusions may be made:

$$P_c = L_n + L_f - L_n L_f = 0.75 \tag{8.24}$$

$$P_m = 1 - (1 - L_t)(1 - L_d) = 0.75 \tag{8.25}$$

Therefore, there is a 75% chance that this is a track continuation with a confidence of 75% that the match is correct. Assuming a perfect sensor, $P_n = 0.25$, since no false detections can be made.

8.8 Cellular Automata Model

Evolving distributed systems have been modeled using CA [16]. A CA is a synchronously interacting set of abstract machines (network nodes). A CA is defined by

- The dimension of the automata, d
- The radius of an element of the automata, r
- The transition rule of the automata
- The set of states of an element of the automata, s

An element's (node's) behavior is a function of its internal state and those of neighboring nodes as defined by δ. The simplest instance of a CA has a dimension of 1, a radius of 1, and a binary set of states, and all elements are uniform. In this case, for each individual cell, there are a total of 2^3 possible configurations of a node's neighborhood at any time step if the cell itself is considered part of its own neighborhood. Each configuration is expressed as an integer v:

$$v = \sum_{i=-1}^{1} 2^{j_i + 1} \tag{8.26}$$

where
  i is the relative position of the cell in the neighborhood (left = −1, current position = 0, right = 1)
  j_i is the binary value of the state of cell i

Each transition rule can therefore be expressed as a single integer r known as its Wolfram number [17]:

$$r = \sum_{v=1}^{8} 2^{j_v + 2^v} \tag{8.27}$$

where j_v is the binary state value for the cell at the next time step if the current configuration is v. This is the most widely studied type of CA. It is a very simple many-to-one mapping for each individual cell.

The four complexity classes shown in Figure 8.16 have been defined for these models. In the uniform class, all cells eventually evolve to the same state. In the periodic class, cells evolve to a periodic fixed structure. The chaotic class evolves to a fractal-like structure. The final class shows an interesting ability to self-organize into regions of local stability. This ability of CA models to capture emergent self-organization in distributed systems is crucial to our study.

We use more complex models than the ones given in Equations 8.1 and 8.2. CA models have been used successfully to study traffic systems and mimic qualitative aspects of many problems found in vehicular traffic flow [16]. For example, they can illustrate how traffic jams propagate through road systems. By modifying system constraints, it is possible to create systems where traffic jams propagate either opposed to or along the direction of traffic flow. This has allowed physicists to empirically study how highway system designs influence the flow of traffic.

Many of the CA models are called "particle-hopping" models. The most widespread particle-hopping CA model is the Nagel–Schreckenberg model [16]. This is a variation of the one-dimensional CA model [18] expressed by Equations 8.1 and 8.2. This approach mainly considers stretches of highway as a one-dimensional CA. It typically models one lane of a highway. The highway is divided into sections, which are typically uniform. Each section of the highway is a cell. The sizes of the cells are such that the state of a cell is defined by the presence or lack of an automobile in the cell. All automobiles move in the same direction. With each time step, every cell's state is probabilistically defined based on the states of its neighbors. Nagel–Schreckenberg's CA is based on mimicking the motion of an automobile. Only one automobile can be in a cell at a time, since two automobiles simultaneously occupying the same space cause a collision. If an automobile occupies a cell, the probability of the automobile moving to the next cell in the direction of travel is determined by the speed of the automobile. The speed of the automobile depends on the amount of free space in front of the automobile, which is defined by the number of vacant cells in front of the automobile. In the absence of other automobiles (particles), an

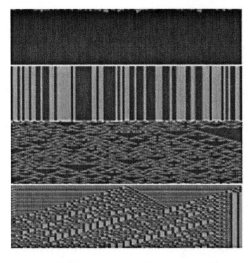

FIGURE 8.16 Examples of the four complexity classes of CA. Four sections from the top to bottom: Uniform, periodic, chaotic, and interesting.

FIGURE 8.17 Example of output from a particle hopping CA. Lighter shades of gray signal higher packet density. This is a one-dimensional example. The *x* dimension is space. Each row is a time step. Time evolves from top to bottom. Black diagonal stripes from top-left to bottom-right show caravan formation. Light stripes from right to left at the top of the image show traffic jam propagation.

automobile moves at maximum speed along the highway by hopping from cell to cell. As more automobiles enter, the highway congestion occurs. The distance between particles decreases and consequently speed decreases. Figure 8.17 shows the evolution of a particle-hopping CA over time.

We adapt this approach to modeling sensor networks. Instead of particles representing automobiles moving along a highway, they represent packets in a multi-hop network moving from node to node. Each cell represents a network node rather than a segment of a highway lane. Since we are considering a two-dimensional surface covered with sensor nodes, we need a two-dimensional CA. The cells are laid out in a regular matrix. A node's neighborhood consists of the eight nodes adjoining it to the north, south, east, west, northwest, northeast, southwest, and southeast directions. For this chapter, we assume that nodes are fixed geographically, for example, nonmobile. A packet can move from a node to any of its neighbors. The number of packets in the cell's node defines the cell's state. Each node has a finite queue length. A packet's speed does not depend on empty cells in its vicinity. It depends on the node's queue length. Cell state is no longer a binary variable; it is an integer value between 0 and 10 (chosen arbitrarily as the maximum value). As with Nagel–Schreckenberg, particle (packet) movement from one cell to another is probabilistic. This mirrors the reality that wireless data transmission is not 100% reliable. Atmospheric and environmental effects, such as sunspots, weather, and jamming, can cause packets to be garbled during transmission. For our initial tests, we have chosen the information sink to be at the center of the bottom edge of the sensor field. Routing is done by sending packets along the shortest viable path from the sensor source to the information sink, which can be determined using local information. Paths are not viable when nodes in the path can no longer receive packets. This may happen when a node's battery is exhausted, or its queue is full.

This adaptation of particle-hopping models is suitable for modeling the information flow in the network; however, it does not adequately express sensing scenarios where a target traverses the sensor field. To express scenarios, we have included "free agents in a cellular space" (FACS) concepts from Ref. [17].

In Ref. [17], Portugali uses ideas from Synergetics and a CA including agents to study the evolution of ethnic distributions in Israeli urban neighborhoods. In the FACS model, agents are free to move from cell to cell in the CA. The presence of an agent modifies the behavior of the cell, and the state of a cell affects the behavior of an agent.

In our experiments, entities traversing the sensor field are free agents. They are free to follow their own trajectory through the field. Detection of an entity by a sensor node (cell) triggers one of the entity-tracking

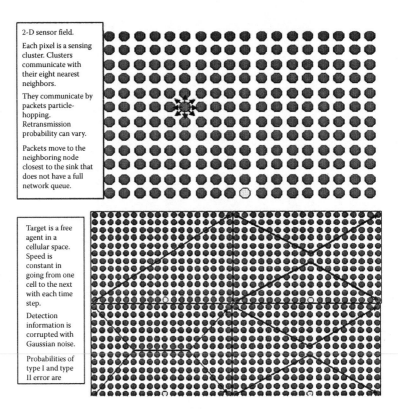

FIGURE 8.18 Top diagram explains the CA model of the sensor field. The bottom diagrams show the four target trajectories used in our simulation scenarios.

algorithms. This causes track information to be transmitted to other nodes and to the information sink. Figure 8.18 describes scenarios we use in this chapter to compare the three tracking approaches proposed.

8.9 Cellular Automata Results

In this section, we present qualitative and quantitative results of CA simulations and a brief summary of the modeling techniques used. Our CA model is designed to mimic higher-level behaviors of clusters of sensor nodes. Each cell corresponds to a localized set of sensor nodes. A traffic sink is present. It connects our sensor grid to the outside world. Traffic is modeled at the cluster level. Simplifying assumptions facilitate the implementation of target tracking. We first present the results of target-tracking algorithms and follow them with network traffic analysis.

8.9.1 Linear Tracks

All three tracking algorithms adequately handle linear tracks. Figure 8.19 shows the track formed using pheromone tracking. This image shows the maximum pheromone concentrations achieved by cells in the sensor grid. At any point in time, the pheromone concentrations are different since the abstract pheromones decay over time.

Figure 8.20 shows the same problem when EKF tracking is used. The fidelity of the model results in a near constant covariance matrix P being computed at run time. The model uses constant cell sizes and Gaussian noise of uniform variance for the results of local collaboration. These simulations are used for a high-level comparison of algorithms and their associated resource consumption. The simplifications are appropriate for this. The expected value for vehicle position is the location of the cell receiving the detection. Speed is

FIGURE 8.19 Track created by the pheromone tracker when an entity crosses the terrain in a straight line from the upper left corner to the lower right corner.

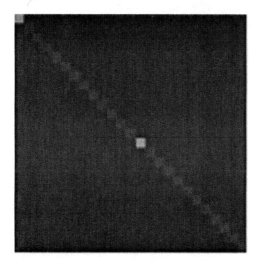

FIGURE 8.20 Results of EKF tracking for the same scenario.

constant with Gaussian noise added. Gray cells indicate a track initiation; dark gray cells indicate a track continuation. In this case, an initiation occurs after the first observation of the agent. A second, erroneous initiation occurs later on as a result of noise. Figure 8.21 presents the results from Bayesian net tracking. Gray squares show track initiation. Light gray squares indicate track continuation. Static conditional probabilities were used for each path through the net. EKF tracking performs best when tracks are not ambiguous. Pheromone routing algorithm performs equally well; however, the track it constructs is significantly wider than the track produced by either the Bayesian net or EKF trackers. The track constructed by the Bayesian net algorithm contains gaps because of errors made in associating detections to tracks.

8.9.2 Crossing Tracks

When two tracks cross, track interpretation can be ambiguous. Unless vehicles can be classified into distinct classes, it is difficult to construct reliable tracks. Figure 8.22 demonstrates this using the pheromone tracker. Gray cells contain two vehicle pheromone trails contrasted to the dark and

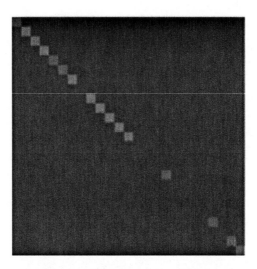

FIGURE 8.21   Results of the Bayesian belief network tracker.

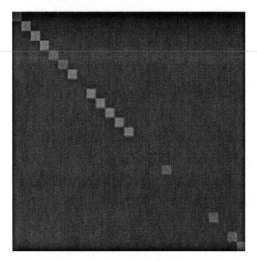

FIGURE 8.22   Results when pheromones track crossing entities.

light gray cells that have only one vehicle. The track information is ambiguous when the vehicles deposit identical pheromones. Figure 8.23 shows tracks formed using the EKF tracker. The target beginning in the lower left-hand corner was successfully tracked until it reached the crossing point. Here, the EKF algorithm was unable to successfully identify the target and began a new track shown by the gray cell. The second track was also followed successfully until the crossing point. After this point, the algorithm consistently propagated incorrect track information. This propagation is a result of the ambiguity in the track crossing. If an incorrect track is matched during disambiguation, the error can be propagated forward for the rest of the scenario. In Figure 8.23 as in the other EKF images,

- Gray pixels signal track initiation
- Dark gray pixels indicate correct track continuation
- Light gray pixels signal incorrect track continuation

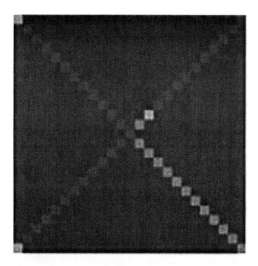

FIGURE 8.23   Results from EKF tracking of two crossing targets.

FIGURE 8.24   Bayesian belief net tracker applied to crossing tracks.

In Figure 8.24, the central region of the sensor field using the Bayesian network approach image continued tracks in both directions. In these tests, we did not provide the Bayesian net with an appropriate disambiguation method. The network only knows that two vehicles passed forming two crossing tracks. It did not estimate which vehicle went in which direction.

Each algorithm constructed tracks representing the shape of the vehicle path. Target to track matching suffered in the EKF most likely as a result of model fidelity. The Bayesian net track formation algorithm performed adequately and comparably to the pheromone-tracking model. Pheromone tracking was able to successfully construct both tracks additionally; different pheromone trails proved to be a powerful device in differentiating between vehicle tracks. Unfortunately, using different "digital pheromones" for each target type will differentiate crossed tracks only when the two vehicles are different. Creating a different pheromone for each track would currently be possible for different target classes when robust classification methods exist. Additional research is desirable to find applicable methods when this is not the case.

8.9.3 Nonlinear Crossing Tracks

We present three cases where the sensor network tracks nonlinear agent motion across the cellular grid. Figure 8.25 displays the pheromone trail of two vehicles following curved paths across the cellular grid. Contrast it with Figures 8.26 and 8.27, which show EKF (Bayesian net) track formation. There is not much difference between the Bayesian net results for nonlinear tracks and that for linear crossing tracks. Pheromone results differ because two distinct pheromones are shown. If both vehicles had equivalent pheromones, the two pheromone history plots would be looked identical. Pheromone concentration can indicate the potential presence of multiple entities. Figure 8.28 shows a plot of pheromone concentration over time around the track intersection. Regions containing multiple entities have higher pheromone concentration levels. Bayesian net formation, however, constructs a more crisp view of the target paths. Notice the ambiguity at the center, however. Here, we see a single-track continuation moving from bottom left to top right and two track initiations. These results indicate the inherent ambiguity of the problem.

FIGURE 8.25   Pheromone trails of two entities that follow nonlinear crossing paths.

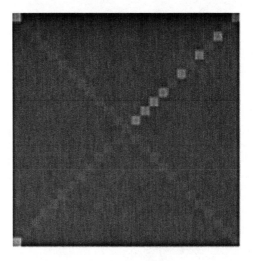

FIGURE 8.26   EKF tracks of two entities in the same scenario.

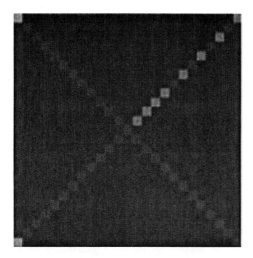

FIGURE 8.27 Bayesian belief net tracks of crossing vehicles taking curved paths.

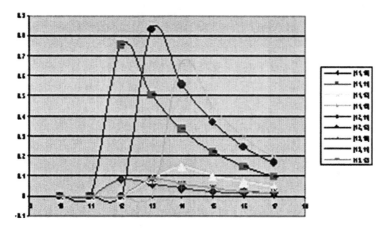

FIGURE 8.28 Pheromone concentration over time.

8.9.4 Intersecting Tracks

Ambiguity increases when two tracks come together for a short time and then split. Figure 8.29 shows one such track formation. The middle section of the track would be ambiguous to the CA pheromone tracking algorithm if both vehicles were mapped to the same pheromone. Minor discontinuities occur in the individual tracks as a result of the agent's path through the cellular grid. The only information available is the existence of the two tracks leaving a different pheromone. Figure 8.30 shows a plot of the pheromone levels through time. Clearly, it is possible to use pheromone concentration as a crude estimate for the number of collocated targets in a given region. Moreover, it may be possible to use the deteriorating nature of pheromone trails to construct a precise history for tracks in a given region.

In Figure 8.30, the central region in Figure 8.29 extends above the other peaks. This would indicate the presence of multiple vehicles. Figure 8.30 shows the track produced by the EKF routing algorithm using the same agent path.

In our simulation, the ability of the EKF to manage this scenario depends on the size of the neighborhood for each cell. The curved path taken by the agent was imposed on a discrete grid. Doing so meant that detections did not always occur in contiguous cells. At this point, it is not clear whether this error is a result of the low fidelity of the CA model or indicative of issues that will occur in the field. Our initial

FIGURE 8.29   Paths of two vehicles intersect, merge for a while, and then diverge. In the absence of classification information, it is impossible to differentiate between two valid interpretations.

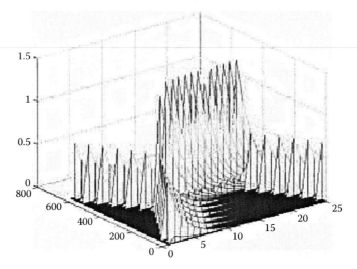

FIGURE 8.30   Pheromone concentrations produced in the scenario shown in Figure 8.29.

interpretation is that this error is significant and should be considered in determining sensor coverage. Ambiguity arises in the region where the tracks merge because both entities have nearly identical state vectors. Cells may choose one or the other with no deleterious effects to track formation. However, ground truth as it is represented in Figure 8.31 can only show that at least one of the two cells selected the incorrect track for continuation. This may also be a residual effect of the synchronous behaviors of the agents as they traverse the cellular grid. Bayesian net track formation had the same problem with contiguous cells as the EKF tracker. Its performance was even more dependent on the ability of the system to provide continuous detections. If an agent activates two non-neighboring cells, the probability of track continuation is zero, because no initial vehicle information was passed between the two nodes (Figure 8.32).

8.9.5 Track Formation Effect on Network Traffic

Network traffic is a nonlinear phenomenon. Our model integrates network traffic analysis into the tracking algorithm simulations. This includes propagating traffic jams as a function of sensor network design.

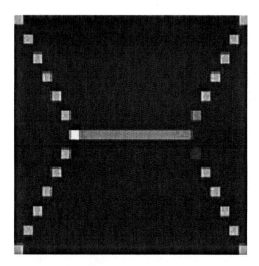

FIGURE 8.31 EKF tracks formed in the same scenario as Figure 8.29.

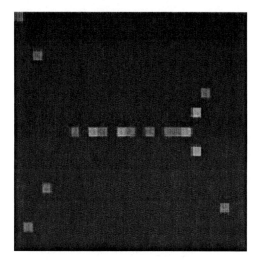

FIGURE 8.32 Results of belief net tracker for this scenario.

Figure 8.33 shows a sensor network randomly generating data packets. Traffic backups occur as data flow to the network sink. These simulations have data packets taking the shortest available route to the sink. When cell packet queues reach their maximum size, cells become unavailable. Packets detour around unavailable cells. Figure 8.34 shows the formation of a traffic jam. Figures 8.35 and 8.36 plot packet density in a region surrounding the sink. The legend indicates the (row–column) position of the cell generating the depicted queue-length history. The existence and rate of growth of traffic jams around the sink is a function of the rate of information detection and the probability of successful data transmission. Consider false detections in the sensor grid, where p is the false alarm probability. For small p, no traffic jams form. If p increases beyond a threshold p_c, traffic jams form around the sink. The value of p_c appears to be unique to each network. In our model, it appears to be unique to each set of CA transition rules. This result is consistent with queuing theory analysis where maximum queue length tends to infinity when the volume of requests for service is greater than the system's capacity to process requests (Figure 8.37). When detections occur, data packets are passed to neighboring cells in the direction the entity is traveling. Neighboring cells store packets and use these data for track formation. Packets are also

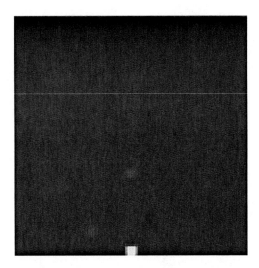

FIGURE 8.33   CA with random detections.

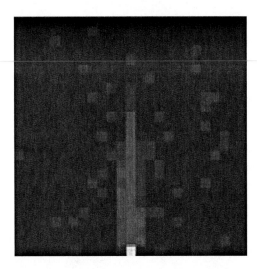

FIGURE 8.34   Traffic jam formation around the data sink. Data packets are generated randomly throughout the network. Light gray cells have maximum queue length. Black cells are empty. Darker shades of gray have shorter queue length than lighter shades.

sent to the data sink along the shortest path. This simple routing algorithm causes traffic jams to form around the network sink. A vertical path above the sink forms, causing a small traffic jam (Figure 8.38). The image in Figure 8.39 shows the queue length of the column in 54 steps. The first 10 rows of data have been discarded. The remaining rows illustrate the traffic jam seen in gray scale in Figure 8.38.

Traffic flux through the sink is proportional to the number of tracks being monitored, the probability of false detection, and the probability of a successful transmission. Assuming a perfect network and nonambiguous tracks, this relationship is linear, for example,

$$\Psi_{\sin k} = kT \tag{8.28}$$

where
     T is the number of tracks
     Ψ is the flux

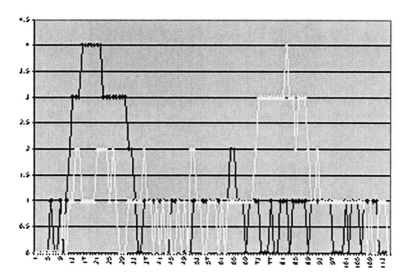

FIGURE 8.35 Average queue length versus time for nodes surrounding the data sink, when probability of false alarm is below the critical value.

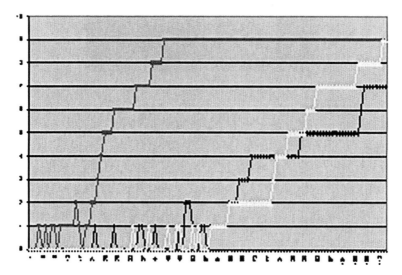

FIGURE 8.36 Average queue length versus time for nodes surrounding the data sink, when probability of false alarm is above the critical value.

Track ambiguities and networking imperfections cause deviations from this linear structure. The exact nature of the distortion depends directly on the CA transition rule and the type of track uncertainty.

8.9.6 Effects of Network Pathologies

Sensing and communication are imperfect in the real world. In this section, we analyze the effects of imperfections on track formation. We analyze the case where false positives occur with a probability of 0.001 per cell per time step. Figure 8.40 shows a pheromone track constructed in the presence of false

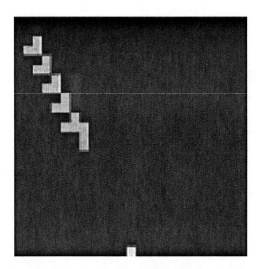

FIGURE 8.37   Data packet propagation in the target-tracking scenario shown in Figures 8.19 through 8.21.

FIGURE 8.38   Formation of a traffic jam above the data sink.

positives. Figure 8.41 illustrates how the existence of false positives degrades the performance of the pheromone tracker. The first peak is false detection. The peak just below this shows the spread of the pheromone. Both peaks decay until an actual detection is made. The first peak could be interpreted as the beginning of track. This misinterpretation is minor in this instance, but these errors could be significant in other examples.

It may be possible to use pheromone decay to modify track continuation probabilities in these cases. If a pheromone has decayed beyond a certain point, it could be assumed that no track was created. In the example, the false detection decayed below a concentration of 0.2 pheromone units before the jump due to the actual sensor detection. If 0.2 were the cut off for track continuation, the node located at cell grid (8,4) would have constructed a track continuation of the true track, not the false one. Further studies would help us determine the proper rates for pheromone diffusion and decay.

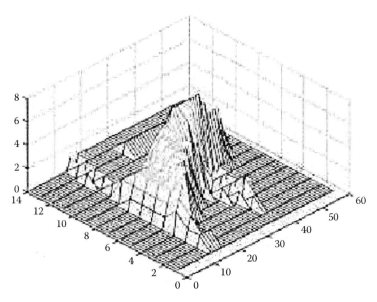

FIGURE 8.39 Three-dimensional view of traffic jam formation in the 12th column of the CA grid.

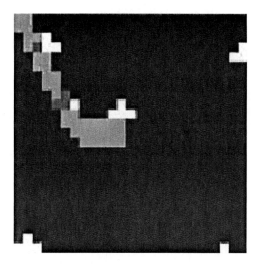

FIGURE 8.40 Pheromone track formed with false-positive inputs. White areas indicate false alarms; darker gray areas indicate regions where two vehicles were detected.

Figure 8.42 shows the track formed using the EKF approach in the presence of false positives. The algorithm is relatively robust to these types of errors. However, as was shown in Figure 8.31, lack of contiguous target sightings plays a significant role in degrading track fidelity. Of the three algorithms studied, the pheromone approach formation is most robust to the presence of false positives. The decay of pheromones over time allows the network to isolate local errors in space and time. The confidence level of information in pheromone systems is proportional to the concentration of the pheromone itself. Thus, as pheromones diffuse through the grid, their concentration and thus confidence decreases. Once pheromone concentration drops below a certain threshold, its value is truncated to zero. EKF and

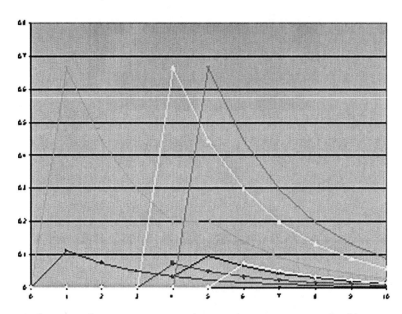

FIGURE 8.41 Ambiguity in pheromone quantities when a track crosses an area with a false positive.

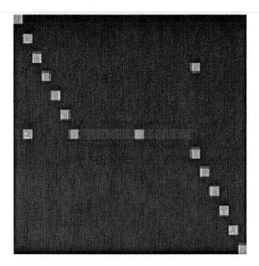

FIGURE 8.42 EKF filter tolerates false positives but is sensitive to the lack of continuous sensor coverage.

Bayesian net track formation is dependent on the spread of information to create track continuations. If a false detection is made near an existing track, it causes errors to occur in initial track formation, which then propagates throughout the network (Figure 8.43).

Network traffic is also affected by the existence of false positives. All detections are transmitted to the central sink for processing. Track formation information is also transmitted to surrounding nodes. As suggested by our empirical analysis, when the false-positive probability is higher than 0.002 for this particular CA, traffic jams form around the sink. This is aggravated by the propagation of false positives. Figure 8.44 displays the relationship between false-positive detection and flux through the sink.

The Bayesian net generates fewer data packets than the other two. The algorithm is designed to disregard some positive readings as false. The others do not. The EKF assumes Gaussian noise. Pheromones propagate uncertain data to be reinforced by others.

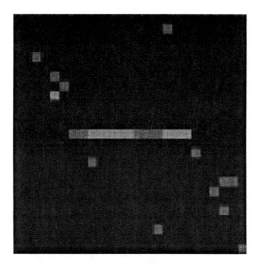

FIGURE 8.43 Bayesian net tracking in the presence of false positives.

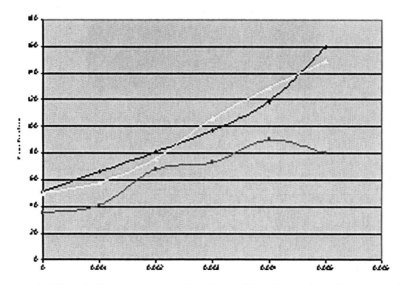

FIGURE 8.44 Probability of a false positive versus the volume of data flowing through the data sink.

Imperfect data transmission also affects sensor networks. Figures 8.45 through 8.47 display the tracks formed by the three track formation algorithms when the probability of intercell communication is reduced to 75%. This means that packets frequently need to be retransmitted. Tracking performance is not badly affected by the lack of timely information. The pheromone track is the least affected by the transmission loss. The fidelity of the track is slightly worse than observed in a perfect network. The track formed by the EKF algorithm is more severely affected. When information is not passed adequately, track continuation is not possible. This leads to a number of track initiations. A similar effect is noted in the Bayesian net track. It is clear that pheromone track formation is the most resilient to the lack of punctual data because the track does not rely on the sharing of information for track continuation. The other two algorithms rely on information from neighbors to continue tracks.

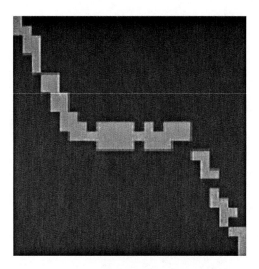

FIGURE 8.45 Pheromone track formed in the presence of frequent data retransmissions.

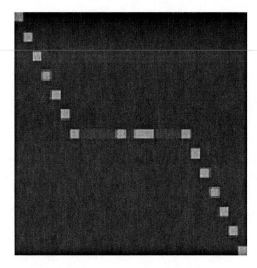

FIGURE 8.46 EKF track in the same conditions.

8.10 Collaborative Tracking Network

The ColTraNe is a fully distributed target-tracking system. It is a prototype implementation of the theoretical intercluster-distributed tracker presented earlier. ColTraNe was implemented and tested as part of a larger program. Sensoria Corporation constructed the sensor nodes used. Individual nodes use SH4 processors running Linux and are battery powered. Wireless communication for ColTraNe uses time division multiplexing. Data routing is done via the diffusion routing approach described in Ref. [19], which supports communication based on data attributes instead of node network addresses. Communication can be directed to geographic locations or regions.

Each node had three sensor inputs: acoustic, seismic, and passive infrared (PIR). Acoustic and seismic sensors are omnidirectional and return time-series data. The PIR sensor is a 2 pixel imager. It detects motion and is directional. Software provided by BAE Systems in Austin, Texas, handles

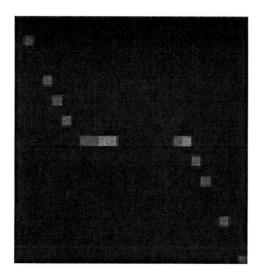

FIGURE 8.47 Bayesian track formed in an imperfect network.

target detection. The software detects and returns CPA events. CPA is a robust, easily detected statistic. A CPA event occurs when the signal intensity received by a sensor starts to decrease. Using CPA events from all sensor types makes combining information from different sensing modes easy. Combining sensory modes makes the system less affected by many types of environmental noise [11].

We summarize the specific application of Figure 8.2 to ColTraNe as follows:

1. Each node waits for CPA events to be triggered by one or more of its sensors. The node also continuously receives information about target tracks heading toward it.
2. When a CPA event occurs, relevant information (node position, CPA time, target class, signal intensity, etc.) is broadcast to nodes in the immediate vicinity.
3. The node with the most intense signal in its immediate neighborhood and current time slice is chosen as the local "clump head." The clump head calculates the geometric centroid of the contributing node's positions, weighted by signal strength. This estimates the target position. Linear regression is used to determine target heading and velocity.
4. The clump head attempts to fit the information from step 3 to the track information received in step 1. We currently use a Euclidean metric for this comparison.
5. If the smallest such track fit is too large, or no incoming track information is found, a new track record is generated with the data from step 3. Otherwise, the current information from step 3 is combined with the information from the track record with the closest track fit to create an updated track record.
6. The record from step 5 is transmitted to the user community.
7. A region is defined containing the likely trajectories of the target; the track record from step 5 is transmitted to all nodes within that region.

Of the three track maintenance techniques evaluated earlier (pheromone, EKF, and Bayesian), field tests showed that the EKF concept was feasible. Processing and networking latency was minimal and allowed the system to track targets in real time.

Distributing logic throughout the network had unexpected advantages in our field test at Twentynine Palms in November 2001. During the test, hardware and environmental conditions caused 55% of the CPA events to be false positives. The tracks initiated by erroneous CPA events were determined by step 3 to have target heading and velocity of 0.0, thereby preventing their propagation to the rest of the nodes in step 7. Thus, ColTraNe automatically filtered this clutter from the data presented to the user.

Problems with the Twentynine Palms implementation were discovered as well:

- Implementation schedule did not allow the EKF version of step 5 to be tested.
- Velocity estimation worked well, but position estimation relied on the position of the clump head.
- Tracks tended to branch, making the results difficult to decipher (see Figure 8.48a).
- Tracking was limited to one target at a time.

Continued development has alleviated these problems. The EKF was integrated into the system. It improves the quality of both track and target position estimates as tracks progress.

An angle gate, which automatically excludes continuations of tracks when velocity estimates show targets are moving in radically different directions, has been inserted into the track-matching metric. This reduces the tendency of tracks to branch, as shown in Figure 8.48b. We constructed a technique for reducing the tendency of tracks to branch. We call this technique lateral inhibition. Before continuing a track, nodes whose current readings match a candidate track broadcast their intention to continue the track. They then wait for a period of time proportional to the log of their goodness-of-fit value. During this time, they can receive messages from other nodes that fit the candidate track better. If better fits are received, they drop their continuations. If no one else reports a better fit within the time-out period, the node continues the track. Figure 8.48d shows a target track with lateral inhibition.

The target position is now estimated as the geometric centroid of local target detections with signal intensity used as the weight. Our tests indicate this is more effective in improving the position estimate than the EKF. The geometric centroid approach was used in the angle filter and lateral inhibition test runs shown in Figure 8.48a through d and Table 8.1.

Differences between the techniques can be seen in the tracks in Figure 8.48a through d. The tracks all use data from a field test with military vehicles at Twentynine Palms Marine Training Ground. Sensor nodes were placed along a road and at an intersection. In the test run depicted in Figure 8.48a through d, the vehicle traversed the sensor field along a road going from the bottom of the diagram to the top. The faint dotted line shows the position of the center of the road. Figure 8.48a shows results from our original implementation. The other diagrams use our improved techniques and the same sensor data. Figure 8.48a illustrates the deficiencies of our original approach. The tracking process works, but many track branches form and the results are difficult to interpret. Introducing a 45° angle gate (Figure 8.48b) reduces track branching. It also helps the system correctly continue the track further than our original approach. Estimating the target position by using the geometric centroid greatly improves the accuracy of the track. This approach works well because it assumes that targets turn slowly and in this test the road section is nearly straight.

Using the EKF (Figure 8.48c) also provides a more accurate and understandable set of tracks. Branching still occurs but is limited to a region that is very close to the actual trajectory of the target. The EKF performs its own computation of the target position. Like the angle filter, the EKF imposes a linear model on the data, and hence works well with the data from the straight road.

The lateral inhibition results (Figure 8.48d) have the least amount of track branching. This track is the most easily understood of all the methods shown. It is nonparametric and does not assume linearity in the data. As with the angle gate, the geometric centroid is a good estimate of the target position.

We have also tested a combination of EKF and lateral inhibition. The results of that approach are worse than either the EKF or lateral inhibition approaches in isolation. Our discussion of the track data is supported by the error data summarized in Table 8.1. Each cell shows the area between the track formed by the approach and the actual target trajectory. The top portion of the table is data from all the tracks taken on November 8, 2001. The bottom portion of the table is from the track shown in

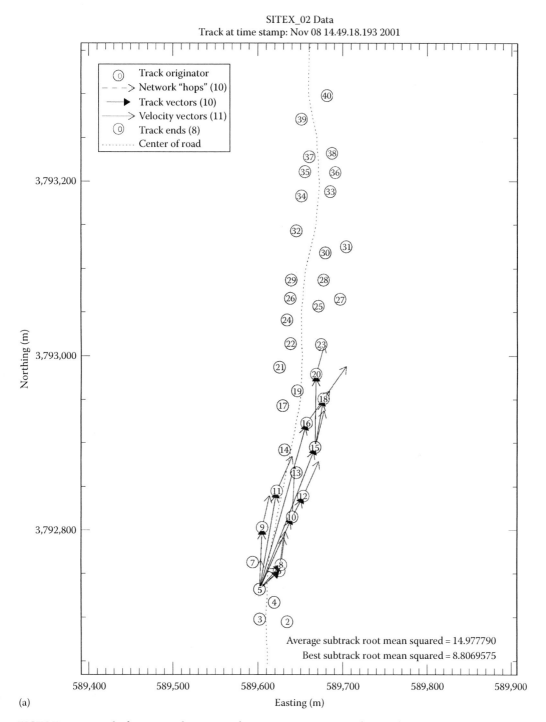

FIGURE 8.48 Tracks from a sample target tracking run at Twentynine Palms. Both axes are UTM coordinates. Circles are sensor nodes. The faint curve through the nodes is the middle of the road. Dark arrows are the reported target tracks. Dotted arrows connect the clump heads that formed the tracks. Filtering not only reduced the systems tendency to branch but also increased the track length: (a) no filtering.

(*continued*)

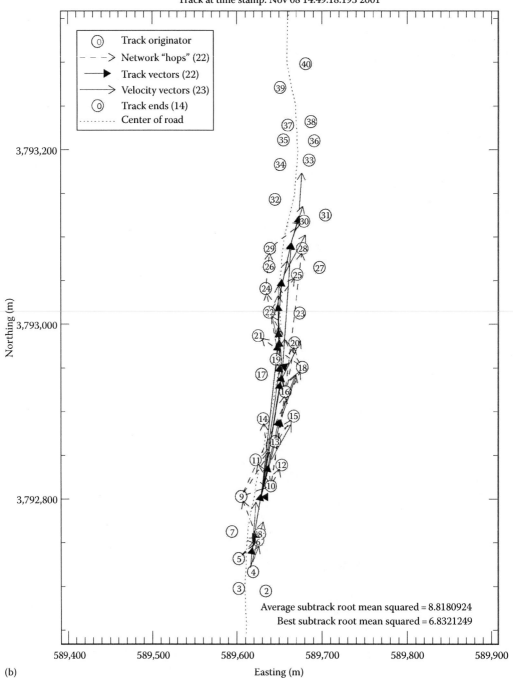

(b)

FIGURE 8.48 (continued)   (b) 45° angle filter.

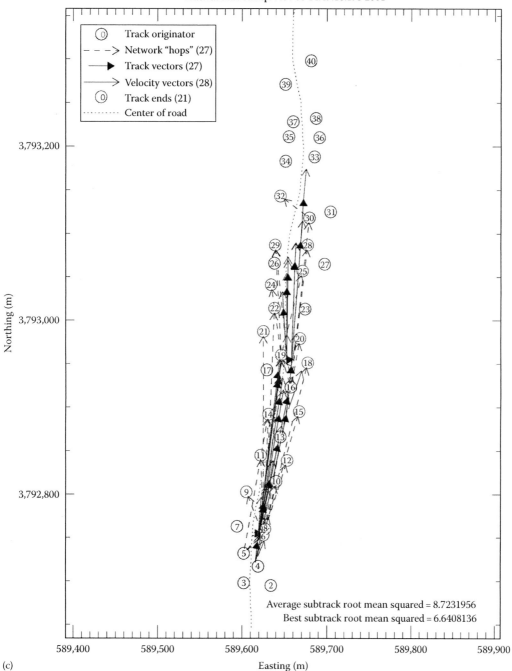

FIGURE 8.48 (continued)   (c) EKF.

(*continued*)

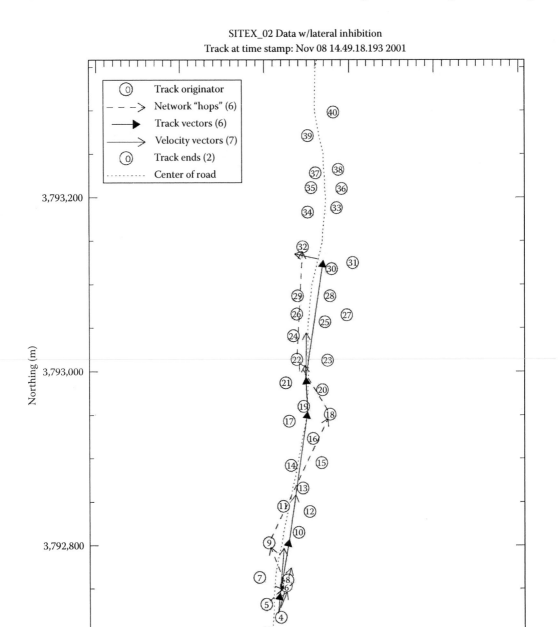

(d)

FIGURE 8.48 (continued)   (d) lateral inhibition.

TABLE 8.1 Root Mean Square Error Comparison for the Data Association
Techniques Discussed

	Live Data	Angle 45	EKF	Lateral Inhibition	EKF and Lateral Inhibition
RMS for tracks from November 8, 2001					
Averaged	18.108328	9.533245	8.877021	9.361643	11.306236
Track summed	81.456893	54.527057	52.775338	13.535534	26.738410
RMS for tracks beginning at Nov-08-14.49.18.193-2001					
Averaged	14.977790	8.818092	8.723196	9.361643	8.979453
Track summed	119.822320	123.453290	183.187110	18.723287	35.917832

The top set of numbers is for all target tracks collected on the day of November 8. The bottom set of numbers is for one specific target run. In each set, the top row is the average error for all tracks made by the target during the run. The bottom row sums the error over the tracks. Since these tests were of a target following a road, the angle and EKF filters have an advantage. They assume a linear trajectory. Lateral inhibition still performs well, although it is nonparametric.

Figure 8.48a through d. In both portions, the top row is the average error for all the tracks formed by a target. The bottom row is the sum of all the errors for all the tracks formed by a target.

If one considers only the average track error, the EKF provides the best results. The original approach provides the worst results. The other three approaches considered are roughly equivalent.

Summing the error of all the tracks formed for a single target penalizes approaches where multiple track branches form. When this is done, lateral inhibition has the most promising results. The second best results are provided by the combination of lateral inhibition and EKF. The other approaches are roughly equivalent.

These results show that the internode coordination provided by lateral inhibition is a promising technique. Since it is nonparametric, it makes very few assumptions about the target trajectory. Geometric centroid is a robust position estimator. Robust local parameter estimation provides a reliable estimate of the target's position and heading. Lateral inhibition reduces the tendency of our original tracking implementation to produce confusing interpretations of the data inputs. The system finds the track continuation that is the best continuation of the last known target position. In combination, both methods track targets moving through a sensor field more clearly.

The distributed nature of this approach makes it very robust to node failure. It also makes multiple target-tracking problems easy to solve when targets are much more sparsely distributed than the sensors. Multiple target tracking becomes a disjoint set of single target-tracking problems.

Multiple target-tracking conflicts arise only when target trajectories cross each other or approach each other too closely. When tracks cross or approach each other closely, linear regression breaks down since CPA events from multiple targets will be used in the same computation. The results will tend to match neither track. The tracks will be continued once the targets no longer interfere with each other.

Classification algorithms will be useful for tracking closely spaced targets. If crossing targets are of different classes and class information is transmitted as part of the CPA event, then the linear regression could be done on events grouped by target class. In which case, target crossing becomes even less of a concern.

Table 8.2 compares the network traffic incurred by the approaches shown in Figure 8.48a through d with the bandwidth required for a centralized approach using CPA data. CPA packets had 40 bytes, and the lateral inhibition packets had 56 bytes. Track data packets vary in size, since the EKF required three data points and a covariance matrix. Table 8.2 shows that lateral inhibition requires the least network bandwidth due to reduced track divergence.

TABLE 8.2 Data Transmission Requirements for the Different Data Association Techniques

	Track Packets	Track Pack Size	CPA Packets	Inhibition/ Packets	Total
EKF	852	296	59	0	254,552
Lateral inhibition	217	56	59	130	21,792
EKF and lateral inhibition	204	296	59	114	69,128
Centralized	0	0	240	0	9,600

The total is the number of bytes sent over the network. The EKF requires covariance data and previous data points. Angle gating and lateral inhibition require less data in the track record. Data are from the tracking period shown in Figure 8.48a through d.

Note from Table 8.2 that in this case centralized tracking required less than half as many bytes as lateral inhibition. These data are somewhat misleading. The data shown are from a network of 40 nodes with an Internet gateway in the middle. As the number of nodes and the distance to the gateway increases, the number of packet transmissions will increase for the centralized case. For the other techniques, the number of packets transmitted will remain constant. Recall the occurrence of tracking filter false positives in the network, which was more than 50% of the CPA's during this test. Reasonably, under those conditions, the centralized data volume would more than double over time and be comparable to the lateral inhibition volume. Note as well that centralized data association would involve as many as 24–30 CPA's for every detection event in our method. When association requires $O(n^2)$ comparisons [20], this becomes an issue.

8.11 Dependability Analysis

Our technology allows the clump head that combines readings to be chosen on the fly. This significantly increases system robustness by allowing the system to adapt to the failure of individual nodes. The nodes that remain exchange readings and find answers.

Since our heading and velocity estimation approach uses triangulation [2], at least three sensor readings are needed to get an answer. In the following, we assume all nodes have an equal probability of failure q. In a nonadaptive system, when the "cluster head" fails, the system fails. The cluster has a probability of failure q no matter how many nodes are in the cluster. In the adaptive case, the system fails only when the number of nodes functioning is three or less. Figure 8.49a and b illustrate the difference in dependability between adaptive and nonadaptive tasking. These figures assume an exponential distribution of independent failure events, which is standard in dependability literature. The probability of failure is constant across time. We assume that all participating nodes have the same probability of failure. This does not account for errors due to loss of power.

In Figure 8.49a, the top line is the probability of failure for a nonadaptive cluster. Since one node is the designated cluster head, when it fails the cluster fails. By definition, this probability of failure is constant. The lower line is the probability of failure of an adaptive cluster as a function of the number of nodes. This is the probability that less than three nodes will be available at any point in time. All individual nodes have the same failure probability, which is the value shown by the top line. The probability of failure of the adaptive cluster drops off exponentially with the number of nodes.

Figure 8.49b shows this same probability of failure as a function of both the number of nodes and the individual node's probability of failure.

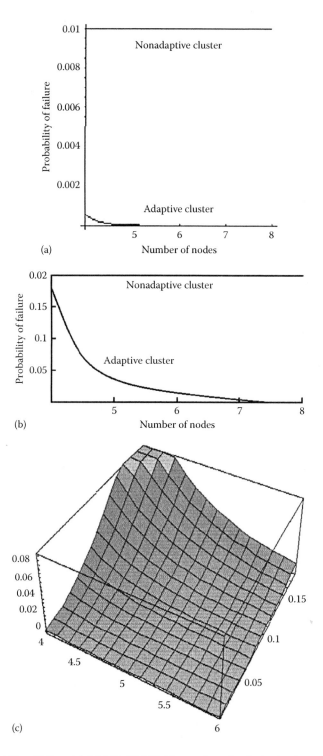

FIGURE 8.49 Probability of failure (q): (a) 0.01, (b) 0.02. The number of nodes in the cluster is varied from 4 to 8, and (c) The surface shows probability of failure (z-axis) for an adaptive cluster as the probability of failure for a single node q varies from 0.01 to 0.2 (side axis), and the number of nodes in the cluster varies from 4 to 6 (front axis).

8.12 Resource Parsimony

We also constructed simulations to analyze the performance and resource consumption of ColTraNe. It is compared to a beamforming algorithm from Ref. [21]. Each approach was used on the same set of target tracks with the same sensor configurations. Simulated target tracks were constructed according to the following equation:

$$y_t = \alpha(x+4) + (1-\alpha)\left(\frac{x^2}{4}\right) \tag{8.29}$$

The sensors were arrayed in a grid over the square area of $x \in [-4, 4]$ and $y \in [0, 8]$. Four configurations consisting of 16, 36, 64, and 100 sensors were constructed in order to examine the effects of sensor density on the results. For each density, five simulations were run for $\alpha = 0, 0.25, 0.5, 0.75,$ and 1, each of which relied on simulated time-series data (in the case of beamforming) and simulated detections (in the case of the CPA-based method).

Parameters measured included average error rate, execution time, bandwidth consumed, and power (or more properly, energy) consumed. Average error is a measure of how much the average estimated target position deviated from the true target position. Bandwidth consumption corresponded to the total amount of data exchanged over all sensor nodes throughout the lifetime of the target track. Power consumption was measured taking into account both the power required by the CPU for computation and the power required by the network to transmit the data to another node.

The resulting graphs are displayed in Figures 8.50 and 8.51. The results for beamforming in Figure 8.50 show that it is possible to reduce power consumption considerably without significantly influencing average error. In the case of ColTraNe, the lowest error resulted from expending power somewhere between the highest and lowest amounts of consumption. Comparing the two algorithms, beamforming produced better results on average but consumed from 100 to 1000 times as much power as the CPA-based method depending on the density of the sensor network.

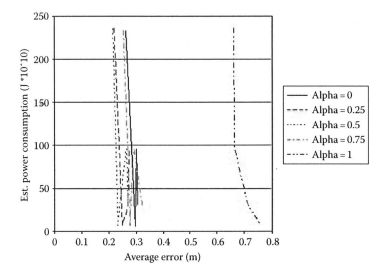

FIGURE 8.50   Beamforming, power consumption versus average error.

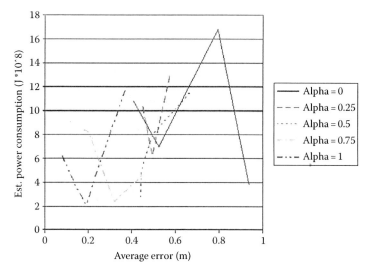

FIGURE 8.51 CPA, power consumption versus average error.

8.13 Multiple Target Tracking

To analyze the ability of ColTraNe to track multiple targets, we performed the following experiment. Two simulated targets were sent through a simulated sensor node field comprised of 400 nodes arranged in a rectangular grid measuring 8 × 8 m. Two different scenarios were used for this simulation:

1. *X path*: Two targets enter the field at the upper and lower left corners, traverse the field crossing each other in the center of the field, and exit at the opposite corners (see Figure 8.52).
2. *Bowtie*: Two targets enter the field at the upper and lower left corner, traverse the field along hyperbolic paths that nearly intersect in the center of the field, and then exit the field at the upper and lower right corners (see Figure 8.52b).

Calculation of the tracking errors was accomplished by determining the area under the curve between a track plot and the target path to which it was related.

The ColTraNe performed very well in the X pattern tests due to the linear nature of our track continuations. Tracks seldom jumped to the opposite target path and almost always tracked both targets separately. Bowtie tracking, however, turned out to be more complex (see Figures 8.53a and b and 8.54a and b).

Bowtie target paths that approach each other too closely at point of nearest approach (conjunction) tend to cause the tracks to cross over to the opposite target path as if the targets' paths had crossed each other (Figure 8.53a and b). Again, this is due to the linear nature of the track continuations.

As conjunction distance increases beyond a certain point (critical conjunction), the incidence of crossover decreases dramatically (Figure 8.54a and b). Minimum effective conjunction is the smallest value of conjunction where the incidence of crossover begins to decrease to acceptable levels. According to our analysis as shown in Figure 8.55, if a clump range equal to the node separation is used, critical conjunction is equal to the node separation and minimum effective conjunction is approximately 1.5 times node separation. If clump range is equal to 2 times node separation, critical conjunction is equal to 2.5 times node separation and minimum effective conjunction is approximately 3 times node separation or 1.5 times clump range. The significant result of this analysis seems to be that the minimum effective conjunction is equal to 1.5 times clump Range. This means that ColTraNe should be able to independently track multiple targets provided they are separated by at

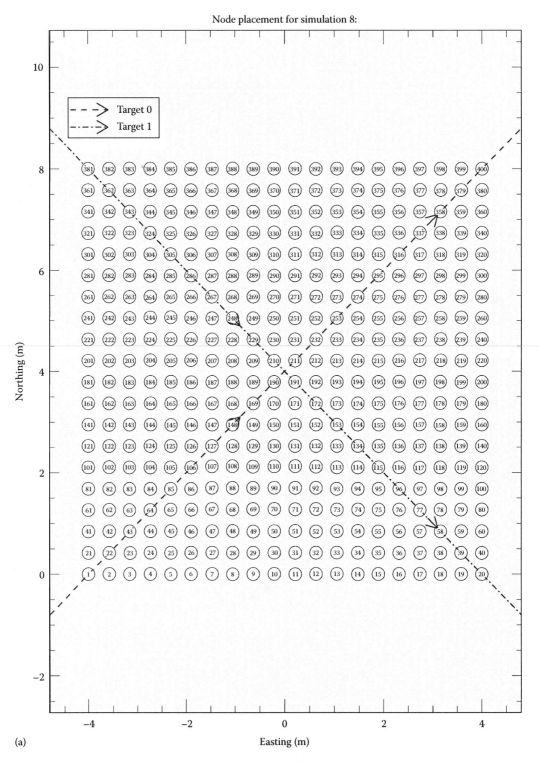

(a)

FIGURE 8.52 Comparison of the two multiple target-tracking simulation scenarios. Circles are sensor nodes. The faint lines crossing the node field are the target paths. (a) X path simulation.

Node placement for simulation 8:

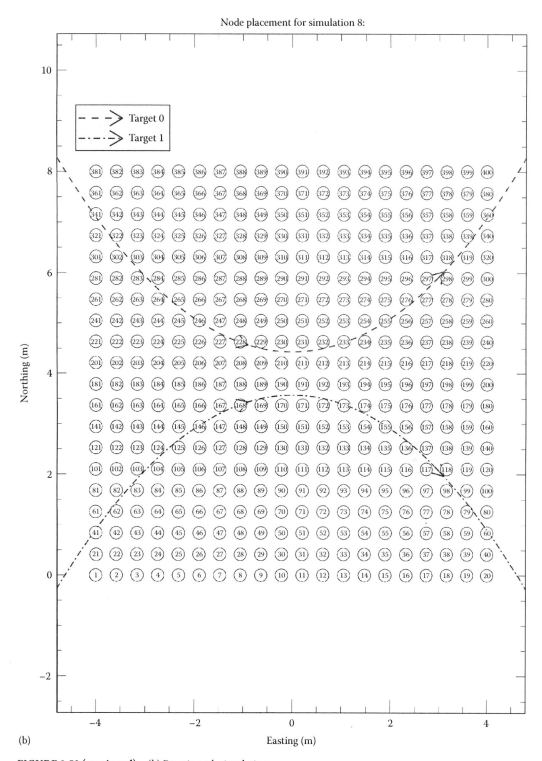

FIGURE 8.52 (continued) (b) Bowtie path simulation.

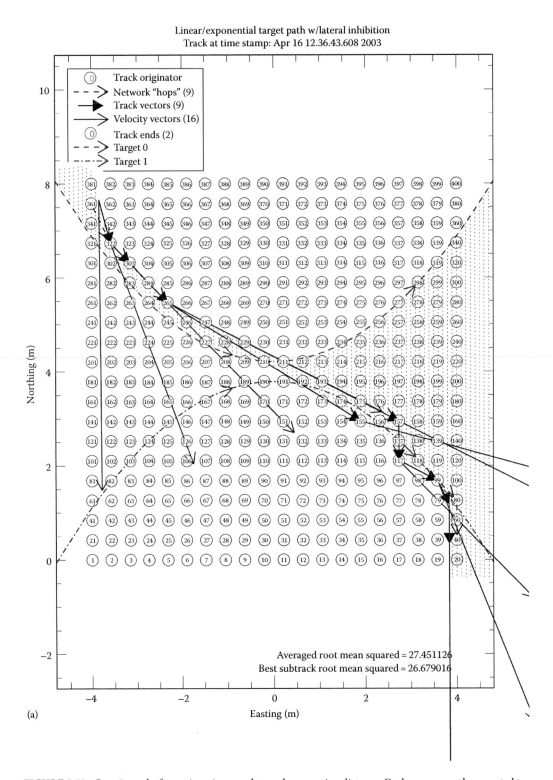

FIGURE 8.53 Bowtie tracks for conjunction equal to node separation distance. Dark arrows are the reported target tracks. Lighter arrows are the calculated velocity vectors. Shaded areas are the area between the curves used to determine track error. Other notations are same as in Figure 8.52a and b. (a) Track for Target 1.

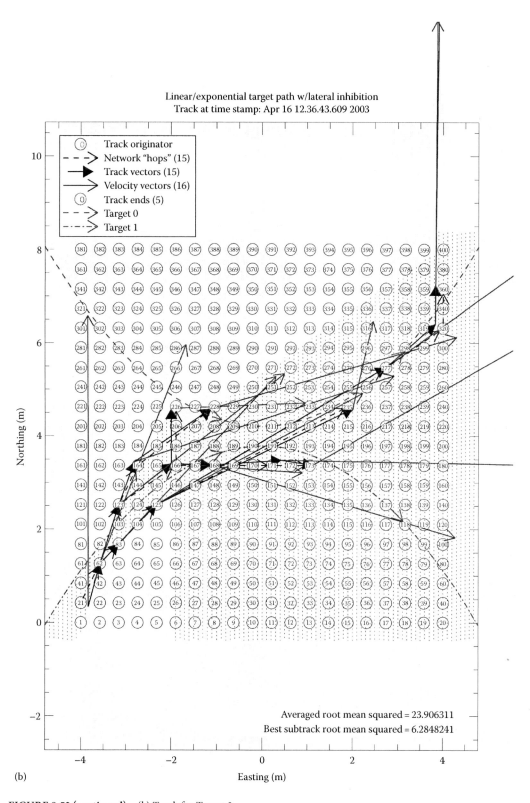

FIGURE 8.53 (continued) (b) Track for Target 2.

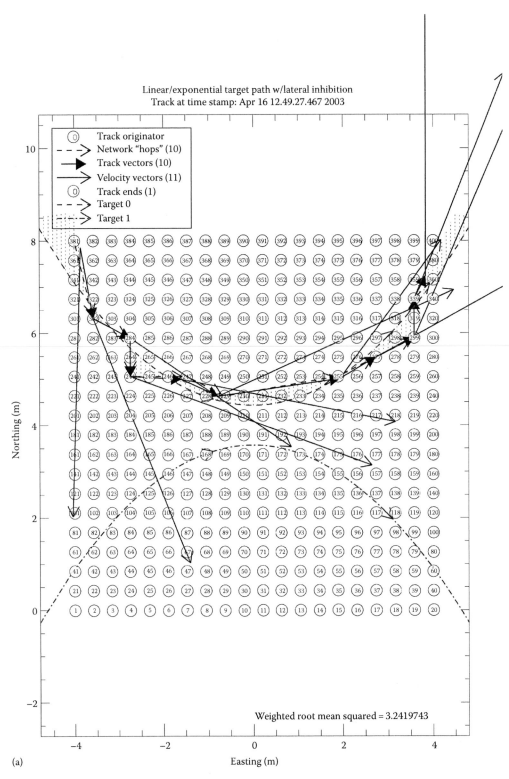

(a)

FIGURE 8.54 Bowtie tracks for conjunction equal to two times node separation distance. Notations are same as in Figure 8.53a and b. (a) Track for Target 1.

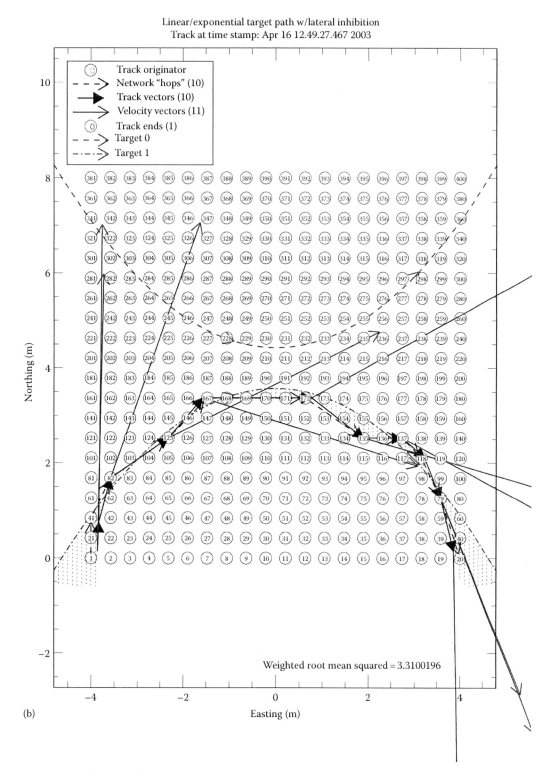

FIGURE 8.54 (continued) (b) Track for Target 2.

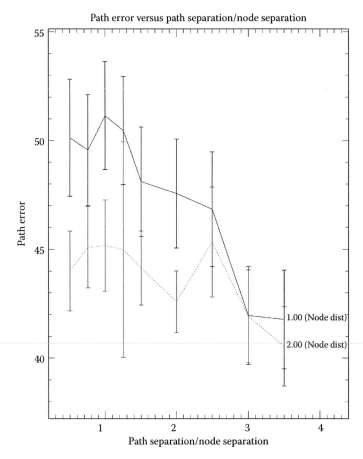

FIGURE 8.55 Finding critical conjunction experimentally. The darker upper line displays the results when the clump range is equal to the node separation distance. The lighter lower line displays the results when clump range is equal to two times the node separation distance.

least 1.5 times the clump Range. This appears to be related to fundamental sampling limitations based on Nyquist sampling theory [22].

8.14 Conclusion

This chapter presents a distributed entity-tracking framework, which embeds the tracking logic in a self-organized distributed network. Tracking is preformed by solving the subproblems of detection, fusion, association, track formation, and track extrapolation. Local collaboration determines the values of parameters such as position, velocity, and entity type at points along the track. These variables become state estimates that are used in track formation and data association. Local processing reduces the amount of information that is global and reduces power consumption.

This approach allows us to study tracking as a distributed computation problem. We use in-house CA to construct models based on the interaction of autonomous nodes. These models include system faults and network traffic. We posit that this type of analysis is important for the design of robust distributed systems, like autonomous sensor networks. Simple CA can be classified into four equivalence classes. Two-dimensional traffic modeling CA are more difficult to classify. The cellular behavior may be periodic, stable, and chaotic in different regions of the CA in question. Exact classification may be impossible or inappropriate.

We have shown that for certain probabilities of false positives, stable traffic jams will form around the sink location, while for other values, unstable traffic jams form. These are traffic jams that continue to form, disappear, and reform. This oscillatory behavior is typical of periodic behavior of CA. It is possible to have a stable traffic jam with an unstable boundary.

In the target-tracking context, we have established strong and weak points of the algorithms used. Pheromones appear to be robust but transmit more data than the other algorithms. They can also be fooled by false positives. The Bayesian network is effective for reducing the transmission of false positives but has difficulty in maintaining track continuation. Most likely, further work is required to tune the probabilities used.

EKF tracking may not be appropriate for this level of analysis, since it is designed to overcome Gaussian noise. At this level of fidelity, that type of noise is less important. The CA model is discrete and the EKF is meant for use with continuous data.

Hybrid approaches may be possible and desirable. One possible avenue to consider is using the Bayesian logic to restrict the propagation of pheromones or to analyze the strength of the pheromone concentration present.

Our tracking algorithm development continues by porting these algorithms to a prototype implementation that has been tested in the field. CPA target detections and EKF track continuations are used to track targets through the field with minimal interference from environmental noise. Lateral inhibition is used to enforce some consistency among track association decisions.

Our work indicates to us that performing target tracking in a distributed manner greatly simplifies the multi-target-tracking problem. If sensor nodes are dense enough and targets are sparse enough, the multi-target tracking is a disjoint set of single target-tracking problems. Centralized approaches will also become untenable as target density increases.

A power analysis of our approach versus a centralized approach such as beamforming was presented. The power analysis shows that ColTraNe is much more efficient than beamforming for distributed sensing. This is because ColTraNe extracts relevant information from time series locally. It also limits information transmission to the regions that absolutely require the information.

The chapter concluded with an analysis of the distributed tracker to distinguish multiple targets in a simulated environment. Analysis shows that ColTraNe can effectively track multiple targets, provided there are at least two nodes between the target paths at all points, as predicted by Nyquist. We are continuing our research in distributed sensing applications. Among the topics of interest are

- Power aware methods of assigning system resources
- Hybrid tracking methods
- Use of symbolic dynamics for inferring target behavior classes
- Development of other peer-to-peer distributed behaviors, such as ColTraNe, that are resistant to hardware failures

Acknowledgments

Efforts are sponsored by the Defense Advanced Research Projects Agency (DARPA) Air Force Research Laboratory, Air Force Materiel Command, USAF, under agreement number F30602-99-2-0520 (Reactive Sensor Network) and DARPA and the Space and Naval Warfare Systems Center, San Diego, under grant N66001-00-G8947 (Semantic Information Fusion). The U.S. Government is authorized to reproduce and distribute reprints for governmental purposes notwithstanding any copyright annotation thereon. The views and conclusions contained herein are those of the authors and should not be interpreted to necessarily represent the official policies or endorsements, either expressed or implied, of the DARPA, the Air Force Research Laboratory, the U.S. Navy, or the U.S. Government.

This material is based upon work supported in part by the Air Force Office of Scientific Research contract/grant number FA9550-09-1-0173, NSF grant EAGER-GENI Experiments on Network Security and

Traffic Analysis contract/grant number CNS-1049765 and NSF-OCI 1064230 EAGER: Collaborative Research: A Peer-to-Peer based Storage System for High-End Computing. Opinions expressed are those of the author and neither the National Science Foundation nor the U.S. Department of Defense.

References

1. C. Intanagonwiwat, R. Govindan, and D. Estrin. Directed diffusion: A scalable and robust communication paradigm for sensor networks. In *Proceedings of the 6th Annual International Conference on Mobile Computing and Networking, MobiCom '00*, Boston, MA, pp. 56–67, ACM, New York, 2000.
2. R.R. Brooks, E. Grele, W. Klimkiewicz, J. Moore, C. Griffin, B. Kovak, and J. Koch. Reactive sensor networks. In *Distributed Autonomous Robotic Systems (DARS)*, Tokyo, Japan, pp. 471–472, 2000.
3. D. Smith and S. Singh. Approaches to multisensor data fusion in target tracking: A survey. *IEEE Transactions on Knowledge and Data Engineering*, 18(12):1696–1710, December 2006.
4. N. Kabaoglu. Target tracking using particle filters with support vector regression. *IEEE Transactions on Vehicular Technology*, 58(5):2569–2573, June 2009.
5. R. Brooks, P. Ramanathan, and A. Sayeed, Distributed target classification and tracking in sensor networks. *Proceedings of the IEEE*, 91:1163–1171, August 2003.
6. S. Aeron, V. Saligrama, and D.A. Castanon. Energy efficient policies for distributed target tracking in multihop sensor networks. In *45th IEEE Conference on Decision and Control*, San Diego, CA, pp. 380–385, December 2006.
7. D.G. Allegretti, G.T. Kenyon, and W.C. Priedhorsky. Cellular automata for distributed sensor networks. *International Journal of High Performance Computing Applications*, 22(2):167–176, 2008.
8. V. Crespi, G. Cybenko, and G. Jiang. The theory of trackability with applications to sensor networks. *Transactions on Sensor Networks*, 4:16-1–16-42, June 2008.
9. Satish Kumar and Ahmed helmy. USC/ISI, Xerox PARC, LBNL, and UCB, Virtual Internet Testbed. Original: September 1996, Last Modified: October 19, 1997. Project Website: http://www.isi.edu/nsnam/vint/
10. S.S. Blackman and T.J. Broida. Multiple sensor data association and fusion in aerospace applications. *Journal of Robotic Systems*, 7(3):445–485, 1990.
11. R.R. Brooks and S.S. Iyengar. *Multi-Sensor Fusion: Fundamentals and Applications with Software*. Prentice-Hall, Inc., Upper Saddle River, NJ, 1998.
12. D.S. Friedlander and S. Phoha. Semantic information fusion for coordinated signal processing in mobile sensor networks. *International Journal of High Performance Computing Applications*, 16(3):235–241, 2002.
13. S.A. Brückner. Return from the ant—Synthetic ecosystems for manufacturing control. PhD thesis, Humboldt University, Berlin, Germany.
14. W.H. Press, S.A. Teukolsky, W.T.F. Vetterling, and B.P. Brooks. *Numerical Recipes in C*. Cambridge University Press, London, U.K., 1997.
15. J. Pearl. Fusion, propagation, and structuring in belief networks. *Artificial Intelligence*, 29(3):241–288, 1986.
16. D. Chowdhury, L. Santen, and A. Schadschneider. Simulation of vehicular traffic: A statistical physics perspective. *Computing in Science Engineering*, 2(5):80–87, September/October 2000.
17. M. Delorme and J. Mazoyer (eds.). *Cellular Automata: A Parallel Model*. Kluwer Academic PTR, Dordrecht, the Netherlands.
18. S. Wolfram. *Cellular Automata and Complexity*, Addison-Wesley Pub.Co., New York, 1994, ISBN: 0201626640.
19. J. Heidemann, F. Silva, C. Intanagonwiwat, R. Govindan, D. Estrin, and D. Ganesan. Building efficient wireless sensor networks with low-level naming. In *Proceedings of the 18th ACM Symposium on Operating Systems Principles, SOSP '01*, Banff, AB, Canada, pp. 146–159, ACM, New York, 2001.

20. Y. Bar-Shalom and X.-R. Li. *Estimation and Tracking: Principles, Techniques, and Software*. Artech House, Boston, MA, 1993.
21. K. Yao, R.E. Hudson, C.W. Reed, D. Chen, and F. Lorenzelli. Blind beamforming on a randomly distributed sensor array system. *IEEE Journal on Selected Areas in Communications*, 16(8):1555–1567, October 1998.
22. N. Jacobson. Target parameter estimation in a distributed acoustic network. Honors thesis, Pennsylvania State University, University Park, PA, 2003.

9

Collaborative Signal and Information Processing: An Information-Directed Approach

Feng Zhao
Microsoft Research Asia

Jie Liu
Microsoft Research

Juan Liu
Palo Alto Research Center

Leonidas Guibas
Stanford University

James Reich
Massachusetts Institute of Technology

9.1 Sensor Network Applications, Constraints, and Challenges..... 191
9.2 Tracking as a Canonical Problem for CSIP 192
    A Tracking Scenario • Design Desiderata in Distributed Tracking
9.3 Information-Driven Sensor Query: A CSIP Approach to
    Target Tracking 195
    Tracking Individual Targets • Information-Based Approaches
9.4 Combinatorial Tracking Problems 199
    Counting the Number of Targets • Contour Tracking • Shadow
    Edge Tracking
9.5 Discussion 203
9.6 Conclusion 204
Acknowledgments 205
References 205

9.1 Sensor Network Applications, Constraints, and Challenges

Networked sensing offers unique advantages over traditional centralized approaches. Dense networks of distributed networked sensors can improve perceived signal-to-noise ratio (SNR) by decreasing average distances from sensor to target. Increased energy efficiency in communications is enabled by the multi-hop topology of the network [1]. Moreover, additional relevant information from other sensors can be aggregated during this multi-hop transmission through in-network processing [2]. But perhaps the greatest advantages of networked sensing are in improved robustness and scalability. A decentralized sensing system is inherently more robust against individual sensor node or link failures, because of redundancy in the network. Decentralized algorithms are also far more scalable in practical deployment, and may be the only way to achieve the large scales needed for some applications.

A sensor network is designed to perform a set of high-level information processing tasks, such as detection, tracking, or classification. Measures of performance for these tasks are well defined, including detection, false alarms or misses, classification errors, and track quality. Commercial and military applications include environmental monitoring (e.g. traffic, habitat, security), industrial sensing and diagnostics (e.g. factory, appliances), infrastructure protection (e.g. power grid, water distributions), and battlefield awareness (e.g. multi-target tracking).

Unlike a centralized system, however, a sensor network is subject to a unique set of resource constraints, such as limited on-board battery power and limited network communication bandwidth. In a typical sensor network, each sensor node operates untethered and has a microprocessor and limited amount of memory for signal processing and task scheduling. Each node is also equipped with one or more of acoustic microphone arrays, video or still cameras, IR, seismic, or magnetic sensing devices. Each sensor node communicates wirelessly with a small number of local nodes within the radio communication range.

The current generation of wireless sensor hardware ranges from the shoe-box-sized Sensoria WINS NG sensors [3] with an SH-4 microprocessor to the matchbox-sized Berkeley motes with an 8 bit microcontroller [4]. It is well known that communicating 1 bit over the wireless medium consumes far more energy than processing the bit. For the Sensoria sensors and Berkeley motes, the ratio of energy consumption for communication and computation is in the range of 1,000–10,000. Despite the advances in silicon fabrication technologies, wireless communication will continue to dominate the energy consumption of embedded networked systems for the foreseeable future [5]. Thus, minimizing the amount and range of communication as much as possible, e.g. through local collaboration, data compression, or invoking only the nodes that are relevant to a given task, can significantly prolong the lifetime of a sensor network and leave nodes free to support multi-user operations.

Traditional signal-processing approaches have focused on optimizing estimation quality for a fixed set of available resources. However, for power-limited and multi-user decentralized systems, it becomes critical to carefully select the embedded sensor nodes that participate in the sensor collaboration, balancing the information contribution of each against its resource consumption or potential utility for other users. This approach is especially important in dense networks, where many measurements may be highly redundant, and communication throughput severely limited. We use the term "collaborative signal and information processing" (CSIP) to refer to signal and information processing problems dominated by this issue of selecting embedded sensors to participate in estimation.

This chapter uses tracking as a representative problem to expose the key issues for CSIP—how to determine what needs to be sensed dynamically, who should sense, how often the information must be communicated, and to whom. The rest of the chapter is organized as follows. Section 9.2 will introduce the tracking problem and present a set of design considerations for CSIP applications. Sections 9.3 and 9.4 will analyze a range of tracking problems that differ in the nature of the information being extracted, and describe and compare several recent contributions that adopted information-based approaches. Section 9.5 will discuss future directions for CSIP research.

9.2 Tracking as a Canonical Problem for CSIP

Tracking is an essential capability in many sensor network applications, and is an excellent vehicle to study information organization problems in CSIP. It is especially useful for illustrating a central problem of CSIP: dynamically defining and forming sensor groups based on task requirements and resource availability.

From a sensing and information processing point of view, we define a sensor network as a tuple, $Sn = \langle V, E, P_V, P_E \rangle$. V and E specify a network graph, with its nodes V, and link connectivity $E \subseteq V \times V$. P_V is a set of functions which characterizes the properties of each node in V, including its location, computational capability, sensing modality, sensor output type, energy reserve, and so on. Possible sensing modalities includes acoustic, seismic, magnetic, IR, temperature, or light. Possible output types include information about signal amplitude, source direction-of-arrival (DOA), target range, or target classification label. Similarly, P_E specifies properties for each link such as link capacity and quality.

A tracking task can be formulated as a constrained optimization problem $Tr = \langle Sn, Tg, Sm, Q, O, C \rangle$. Sn is the sensor network specified above. Tg is a set of targets, specifying for each target the location, shape (if not a point source), and signal source type. Sm is a signal model for how the target signals propagate and attenuate in the physical medium. For example, a possible power attenuation model for an acoustic signal is the inverse distance squared model. Q is a set of user queries, specifying query instances and query entry points into the network. A sample query is "Count the number of targets in region R." O is an objective function, defined by task requirements. For example, for a target localization task, the objective function could be the localization accuracy, expressed as the trace of the covariance matrix for the position estimate. $C = \{C_1, C_2, \ldots\}$ specifies a set of constraints. An example is localizing an object within a certain amount of time and using no more than a certain quantity of energy. The constrained optimization finds a set of feasible sensing and communication solutions for the problem that satisfies the given set of constraints. For example, a solution to the localization problem above could be a set of sensor nodes on a path that gathers and combines data and routes the result back to the querying node.

In wireless sensor networks, some of the information defining the objective function and/or constraints is only available at run time. Furthermore, the optimization problem may have to be solved in a decentralized way. In addition, anytime algorithms are desirable, because constraints and resource availability may change dynamically.

9.2.1 A Tracking Scenario

We use the following tracking scenario (Figure 9.1) to bring out key CSIP issues. As a target X moves from left to right, a number of activities occur in the network:

1. *Discovery*: Node a detects X and initiates tracking.
2. *Query processing*: A user query Q enters the network and is routed toward regions of interest, in this case the region around node a. It should be noted that other types of query, such as long-running query that dwell in a network over a period of time, are also possible.

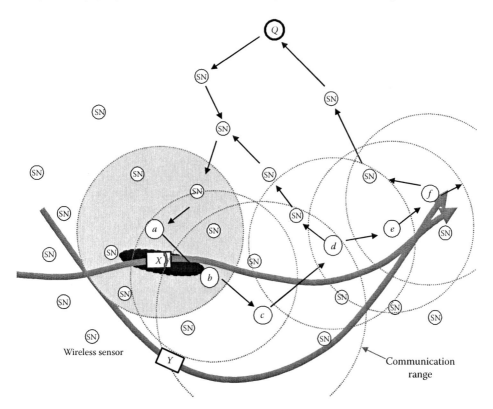

FIGURE 9.1 A tracking scenario, showing two moving targets, X and Y, in a field of sensors. Large circles represent the range of radio communication from each node.

3: *Collaborative processing*: Node *a* estimates the target location, possibly with help from neighboring nodes.

4: *Communication*: Node *a* may hand off data to node *b*, *b* to *c*, etc.

5: *Reporting*: Node *d* or *f* summarizes track data and sends it back to the querying node.

Let us now assume another target, *Y*, enters the region around the same time. The network will have to handle multiple tasks in order to track both targets simultaneously. When the two targets move close to each other, the problem of properly associating a measurement to a target track, the so-called *data association problem*, becomes tricky. In addition, collaborative sensor groups, as defined earlier, must be selected carefully, since multiple groups might need to share the same physical hardware [6].

This tracking scenario raises a number of fundamental information-processing problems in distributed information discovery, representation, communication, storage, and querying: (1) in collaborative processing, the issues of target detection, localization, tracking, and sensor tasking and control; (2) in networking, the issues of data naming, aggregation, and routing; (3) in databases, the issues of data abstraction and query optimization; (4) in human–computer interface, the issues of data browsing, search, and visualization; (5) in software services, the issues of network initialization and discovery, time and location services, fault management, and security. In the rest of the chapter, we will focus on the collaborative processing aspects and touch on other issues only as necessary.

A common task for a sensor network is to gather information from the environment. Doing this under the resource constraints of a sensor network may require data-centric routing and aggregation techniques which differ considerably from TCP/IP end-to-end communication. Consequently, the research community has been searching for the right "sensor net stack" that can provide suitable abstractions over networking and hardware resources. While defining a unifying architecture for sensor networks is still an open problem, we believe a key element of such an architecture is the *principled interaction between the application and networking layers*. For example, Section 9.3 will describe an approach that expresses application requirements as a set of information and cost constraints so that an ad hoc networking layer using, for example, the diffusion routing protocol [2], can effectively support the application.

9.2.2 Design Desiderata in Distributed Tracking

In essence, a tracking system attempts to recover the state of a target (or targets) from observations. Informally, we refer to the information about the target state distilled from measurement data as a belief or belief state. An example is the posterior probability distribution of target state, as discussed in Section 9.3. As more observation data are available, the belief may be refined and updated.

In sensor networks, the belief state can be stored centrally at a fixed node, at a sequence of nodes through successive hand-offs, or at a set of nodes concurrently. In the first case (Figure 9.2a), a fixed node is designated to receive measurements from other relevant sensors through communication. This simpler tracker design is obtained at the cost of potentially excessive communication and reduced robustness to node failure. It is feasible only for tracking nearly stationary targets, and is in general neither efficient nor scalable.

In the second case (Figure 9.2b), the belief is stored at a node called the leader node, which collects data from nearby, relevant sensors. As the phenomenon of interest moves or environmental conditions vary, the leadership may change hands among sensor nodes. Since the changes in physical conditions are often continuous in nature, these handoffs often occur within a local geographic neighborhood. This moving leader design localizes communication, reducing overall communication and increasing the lifetime of the network. The robustness of this method may suffer from potential leader node attrition, but this can be mitigated by maintaining copies of the belief in nearby nodes and detecting and responding to leader failure. The key research challenge for this design is to define an effective selection criterion for sensor leaders, to be addressed in Section 9.3.

Finally, the belief state can be completely distributed across multiple sensor nodes (Figure 9.2c). The inference from observation data is accomplished nodewise, thus localizing the communication. This is attractive from the robustness point of view. The major design challenge is to infer global properties about targets

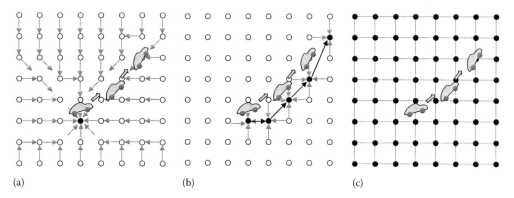

FIGURE 9.2 Storage and communication of target state information in a networked distributed tracker. Circles on the grid represent sensor nodes, and some of the nodes, denoted by solid circles, store target state information. Thin, faded arrows or lines denote communication paths among the neighbor nodes. Thin, dark arrows denote sensor hand-offs. A target moves through the sensor field, indicated by thick arrows. (a) A fixed single leader node has the target state. (b) A succession of leader nodes is selected according to information such as vehicle movement. (c) Every node in the network stores and updates target state information.

efficiently, some of which may be discrete and abstract, from partial, local information, and to maintain information consistency across multiple nodes. Section 9.4 addresses the challenge. Many issues about leaderless distributed trackers are still open and deserve much attention from the research community.

9.3 Information-Driven Sensor Query: A CSIP Approach to Target Tracking

Distributed tracking is a very active field, and it is beyond the scope of this chapter to provide a comprehensive survey. Instead, we will focus on the information processing aspect of the tracking problems, answering questions such as what information is collected by the sensors, how that information is aggregated in the network, and what high-level user queries are answered. This section describes an information-driven sensor query (IDSQ), a set of information-based approaches to tracking individual targets, and discusses major issues in designing CSIP solutions. Next, Section 9.4 presents approaches to other tracking problems, where the focus is more on uncovering abstract and discrete target properties, such as target density, rather than just their locations.

9.3.1 Tracking Individual Targets

The basic task of tracking a moving target in a sensor field is to determine and report the underlying target state $x^{(t)}$, such as its position and velocity, based on the sensor measurements up to time t, denoted as $\overline{z^{(t)}} = \{z^{(0)}, z^{(1)}, \ldots, z^{(t)}\}$. Many approaches have been developed over the last half century. These include Kalman filters, which assume a Gaussian observation model and linear state dynamics, and, more generally, sequential Bayesian filtering, which computes the posterior belief at time $t + 1$ based on the new measurement $z^{(t+1)}$ and the belief $p(x^{(t)} \mid \overline{z^{(t)}})$ inherited from time t:

$$p(x^{(t+1)} \mid \overline{z^{(t+1)}}) \alpha p(z^{(t+1)} \mid x^{(t+1)} \cdot \int p(x^{(t+1)} \mid x^{(t)}) \cdot p(x^{(t)} \mid \overline{z^{(t)}}) dx^{(t)}$$

where
  $p(z^{(t+1)}|x^{(t+1)})$ denotes the observation model
  $p(x(t + 1)|x^{(t)})$ the state dynamics model

As more data are gathered over time, the belief $p(x^{(t)} \mid \overline{z^{(t)}})$ is successively refined.

Kalman filters and many practical forms of Bayesian filter assume that the measurement noise across multiple sensors is independent, which is not always the case. Algorithms, such as covariance intersection, have been proposed to combine data from sensors with correlated information. Although these methods have been successfully implemented in applications, they were primarily designed for centralized platforms. Relatively little consideration was given to the fundamental problems of moving data across sensor nodes in order to combine data and update track information. There was no cost model for communication in the tracker. Furthermore, owing to communication delays, sensor data may arrive at a tracking node out of order compared with the original time sequence of the measurements. Kalman or Bayesian filters assume a strict temporal order on the data during the sequential update, and may have to roll back the tracker in order to incorporate "past" measurements, or throw away the data entirely.

For multi-target tracking, methods such as multiple hypothesis tracking (MHT) [7] and joint probabilistic data association (JPDA) [8] have been proposed. They addressed the key problem of data association, of pairing sensor data with targets, thus creating association hypotheses. MHT forms and maintains multiple association hypotheses. For each hypothesis, it computes the probability that it is correct. On the other hand, JPDA evaluates the association probabilities and combines them to compute the state estimate. Straightforward applications of MHT and JPDA suffer from a combinatorial explosion in data association. Knowledge about targets, environment, and sensors can be exploited to rank and prune hypotheses [9,10].

9.3.2 Information-Based Approaches

The main idea of information-based approaches is to base sensor collaboration decisions on information content, as well as constraints on resource consumption, latency, and other costs. Using information utility measures, sensors in a network can exploit the information content of data already received to optimize the utility of future sensing actions, thereby efficiently managing scarce communication and processing resources. The distributed information filter, as described by Manyika and Durrant-Whyte [11], is a global method requiring each sensor node to communicate its measurement to a central node where estimation and tracking are carried out. In this method, sensing is distributed and tracking is centralized. Directed-diffusion routes sensor data in a network to minimize communication distance between data sources and data sinks [2,12]. This is an interesting way of organizing a network to allow publish-and-subscribe to occur at a very fine grained level. A prediction-based tracking algorithm is described by Brooks et al. [13] which uses estimates of target velocity to select which sensors to query. An IDSQ [14,15] formulates the tracking problem as a more general distributed constrained optimization that maximizes information gain of sensors while minimizing communication and resource usage. We describe the main elements of an IDSQ here.

Given the current belief state, we wish to update the belief incrementally by incorporating the measurements of other nearby sensors. However, not all available sensors in the network provide useful information that improves the estimate. Furthermore, some information may be redundant. The task is to select an optimal subset and an optimal order of incorporating these measurements into our belief update. Note that, in order to avoid prohibitive communication costs, this selection must be done without explicit knowledge of measurements residing at other sensors. The decision must be made solely based upon known characteristics of other sensors, such as their position and sensing modality, and predictions of their contributions, given the current belief.

Figure 9.3 illustrates the basic idea of optimal sensor selection. The illustration is based upon the assumption that estimation uncertainty can be effectively approximated by a Gaussian distribution, illustrated by uncertainty ellipsoids in the state space. In the figure, the solid ellipsoid indicates the belief state at time t, and the dashed ellipsoids are the incrementally updated belief after incorporating an additional measurement from a sensor, S1 or S2, at the next time step. Although in both cases, S1 and S2, the area of high uncertainty is reduced by 50%, the residual uncertainty of the S2 case is not reduced along the long principal axis of the ellipse. If we were to decide between the two sensors, then we might favor case S1 over case S2, based upon the underlying measurement task.

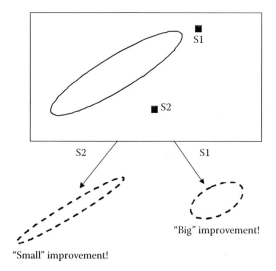

FIGURE 9.3 Sensor selection based on information gain of individual sensor contributions. The information gain is measured by the reduction in the error ellipsoid. In the figure, reduction along the longest axis of the error ellipsoid produces a larger improvement in reducing uncertainty. Sensor placement geometry and sensing modality can be used to compare the possible information gain from each possible sensor selection, S1 or S2.

In distributed sensor network systems we must balance the information contribution of individual sensors against the cost of communicating with them. For example, consider the task of selecting among K sensors with measurements $\{\mathbf{z}_i\}_{i=1}^{K}$. Given the current belief $p(\mathrm{x}\,|\,\{z_i\}i \in U)$, where $U \subset \{1, \ldots, K\}$ is the subset of sensors whose measurement has already been incorporated, the task is to choose which sensor to query among the remaining unincorporated set $A = \{1, \ldots, K\}\backslash U$. For this task, an objective function as a mixture of information and cost is designed in Ref. [15]:

$$O\left(p\left(x\,|\,\overline{z_j^{(t)}}\right)\right) = \alpha\phi\left(p\left(x\,|\,\overline{z_{j-1}^{(t)}},\, z_j^{(t)}\right)\right) - (1-\alpha)\psi\left(z_j^{(t)}\right) \qquad (9.1)$$

where

ϕ measures the information utility of incorporating the measurement $z_j^{(t)}$ from sensor j
ψ is the cost of communication and other resources
α is the relative weighting of the utility and cost

With this objective function, the sensor selection criterion takes the form

$$\hat{j} = \arg\max_{j \in A} O(p(x\,|\,\{z_i\}_{i \in U} \cup \{z_j\})) \qquad (9.2)$$

This strategy selects the best sensor given the current state $p(x\,|\,\{z_i\}_{i \in U})$. A less greedy algorithm has been proposed by Liu et al. [16], extending the sensor selection over a finite look-ahead horizon.

Metrics of information utility ϕ and cost ψ may take various forms, depending on the application and assumptions [14]. For example, Chu et al. [15] considered the query routing problem: assuming a query has entered from a fixed node, denoted by "?" in Figure 9.4, the task is to route the query to the target vicinity, collect information along an optimal path, and report back to the querying node. Assuming the belief state is well approximated by a Gaussian distribution, the usefulness of the sensor data (in this case, range data) ϕ is measured by how close the sensor is to the mean of the belief state under a Mahalanobis metric, assuming that close-by sensors provide more discriminating information.

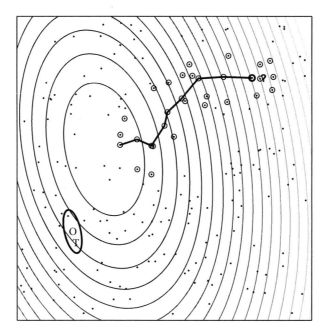

FIGURE 9.4 Sensor querying and data routing by optimizing an objective function of information gain and communication cost, whose iso-contours are shown as the set of concentric ellipses. The circled dots are the sensors being queried for data along the querying path. "T" represents the target position and "?" denotes the position of the query origin.

The cost ψ is given here by the squared Euclidean distance from the sensor to the current leader, a simplified model of the energy expense of radio transmission for some environments. The optimal path results from the tradeoff between these two terms. Figure 9.4 plots such a sample path. Note that the belief is updated incrementally along the information collection path. The ellipses in Figure 9.4 show a snapshot of the objective function that an active leader node evaluates locally at a given time step.

For multi-modal non-Gaussian distributions, a mutual information-based sensor selection criterion has been developed and successfully tested on real data [17]. The problem is as follows: assuming that a leader node holds the current belief $p(x^{(t)} \mid \overline{\mathbf{z}}^{(t)})$, and the cost to query any sensor in its neighborhood N is identical (e.g. over a wired network or using a fixed power-level radio), the leader selects from N the most informative sensor to track the moving target. In this scenario, the selection criterion of Equation 9.2 takes the form

$$\hat{j}_{\text{IDSQ}} = \arg\max_{j \in N} I\left(X^{(t+1)}; Z_j^{(t+1)} \mid \overline{Z^{(t)}} = \overline{\mathbf{z}^{(t)}}\right) \qquad (9.3)$$

where $I\,(.;\,.)$ measures the mutual information in bits between two random variables. Essentially, this criterion selects a sensor whose measurement $\mathbf{z}_j^{(t+1)}$, combined with the current measurement history $\overline{\mathbf{z}}^{(t)}$, would provide the greatest amount of information about the target location $x^{(t+1)}$. The mutual information can be interpreted as Kullback–Leibler divergence between the beliefs after and before applying the new measurement $\mathbf{z}_j^{(t+1)}$ Therefore, this criterion favors the sensor which, on average, gives the greatest change to the current belief.

To analyze the performance of the IDSQ tracker, we measure how the tracking error varies with sensor density through simulation. Figure 9.5 shows that, as the sensor density increases, the tracking error, expressed as the mean error of the location estimate, decreases, as one would expect, and tends to a floor dominated by sensor noise. This indicates that there is a maximum density beyond which using more sensors gains very little in tracking accuracy.

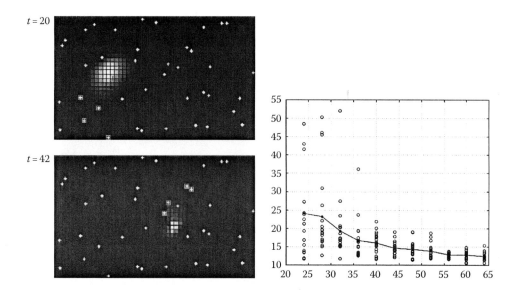

FIGURE 9.5 Experimental results (right figure) show how the tracking error (vertical axis), defined as the mean error of estimated target positions, varies with the sensor density (horizontal axis), defined as the number of sensors in the sensor field. The left figure shows snapshots of a belief "cloud"—the probability density function of the location estimate—for different local sensor densities.

The IDSQ tracker has been successfully tested in a DARPA tracking experiment at Twentynine Palms, November 2001. In the experiment, 21 Sensoria WINS NG wireless sensors were used to collect acoustic data from moving vehicles. Details of the results can be found in Ref. [17].

9.4 Combinatorial Tracking Problems

The discussion of tracking so far has focused on localizing targets over time. In many applications, however, the phenomenon of interest may not be the exact locations of individual objects, but global properties regarding a collection of objects, e.g. the number of targets, their regions of influence, or their boundaries. The information to be extracted in this case may be more discrete and abstract, and may be used to answer high-level queries about the world-state or to make strategic decisions about actions to take.

An expensive way to compute such global class properties of objects is to locate and identify each object in the collection, determine its individual properties, and combine the individual information to form the global answer, such as the total number of objects in the collection. However, in many cases, these class properties can be inferred without accurate localization or identification of all the objects in question. For example, it may be possible to focus on attributes or relations that can be directly sensed by the sensors. This may both make the tracking results more robust to noise and may simplify the algorithms to the point where they can be implemented on less powerful sensor nodes. We call these approaches *combinatorial tracking*.

9.4.1 Counting the Number of Targets

Target counting is an attempt to keep track of the number of distinct targets in a sensor field, even as they move, cross-over, merge, or split. It is representative of a class of applications that need to monitor intensity of activities in an area. To describe the problem, let us consider counting multiple targets in a two-dimensional sensor field, as shown in Figure 9.6. We assume that targets are point-source acoustic signals and can

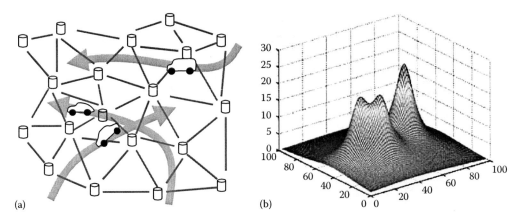

FIGURE 9.6 Target counting scenario, showing three targets in a sensor field (a). The goal is to count and report the number of distinct targets. With the signal field plotted in (b), the target counting becomes a peak counting problem.

be stationary or moving at any time, independent of the state of other targets. Sensors measure acoustic power and are time synchronized to a global clock. We assume that signals from two targets simply add at a receiving sensor, which is reasonable for noncoherent interference between acoustic sources.

The task here is to determine the number of targets in the region. One way to solve the problem is to compute an initial count and then update the count as targets move, enter, or leave the region. Here, we describe a leader-based counting approach, where a sensor leader is elected for each distinct target. A leader is initialized when a target moves into the field. As the target moves, the leadership may switch between sensor nodes to reflect the state change. When a target moves out of the region, the corresponding leader node is deactivated. Note here that the leader election does not rely on accurate target localization, as will be discussed later. The target count is obtained by noting the number of active leader nodes in the network (and the number of targets each is responsible for). Here, we will focus on the leader election process, omitting details of signal and query processing.

Since the sensors in the network only sense signal energy, we need to examine the spatial characteristics of target signals when multiple targets are in close proximity to each other. In Figure 9.6b, the three-dimensional surface shown represents total target signal energy. Three targets are plotted, with two targets near each other and one target well separated from the rest of the group.

There are several interesting observations to make here:

1. Call the set of sensors that can "hear" a target the target influence area. When targets' influence areas are well separated, target counting can be considered as a clustering and a cluster leader election problem. Otherwise, it becomes a peak counting problem.
2. The target signal propagation model has a large impact on target "resolution." The faster the signal attenuates with distance from the source, the easier it is to discern targets from neighboring targets based on the energy of signals they emit.
3. Sensor spacing is also critical in obtaining correct target count. Sensor density has to be sufficient to capture the peaks and valleys of the underlying energy field, yet very densely packed sensors are often redundant, wasting resources.

A decentralized algorithm was introduced for the target counting task [10]. This algorithm forms equivalence classes among sensors, elects a leader node for each class based on the relative power detected at each sensor, and counts the number of such leaders. The algorithm comprises a decision predicate P which, for each node i, tests if it should participate in an equivalence class and a message exchange schema M about how the predicate P is applied to nodes. A node determines whether it belongs to an equivalence class based on the result of applying the predicate to the data of the node, as well as on

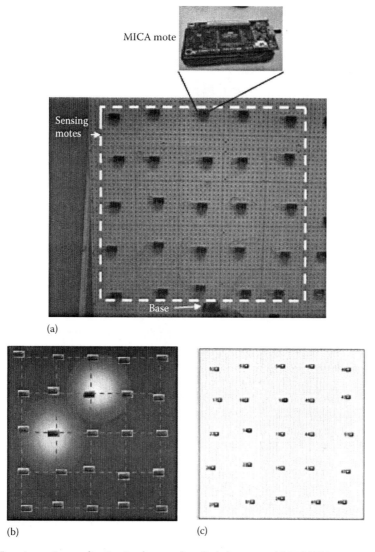

FIGURE 9.7 Target counting application implemented on Berkeley motes: (a) 25 MICA motes with light sensors are placed on a perturbed grid in a dark room; (b) two light blobs emulating $1/r^2$ signal attenuation are projected onto the mote board; (c) the leader of each collaboration group sends its location back to a base station GUI.

information from other nearby nodes. Equivalence classes are formed when the process converges. This protocol finds equivalence classes even when multiple targets interfere.

This leader election protocol is very powerful, yet it is lightweight enough to be implemented on sensor nodes such as the Berkeley motes. Figure 9.7 shows an experiment consists of 25 MICA motes with light sensors. The entire application, including code for collaborative leader election and multi-hop communication to send the leader information back to the base station, takes about 10 kB memory space on a mote.

9.4.2 Contour Tracking

Contour tracking is another example of finding the influence regions of targets without locating them. For a given signal strength, the tracking results are a set of contours, each of which contains one or more targets.

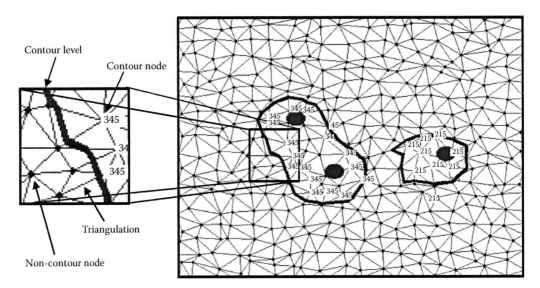

FIGURE 9.8 Simulation result showing contours for three point targets in a sensor field. The contours are constructed using a distributed marching-squares-like algorithm and are updated as targets move.

As in the target counting scenario, let us consider a two-dimensions sensor field and point-source targets. One way of determining the contours is by building a mesh over distributed sensor nodes via a Delaunay triangulation or a similar algorithm. The triangulation can be computed offline when setting up the network. Nodes that are connected by an edge of a triangle are called direct neighbors. Given a measurement threshold σ, which defines a σ-contour, a node is called a contour node if it has a sensor reading above σ and at least one of its direct neighbors has a sensor reading below σ. For a sufficiently smooth contour and dense sensor network, a contour can be assumed to intersect an edge only once, and a triangle at exactly two edges, as shown in Figure 9.8. By following this observation, we can traverse the contour by "walking" along the contour nodes. Again, purely local algorithms exist to maintain these contours as the targets move.

9.4.3 Shadow Edge Tracking

Contour tracking can be viewed as a way to determine the boundary of a group of targets. In an extreme case, the group of targets can be a continuum over space, where no single sensor alone can determine the global information from its local measurement. An example of this is to determine and track the boundary of a large object moving in a sensor field, where each sensor only "sees" a portion of the object. One such application is tracking a moving chemical plume over an extended area using airborne and ground chemical sensors.

We assume the boundary of the object is a polygon made of line segments. Our approach is to convert the problem of estimating and tracking a nonlocal (possibly very long) line segment into a local problem using a dual-space transformation [19]. Just as a Fourier transform maps a global property of a signal, such as periodicity in the time domain, to a local feature in the frequency domain, the dual-space transform maps a line in the primal space into a point in the dual space, and vice versa (Figure 9.9). Using a primal–dual transformation, each edge of a polygonal object can be tracked as a point in the dual space. A tracking algorithm has been developed based on the dual-space analysis and implemented on the Berkeley motes [19]. A key feature of this algorithm is that it allows us to put to sleep all sensor nodes except those in the vicinity of the object boundary, yielding significant energy savings.

Tracking relations among a set of objects is another form of global, discrete analysis of a collection of objects, as described by Guibas [20]. An example is determining whether a friendly vehicle is surrounded by a number of enemy tanks. Just as in the target counting problem, the "am I surrounded" relation can be resolved without having to solve the local problems of localizing all individual objects first.

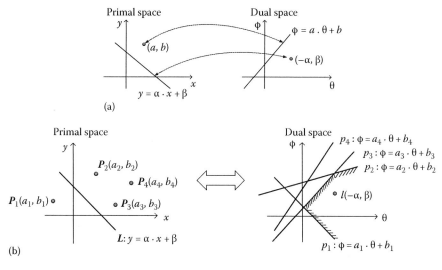

FIGURE 9.9 Primal–dual transformation, a one-to-one mapping where a point maps to a line and a line maps to a point (a). The image of a half-place shadow edge in the dual space is a point located in a cell formed by the duals of the sensor nodes (b).

9.5 Discussion

We have used the tracking problem as a vehicle to discuss sensor network CSIP design. We have focused on the estimation and tracking aspects and skipped over other important details, such as target detection and classification, for space reasons.

Detection is an important capability for a sensor network, as a tracker must rely on detection to initialize itself as new events emerge [1,22]. Traditional detection methods have focused on minimizing false alarms or the miss rate. In a distributed sensor network, the more challenging problem for detection is the proper allocation of sensing and communication resources to multiple competing detection tasks spawned by emerging stimuli. This dynamic allocation and focusing of resources in response to external events is somewhat analogous to attentional mechanisms in human vision systems, and is clearly a future research direction. More research should also be directed to the information architecture of distributed detection and tracking, and to addressing the problems of "information double-counting" and data association in a distributed network [6,23].

Optimizing resources for a given task, as for example in IDSQ, relies on accurate models of information gain and cost. To apply the information-driven approach to tracking problems involving other sensing modalities, or to problems other than tracking, we will need to generalize our models for sensing and estimation quality and our models of the tradeoff between resource use and quality. For example, what is the expected information gain per unit energy consumption in a network? One must make assumptions about the network, stimuli, and tasks in order to build such models. Another interesting problem for future research is to consider routing and sensing simultaneously and optimize for the overall gain of information.

We have not yet touched upon the programming issues in sensor networks. The complexity of the applications, the collaborative nature of the algorithms, and the plurality and diversity of resource constraints demand novel ways to construct, configure, test, and debug the system, especially the software. This is more challenging than traditional collection-based computation in parallel-processing research because sensor group management is typically dynamic and driven by physical events. In addition, the existing development and optimization techniques for embedded software are largely at the assembly level and do not scale to collaborative algorithms for large-scale distributed sensor networks.

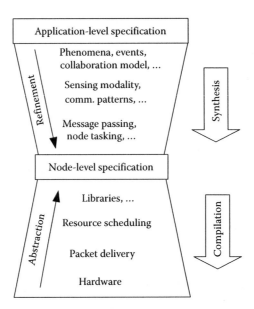

FIGURE 9.10   A programming methodology for deeply embedded systems.

We need high-level system organizational principles, programming models, data structures, and processing primitives to express and reason about system properties, physical data, and their aggregation and abstraction, without losing relevant physical and resource constraints.

A possible programming methodology for distributed embedded sensing systems is shown in Figure 9.10. Given a specification at a collaborative behavioral level, software tools automatically generate the interactions of algorithm components and map them onto the physical hardware of sensor networks.

At the top level, the programming model should be expressive enough to describe application-level concerns, e.g. physical phenomena to be sensed, user interaction, and collaborative processing algorithms, without the need to manage node-level interactions. The programming model may be domain specific. For example, SAL [24] is a language for expressing and reasoning about geometries of physical data in distributed sensing and control applications; various biologically inspired computational models [25,26] study how complex collaborative behaviors can be built from simple components. The programming model should be structural enough to allow synthesis algorithms to exploit commonly occurring patterns and generate efficient code. TinyGALS [27] is an example of a synthesizable programming model for event-driven embedded software.

Automated software synthesis is a critical step in achieving the scalability of sensor network programming. Hardware-oriented concerns, such as timing and location, may be introduced gradually by refinement and configuration processes. The final outputs of software synthesis are operational code for each node, typically in forms of imperative languages, from which the more classical operating system, networking, and compiler technologies can be applied to produce executables. The libraries supporting node-level specifications need to abstract away hardware idiosyncrasy across different platforms, but still expose enough low-level features for applications to take advantage of.

9.6 Conclusion

This chapter has focused on the CSIP issues in designing and analyzing sensor network applications. In particular, we have used tracking as a canonical problem to expose important constraints in designing, scaling, and deploying these sensor networks, and have described approaches to several tracking problems that are at progressively higher levels with respect to the nature of information being extracted.

From the discussions, it is clear that, for resource-limited sensor networks, one must take a more holistic approach and break the traditional barrier between the application and networking layers. The challenge is to define the constraints from an application in a general way so that the networking layers can exploit, and vice versa. An important contribution of the approaches described in this chapter is the formulation of application requirements and network resources as a set of generic constraints so that target tracking and data routing can be jointly optimized.

Acknowledgments

This work is supported in part by the Defense Advanced Research Projects Agency (DARPA) under contract number F30602-00-C-0139 through the Sensor Information Technology Program. The views and conclusions contained herein are those of the authors and should note be interpreted as representing the official policies, either expressed or implied, of the Defense Advanced Research Projects Agency or the U.S. Government.

This chapter was originally published in the August 2003 issue of the *Proceedings of the IEEE*. It is reprinted here with permission.

The algorithm and experiment for the target counting problem were designed and carried out in collaboration with Qing Fang, Judy Liebman, and Elaine Cheong. The contour tracking algorithm and simulation were jointly developed with Krishnan Eswaran.

Patrich Cheung designed, prototyped, and calibrated the PARC sensor network testbeds and supported the laboratory and field experiments for the algorithms and software described in this chapter.

References

1. Pottie, G.J. and Kaiser, W.J., Wireless integrated network sensors, *Communications of the ACM*, 43(5), 51, 2000.
2. Intanagonwiwat, C. et al., Directed diffusion: A scalable and robust communication paradigm for sensor networks, in *Proceedings of ACM MobiCOM*, Boston, MA, August 2000.
3. Merrill, W.M. et al., Open standard development platforms for distributed sensor networks, in *Proceedings of SPIE, Unattended Ground Sensor Technologies and Applications IV*, AeroSense 2002, Vol. 4743, Orlando, FL, April 2–5, 2002, p. 327.
4. Hill, J. et al., System architecture directions for networked sensors, in *The International Conference on Architectural Support for Programming Languages and Operating Systems (ASPLOS)*, Cambridge, MA, 2000.
5. Doherty, L. et al., Energy and performance considerations for smart dust, *International Journal of Parallel Distributed Systems and Networks*, 4(3), 121, 2001.
6. Liu, J.J. et al., Distributed group management for track initiation and maintenance in target localization applications, in *Proceedings of 2nd International Workshop on Information Processing in Sensor Networks (IPSN)*, Palo Alto, CA, April 2003.
7. Reid, D.B., An algorithm for tracking multiple targets, *IEEE Transactions on Automatic Control*, 24, 6, 1979.
8. Bar-Shalom, Y. and Li, X.R., *Multitarget–Multisensor Tracking: Principles and Techniques*, YBS Publishing, Storrs, CT, 1995.
9. Cox, I.J. and Hingorani, S.L., An efficient implementation of Reid's multiple hypothesis tracking algorithm and its evaluation for the purpose of visual tracking, *IEEE Transactions on Pattern Analysis and Machine Intelligence*, 18(2), 138, 1996.
10. Poore, A.B., Multidimensional assignment formulation of data association problems arising from multitarget and multisensor tracking, *Computational Optimization and Applications*, 3, 27, 1994.
11. Manyika, J. and Durrant-Whyte, H., *Data Fusion and Sensor Management: A Decentralized Information-Theoretic Approach*, Ellis Horwood, New York, 1994.

12. Estrin, D. et al., Next century challenges: Scalable coordination in sensor networks, in *Proceedings of the Fifth Annual International Conference on Mobile Computing and Networks (MobiCOM'99)*, Seattle, Washington, DC, August 1999.
13. Brooks, R.R. et al., Self-organized distributed sensor network entity tracking, *International Journal of High-Performance Computing Applications*, 16(3), 207, 2002.
14. Zhao, F. et al., Information-driven dynamic sensor collaboration, *IEEE Signal Processing Magazine*, 19(2), 61, 2002.
15. Chu, M. et al., Scalable information-driven sensor querying and routing for *ad hoc* heterogeneous sensor networks, *International Journal of High-Performance Computing Applications*, 16(3), 90, 2002.
16. Liu, J.J. et al., Multi-step information-directed sensor querying in distributed sensor networks, in *Proceedings of IEEE International Conference on Acoustics, Speech, and Signal Processing (ICASSP)*, Hong Kong, China, April 2003.
17. Liu, J.J. et al., Collaborative in-network processing for target tracking, *EURASIP Journal of Applied Signal Processing*, 2003(4), 379, 2003.
18. Fang, Q. et al., Lightweight sensing and communication protocols for target enumeration and aggregation, in *ACM Symposium on Mobile Ad Hoc Networking and Computing (MobiHoc)*, Annapolis, MD, 2003.
19. Liu, J. et al., A dual-space approach to tracking and sensor management in wireless sensor networks, in *Proceedings of 1st ACM International Workshop on Wireless Sensor Networks and Applications*, Atlanta, GA, April, 2002, p. 131.
20. Guibas, L., Sensing, tracking, and reasoning with relations, *IEEE Signal Processing Magazine*, 19(2), 73, 2002.
21. Tenney, R.R. and Sandell, N.R. Jr., Detection with distributed sensors, *IEEE Transactions on Aerospace and Electronic Systems*, 17, 501, 1981.
22. Li, D. et al., Detection, classification and tracking of targets in distributed sensor networks, *IEEE Signal Processing Magazine*, 19(2), 17, 2002.
23. Shin, J. et al., A distributed algorithm for managing multi-target identities in wireless *ad-hoc* sensor networks, in *Proceedings of 2nd International Workshop on Information Processing in Sensor Networks (IPSN)*, Palo Alto, CA, April 2003.
24. Zhao, F. et al., Physics-based encapsulation in embedded software for distributed sensing and control applications, *Proceedings of the IEEE*, 91(1), 40, 2003.
25. Abelson, H. et al., Amorphous computing, *Communications of the ACM*, 43(5), 74, 2001.
26. Calude, C. et al. (eds.), *Unconventional Models of Computation*, LNCS 2509, Springer, New York, 2002.
27. Cheong, E. et al., TinyGALS: A programming model for event-driven embedded systems, in *18th ACM Symposium on Applied Computing*, Melbourne, FL, March 2003, p. 698.

10

Environmental Effects

10.1 Introduction 207
10.2 Sensor Statistical Confidence Metrics 207
10.3 Atmospheric Dynamics 208
    Acoustic Environmental Effects • Seismic Environmental
    Effects • EM Environmental Effects • Optical Environmental
    Effects • Environmental Effects on Chemical and Biological
    Detection and Plumes Tracking
10.4 Propagation of Sound Waves 214
References 217

David C. Swanson
*The Pennsylvania
State University*

10.1 Introduction

Sensor networks can be significantly impacted by environmental effects from electromagnetic (EM) fields, temperature, humidity, background noise, obscuration, and for acoustic sensors outdoors, the effects of wind, turbulence, and temperature gradients. The prudent design strategy for intelligent sensors is to, at a minimum, characterize and measure the environmental effect on the sensor information reported while also using best practices to minimize any negative environmental effects. This approach can be seen as essential *information reporting* by sensors, rather than simple data reporting. Information reporting by sensors allows the data to be put into the context of the environmental and sensor system conditions, which translates into a confidence metric for proper use of the information. Reporting information, rather than data, is consistent with data fusion hierarchies and global situational awareness goals of the sensor network.

Sensor signals can be related to known signal patterns when the signal-to-noise ratio (SNR) is high, as well as when known environmental effects are occurring. A microprocessor is used to record and transmit the sensor signal, so it is a straightforward process to evaluate the sensor signal relative to known patterns. Does the signal fit the expected pattern? Are there also measured environmental parameters that indicate a possible bias or noise problem? Should the sensor measurement process adapt to this environmental condition? These questions can be incorporated into the sensor node's program flow chart to report the best possible sensor information, including any relevant environmental effects, or any unexplained environmental effect on SNR. But, in order to put these effects into an objective context, we first must establish a straightforward confidence metric based on statistical moments.

10.2 Sensor Statistical Confidence Metrics

Almost all sensors in use today have some performance impact from environmental factors that can be statistically measured. The most common environmental factors cited for electronic sensors are temperature and humidity, which can impact "error bars" for bias and/or random sensor error. If the environmental effect is a repeatable bias, then it can be removed by a calibration algorithm, such as in the corrections applied to J-type thermocouples as a function of temperature. If the bias is random from sensor to sensor (say due to manufacturing or age effects), then it can be removed via periodic calibration of each specific sensor in the network. But, if the measurement error is random due to low SNR or background noise interference, then we should measure the sensor signal statistically (mean and standard deviation). We should also report how the statistical estimate was measured in terms of numbers of observation samples and time intervals of observations. This also allows an estimate of the confidence of the variance to be reported using the Cramer–Rao lower bound [1]. If we assume the signal distribution has mean m and variance σ^2, and we observe N observations to estimate the mean and variance, our estimate has fundamental limitations on its accuracy. The estimated mean from N observations, m_N, is

$$m_N = m \pm \frac{\sigma}{\sqrt{N}} \tag{10.1}$$

The estimated variance σ_N^2 is

$$\sigma_N^2 = \sigma^2 \pm \frac{\sigma^2 \sqrt{2}}{\sqrt{N}} \tag{10.2}$$

As Equations 10.1 and 10.2 show, only as N becomes large can the estimated mean and variance be assumed to match the actual mean and variance. For N observations, the mean and variance estimates cannot be more accurate than that seen in Equations 10.1 and 10.2 respectively. Reporting the estimated mean, standard deviation, the number of observations, and the associated time interval with those observations is essential to assembling the full picture of the sensor information.

For example, a temperature sensor's output in the absence of a daytime wind may not reflect air temperature only, but rather solar loading. To include this in the temperature information output one most have wind and insolation (solar heat flux) sensors and a physical algorithm to include these effects to remove bias in the temperature information. If the temperature is fluctuating, then it could be electrical noise or interference, but it could also be a real fluctuation due to air turbulence, partly cloudy skies during a sunny day, or cold rain drops. The temporal response of a temperature sensor provides a physical basis for classifying some fluctuations as electrical noise and others as possibly real atmospheric dynamics. For the electronic thermometer or relative humidity sensor, fluctuations faster than around 1 s can be seen as likely electrical noise. However, this is not necessarily the case for wind, barometer, and solar flux sensors.

10.3 Atmospheric Dynamics

The surface layer of the atmosphere is driven by the heat flux from the sun (and nighttime re-radiation into space), the latent heat contained in water vapor, the forces of gravity, and the forces of the prevailing geotropic wind. The physical details of the surface-layer are well described in an excellent introductory text by Stull [2]. Here, we describe the surface-layer dynamics as they pertain to unattended ground sensor (UGS) networks and the impact these atmospheric effects can have on acoustic, seismic, EM, and optical image data. However, we should also keep in mind that atmospheric parameters are physically interrelated and that the confidence and bias in a given sensor's reported data can be related physically to a calibration model with a broad range of environmental inputs.

Propagation outdoors can be categorized into four main wave types: acoustic, seismic, EM, and optical. While seismic propagation varies seasonally (as does underwater sound propagation), acoustic, optical, and EM wave propagation varies diurnally, and as fast as by the minute when one includes weather effects. The diurnal cycle starts with stable cold air near the ground in the early morning. As the sun heats the ground, the ground heats the air parcels into unstable thermal plumes, which rise upwards and draw cooler upper air parcels to the surface. The thermal plumes lead to turbulence and eventually to an increase in surface winds. Once the sun sets, the ground heating turns to radiation into space. The lack of solar heating stops the thermal plumes from forming. Colder parcels of air settle by gravity to the surface, forming a stable nocturnal boundary layer. The prevailing winds tend to be elevated over this stable cold air layer. Eventually, the cold air starts to drain downhill into the valleys and low-lying areas, in what are called katabatic winds. These nocturnal katabatic winds are very light and quite independent of the upper atmosphere prevailing geotropic winds. If there is a cold or warm front or storm moving across the area, then the wind and temperature tend to be very turbulent and fluctuating. It is important to keep in mind that these atmospheric effects on the surface are very local and have a significant effect on the ground sensor data. We will discuss these effects by wave type below.

10.3.1 Acoustic Environmental Effects

Acoustic waves outdoors have a great sensitivity to the local weather, especially wind. The sound wave speed in air is relatively slow at about 344 m/s at room temperature relative to wind speed, which can routinely approach tens of meters per second. The sound wave travels faster in downwind directions than upwind directions. Since the wind speed increases as one moves up in elevation (due to surface turbulence and drag), sound rays from a source on the ground tend to refract upward in the upwind direction and downward in the downwind direction. This means that a acoustic UGS will detect sound at much greater distances in the direction the wind is coming from. However, the wind will also generate acoustic noise from the environment and from the microphone. Wind noise is perhaps the most significant detection performance limitation for UGS networks. Turbulence in the atmosphere will tend to scatter and refract sound rays randomly. When the UGS is upwind of a sound source, this scattering will tend to make some of the upward refracting sound detectable by the UGS. The received spectra of the sound source in a turbulent atmosphere will fluctuate in amplitude randomly for each frequency due to the effects of multipaths. However, the effects of wind and turbulence on, and bearing measurement by, a UGS are usually quite small, since the UGS microphone array is typically only a meter or less. Building UGS arrays larger than about 2 m becomes mechanically complex, is subject to wind damage, and does not benefit from signal coherence and noise independence like EM or underwater arrays. This is because the sound speed is slow compared with the wind speed, making the acoustic signal spatially coherent over shorter distances in air.

When the wind is very light, temperature effects on sound waves tend to dominate. Sound travels faster in warmer air. During a sunny day the air temperature at the surface can be significantly warmer than that just a few meters above. On sunny days with light winds, the sound tends to refract upwards, making detection by a UGS more difficult. At night the opposite occurs, where colder air settles near the surface and the air above is warmer. Like the downwind propagation case, the higher elevation part of the wave outruns the slower wave near the ground. Thus, downward refraction occurs in all directions on a near windless night. This case makes detection of distant sources by a UGS much easier, especially because there is little wind noise. In general, detection ranges for a UGS at night can be two orders of magnitude better (yes, 100 times longer detection distances for the same source level). This performance characteristic is so significant that the UGS networks should be described operationally as nocturnal sensors. Figure 10.1 shows the measured sound from a controlled loudspeaker monitored continuously over a 3 day period.

Humidity effects on sound propagation are quite small, but still significant when considering long-range sound propagation. Water vapor in the air changes the molecular relaxation, thus affecting the energy absorption of sound into the atmosphere [3]. However, absorption is greatest in hot, dry air and at ultrasonic frequencies. For audible frequencies, absorption of sound by the atmosphere can be a few decibels per kilometer of propagation. When the air is saturated with water vapor, the relative humidity is 100% and typically fog forms from aerosols of water droplets. The saturation of water vapor in air h in grams per cubic meter can be approximated by

$$h\,(\text{g/m}^3) = 0.94T + 0.345 \tag{10.3}$$

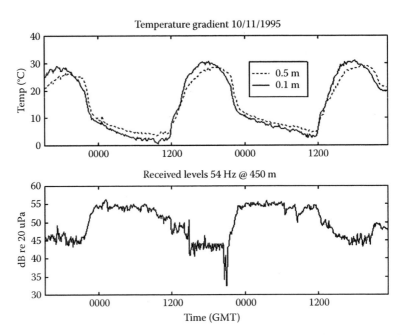

FIGURE 10.1   Air temperature at two heights (top) and received sound at 54 Hz from a controlled loudspeaker 450 m away over a 3 days period showing a 10 dB increase in sound during nighttime temperature inversions.

where T is in degrees centigrade. If the sensor reports the percent relative humidity RH at a given temperature T, one can estimate the dewpoint T_{dp}, or temperature where saturation occurs, by

$$T_{dp} = \frac{(0.94T + 0.345) \times \text{RH}/100 - 0.345}{0.94} \qquad (10.4)$$

One can very simply approximate the saturation density h by the current temperature in centigrade and get the dewpoint temperature by multiplying by the relative humidity fraction. For example, if the temperature is 18°C there is roughly 18 g/m³ of water vapor if the air is 100% saturated. If the RH is 20%, then there is roughly 3.6 g/m³ of water vapor and the dewpoint is approximately 3.6°C. The actual number using Equations 10.3 and 10.4 are 17.3 g/m³ for saturation, 3.46 g/m³ for 20% RH, and a dewpoint of 3.3°C. Knowledge of the humidity is useful in particular for chemical or biological aerosol hazards, EM propagation, and optical propagation, but it does not significantly affect sound propagation. Humidity sensors generally are only accurate to a few percent, unless they are the expensive "chilled mirror" type of optical humidity sensor. There are also more detailed relative humidity models for estimating dew-point and frost-point temperatures in the literature. Equations 10.3 and 10.4 are a useful and practical approximation for UGS networks.

10.3.2 Seismic Environmental Effects

Seismic waves are relatively immune to the weather, except for cases where ground water changes or freezes. However, seismic propagation is significantly dependent on the subterranean rock structures and material. Solid materials, such as a dry lake bed, form ideal seismic wave propagation areas. Rock fissures, water, and back-filled areas of earth tend to block seismic waves. If one knows the seismic propagation details for a given area, then seismic arrays can make very effective UGS networks. If the source is on or near the surface, two types of wave are typically generated, a spherically radiating pressure wave, or p-wave, and a circularly radiating surface shear wave, or s-wave. The p-wave can carry a lot of energy at very fast speeds due to the compressional stiffness of the ground. However, since it is spherically spreading (approximately) its amplitude (dB) decays with distance R by 20 log R, or 60 dB in the first 1 km.

The s-wave speed depends on the shear stiffness of the ground and the frequency, where high frequencies travel faster than low frequencies. Since the s-wave spreads circularly on the surface, its approximate amplitude dependence (dB) with distance R is 10 log R, or only 30 dB in the first 1 km. A UGS network detecting seismic waves from ground vehicles is predominately detecting s-waves, the propagation of which is highly dependent on the ground structure. In addition, there will always be a narrow frequency range where the s-wave speed is very close to the acoustic wave speed in the air. This band in the seismic spectrum will detect acoustic sources as well as seismic sources. Seismic sensors (typically geophones) will also detect wind noise through tree roots and structure foundations. If the UGS sensor is near a surf zone, rapids, airport, highway, or railroad, it will also detect these sources of noise or signal, depending on the UGS application. When the ground freezes, frozen water will tend to make the surface stiffer and the s-waves faster. Snow cover will tend to insolate the ground from wind noise. Changes in soil moisture can also effect seismic propagation, but in complicated ways depending on the composition of the soil.

10.3.3 EM Environmental Effects

Environmental effects on EM waves include the effect of the sun's radiation, the ionosphere, and, most important to UGS communications, the humidity and moisture on the ground. The water vapor density in the air, if not uniform, has the effect of changing the EM impedance, which can refract EM waves. When the air is saturated, condensation in the form of aerosols can also weaken EM wave propagation through scattering, although this effect is fairly small at frequencies below 60 GHz (the wavelength at 60 GHz is 5 mm). In the hundreds of Megahertz range, propagation is basically line of sight except for the first couple of reflections from large objects, such as buildings (the wavelength at 300 MHz is 1 m). Below 1 MHz, the charged particles of the Earth's ionosphere begin to play a significant role in the EM wave propagation. The ground and the ionosphere create a waveguide, allowing long-range "over the horizon" wave propagation. The wavelength at 300 kHz is 1 km, so there are little environmental effects from manmade objects in the propagation path, provided one has a large enough antenna to radiate such a long wavelength. In addition, EM radiation by the sun raises background noise during the daytime.

For all ground-to-ground EM waves, the problem for UGS networks sending and receiving is the practical fact that the antenna needs to be small and cannot have a significant height above the ground plane. This propagation problem is crippling when the dewpoint temperature (or frost point) is reached, which effectively can raise the ground plane well above a practical UGS antenna height, rendering the antenna efficiency to minimal levels. Unfortunately, there are no answers for communication attempts using small antennas near the ground in high humidity. Vegetation, rough terrain, limited line of sight, and especially dew- or frost-covered environments are the design weak point of UGS networks. This can be practically managed by vertical propagation to satellites or air vehicles. Knowledge of humidity and temperature can be very useful in assessing the required power levels for EM transmission, as well as for managing communications during problem environmental conditions.

10.3.4 Optical Environmental Effects

Optical waves are also EM waves, but with wavelengths in the 0.001 mm (infrared) to a few hundred nanometers. The visual range is from around 700 nm (red) to 400 nm (violet) wavelengths. These small wavelengths are affected by dust particles, and even the molecular absorption by the atmospheric gases. Scattering is stronger at shorter wavelengths, which is why the sky is blue during the day. At sunrise and sunset, the sunlight reaches us by passing through more of the atmosphere, this scattering that blue light, leaving red, orange, and yellow. Large amounts of pollutants, such as smoke, ozone, hydrocarbons, and sulfur dioxide, can also absorb and scatter light, obscuring optical image quality. Another obvious environmental effect for imagery is obscuration by rain or snow. However, thermal plumes and temperature gradients also cause local changes in air density and the EM index of refraction, which cause fluctuations in images.

Measuring the environmental effects directly can provide an information context for image features and automatic target recognition. The syntax for logically discounting or enhancing the weight of some features in response to the environment creates a very sophisticated and environmentally robust UGS. More importantly, it provides a scientific strategy for controlling false alarms due to environmental effects.

10.3.5 Environmental Effects on Chemical and Biological Detection and Plumes Tracking

Perhaps one of the most challenging and valuable tasks of a UGS network on the battlefield is to provide real-time guidance on the detection and tracking of plumes of harmful chemical vapors, aerosols, or biological weapons. The environment in general, and in particular the temperature and humidity, have an unfortunate direct effect on the performance of many chemical and biological sensors. Putting the chemical and biological sensor performance aside, the movement and dispersion of a detected chem/bio plume is of immediate importance once it is detected; thus, this capability is a major added value for UGS networks.

Liquid chemical and biological aerosols will evaporate based on their vapor pressures at a particular temperature and the partial pressures of the other gases, most notably water, in the atmosphere. Once vaporized, the chemicals will diffuse at nearly the speed of sound and the concentration will decrease rapidly. Since vaporization and diffusion are highly dependent on temperature, the local environmental conditions play a dominant role in how fast a chemical threat will diffuse and which direction the threat will move. To maximize the threat of a chemical weapon, one would design the material to be a power of low vaporization aerosol to maintain high concentration for as long as possible in a given area [4]. The environmental condition most threatening to people is when a chemical or biological weapon is deployed in a cold, wet fog or drizzle where little wind or rain is present to help disperse the threat. During such conditions, temperature inversions are present where cold stable air masses remain at the surface, maximizing exposure to the threat. A UGS network can measure simple parameters such as temperature gradients, humidity, and wind to form physical features, such as the bulk Richardson index R_B, to indicate in a single number the stability of the atmosphere and the probability of turbulence, as seen in Equation 10.5 and Figure 10.2 [5]:

$$R_B = \frac{g\Delta T \Delta z}{T(\Delta U^2 + \Delta V^2)} \tag{10.5}$$

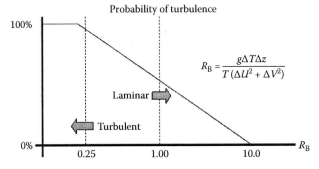

FIGURE 10.2 The bulk Richardson index provides a physical atmospheric parameter representing the likelihood of turbulence given the wind and temperature measured by a UGS.

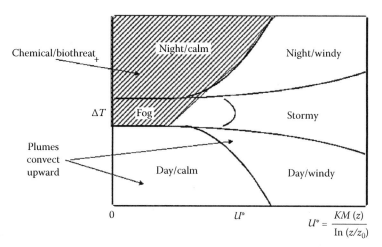

FIGURE 10.3 Using temperature gradient and wind *M*, we can show a simplified atmospheric state space that illustrates the conditions for an elevated chemical or biological threat ($k = 0.4$).

In Equation 10.5, the parameters ΔU and ΔV represent the horizontal wind gradient components, g is the acceleration due to gravity, ΔT is the temperature gradient (top minus bottom), and Δz is the separation of the temperature readings. If the bottom is the ground surface, then one can assume that the wind is zero. Clearly, this is a good example of a natural physical feature for the state of the environment. When $R_B > 1$, stable air near the ground enhances the exposure of chemical or biological weapons. This problem arises from the lack of turbulence and mixing that helps disperse aerosols.

For measurements such as wind, it not only makes sense to report the mean wind speed, but also the standard deviation, sample interval, and number of samples in the estimate. Information on the sample set size is used in Equations 10.1 and 10.2 to provide information confidence bounds, rather than simply data. Combining wind and temperature data we can devise a chart for atmospheric state, as seen in Figure 10.3.

A UGS can extract another important turbulence factor from the wind called the dissipation rate of the Kolmogorov spectrum [2]. The Kolmogorov spectrum represents the wave structure of the turbulence. It is useful for sound propagation models and for chemical and biological plume transport models. One calculates the mean wind speed and the Fourier transform of a regularly time-sampled series of wind-speed measurements. The mean wind speed is used to convert the time samples to special samples such that the Fourier spectrum represents a wavenumber spectrum. This is consistent with Taylor's hypothesis [2] of spatially frozen turbulence, meaning that the spatial turbulence structure remains essentially the same as it drifts with the wind. Figure 10.4 shows the Kolmogorov spectrum for over 21 h of wind measurements every 5 min. The physical model represented by the Kolmogorov spectrum has importance in meteorology, as well as in sound-propagation modeling, where turbulence cause random variations in sound speed.

The surface layer of the atmosphere is of great interest to the meteorologist because much of the heat energy transport occurs there. There is the obvious heating by the sun and cooling by the night sky, both of which are significantly impacted by moisture and wind. There is also a latent heat flux associated with water vapor given off by vegetation, the soil, and sources of water. While atmospheric predictive models such as MM5 can provide multi-elevation weather data at grid points as close as 20 km, having actual measurements on the surface is always of some value, especially if one is most interested in the weather in the immediate vicinity. These local surface inputs to the large-scale weather models are automated using many fixed sites and mobile sensors (such as on freight trucks) that provide both local and national surface-layer measurements.

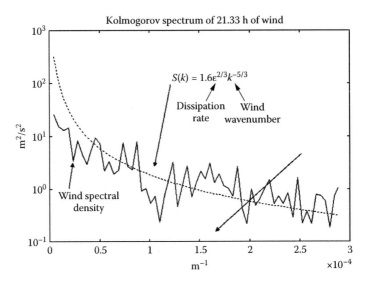

FIGURE 10.4 The Kolmogorov spectrum provides a means to characterize the turbulent structure of the wind using a single parameter called the dissipation rate ε.

10.4 Propagation of Sound Waves

Sound will travel faster in downwind directions and in hotter air, whereas aerosols will disperse slowly in the still cool air of a nocturnal boundary layer. The atmospheric condition of concern for both noise and air pollution is the case of a cool, still, surface boundary layer, as this traps pollutants and downward-refracts sound waves near the ground and often occurs at night. The local terrain is very important to the formation of a nocturnal boundary layer, as low-lying and riverine areas will tend to collect cold parcels of heavy air. Slow katabatic winds form from the "draining" of these cool air parcels downhill. Local sensors can detect the environmental conditions leading to the formation of nocturnal boundary layers in these local areas to help avoid problems from local noise and chemical pollution.

Given the weather information, we can construct a volume of elements each with a sound speed plus a wind vector. The wind speed adds to the sound speed in the direction of propagation, which is calculated as a dot product of the wind vector and the propagation direction vector, plus the sound speed scalar. We will develop a general propagation algorithm by first starting with a simple point source (small compared with wavelength) and noting the pressure field created:

$$p(r,k) = \frac{A}{R} e^{j(\omega t - kr)} \qquad (10.6)$$

The pressure p in Equation 10.6 decays with distance r for frequency ω and wavenumber k. Equation 10.6 describes an outgoing wave; as can be seen, as the time t increases, so the distance r must also increase to keep the phase constant. Since we are modeling wave propagation in one direction only, we can adopt a cylindrical coordinate system and concern our model with a particular direction only. However, this requires that we factor a square root of distance out to account for cylindrical verse spherical spreading.

$$\psi_{2D}(r,k) = \sqrt{r}\, p_{3D}(r,k) \qquad (10.7)$$

We can now decompose the wavenumber k into its r (distance) and z (elevation) components. Since $k = \omega/c$, where ω is radian frequency and c is the sound speed, the meteorological variations in sound speed will impact the wavenumber:

$$k = \sqrt{k_r^2 + k_z^2} \tag{10.8}$$

The pressure wavenumber spectrum of the field at some distance r is the Fourier transform of a slice of the field along the z-direction:

$$\psi(r,k_z) = \int_{-\infty}^{+\infty} \psi(r,z)e^{-jk_z z}\,dz \tag{10.9}$$

Using wavenumber spectra, we can write the spectrum at a distance $r + \Delta r$ in terms of the spectrum at r:

$$\psi(r + \Delta r, k_z) = e^{j\Delta_r \sqrt{k^2 - k_z^2}} \psi(r,k_z) \tag{10.10}$$

Equation 10.10 may not seem significant, but one cannot relate the pressure fields at r and $r + \Delta r$ directly in this manner.

We now consider variations in the total wavenumber k, due to small variations in sound speed, as the horizontal wavenumber at the surface $k_r(0)$ plus a small variation in wavenumber due to the environmental sound speed:

$$k^2 = k_r^2(0) + \delta k^2(z) \tag{10.11}$$

If the wavenumber variation is small, say on the order of a few percent or less, k_r can be approximated by

$$k_r \approx \sqrt{k_r^2(0) - k_z^2(r)} + \frac{\delta k^2(r,z)}{2k_r(0)} \tag{10.12}$$

Equation 10.10 is now rewritten as

$$\psi(r + \Delta r, k_z) = e^{j\Delta r \frac{\delta k^2(r,z)}{2k_r(0)} e^{j\Delta r \sqrt{k_r^2(0) - k_z^2(r)}}} \psi(r,k_z) \tag{10.13}$$

and using Fourier transforms can be seen as

$$\psi(r + \Delta r, z) = e^{j\Delta r \frac{\delta k^2(r,z)}{2k_r(0)} \frac{1}{2\pi} \int_{-\infty}^{+\infty} \psi(r,k_z)e^{j\Delta r \sqrt{k_r^2(0) - k_z^2(r)}} e^{jk_z z}\,dz} \tag{10.14}$$

so that we have a process cycle of calculating a Fourier transform of the acoustic pressure, multiplying by a low-pass filter, inverse Fourier transforming, and finally multiplying the result by a phase variation with height for the particular range step. In this process, the acoustic wave fronts will diffract according to the sound speed variations and we can efficiently calculate good results along the r-direction of propagation.

For real outdoor sound propagation environments, we must also include the effects of the ground. Following the developments of Gilbert and Di [6], we include a normalized ground impedance $Z_g(r)$ with respect to $\rho c = 415$ Rayls for the impedance of air. This requires a ground reflection factor

$$R(k_z) = \frac{k_z(r)Z_g(r) - k_r(0)}{k_z(r)Z_g(r) + k_r(0)} \tag{10.15}$$

and a surface complex wavenumber $\beta = k_r(0)/Z_g(r)$, which accounts for soft grounds. The complete solution is

$$\psi(r + \Delta r, z) = e^{j\Delta r \frac{\delta k^2(r,z)}{2k_r(0)}} \left[\begin{array}{l} \left\{ \dfrac{1}{2\pi} \displaystyle\int_{-\infty}^{+\infty} \left[\psi(r, k_z) e^{j\Delta r \sqrt{k_r^2(0) - k_z^2(r)}} \right] e^{+jk_z z} dz \right\} \\[2ex] + \left\{ \dfrac{1}{2\pi} \displaystyle\int_{-\infty}^{+\infty} \left[\psi(r, -k_z) R(k_z) e^{j\Delta r \sqrt{k_r^2(0) - k_z^2(r)}} \right] e^{+jk_z z} dz \right\} \\[2ex] + 2j\beta e^{-j\beta(r)z} e^{j\Delta r \sqrt{k_r^2(0) - \beta^2(r)}} \psi(r, \beta) \end{array} \right] \tag{10.16}$$

where the first term in Equation 10.16 is the direct wave, the middle term is the reflected wave, and the last term is the surface wave. For rough surfaces, one can step up or down the elevation as a phase shift in the wavenumber domain. Figure 10.5 shows some interesting results using a Green's function parabolic equation model.

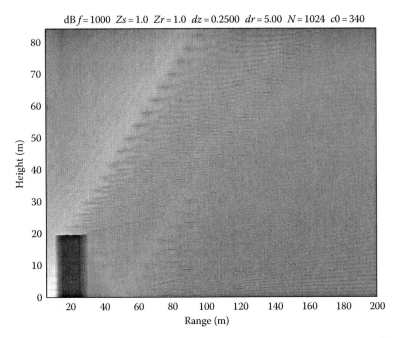

FIGURE 10.5 Propagation model results with a source to the left of a large wall showing diffraction, ground reflection, and atmospheric refraction.

References

1. Swanson, D.C., *Signal Processing for Intelligent Sensor Systems*, Marcel-Dekker, New York, 2000, p. 362.
2. Stull, R., *An Introduction to Boundary Layer Meteorology*, Kluwer Academic Publishers, Boston, MA, 1991.
3. Kinsler, L.E. and Frey, A.R., *Fundamentals of Acoustics*, Wiley, New York, 1962, p. 227.
4. Ali, J. et al., *US Chemical–Biological Defense Guidebook*, Jane's Information Group, Alexandria, VA, 1997, p. 150.
5. Garratt, J.R., *The Atmospheric Boundary Layer*, Cambridge University Press, New York, 1992, p. 32.
6. Gilbert, K.E. and Di, X., A fast Green's function method for one-way sound propagation in the atmosphere, *Journal of the Acoustical Society of America*, 94(4), 2343, 1993.

11

Detecting and Counteracting Atmospheric Effects

11.1 Motivation: The Problem 219
    What Is Atmosphere?
11.2 Sensor-Specific Issues 224
    Visible Spectrum Cameras • Infrared Sensors • MMW
    Radar Sensors • LADAR Sensors • Multispectral Sensors •
    Sonar Sensors
11.3 Physics-Based Solutions 226
11.4 Heuristics and Non-Physics-Based Solutions 234
11.5 Conclusion 240
References 240

Lynne L. Grewe
California State University

11.1 Motivation: The Problem

With the increasing use of vision systems in uncontrolled environments, reducing the effect of atmospheric conditions such as fog in images has become a significant problem. Many applications (e.g., surveillance) rely on accurate images of the objects under scrutiny [Brooks 98]. Poor visibility is an issue in aircraft navigation [Huxtable 97, Moller 94, Oakley 96, Sweet 96], highway monitoring [Arya 97], and commercial and military vehicles [Barducci 95, Pencikowski 96]. Figure 11.1 shows an example scene from a visible spectrum camera where the presence of fog severely impedes the ability to recognize objects in the scene.

Before discussing how particular sensors respond to atmospheric conditions and algorithms to improve the resulting images, let us discuss what is meant by atmosphere.

11.1.1 What Is Atmosphere?

Atmosphere whether caused by fog, rain, or even smoke involves the presence of particles in the air. On a clear day, the particles present, which make up "air" like oxygen, are so small that sensors are not impaired in their capture of the scene elements. However, this is not true with the presence of atmospheric conditions like fog because these particles are larger in size and impede the transmission of light (electromagnetic radiation) from the object to the sensor through scattering and absorption.

There have been many models proposed to understand how these particles interact with electromagnetic radiation. These models are referred to as scattering theories. Models consider parameters like

FIGURE 11.1 Building in foggy conditions, visibility of objects is impaired.

radius of particle size, wavelength of light, density of particle material, shape of particles, etc. Selection of the appropriate model is a function of the ratio of the particle radius to the wavelength of light being considered as well what parameters are known.

When this ratio of particle size to wavelength is near one, most theories used are derived from the "Mie scattering theory" [Mie 08]. This theory is the result of solving the Maxwell equations for the interaction of an electromagnetic wave with a spherical particle. This theory takes into account absorption and the refractive index (related to the angle to which light is bent going through the particle).

Other theories exist that do not assume the particle has a spherical shape and alter the relationship between size, distribution, reflection, and refraction. Size and shape of these particles vary. For example, water-based particles range from some microns and perfect spheres in case of liquid cloud droplets to large raindrops (up to 9 mm diameter), which are known to be highly distorted. Ice particles have a variety of nonspherical shapes. Their size can be some millimeters in case of hail and snow but also extends to the regime of cirrus particles, forming needles or plates of ten to hundreds of micrometers.

Most scattering theories use the following basic formula that relates the incident (incoming) spectral irradiance, $E(\lambda)$, to the outgoing radiance in the direction of the viewer, $I(\theta, \lambda)$, and the angular scattering function, $\beta(\theta, \lambda)$:

$$I(\theta,\lambda) = \beta(\theta,\lambda)E(\lambda) \tag{11.1}$$

where
  λ is the wavelength of the light
  θ is the angle the viewer is at with regard to the normal to the surface of the atmospheric patch the incident light is hitting

What is different among the various models is $\beta(\theta, \lambda)$. An observed phenomenon regarding physics-based models is that light energy at most wavelengths (λ) is attenuated as it travels through the atmosphere. In general, as it travels through more of the atmosphere, the signal will be diminished as shown

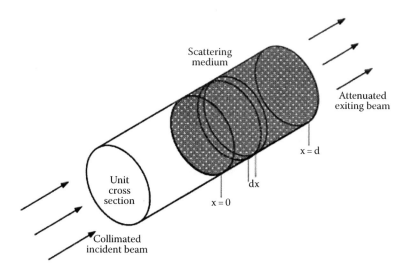

FIGURE 11.2 Demonstration of light attenuating as it travels through the atmosphere. (From Narasimhan, S. and Nayar, S., *Int. J. Comput. Vision*, 48(3), 233, 2002.)

in Figure 11.2. This is why as humans we can only see objects close to us in heavy fog. This is described as the change in the incoming light as follows (from [McCartney 75]):

$$\frac{dE(x,\lambda)}{E(x,\lambda)} = -\beta(\lambda)dx \tag{11.2}$$

where
   x is the travel direction
   dx is the distance traveled along that direction

Note that $\beta(\lambda)$ called the total scattering function is simply $\beta(\theta, \lambda)$ integrated over all angles. If we travel from the starting point of x = 0 to a distance d, the following will be the irradiance at d:

$$E(d,\lambda) = Eo(\lambda)\exp(-\beta(\lambda)d) \tag{11.3}$$

where $Eo(\lambda)$ is the initial irradiance at x = 0.

Many of the scattering theories treat each particle as independent of other particles because their separation distance is many times greater than their own diameter. However, multiple scatterings from one particle to the next of the original irradiance occur. This phenomenon when produced from environmental illumination like direct sunlight was coined "airlight" by [Koschmieder 24]. In this case, unlike the previously described attenuation phenomenon, there is an increase in the radiation as the particles interscatter the energy. Humans experience this as foggy areas being "white" or "bright." Figure 11.3 illustrates this process [Narasimhan 02]. [Narasimhan 02] develop a model that integrates the scattering equation of Equation 11.1 with the "airlight" phenomenon using an overcast environmental illumination model present on most foggy days.

The reader is referred to [Curry 02], [Day 98], [Kyle 91], and [Bohren 89] for details on specific atmosphere scattering theories and their equations. In [Narasimhan 03c], the authors look at the multiple scatter effects in atmospheric conditions that yield effects like glow of light sources.

Figure 11.4 shows a system [Hautiere 07] to reduce fog effects in vehicle-mounted camera images and illustrates the idea that the light received by the sensor is a combination of direct light from the object

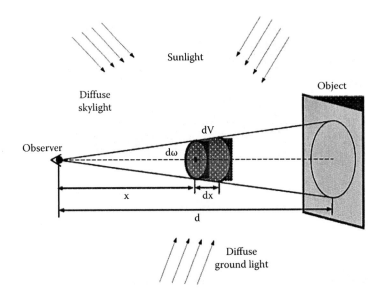

FIGURE 11.3   Cone of atmosphere between an observer and object scatters environmental illumination in the direction of observer. Thus, acting like a light source, called "airlight," whose brightness increases with path length d. (From Narasimhan, S. and Nayar, S., *Int. J. Comput. Vision*, 48(3), 233, 2002.)

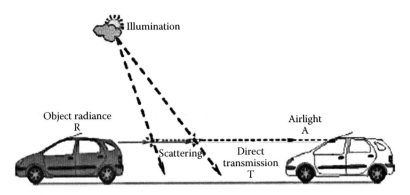

FIGURE 11.4   Light received by white car's camera is a function of the light reflected from the grey car, T, and the scattered light from the atmospheric particles called "airlight," A. (From Hautiere, N. et al., Towards fog-free in-vehicle vision systems through contrast restoration, *IEEE Conference on Computer Vision and Pattern Recognition, CVPR'07*, pp. 1–8, 2007.)

(green car) transmitted to the sensor (white car) plus the "airlight," which is the light scattered by the atmospheric particles. Some systems simplify the model by assuming a linear combination of these two elements that is dependent on the light wavelength and the depth of the object point in the scene from the camera. Some systems further simply by assuming this combination is independent of wavelength. Equation 11.4 shows the linear combination that is a common model used in current research. $I(x)$ represents a point in the image. The first term $L_0 \exp(-\beta(\lambda)d(x))$ represents the direct transmission light, T, in Figure 11.4. L_0 represents the luminance measured close to the object. Hence, the exponential factor represents the degradation with distance in the scene, $d(x)$. The second term $L_{inf}(1 - \exp(-\beta(\lambda) d(x)))$ represents the "airlight" component, A in Figure 11.4. $\beta(\lambda)$ is called the atmosphere extinction coefficient and can change with wavelength though as we will see in some of the research it is assumed to be constant over wavelength. It is important to note that the second term $L_{inf}(1 - \exp(-\beta(\lambda)d(x)))$

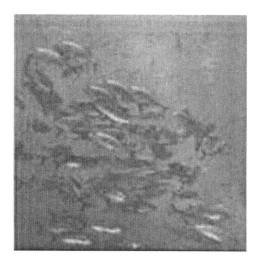

FIGURE 11.5 Operating underwater environment. (From Iqbal, K. et al., Enhancing the low quality images using unsupervised colour correction method, *2010 IEEE International Conference on Systems, Man and Cybernetics*, Istanbul, Turkey, pp. 1703–1709, 2010.)

increases in value with increasing scene depth. This corresponds to the observation that in foggy conditions the further away an object is the more "obscured" it is in the image appearing light indicating the bright fog. Linf is commonly called the atmospheric luminance:

$$I(x) = Lo \exp(-\beta(\lambda)d(x)) + Linf(1 - \exp(-\beta(\lambda)d(x))) \tag{11.4}$$

Another kind of "atmosphere" to consider is in underwater environments. Figure 11.5 shows an underwater environment with a traditional optical camera [Iqbal 10]. In [Sulzberger 09], the authors search for sea mines using both magnetic and electro-optic sensors under water. Figure 11.6 shows images from their sonar and optical sensor in underwater environments [Sulzberger 09].

The optical qualities of water both in reflection and refraction create differences in images produced by traditional optical cameras as evidenced in Figures 11.5 and 11.6. Bad weather can further increase difficulties by adding more particulates [Ahlen 05], and water movement can mimic problems seen in heavy smoke and fog conditions. There are some unique systems like maritime patrol and anti-submarine warfare Aircraft that deal with atmospheric conditions both in air and sea. In strictly underwater environments, sensors that can penetrate the water atmosphere like sonar are often used. In this chapter, we will concentrate on the techniques used for air/space atmospheric correction knowing that

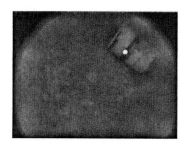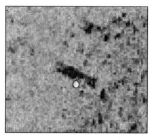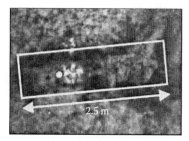

FIGURE 11.6 On the left and right, optical camera images in different underwater "atmospheres" and in the center a sonar image corresponding to right optical image. (From Sulzberger, G. et al., Hunting sea mines with UUV-based magnetic and electro-optic sensors, *OCEANS 2009*, Biloxi, MS, pp. 1–5, 2009.)

many of these techniques will also be useful for underwater "atmospheric" correction. The reader is referred to [Garcia 02], [Chambah 04], [Schechner 04], [Clem 02], [Clem 03], and [Kumar 06] for just a few selections on research with underwater sensors.

11.2 Sensor-Specific Issues

Distributed sensor networks employ many kinds of sensors. Each kind of sensor responds differently to the presence of atmosphere. While some sensors operate much better than others in such conditions, it is true that in most cases there will be some degradation to the image. We highlight a few types of sensors and discuss how they are affected.

The system objectives including the environmental operating conditions should select the sensors used in a sensor network. An interesting work that compares the response of visible, near-infrared, and thermal-infrared sensors in terms of target detection is discussed in [Sadot 95].

11.2.1 Visible Spectrum Cameras

Visual spectrum, photometric images are very susceptible to atmospheric effects. As shown in Figure 11.1, the presence of atmosphere causes the image quality to degrade. Specifically, images can appear fuzzy, out of focus, objects may be obscured behind the visual blanket of the atmosphere, and other artifacts may occur. As discussed in the previous section, in fog and haze, images appear brighter, the atmosphere itself often being near-white, and farther scene objects are not or fuzzy.

Figure 11.7 shows a visible spectrum image and the corresponding infrared (IR) and millimeter wave (MMW) images of the same scene. Figure 11.8 shows another visible spectrum image in foggy conditions.

11.2.2 Infrared Sensors

IR light can be absorbed by water-based molecules in the atmosphere but imaging using IR sensors yields better results than visible spectrum cameras. This is demonstrated in Figure 11.8. As a consequence, IR sensors have been used for imaging in smoke and adverse atmospheric conditions. For example, IR cameras are used by firefighters to find people and animals in smoke-filled buildings. IR as discussed in Chapter 5

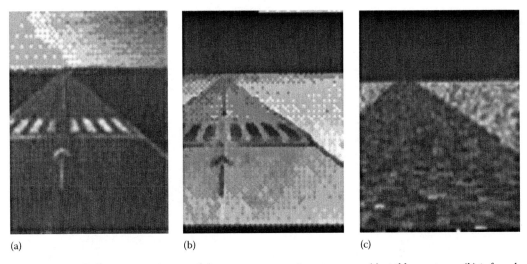

(a)   (b)   (c)

FIGURE 11.7   Different sensor images of the same scene, an airport runway: (a) visible spectrum, (b) infrared, and (c) MMW.

(a)

(b)

Visible   Infrared

(c)

(d)

FIGURE 11.8 IR is better in imaging through fog than visible spectrum camera: (a) visible spectrum image (color) on clear day, (b) grayscale visible spectrum image on a foggy day, and (c) IR image on same foggy day. (a–c: From Hosgood, B., Some examples of thermal infrared applications, IPSC website, http://humanitarian-security. jrc.it/demining/infrared_files/IR_show4/sld006.htm, 2003.) (d) Visible and IR images of a runway. (From IPAC, NASA, What is infrared? http://sirtf.caltech.edu/EPO/Kidszone/infrared.html, 2003.)

often is associated with thermal imaging and is used in night vision systems. Of course some "cold" objects, which do not put out IR energy, cannot be sensed by an IR sensor and hence in comparison to a visible spectrum image of the scene may lack information or desired detail as well as introduce unwanted detail. This lack of detail can be observed by comparing Figure 11.8a and c and noting that the IR image would not change much in clear conditions. Figure 11.11 also shows another pair of IR and visible spectrum images.

11.2.3 MMW Radar Sensors

MMW radar is another good sensor for imaging in atmospheric conditions. One reason is that compared to many sensors, it can sense at a relatively long range. MMW radar works by emitting a beam of electromagnetic waves. The beam is scanned over the scene and the reflected intensity of radiation is recorded as a function of return time. The return time is correlated with range, and this information is used to create a range image.

In MMW radar, the frequency of the beam allows it in comparison to many other sensors pass by atmospheric particles because they are too small to affect the beam. MMW radar can thus "penetrate" atmospheric layers like smoke, fog, etc. However, because of the frequency, the resolution of the image produced compared to many other sensors is poorer as evidenced by Figure 11.7c, which shows the MMW radar image produced along side of the photometric and IR images of the same scene. In addition, it is measuring range and not reflected color or other parts of the spectrum that may be required for the task at hand.

11.2.4 LADAR Sensors

Another kind of radar system is that of laser radar or LADAR. LADAR sensors uses shorter wavelengths than other radar systems (e.g., MMW) and can thus achieve better resolution. Depending on the application, this resolution may be a requirement. A LADAR sensor sends out a laser beam and the time of return in the reflection measures the distance from the scene point. Through scanning of the scene, a range image is created. As the laser beam propagates through the atmosphere, fog droplets and raindrops cause image degradation, which are manifested as either dropouts (meaning not enough reflection is returned to be registered) or false returns (the beam is returned from the atmospheric particle itself). [Campbell 98] studies this phenomenon for various weather conditions yielding a number of performance plots. One conclusion was that for false returns to occur, the atmospheric moisture (rain) droplets had to be at least 3 mm in diameter. This work was done with a 1.06 μm wavelength LADAR sensor. Results would change with changing wavelength. The produced performance plots can be used as thresholds in determining predicted performance of the system given current atmospheric conditions and could potentially be used to select image-processing algorithms to improve image quality.

11.2.5 Multispectral Sensors

Satellite imagery systems often encounter problems with analysis and clarity of original images due to atmospheric conditions. Satellites must penetrate through many layers of the Earth's atmosphere in order to view ground or near-ground scenes. Hence, this problem has been given great attention in research. Most of this work involves various remote sensing applications like terrain mapping, vegetation monitoring, and weather prediction and monitoring systems. However, with high spatial resolution satellites, other less "remote" applications like surveillance and detailed mapping can suffer from atmospheric effects.

11.2.6 Sonar Sensors

Sonar sensors utilize sound in its creation of measurements. In this case, the atmospheric effects that may be corrected can include noise in the environment but physical atmospheric elements also can affect the filtration of sound before it reaches the sonar sensors. In underwater environments where sonar is popular, sounds from boats and other vehicles, pipes and other cables, can influence reception. In this case, even the amount of salt in the water can influence the speed of sound slightly. Small particles in the water can cause sonar scattering and is analogous to the effects of atmospheric fog in above water (air) environments.

11.3 Physics-Based Solutions

As discussed earlier, physics-based algorithm is one of the two classifications that atmospheric detection and correction systems can be grouped into. In this section, we describe a few of the systems that use physics-based models. Most of these systems consider only one kind of sensor. Hence, in heterogeneous distributed sensor networks, one would have to apply the appropriate technique for each different kind of sensor used. The reader should read Section 11.1.1 for a discussion on atmosphere and physics-based scattering before reading this section.

[Narasimhan 02] discuss in detail the issue of atmosphere and visible spectrum images. In this work, they develop a dichromatic atmospheric scattering model that tracks how images are affected by atmospheric particles as a function of color, distance, and environmental lighting interactions ("airlight" phenomenon, see Section 11.1.1). In general, when particle sizes are comparable to the wavelength of reflected object light, the transmitted light through the particle will have its spectral composition altered. In [Narasimhan O2] they consider fog and dense haze; the shifts in the spectral composition for the visible spectrum are minimal and hence they assume that the hue of the transmitted light to be independent of the depth of the atmosphere. They postulate that with an overcast sky the hue of "airlight" depends on the particle size distribution and tends to be gray or light blue in the case of haze and fog. Recall, "airlight" is the transmitted light that originated directly from environmental light like direct sunlight. Their model for the spectral distribution of light received by the observer is the sum of the distribution from the scene objects' reflected light, taking into account the attenuation phenomenon (see Section 11.1.1) and airlight. It is similar to the dichromatic reflectance model in [Shafer 85] that describes the spectral effects of diffuse and specular surface reflections.

[Narasimhan 02] reduce their wavelength-based equations to the red, green, and blue color space and in doing so show that equation relating the received color remains a linear combination of the scene object transmitted light color and the environmental "airlight" color. Narasimhan and Nayar (2002) hypothesize that a simple way of reducing the effect of atmosphere on an image would be to subtract the "airlight" color component from the image.

Hence, [Narasimhan 02] discuss how to measure the "airlight" color component. A color component (like "airlight") is described by a color unit vector and its magnitude. The color unit vector for "airlight" can be estimated as the unit vector of the average color in an area of the image that should be registered as black. However, this kind of calibration may not be possible and they discuss a method of computing it using all of the color pixels values in an image formulated as an optimization problem.

Using the estimate of the "airlight" color component's unit vector and given the magnitude of this component at just one point in an image, as well as two perfectly registered images of the scene, [Narasimhan 02] are able to calculate the "airlight" component at each pixel. Subtracting the "airlight" color component at each pixel value yields an atmospheric-corrected image. It is also sufficient to know the true transmission color component at one point to perform this process. Figure 11.9 shows the results of this process. This system requires multiple registered images of the scene as well as true color information at one pixel in the image. If this is not possible, a different technique should be applied.

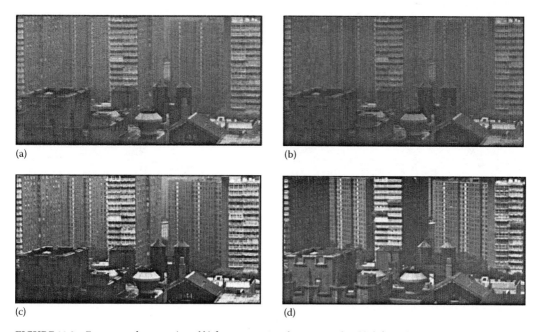

(a)   (b)

(c)   (d)

FIGURE 11.9   Fog removal system: (a and b) foggy images under overcast day, (c) defogged image, and (d) image taken on clear day under partly cloudy sky. (From Narasimhan, S. and Nayar, S., *Int. J. Comput. Vision*, 48(3), 233, 2002.)

Obviously, this technique works for visible spectrum images and under the assumption that the atmospheric condition does not (significantly) alter the color of the received light.

Scattering can change for some aerosols with wavelength and the work in [Narasimhan 03a] is an extension of the system in [Narasimhan 02] that addresses this issue. Another limitation of the previous system [Narasimhan 02] is that the model makes it ambiguous for scene points whose color matches the fog color. In [Narashimhan 03b], the idea that parts of the image at the same scene depth should be enhanced separately from other parts of the image at different scene depths is explored. This relates to the modeling of "airlight" as being more intense with greater scene depth as illustrated by Equation 11.4. The authors describe techniques including detection of depth and reflectance edges in the scene to guide scene depth image map creation. Figure 11.10 shows some results where (a) and (b) show the two images of the scene to be processed, (c) is the calculated depth image map, and (d) the resulting restored image after removal of the depth-related estimate of "airlight."

[Narashimhan 03b] discuss an approach where the user is in loop and specifies information that guides the system in how to perform atmospheric reduction. The addition of the user information illuminates the need for two registered images and this system only uses a single image. The user specifies an initial area in the image that is not corrupted by atmosphere and also an area that contains the color of the atmosphere (an area filled with fog). This is followed by the user specifying some depth information including a vanishing point and a minimum and maximum distance in the scene. From all of this user-specified information, the airlight component is calculated and the image corrected. This technique is heavily reliant on a not-insignificant amount of information for each image processed and is only viable in situations when this is possible.

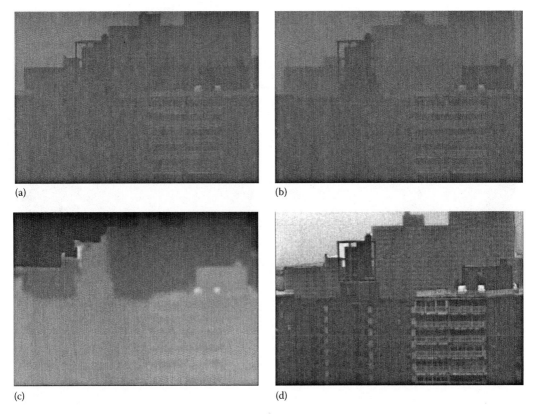

(a) (b) (c) (d)

FIGURE 11.10 Depth-related fog removal: (a and b) two original images of same scene, (c) calculated depth image map, and (d) restored image. (From Narasimhan, S. and Nayar, S., *Pattern Anal. Mach. Intell.*, 25(6), 713, 2003a.)

Another system that has the user in the loop is in [Hautiere 07] where an initial calculation of a horizontal line in an image from a car camera is taken using a calibration process involving an inertial sensor and pitching of the car. By doing this (given the more restricted environment of a car camera), the authors [Hautiere 07] reduce the amount of user input necessary over the system of [Narashimhan 03b].

The system in [Hautiere 07] is also interesting in that it has an iterative correction algorithm that includes an object detection subsystem that looks to find regions containing cars. This is an example of a system that looks to correct the effects of atmosphere differently in object and nonobject regions. This is an intriguing idea but only makes sense when there are objects of interest for the system in the scene. Figure 11.11 shows the results of this algorithm, which does initial restoration on the entire image and

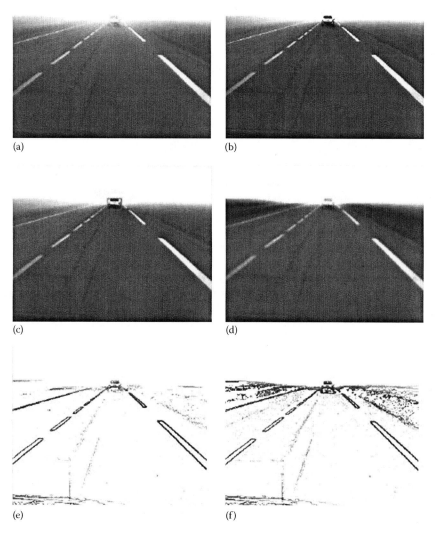

(a)   (b)

(c)   (d)

(e)   (f)

FIGURE 11.11 System to reduce effects of fog in vehicle images using an iterative contrast restoration scheme that utilizes, in part, the detection of vehicles in the scene: (a) original image, (b) restored image after initial restoration, (c) vehicle detected using image in (b), (d) final restored image with vehicle properly restored, (e) edge image of original image, and (f) edge image of restored image (d) showing improved contrast in image. (From Hautiere, N. et al., Towards fog-free in-vehicle vision systems through contrast restoration, *IEEE Conference on Computer Vision and Pattern Recognition, CVPR'07*, pp. 1–8, 2007.)

then detects vehicles by putting bounding boxes on them. The restoration continues but is different in nonvehicle areas of the image over vehicle areas. The vehicle detection algorithm is applied potentially multiple times in the iteration loop of restoration, detection and repeat. Some question as to how to control this process and when to stop are not clearly addressed.

Another system that utilizes only a single image is [Tan 07] system, which estimates the "airlight" component using color and intensity information of only a single image. An interesting point made in this chapter is that estimating "airlight" from a single image is ill posed and their goal is not the absolute measurement of "airlight" but the improvement of visibility in the image. The system has two parts: first is the normalization of the pixel values by the color of the scene illumination and second the estimation of "airlight" and its removal.

The [Tan 07] system begins by estimating the scene illumination color, which, they show, can be represented as the linear intercept in a special transform space called the inverse intensity chromaticity space for all pixels that come from "airlight." Hence, the top 20% intensity pixels are assumed to belong to the fog (airlight) and are mapped into the inverse intensity chromaticity space. The Hough transform is used to find the intercept in this space that is used to estimate the scene illumination color. For theoretical details of the equations relating the scene illumination color and the intercept of the inverse chromaticity space, the reader is referred to [Tan 07]. Of course, one difficulty of this part of the system is the accuracy of the intercept detection as well as the empirical choice of the top 20% intense pixels corresponding to fog pixels. The system normalizes the pixels by this estimated scene illumination color. The result is hypothesized to be an image illuminated by a "white" illumination source.

In a second part of the [Tan 07] system, the "airlight" component at each pixel is estimated using the Y (brightness) metric from the YIQ transform. In this method, the maximum intensity pixel found in the image is assumed to represent the intensity of the scene illumination. Note that the brightest pixel is assumed to be furthest away in the scene (i.e., the sky or similar distant point). Using the Y value at each pixel and the scene illumination intensity (maximum value in image), the relative distance (the scene illumination intensity represents maximum distance) of a pixel's corresponding scene point can be estimated. The "airlight" component at a pixel is calculated and subtracted from the pixel value yielding a restored image. Figure 11.12 shows some results in heavy fog. This method relies on the assumption that the Y value of each pixel represents the distant proportional "airlight" component of this pixel. As the authors state, finding the absolute value of the "airlight" component is not possible in single-image analysis but rather the improved visibility is the goal.

(a)   (b)

FIGURE 11.12 Results shown in heavy fog: (a) original image and (b) restored image. (From Tan, R. et al., Visibility enhancement for roads with foggy or hazy scenes, *IEEE Intelligent Vehicles Symposium*, Istanbul, Turkey, pp. 19–24, 2007.)

Because solving Equation 11.4 with only one image as discussed in [Tan 07] is ill posed, the single-image systems we have seen impose some constraints or assumptions to solve this problem or in systems with a user in the loop rely on the user to provide additional information. Sometimes the constraints will be assumptions about the depth of the scene d(x) or atmospheric luminance Linf or both. These constraints as discussed in [Tan 07] yield not the ideal results but can improve results.

A question that naturally arises is what system is best. Unfortunately, there is no theoretically best solution. Beyond an understanding of what a candidate system does, there is the comparison of results. This can be even more critical for single-image systems where typically there are a greater number of constraints or assumptions that must be taken. [Yu 10] shows a comparison of their system on some images with that of [Tarel 09] as shown in Figure 11.13. It is not clear from the upper row, middle image of Figure 11.13 if it is better to recover the water behind, leaving an unusual clumping of fog around the palm trees, or to leave the fog in front of the water as shown in the images on the right. The answer to the question may be a function of what your system is trying to do.

For other variations on the use of single images in physics-based atmospheric reduction, see [Tan 08], [Fattal 08], and [Guo 10].

In [Richter 98], a method for correcting satellite imagery taken over mountainous terrain has been developed to remove atmospheric and topographic effects. The algorithm accounts for horizontally varying atmospheric conditions and also includes the height dependence of the atmospheric radiance and transmittance functions to simulate the simplified properties of a three-dimensional (3D) atmosphere. A database was compiled that contains the results of radiative transfer calculations for a wide range of weather conditions. A digital elevation model is used to obtain information about the surface elevation, slope, and orientation. Based on Lambertian assumptions, the surface reflectance in rugged terrain is calculated for the specified atmospheric conditions. Regions with extreme illumination

FIGURE 11.13 Images on left are original. Images in middle are results from [Tarel 09] system and images on right are results from [Yu 10] system. (From Tarel, J. and Hauti, N., *Fast visibility restoration from a single color or gray level image, Proceedings of the IEEE International Conference on Computer Vision*, Kyoto, Japan, 2009; Yu, J. et al., Physics-based fast single image fog removal, *IEEE 10th International Conference on Signal Processing (ICSP)*, Beijing, China, pp. 1048–1052, 2010.)

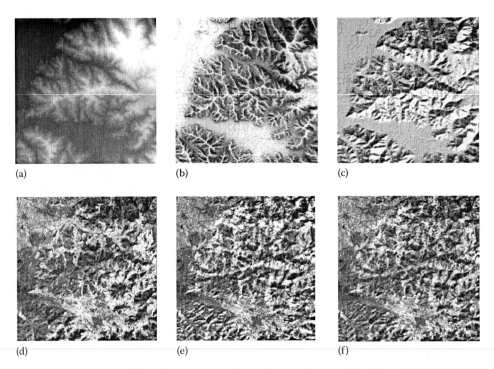

(a)   (b)   (c)

(d)   (e)   (f)

FIGURE 11.14 System to model and remove atmospheric effects and topographic effects: (a) DEM-digital elevation model, (b) sky view factor, (c) illumination image, (d) original TM band 4 image, (e) reflectance image without processing of low illumination areas, and (f) reflectance image with processing of low illumination areas. (From Richter, R., *Appl. Opt.*, 37(18), 4004, 1998.)

geometries sensitive to bidirectional reflectance distribution function (BRDF) effects can be processed separately. This method works for high spatial resolution satellite sensor data with small swath angles.

Figure 11.14 shows the results of the method in [Richter 98] on satellite imagery.

While our previous discussion of sensors focused on the issue of how atmosphere affects them, in [Schechner 01] the use of a rotating polarizer attached to the lens of a camera is used to reduce the effects of fog. This is done by "dehazing" the images using polarization. It works on the premise that light scattered by atmospheric particles (like fog) is partially polarized. The approach assumes that the "foggy/hazy" image is composed of a combination of light from the objects in the scene and "airlight," which is the light that is coming from scattering effect on the atmospheric particles. To measure both and attempt to remove the "airlight," two different polarized images are gathered of the scene, and, given the "airlight" will be partially polarized, it is detected and removed from the image. See [Schechner 01] for derivations. An interesting by-product of these calculations from the two images is the creation of a 3D range map. It should be noted that as the system depends on polarization, as the light becomes less polarized, the system's performance will degrade. This can occur in situations of overcast lighting and in situations of heavy fog and haze. However, rain is heavily polarizing and this system may work well in that condition (see [Shwartz 06] for a continuation of this work).

Some work looks at the detection of particular kinds of atmosphere, which can potentially yield to its removal or correction. In [Halimeh 09], a system to detect raindrops in vehicle cameras is discussed. Figure 11.15a shows the geometry involved in the geometric–photometric model of a raindrop the authors create. This model is used in the detection of raindrops illustrated from [Halimeh 09] in the remaining images of Figure 11.15. Figure 11.15c shows input images and the results of using the model. It contains a sequence of images in two environments: the upper sequence is for an artificial background and the lower is in a real environment. Note that the top images in a sequence represent the

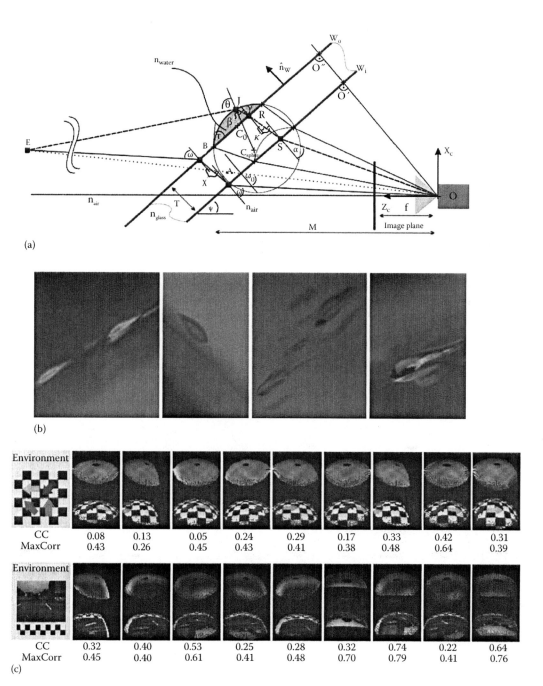

FIGURE 11.15 System to detect raindrops in images from vehicle cameras: (a) geometry behind geometric–photometric model, (b) sample images of raindrops on car windshield, and (c) sequence of images with an artificial background and a real environment background; the upper images are the observed (real) and lower are constructed by the model. (From Halimeh, J. and Roser, M., Raindrop detection on car windshields using geometric–photometric environment construction and intensity-based correlation, *2009 IEEE Intelligent Vehicles Symposium*, Xi'an, China, pp. 610–615, 2009.)

FIGURE 11.16 Results for fog reduction in images. (From Desai, N. et al., A fuzzy logic based approach to de-weather fog-degraded images, *Sixth International Conference on Computer Graphics, Imaging and Visualization,* Tianjin, China, pp. 383–387, 2009.)

real, observed data and the bottom contains the results using the model developed by the authors. The parameters CC and MaxCC represent correlation coefficient measurements between the real, observed data and the model's data. Note that even in the best (highest) correlations the model produces "crisper" results. This is, as the authors' state, due to the fact that they have not captured the lens type properties of the raindrop in the focusing of light. However, they conjecture that even so their system produces good results. Another system that looks at raindrop detection can be found in [Görmer 09].

Straddling somewhere between a physics-based approach and a heuristic/non-physics-based approach (topic of the next section) is the work presented in [Desai 09]. The authors take the point of view that many of the physics-based solutions are too complex and other methods that require multiple images again take too much time for some real-time applications like driving. They propose their "simpler" physics-based solution that uses fuzzy logic on single images as a good approach for real-time applications for fog removal. The light scattering caused by fog is as we saw previously mathematically modeled as the linear combination of directly transmitted light (no effect from fog) plus "airlight," which is the light received that is scattered on the fog particles in the air.

Recall, the linear model is such that the effect from "airlight" (fog) increases as the distance to the object increases. Thus, they attempt to measure the "airlight" in the scene by image points corresponding to distant scene points. This system attempts to find "sky" and assumes that the image will contain this as a component (note this is a restriction of this system). Utilizing the idea that sky is at the top of the image (again a restriction—but reasonable in most images containing sky), an image region is mapped to one of three fuzzy sets top, middle, or bottom. Based on pixel information, pixels are further mapped to fuzzy sets reflecting how much "airlight" component is in it as "high," "moderate," or "little." The system estimates "airlight" in this way and globally reduces this effect by dividing the image intensity by the average light estimate in the "airlight" areas.

The next stage of the system is to perform color correction on each pixel that is dependent on its nearby neighbors. Figure 11.16 shows some results of this system.

11.4 Heuristics and Non-Physics-Based Solutions

Heuristic and non-physics-based approaches represent the other paradigm of atmospheric detection and correction algorithms. These algorithms do not attempt to directly model the physics behind the atmosphere. There is a wide range of algorithms: some based on empirical data, others on observations, and others that alter the sensing system itself.

Work has been done to specifically detect image areas where various atmospheric conditions are present with the idea that further processing to enhance the image in these areas could then be done.

TABLE 11.1 MTI Bands

Band	Band No.	Range (μm)	Resolution (m)	Function
A	1	0.45–0.52	5	Blue, true color
B	2	0.52–0.60	5	Green, true color
C	3	0.62–0.68	5	Red, true color
D	4	0.76–0.86	5	Vegetation
E	5	0.86–0.89	20	Water vapor
F	6	0.91–0.97	20	Water vapor
G	7	0.99–1.04	20	Water vapor
H	8	1.36–1.39	20	Cirrus
I	9	1.54–1.75	20	Surface
O	10	2.08–2.37	20	Surface
J	11	3.49–4.10	20	Atmosphere
K	12	4.85–5.05	20	Atmosphere
L	13	8.01–8.39	20	Surface
M	14	8.42–8.83	20	Surface
N	15	10.15–10.7	20	Surface

In particular, there is a body of research on cloud detection. After detection, some of these systems attempt to eliminate the clouds while most use this information for weather analysis.

In [Rohde 01], a system is discussed to detect dense clouds in daytime multispectral thermal imager (MTI) satellite images. The [Rohde 01] system uses 15 spectral bands (images) shown in Table 11.1 that ranges from visible wavelengths to longwave IR. Rhode et al. hypothesize that clouds in the visible wavelength spectrum are bright and evenly reflect much of the wavelengths from visible to near-IR. Also, clouds appear higher in the atmosphere than most image features and are hence colder and drier than other features. Recall that the IR spectrum is a measure of temperature. Also, for the spatial resolution of their images, clouds cover large areas of the image. Given these characteristics, they have come up with a number of parameters that can be used to classify pixels as belonging to a cloud or not.

First, the pixels are subjected to a thresholding technique in both the visible and IR ranges that threshold on "clouds being bright" and "even-reflection" properties. A minimum brightness value is selected, and, in the visible range (band C), any pixel above this brightness value is retained as a potential cloud pixel; all others are rejected. In the IR band N, an upper limit on temperature is given and used to reject pixels as potential cloud pixels.

Next, the "whiteness" property is tested by using a ratio of the difference of the E and C bands to their sum. Evenly reflected or "white" pixels will have a ratio around zero. Only pixels with ratios near zero will be retained as potential cloud pixels. The last thresholding operation is done using CIBR, which stands for continuum interpolated band ratio (see [Gao 90] for details). CIBR can be used as a measure of the "wetness" (and thus "dryness") in the IR spectrum. CIBR is a ratio of band F to a linear combination of bands E and G.

At this point, the system has a set of pixels that conform to being "bright," "white," "cold," and "dry." What is left is to group nearby pixels and test if they form regions large enough to indicate a cloud. This is accomplished through the removal of blobs too small to indicate clouds via a morphological opening operator (see Chapter 5). The final result is a binary "cloud map" where the nonzero pixels represent cloud pixels. The system seems to work well on the presented images, but its performance is tuned to the spatial resolution of the images and is not extensible to non-MTI like satellite imagery. However, combinations of photometric images with IR camera images could use these techniques and there do exist a number of sensor networks with these two kinds of imaging devices.

Another system dealing with satellite imagery is presented in [Hu 09] and [He 10], where a three-step process of haze detection, perfection, and removal takes place using filtering of particular spectra of

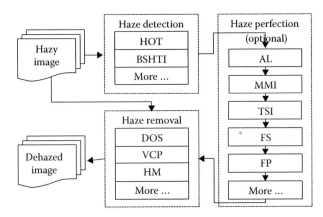

FIGURE 11.17 System takes toolkit approach where user selects different algorithms to apply. (From Hu, H. et al., A haze removal module for multi-spectral satellite imagery, *Proceedings of the Urban Remote Sensing Joint Event*, Shanghai, China, 2009.)

multispectrum images as shown in Figure 11.17. In each module, the authors propose a set of possible filtering algorithms that can be chosen by the user. In this sense, it is a toolkit of possibilities and as the user needs to select the options, the authors suggest further work should be done to automate this selection process. Many of the algorithms mentioned in Figure 11.17 are currently popular for the stages of haze detection, "perfection," and removal in multispectral data and for this point is an interesting review.

The first stage of [Hu 09] is to estimate how much haze is present in the image. This potentially could be useful in the choice and tuning of the algorithm to remove the haze. One algorithm the authors feel is particularly successful in detecting haze is their extension of the haze optimization transformation (HOT) originally described in [Zhang 02], which performs its analysis in the red and blue visual bands based on the assumption that these bands are highly correlated in clear (no haze) conditions. When there is a deviation in this correlation, the hypothesis is that it is due to haze. The authors extend HOT to an algorithm they call AHOT [He 10] that uses spatial information to reduce some of the problems when the land cover is either bare soil or water with the original HOT algorithm. BSHTI (see Figure 11.17) stands for background suppressed haze thickness index and is another alternative to HOT and AHOT in the haze detection toolkit.

The second stage of haze "perfection" is an optional stage the user may choose not to employ. The purpose of this "perfection" is to rectify the detection stage in any poor detection cases caused by some kinds of land covers. A suite of possible algorithms are proposed that use spatial filtering to guide this rectification process including an algorithm to adjust bias of each land cover type in TM imagery (AL in Figure 11.17); masking areas manually followed by a smoothing interpolation (MMI in Figure 11.17); thresholding, segmentation, and interpolation (TSI in Figure 11.17); a fill sink operation that interprets results from previous haze detection algorithm as a digital elevation map (DEM) and fills in low spots in the DEM assuming them to be erroneous extremes (FS in Figure 11.17) [He 10]; and a flatten peak operation that similarly represents pervious haze detection results as a DEM and this time flattens peaks in the DEM if the peak is surrounded by sharp areas of contrast assuming these peaks to be areas of pseudo-haze-like water, which have sharp boundaries where it meets with land (FP in Figure 11.17) [He 10].

The final stage is to remove the haze previously detected and again we see the authors suggesting a toolkit of options including dark object subtraction (DOS in Figure 11.17) [Zhang 02], virtual cloud point removal (VCP in Figure 11.17), and histogram matching (HM in Figure 11.17). Figure 11.18 shows some results: the image on the left is the original and the image on the right has the HOT algorithm used for the detection phase, followed by the fill sink and flatten peak algorithms for the haze "perfection" phase, and finally the virtual cloud point removal algorithm [He 10] to remove the detected haze (see [Chen 09] for another system that uses filtering in multispectral imagery).

FIGURE 11.18 (a) Original. (b) Results processing using HOT + (FS + FP) + VCP algorithms of system shown in Figure 11.17. (From Hu, H. et al., A haze removal module for multi-spectral satellite imagery, *Proceedings of the Urban Remote Sensing Joint Event*, Shanghai, China, 2009.)

FIGURE 11.19 Results of fusion system for atmospheric correction. (a) One of two original foggy images, visible spectrum, (b) fussed image from two foggy images, (c) blowup of portion of an original foggy image, and (d) corresponding region to image (c) for fused image. (From Grewe, L. et al., Atmospheric attenuation through multi-sensor fusion, *SPIE AeroSense: Sensor Fusion: Architectures, Algorithms, and Applications II*, April 1998.)

[Grewe 01] and [Grewe 98] describe the creation of a system for correction of images in atmospheric conditions. [Grewe 98] describes a system that uses multi-image fusion to reduce atmospheric attenuation in visible spectrum images. The premise is that atmospheric conditions like fog, rain, and smoke are transient. Airflow moves the atmospheric particles in time such that at one moment certain areas of a scene are clearly imaged, while at other moments other areas are visible. By fusing multiple images taken at different times, we may be able to improve the quality. Figure 11.19 shows typical results of the system when only two images are fused. It was discovered that the fusion engine parameters are a function of the type and level of atmosphere in the scene. Hence, this system first detects the level and type of atmospheric conditions in the image [Grewe 01] and this is used to select which fusion engine to apply to the image set. The system uses a wavelet transform to both [Brooks 01] detect atmospheric conditions as well as fuse multiple images of the same scene to reduce these effects. A neural network is trained to detect the level and type of atmosphere using a combination of wavelet and spatial features like brightness, focus, and scale changes. Typical results are shown in Figure 11.20.

In [Honda 02], a system is developed to detect moving objects in a time-series sequence of images that is applied to cloud. Again the fact that clouds move through time is taken advantage of. Here a motion detection algorithm is applied. This work only looks at extracting these clouds for further processing for weather applications; no correction phase takes place.

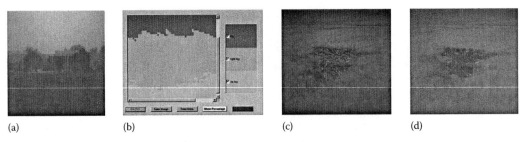

(a)   (b)   (c)   (d)

FIGURE 11.20 Detection of level and type of atmospheric conditions. (a) Original image; (b) application showing pseudo-shaded (original colored), light gray, no fog; medium gray, light fog; dark gray, fog; (c) original image with cloud cover; (d) superimposed pseudo-shaded (original colored) detection map where light gray represents heavy atmosphere (cloud) and gray represents no atmosphere. (From Grewe, L., Detection of atmospheric conditions in images, *Proceedings of the SPIE AeroSense: Signal Processing, Sensor Fusion and Target Recognition*, April 2001.)

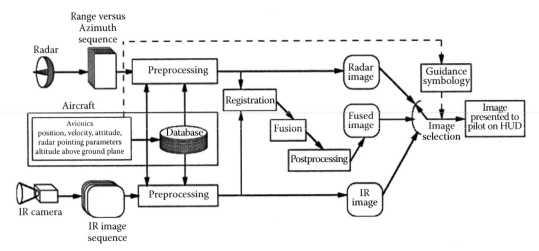

FIGURE 11.21 Components of a sensor system using IR and MMW radar to aid in landing aircraft in low-visibility conditions like fog. (From Sweet, B. and Tiana, C., Image processing and fusion for landing guidance, *Proceedings of SPIE*, Vol. 2736, pp. 84–95, 1996.)

There are also systems that use sensor selection to avoid atmospheric distortions. Sweet and Tiana discuss a system in which the sensors are selected to penetrate obscuring visual phenomena such as fog, snow, and smoke for the application of enabling aircraft landings in low-visibility conditions [Sweet 96]. In particular, the use of IR and MMW imaging radar is investigated.

Figure 11.21 shows the components of the [Sweet 96] system, and Figure 11.22 shows some images from that system. Both the IR and MMW radar images go through a preprocessing stage followed by registration and fusion. The pilot through a head-up display is able to select the IR, MMW, or fused images to view for assistance in landing.

[Burt 93] discuss a system similar to [Sweet 96], but MMW images are not used in the fusion process. Also, in [Pucher 10], a system is developed that fuses video and sound for highway monitoring, which the authors suggest may be an improvement over video only in fog conditions.

Another possible solution to the atmospheric correction problem is to treat it more generically as an image restoration problem. In this case, there is some similarity with blurred images and noisy images. There are a number of techniques, some discussed in Chapter 5, like deblurring or sharpening filters that could improve image quality. See also [Banham 96] and [Kundur 96] for algorithms for deblurring images.

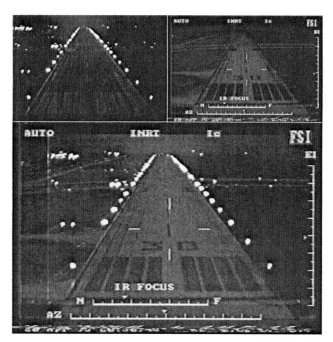

FIGURE 11.22 (Upper left) visible spectrum image, (upper right) IR image, and (lower) fused IR and visible spectrum images. (From Sweet, B. and Tiana, C., Image processing and fusion for landing guidance, *Proceedings of SPIE*, Vol. 2736, pp. 84–95, 1996.)

One system that treats atmospheric correction as an image restoration problem is [Joshi 10] where they discuss the "denoising" of the image, particularly with long-range scene points through the use of image restoration techniques. They discus a local weighted averaging method based on ideas of "lucky imaging" (an astronomy technique) to minimize blur and alignment errors evident in the haze/fog-filled images. An interesting point of this work is they consider not only the atmospheric attenuation effects but also the blurring that is caused and, in the case of multiple images taken over a time sequence, the misalignment that can occur. The alignment issue is taken care of with a simple registration process that is assisted by the fact that the images under consideration are of far-distant objects, like the mountain image shown in Figure 11.23.

After registration of the multiple images, a single new "restored" image is created through a special weighting of the images, which favors pixels in an image that contain a higher value of local sharpness

(a)   (b)

FIGURE 11.23 (a) One of the original images in the temporal input set. (b) Restored image. (From Joshi, N. and Cohen, M., Seeing Mt. Rainier: Lucky imaging for multi-image denoising, sharpening, and haze removal, *IEEE Conference on Computational Photography*, Cambridge, MA, March 28–30, 2010.)

that is measured with the Laplacian function (see [Joshi 10] for further details). Figure 11.23 shows some results of this system.

Related to the work in [Grewe 01] is the topic of high dynamic range imaging [Reinhard 05], which is discussed more in Chapter 5. It relates to the fusion of images of different exposures. This could be an extension to the [Grewe 01] that could be potentially applied for use in atmospheric correction and not just variation in lighting conditions.

11.5 Conclusion

Unless you can effectively model the atmosphere, a very challenging problem, heuristics and non-physics-based solutions may be the only viable solutions for a distributed sensor network. Even in the case of accurate modeling, the model would need to be dynamic and alter with the temporal changes to the atmosphere.

Very little work has been done in comparing techniques. However, in [Nikolakopoulos 02], the authors compare two algorithms: one that explicitly models the atmosphere using a number of environmental parameters and another that uses a heuristic approach. The tests were done on multispectral data, and they found superior results for the model-based technique. Unfortunately, this work only compares two very specific algorithms and the heuristic approach is simplistic as it involves only histogram shifting. So, this should not be taken as an indication that model-based techniques are superior. If the assumptions and measurements made in the model are accurate, it is reasonable to assume that model-based techniques will give good results. The information you have, the environments you wish your system to operate in, and, finally, empirical testing is what is required to select the best atmospheric correction technique for your system.

References

[Ahlen 05] Ahlen, J., D. Sundren, T. Lindell, and E. Bengtsson, Dissolved organic matters impact on colour reconstruction in underwater images, *Proceedings of the 14th Scandinavian Conference on Image Analysis*, Joensuu, Finland, pp. 1148–1156, June 19–22, 2005.

[Arya 97] Arya, V., P. Duncan, M. Devries, and R. Claus, Optical fiber sensors for monitoring visibility, *Proceedings of the SPIE Transportation Sensors and Controls: Collisions Avoidance, Traffic Management, and ITS*, pp. 212–218, February 1997. Available at: http://spie.org/x648.html? product_id=267147.

[Banham 96] Banham, M. and A. Katasaggelos, Spatially adaptive wavelet-based multiscale image restoration, *IEEE Transactions on Image Processing*, 5(4), 619–634, 1996.

[Barducci 95] Barducci, A. and I. Pippi, Retrieval of atmospheric parameters from hyperspectral image data, *International Geoscience and Remote Sensing Symposium*, Florence, Italy, pp. 138–140, July 1995.

[Bohren 89] Bohren, C., Selected papers on scattering in the atmosphere, Milestone series Vol. MS07, October 1, 1989, ISBN-10:0819402400; ISBN-13:978-0819402400.

[Brooks 01] Brooks, R., L. Grewe, and S. Iyengar, Recognition in the wavelet domain, *Journal of Electronic Imaging*, 10(03), pp. 757–784, July 2001.

[Brooks 98] Brooks, R. R. and S. S. Iyengar, *Multi-Sensor Fusion: Fundamentals and Applications with Software*, Prentice Hall PTR, Saddle River, NJ, 1998.

[Burt 93] Burt, P. and R. Kolezynski, Enhanced image capture through fusion, *IEEE 4th International Conference on Computer Vision*, Berlin, Germany, pp. 173–182, 1993.

[Campbell 98] Campbell, K., Performance of imaging laser radar in rain and fog, Master's thesis, Wright-Patterson AFB, Dayton, OH, 1998.

[Chambah 04] Chambah, M., A. Renouf, D. Semani, P. Courtellemont, and A. Rizzi, Underwater colour constancy: Enhancement of automatic live fish recognition, *Electronic Imaging*, 5293, 157–168, 2004.

[Chen 09] Chen, W., Z. Zhang, Y. Wang, and X. Wen, Atmospheric correction of SPOT5 land surface imagery, *CISP '09, 2nd International Congress on Image and Signal Processing*, Tianjing, China, pp. 1–5, 2009.

[Clem 02] Clem, T., Sensor technologies for hunting buried sea mines, *Proceedings of the MTS/IEEE Oceans*, Biloxi, MS, pp. 29–31, 2002.

[Clem 03] Clem, T. and J. Lopes, Progress in the development of buried minehunting systems, *Proceedings of the MTS/IEEE Oceans*, San Diego, CA, pp. 22–26, 2003.

[Curry 02] Curry, J., J. Pyle, and J. Holton, *Encyclopedia of Atmospheric Sciences*, Academic Press, Waltham, MA, 2002.

[Day 98] Day, J., V. Schaefer, and C. Day, *A Field Guide to the Atmosphere*, Houghton Mifflin Co., Boston, MA, 1998.

[Desai 09] Desai, N., A. Chatterjee, S. Mishra, D. Chudasama, S. Choudhary, and S. Barai, A fuzzy logic based approach to de-weather fog-degraded images, *Sixth International Conference on Computer Graphics, Imaging and Visualization*, Tianjin, China, pp. 383–387, 2009.

[Fattal 08] Fattal, R., Single image dehazing, *ACM Transactions on Graphics*, 27(3), pp. 1–9, August 2008.

[Gao 90] Gao, B. and A. Goetz, Column atmospheric water vapor and vegetation liquid water retrievals from airborne imaging spectrometer data, *Journal of Geophysical Research*, 95, 3549–3564, 1990.

[Garcia 02] Garcia, R., T. Nicosevici, and X. Cull, On the way to solve lighting problems in underwater imaging, *Proceedings of the IEEE OCEANS Conference*, pp. 1018–1024, 2002.

[Görmer 09] Görmer, S., A. Kummert, S. Park, and P. Egbert, Vision-based rain sensing with an in-vehicle camera, *IEEE Intelligent Vehicle Symposium*, Xi'an, China, 2009.

[Grewe 01] Grewe, L., Detection of atmospheric conditions in images, *Proceedings of the SPIE AeroSense: Signal Processing, Sensor Fusion and Target Recognition*, Orlando, FL, pp 757–784, April 2001.

[Grewe 98] Grewe, L., R. Brooks, and K. Scott, Atmospheric attenuation through multi-sensor fusion, *SPIE AeroSense: Sensor Fusion: Architectures, Algorithms, and Applications II*, Orlando, FL, pp. 102–109, April 1998.

[Guo 10] Guo, F., Z. Cai, B. Xie, and J. Tang, Automatic image haze removal based on luminance component, *International Conference on Wireless Communications Networking and Mobile Computing (WiCOM)*, Chengdu pp. 1–4, China, 2010.

[Halimeh 09] Halimeh, J. and M. Roser, Raindrop detection on car windshields using geometric-photometric environment construction and intensity-based correlation, *2009 IEEE Intelligent Vehicles Symposium*, Xi'an, China, pp. 610–615, 2009.

[Hautiere 07] Hautiere, N., J. Tarel, and D. Aubert, Towards fog-free in-vehicle vision systems through contrast restoration, *IEEE Conference on Computer Vision and Pattern Recognition, CVPR'07*, Minneapolis, Minnesota, pp. 1–8, 2007.

[He 10] He, X., J. Hu, W. Chen, and X. Li, Haze removal based on advanced haze optimized transformation (AHOT) for multispectral imagery, *International Journal of Remote Sensing*, 31(20), 5331–5348, 2010.

[Honda 02] Honda, R., S. Wang, T. Kikuchi, and O. Konishi, Mining of moving objects from time-series images and its application to satellite weather imagery, *Journal of Intelligent Information Systems*, 19(1), 79–93, 2002.

[Hosgood 03] Hosgood, B., Some examples of thermal infrared applications, IPSC website, http://humanitarian-security.jrc.it/demining/infrared_files/IR_show4/sld006.htm, 2003.

[Hu 09] Hu, H., W. Chen, L. Xiaoyu, and X. He, A haze removal module for multi-spectral satellite imagery, *Proceedings of the Urban Remote Sensing Joint Event*, Shanghai, pp. 1–4, China, 2009.

[Huxtable 97] Huxtable, B. et al., A synthetic aperture radar processing system for search and rescue, *Proceedings of the SPIE Automatic Target Recognition VII*, Orlando, FL, pp. 185–192, April 1997.

[NASA 03] IPAC, NASA, What is infrared? http://sirtf.caltech.edu/EPO/Kidszone/infrared.html, 2003.

[Iqbal 10] Iqbal, K., M. Odetayo, A. James, R. Salam, and A. Talib, Enhancing the low quality images using unsupervised colour correction method, *2010 IEEE International Conference on Systems, Man and Cybernetics*, Istanbul, Turkey, pp. 1703–1709, 2010.

[Joshi 10] Joshi, N. and M. Cohen, Seeing Mt. Rainier: Lucky imaging for multi-image denoising, sharpening, and haze removal, *IEEE Conference on Computational Photography*, Cambridge, MA, March 28–30, 2010.

[Koschmieder 24] Koschmieder, H., Theorie der horizontalen sichtweite, (*Beitr. Zur Phys. eL freien Atm.*) 12(33–53), 171–181, 1924.

[Kumar 06] Kumar, S., D. Skvoretz, M. Elbert, C. Moeller, R. Ostrom, A. Perry, A. Tzouris, S. Bennett, and P. Czipott, Real-time tracking gradiometer for use in an autonomous underwater vehicle for buried minehunting, *Proceedings of the MTS/IEEE Oceans*, Boston, pp. 2108–2111; MA, 2006.

[Kundur 96] Kundur, D. and D. Hatzinakos, Blind image deconvolution, *IEEE Signal Processing Magazine*, 13(3), 43–64, 1996.

[Kyle 91] Kyle, T., *Atmospheric Transmission, Emission and Scattering*, Pergamon Press, Oxford, U.K., 1991.

[McCartney 75] McCartney, E., *Optics of the Atmosphere: Scattering by Molecules and Particles*, John Wiley & Sons, New York, 1975.

[Mie 08] Mie, G., A contribution to the optics of turbid media, especially colloidal metallic suspensions, *Annals of Physics*, 25(4), 377–445, 1908.

[Moller 94] Moller, H. and G. Sachs, Synthetic vision for enhancing poor visibility flight operations, *IEEE AES Systems Magazine*, 9(3), 27–33, March 1994.

[Narasimhan 02] Narasimhan, S. and S. Nayar, Vision and the atmosphere, *International Journal of Computer Vision*, 48(3), 233–254, 2002.

[Narasimhan 03] Narasimhan, S. and S. Nayar, Contrast restoration of weather degraded images, *Pattern Analysis and Machine Intelligence*, 25(6), 713–724, 2003a.

[Narasimhan 03b] Narasimhan, S. G. and S. K. Nayar, Interactive de-weathering of an image using physical models, *IEEE Workshop on Color and Photometric Methods in Computer Vision (in Conjunction with ICCV)*, pp. 528–535, Nice, France, October 2003b.

[Narasimhan 03c] Narasimhan, S. G. and S. Nayar, Shedding light on the weather, *IEEE Conference on Computer Vision and Pattern Recognition*, Madison, Wisconsin, 2003c.

[Nikolakopoulos 02] Nikolakopoulos, K., D. Vaiopoulos, and G. Skiani, A comparative study of different atmospheric correction algorithms over an area with complex geomorphology in Western Peloponnese, Greece, *IEEE International Geoscience and Remote Sensing Symposium Proceedings*, Rome, Italy, Vol. 4, pp. 2492–2494, 2002.

[Oakley 96] Oakley, J., B. Satherley, C. Harrixon, and C. Xydeas, Enhancement of image sequences from a forward-looking airborne camera, *SPIE Image and Video Processing IV*, San Jose, CA, pp. 266–276, February 1996.

[Pencikoswki 96] Pencikowski, P., A low cost vehicle-mounted enhanced vision system comprised of a laser illuminator and range-gated camera, *SPIE Enhanced and Synthetic Vision*, pp. 222–227, April 1996.

[Pucher 10] Pucher, M., D. Schabus, P. Schallauer, P. Lypetskyy, Y. Graf, Fl. Rainer, H. Stadtschnitzer et al., Multimodal highway monitoring for robust incident detection, *IEEE Annual Conference on Intelligent Transportation Systems*, Madeira Island, Portugal, pp. 837–842, 2010.

[Reinhard 05] Reinhard, E., G. Ward, S. Pattanaik, and P. Debevec, *High Dynamic Range Imaging, Acquisition, Manipulation and Display*, Morgan Kaufmann, Waltham, MA, 2005.

[Richter 98] Richter, R., Correction of satellite imagery over mountainous terrain, *Applied Optics*, 37(18), 4004–4015, 1998.

[Rohde 01] Rohde, C., K. Hirsch, and A. Davis, Performance of the interactive procedures for daytime detection of dense clouds in the MTI pipeline, *Algorithms for Multispectral, Hyperspectral, and Ultraspectral Imagery VII*, SPIE, Orlando, FL, Vol. 4381, pp. 204–212, 2001.

[Sadot 95] Sadot, D., N. Kopeika, and S. Rotman, Target acquisition modeling for contrast-limited imaging: Effects of atmospheric blur and image restoration, *Journal of the Optical Society of America*, 12(11), 2401–2414, 1995.

[Schechner 04] Schechner, Y. and N. Karpel, Clear underwater vision, *Proceedings of the IEEE CVPR*, Washington, DC, Vol. 1, pp. 536–543, 2004.

[Schechner 01] Schechner, Y., S. Narasimhan, and S. Nayar, Instant dehazing of images using polarization, *IEEE Conference on Computer Vision and Pattern Recognition*, Kauai, HI pp. 325–332, 2001.

[Shafer 85] Shafer, S., Using color to separate reflection components, *Color Research and Applications,* 10(4), 210–218, 1985.

[Shwartz 06] Shwartz, S., E. Namer, and Y. Schechner, Blind haze separation, *IEEE Conference on Computer Vision and Pattern Recognition,* New York, NY, pp. 1984–1991, 2006.

[Sulzberger 09] Sulzberger, G., J. Bono, R. Manley, T. Clem, L. Vaizer, and R. Holtzapple, Hunting sea mines with UUV-based magnetic and electro-optic sensors, *OCEANS 2009,* Biloxi, MS, pp. 1–5, 2009.

[Sweet 96] Sweet, B. and C. Tiana, Image processing and fusion for landing guidance, *Proceedings of SPIE,* Orlando, FL, Vol. 2736, pp. 84–95, 1996.

[Tan 08] Tan, R., Visibility in bad weather from a single image, *IEEE Conference on Computer Vision and Pattern Recognition,* Ancorage, Alaska, pp. 1–8, 2008.

[Tan 07] Tan, R., N. Petersson, and L. Petersson, Visibility enhancement for roads with foggy or hazy scenes, *IEEE Intelligent Vehicles Symposium,* Istanbul, Turkey, pp. 19–24, 2007.

[Tarel 09] Tarel, J. and N. Hauti, Fast visibility restoration from a single color or gray level image, *Proceedings of the IEEE International Conference on Computer Vision,* pp. 2201–2208. Kyoto, Japan, 2009.

[Yu 10] Yu, J., C. Xiao, and D. Li, Physics-based fast single image fog removal, *IEEE 10th International Conference on Signal Processing (ICSP),* Beijing, China, pp. 1048–1052, 2010.

[Zhang 02] Zhang, Y., B. Guindon, and J. Cihlar, An image transform to characterize and compensate for spatial variations in thin cloud contamination of Landsat images, *Remote Sensing of Environment,* 82(2–3), 173–187, October 2002.

12

Signal Processing and Propagation for Aeroacoustic Sensor Networks

Richard J. Kozick
Bucknell University

Brian M. Sadler
U.S. Army Research Laboratory

D. Keith Wilson
U.S. Army Cold Regions Research and Engineering Laboratory

12.1 Introduction 245
12.2 Models for Source Signals and Propagation............................247
 Basic Considerations • Narrowband Model with No
 Scattering • Narrowband Model with Scattering • Model for
 Extinction Coefficients • Multiple Frequencies and Sources
12.3 Signal Processing 261
 Angle of Arrival Estimation • Localization with Distributed Sensor
 Arrays • Tracking Moving Sources • Detection and Classification
12.4 Concluding Remarks..284
Acknowledgments...285
References...285

12.1 Introduction

Passive sensing of acoustic sources is attractive in many respects, including the relatively low signal bandwidth of sound waves, the loudness of most sources of interest, and the inherent difficulty of disguising or concealing emitted acoustic signals. The availability of inexpensive, low-power sensing and signal processing hardware enables application of sophisticated real-time signal processing. Among the many applications of aeroacoustic sensors, we focus in this chapter on detection and localization of ground and air (both jet and rotary) vehicles from ground-based sensor networks. Tracking and classification are briefly considered as well.

Elaborate, aeroacoustic systems for passive vehicle detection were developed as early as World War I [1]. Despite this early start, interest in aeroacoustic sensing has generally lagged other technologies until the recent packaging of small microphones, digital signal processing, and wireless communications into compact, unattended systems. An overview of modern outdoor acoustic sensing is presented by Becker and Güdesen [2]. Experiments in the early 1990s, such as those described by Srour and Robertson [3], demonstrated the feasibility of network detection, array processing, localization, and multiple target tracking via Kalman filtering. Many of the fundamental issues and challenges described by Srour and Robertson remain relevant today.

Except at very close range, the typical operating frequency range we consider is roughly 30–250 Hz. Below 30 Hz (the infrasonic regime), wavelengths are greater than 10 m so that rather large arrays may be required. Furthermore, wind noise (random pressure fluctuations induced by atmospheric turbulence) reduces the observed signal-to-noise ratio (SNR) [2]. At frequencies above several hundred Hz, molecular absorption of sound and interference between direct and ground-reflected waves attenuate received signals significantly [4]. In effect, the propagation environment acts as a low pass filter; this is particularly evident at longer ranges.

Aeroacoustics is inherently an ultra-wideband array processing problem, e.g., operating in [30,250] Hz yields a 157% fractional bandwidth centered at 140 Hz. To process under the narrow band array assumptions will require the fractional bandwidth to be on the order of a few percent or less, limiting the bandwidth to perhaps a few Hz in this example. The wide bandwidth significantly complicates the array signal processing, including angle-of-arrival (AOA) estimation, wideband Doppler compensation, beamforming, and blind source separation (which becomes convolutional).

The typical source of interest here has a primary contribution due to rotating machinery (engines), and may include tire and/or exhaust noise, vibrating surfaces, and other contributions. Internal combustion engines typically exhibit a strong sum of harmonics acoustic signature tied to the cylinder firing rate, a feature that can be exploited in virtually all phases of signal processing. Tracked vehicles also exhibit tread slap, which can produce very strong spectral lines, while helicopters produce strong harmonic sets related to the blade rotation rates. Turbine engines, on the other hand, exhibit a much more smoothly broad spectrum and consequently call for different algorithmic approaches in some cases. Many heavy vehicles and aircraft are quite loud and can be detected from ranges of several kilometers or more. Ground vehicles may also produce significant seismic waves, although we do not consider multi-modal sensing or sensor fusion here.

The problem is also complicated by time-varying factors that are difficult to model, such as source signature variations resulting from acceleration/deceleration of vehicles, changing meteorological conditions, multiple soft and loud sources, aspect angle source signature dependency, Doppler shifts (with 1 Hz shifts at a 100 Hz center frequency not unusual), multipath, and so on. Fortunately, at least for many sources of interest, a piecewise stationary model is reasonable on time scales of 1 s or less, although fast moving sources may require some form of time-varying model.

Sensor networks of interest are generally connected with wireless links, and are battery powered. Consequently, the node power budget may be dominated by the communications (radio). Therefore, a fundamental design question is how to perform distributed processing in order to reduce communication bandwidth, while achieving near optimal detection, estimation, and classification performance. We focus on this question, taking the aeroacoustic environment into account.

In particular, we consider the impact of random atmospheric inhomogeneities (primarily thermal and wind variations caused by turbulence) on the ability of an aeroacoustic sensor network to localize sources. Given that turbulence induces acoustical index-of-refraction variations several orders of magnitude greater than corresponding electromagnetic variations [5], this impact is quite significant. Turbulent scattering of sound waves causes random fluctuations in signals as observed at a single sensor, with variations occurring on time scales from roughly one to hundreds of seconds in our frequency range of interest [6–8]. Scattering is also responsible for losses in the observed spatial coherence measured between two sensors [9–11]. The scattering may be weak or strong, which are analogous to Rician and Rayleigh fading in radio propagation, respectively.

The impact of spatial coherence loss is significant, and generally becomes worse with increasing distance between sensors. This effect, as well as practical size constraints, limits individual sensor node array apertures to perhaps a few meters. At the same time, the acoustic wavelengths λ of interest are about 1–10 m ($\lambda = (330 \, \text{m/s})/(30 \, \text{Hz}) = 11 \, \text{m}$ at 30 Hz, and $\lambda = 1.32 \, \text{m}$ at 250 Hz). Thus, the typical array aperture will only span a fraction of a wavelength, and accurate AOA estimation requires wideband superresolution methods. The source may generally be considered to be in the far field of these small arrays. Indeed, if it is in the near field, then the rate of change of the AOA as the source moves past the array must be considered.

The signal-coherence characteristics suggest deployment of multiple, small-baseline arrays as nodes within an overall large-baseline array (see Figure 12.7 in Section 12.3.2). The source is intended to be in the near-field of the large-baseline array. Exploitation of this larger baseline is highly desirable, as it potentially leads to very accurate localization. We characterize this problem in terms of the atmosphere-induced spatial coherence loss, and show fundamental bounds on the ability to localize a source in such conditions. This leads to a family of localization approaches, spanning triangulation (which minimizes inter-node communication), to time-delay estimation (TDE), to fully centralized processing (which maximizes communication use and is therefore undesirable). The achievable localization accuracy depends on both the propagation conditions and the time-bandwidth product of the source.

The chapter is organized as follows. In Section 12.2 we introduce the wideband source array signal processing model, develop the atmospheric scattering model, and incorporate the scattering into the array model. We consider array signal processing in Section 12.3, including narrowband AOA estimation with scattering present. We review wideband AOA estimation techniques, and highlight various aeroacoustic wideband AOA experiments. Next, we consider localization with multiple nodes (arrays) in the presence of scattering. We develop fundamental and tight performance bounds on TDE in the turbulent atmosphere, as well as bounds on localization. Localization performance is illustrated via simulation and experiments. We then briefly consider the propagation impact on detection and classification. Finally, in Section 12.4 we consider some emerging aspects and open questions.

12.2 Models for Source Signals and Propagation

In this section, we present a general model for the signals received by an aeroacoustic sensor array. We begin by briefly considering models for the signals emitted by ground vehicles and aircraft in Section 12.2.1. Atmospheric phenomena affecting propagation of the signal are also summarized. In Section 12.2.2, we consider the simplest possible case for the received signals: a single nonmoving source emits a sinusoidal waveform, and the atmosphere induces no scattering (randomization of the signal). Then in Section 12.2.3, we extend the model to include the effects of scattering, and in Section 12.2.4, approximate models for the scattering as a function of source range, frequency, and atmospheric conditions are presented. The model is extended to multiple sources and multiple frequencies (wideband) in Section 12.2.5.

12.2.1 Basic Considerations

As we noted in Section 12.1, the sources of interest typically have spectra that are harmonic lines, or have relatively continuous broadband spectra, or some combination. The signal processing for detection, localization, and classification is highly dependent on whether the source spectrum is harmonic or broadband. For example, broadband sources allow time-difference of arrival processing for localization, while harmonic sources allow differential Doppler estimation.

Various deterministic and random source models may be employed. Autoregressive (AR) processes are well suited to modeling sums of harmonics, at least for the case of a single source, and may be used for detection, Doppler estimation, filtering, AOA estimation, and so on [12–14]. Sum of harmonic models, with unknown harmonic structure, lead naturally to detection tests in the frequency domain [15].

More generally, a Gaussian random process model may be employed to describe both harmonic sets and wideband sources [16]; we adopt such a point of view here. We also assume a piecewise stationary (quasi-static) viewpoint: although the source may actually be moving, the processing interval is assumed to be short enough that the signal characteristics are nearly constant.

Four phenomena are primarily responsible for modifying the source signal to produce the actual signal observed at the sensor array:

1. The propagation delay from the source to the sensors
2. Random fluctuations in the amplitude and phase of the signals caused by scattering from random inhomogeneities in the atmosphere such as turbulence
3. Additive noise at the sensors caused by thermal noise, wind noise, and directional interference
4. Transmission loss caused by spreading of the wavefronts, refraction by wind and temperature gradients, ground interactions, and molecular absorption of sound energy

Thermal noise at the sensors is typically independent from sensor to sensor. In contrast, interference from an undesired source produces additive noise that is (spatially) correlated from sensor to sensor. Wind noise, which consists of low-frequency turbulent pressure fluctuations intrinsic to the atmospheric flow (and, to a lesser extent, flow distortions induced by the microphone itself [2,17]), exhibits high spatial correlation over distances of several meters [18].

The transmission loss (TL) is defined as the diminishment in sound energy from a reference value S_{ref}, which would hypothetically be observed in free space at 1 m from the source, to the actual value observed at the sensor S. To a first approximation, the sound energy spreads spherically; that is, it diminishes as the inverse of the squared distance from the source. In actuality the TL for sound wave propagating near the ground involves many complex, interacting phenomena, so that the spherical spreading condition is rarely observed in practice, except perhaps within the first 10–30 m [4]. Fortunately, several well refined and accurate numerical procedures for calculating TL have been developed [19]. For simplicity, here we model S as a deterministic parameter, which is reasonable when the state of the atmosphere does not change dramatically during the data collection.

Particularly significant to the present discussion is the second phenomenon in the preceding list, namely scattering by turbulence. The turbulence consists of random atmospheric motions occurring on time scales from seconds to several minutes. Scattering from these motions causes random fluctuations in the complex signals at the individual sensors and diminishes the cross coherence of signals between sensors. The effects of scattering on array performance will be analyzed in Sections 12.2.2 and 12.2.4.

The sinusoidal source signal that is measured at the reference distance of 1 m from the source is written as

$$s_{ref}(t) = \sqrt{S_{ref}}\,\cos(2\pi f_o t + \chi), \tag{12.1}$$

where
    The frequency of the tone is $f_o = \omega_o/(2\pi)$ Hz
    The period is T_o s
    The phase is χ
    The amplitude is $\sqrt{S_{ref}}$

The sound waves propagate with wavelength $\lambda = c/f_o$, where c is the speed of sound. The wavenumber is $k = 2\pi/\lambda = \omega_o/c$. We will represent sinusoidal and narrowband signals by their complex envelope, which may be defined in two ways, as in (12.2):

$$\mathcal{C}\{s_{ref}(t)\} = \tilde{s}_{ref}(t) = s_{ref}^{(I)}(t) + j s_{ref}^{(Q)}(t) = \left[s_{ref}(t) + j\mathcal{H}\{s_{ref}(t)\} \right] \exp(-j2\pi f_o t) \tag{12.2}$$

$$= \sqrt{S_{ref}}\,\exp(j\chi). \tag{12.3}$$

We will represent the complex envelope of a quantity with the notation $\mathcal{C}\{\cdot\}$ or $(\tilde{\cdot})$ the in-phase component with $(\cdot)^{(I)}$, the quadrature component with $(\cdot)^{(Q)}$, and the Hilbert transform with $\mathcal{H}\{\cdot\}$. The in-phase (I) and quadrature (Q) components of a signal are obtained by the processing in Figure 12.2. The FFT is often used to approximate the processing in Figure 12.2 for a finite block of data, where the real and imaginary parts of the FFT coefficient at frequency f_o are proportional to the I and Q components, respectively. The complex envelope of the sinusoid in (12.1) is given by (12.3), which is not time-varying, so the average power is $|\tilde{s}_{ref}(t)|^2 = S_{ref}$.

It is easy to see for the sinusoidal signal (12.1) that shifting $s_{ref}(t)$ in time causes a phase shift in the corresponding complex envelope, i.e., $\mathcal{C}\{s_{ref}(t - \tau_o)\} = \exp(-j2\pi f_o \tau_o)\tilde{s}_{ref}(t)$. A similar property is true for *narrowband* signals whose frequency spectrum is confined to a bandwidth B Hz around a center

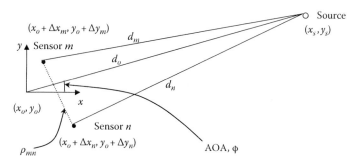

FIGURE 12.1 Geometry of source and sensor locations.

frequency f_o Hz, where $B \ll f_o$. For a narrowband signal $z(t)$ with complex envelope $\tilde{z}(t)$, a shift in time is well-approximated by a phase shift in the corresponding complex envelope:

$$\mathcal{C}\{z(t - \tau_o)\} \approx \exp(-j2\pi f_o \tau_o)\tilde{z}(t) \quad \text{(narrowband approximation)}. \tag{12.4}$$

Equation 12.4 is the well-known Fourier transform relationship between shifts in time and phase shifts that are linearly proportional to frequency. The approximation is accurate when the frequency band is narrow enough so that the linearly increasing phase shift is close to $\exp(-j2\pi f_o \tau_o)$ over the band.

The source and array geometry is illustrated in Figure 12.1. The source is located at coordinates (x_s, y_s) in the (x, y) plane. The array contains N sensors, with sensor n located at $(x_o + \Delta x_n, y_o + \Delta y_n)$, where (x_o, y_o) is the center of the array and $(\Delta x_n, \Delta y_n)$ is the relative sensor location. The propagation time from the source to the array center is

$$\tau_o = \frac{d_o}{c} = \frac{1}{c}\left[(x_s - x_o)^2 + (y_s - y_o)^2\right]^{1/2}, \tag{12.5}$$

where d_o is the distance from the source to the array center. The propagation time from the source to sensor n is

$$\tau_n = \frac{d_n}{c} = \frac{1}{c}\left[(x_s - x_o - \Delta x_n)^2 + (y_s - y_o - \Delta y_n)^2\right]^{1/2}. \tag{12.6}$$

Let us denote the array diameter by $L = \max\{\rho_{mn}\}$, where ρ_{mn} is the separation between sensors m and n, as shown in Figure 12.1. The source is in the *far-field* of the array when the source distance satisfies $d_o \gg L^2/\lambda$, in which case (12.6) may be approximated with the first term in the Taylor series $(1 + u)^{1/2} \approx 1 + u/2$. Then $\tau_n \approx \tau_o + \tau_{o,n}$ with error that is much smaller than the source period, T_o, where

$$\tau_{o,n} = -\frac{1}{c}\left[\frac{x_s - x_o}{d_o}\Delta x_n + \frac{y_s - y_o}{d_o}\Delta y_n\right] = -\frac{1}{c}\left[(\cos\phi)\Delta x_n + (\sin\phi)\Delta y_n\right]. \tag{12.7}$$

The angle ϕ is the azimuth bearing, or AOA, as shown in Figure 12.1. In the far-field, the spherical wavefront is approximated as a plane wave over the array aperture, so the bearing ϕ contains the available information about the source location. For array diameters $L < 2$ m and tone frequencies $f_o < 200$ Hz so that $\lambda > 1.5$ m, the quantity $L^2/\lambda < 2.7$ m. Thus the far-field is valid for source distances on the order of tens of meters. For smaller source distances and/or larger array apertures, the curvature of the wavefront over the array aperture must be included in τ_n according to (12.6). We develop the model for the far-field case in the next section. However, the extension to the near-field is easily accomplished by redefining the array response vector (**a** in (12.20)) to include the wavefront curvature with $a_n = \exp(-j2\pi f_o \tau_n)$.

12.2.2 Narrowband Model with No Scattering

Here we present the model for the signals impinging on the sensor array when there is no scattering. Using the far-field approximation, the noisy measurements at the sensors are

$$z_n(t) = s_n(t - \tau_o - \tau_{o,n}) + w_n(t), \quad n = 1, \ldots, N. \tag{12.8}$$

In the absence of scattering, the signal components are pure sinusoids:

$$s_n(t) = \sqrt{S} \cos(2\pi f_o t + \chi). \tag{12.9}$$

The $w_n(t)$ are additive, white, Gaussian noise (AWGN) processes that are real-valued, continuous-time, zero-mean, jointly wide-sense stationary, and mutually uncorrelated at distinct sensors with power spectral density (PSD) $(\mathcal{N}_o/2)$ W/Hz. That is, the noise correlation properties are

$$E\{w_n(t)\} = 0, \quad -\infty < t < \infty, \quad n = 1, \ldots, N, \tag{12.10}$$

$$r_{w,mn}(\xi) = E\{w_m(t + \xi)w_n(t)\} = r_w(\xi)\delta_{mn}, \tag{12.11}$$

where
  $E\{\cdot\}$ denotes expectation
  $r_w(\xi) = (\mathcal{N}_o/2)\delta(\xi)$ is the noise autocorrelation function that is common at all sensors

The Dirac delta function is $\delta(\cdot)$, and the Kronecker delta function is $\delta_{mn} = 1$ if $m = n$ and 0 otherwise. As noted above, modeling the noise as spatially white may be inaccurate if wind noise or interfering sources are present in the environment. The noise PSD is

$$G_w(f) = \mathcal{F}\{r_w(\xi)\} = \frac{\mathcal{N}_o}{2}, \tag{12.12}$$

where $\mathcal{F}\{\}$ denotes Fourier transform. With no scattering, the complex envelope of $z_n(t)$ in (12.8) and (12.9) is, using (12.4),

$$\tilde{z}_n(t) = \exp[-j(\omega_o \tau_o + \omega_o \tau_{o,n})]\tilde{s}_n(t) + \tilde{w}_n(t)$$

$$= \sqrt{S} \exp[j(\chi - \omega_o \tau_o)]\exp[-j\omega_o \tau_{o,n}] + \tilde{w}_n(t), \tag{12.13}$$

where the complex envelope of the narrowband source component is

$$\tilde{s}_n(t) = \sqrt{S}e^{j\chi}, \quad n = 1, \ldots, N \text{ (no scattering)}. \tag{12.14}$$

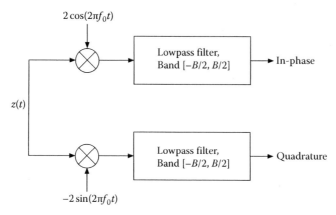

$2\cos(2\pi f_0 t)$

Lowpass filter, Band $[-B/2, B/2]$ → In-phase

$z(t)$

Lowpass filter, Band $[-B/2, B/2]$ → Quadrature

$-2\sin(2\pi f_0 t)$

FIGURE 12.2 Processing to obtain in-phase and quadrature components, $z^{(I)}(t)$ and $z^{(Q)}(t)$.

We assume that the complex envelope is lowpass filtered with bandwidth from $[-B/2, B/2]$ Hz, e.g., as in Figure 12.2. Assuming that the lowpass filter is ideal, the complex envelope of the noise, $\tilde{w}_n(t)$, has PSD and correlation

$$G_{\tilde{w}}(f) = (2\mathcal{N}_o)\text{rect}\left(\frac{f}{B}\right), \tag{12.15}$$

$$r_{\tilde{w}}(\xi) = E\{\tilde{w}_n(t+\xi)\tilde{w}_n(t)^*\} = \mathcal{F}^{-1}\{G_{\tilde{w}}(f)\} = (2\mathcal{N}_oB)\text{sinc}(B\xi), \tag{12.16}$$

$$r_{\tilde{w},mn}(\xi) = E\{\tilde{w}_m(t+\xi)\tilde{w}_n(t)^*\} = r_{\tilde{w}}(\xi)\delta_{mn}, \tag{12.17}$$

where

$(\cdot)^*$ denotes complex conjugate

rect $(u) = 1$ for $-1/2 < u < 1/2$ and 0 otherwise

sinc $(u) = \sin(\pi u)/(\pi u)$

Note that the noise samples are uncorrelated (and independent since Gaussian) at sample times spaced by $1/B$ s. In practice, the noise PSD $G_{\tilde{w}}(f)$ is neither flat nor perfectly band-limited as in (12.15). However, the lowpass filtering to bandwidth B Hz implies that the noise samples have decreasing correlation for time spacing greater than $1/B$ s.

Let us define the vectors

$$\tilde{\mathbf{z}}(t) = \begin{bmatrix} \tilde{z}_1(t) \\ \vdots \\ \tilde{z}_N(t) \end{bmatrix}, \quad \tilde{\mathbf{s}}(t) = \begin{bmatrix} \tilde{s}_1(t) \\ \vdots \\ \tilde{s}_N(t) \end{bmatrix}, \quad \tilde{\mathbf{w}}(t) = \begin{bmatrix} \tilde{w}_1(t) \\ \vdots \\ \tilde{w}_N(t) \end{bmatrix}. \tag{12.18}$$

Then using (12.13) with (12.7),

$$\tilde{\mathbf{z}}(t) = \sqrt{S}\exp\left[j(\chi - \omega_o\tau_o)\right]\mathbf{a} + \tilde{\mathbf{w}}(t) = \sqrt{S}e^{j\theta}\mathbf{a} + \tilde{\mathbf{w}}(t), \tag{12.19}$$

where **a** is the array steering vector (or array manifold)

$$\mathbf{a} = \begin{bmatrix} \exp[jk((\cos\phi)\Delta x_1 + (\sin\phi)\Delta y_1)] \\ \vdots \\ \exp[jk((\cos\phi)\Delta x_N + (\sin\phi)\Delta y_N)] \end{bmatrix} \qquad (12.20)$$

with $k = \omega_o/c$. Note that the steering vector, **a**, depends on the frequency ω_o, the sensor locations $(\Delta x_n, \Delta y_n)$, and the source bearing ϕ. The common phase factor at all of the sensors, $\exp[j(\chi - \omega_o\tau_o)] = \exp[j(\chi - kd_o)]$, depends on the phase of the signal emitted by the source (χ) and the propagation distance to the center of the array (kd_o). We simplify the notation and define

$$\theta \triangleq \chi - kd_o, \qquad (12.21)$$

which is a deterministic parameter.

In preparation for the introduction of scattering into the model, let us write expressions for the first- and second-order moments of the vectors $\tilde{\mathbf{s}}(t)$ and $\tilde{\mathbf{z}}(t)$. Let **1** be an $N \times 1$ vector of ones, $\mathbf{R}_{\tilde{z}}(\xi) = E\{\tilde{\mathbf{z}}(t+\xi)\tilde{\mathbf{z}}(t)^\dagger\}$ be the $N \times N$ cross-correlation function matrix with (m, n) element $r_{\tilde{z},mn}(\xi) = E\{\tilde{z}_m(t+\xi)\tilde{z}_n(t)^*\}$, and $\mathbf{G}_{\tilde{z}}(f) = \mathcal{F}\{\mathbf{R}_{\tilde{z}}(\xi)\}$ be the cross-spectral density (CSD) matrix, then

$$E\{\tilde{\mathbf{s}}(t)\} = \sqrt{S}e^{j\chi}\mathbf{1} \quad E\{\tilde{\mathbf{z}}(t)\} = \sqrt{S}e^{j\theta}\mathbf{a}, \qquad (12.22)$$

$$\mathbf{R}_{\tilde{s}}(\xi) = S\mathbf{1}\mathbf{1}^T \quad \mathbf{R}_{\tilde{z}}(\xi) = S\mathbf{a}\mathbf{a}^\dagger + r_{\tilde{w}}(\xi)\mathbf{I}, \qquad (12.23)$$

$$\mathbf{G}_{\tilde{s}}(f) = S\mathbf{1}\mathbf{1}^T\delta(f) \quad \mathbf{G}_{\tilde{z}}(f) = S\mathbf{a}\mathbf{a}^\dagger\delta(f) + G_{\tilde{w}}(f)\mathbf{I}, \qquad (12.24)$$

$$E\{\tilde{\mathbf{s}}(t)\tilde{\mathbf{s}}(t)^\dagger\} = \mathbf{R}_{\tilde{s}}(0) = S\mathbf{1}\mathbf{1}^T \quad E\{\tilde{\mathbf{z}}(t)\tilde{\mathbf{z}}(t)^\dagger\} = \mathbf{R}_{\tilde{z}}(0) = S\mathbf{a}\mathbf{a}^\dagger + \sigma_{\tilde{w}}^2\mathbf{I}, \qquad (12.25)$$

where

$(\cdot)^T$ denotes transpose
$(\cdot)^*$ denotes complex conjugate
$(\cdot)^\dagger$ denotes complex conjugate transpose
I is the $N \times N$ identity matrix
$\sigma_{\tilde{w}}^2$ is the variance of the noise samples

$$\sigma_{\tilde{w}}^2 = E\{|\tilde{\mathbf{w}}(t)|^2\} = r_{\tilde{w}}(0) = 2\mathcal{N}_oB. \qquad (12.26)$$

Note from (12.24) that the PSD at each sensor contains a spectral line since the source signal is sinusoidal. Note from (12.25) that at each sensor, the average power of the signal component is S, so the SNR at each sensor is

$$\text{SNR} = \frac{S}{\sigma_{\tilde{w}}^2} = \frac{S}{2\mathcal{N}_oB}. \qquad (12.27)$$

The complex envelope vector $\tilde{\mathbf{z}}(t)$ is typically sampled at a rate $f_s = B$ samples/s, so the samples are spaced by $T_s = 1/f_s = 1/B$ s:

$$\tilde{\mathbf{z}}(iT_s) = \sqrt{S}e^{j\theta}\mathbf{a} + \tilde{\mathbf{w}}(iT_s), \quad i = 0,\dots,T-1. \qquad (12.28)$$

According to (12.17), the noise samples are spatially independent as well as temporally independent, since $r_{\tilde{w}}(iT_s) = r_{\tilde{w}}(i/B) = 0$. Thus the vectors $\tilde{\mathbf{z}}(0), \tilde{\mathbf{z}}(T_s), \ldots, \tilde{\mathbf{z}}((T-1)T_s)$ in (12.28) are independent and identically distributed (iid) with complex normal distribution, which we denote by $\tilde{\mathbf{z}}(iT_s) \sim \mathrm{CN}(\mathbf{m}_{\tilde{z}}, \mathbf{C}_{\tilde{z}})$, with mean and covariance matrix:

$$\mathbf{m}_{\tilde{z}} = \sqrt{S}e^{j\theta}\mathbf{a} \quad \text{and} \quad \mathbf{C}_{\tilde{z}} = \sigma_{\tilde{w}}^2\mathbf{I} \quad \text{(no scattering)}. \tag{12.29}$$

The joint probability density function for $\mathrm{CN}(\mathbf{m}_{\tilde{z}}, \mathbf{C}_{\tilde{z}})$ is given by [20]

$$f(\tilde{\mathbf{z}}) = \frac{1}{\pi^N \det(\mathbf{C}_{\tilde{z}})}\exp[-(\tilde{\mathbf{z}} - \mathbf{m}_{\tilde{z}})^\dagger \mathbf{C}_{\tilde{z}}^{-1}(\tilde{\mathbf{z}} - \mathbf{m}_{\tilde{z}})], \tag{12.30}$$

where "det" denotes determinant. In the absence of scattering, the information about the source location (bearing) is contained in the mean of the sensor observations. If the T time samples in (12.28) are coherently averaged, then the resulting SNR per sensor is T times that in (12.27), so $\mathrm{SNR}' = T \cdot (S/\sigma_{\tilde{w}}^2) = T \cdot [S/(2\mathcal{N}_o/T_s)] = \mathcal{T} \cdot S/(2\mathcal{N}_o)$, where $\mathcal{T} = T \cdot T_s$ is the total observation time, in seconds.

12.2.3 Narrowband Model with Scattering

Next, we include the effects of scattering by atmospheric turbulence in the model for the signals measured at the sensors in the array. As mentioned earlier, the scattering introduces random fluctuations in the signals and diminishes the cross coherence between the array elements. The formulation we present for the scattering effects was developed by Wilson et al. [11,21–26]. The reader may refer to these studies for details about the physical modeling and references to additional primary source material. Several assumptions and simplifications are involved in the formulation: (12.1) the propagation is line-of-sight (no multipath), (12.2) the additive noise is independent from sensor to sensor, and (12.3) the random fluctuations caused by scattering are complex, circular, Gaussian random processes with partial correlation between the sensors.

The line-of-sight propagation assumption is consistent with Section 12.2.2 and is reasonable for propagation over fairly flat, open terrain in the frequency range of interest here (below several hundred Hz). Significant acoustic multipath may result from reflections off hard objects such as buildings, trees, and (sometimes) the ground. Multipath can also result from refraction of sound waves by vertical gradients in the wind and temperature.

By assuming independent, additive noise, we ignore the potential spatial correlation of wind noise and interference from other undesired sources. This restriction may be averted by extending the models to include spatially-correlated additive noise, although the signal processing may be more complicated in this case.

Modeling of the scattered signals as complex, circular, Gaussian random processes is a substantial improvement on the constant signal model (Section 12.2.2), but it is, nonetheless, rather idealized. Waves that have propagated through a random medium can exhibit a variety of statistical behaviors, depending on such factors as the strength of the turbulence, the propagation distance, and the ratio of the wavelength to the predominant eddy size [5,27]. Experimental studies [8,28,29] conducted over short horizontal propagation distances with frequencies below 1000 Hz demonstrate that the effect of turbulence is highly significant, with phase variations much larger than 2π rad and deep fades in amplitude often developing. The measurements demonstrate that the Gaussian model is valid in many conditions, although non-Gaussian scattering characterized by large phase but small amplitude variations is observed at some frequencies and propagation distances. The Gaussian model applies in many cases of interest, and we apply it in this chapter. The effect of non-Gaussian signal scattering on aeroacoustic array performance remains to be determined.

The scattering modifies the complex envelope of the signals at the array by spreading a portion of the power from the (deterministic) mean component into a zero-mean random process with a PSD centered at 0 Hz. We assume that the bandwidth of the scattered signal, which we denote by B, is much smaller than the tone frequency, f_o. The *saturation* parameter [25,26], denoted by $\Omega \in [0, 1]$, defines the fraction of average signal power that is scattered from the mean into the random component. The scattering may be weak ($\Omega \approx 0$) or *strong* ($\Omega \approx 1$), which are analogous to Rician and Rayleigh fading in the radio propagation literature. The modification of (12.8), (12.9), (12.13), and (12.14) to include scattering is as follows, where $\tilde{z}_n(t)$ is the signal measured at sensor n:

$$\tilde{z}_n(t) = \exp\left[-j(\omega_o \tau_o + \omega_o \tau_{o,n})\right] \tilde{s}_n(t) + \tilde{w}_n(t), \tag{12.31}$$

$$\tilde{s}_n(t) = \sqrt{(1-\Omega)S}e^{j\chi} + \tilde{v}_n(t)e^{j\chi}, \quad n = 1, \ldots, N \text{ (with scattering).} \tag{12.32}$$

In order to satisfy conservation of energy with $E\{|\tilde{s}_n(t)|^2\} = S$, the average power of the scattered component must be $E\{|\tilde{v}_n(t)|^2\} = \Omega S$. The value of the saturation Ω and the correlation properties of the vector of scattered processes, $\tilde{\mathbf{v}}(t) = [\tilde{v}_1(t), \ldots, \tilde{v}_N(t)]^T$, depend on the source distance (d_o) and the meteorological conditions. The vector of scattered processes $\tilde{\mathbf{v}}(t)$ and the additive noise vector $\tilde{\mathbf{w}}(t)$ contain zero-mean, jointly wide-sense stationary, complex, circular Gaussian random processes. The scattered processes and the noise are modeled as independent, $E\{\tilde{\mathbf{v}}(t + \xi)\tilde{\mathbf{w}}(t)^\dagger\} = \mathbf{0}$. The noise is described by (12.15) through (12.17), while the saturation Ω and statistics of $\tilde{\mathbf{v}}(t)$ are determined by the "extinction coefficients" of the first and second moments of $\tilde{s}(t)$. As will be discussed in Section 12.2.4, approximate analytical models for the extinction coefficients are available from physical modeling of the turbulence in the atmosphere. In the remainder of this section, we define the extinction coefficients and relate them to Ω and the statistics of $\tilde{\mathbf{v}}(t)$, thereby providing models for the sensor array data that include turbulent scattering by the atmosphere.

We denote the extinction coefficients for the first and second moments of $\tilde{s}(t)$ by μ and $\nu(\rho_{mn})$, respectively, where ρ_{mn} is the distance between sensors m and n (see Figure 12.1). The extinction coefficients are implicitly defined as follows:

$$E\{\tilde{s}_n(t)\} = \sqrt{(1-\Omega)S}e^{j\chi} \triangleq \sqrt{S}e^{j\chi}e^{-\mu d_o}, \tag{12.33}$$

$$r_{\tilde{s},mn}(0) = E\{\tilde{s}_m(t)\tilde{s}_n(t)^*\} = (1-\Omega)S + r_{\tilde{v},mn}(0) \triangleq Se^{-\nu(\rho_{mn})d_o}, \tag{12.34}$$

where

$$r_{\tilde{s},mn}(\xi) = E\{\tilde{s}_m(t+\xi)\tilde{s}_n(t)^*\} = (1-\Omega)S + r_{\tilde{v},mn}(\xi). \tag{12.35}$$

The right sides of (12.33) and (12.34) are the first and second moments *without* scattering, from (12.22) and (12.23), respectively, multiplied by a factor that decays exponentially with increasing distance d_o from the source. From (12.33), we obtain

$$\sqrt{(1-\Omega)} = e^{-\mu d_o} \quad \text{and} \quad \Omega = 1 - e^{-2\mu d_o}. \tag{12.36}$$

Also, by conservation of energy with $m = n$ in (12.34), adding the average powers in the unscattered and scattered components of $\tilde{s}_n(t)$ must equal S, so

$$r_{\tilde{s}}(0) = E\left\{\left|\tilde{s}_n(t)\right|\right\} = e^{-2\mu d_o} S + r_{\tilde{v}}(0) = S, \tag{12.37}$$

$$\Rightarrow r_{\tilde{v}}(0) = E\left\{\left|\tilde{v}_n(t)\right|\right\} = \int_{-\infty}^{\infty} G_{\tilde{v}}(f)df = (1 - e^{-2\mu d_o})S = \Omega S, \tag{12.38}$$

where

$r_{\tilde{v}}(\xi) = E\{\tilde{v}_n(t+\xi)\tilde{v}_n(t)^*\}$ is the autocorrelation function (which is the same for all n)
$G_{\tilde{v}}(f)$ is the corresponding PSD

Therefore for source distances $d_o \ll 1/(2\mu)$, the saturation $\Omega \approx 0$ and most of the energy from the source arrives at the sensor in the unscattered (deterministic mean) component of $\tilde{s}_n(t)$. For source distances $d_o \gg 1/(2\mu)$, the saturation $\Omega \approx 1$ and most of the energy arrives in the scattered (random) component.

Next, we use (12.34) to relate the correlation of the scattered signals at sensors m and n, $r_{\tilde{v},mn}(\xi)$, to the second moment extinction coefficient, $v(\rho_{mn})$. Since the autocorrelation of $\tilde{v}_n(t)$ is identical at each sensor n and equal to $r_{\tilde{v}}(\xi)$, and assuming that the PSD $G_{\tilde{v}}(f)$ occupies a narrow bandwidth centered at $0\,\text{Hz}$, the cross-correlation and CSD satisfy

$$r_{\tilde{v},mn}(\xi) = \gamma_{mn} r_{\tilde{v}}(\xi) \quad \text{and} \quad G_{\tilde{v},mn}(f) = \mathcal{F}\{r_{\tilde{v},mn}(\xi)\} = \gamma_{mn} G_{\tilde{v}}(f), \tag{12.39}$$

where $|\gamma_{mn}| \leq 1$ is a measure of the coherence between $\tilde{v}_m(t)$ and $\tilde{v}_n(t)$. The definition of γ_{mn} as a *constant* includes an approximation that the coherence does not vary with frequency, which is reasonable when the bandwidth of $G_{\tilde{v}}(f)$ is narrow. Although systematic studies of the coherence time of narrowband acoustic signals have not been made, data and theoretical considerations (such as in Ref. [27, Sec. 8.4]) are consistent with values ranging from tens of seconds to several minutes in the frequency range [50, 250] Hz. Therefore the bandwidth of $G_{\tilde{v}}(f)$ may be expected to be less than 1 Hz. The bandwidth B in the lowpass filters for the complex amplitude in Figure 12.2 should be chosen to be equal to the bandwidth of $G_{\tilde{v}}(f)$. We assume that γ_{mn} in (12.39) is real-valued and non-negative, which implies that phase fluctuations at sensor pairs are not biased toward positive or negative values. Then using (12.39) with (12.38) and (12.36) in (12.34) yields the following relation between γ_{mn} and μ, v:

$$\gamma_{mn} = \frac{e^{-v(\rho_{mn})d_o} - e^{-2\mu d_o}}{1 - e^{-2\mu d_o}}, \quad m, n = 1, \ldots, N. \tag{12.40}$$

We define Γ as the $N \times N$ matrix with elements γ_{mn}. The second moment extinction coefficient $v(\rho_{mn})$ is a monotonically increasing function, with $v(0) = 0$ and $v(\infty) = 2\mu$, so $\gamma_{mn} \in [0, 1]$.

Combining (12.31) and (12.32) into vectors, and using (12.36) yields

$$\tilde{\mathbf{z}}(t) = \sqrt{S}e^{j\theta}e^{-\mu d_o}\mathbf{a} + e^{j\theta}\mathbf{a}\tilde{\mathbf{v}}(t) + \tilde{\mathbf{w}}(t), \tag{12.41}$$

where

θ is defined in (12.21)
$\mathbf{a}$ is the array steering vector in (12.20)

We define the matrix **B** with elements

$$B_{mn} = \exp\left[-v(\rho_{mn})d_o\right], \tag{12.42}$$

and then we can extend the second-order moments in (12.22) through (12.25) to the case with scattering as

$$E\{\tilde{\mathbf{z}}(t)\} = e^{-\mu d_o}\sqrt{S}e^{j\theta}\mathbf{a} \triangleq \mathbf{m}_{\tilde{z}}, \tag{12.43}$$

$$\mathbf{R}_{\tilde{z}}(\xi) = e^{-2\mu d_o}S\mathbf{a}\mathbf{a}^\dagger + S[\mathbf{B}\circ(\mathbf{a}\mathbf{a}^\dagger) - e^{-2\mu d_o}\mathbf{a}\mathbf{a}^\dagger]\frac{r_{\tilde{v}}(\xi)}{S(1-e^{-2\mu d_o})} + r_{\tilde{w}}(\xi)\mathbf{I}, \tag{12.44}$$

$$\mathbf{G}_{\tilde{z}}(f) = e^{-2\mu d_o}S\mathbf{a}\mathbf{a}^\dagger\delta(f) + S[\mathbf{B}\circ(\mathbf{a}\mathbf{a}^\dagger) - e^{-2\mu d_o}\mathbf{a}\mathbf{a}^\dagger]\frac{G_{\tilde{v}}(f)}{S(1-e^{-2\mu d_o})} + G_{\tilde{w}}(f)\mathbf{I}, \tag{12.45}$$

$$E\{\tilde{\mathbf{z}}(t)\tilde{\mathbf{z}}(t)^\dagger\} = \mathbf{R}_{\tilde{z}}(0) = S\mathbf{B}\circ(\mathbf{a}\mathbf{a}^\dagger) + \sigma_w^2\mathbf{I} \triangleq \mathbf{C}_{\tilde{z}} + \mathbf{m}_{\tilde{z}}\mathbf{m}_{\tilde{z}}^\dagger, \tag{12.46}$$

where ∘ denotes element-wise product between matrices. The normalizing quantity $S(1-e^{-2\mu d_o})$ that divides the autocorrelation $r_{\tilde{v}}(\xi)$ and the PSD $G_{\tilde{v}}(f)$ in (12.44) and (12.45) is equal to $r_{\tilde{v}}(0) = \int G_{\tilde{v}}(f)df$. Therefore the maximum of the normalized autocorrelation is 1, and the area under the normalized PSD is 1. The complex envelope samples $\tilde{z}(t)$ have the complex normal distribution $CN(\mathbf{m}_{\tilde{z}},\mathbf{C}_{\tilde{z}})$, which is defined in (12.30). The mean vector and covariance matrix are given in (12.43) and (12.46), but we repeat them below for comparison with (12.29):

$$\mathbf{m}_{\tilde{z}} = e^{-\mu d_o}\sqrt{S}e^{j\theta}\mathbf{a} \quad \text{(with scattering)}, \tag{12.47}$$

$$\mathbf{C}_{\tilde{z}} = S[\mathbf{B}\circ(\mathbf{a}\mathbf{a}^\dagger) - e^{-2\mu d_o}\mathbf{a}\mathbf{a}^\dagger] + \sigma_w^2\mathbf{I} \quad \text{(with scattering)}. \tag{12.48}$$

Note that the scattering is negligible if $d_o \ll 1/(2\mu)$, in which case $e^{-2\mu d_o} \approx 1$ and $\Omega \approx 0$. Then most of the signal energy is in the mean, with $B \approx \mathbf{1}\mathbf{1}^T$ and $\gamma_{mn} \approx 1$ in (12.40), since $v(\rho_{mn}) < 2\mu$. For larger values of the source range d_o, more of the signal energy is scattered, and B may deviate from $\mathbf{1}\mathbf{1}^T$ (and $\gamma_{mn} < 1$ for $m \neq n$) due to coherence losses between the sensors. At full saturation ($\Omega = 1$), $\mathbf{B} = \Gamma$.

The scattering model in (12.41) may be formulated as multiplicative noise on the steering vector:

$$\tilde{\mathbf{z}}(t) = \sqrt{S}e^{j\theta}\mathbf{a}\circ\left[e^{-\mu d_o}\mathbf{1} + \frac{\tilde{\mathbf{v}}(t)}{\sqrt{S}}\right] + \tilde{\mathbf{w}}(t) \triangleq \sqrt{S}e^{j\theta}(\mathbf{a}\circ\tilde{\mathbf{u}}(t)) + \tilde{\mathbf{w}}(t). \tag{12.49}$$

The multiplicative noise process, $\tilde{\mathbf{u}}(t)$, is complex normal with $\mathbf{m}_{\tilde{u}} = E\{\tilde{\mathbf{u}}(t)\} = e^{-\mu d_o}\mathbf{1}$ and $E\{\tilde{\mathbf{u}}(t)\tilde{\mathbf{u}}(t)^\dagger\} = \mathbf{B}$, so the covariance matrix is $\mathbf{C}_{\tilde{u}} = \mathbf{B} - e^{-2\mu d_o}\mathbf{1}\mathbf{1}^T = \Omega\Gamma$, where Γ has elements γ_{mn} in (12.40). The mean vector and covariance matrix in (12.47) and (12.48) may be represented as $\mathbf{m}_{\tilde{z}} = \sqrt{S}e^{j\theta}(\mathbf{a}\circ\mathbf{m}_{\tilde{u}})$ and $\mathbf{C}_{\tilde{z}} = S[(\mathbf{a}\mathbf{a}^\dagger)\circ\mathbf{C}_{\tilde{u}}] + \sigma_w^2\mathbf{I}$.

12.2.4 Model for Extinction Coefficients

During the past several decades, considerable effort has been devoted to the modeling of wave propagation through random media. Theoretical models have been developed for the extinction coefficients of the first and second moments, μ and $v(\rho)$, along nearly line-of-sight paths. For general background, we refer the reader to Refs. [5,10,27,30]. Here we consider some specific results relevant to turbulence effects on aeroacoustic arrays.

The extent that scattering affects array performance depends on many factors, including the wavelength of the sound, the propagation distance from the source to the sensor array, the spacing between the sensors, the strength of the turbulence (as characterized by the variance of the temperature and wind-velocity fluctuations), and the size range of the turbulent eddies. Turbulence in the near-ground atmosphere spans a vast range of spatial scales, from millimeters to hundreds of meters. If the sensor spacing ρ is small compared to the size ℓ of the smallest eddies (a case highly relevant to optics but not low-frequency acoustics), $v(\rho)$ is proportional to $k^2\rho^2$, where $k = \omega/c_0$ is the wavenumber of the sound and c_0 the ambient sound speed [27]. In this situation, the loss in coherence between sensors results entirely from turbulence-induced variability in the AOA. Of greater practical importance in acoustics are situations where $\rho \gg \ell$. The spacing ρ may be smaller or larger than $\mathcal{L}$, the size of the largest eddies.

When $\rho \gg \ell$ and $\rho \ll \mathcal{L}$, the sensor spacing resides in the *inertial subrange* of the turbulence [5]. Because the strength of turbulence increases with the size of the eddies, this case has qualitative similarities to $\rho \ll \ell$. The wavefronts impinging on the array have a roughly constant AOA over the aperture and the apparent bearing of the source varies randomly about the actual bearing. Increasing the separation between sensors can dramatically decrease the coherence. In contrast, when $\rho \gg \mathcal{L}$ is large, the wavefront distortions induced by the turbulence produce nearly uncorrelated signal variations at the sensors. In this case, further increasing separation does not affect coherence: it is "saturated" at a value determined by the strength of the turbulence, and therefore has an effect similar to additive, uncorrelated noise. These two extreme cases are illustrated in Figure 12.3. The resulting behavior of $v(\rho)$ and B_{mn} (Equation 12.42) are shown in Figure 12.4.

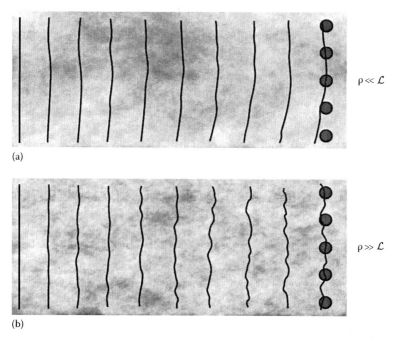

(a)

(b)

FIGURE 12.3 Turbulence-induced distortions of acoustic wavefronts impinging on an array. The wavefronts are initially smooth (left) and become progressively more distorted until they arrive at the array (right). (a) Sensor separations within the inertial subrange of the turbulence ($\rho \gg \ell$ and $\rho \ll \mathcal{L}$). The wavefronts are fairly smooth but the AOA (and therefore the apparent source bearing) varies. (b) Sensor separations much larger than the scale of the largest turbulent eddies ($\rho \gg \mathcal{L}$). The wavefronts have a very rough appearance and the effect of the scattering is similar to uncorrelated noise.

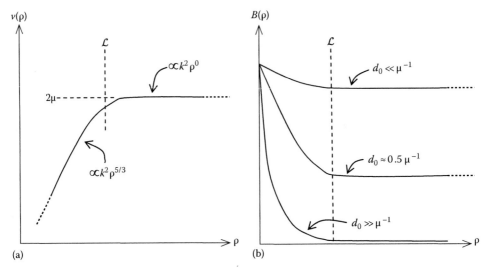

FIGURE 12.4 (a) Characteristic behavior of the second-moment extinction coefficient, $v(\rho)$. It initially increases with increasing sensor separation ρ, and then saturates at a fixed value 2μ (where μ is the first-moment extinction coefficient) when ρ is large compared to the size of the largest turbulent eddies. (b) Resulting behavior of the total signal coherence, B_{mn}, (12.42), for several values of the propagation distance d_o.

The general results for the extinction coefficients of a spherically propagating wave, derived with the parabolic (narrow-angle) and Markov approximations, and assuming $\rho \gg \ell$, are (Ref. [10], Equations 7.60 and 7.71; Ref. [30], Equations 20 through 28)

$$\mu = \frac{\pi^2 k^2}{2} \int_0^\infty dK_\perp K_\perp \Phi_{eff}(K_\parallel = 0, K_\perp) = \frac{k^2 \sigma_{eff}^2 \mathcal{L}_{eff}}{4}, \tag{12.50}$$

$$v(\rho) = \pi^2 k^2 \int_0^1 dt \int_0^\infty dK_\perp K_\perp \left[1 - J_0(K_\perp \rho t)\right] \Phi_{eff}(K_\parallel = 0, K_\perp), \tag{12.51}$$

in which J_0 is the zeroth-order Bessel function of the first kind and $\mathbf{K} = K_\parallel + K_\perp$ is the turbulence wavenumber vector decomposed into components parallel and perpendicular to the propagation path. The quantities $\Phi_{eff}(\mathbf{K})$, σ_{eff}, and $\mathcal{L}_{eff}$ are the effective turbulence spectrum, effective variance, and effective integral length scale. (The integral length scale is a quantitative measure of the size of the largest eddies.) The spectrum is defined as

$$\Phi_{eff}(\mathbf{K}) = \frac{\Phi_T(\mathbf{K})}{T_0^2} + \frac{4\Phi_v(\mathbf{K})}{c_0^2}, \tag{12.52}$$

where

T_0 is the ambient temperature

The subscripts T and v indicate the temperature and wind-velocity fields, respectively

The definition of the effective variance is the same, except with σ^2 replacing $\Phi(\mathbf{K})$. The effective integral length scale is defined as

$$\mathcal{L}_{eff} = \frac{1}{\sigma_{eff}} \left(\mathcal{L}_T \frac{\sigma_T^2}{T_0^2} + \mathcal{L}_v \frac{4\sigma_v^2}{c_0^2} \right). \tag{12.53}$$

For the case $\rho/\mathcal{L}_{\text{eff}} \gg 1$, the contribution from the term in (12.51) involving the Bessel function is small and one has $v(\rho) \rightarrow 2\mu$, as anticipated from the discussion after (12.40). When $\rho/\mathcal{L}_{\text{eff}} \ll 1$, the inertial-subrange properties of the turbulence come into play and one finds (Ref. [10], Equation 7.87)

$$v(\rho) = 0.137\left(\frac{C_T^2}{T_0^2} + \frac{22}{3}\frac{C_v^2}{c_0^2}\right)k^2\rho^{5/3},\qquad(12.54)$$

where C_T^2 and C_v^2 are the structure-function parameters for the temperature and wind fields, respectively. The structure-function parameters represent the strength of the turbulence in the inertial subrange.

Note that the extinction coefficients for both moments depend quadratically on the frequency of the tone, regardless of the separation between the sensors. The quantities μ, C_T^2, C_v^2, and $\mathcal{L}_{\text{eff}}$ each depend strongly on atmospheric conditions. Table 12.1 provides estimated values for typical atmospheric conditions based on the turbulence models in Refs. [11,24]. These calculations were performed for a propagation path height of 2 m.

It is evident from Table 12.1 that the entire range of saturation parameter values from $\Omega \approx 0$ to may be encountered in aeroacoustic applications, which typically have source ranges from meters to kilometers. Also, saturation occurs at distances several times closer to the source in sunny conditions than in cloudy ones. In a typical scenario in aeroacoustics involving a sensor standoff distance of several hundred meters, saturation will be small only for frequencies of about 100 Hz and lower. At frequencies above 200 Hz or so, the signal is generally saturated and random fluctuations dominate.

Based on the values for C_T^2 and C_v^2 in Table 12.1, coherence of signals is determined primarily by wind-velocity fluctuations (as opposed to temperature), except for mostly sunny, light wind conditions. It may at first seem a contradiction that the first-moment extinction coefficient μ is determined mainly by cloud cover (which affects solar heating of the ground), as opposed to the wind speed. Indeed, the source distance d_0 at which a given value of Ω is obtain is several times longer in cloudy conditions than in sunny ones. This can be understood from the fact that cloud cover damps strong thermal plumes (such as those used by hang gliders and seagulls to stay aloft), which are responsible for wind-velocity fluctuations that strongly affect acoustic signals.

Interestingly, the effective integral length scale for the sound field usually takes on a value intermediate between the microphone separations within small arrays (around 1 m) and the spacing between typical network nodes (which may be 100 m or more). As a result, high coherence can be expected

TABLE 12.1 Modeled Turbulence Quantities and Inverse Extinction Coefficients for Various Atmospheric Conditions

Atmospheric Condition	μ^{-1} (m) at 50 Hz	μ^{-1} (m) at 200 Hz	C_T^2/T_0^2 (m$^{-2/3}$)	$(22/3)\,C_v^2/c_0^2$ (m$^{-2/3}$)	$\mathcal{L}_{\text{eff}}$ (m)
Mostly sunny, light wind	990	62	2.0×10^{-5}	8.0×10^{-6}	100
Mostly sunny, moderate wind	980	61	7.6×10^{-6}	2.8×10^{-5}	91
Mostly sunny, strong wind	950	59	2.4×10^{-6}	1.3×10^{-4}	55
Mostly cloudy, light wind	2900	180	1.5×10^{-6}	4.4×10^{-6}	110
Mostly cloudy, moderate wind	2800	180	4.5×10^{-7}	2.4×10^{-5}	75
Mostly cloudy, strong wind	2600	160	1.1×10^{-7}	1.2×10^{-4}	28

The atmospheric conditions are described quantitatively in Ref. [24]. The second and third columns give the inverse extinction coefficients at 50 and 200 Hz, respectively. These values indicate the distance at which random fluctuations in the complex signal become strong. The fourth and fifth columns represent the relative contributions of temperature and wind fluctuations to the field coherence. The sixth column is the effective integral length scale for the scattered sound field; at sensor separations greater than this value, the coherence is "saturated."

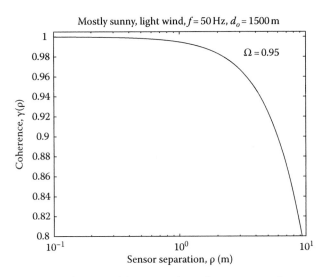

FIGURE 12.5   Evaluation of the coherence of the scattered signals at sensors with separation ρ, using $f = 50\,\text{Hz}$, $d_o = 1500\,\text{m}$, mostly sunny, light wind conditions (Table 12.1), $v(\rho)$ is computed with (12.54), and the coherence, $\gamma(\rho)$, is computed with (12.40).

within small arrays. However, coherence between nodes in a widely spaced network will be quite small, particularly at frequencies above 200 Hz or so.

Figure 12.5 illustrates the coherence of the scattered signals, γ_{mn} in (12.40), as a function of the sensor separation, ρ. The extinction coefficient in (12.54) is computed at frequency $f = 50\,\text{Hz}$ and source range $d_o = 1500\,\text{m}$, with mostly sunny, light wind conditions from Table 12.1, so $\Omega = 0.95$. Note the coherence is nearly perfect for sensor separations $\rho < 1\,\text{m}$, then the coherence declines steeply for larger separations.

12.2.5 Multiple Frequencies and Sources

The model in (12.49) is for a single source that emits a single frequency, $\omega = 2\pi f_o$ rad/s. The complex envelope processing in (12.2) and Figure 12.2 is a function of the source frequency. We can extend the model in (12.49) to the case of K sources that emit tones at L frequencies $\omega_1, \ldots, \omega_L$, as follows:

$$\tilde{\mathbf{z}}(iT_s; \omega_l) = \sum_{k=1}^{K} \sqrt{S_k(\omega_l)}\, e^{j\theta_{k,l}}\left(\mathbf{a}_k(\omega_l) \circ \tilde{\mathbf{u}}_k(iT_s; \omega_l)\right) + \tilde{\mathbf{w}}(iT_s; \omega_l), \quad \begin{array}{l} i = 1, \ldots, T \\ l = 1, \ldots, L \end{array} \tag{12.55}$$

$$= \left([\mathbf{a}_1(\omega_l) \ldots \mathbf{a}_K(\omega_l)] \circ [\tilde{\mathbf{u}}_1(iT_s; \omega_l) \ldots \tilde{\mathbf{u}}_K(iT_s; \omega_l)]\right) \begin{bmatrix} \sqrt{S_1(\omega_l)}\, e^{j\theta_{1,l}} \\ \vdots \\ \sqrt{S_K(\omega_l)}\, e^{j\theta_{K,l}} \end{bmatrix} + \tilde{\mathbf{w}}(iT_s; \omega_l)$$

$$\triangleq \left(\mathbf{A}(\omega_l) \circ \tilde{\mathbf{U}}(iT_s; \omega_l)\right)\tilde{\mathbf{p}}(\omega_l) + \tilde{\mathbf{w}}(iT_s; \omega_l). \tag{12.56}$$

In (12.55), $S_k(\omega_l)$ is the average power of source k at frequency ω_l, $\mathbf{a}_k(\omega_l)$ is the steering vector for source k at frequency ω_l as in (12.20), $\tilde{\mathbf{u}}_k(iT_s; \omega_l)$ is the scattering of source k at frequency ω_l at time sample i, and T is the number of time samples. In (12.56), the steering vector matrices $A(\omega_l)$, the scattering matrices $\tilde{\mathbf{U}}(iT_s; \omega_l)$, and the source amplitude vectors $\tilde{\mathbf{p}}(\omega_l)$ for $l = 1, \ldots, L$ and $i = 1, \ldots, T$, are defined by the context. If the sample spacing T_s is chosen appropriately, then the samples at a given frequency ω_l are independent in time. We will also model the scattered signals at different frequencies as independent. Cross-frequency coherence has been previously studied theoretically and experimentally, with Refs. [8,31] presenting experimental studies in the atmosphere. However, models for cross-frequency coherence in the atmosphere are at a very preliminary stage. It may be possible to revise the assumption of independent scattering at different frequencies as better models become available.

The covariance matrix at frequency ω_l is, by extending the discussion following (12.49),

$$\mathbf{C}_{\tilde{z}}(\omega_l) = \sum_{k=1}^{K} S_k(\omega_l)\Omega_k(\omega_l)\left[\Gamma_k(\omega_l)\circ\left(\mathbf{a}_k(\omega_l)\mathbf{a}_k(\omega_l)^\dagger\right)\right] + \sigma_{\tilde{w}}(\omega_l)^2\mathbf{I}, \quad (12.57)$$

where the scattered signals from different sources are assumed to be independent. If we assume full saturation ($\Omega_k(\omega_l) = 1$) and negligible coherence loss across the array aperture ($\Gamma_k(\omega_l) = 11^T$), then the sensor signals in (12.55) have zero mean, and the covariance matrix in (12.57) reduces to the familiar correlation matrix of the form

$$\mathbf{R}_{\tilde{z}}(0;\omega_l) = E\left\{\tilde{\mathbf{z}}(iT_s;\omega_l)\tilde{\mathbf{z}}(iT_s;\omega_l)^\dagger\right\}$$
$$= \mathbf{A}(\omega_l)\mathbf{S}(\omega_l)\mathbf{A}(\omega_l)^\dagger + \sigma_{\tilde{w}}(\omega_l)^2\mathbf{I} \quad (\Omega_k(\omega_l)=1 \text{ and no coherence loss}), \quad (12.58)$$

where $\mathbf{S}(\omega_l)$ is a diagonal matrix with $S_1(\omega_l),\ldots,S_K(\omega_l)$ along the diagonal.*

12.3 Signal Processing

In this section, we discuss signal processing methods for aeroacoustic sensor networks. The signal processing takes into account the source and propagation models presented in the previous section, as well as minimization of the communication bandwidth between sensor nodes connected by a wireless link. We begin with AOA estimation using a single sensor array in Section 12.3.1. Then we discuss source localization with multiple sensor arrays in Section 12.3.2, and we briefly describe implications for tracking, detection, and classification algorithms in Sections 12.3.3 and 12.3.4.

12.3.1 Angle of Arrival Estimation

We discuss narrowband AOA estimation with scattering in Section 12.3.1.1, and then we discuss wideband AOA estimation without scattering in Section 12.3.1.2.

12.3.1.1 Narrowband AOA Estimation with Scattering

In this section, we review some performance analyses and algorithms that have been investigated for narrowband AOA estimation with scattering. Most of the methods are based on scattering models that are similar to the single-source model in Section 12.2.3 or the multiple-source model in Section 12.2.5 at a single frequency. Many of the references cited below are formulated for radio frequency (RF) channels, so the equivalent channel effect is caused by multipath propagation and Doppler. The models for the RF case are similar to those presented in Section 12.2.

---

* For the fully saturated case with no coherence loss, we can relax the assumption that the scattered signals from different sources are independent by replacing the diagonal matrix $\mathbf{S}(\omega_l)$ in (12.58) with a positive semidefinite matrix with (m, n) element $\sqrt{S_m(\omega_l)S_n(\omega_l)}\cdot E\{\tilde{u}_m(iT_s;\omega_l)\,\tilde{u}_n(iT_s;\omega_l)^\star\}$, where $\tilde{u}_m(iT_s;\omega_l)$ is the scattered signal for source m.

262 *Image and Sensor Signal Processing*

Wilson [21] analyzed the Cramér-Rao bound (CRB) on AOA estimation for a single source using several models for atmospheric turbulence. Rayleigh signal fading was assumed. Collier and Wilson extended the work [22,23] to include unknown turbulence parameters in the CRB, along with the source AOA. Their CRB analysis provides insight into the combinations of atmospheric conditions, array geometry, and source location that are favorable for accurate AOA estimation. They note that refraction effects make it difficult to accurately estimate the elevation angle when the source and sensors are near the ground, so aeroacoustic sensor arrays are most effective for azimuth estimation.

Other researchers that have investigated the problem of imperfect spatial coherence in the context of narrowband AOA estimation include Refs. [32–40]. Paulraj and Kailath [32] presented a MUSIC algorithm that incorporates nonideal spatial coherence, assuming that the coherence losses are known. Song and Ritcey [33] provided maximum-likelihood (ML) methods for estimating the angles of arrival and the parameters in a coherence model. Gershman et al. [34] provided a procedure to jointly estimate the spatial coherence loss and the angles of arrival. In the series of papers [35–38], stochastic and deterministic models were studied for imperfect spatial coherence, and the performance of various AOA estimators was analyzed. Ghogho et al. [39] presented an algorithm for AOA estimation with multiple sources in the fully-saturated case. Their algorithm exploits the Toeplitz structure of the **B** matrix in (12.42) for a uniform linear array (ULA).

None of the Refs. [32–39] handle range of scattering scenarios from weak ($\Omega = 0$) to strong ($\Omega = 1$). Fuks et al. [40] treat the case of Rician scattering on RF channels, so this approach does include the entire range from weak to strong scattering. Indeed, the "Rice factor" in the Rician fading model is related to the saturation parameter through $(1 - \Omega)/\Omega$. The main focus in Ref. [40] is on CRBs for AOA estimation.

12.3.1.2 Wideband AOA Estimation without Scattering

Narrow band processing in the aeroacoustic context will limit the bandwidth to perhaps a few hertz, and the large fractional bandwidth encountered in aeroacoustics significantly complicates the array signal processing. A variety of methods are available for wideband AOA estimation, with varying complexity and applicability. Application of these to specific practical problems leads to a complicated task of appropriate procedure choice. We outline some of these methods and various tradeoffs, and describe some experimental results. Basic approaches include: classical delay-and-sum beamformer, incoherent averaging over narrow band spatial spectra, maximum likelihood, coherent signal subspace methods, steered matrix techniques, spatial resampling (array interpolation), and frequency-invariant beamforming. Useful overviews include Boehme [41], and Van Trees [42]. Significant progress in this area has occurred in the previous 15 years or so; major earlier efforts include the underwater acoustics area, e.g., see Owsley [43].

Using frequency decomposition at each sensor, we obtained the array data model in (12.55). For our discussion of wideband AOA methods, we will ignore the scattering, and so assume the spatial covariance can be written as in (12.58). Equation 12.58 may be interpreted as the covariance matrix of the Fourier-transformed (narrowband) observations (12.55). The noise is typically assumed to be Gaussian and spatially white, although generalizations to spatially correlated noise are also possible, which can be useful for modeling unknown spatial interference.

Working with an estimate $\hat{\mathbf{R}}_{\tilde{z}}(0;\omega_l)$, we may apply covariance-based high resolution AOA estimators (MUSIC, MLE, etc.), although this results in many frequency-dependent angle estimates that must be associated in some way for each source. A simple approach is to sum the resulting narrowband spatial spectra, e.g., see Ref. [44]; this is referred to as noncoherent averaging. This approach has the advantages of straightforward extension of narrowband methods and relatively low complexity, but can produce artifacts. And, noncoherent averaging requires that the SNRs after channelization be adequate to support the chosen narrow band AOA estimator; in effect the method does not take strong advantage of the wideband nature of the signal. However, loud harmonic sources can be processed in this manner with success.

A more general approach was first developed by Wang and Kaveh [45], based on the following additive composition of transformed narrowband covariance matrices:

$$\mathbf{R}_{scm}(\phi_i) = \sum_l \mathbf{T}(\phi_i, \omega_l)\mathbf{R}_{\tilde{z}}(0; \omega_l)\mathbf{T}(\phi_i, \omega_l)^\dagger, \tag{12.59}$$

where ϕ_i is the ith AOA. $\mathbf{R}_{scm}(\phi_i)$ is referred to as the *steered covariance matrix* or the *focused wideband covariance matrix*. The transformation matrix $\mathbf{T}(\phi_i, \omega_l)$, sometimes called the *focusing matrix*, can be viewed as selecting delays to coincide with delay-sum beamforming, so that the transformation depends on both AOA and frequency. Viewed in another way, the transformation matrix acts to align the signal subspaces, so that the resulting matrix $\mathbf{R}_{scm}(\phi_i)$ has a rank one contribution from a wideband source at angle ϕ_i. Now, narrowband covariance-based AOA estimation methods may be applied to the matrix $\mathbf{R}_{scm}(\phi_i)$. This approach is generally referred to as the *coherent subspace method* (CSM). CSM has significant advantages: it can handle correlated sources (due to the averaging over frequencies), it averages over the entire source bandwidth, and has good statistical stability. On the other hand, it requires significant complexity, and as originally proposed requires pre-estimation of the AOAs which can lead to biased estimates [46]. (Valaee and Kabal [47] present an alternative formulation of focusing matrices for CSM using a two-sided transformation, attempting to reduce the bias associated with CSM.)

A major drawback to CSM is the dependence of $\mathbf{T}$ on the AOA. The most general form requires generation and eigendecomposition of $\mathbf{R}_{scm}(X\phi_i)$ for each look angle; this is clearly undesirable from a computational standpoint.[*] The dependence of $\mathbf{T}$ on ϕ_i can be removed in some cases by incorporating spatial interpolation, thereby greatly reducing the complexity. The basic ideas are established by Krolik and Swingler in Ref. [48]; for an overview (including CSM methods) see Krolik [49].

As an example, consider a ULA [48,49], with $d = \lambda_i/2$ spacing. In order to process over another wavelength choice $\lambda_j (\lambda_j > \lambda_i)$, we could spatially interpolate the physical array to a virtual array with the desired spacing ($d_j = \lambda_j/2$). The spatial resampling approach adjusts the spatial sampling interval d as a function of source wavelength λ_j. The result is a simplification of (12.59) to

$$\mathbf{R}_{sr} = \sum_l \mathbf{T}(\omega_l)\mathbf{R}_{\tilde{z}}(0; \omega_l)\mathbf{T}(\omega_l)^\dagger, \tag{12.60}$$

where the angular dependence is now removed. The resampling acts to align the signal subspace contributions over frequency, so that a single wideband source results in a rank one contribution to $\mathbf{R}_{sr}$. Note that the spatial resampling is implicit in (12.60) via the matrices $\mathbf{T}(\omega_l)$. Conventional narrow band AOA estimation methods may now be applied to $\mathbf{R}_{sr}$ and, in contrast to CSM, this operation is conducted once for all angles.

Extensions of Ref. [48] from ULAs to arbitrary array geometries can be undertaken, but the dependence on look angle returns, and the resulting complexity is then similar to the CSM approaches. To avoid this, Friedlander and Weiss considered spatial interpolation of an arbitrary physical array to virtual arrays that are uniform and linear [50], thereby returning to a formulation like (12.60). Doron et al. [51] developed a spatial interpolation method for forming a focused covariance matrix with arbitrary arrays. The formulation relies on a truncated series expansion of plane waves in polar coordinates. The array manifold vector is now separable, allowing focusing matrices that are not a function of angle. The specific case of a *circular* array leads to an FFT-based implementation that is appealing due to its relatively low complexity.

While the spatial resampling methods are clearly desirable from a complexity standpoint, experiments indicate that they break down as the fractional bandwidth grows (see the examples that follow).

---

[*] In their original work, Wang and Kaveh relied on pre-estimates of the AOAs to lower the computational burden [45].

This depends on the particular method, and the original array geometry. This may be due to accumulated interpolation error, undersampling, and calibration error. As we have noted, and show in our examples, fractional bandwidths of interest in aeroacoustics may easily exceed 100%. Thus, the spatial resampling methods should be applied with some caution in cases of large fractional bandwidth.

Alternatives to the CSM approach are also available. Many of these methods incorporate time domain processing, and so may avoid the frequency decomposition (DFT) associated with CSM. Buckley and Griffiths [52], and Agrawal and Prasad [53], have developed methods based on wideband correlation matrices. (The work of Agrawal and Prasad [53] generally relies on a white or near-white source spectrum assumption, and so might not be appropriate for harmonic sources.) Sivanand et al. [54–56] have shown that the CSM focusing can be achieved in the time domain, and treat the problem from a multichannel FIR filtering perspective. Another FIR based method employs frequency invariant beamforming, e.g., see Ward et al. [57] and references therein.

12.3.1.3 Performance Analysis and Wideband Beamforming

Cramer-Rao bounds (CRBs) on wideband AOA estimation can be established using either a deterministic or random Gaussian source model, in additive Gaussian noise. The basic results were shown by Bangs [58]; see also Swingler [59]. The deterministic source case in (possibly colored) Gaussian noise is described in Kay [20]. Performance analysis of spatial resampling methods is considered by Friedlander and Weiss, who also provide CRBs, as well as a description of ML wideband AOA estimation [50].

These CRBs typically require known source statistics, apply to unbiased estimates, and assume no scattering, whereas prior spectrum knowledge is usually not available, and the above wideband methods may result in biased estimates. Nevertheless, the CRB provides a valuable fundamental performance bound.

Basic extensions of narrow band beamforming methods are reviewed in Van Trees [42, Chapter 6], including delay-sum and wideband minimum variance distortionless response (MVDR) techniques. The CSM techniques also extend to wideband beamforming, e.g., see Yang and Kaveh [60].

12.3.1.4 AOA Experiments

Next, we highlight some experimental examples and results, based on extensive aeroacoustic experiments carried out since the early 1990s [3,61–66]. These experiments were designed to test wideband superresolution AOA estimation algorithms based on array apertures of a few meters or less. The arrays were typically only approximately calibrated, roughly operating in [50,250] Hz, primarily circular in geometry, and planar (on the ground). Testing focused on military vehicles, and low flying rotary and fixed wing aircraft, and ground truth was typically obtained from GPS receivers on the sources.

Early results showed that superresolution AOA estimates could be achieved at ranges of 1–2 km [61], depending on the various propagation conditions and source loudness, and that non-coherent summation of narrowband MUSIC spatial signatures significantly outperforms conventional wideband delay-sum beamforming [62]. When the sources had strong harmonic structure, it was a straightforward matter to select the spectral peaks for narrowband AOA estimation. These experiments also verified that a piece-wise stationary assumption was valid over intervals approximately below 1 s, that the observed spatial coherence was good over apertures of a few meters or less, and that only rough calibration was required with relatively inexpensive microphones. Outlier AOA estimates were also observed, even in apparently high SNR and good propagation conditions. In some cases outliers composed 10% of the AOA estimates, but these were infrequent enough that a robust tracking algorithm can reject them.

Tests of the CSM method (CSM-MUSIC) were conducted with diesel engine vehicles exhibiting strong harmonic signatures [63], as well as turbine engines exhibiting broad, relatively flat spectral signatures [64]. The CSM-MUSIC approach was contrasted with noncoherent MUSIC. In both cases the M largest spectral bins were selected adaptively for each data block. CSM-MUSIC was implemented with focusing matrix $\mathbf{T}$ diagonal. For harmonic source signatures, the noncoherent MUSIC method was

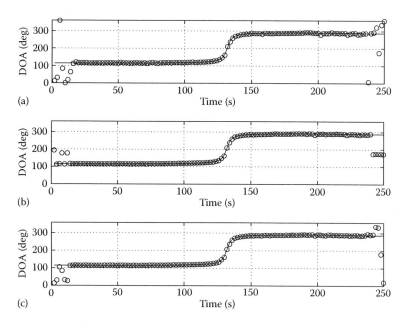

FIGURE 12.6 Experimental wideband AOA estimation over 250 s, covering a range of approximately ±1 km. Three methods are depicted with M highest SNR frequency bins: (a) narrowband MUSIC ($M = 1$), (b) incoherent MUSIC ($M = 20$), and (c) CSM-MUSIC ($M = 20$). Solid lines depict GPS-derived AOA ground truth.

shown to outperform CSM-MUSIC in many cases, generally depending on the observed narrowband SNRs [63]. On the other hand, the CSM-MUSIC method displays good statistical stability at a higher computational cost. And, inclusion of lower SNR frequency bins in noncoherent MUSIC can lead to artifacts in the resulting spatial spectrum.

For the broadband turbine source, the CSM-MUSIC approach generally performed better than noncoherent MUSIC, due to the ability of CSM to capture the broad spectral spread of the source energy [64]. Figure 12.6 depicts a typical experiment with a turbine vehicle, showing AOA estimates over a 250 s span, where the vehicle traverses approximately a ±1 km path past the array. The largest $M = 20$ frequency bins were selected for each estimate. The AOA estimates (circles) are overlaid on GPS ground truth (solid line). The AOA estimators break down at the farthest ranges (the beginning and end of the data). Numerical comparison with the GPS-derived AOA's reveals the CSM-MUSIC to have slightly lower mean square error. While the three AOA estimators shown in Figure 12.6 for this single source case have roughly the same performance, we emphasize that examination of the beam patterns reveals that the CSM-MUSIC method exhibits the best statistical stability and lower sidelobe behavior over the entire data set [64]. In addition, the CSM-MUSIC approach exhibited better performance in multiple source testing.

Experiments with the spatial resampling approaches reveal that they require spatial oversampling to handle large fractional bandwidths [65,66]. For example, the array manifold interpolation (AMI) method of Doron et al. [51] was tested experimentally and via simulation using a 12-element uniform circular array. While the CSM-MUSIC approach was asymptotically efficient in simulation, the AMI technique did not achieve the CRB. The AMI algorithm performance degraded as the fractional bandwidth was increased for a fixed spatial sampling rate. While the AMI approach is appealing from a complexity standpoint, effective application of AMI requires careful attention to the fractional bandwidth, maximum source frequency, array aperture, and degree of oversampling. Generally, the AMI approach required higher spatial sampling when compared to CSM type methods, and so AMI lost some of its potential complexity savings in both hardware and software.

12.3.2 Localization with Distributed Sensor Arrays

The previous section was concerned with AOA estimation using a single sensor array. The (x, y) location of a source in the plane may be estimated efficiently using multiple sensor arrays that are distributed over a wide area. We consider source localization in this section using a network of sensors that are placed in an "array of arrays" configuration, as illustrated in Figure 12.7. Each array contains local processing capability and a wireless communication link with a fusion center. A standard approach for estimating the source locations involves AOA estimation at the individual arrays, communication of the bearings to the fusion center, and triangulation of the bearing estimates at the fusion center (e.g., see Refs. [67–71]). This approach is characterized by low communication bandwidth and low complexity, but the localization accuracy is generally inferior to the optimal solution in which the fusion center jointly processes all of the sensor data. The optimal solution requires high communication bandwidth and high processing complexity. The amount of improvement in localization accuracy that is enabled by greater communication bandwidth and processing complexity is dependent on the scenario, which we characterize in terms of the power spectra (and bandwidth) of the signals and noise at the sensors, the coherence between the source signals received at widely separated sensors, and the observation time (amount of data).

   We have studied this scenario in Ref. [16], where a framework is presented to identify situations that have the potential for improved localization accuracy relative to the standard bearings-only triangulation method. We proposed an algorithm that is bandwidth-efficient and nearly optimal that uses beamforming at small-aperture sensor arrays and TDE between widely-separated sensors. Accurate TD estimates using widely-separated sensors are utilized to achieve improved localization accuracy relative to bearings-only triangulation, and the scattering of acoustic signals by the atmosphere significantly impacts the accuracy of TDE. We provide a detailed study of TDE with scattered signals that are *partially* coherent at widely-spaced sensors in Ref. [16]. Our results quantify the scenarios in which TDE is feasible as a function of signal coherence, SNR per sensor, fractional bandwidth of the signal, and time-bandwidth product of the observed data. The basic result is that for a given SNR, fractional bandwidth, and time-bandwidth product, there exists a "threshold coherence" value that must be exceeded in order for TDE to achieve the CRB. The analysis is based on Ziv-Zakai bounds for TDE, expanding upon the results in Refs. [72,73]. Time synchronization is required between the arrays for TDE.

   Previous work on source localization with aeroacoustic arrays has focused on AOA estimation with a *single* array, e.g., Refs. [61–66,74,75], as discussed in Section 12.3.1. The problem of imperfect spatial coherence in the context of narrowband AOA estimation with a single array was studied in Refs. [21–23,32–40], as discussed in Section 12.3.1.1. The problem of decentralized array processing was studied in Refs. [76,77]. Wax and Kailath [76] presented subspace algorithms for narrowband

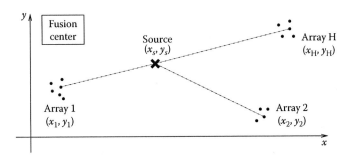

FIGURE 12.7 Geometry of non-moving source location and an array of arrays. A communication link is available between each array and the fusion center. (Originally published in Kozick, R.J. and Sadler, B.M., *IEEE Trans. Signal Process.*, © 2003 IEEE, reprinted with permission.)

signals and distributed arrays, assuming perfect spatial coherence across each array but neglecting any spatial coherence that may exist between arrays. Stoica et al. [77] considered maximum likelihood AOA estimation with a large, perfectly coherent array that is partitioned into subarrays. Weinstein [78] presented performance analysis for pairwise processing of the wideband sensor signals from a single array, and he showed that pairwise processing is nearly optimal when the SNR is high. In Ref. [79], Moses and Patterson studied autocalibration of sensor arrays, where for aeroacoustic arrays the loss of signal coherence at widely-separated sensors impacts the performance of autocalibration.

The results in Ref. [16] are distinguished from those cited in the previous paragraph in that the primary focus is a performance analysis that explicitly models partial spatial coherence in the signals at *different* sensor arrays in an array of arrays configuration, along with an analysis of decentralized processing schemes for this model. The previous works have considered wideband processing of aeroacoustic signals using a single array with perfect spatial coherence [61–66,74,75], imperfect spatial coherence across a single array aperture [21–23,32–40], and decentralized processing with either zero coherence between distributed arrays [76] or full coherence between all sensors [77,78]. We summarize the key results from Ref. [16] in Sections 12.3.2.1 through 12.3.2.3.

Source localization using the method of travel-time tomography is described in Refs. [80,81]. In this type of tomography, TDEs are formed by cross correlating signals from widely spaced sensors. The TDEs are incorporated into a general inverse procedure that provides information on the atmospheric wind and temperature fields in addition to the source location. The tomography thereby adapts to time delay shifts that result from the intervening atmospheric structure.

Ferguson [82] describes localization of small-arms fire using the near-field wavefront curvature. The range and bearing of the source are estimated from two adjacent sensors. Ferguson's experimental results clearly illustrate random localization errors induced by atmospheric turbulence. In a separate article, Ferguson [83] discusses time-scale compression to compensate TDEs for differential Doppler resulting from fast-moving sources.

12.3.2.1 Model for Array of Arrays

Our model for the array of arrays scenario in Figure 12.7 is a wideband extension of the single-array, narrowband model in Section 12.2. Our array of arrays model includes two key assumptions:

1. The distance from the source to each array is sufficiently large so that the signals are fully saturated, i.e., $\Omega^{(h)}(\omega) \approx 1$ for $h = 1, ..., H$ and all ω. Therefore according to the model in Section 12.2.3, the sensor signals have zero mean.
2. Each array aperture is sufficiently small so that the coherence loss is negligible between sensor pairs in the array. For the example in Figure 12.5, this approximation is valid for array apertures less than 1 m.

It may be useful to relax these assumptions in order to consider the effects of nonzero mean signals and coherence losses across individual arrays. However, these assumptions allow us to focus on the impact of coherence losses in the signals at *different* arrays.

As in Section 12.2.1, we let (x_s, y_s) denote the coordinates of a single non-moving source, and we consider H arrays that are distributed in the same plane, as illustrated in Figure 12.7. Each array $h \in \{1, ..., H\}$ contains N_h sensors and has a reference sensor located at coordinates (x_h, y_h). The location of sensor $n \in \{1, ..., N_h\}$ is at $(x_h + \Delta x_{hn}, y_h + \Delta y_{hn})$, where $(\Delta x_{hn}, \Delta y_{hn})$ is the relative location with respect to the reference sensor. If c is the speed of propagation, then the propagation time from the source to the reference sensor on array h is

$$\tau_h = \frac{d_h}{c} = \frac{1}{c}\left[(x_s - x_h)^2 + (y_s - y_h)^2\right]^{1/2}, \tag{12.61}$$

where d_h is the distance from the source to array h, as in (12.5). We model the wavefronts over individual array apertures as perfectly coherent plane waves, so in the far-field approximation, the propagation time from the source to sensor n on array h is expressed by $\tau_h + \tau_{hn}$, where

$$\tau_{hn} \approx -\frac{1}{c}\left[\frac{x_s - x_h}{d_h}\Delta x_{hn} + \frac{y_s - y_h}{d_h}\Delta y_{hn}\right] = -\frac{1}{c}\left[(\cos\phi_h)\Delta x_{hn} + (\sin\phi_h)\Delta y_{hn}\right] \tag{12.62}$$

is the propagation time from the reference sensor on array h to sensor n on array h, and ϕ_h is the bearing of the source with respect to array h. Note that while the far-field approximation (12.62) is reasonable over individual array apertures, the wavefront curvature that is inherent in (12.61) must be retained in order to model wide separations between arrays.

The time signal received at sensor n on array h due to the source will be denoted as $s_h(t - \tau_h - \tau_{hn})$, where the vector $\mathbf{s}(t) = [s_1(t), \ldots, s_H(t)]^T$ contains the signals received at the reference sensors on the H arrays. The elements of $\mathbf{s}(t)$ are modeled as real-valued, continuous-time, zero-mean, jointly wide-sense stationary, Gaussian random processes with $-\infty < t < \infty$. These processes are fully specified by the $H \times H$ cross-correlation matrix

$$\mathbf{R}_s(\xi) = E\{\mathbf{s}(t + \xi)\mathbf{s}(t)^T\}. \tag{12.63}$$

The (g, h) element in (12.63) is the cross-correlation function

$$r_{s,gh}(\xi) = E\{s_g(t + \xi)s_h(t)\} \tag{12.64}$$

between the signals received at arrays g and h. The correlation functions (12.63) and (12.64) are equivalently characterized by their Fourier transforms, which are the CSD functions in (12.65) and CSD matrix in (12.66):

$$G_{s,gh}(\omega) = \mathcal{F}\{r_{s,gh}(\xi)\} = \int_{-\infty}^{\infty} r_{s,gh}(\xi)\exp(-j\omega\xi)d\xi \tag{12.65}$$

$$\mathbf{G}_s(\omega) = \mathcal{F}\{\mathbf{R}_s(\xi)\}. \tag{12.66}$$

The diagonal elements $G_{s,hh}(\omega)$ of (12.66) are the PSD functions of the signals $s_h(t)$, and hence they describe the distribution of average signal power with frequency. The model allows the PSD to vary from one array to another to reflect differences in transmission loss and source aspect angle.

The off-diagonal elements of (12.66), $G_{s,gh}(\omega)$, are the CSD functions for the signals $s_g(t)$ and $s_h(t)$ received at distinct arrays $g \neq h$. In general, the CSD functions have the form

$$G_{s,gh}(\omega) = \gamma_{s,gh}(\omega)\left[G_{s,gg}(\omega)G_{s,hh}(\omega)\right]^{1/2}, \tag{12.67}$$

where $\gamma_{s,gh}(\omega)$ is the spectral coherence function for the signals, which has the property $0 \leq |\gamma_{s,gh}(\omega)| \leq 1$. Coherence magnitude $|\gamma_{s,gh}(\omega)| = 1$ corresponds to perfect correlation between the signals at arrays g and h, while the partially coherent case $|\gamma_{s,gh}(\omega)| < 1$ models random scattering in the propagation paths from the source to arrays g and h. Note that our assumption of perfect spatial coherence across individual arrays implies that the scattering has negligible impact on the intra-array delays τ_{hn} in (12.62) and the bearings $\phi_1, \ldots, \phi_H$. The coherence $\gamma_{s,gh}(\omega)$ in (12.67) is an extension of the narrowband, short-baseline coherence γ_{mn} in (12.39). However, the relation to extinction coefficients in (12.40) is not necessarily valid for very large sensor separations.

The signal received at sensor n on array h is the delayed source signal plus noise,

$$z_{hn}(t) = s_h(t - \tau_h - \tau_{hn}) + w_{hn}(t), \tag{12.68}$$

where the noise signals $w_{hn}(t)$ are modeled as real-valued, continuous-time, zero-mean, jointly wide-sense stationary, Gaussian random processes that are mutually uncorrelated at distinct sensors, and are uncorrelated from the signals. That is, the noise correlation properties are

$$E\{w_{gm}(t + \xi)w_{hn}(t)\} = r_w(\xi)\delta_{gh}\delta_{mn} \quad \text{and} \quad E\{w_{gm}(t + \xi)s_h(t)\} = 0, \tag{12.69}$$

where $r_w(\xi)$ is the noise autocorrelation function and the noise PSD is $G_w(\omega) = \mathcal{F}\{r_w(\xi)\}$.

We then collect the observations at each array h into $N_h \times 1$ vectors $z_h(t) = [z_{h1}(t), \ldots, Z_{h,N_h}(t)]^T$ for $h = 1, \ldots, H$, and we further collect the observations from the H arrays into a vector

$$\mathbf{Z}(t) = \left[\mathbf{z}_1(t)^T \quad \cdots \quad \mathbf{z}_H(t)^T \right]^T. \tag{12.70}$$

The elements of $\mathbf{Z}(t)$ in (12.70) are zero-mean, jointly wide-sense stationary, Gaussian random processes. We can express the CSD matrix of $\mathbf{Z}(t)$ in a convenient form with the following definitions. We denote the array steering vector for array h at frequency ω as

$$\mathbf{a}^{(h)}(\omega) = \begin{bmatrix} \exp(-j\omega\tau_{h1}) \\ \vdots \\ \exp(-j\omega\tau_{h,N_h}) \end{bmatrix} = \begin{bmatrix} \exp\left[j\dfrac{\omega}{c}((\cos\phi_h)\Delta x_{h1} + (\sin\phi_h)\Delta y_{h1}) \right] \\ \vdots \\ \exp\left[j\dfrac{\omega}{c}((\cos\phi_h)\Delta x_{h,N_h} + (\sin\phi_h)\Delta y_{h,N_h}) \right] \end{bmatrix}, \tag{12.71}$$

using τ_{hn} from (12.62) and assuming that the sensors have omnidirectional response. Let us define the relative time delay of the signal at arrays g and h as

$$D_{gh} = \tau_g - \tau_h, \tag{12.72}$$

where τ_h is defined in (12.61). Then the CSD matrix of $\mathbf{Z}(t)$ in (12.70) has the form

$$\mathbf{G_Z}(\omega) = \begin{bmatrix} \mathbf{a}^{(1)}(\omega)\mathbf{a}^{(1)}(\omega)^\dagger G_{s,11}(\omega) & \cdots & \mathbf{a}^{(1)}(\omega)\mathbf{a}^{(H)}(\omega)^\dagger \exp(-j\omega D_{1H})G_{s,1H}(\omega) \\ \vdots & \ddots & \vdots \\ \mathbf{a}^{(H)}(\omega)\mathbf{a}^{(1)}(\omega)^\dagger \exp(+j\omega D_{1H})G_{s,1H}(\omega)^* & \cdots & \mathbf{a}^{(H)}(\omega)\mathbf{a}^{(H)}(\omega)^\dagger G_{s,HH}(\omega) \end{bmatrix} + G_w(\omega)\mathbf{I}$$

$$\tag{12.73}$$

Recall that the source CSD functions $G_{s,gh}(\omega)$ in (12.73) depend on the signal PSDs and spectral coherence $\gamma_{s,gh}(\omega)$ according to (12.67). Note that (12.73) depends on the source location parameters (x_s, y_s) through the bearings ϕ_h in $\mathbf{a}^{(h)}(\omega)$ and the pairwise time-delay differences D_{gh}.

12.3.2.2 Cramér-Rao Bounds and Examples

The problem of interest is estimation of the source location parameter vector $\Theta = [x_s, y_s]^T$ using T independent samples of the sensor signals $\mathbf{Z}(0)$, $\mathbf{Z}(T_s)$, ..., $\mathbf{Z}((T-1) \ldots T_s)$, where T_s is the sampling period. The total observation time is $\mathcal{T} = T \cdot T_s$, and the sampling rate is $f_s = 1/T_s$ and $\omega_s = 2\pi f_s$. We will assume that the continuous-time random processes $Z(t)$ are band-limited, and that the sampling rate f_s is greater than twice the bandwidth of the processes. Then it has been shown [84,85] that the Fisher information matrix (FIM) $\mathbf{J}$ for the parameters Θ based on the samples $\mathbf{Z}(0)$, $\mathbf{Z}(T_s)$, ..., $\mathbf{Z}((T-1) \ldots T_s)$ has elements

$$J_{ij} = \frac{\mathcal{T}}{4\pi} \int_0^{\omega_s} \text{tr} \left\{ \frac{\partial \mathbf{G}_Z(\omega)}{\partial \theta_i} \mathbf{G}_Z(\omega)^{-1} \frac{\partial \mathbf{G}_Z(\omega)}{\partial \theta_j} \mathbf{G}_Z(\omega)^{-1} \right\} d\omega, \quad i,j = 1,2, \tag{12.74}$$

where "tr" denotes the trace of the matrix. The CRB matrix $\mathbf{C} = \mathbf{J}^{-1}$ then has the property that the covariance matrix of any unbiased estimator $\hat{\Theta}$ satisfies $\text{Cov}(\hat{\Theta}) - \mathbf{C} \geq 0$, where ≥ 0 means that $\text{Cov}(\hat{\Theta}) - \mathbf{C}$ is positive semidefinite. Equation 12.74 provides a convenient way to compute the FIM for the array of arrays model as a function of the signal coherence between distributed arrays, the signal and noise bandwidth and power spectra, and the sensor placement geometry.

The CRB presented in (12.74) provides a performance bound on source location estimation methods that *jointly* process all the data from all the sensors. Such processing provides the best attainable results, but also requires significant communication bandwidth to transmit data from the individual arrays to the fusion center. Next we develop approximate performance bounds on schemes that perform bearing estimation at the individual arrays in order to reduce the required communication bandwidth to the fusion center. These CRBs facilitate a study of the tradeoff between source location accuracy and communication bandwidth between the arrays and the fusion center. The methods that we consider are summarized as follows:

1. Each array estimates the source bearing, transmits the bearing estimate to the fusion center, and the fusion processor triangulates the bearings to estimate the source location. This approach does not exploit wavefront coherence between the distributed arrays, but it greatly reduces the communication bandwidth to the fusion center.
2. The raw data from all sensors is jointly processed to estimate the source location. This is the optimum approach that fully utilizes the coherence between distributed arrays, but it requires large communication bandwidth.
3. Combination of methods 1 and 2, where each array estimates the source bearing and transmits the bearing estimate to the fusion center. In addition, the raw data from one sensor in each array is transmitted to the fusion center. The fusion center estimates the propagation time delay between pairs of distributed arrays, and processes these time delay estimates with the bearing estimates to localize the source.

Next we evaluate CRBs for the three schemes for a narrowband source and a wideband source. Consider $H = 3$ identical arrays, each of which contains $N_1 = \ldots = N_H = 7$ sensors. Each array is circular with 4 ft radius, and six sensors are equally spaced around the perimeter and one sensor is in the center. We first evaluate the CRB for a narrowband source with a 1 Hz bandwidth centered at 50 Hz and SNR = 10 dB at each sensor. That is, $G_{s,hh}(\omega)/G_w(\omega) = 10$ for $h = 1, \ldots, H$ and $2\pi(49.5) < \omega < 2\pi(50.5)$ rad/s. The signal coherence $\gamma_{s,gh}(\omega) = \gamma_s(\omega)$ is varied between 0 and 1. We assume that $T = 4000$ time samples are obtained at each sensor with sampling rate $f_s = 2000$ samples/s. The source localization performance is evaluated by computing the ellipse in (x, y) coordinates that satisfies the expression $[x \quad y]\mathbf{J}\begin{bmatrix} x \\ y \end{bmatrix} = 1$, where $\mathbf{J}$ is the FIM in (12.74). If the errors in (x, y) localization are jointly Gaussian distributed, then the ellipse represents the contour at one standard deviation in root-mean-square (RMS) error. The error ellipse for any unbiased estimator of source location cannot be smaller than this ellipse derived from the FIM.

The $H = 3$ arrays are located at coordinates $(x_1, y_1) = (0, 0)$, $(x_2, y_2) = (400, 400)$, and $(x_3, y_3) = (100, 0)$, where the units are meters. One source is located at $(x_s, y_s) = (200, 300)$, as illustrated in Figure 12.8a. The RMS error ellipses for joint processing of all sensor data for coherence values $\gamma_s(\omega) = 0, 0.5$, and 1 are also shown in Figure 12.8a. The coherence between all pairs of arrays is assumed to be identical, i.e., $\gamma_{s,gh}(\omega) = \gamma_s(\omega)$ for $(g, h) = (1, 2), (1, 3), (2, 3)$. The largest ellipse in Figure 12.8a corresponds to incoherent signals, i.e., $\gamma_s(\omega) = 0$, and characterizes the performance of the simple method of triangulation using the bearing estimates from the three arrays. Figure 12.8b shows the ellipse radius = [(major axis)^2 + (minor axis)^2]$^{1/2}$ for various values of the signal coherence $\gamma_s(\omega)$. The ellipses for $\gamma_s(\omega) = 0.5$ and 1 are difficult to see in Figure 12.8a because they fall on the lines of the $\times$ that marks the source location, illustrating that signal coherence between the arrays significantly improves the CRB on source localization accuracy. Note also that for this scenario, the localization scheme based on bearing estimation with each array and TDE using one sensor from each array has the same CRB as the optimum, joint processing scheme.

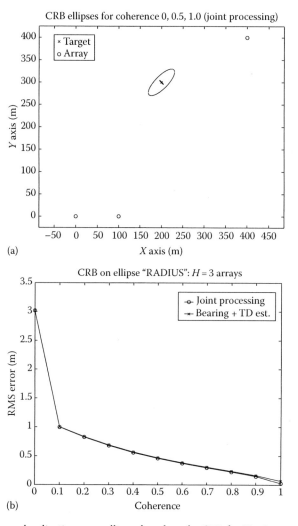

(a)

(b)

FIGURE 12.8 RMS source localization error ellipses based on the CRB for $H = 3$ arrays and one narrowband source in (a–c) and one wideband source in (d–f). (Originally published in Kozick, R.J. and Sadler, B.M., *IEEE Trans. Signal Process.*, © 2003 IEEE, reprinted with permission.)

(continued)

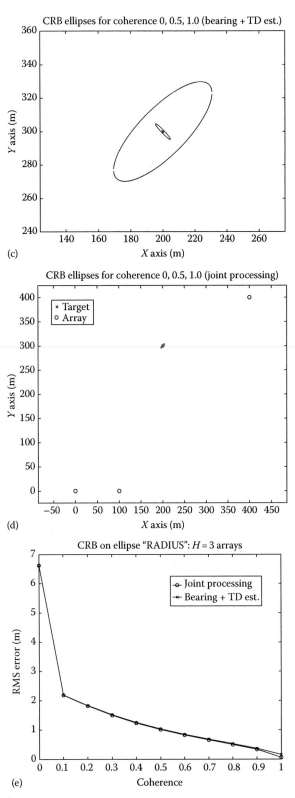

FIGURE 12.8 (continued)

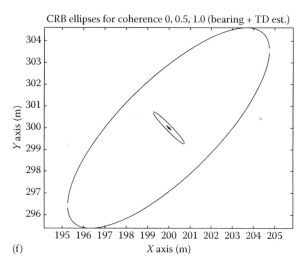

(f)

FIGURE 12.8 (continued)

Figure 12.8c shows a closer view of the error ellipses for the scheme of bearing estimation plus TDE with one sensor from each array. The ellipses are identical to those in Figure 12.8a for joint processing.

Figure 12.8d through f present corresponding results for a wideband source with bandwidth 20 Hz centered at 50 Hz and SNR 16 dB. That is, $G_{s,hh}/G_w = 40$ for $2\pi(40) < \omega < 2\pi(60)$ rad/s, $h = 1, \ldots, H$. $T = 2000$ time samples are obtained at each sensor with sampling rate $f_s = 2000$ samples/s, so the observation time is 1 s. As in the narrowband case in Figure 12.8a through c, joint processing reduces the CRB compared with bearings-only triangulation, and bearing plus TDE is nearly optimum.

The CRB provides a lower bound on the variance of unbiased estimates, so an important question is whether an estimator can *achieve* the CRB. We show next in Section 12.3.2.3 that the coherent processing CRBs for the narrowband scenario illustrated in Figure 12.8a through c are achievable only when the coherence is perfect, i.e. $\gamma_s = 1$. Therefore for that scenario, bearings-only triangulation is optimum in the presence of even small coherence losses. However, for the wideband scenario illustrated in Figure 12.8d through f, the coherent processing CRBs are achievable for coherence values $\gamma_s \gtrsim 0.75$.

12.3.2.3 TDE and Examples

The CRB results presented in Section 12.3.2.2 indicate that TDE between widely-spaced sensors may be an effective way to improve the source localization accuracy with joint processing. Fundamental performance limits for passive time delay and Doppler estimation have been studied extensively for several decades, e.g., see the collection of papers in Ref. [86]. The fundamental limits are usually parameterized in terms of the SNR at each sensor, the spectral support of the signals (fractional bandwidth), and the time-bandwidth product of the observations. However, the effect of coherence loss on TDE accuracy has not been explicitly considered.

In this section, we quantify the effect of partial signal coherence on TDE. We present Cramér-Rao and Ziv-Zakai bounds that are explicitly parameterized by the signal coherence, along with the traditional parameters of SNR, fractional bandwidth, and time-bandwidth product. This analysis of TDE is relevant to method 3 in Section 12.3.2.2. We focus on the case of $H = 2$ sensors here. The extension to $H > 2$ sensors is outlined in Ref. [16].

Let us specialize (12.68) to the case of two sensors, with $H = 2$ and $N_1 = N_2 = 1$, so

$$z_1(t) = s_1(t) + w_1(t) \quad \text{and} \quad z_2(t) = s_2(t - D) + w_2(t), \tag{12.75}$$

where $D = D_{21}$ is the differential time delay. Following (12.73), the CSD matrix is

$$\text{CSD}\begin{bmatrix} z_1(t) \\ z_2(t) \end{bmatrix} = \mathbf{G_Z}(\omega) = \begin{bmatrix} G_{s,11}(\omega) + G_w(\omega) & e^{+j\omega D}\gamma_{s,12}(\omega)[G_{s,11}(\omega)G_{s,22}(\omega)]^{1/2} \\ e^{-j\omega D}\gamma_{s,12}(\omega)^*[G_{s,11}(\omega)G_{s,22}(\omega)]^{1/2} & G_{s,22}(\omega) + G_w(\omega) \end{bmatrix}.$$

$$\text{(12.76)}$$

The signal coherence function $\gamma_{s,12}(\omega)$ describes the degree of correlation that remains in the signal emitted by the source at each frequency ω after propagating to sensors 1 and 2.

We consider the following simplified scenario. The signal and noise spectra are flat over a bandwidth of $\Delta\omega$ rad/s centered at ω_0 rad/s, the observation time is T s, and the propagation is fully saturated, so the signal mean is zero. Further, the signal PSDs are identical at each sensor, and we define the following constants for notational simplicity:

$$G_{s,11}(\omega_0) = G_{s,22}(\omega_0) = G_s, \quad G_w(\omega_0) = G_w, \quad \text{and} \quad \gamma_{s,12}(\omega_0) = \gamma_s. \qquad \text{(12.77)}$$

Then we can use (12.76) in (12.74) to find the CRB for TDE with $H = 2$ sensors, yielding

$$\text{CRB}(D) = \frac{1}{2\omega_0^2\left(\Delta\omega T/2\pi\right)\left[1 + \frac{1}{12}\left(\Delta\omega/\omega_0\right)^2\right]}\left[\frac{1}{|\gamma_s|^2}\left(1 + \frac{1}{(G_s/G_w)}\right)^2 - 1\right] \qquad \text{(12.78)}$$

$$> \frac{1}{2\omega_0^2\left(\Delta\omega T/2\pi\right)\left[1 + \frac{1}{12}\left(\Delta\omega/\omega_0\right)^2\right]}\left[\frac{1}{|\gamma_s|^2} - 1\right]. \qquad \text{(12.79)}$$

The quantity $(\Delta\omega T/2\pi)$ is the time-bandwidth product of the observations, $(\Delta\omega/\omega_0)$ is the fractional bandwidth of the signal, and G_s/G_w is the SNR at each sensor. Note from the high-SNR limit in (12.79) that when the signals are partially coherent, so that $|\gamma_s| < 1$, increased source power does not reduce the CRB. Improved TDE accuracy is obtained with partially coherent signals by increasing the observation time T or changing the spectral support of the signal, which is $[\omega_0 - \Delta\omega/2, \omega_0 + \Delta\omega/2]$. The spectral support of the signal is not controllable in passive TDE applications, so increased observation time is the only means for improving the TDE accuracy with partially coherent signals. Source motion becomes more important during long observation times, as we discuss in Section 12.3.3.

We have shown in Ref. [16] that the CRB on TDE is achievable only when the coherence, γ_s, exceeds a threshold. The analysis is based on Ziv-Zakai bounds as in Refs. [72,73], and the result is that the coherence must satisfy the following inequality in order for the CRB on TDE in (12.78) to be achievable:

$$|\gamma_s|^2 \geq \frac{\left(1 + \frac{1}{(G_s/G_w)}\right)^2}{1 + (1/\text{SNR}_{thresh})}, \quad \text{so} \quad |\gamma_s|^2 \geq \frac{1}{1 + (1/\text{SNR}_{thresh})} \quad \text{as} \quad \frac{G_s}{G_w} \to \infty. \qquad \text{(12.80)}$$

The quantity SNR_{thresh} is

$$\text{SNR}_{thresh} = \frac{6}{\pi^2\left(\Delta\omega T/2\pi\right)}\left(\frac{\omega_0}{\Delta\omega}\right)^2\left[\varphi^{-1}\left(\frac{1}{24}\left(\frac{\Delta\omega}{\omega_0}\right)^2\right)\right]^2 \qquad \text{(12.81)}$$

where $\varphi(y) = 1/\sqrt{2\pi} \int_y^\infty \exp(-t^2/2)dt$. Since $|\gamma_s|^2 \leq 1$, (12.80) is useful only if $G_s/G_w > \text{SNR}_{thresh}$. Note that the threshold coherence value in (12.80) is a function of the time-bandwidth product, $(\Delta\omega \cdot T/2\pi)$, and the fractional bandwidth, $(\Delta\omega/\omega_0)$, through the formula for SNR_{thresh} in (12.81).

Figure 12.9a contains a plot of (12.80) for a particular case in which the signals are in a band centered at $\omega_0 = 2\pi 50$ rad/s and the time duration is $T = 2$ s. Figure 12.9a shows the variation in threshold coherence as a function of signal bandwidth, $\Delta\omega$. Note that nearly perfect coherence is required when the signal bandwidth is less than 5 Hz (or 10% fractional bandwidth). The threshold coherence drops sharply for values of signal bandwidth greater than 10 Hz (20% fractional bandwidth). Thus for sufficiently

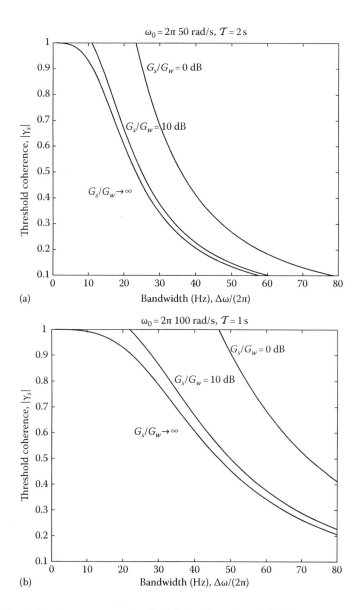

FIGURE 12.9 Threshold coherence versus bandwidth based on (12.80) for (a) $\omega_0 = 2\pi 50$ rad/s, $T = 2$ s and (b) $\omega_0 = 2\pi 100$ rad/s, $T = 1$ s for SNRs $G_s/G_w = 0$, 10, and ∞ dB.

(continued)

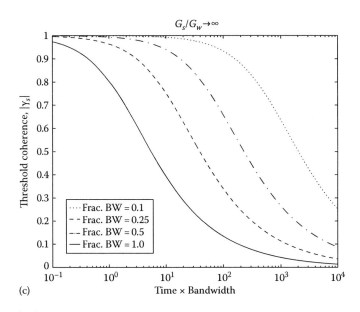

$$G_s/G_w \to \infty$$

FIGURE 12.9 (continued) (c) Threshold coherence value from (12.80) versus time-bandwidth product ($\Delta\omega \cdot \mathcal{T}/2\pi$) for several values of fractional bandwidth ($\Delta\omega/\omega_0$) and high SNR, $G_s/G_w \to \infty$. (Originally published in Kozick, R.J. and Sadler, B.M., *IEEE Trans. Signal Process.*, © 2003 IEEE, reprinted with permission.)

wideband signals, e.g., $\Delta\omega \geq 2\pi10$ rad/s, a certain amount of coherence loss can be tolerated while still allowing unambiguous TDE. Figure 12.9b shows corresponding results for a case with twice the center frequency and half the observation time. Figure 12.9c shows the threshold coherence as a function of the time-bandwidth product and the fractional bandwidth for large SNR, $G_s/G_w \to \infty$. Note that very large time-bandwidth product is required to overcome coherence loss when the fractional bandwidth is small. For example, if the fractional bandwidth is 0.1, then the time-bandwidth product must exceed 100 if the coherence is 0.9. For threshold coherence values in the range from about 0.1 to 0.9, each doubling of the fractional bandwidth reduces the required time-bandwidth product by a factor of 10.

Let us examine a scenario that is typical in aeroacoustics, with center frequency $f_o = \omega_o/(2\pi) = 50$ Hz and bandwidth $\Delta f = \Delta\omega/(2\pi) = 5$ Hz, so the fractional bandwidth is $\Delta f/f_o = 0.1$. From Figure 12.9c, signal coherence $|\gamma_s| = 0.8$ requires time-bandwidth product $\Delta f \cdot \mathcal{T} > 200$, so the necessary time duration $\mathcal{T} = 40$ s for TDE is impractical for moving sources.

Larger time-bandwidth products of the observed signals are required in order to make TDE feasible in environments with signal coherence loss. As discussed previously, only the observation time is controllable in passive applications, thus leading us to consider source motion models in Section 12.3.3 for use during long observation intervals.

We can evaluate the threshold coherence for the narrowband and wideband scenarios considered in Section 12.3.2.2 for the CRB examples in Figure 12.8. The results are as follows, using (12.80) and (12.81):

- Narrowband case: $G_s/G_w = 10$, $\omega_0 = 2\pi50$ rad/s, $\Delta\omega = 2\pi$ rad/s, $\mathcal{T} = 2$ s $\Rightarrow$ Threshold coherence ≈ 1
- Wideband case: $G_s/G_w = 40$, $\omega_0 = 2\pi50$ rad/s, $\Delta\omega = 2\pi20$ rad/s, $\mathcal{T} = 1$ s $\Rightarrow$ Threshold coherence ≈ 0.75

Therefore for the narrowband case, joint processing of the data from different arrays will not achieve the CRBs in Figure 12.8a through c when there is any loss in signal coherence. For the wideband case, joint processing can achieve the CRBs in Figure 12.8d through f for coherence values ≥ 0.75.

We have presented simulation examples in Ref. [16] that confirm the accuracy of the CRB in (12.78) and threshold coherence in (12.80). In particular, the simulations show that TDE based on cross-correlation processing achieves the CRB only when the threshold coherence is exceeded.

We conclude this section with a TDE example based on data that was measured by BAE Systems using a synthetically-generated, non-moving, wideband acoustic source. The source bandwidth is approximately 50 Hz with center frequency 100 Hz, so the fractional bandwidth is 0.5. Four nodes are labeled and placed in the locations shown in Figure 12.9a. The nodes are arranged in a triangle, with nodes on opposite vertices separated by about 330 ft, and adjacent vertices separated by about 230 ft. The source is at node 0, and receiving sensors are located at nodes 1, 2, and 3.

The PSDs estimated at sensors 1 and 3 are shown in Figure 12.10b, and the estimated coherence magnitude between sensors 1 and 3 is shown in Figure 12.10c. The PSDs and coherence are estimated using data segments of duration 1 s. Note that the PSDs are not identical due to differences in the

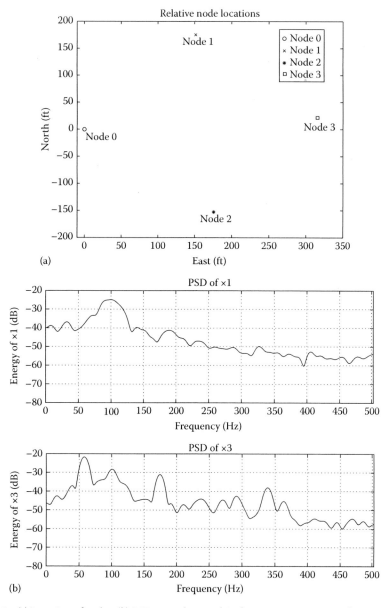

FIGURE 12.10   (a) Location of nodes. (b) PSDs at nodes 1 and 3 when transmitter is at node 0.

(*continued*)

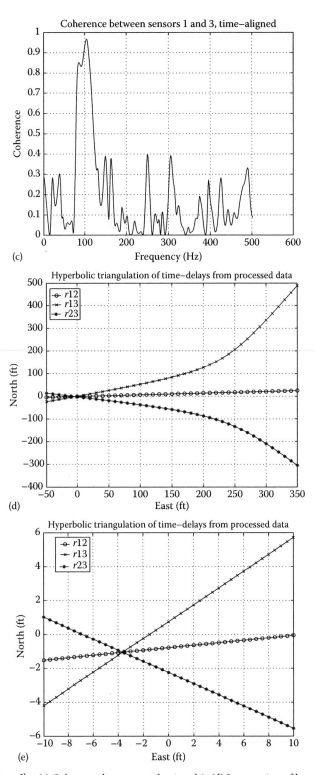

FIGURE 12.10 (continued)   (c) Coherence between nodes 1 and 3. (d) Intersection of hyperbolas obtained from differential time delays estimated at nodes 1, 2, and 3. (e) Expanded view of part (d). (Originally published in Kozick, R.J. and Sadler, B.M., *IEEE Trans. Signal Process.*, © 2003 IEEE, reprinted with permission.)

propagation paths. The coherence magnitude exceeds 0.8 over an appreciable band centered at 100 Hz. The threshold coherence value from (12.80) for the parameters in this experiment is 0.5, so the actual coherence of 0.8 exceeds the threshold. Thus accurate TDE should be feasible, and indeed, we found that generalized cross-correlation yielded accurated TDE estimates. Differential time delays were estimated using the signals measured at nodes 1, 2, and 3, and the TDEs were hyperbolically triangulated to estimate the location of the source (which is at node 0). Figure 12.10d shows the hyperbolas obtained from the three differential time delay estimates, and Figure 12.10e shows an expanded view near the intersection point. The triangulated location is within 1 ft of the true source location, which is at (−3, 0) ft.

This example shows the feasibility of TDE with acoustic signals measured at widely-separated sensors, provided that the SNR, fractional bandwidth, time-bandwidth product, and coherence meet the required thresholds. If the signal properties do not satisfy the thresholds, then accurate TDE is not feasible and triangulation of AOAs is optimum.

12.3.3 Tracking Moving Sources

In this section, we summarize past work and key issues for tracking moving sources. A widely-studied approach for estimating the locations of moving sources with an array of arrays involves bearing estimation at the individual arrays, communication of the bearings to the fusion center, and processing of the bearing estimates at the fusion center with a tracking algorithm (e.g., see Refs. [67–71]).

As discussed in Section 12.3.2, jointly processing data from widely-spaced sensors has the potential for improved source localization accuracy, compared with incoherent triangulation/tracking of bearing estimates. The potential for improved accuracy depends directly on TDE between the sensors, which is feasible only with an increased time-bandwidth product of the sensor signals. This leads to a constraint on the minimum observation time, T, in passive applications where the signal bandwidth is fixed. If the source is moving, then approximating it as non-moving becomes poorer as T increases, so modeling the source motion becomes more important.

Approximate bounds are known [87,88] that specify the maximum time interval over which moving sources are may be approximated as nonmoving for TDE. We have applied the bounds to a typical scenario in aeroacoustics [89]. Let us consider $H = 2$ sensors, and a vehicle moving at 15 m/s (about 5% the speed of sound), with radial motion that is in opposite directions at the two sensors. If the highest frequency of interest is 100 Hz, then the time interval over which the source is well-approximated as nonmoving is $T \ll 0.1$ s. According to the TDE analysis in Section 12.3.2, this yields insufficient time-bandwidth product for partially coherent signals that are typically encountered. Thus motion modeling and Doppler estimation/compensation are critical, even for aeroacoustic sources that move more slowly than in this example.

We have extended the model for a nonmoving source presented in Section 12.3.2 to a moving source with a first-order motion model in Ref. [89]. We have also presented an algorithm for estimating the motion parameters for multiple moving sources in Ref. [89], and the algorithm is tested with measured aeroacoustic data. The algorithm is initialized using the local polynomial approximation (LPA) beamformer [90] at each array to estimate the bearings and bearing rates. If the signals have sufficient coherence and bandwidth at the arrays, then the differential TDEs and Doppler shifts may be estimated. The maximum likelihood solution involves wideband ambiguity function search over Doppler and TDE [87], but computationally simpler alternatives have been investigated [91]. If TDE is not feasible, then the source may be localized by triangulating bearing, bearing rate, and differential Doppler. Interestingly, differential Doppler provides sufficient information for source localization, even without TDE, as long as five or more sensors are available [92]. Thus the source motion may be exploited via Doppler estimation in scenarios where TDE is not feasible, such as narrowband or harmonic signals.

Recent work on tracking multiple sources with aeroacoustic sensors include the penalized maximum likelihood approach in Ref. [75], and the $\alpha - \beta$/Kalman tracking algorithms in Ref. [93]. It may be feasible to use source aspect angle differences and Doppler estimation to help solve the data association problem in multiple target tracking based on data from multiple sensor arrays.

12.3.4 Detection and Classification

It is necessary to detect the presence of a source before carrying out the localization processing discussed in Sections 12.3.1 through 12.3.3. Detection is typically performed by comparing the energy at a sensor with a threshold. The acoustic propagation model presented in Section 12.2 implies that the energy fluctuates due to scattering, so the scattering has a significant impact on detection algorithms and their performance.

In addition to detecting a source and localizing its position, it is desirable to identify (or classify) the type of vehicle from its acoustic signature. The objective is to broadly classify into categories such as "ground, tracked," "ground, wheeled," "airborne, fixed wing," "airborne, rotary wing," and to further identify the particular vehicle type within these categories. Most classification algorithms that have been developed for this problem use the relative amplitudes of harmonic components in the acoustic signal as features to distinguish between vehicle types [94–101]. However, the harmonic amplitudes for a given source may vary significantly due to several factors. The scattering model presented in Section 12.2 implies that the energy in each harmonic will randomly fluctuate due to scattering, and the fluctuations will be stronger at higher frequencies. The harmonic amplitudes may also vary with engine speed and the orientation of the source with respect to the sensor (aspect angle).

In this section, we specialize the scattering model from Section 12.2 to describe the probability distribution for the energy at a single sensor for a source with a harmonic spectrum. We then discuss the implications for detection and classification performance. More detailed discussions may be found in Ref. [25] for detection and [102] for classification.

The source spectrum is assumed to be harmonic, with energy at frequencies $\omega_1, \ldots, \omega_L$. Following the notation in Section 12.2.5 and specializing to the case of one source and one sensor, $S(\omega_l)$, $\Omega(\omega_l)$, and $\sigma_w^2(\omega_l)$ represent the average source power, the saturation, and the average noise power at frequency ω_l, respectively. The complex envelope samples at each frequency ω_l are then modeled with the first element of the vector in (12.55) with $K = 1$ source, and they have a complex Gaussian distribution,

$$\tilde{z}(iT_s; \omega_l) \sim CN\left(\sqrt{\left[1 - \Omega(\omega_l)\right]S(\omega_l)}e^{j\theta(i;\omega_l)}, \Omega(\omega_l)S(\omega_l) + \sigma_w^2(\omega_l)\right), \quad \begin{matrix} i = 1, \ldots, T \\ l = 1, \ldots, L \end{matrix}. \tag{12.82}$$

The number of samples is T, and the phase $\theta(i; \omega_l)$ is defined in (12.21) and depends on the source phase and distance. We allow $\theta(i; \omega_l)$ to vary with the time sample index, i, in case the source phase (χ) or the source distance (d_o) changes. As discussed in Section 12.2.5, we model the complex Gaussian random variables in (12.82) as independent.

As discussed in Sections 12.2.3 and 12.2.4, the saturation (Ω) is related to the extinction coefficient of the first moment (μ) according to $\Omega(\omega_l) = 1 - \exp(-2\mu(\omega_l)d_o)$, where d_o is the distance from the source to the sensor. The dependence of the saturation on frequency and weather conditions is modeled by the following approximate formula for μ:

$$\mu(\omega) \approx \begin{cases} 4.03 \times 10^{-7}\left(\dfrac{\omega}{2\pi}\right)^2, & \text{mostly sunny} \\ \\ 1.42 \times 10^{-7}\left(\dfrac{\omega}{2\pi}\right)^2, & \text{mostly cloudy} \end{cases}, \quad \dfrac{\omega}{2\pi} \in [30, 200]\,\text{Hz}, \tag{12.83}$$

which is obtained by fitting (12.50) to the values for μ^{-1} in Table 12.1. A contour plot of the saturation as a function of frequency and source range is shown in Figure 12.11a using (12.83) for mostly sunny conditions. Note that the saturation varies significantly with frequency for ranges >100 m. Larger saturation values imply more scattering, so the energy in the higher harmonics will fluctuate more widely than the lower harmonics.

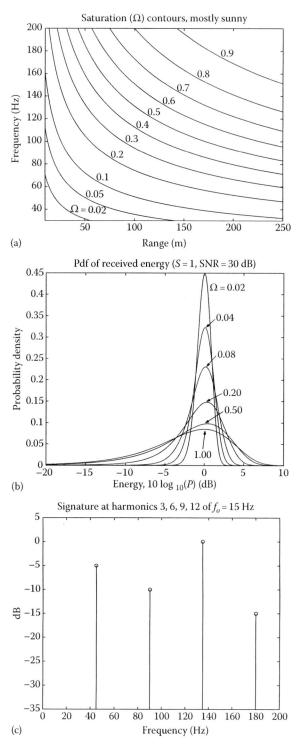

(a)

(b)

(c)

FIGURE 12.11 (a) Variation of saturation Ω with frequency f and range d_o. (b) Probability density function (pdf) of average power $10\log_{10}(P)$ measured at the sensor for $T = 1$ sample of a signal with $S = 1$ (0 dB), $\text{SNR} = 1/\sigma_w^2 = 10^3 = 30\,\text{dB}$, and various values of the saturation, Ω. (c) Harmonic signature with no scattering.

(*continued*)

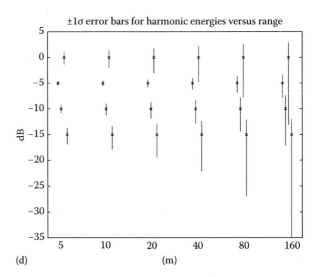

(d)   (m)

FIGURE 12.11 (continued)   (d) Error bars for harmonic signatures ±1 standard deviation caused by scattering at different source ranges.

We will let $P(\omega_1)$, ..., $P(\omega_L)$ denote the estimated energy at each frequency. The energy may be estimated from the complex envelope samples in (12.82) by coherent or incoherent combining:

$$P_C(\omega_l) = \left| \frac{1}{T} \sum_{i=1}^{T} \tilde{z}(iT_s; \omega_l) e^{-j\theta(i;\omega_l)} \right|^2, \quad l = 1, \ldots, L \qquad (12.84)$$

$$P_I(\omega_l) = \frac{1}{T} \sum_{i=1}^{T} \left| \tilde{z}(iT_s; \omega_l) \right|^2, \quad l = 1, \ldots, L. \qquad (12.85)$$

(12.85) Coherent combining is feasible only if the phase shifts $\theta(i; \omega_l)$ are known or are constant with i. Our assumptions imply that the random variables in (12.84) are independent over l, as are the random variables in (12.85). The probability distribution functions (pdfs) for P_C and P_I are noncentral chi-squared distributions.* We let $\chi^2(D, \delta)$ denote the standard noncentral chi-squared distribution with D degrees of freedom and noncentrality parameter δ. Then the random variables in (12.84) and (12.85) may be scaled so that their pdfs are standard noncentral chi-squared distributions

$$\frac{P_C(\omega_l)}{\left[\Omega(\omega_l)S(\omega_l) + \sigma_{\tilde{w}}^2(\omega_l) \right]/(2T)} \sim \chi^2(2, \delta(\omega_l)), \qquad (12.86)$$

$$\frac{P_I(\omega_l)}{\left[\Omega(\omega_l)S(\omega_l) + \sigma_{\tilde{w}}^2(\omega_l) \right]/(2T)} \sim \chi^2(2T, \delta(\omega_l)), \qquad (12.87)$$

---

* The random variable $\sqrt{P_C}$ in (12.84) has a Rician distribution, which is widely used to model fading RF communication channels.

where the noncentrality parameter is

$$\delta(\omega_l) = \frac{\left[1 - \Omega(\omega_l)\right]S(\omega_l)}{\left[\Omega(\omega_l)S(\omega_l) + \sigma_w^2(\omega_l)\right]/(2T)}. \tag{12.88}$$

The only difference in the pdfs for coherent and incoherent combining is the number of degrees of freedom in the noncentral chi-squared pdf: 2 degrees of freedom for coherent, and $2T$ degrees of freedom for incoherent.

The noncentral chi-squared pdf is readily available in analytical form and in statistical software packages, so the performance of detection algorithms may be evaluated as a function of SNR $= S/\sigma_w^2$ and saturation Ω. To illustrate the impact of Ω on the energy fluctuations, Figure 12.11b shows plots of the pdf of $10\log_{10}(P)$ for $T = 1$ sample (so coherent and incoherent are identical), $S = 1$, and SNR $= 1/\sigma_w^2 = 10^3 = 30$ dB. Note that a small deviation in the saturation from $\Omega = 0$ causes a significant spread in the distribution of P around the unscattered signal power, $S = 1$ (0 dB). This variation in P affects detection performance and limits the performance of classification algorithms that use P as a feature.

Figure 12.12 illustrates signal saturation effects on detection probabilities. In this example, the Neyman–Pearson detection criterion [103] with false-alarm probability of 0.01 was used. The noise is zero-mean Gaussian, as in Section 12.2.2. When $\Omega = 0$, the detection probability is nearly zero for SNR $= 2$ dB but quickly changes to one when the SNR increases by about 6 dB. When $\Omega = 1$, however, the transition is much more gradual: even at SNR $= 15$ dB, the detection probability is less than 0.9.

The impact of scattering on classification performance can be illustrated by comparing the fluctuations in the measured harmonic signature, $\mathbf{P} = [P(\omega_1), ..., P(\omega_L)]^T$, with the "true" signature, $\mathbf{S} = [S(\omega_1), ..., S(\omega_L)]^T$, that would be measured in the absence of scattering and additive noise. Figure 12.11c and d illustrate this variability in the harmonic signature as the range to the target increases. Figure 12.11c shows the "ideal" harmonic signature for this example (no scattering and no noise). Figure 12.11d shows ±1 standard deviation error bars on the harmonics for ranges 5, 10, 20, 40, 80, 160 m under "mostly sunny" conditions, using (12.83). For ranges beyond 80 m, the harmonic components display significant variations, and rank ordering of the harmonic amplitudes would exhibit variations also. The higher frequency

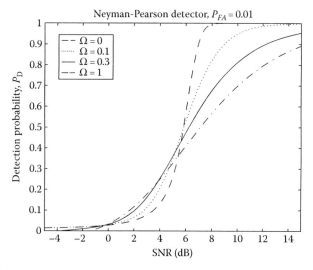

FIGURE 12.12 Probability of detection as a function of SNR for several values of the saturation parameter Ω. The Neyman–Pearson criterion is used with probability of false-alarm $P_{FA} = 0.01$.

harmonics experience larger variations, as expected. Classification based on relative harmonic amplitudes may experience significant performance degradations at these ranges, particularly for sources that have similar harmonic signatures.

12.4 Concluding Remarks

Aeroacoustics has a demonstrated capability for sensor networking applications, providing a low bandwidth sensing modality that leads to relatively low cost nodes. In battery operated conditions, where long lifetime in the field is expected, the node power budget is dominated by the cost of the communications. Consequently, the interplay between the communications and distributed signal processing is critical. We seek optimal network performance while minimizing the communication overhead.

We have considered the impact of the propagation phenomena on our ability to detect, localize, track, and classify acoustic sources. The strengths and limitations of acoustic sensing become clear in this light. Detection ranges and localization accuracy may be reasonably predicted. The turbulent atmosphere introduces spatial coherence losses that impact the ability to exploit large baselines between nodes for increased localization accuracy. The induced statistical fluctuations in amplitude place limits on the ability to classify sources at longer ranges. Very good performance has been demonstrated in many experiments; the analysis and experiments described here and elsewhere bound the problem and its solution space.

Because it is passive, and depends on the current atmospheric conditions, acoustic sensing may be strongly degraded in some cases. Passive sensing with high performance in all conditions will very likely require multiple sensing modalities, as well as hierarchical networks. This leads to interesting problems in fusion, sensor density and placement, as well as distributed processing and communications. For example, when very simple acoustic nodes with the limited capability of measuring loudness are densely deployed, they provide inherent localization capability [104,105]. Such a system, operating at relatively short ranges, provides significant robustness to many of the limitations described here, and may act to queue other sensing modalities for classification or even identification.

Localization based on accurate AOA estimation with short baseline arrays has been carefully analyzed, leading to well known triangulation strategies. Much more accurate localization, based on cooperative nodes, is possible in some conditions. These conditions depend fundamentally on the time-bandwidth of the observed signal, as well as the spatial coherence. For moving harmonic sources, these conditions are not likely to be supported, whereas sources that are more continuously broadband may be handled in at least some cases. It is important to note that the spatial coherence over a long baseline may be passively estimated in a straightforward way, leading to adaptive approaches that exploit the coherence when it is present. Localization updates, coupled with tracking, leads to an accurate picture of the nonstationary source environment.

Acoustic-based classification is the most challenging signal processing task, due to the source non-stationarities, inherent similarities between the sources, and propagation-induced statistical fluctuations. While the propagation places range limitations on present algorithms, it appears that the source similarities and nonstationarities may be the ultimate limiting factors in acoustic classification. Highly accurate classification will likely require the incorporation of other sensing modalities because of the challenging source characteristics.

Other interesting signal acoustic signal processing includes exploitation of Doppler, hierarchical and multi-modal processing, and handling multipath effects. Complex environments, such as indoor, urban, and forest, create multipath and diffraction that greatly complicate sensor signal processing and performance modeling. Improved understanding of the impact of these effects, and robust techniques for overcoming them, are needed. Exploitation of the very long range propagation distances possible with infrasound (frequencies below 20 Hz) [106] also requires further study and experimentation. Finally, we note that strong linkages between the communications network and the sensor signal processing are very important for overall resource utilization, especially including the MAC networking layer.

Acknowledgments

The authors thank Tien Pham of the Army Research Laboratory for contributions to the wideband AOA estimation material in this chapter, and we thank Sandra Collier of the Army Research Laboratory for many helpful discussions on beamforming in random media.

References

1. M. V. Namorato, A concise history of acoustics in warfare, *Appl. Acoust.*, 59: 101–135, 2000.
2. G. Becker and A. Güdesen, Passive sensing with acoustics on the battlefield, *Appl. Acoust.*, 59: 149–178, 2000.
3. N. Srour and J. Robertson, Remote netted acoustic detection system, Army Research Laboratory Technical Report, ARL-TR-706, May 1995.
4. T. F. W. Embleton, Tutorial on sound propagation outdoors, *J. Acoust. Soc. Am.*, 100: 31–48, 1996.
5. V. I. Tatarskii, *The Effects of the Turbulent Atmosphere on Wave Propagation* (Keter Press, Jerusalem, Israel, 1971).
6. J. M. Noble, H. E. Bass, and R. Raspet, The effect of large-scale atmospheric inhomogeneities on acoustic propagation, *J. Acoust. Soc. Am.*, 92: 1040–1046, 1992.
7. D. K. Wilson and D. W. Thomson, Acoustic propagation through anisotropic, surface-layer turbulence, *J. Acoust. Soc. Am.*, 96: 1080–1095, 1994.
8. D. E. Norris, D. K. Wilson, and D. W. Thomson, Correlations between acoustic travel-time fluctuations and turbulence in the atmospheric surface layer, *Acta Acust.*, 87: 677–684, 2001.
9. G. A. Daigle, T. F. W. Embleton, and J. E. Piercy, Propagation of sound in the presence of gradients and turbulence near the ground, *J. Acoust. Soc. Am.*, 79: 613–627, 1986.
10. V. E. Ostashev, *Acoustics in Moving Inhomogeneous Media* (E & FN Spon, London, U.K., 1997).
11. D. K. Wilson, A turbulence spectral model for sound propagation in the atmosphere that incorporates shear and buoyancy forcings, *J. Acoust. Soc. Am.*, 108(5), Pt. 1: 2021–2038, November 2000.
12. S. M. Kay, V. Nagesha, and J. Salisbury, Broad-band detection based on two-dimensional mixed autoregressive models, *IEEE Trans. Signal Process.*, 41(7): 2413–2428, July 1993.
13. M. Agrawal and S. Prasad, DOA estimation of wideband sources using a harmonic source model and uniform linear array, *IEEE Trans. Signal Process.*, 47(3): 619–629, March 1999.
14. M. Feder, Parameter estimation and extraction of helicopter signals observed with a wide-band interference, *IEEE Trans. Signal Process.*, 41(1): 232–244, January 1993.
15. M. Zeytinoglu and K. M. Wong, Detection of harmonic sets, *IEEE Trans. Signal Process.*, 43(11): 2618–2630, November 1995.
16. R. J. Kozick and B. M. Sadler, Source localization with distributed sensor arrays and partial spatial coherence, to appear in *IEEE Trans. Signal Process.*, 2003.
17. S. Morgan and R. Raspet, Investigation of the mechanisms of low-frequency wind noise generation outdoors, *J. Acoust. Soc. Am.*, 92: 1180–1183, 1992.
18. H. E. Bass, R. Raspet, and J. O. Messer, Experimental determination of wind speed and direction using a three microphone array, *J. Acoust. Soc. Am.*, 97: 695–696, 1995.
19. E. M. Salomons, *Computational Atmospheric Acoustics* (Kluwer, Dordrecht, the Netherlands, 2001).
20. S. M. Kay, *Fundamentals of Statistical Signal Processing, Estimation Theory* (Prentice-Hall, Englewood Cliffs, NJ, 1993).
21. D. K. Wilson, Performance bounds for acoustic direction-of-arrival arrays operating in atmospheric turbulence, *J. Acoust. Soc. Am.*, 103(3): 1306–1319, March 1998.
22. S. L. Collier and D. K. Wilson, Performance bounds for passive arrays operating in a turbulent medium: Plane-wave analysis, *J. Acoust. Soc. Am.*, 113(5): 2704–2718, May 2003.
23. S. L. Collier and D. K. Wilson, Performance bounds for passive sensor arrays operating in a turbulent medium II: Spherical-wave analysis, in review at *J. Acoust. Soc. Am.*, June 2003.

24. V. E. Ostashev and D. K. Wilson, Relative contributions from temperature and wind velocity fluctuations to the statistical moments of a sound field in a turbulent atmosphere, *Acta Acust.*, 86: 260–268, 2000.

25. D. K. Wilson, B. M. Sadler, and T. Pham, Simulation of detection and beamforming with acoustical ground sensors, *Proceedings of the SPIE 2002 AeroSense Symposium*, Orlando, FL, April 1–5, 2002, pp. 50–61.

26. D. E. Norris, D. K. Wilson, and D. W. Thomson, Atmospheric scattering for varying degrees of saturation and turbulent intermittency, *J. Acoust. Soc. Am.*, 109: 1871–1880, 2001.

27. S. M. Flatté, R. Dashen, W. H. Munk, K. M. Watson, and F. Zachariasen, *Sound Transmission Through a Fluctuating Ocean* (Cambridge University Press, Cambridge, U.K., 1979).

28. G. A. Daigle, J. E. Piercy, and T. F. W. Embleton, Line-of-sight propagation through atmospheric turbulence near the ground, *J. Acoust. Soc. Am.*, 74: 1505–1513, 1983.

29. H. E. Bass, L. N. Bolen, R. Raspet, W. McBride, and J. Noble, Acoustic propagation through a turbulent atmosphere: Experimental characterization, *J. Acoust. Soc. Am.*, 90: 3307–3313, 1991.

30. A. Ishimaru, *Wave Propagation and Scattering in Random Media* (IEEE Press, New York, 1997).

31. D. I. Havelock, M. R. Stinson, and G. A. Daigle, Measurements of the two-frequency mutual coherence function for sound propagation through a turbulent atmosphere, *J. Acoust. Soc. Am.*, 104(1): 91–99, July 1998.

32. A. Paulraj and T. Kailath, Direction of arrival estimation by eigenstructure methods with imperfect spatial coherence of wavefronts, *J. Acoust. Soc. Am.*, 83: 1034–1040, March 1988.

33. B.-G. Song and J. A. Ritcey, Angle of arrival estimation of plane waves propagating in random media, *J. Acoust. Soc. Am.*, 99(3): 1370–1379, March 1996.

34. A. B. Gershman, C. F. Mecklenbrauker, and J. F. Bohme, Matrix fitting approach to direction of arrival estimation with imperfect spatial coherence, *IEEE Trans. Signal Process.*, 45(7): 1894–1899, July 1997.

35. O. Besson, F. Vincent, P. Stoica, and A. B. Gershman, Approximate maximum likelihood estimators for array processing in multiplicative noise environments, *IEEE Trans. Signal Process.*, 48(9): 2506–2518, September 2000.

36. J. Ringelstein, A. B. Gershman, and J. F. Bohme, Direction finding in random inhomogeneous media in the presence of multiplicative noise, *IEEE Signal Process. Lett.*, 7(10): 269–272, October 2000.

37. P. Stoica, O. Besson, and A. B. Gershman, Direction-of-arrival estimation of an amplitude-distorted wavefront, *IEEE Trans. Signal Process.*, 49(2): 269–276, February 2001.

38. O. Besson, P. Stoica, and A. B. Gershman, Simple and accurate direction of arrival estimator in the case of imperfect spatial coherence, *IEEE Trans. Signal Process.*, 49(4): 730–737, April 2001.

39. M. Ghogho, O. Besson, and A. Swami, Estimation of directions of arrival of multiple scattered sources, *IEEE Trans. Signal Process.*, 49(11): 2467–2480, November 2001.

40. G. Fuks, J. Goldberg, and H. Messer, Bearing estimation in a Ricean channel–Part I: Inherent accuracy limitations, *IEEE Trans. Signal Process.*, 49(5): 925–937, May 2001.

41. J. F. Boehme, Array processing, in *Advances in Spectrum Analysis and Array Processing*, Vol. 2, Chapter 1, S. Haykin, ed. (Prentice-Hall, Englewood Cliffs, NJ, 1991).

42. H. L. Van Trees, *Optimum Array Processing* (Wiley, New York, 2002).

43. N. Owsley, Sonar array processing, in *Array Signal Processing*, S. Haykin, ed. (Prentice-Hall, Englewood Cliffs, NJ, 1984).

44. G. Su and M. Morf, Signal subspace approach for multiple wideband emitter location, *IEEE Trans. ASSP*, 31(6): 1502–1522, December 1983.

45. H. Wang and M. Kaveh, Coherent signal-subspace processing for the detection and estimation of angles of arrival of multiple wide-band sources, *IEEE Trans. ASSP*, SSP-33(4): 823–831, August 1985.

46. D. N. Swingler and J. Krolik, Source location bias in the coherently focused high-resolution broadband beamformer, *IEEE Trans. ASSP*, 37(1): 143–145, January 1989.

47. S. Valaee and P. Kabal, Wideband array processing using a two-sided correlation transformation, *IEEE Trans. Signal Process.*, 43(1): 160–172, January 1995.

48. J. Krolik and D. Swingler, Focused wide-band array processing by spatial resampling, *IEEE Trans. ASSP*, 38(2): 356–360, February 1990.

49. J. Krolik, Focused wide-band array processing for spatial spectral estimation, in *Advances in Spectrum Analysis and Array Processing*, Vol. 2, Chapter 6, S. Haykin, ed. (Prentice-Hall, Englewood Cliffs, NJ, 1991).

50. B. Friedlander and A. J. Weiss, Direction finding for wide-band signals using an interpolated array, *IEEE Trans. Signal Process.*, 41(4): 1618–1634, April 1993.

51. M. A. Doron, E. Doron, and A. J. Weiss, Coherent wide-band processing for arbitrary array geometry, *IEEE Trans. Signal Process.*, 41(1): 414–417, January 1993.

52. K. M. Buckley and L. J. Griffiths, Broad-band signal-subspace spatial-spectrum (BASS-ALE) estimation, *IEEE Trans. ASSP*, 36(7): 953–964, July 1988.

53. M. Agrawal and S. Prasad, Broadband DOA estimation using spatial-only modeling of array data, *IEEE Trans. Signal Process.*, 48(3): 663–670, March 2000.

54. S. Sivanand, J. Yang, and M. Kaveh, Focusing filters for wide-band direction finding, *IEEE Trans. Signal Process.*, 39(2): 437–445, February 1991.

55. S. Sivanand and M. Kaveh, Multichannel filtering for wide-band direction finding, *IEEE Trans. Signal Process.*, 39(9): 2128–2132, September 1991.

56. S. Sivanand, On focusing preprocessor for broadband beamforming, *Sixth SP Workshop on Statistical Signal and Array Proceedings*, Victoria, British Columbia, Canada, October 1992, pp. 350–353.

57. D. B. Ward, Z. Ding, and R. A. Kennedy, Broadband DOA estimation using frequency invariant beamforming, *IEEE Trans. Signal Process.*, 46(5): 1463–1469, May 1998.

58. W. J. Bangs, Array processing with generalized beamformers, PhD dissertation, Yale University, New Haven, CT, 1972.

59. D. N. Swingler, An approximate expression for the Cramer-Rao bound on DOA estimates of closely spaced sources in broadband line-array beamforming, *IEEE Trans. Signal Process.*, 42(6): 1540–1543, June 1994.

60. J. Yang and M. Kaveh, Coherent signal-subspace transformation beamformer, *IEE Proc.*, 137, Pt. F(4): 267–275, August 1990.

61. T. Pham and B. M. Sadler, Acoustic tracking of ground vehicles using ESPRIT, *Automatic Object Recognition V—Volume 2485 of Proceedings of SPIE*, SPIE, Orlando, FL, April 1995, pp. 268–274.

62. T. Pham, B. M. Sadler, M. Fong, and D. Messer, High resolution acoustic direction finding algorithm to detect and track ground vehicles, *20th Army Science Conference*, Norfolk, VA, June 1996; see also *Twentieth Army Science Conference, Award Winning Papers* (World Scientific, Singapore, 1997).

63. T. Pham and B. M. Sadler, Adaptive wideband aeroacoustic array processing, *8th IEEE Statistical Signal and Array Processing Workshop*, Corfu, Greece, June 1996, pp. 295–298.

64. T. Pham and B. M. Sadler, Adaptive wideband aeroacoustic array processing, *Proceedings of the 1st Annual Conference of the Sensors and Electron Devices Federated Laboratory Research Program*, College Park, MD, January 1997.

65. T. Pham and B. M. Sadler, Focused wideband array processing algorithms for high-resolution direction finding, *Proceedings of the Meeting of the MSS Specialty Group on Battlefield Acoustics and Seismic Sensing*, Chicago, IL, September 1998.

66. T. Pham and B. M. Sadler, Wideband array processing algorithms for acoustic tracking of ground vehicles, *Proceedings of the 21st Army Science Conference*, Norfolk, VA, 1998.

67. R. R. Tenney and J. R. Delaney, A distributed aeroacoustic tracking algorithm, *Proceedings of the American Control Conference*, San Diego, CA, June 1984, pp. 1440–1450.

68. Y. Bar-Shalom and X.-R. Li, *Multitarget-Multisensor Tracking: Principles and Techniques*, YBS, Storrs, CT, 1995.

69. A. Farina, Target tracking with bearings-only measurements, *Signal Process.*, 78: 61–78, 1999.

70. B. Ristic, S. Arulampalam, and C. Musso, The influence of communication bandwidth on target tracking with angle only measurements from two platforms, *Signal Process.*, 81: 1801–1811, 2001.

71. L. M. Kaplan, P. Molnar, and Q. Le, Bearings-only target localization for an acoustical unattended ground sensor network, *Proceedings of SPIE AeroSense*, Orlando, FL, April 2001.

72. A. J. Weiss and E. Weinstein, Fundamental limitations in passive time delay estimation—Part 1: Narrowband systems, *IEEE Trans. Acoust., Speech, Signal Process.*, ASSP-31(2): 472–485, April 1983.

73. E. Weinstein and A. J. Weiss, Fundamental limitations in passive time delay estimation—Part 2: Wideband systems, *IEEE Trans. Acoust., Speech, Signal Process.*, ASSP-32(5): 1064–1077, October 1984.

74. K. Bell, Wideband direction of arrival (DOA) estimation for multiple aeroacoustic sources, *Proceedings, 2000 Meeting of the MSS Specialty Group on Battlefield Acoustics and Seismics*, Laurel, MD, October 18–20, 2000.

75. K. Bell, Maximum a posteriori (MAP) multitarget tracking for broadband aeroacoustic sources, *Proceedings, 2001 Meeting of the MSS Specialty Group on Battlefield Acoustics and Seismics*, Laurel, MD, October 23–26, 2001.

76. M. Wax and T. Kailath, Decentralized processing in sensor arrays, *IEEE Trans. Acoust., Speech, Signal Process.*, ASSP-33(4): 1123–1129, October 1985.

77. P. Stoica, A. Nehorai, and T. Soderstrom, Decentralized array processing using the MODE algorithm, *Circuits Syst. Signal Process.*, 14(1): 17–38, 1995.

78. E. Weinstein, Decentralization of the Gaussian maximum likelihood estimator and its applications to passive array processing, *IEEE Trans. Acoust., Speech, Signal Process.*, ASSP-29(5): 945–951, October 1981.

79. R. L. Moses and R. Patterson, Self-calibration of sensor networks, *Proceedings of SPIE AeroSense*, Vol. 4743, Orlando, FL, April 2002, pp. 108–119.

80. J. L. Spiesberger, Locating animals from their sounds and tomography of the atmosphere: Experimental demonstration, *J. Acoust. Soc. Am.*, 106: 837–846, 1999.

81. D. K. Wilson, A. Ziemann, V. E. Ostashev, and A. G. Voronovich, An overview of acoustic travel-time tomography in the atmosphere and its potential applications, *Acta Acust.*, 87: 721–730, 2001.

82. B. G. Ferguson, Variability in the passive ranging of acoustic sources in air using a wavefront curvature technique, *J. Acoust. Soc. Am.*, 108(4): 1535–1544, October 2000.

83. B. G. Ferguson, Time-delay estimation techniques applied to the acoustic detection of jet aircraft transits, *J. Acoust. Soc. Am.*, 106(1): 255–264, July 1999.

84. B. Friedlander, On the Cramer-Rao bound for time delay and Doppler estimation, *IEEE Trans. Info. Theory*, IT-30(3): 575–580, May 1984.

85. P. Whittle, The analysis of multiple stationary time series, *J. Roy. Stat. Soc.*, 15: 125–139, 1953.

86. G. C. Carter (ed.), *Coherence and Time Delay Estimation* (Selected Reprint Volume) (IEEE Press, New York, 1993).

87. C. H. Knapp and G. C. Carter, Estimation of time delay in the presence of source or receiver motion, *J. Acoust. Soc. Am.*, 61(6): 1545–1549, June 1977.

88. W. B. Adams, J. P. Kuhn, and W. P. Whyland, Correlator compensation requirements for passive time-delay estimation with moving source or receivers, *IEEE Trans. Acoust., Speech, Signal Process.*, ASSP-28(2): 158–168, April 1980.

89. R. J. Kozick and B. M. Sadler, Tracking moving acoustic sources with a network of sensors, Army Research Laboratory Technical Report ARL-TR-2750, October 2002.

90. V. Katkovnik and A. B. Gershman, A local polynomial approximation based beamforming for source localization and tracking in nonstationary environments, *IEEE Signal Process. Lett.*, 7(1): 3–5, January 2000.

91. J. W. Betz, Comparison of the deskewed short-time correlator and the maximum likelihood correlator, *IEEE Trans. Acoust., Speech, Signal Process.*, ASSP-32(2): 285–294, April 1984.

92. P. M. Schultheiss and E. Weinstein, Estimation of differential Doppler shifts, *J. Acoust. Soc. Am.*, 66(5): 1412–1419, November 1979.

93. T. R. Damarla, T. Pham, J. Gerber, and D. Hillis, Army acoustic tracking algorithm, *Proceedings, 2002 Meeting of the MSS Specialty Group on Battlefield Acoustics and Seismics*, Laurel, MD, September 24–27, 2002.

94. M. Wellman, N. Srour, and D. B. Hillis, Acoustic feature extraction for a neural network classifier, Army Research Laboratory Report ARL-TR-1166, January 1997.

95. N. Srour, D. Lake, and M. Miller, Utilizing acoustic propagation models for robust battlefield target identification, *Proceedings, 1998 Meeting of the IRIS Specialty Group on Acoustic and Seismic Sensing*, Laurel, MD, September 1998.

96. D. Lake, Robust battlefield acoustic target identification, *Proceedings, 1998 Meeting of the IRIS Specialty Group on Acoustic and Seismic Sensing*, Laurel, MD, September 1998.

97. D. Lake, Efficient maximum likelihood estimation for multiple and coupled harmonics, Army Research Laboratory Report ARL-TR-2014, December 1999.

98. D. Lake, Harmonic phase coupling for battlefield acoustic target identification, *Proceedings of IEEE ICASSP*, Seattle, WA, 1998, pp. 2049–2052.

99. H. Hurd and T. Pham, Target association using harmonic frequency tracks, *Proceedings of the Fifth IEEE International Conference on Information Fusion*, Annapolis, MD, 2002, pp. 860–864.

100. H. Wu and J. M. Mendel, Data analysis and feature extraction for ground vehicle identification using acoustic data, *2001 MSS Specialty Group Meeting on Battlefield Acoustics and Seismic Sensing*, Johns Hopkins University, Laurel, MD, October 2001.

101. H. Wu and J. M. Mendel, Classification of ground vehicles from acoustic data using fuzzy logic rule-based classifiers: Early results, *Proceedings of SPIE AeroSense*, pp. 62–72, Orland, FL, April 1–5, 2002.

102. R. J. Kozick and B. M. Sadler, Information sharing between localization, tracking, and identification algorithms, *Proceedings, 2002 Meeting of the MSS Specialty Group on Battlefield Acoustics and Seismics*, Laurel, MD, September 24–27, 2002.

103. S. M. Kay, *Fundamentals of Statistical Signal Processing, Detection Theory* (Prentice-Hall, Englewood Cliffs, NJ, 1998).

104. T. Pham and B. M. Sadler, Energy-based detection and localization of stochastic signals, *2002 Meeting of the MSS Specialty Group on Battlefield Acoustic and Seismic Sensing*, Laurel, MD, September 2002.

105. T. Pham, Localization algorithms for ad-hoc network of disposable sensors, *2003 MSS National Symposium on Sensor and Data Fusion*, San Diego, CA, June 2003.

106. A. J. Bedard and T. M. Georges, Atmospheric infrasound, *Phys. Today*, 53: 32–37, 2000.

13

Distributed Multi-Target Detection in Sensor Networks

Xiaoling Wang
The University of Tennessee

Hairong Qi
The University of Tennessee

Steve Beck
BAE Systems

13.1 Blind Source Separation 292
13.2 Source Number Estimation...293
  Bayesian Source Number Estimation • Sample-Based Source
  Number Estimation • Variational Learning
13.3 Distributed Source Number Estimation 296
  Distributed Hierarchy in Sensor Networks • Posterior Probability
  Fusion Based on Bayes' Theorem
13.4 Performance Evaluation 299
  Evaluation Metrics • Experimental Results • Discussion
13.5 Conclusions...303
References...304

Recent advances in micro electro mechanical systems (MEMS), wireless communication technologies, and digital electronics are responsible for the emergence of sensor networks that deploy thousands of low-cost sensor nodes integrating sensing, processing, and communication capabilities. These sensor networks have been employed in a wide variety of applications, ranging from military surveillance to civilian and environmental monitoring. Examples of such applications include battlefield command, control, and communication [1], target detection, localization, tracking and classification [2–6], transportation monitoring [7], pollution monitoring in the air, soil, and water [8,9], ecosystem monitoring [10], etc.

A fundamental problem addressed by these sensor network applications is target detection in the field of interest. This problem has two primary levels of difficulties: *single target detection* and *multiple target detection*. The single target detection problem can be solved using statistical signal processing methods. For example, a constant false-alarm rate (CFAR) detector on the acoustic signals can determine the presence of a target if the signal energy exceeds an adaptive threshold. On the other hand, the multiple target detection problem is rather challenging and very difficult to solve.

Over the years, researchers have employed different sensing modalities, 1-D or 2-D, to tackle the problem. Take 2-D imagers as an example: through image segmentation, the targets of interest can be separated from the background and later identified using pattern classification methods. However, if these targets appear occluded with each other in a single image frame, or the target pixels are mixed with background clutter, which is almost always the case, detecting these targets can be extremely difficult. In such situations, array processing or distributed collaborative processing of 1-D signals such as the acoustic and seismic signals may offer advantages because of the time delays among the sensors' relative positions as well as the intrinsic correlation among the target signatures. For instance, we can model the acoustic signal received at an individual sensor node as a linear/nonlinear weighted combination of the signals radiated from different targets with the weights determined by the signal propagation model and the distance between the targets and the sensor node.

The problem of detecting multiple targets in sensor networks from their linear/nonlinear mixtures is similar to the traditional blind source separation (BSS) problem [11,12], where different targets in the field are considered as the *sources*. The "blind" qualification of BSS refers to the fact that there is no a priori information available on the number of sources, the distribution of sources, or the mixing model [13]. In general, BSS problems involve two aspects of research: *source number estimation* and *source separation*, where source number estimation is the process of estimating the number of targets and source separation is the process of identifying and extracting target signals from the mixtures. Independent component analysis (ICA) [14–17] has been a widely accepted technique to solve BSS problems, but it has two major drawbacks when applied in the context of sensor networks.

First, for conceptual and computational simplicity, most ICA algorithms make the assumption that the number of sources equals to the number of observations, so that the mixing/unmixing matrix is square and can be easily estimated. However, this equality assumption is generally not the case in sensor network applications since thousands of sensors can be densely deployed within the sensing field and the number of sensors can easily exceed the number of sources. Hence, the number of sources has to be estimated before any further calculations can be done. In Ref. [18], the problem of source number estimation is also referred to as the problem of *model order estimation*. We will use these two terms interchangeably in this section.

The second drawback is that most model order estimation methods developed to date require centralized processing and are derived under the assumption that sufficient data from all the involved sensors are available in order to estimate the most probable number of sources as well as the mixing matrix. However, this assumption is not feasible for real-time processing in sensor networks because of the sheer amount of sensor nodes deployed in the field and the limited power supply on the battery-supported sensor nodes.

In this section, we focus our discussion on model order estimation. We develop a distributed multiple target detection framework for sensor network applications based on centralized blind source estimation techniques. The outline of this section is as follows: We first describe the problem of BSS in Section 13.1 and source number estimation in Section 13.2. Based on the background introduction of these two related topics, we then present in Section 13.3 a distributed source number estimation technique for multiple target detection in sensor networks. In Section 13.4, we conduct experiments to evaluate the performance of the proposed distributed method compared to the existing centralized approach.

13.1 Blind Source Separation

The BSS problem [11,12] considers how to extract source signals from their linear or nonlinear mixtures using a minimum of a priori information. We start our discussion with a very intuitive example, the so-called *cocktail-party problem* [15]. Suppose there are two people speaking simultaneously in a room and two microphones placed in different locations of the room. Let $x_1(t)$ and $x_2(t)$ denote the amplitude of the speech signals *recorded* at the two microphones, and $s_1(t)$ and $s_2(t)$ the amplitude of the speech signals *generated* by the two speakers. We call $x_1(t)$ and $x_2(t)$ the *observed signals* and $s_1(t)$ and $s_2(t)$ the *source signals*. Intuitively, we know that both the observed signals are mixtures of the two source signals. If we assume the mixing process is linear, then we can model it using Equation 13.1, where the observed signals ($x_1(t)$ and $x_2(t)$) are weighted sums of the source signals ($s_1(t)$ and $s_2(t)$), and a_{11}, a_{12}, a_{21}, and a_{22} denote the weights which are normally dependent upon the distances between the microphones and the speakers:

$$x_1(t) = a_{11}s_1(t) + a_{12}s_2(t)$$
$$x_2(t) = a_{21}s_1(t) + a_{22}s_2(t)$$

(13.1)

If the a_{ij}'s are known, the solutions to the linear equations in Equation 13.1 are straightforward; otherwise, the problem is considerably more difficult. A common approach is to adopt some statistical properties of the source signals (e.g., making the assumption that the source signals $s_i(t)$, at each time instant t, are statistically independent) to help estimate the weights a_{ij}.

In sensor networks, sensor nodes are usually densely deployed in the field. If the targets are close to each other, the observation from each individual sensor node is a mixture of the source signals generated by the targets. Therefore, the basic formulation of the BSS problem and its ICA-based solution are applicable to multiple target detection in sensor networks as well.

Suppose there are m targets in the sensor field generating source signals $S_i(t)$, $i = 1, \ldots, m$, and n sensor observations recorded at each sensor node $x_j(t)$, $j = 1, \ldots, n$, where $t = 1, \ldots, T$ indicates the time index of the discrete-time signals. Then the sources and the observed mixtures at t can be denoted as vectors $s(t) = [s_1(t), \ldots, s_m(t)]T$ and $x(t) = [x_1(t), \ldots, x_n(t)]^T$ respectively. Let $\mathbf{X}_{n \times p} = \{x(t)\}$ represent the sensor observation matrix and $\mathbf{S}_{m \times p} = \{s(t)\}$ the unknown source matrix, where p is the number of discrete times.

If we assume the mixing process is linear, then $\mathbf{X}$ can be represented as

$$\mathbf{X} = | \mathbf{AS} \tag{13.2}$$

where $\mathbf{A}_{n \times m}$ is the unknown non-singular scalar mixing matrix.

In order to solve Equation 13.2 using the ICA algorithms, we need to make the following three assumptions. First, the mixing process is instantaneous so that there is no time delay between the source signals and the sensor observations. Second, the source signals $s(t)$ are mutually independent at each time instant t. This assumption is not unrealistic in many cases since the estimation results can provide a good approximation of the real source signals [15]. In this sense, the BSS problem is to determine a constant (weight) matrix $\mathbf{W}$ so that $\hat{\mathbf{S}}$, an estimate of the source matrix, is as independent as possible:

$$\hat{\mathbf{S}} = \mathbf{WX} \tag{13.3}$$

In theory, the unmixing matrix $\mathbf{W}_{m \times n}$ can be solved using the Moore–Penrose pseudo-inverse of the mixing matrix $\mathbf{A}$:

$$\mathbf{W} = (\mathbf{A}^T \mathbf{A})^{-1} \mathbf{A}^T \tag{13.4}$$

Correspondingly, the estimation of one independent component (one row of $\hat{\mathbf{S}}$) can be denoted as $\mathcal{H}_m \hat{S}_i = \mathbf{wX}$, where $\mathbf{w}$ is one row of the unmixing matrix $\mathbf{W}$. Define $z = \mathbf{A}^T \mathbf{w}^T$, then the independent component $\hat{S}_i = \mathbf{wX} = \mathbf{wAS} = z^T S$, which is a linear combination of S_i's with the weights given by z. According to the Central Limit Theorem, the distribution of a sum of independent random variables converges to a Gaussian. Thus, $z^T S$ is more Gaussian than any of the components s_j and becomes least Gaussian when it in fact equals one of the s_j's, that is, when it gives the correct estimation of one of the sources [15]. Therefore, in the context of ICA, nongaussianity indicates independence. Many metrics have been studied to measure the nongaussianity of the independent components, such as the kurtosis [13,19], the mutual information [11,20], and the negentropy [14,21].

The third assumption is related to the independence criterion stated above. Since the mixture of two or more Gaussian sources is still a Gaussian, which makes it impossible to separate them from each other, we need to assume that at most one source signal is normally distributed for the linear mixing/unmixing model [17]. This assumption is reasonable in practice since pure Gaussian processes are rare in real data.

13.2 Source Number Estimation

As we discussed above, BSS problems involve two aspects: source number estimation and source separation. Most ICA-based algorithms assume that the number of sources equals the number of observations to make the mixing matrix $\mathbf{A}$ and the unmixing matrix $\mathbf{W}$ square in order to simplify the problem.

However, this assumption is not feasible in sensor networks due to the sheer amount of sensor nodes deployed. Hence, the number of targets has to be estimated before any further operations can be done.

Several approaches have been introduced to solve the source number estimation problem so far, some heuristic, others based on more principled approaches [22–24]. As discussed recently in Ref. [18], it has become clear that techniques of the latter category are superior, and heuristic methods may be seen at best as approximations to more detailed underlying principles. In this section, we focus on the discussion of principled source number estimation algorithms, which construct multiple hypotheses corresponding to different number of sources. Suppose $\mathcal{H}_m$ denotes the hypothesis on the number of sources m. The goal of principled source number estimation is to find m whose corresponding hypothesis $\mathcal{H}_{\hat{m}}$ maximizes the posterior probability given only the observation matrix $\mathbf{X}$:

$$\hat{m} = \arg\max_{m} P(\mathcal{H}_m \mid \mathbf{X}) \tag{13.5}$$

A brief introduction on different principled source number estimation methods is given next.

13.2.1 Bayesian Source Number Estimation

Roberts proposed a Bayesian source number estimation approach to finding the hypothesis that maximizes the posterior probability $P(\mathcal{H}_m \mid \mathbf{X})$. Interested readers are referred to Ref. [24] for detailed theoretical derivation.

According to Bayes' theorem, the posterior probability of the hypothesis can be written as

$$P(\mathcal{H}_m \mid \mathbf{X}) = \frac{p(\mathbf{X} \mid \mathcal{H}_m)P(\mathcal{H}_m)}{p(\mathbf{X})} \tag{13.6}$$

Assume the hypothesis $\mathcal{H}_m$ of different number of sources m has a uniform distribution, i.e. equal prior probability $P(\mathcal{H}_m)$ and since $p(\mathbf{X})$ is a constant, the measurement of the posterior probability can be simplified to the calculation of likelihood $p(\mathbf{X} \mid \mathcal{H}_m)$. By marginalizing the likelihood over the system parameters space and approximating the marginal integrals using the Laplace approximation method, a log-likelihood function proportional to the posterior probability can be written as

$$L(m) = \log p(\mathbf{x}(t) \mid \mathcal{H}_m)$$

$$= \log \pi(\hat{S}(t)) + \frac{1}{2}(n-m)\log\left(\frac{\hat{\beta}}{2\pi}\right) - \frac{1}{2}\log|\hat{\mathbf{A}}^T\hat{\mathbf{A}}| - \frac{\hat{\beta}}{2}(\mathbf{x}(t) - \hat{\mathbf{A}}\hat{S}(t))^2$$

$$- \left[\frac{mn}{2}\log\left(\frac{\hat{\beta}}{2\pi}\right) + \frac{n}{2}\left(\sum_{j=1}^{m}\log \hat{S}_j(t)^2 + mn\log\gamma\right)\right] \tag{13.7}$$

where
  $\mathbf{x}(t)$ is the sensor observations
  $\hat{\mathbf{A}}$ is the estimate of the mixing matrix
  $\hat{s}(t) = \mathbf{W}\mathbf{x}(t)$ is the estimate of the independent sources, $\mathbf{W} = (\hat{\mathbf{A}}^T\hat{\mathbf{A}})^{-1}\hat{\mathbf{A}}^T$
  $\hat{\beta}$ is the variance of noise component
  γ is a constant
  $\pi(\cdot)$ is the assumed marginal distribution of the source

The Bayesian source number estimation method considers a set of Laplace approximations to infer the posterior probabilities of specific hypotheses. This approach has a solid theoretical background and the objective function is easy to calculate, hence, it provides a practical solution for the source number estimation problem.

13.2.2 Sample-Based Source Number Estimation

Other than the Laplace approximation method, the posterior probabilities of specific hypotheses can also be evaluated using a sample-based approach. In this approach, a reversible-jump Markov chain Monte Carlo (RJ-MCMC) method is proposed to estimate the joint density over the mixing matrix $\mathbf{A}$, the hypothesized number of sources m, and the noise component R_n, which is denoted as $P(\mathbf{A}, m, R_n)$ [18,23]. The basic idea is to construct a Markov chain which generates samples from the hypothesis probability and to use the Monte Carlo method to estimate the posterior probability from the samples. An introduction of Monte Carlo methods can be found in Ref. [25].

RJ-MCMC is actually a random-sweep Metropolis–Hastings method, where the transition probability of the Markov chain from state $(\mathbf{A}, m, R_n)$ to state $(\mathbf{A}', m', R_n')$ is

$$p = \min\left\{1, \frac{P(\mathbf{A}', m', R_n' \mid \mathbf{X})}{P(\mathbf{A}, m, R_n \mid \mathbf{X})} \frac{q(\mathbf{A}, m, R_n \mid \mathbf{X})}{q(\mathbf{A}', m', R_n' \mid \mathbf{X})} J\right\} \tag{13.8}$$

where

$P(\cdot)$ is the posterior probability of the unknown parameters of interest
$q(\cdot)$ is a proposal density for moving from state $(\mathbf{A}, m, R_n)$ to state $(\mathbf{A}', m', R_n')$
J is the ratio of Jacobians for the proposal transition between the two states [18]

More detailed derivation of this method is provided in Ref. [23].

13.2.3 Variational Learning

In recent years, the Bayesian inference problem shown in Equation 13.6 is also tackled using another approximative method known as variational learning [26,27]. In ICA problems, variables are divided into two classes: the visible variables v and the hidden variables h. An example of visible variables is the observation matrix $\mathbf{X}$; examples of hidden variables include an ensemble of the parameters of $\mathbf{A}$, the noise covariance matrix, any parameters in the source density models, and all associated hyperparameters such as the number of sources m [18]. Suppose $q(h)$ denotes the variational approximation to the posterior probability of the hidden variables $P(h|v)$, then the negative variational free energy, F, is defined as

$$F = \int q(h)\ln P(h|\upsilon)\,dh + H[q(h)] \tag{13.9}$$

where $H[q(h)]$ is the differential entropy of $q(h)$. The negative free energy F forms a strict lower bound on the evidence of the model, $\ln p(v) = \int p(\upsilon|h)p(h)\,dh)$. The difference between this variational bound and the true evidence is the Kullback–Leibler (KL) divergence between $q(h)$ and the true posteriors $P(h|v)$ [28]. Therefore, maximizing F is equivalent to minimizing the KL divergence, and this process provides a direct method for source number estimation.

Another promising source number estimation approach using variational learning is the so-called Automatic Relevance Determination (ARD) scheme [28]. The basic idea of ARD is to suppress sources that are unsupported by the data. For example, assume each hypothesized source has a Gaussian prior with separate variances, those sources that do not contribute to modeling the observations tend to have very small variances and the corresponding source models do not move significantly from their priors [18]. After eliminating those unsupported sources, the sustained sources give the true number of sources of interest.

Even though variational learning is a particularly powerful approximative approach, it is yet to be developed into a more mature form. In addition, it presents difficulties to estimate the true number of sources with noisy data.

13.3 Distributed Source Number Estimation

The source number estimation algorithms described in Section 13.2 are all centralized processes in the sense that the observed signals from all sensor nodes are collected at a processing center and estimation needs to be performed on the whole data set. While this assumption works well for small sensor array applications like in speech analysis, it is not necessarily the case for real-time applications in sensor networks due to the large scale of network, as well as the severe resource constraints. In sensor networks, the sensor nodes are usually battery-operated and cannot be recharged in real time, which makes energy the most constrained resource. Since wireless communication consumes the most energy among all the activities conducted on the sensor node [29], the centralized scheme becomes very energy-intensive due to large amount of data transmission and is not cost effective for real-time sensor network applications. On the contrary, when implemented in a distributed manner, data can be processed locally on a cluster of sensor nodes that are close in geographical location and only local decisions need to be transferred for further processing. In this way, the distributed target detection framework can dramatically reduce long-distance network traffic and therefore conserve energy consumed on data transmissions and prolong the lifetime of the sensor network.

13.3.1 Distributed Hierarchy in Sensor Networks

In the context of the proposed distributed solution to the source number estimation problem, we assume a clustering protocol has been applied and the sensor nodes have organized themselves into clusters with each node assigned to one and only one cluster. Nodes can communicate locally within the same cluster, and different clusters communicate through a cluster head specified within each cluster. An example of a clustered sensor network is illustrated in Figure 13.1.

Suppose there are m targets present in the sensor field, and the sensor nodes are divided into L clusters. Each cluster l ($l = 1, ..., L$) can sense the environment independently and generate an observation matrix

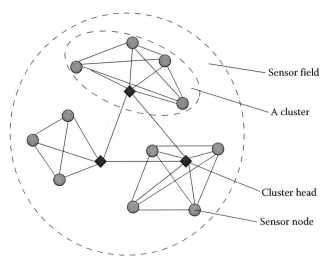

FIGURE 13.1   An example of a clustered sensor network.

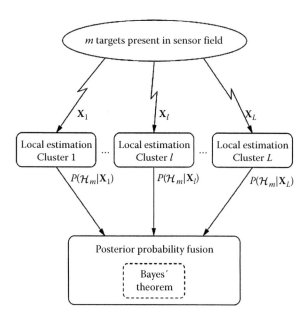

FIGURE 13.2 The structure of the distributed source number estimation hierarchy.

$\mathbf{X}_l$ which consists of mixtures of the source signals generated by the m targets. The distributed estimation hierarchy includes two levels of processing: First, we estimate the posterior probability $P(\mathcal{H}_m \mid \mathbf{X}_l)$ of each hypothesis $\mathcal{H}_m$ given a local observation matrix $\mathbf{X}_l$. The Bayesian source number estimation approach proposed by Roberts [24] is employed in this step. Secondly, we develop a posterior probability fusion algorithm to integrate the decisions from each cluster. The structure of the hierarchy is illustrated in Figure 13.2.

This hierarchy for distributed source number estimation benefits from two areas of research achievements, distributed detection and ICA model order estimation. However, it exhibits some unique features that make it advantageous for multiple target detection in sensor networks from both the theoretical and practical points of view:

- M-ary hypothesis testing. Most distributed detection algorithms are derived under the binary hypothesis assumption, where $\mathcal{H}$ takes on one of two possible values corresponding to the presence and absence of the target [30]. The distributed framework developed here allows the traditional binary hypothesis testing problem to be extended to the M-ary case, where the values of $\mathcal{H}$ correspond to the different numbers of sources.
- Fusion of detection probabilities. Instead of making a crisp decision from local cluster estimates as in the classic distributed detection algorithms, a Bayesian source number estimation algorithm is performed on the observations from each cluster, and the posterior probability for each hypothesis is estimated. These probabilities are then sent to a fusion center where a decision regarding the source number hypothesis is made. This process is also referred to as the fusion of detection probabilities [31] or combination of level of significance [32]. By estimating and fusing the probabilities of hypothesis from each cluster, it is possible for the system to achieve a higher detection accuracy because faults are constrained within the processing of each cluster.
- Distributed structure. Even though the source number estimation technique is usually implemented in a centralized manner, the distributed framework presents several advantages. For example, in the distributed framework, data is processed locally in each cluster and only the estimated hypothesis probabilities are transmitted through the network. Hence, heavy network traffic can be significantly reduced and communication energy conserved. Furthermore, since the estimation process is performed in parallel within each cluster, the computation burden is distributed and computation time reduced.

After local source number estimation is conducted within each cluster, we develop a posterior probability fusion method based on Bayes' theorem to fuse the results from each cluster.

13.3.2 Posterior Probability Fusion Based on Bayes' Theorem

The objective of source number estimation approaches is to find the optimal number of sources m that maximizes the posterior probability $P(\mathcal{H}_m \mid \mathbf{X})$. When implemented in the distributed hierarchy, the local estimation approach calculates the posterior probability corresponding to each hypothesis $\mathcal{H}_m$ from each cluster, $P(\mathcal{H}_m \mid \mathbf{X}_1),\ldots,P(\mathcal{H}_m \mid \mathbf{X}_L)$. According to Bayes' theorem, the fused posterior probability can be written as

$$P(\mathcal{H}_m \mid \mathbf{X}) = \frac{P(\mathbf{X} \mid \mathcal{H}_m)P(\mathcal{H}_m)}{p(\mathbf{X})} \tag{13.10}$$

Assume the clustering of sensor nodes is exclusive, that is, $\mathbf{X} = \mathbf{X}_1 \cup \mathbf{X}_2 \cup \ldots \cup \mathbf{X}_L$ and $\mathbf{X}_l \cap \mathbf{X}_q = \emptyset$ for any $l \neq q, l = 1, \ldots, L, q = 1, \ldots, L$, the posterior probability $P(\mathcal{H}_m \mid \mathbf{X})$ can be represented as

$$P(\mathcal{H}_m \mid \mathbf{X}) = \frac{P(\mathbf{X}_1 \cup \mathbf{X}_2 \cup \cdots \cup \mathbf{X}_L \mid \mathcal{H}_m)P(\mathcal{H}_m)}{p(\mathbf{X}_1 \cup \mathbf{X}_2 \cup \cdots \cup \mathbf{X}_L)} \tag{13.11}$$

Since the observations from different clusters are assumed to be independent, $p(\mathbf{X}_l \cap \mathbf{X}_q) = 0$, for any $l \neq q$, we then have

$$p(\mathbf{X}_1 \cup \mathbf{X}_2 \cup \cdots \cup \mathbf{X}_L \mid \mathcal{H}_m) = \sum_{l=1}^{L} p(\mathbf{X}_l \mid \mathcal{H}_m) - \sum_{l,q=1,l\neq q}^{L} p(\mathbf{X}_l \cap \mathbf{X}_q \mid \mathcal{H}_m)$$

$$= \sum_{l=1}^{L} p(\mathbf{X}_l \mid \mathcal{H}_m) \tag{13.12}$$

Substituting Equation 13.12 into Equation 13.11, the fused posterior probability can be derived as

$$P(\mathcal{H}_m \mid \mathbf{X}) = \frac{\sum_{l=1}^{L} p(\mathbf{X}_l \mid \mathcal{H}_m)P(\mathcal{H}_m)}{\sum_{l=1}^{L} p(\mathbf{X}_l)}$$

$$= \frac{\sum_{l=1}^{L} \dfrac{P(\mathcal{H}_m \mid \mathbf{X}_l)\, p(\mathbf{X}_l)}{P(\mathcal{H}_m)} P(\mathcal{H}_m)}{\sum_{l=1}^{L} p(\mathbf{X}_l)}$$

$$= \sum_{l=1}^{L} \frac{p(\mathbf{X}_l)}{\sum_{q=1}^{L} p(\mathbf{X}_q)} P(\mathcal{H}_m \mid \mathbf{X}_l) \tag{13.13}$$

where
   $P(\mathcal{H}_m \mid \mathbf{X}_l)$ denotes the posterior probability calculated in cluster l
   the term $p(\mathbf{X}_l)/\sum_{q=1}^{L} p(\mathbf{X}_q)$ reflects the physical characteristic from clustering in sensor networks which is application-specific

For example, in the case of distributed multiple target detection using acoustic signals, the propagation of acoustic signals follows the energy decay model where the detected energy is inversely proportional to the square of the distance between the source and the sensor node, i.e., $E_{sensor} \propto (1/d^2)E_{source}$. Therefore, the term $p(\mathbf{X}_l)/\sum_{q=1}^{L} p(\mathbf{X}_q)$ can be considered as the relative detection sensitivity of the sensor nodes in cluster l and is proportional to the average energy captured by the sensor nodes

$$\frac{p(\mathbf{X}_l)}{\sum_{q=1}^{L} p(\mathbf{X}_q)} \propto \frac{1}{K_l}\sum_{k=1}^{K_l} E_k \propto \frac{1}{K_l}\sum_{k=1}^{K_l} \frac{1}{d_k^2} \tag{13.14}$$

where K_l denotes the number of sensor nodes in cluster l.

13.4 Performance Evaluation

We apply the proposed distributed source number estimation hierarchy to detect multiple civilian targets using data collected from a field demo held at BAE Systems, Austin, TX in August, 2002. We also compare the performance between the centralized Bayesian source number estimation algorithm and the distributed hierarchy using the evaluation metrics described next.

13.4.1 Evaluation Metrics

As mentioned before, source number estimation is basically an optimization problem in which an optimal hypothesis $\mathcal{H}_m$ is pursued that maximizes the posterior probability given the observation matrix, $P(\mathcal{H}_m | \mathbf{X}_l)$. The optimization process is affected by the initialization condition and the update procedure of the algorithm itself. To compensate for the randomness and to stabilize the overall performance, the algorithms are iteratively repeated, for example, 20 times in this experiment.

Detection probability $(P_{detection})$ is the most intuitive metric for measuring the accuracy of a detection approach. In our experiment, we define detection probability as the ratio between the correct source number estimates and the total number of estimations, i.e., $P_{detection} = N_{correct}/N_{total}$, where $N_{correct}$ denotes the number of correct estimations and N_{total} is the total number of estimations.

After executing the algorithm multiple times, a *histogram* can be generated that shows the accumulated number of estimations corresponding to different hypotheses of the number of sources. The histogram also represents the reliability of algorithm, the larger the difference of histogram values between the hypothesis of the correct estimate and other hypotheses, the more deterministic and reliable the algorithm. We use kurtosis (β) to extract this characteristic of the histogram. Kurtosis calculates the flatness of the histogram

$$\beta = \frac{1}{C}\sum_{k=1}^{N} k\left(\frac{h_k - \mu}{\theta}\right)^4 - 3 \tag{13.15}$$

where
$\quad h_k$ denotes the value of the kth bin in the histogram
$\quad N$ is the total number of bins
$\quad C = \sum_{k=1}^{N} h_k$
$\quad \mu = \frac{1}{C}\sum_{k=1}^{N} kh_k$ is the mean
$\quad \theta = \sqrt{\frac{1}{C}\sum_{k=1}^{N} k(h_k - \mu)^2}$ is the variance

Intuitively, the larger the kurtosis, the more deterministic the algorithm, and the more reliable the estimation.

Since the source number estimation is designed for real time multiple target detection in sensor networks, the *computation time* is also an important metric for performance evaluation.

13.4.2 Experimental Results

In the field demo, we let two civilian vehicles, a motorcycle and a diesel truck, as shown in Figure 13.3, travel along the N–S road from opposite directions and intersect at the T-junction. There are 15 nodes deployed along the road. For this experiment, we assume 2 clusters of 5 sensor nodes exist for the distributed processing. The sensor network setup is illustrated in Figure 13.4a. We use Sensoria WINS NG-2.0 sensor nodes (as shown in Figure 13.4b), which consist of a dual-issue SH-4 processor running at 167 MHz with 300 MIPS of processing power, RF modem for wireless communication, and up to four channels of sensing modalities, such as acoustic, seismic, and infrared. In this experiment, we perform multiple target detection algorithms on the acoustic signals captured by the microphone on each sensor node. The observations from sensor nodes are preprocessed component-wise to be zero-mean, unit-variance distributed.

First, the centralized Bayesian source number estimation algorithm is performed using all the 10 sensor observations. Secondly, the distributed hierarchy is applied as shown in Figure 13.2, which first calculates the corresponding posterior probabilities of different hypotheses in the two clusters and then fuses the local results using the Bayesian posterior probability fusion method.

Figure 13.5a shows the average value of the log-likelihood function in Equation 13.7 corresponding to different hypothesized numbers of sources estimated over 20 repetitions. Figure 13.5b displays the histogram of the occurrence of the most probable number of sources when the log-likelihood function is evaluated 20 times. Each evaluation randomly initializes the mixing matrix **A** with values drawn from a zero-mean, unit-variance normal distribution. The left column in the figure corresponds to the performance from applying the centralized Bayesian source number estimation approach on all the 10 sensor observations. The right column shows the corresponding performance of the distributed hierarchy with the proposed Bayesian posterior probability fusion method. Based on the average log-likelihood, it is clear that in both approaches the hypothesis with the true number of sources ($m = 2$) has the greatest support. However, the two approaches have different rates of correct estimations and different levels of uncertainty.

(a) (b)

FIGURE 13.3 Vehicles used in the experiment.

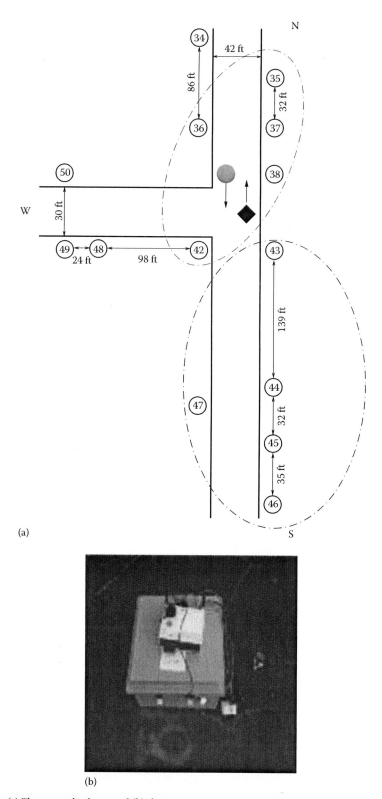

(a)

(b)

FIGURE 13.4 (a) The sensor laydown and (b) the Sensoria sensor node used.

302   *Image and Sensor Signal Processing*

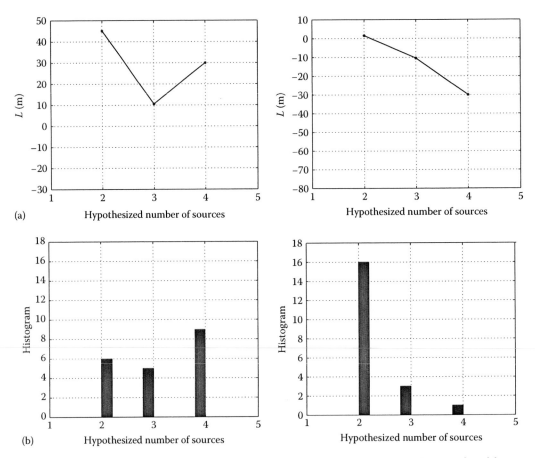

FIGURE 13.5 Performance comparison. (a) Left: Centralized Bayesian approach. Right: Distributed hierarchy with the Bayesian posterior probability fusion. (b) Left: Histogram of centralized scheme for 20 runs. Right: Histogram of distributed hierarchy with Bayes probability fusion for 20 runs.

Figure 13.6a illustrates the kurtosis calculated from the two histograms in Figure 13.5b. We can see that the kurtosis of the distributed approach is eight times higher than that of the centralized approach. The detection probabilities are shown in Figure 13.6b. We observe that the centralized Bayesian algorithm can detect the correct number of sources 30% of the time, while the distributed approach increases the detection probability of centralized scheme by an average of 50%. A comparison of computation times between the centralized scheme and the distributed hierarchy are shown in Figure 13.6c. It is clear that by using the distributed hierarchy, the computation time is generally reduced by a factor of 2.

13.4.3 Discussion

As demonstrated in the experiment as well as the performance evaluation, the distributed hierarchy with the proposed Bayesian posterior probability fusion method has better performance in the sense of providing a higher detection probability and a more deterministic and reliable response. The reasons include: (1) The centralized scheme uses the observations from all the sensor nodes as inputs to the Bayesian source number estimation algorithm. The algorithm is thus sensitive to signal variations due to node failure or environmental noise in each input signal. While in the distributed framework, the source number estimation algorithm is only performed within each cluster, therefore, the effect of signal variations are limited locally and might contribute less in the posterior probability fusion process. (2) In the derivation of the Bayesian posterior probability fusion

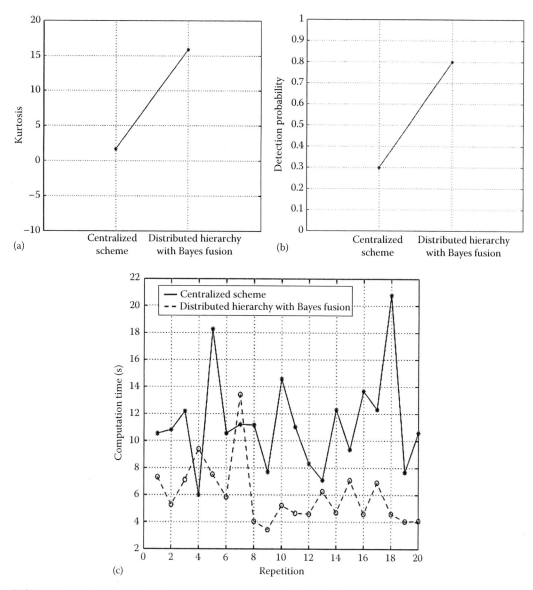

FIGURE 13.6 Comparison: (a) kurtosis, (b) detection probability, and (c) computation time.

method, the physical characteristics of sensor networks, such as the signal energy captured by each sensor node versus its geographical position, are considered, making this method more adaptive to real applications.

Furthermore, the distributed hierarchy is able to reduce the network traffic by avoiding large amount of data transmission, hence conserving energy and providing a scalable solution. The parallel implementation of the estimation algorithm in each cluster can also reduce the computation time by half.

13.5 Conclusions

This work addressed the problem of source number estimation in sensor networks for multiple target detection. This problem is similar to the BSS problem in signal processing and ICA is traditionally the most popular algorithm to solve it. The classical BSS problem includes two aspects of research: source number estimation and source separation. The multiple target detection problem in sensor networks is similar to source number estimation. Based on the description of several centralized source number estimation approaches, we developed a distributed source estimation algorithm that avoids large amounts of long-distance data transmission, and in turn reduces the network traffic and conserves energy. The distributed processing hierarchy consists of two levels: First, a local source number estimation is performed in each cluster using the centralized Bayesian source number estimation approach. Then a posterior probability fusion method is derived based on Bayes' theorem to combine the local estimations and generate a global decision. An experiment is conducted on the detection of multiple civilian vehicles using acoustic signals to evaluate the performance of the two approaches. The distributed hierarchy with the Bayesian posterior probability fusion method is shown to provide better performance in terms of the detection probability and reliability. In addition, the distributed framework can reduce the computation time by half.

References

1. I. F. Akyildiz, W. Su, Y. Sankarasubramaniam, and E. Cayirci, A survey on sensor networks, *IEEE Communications Magazine*, 40(8): 102–114, August 2002.
2. S. Kumar, D. Shepherd, and F. Zhao, Collaborative signal and information processing in microsensor networks, *IEEE Signal Processing Magazine*, 19(2): 13–14, March 2002.
3. D. Li, K. D. Wong, Y. H. Hu, and A. M. Sayeed, Detection, classification, and tracking of targets, *IEEE Signal Processing Magazine*, 19(2): 17–29, March 2002.
4. X. Wang, H. Qi, and S. S. Iyengar, Collaborative multi-modality target classification in distributed sensor networks, in *Proceedings of the Fifth International Conference on Information Fusion*, Annapolis, MD, July 2002, Vol. 1, pp. 285–290.
5. K. Yao, J. C. Chen, and R. E. Hudson, Maximum-likelihood acoustic source localization: Experimental results, in *Proceedings of IEEE International Conference on Acoustics, Speech, and Signal Processing*, Salt Lake City, UT, 2002, Vol. 3, pp. 2949–2952.
6. F. Zhao, J. Shin, and J. Reich, Information-driven dynamic sensor collaboration, *IEEE Signal Processing Magazine*, 19(2): 61–72, March 2002.
7. A. N. Knaian, A wireless sensor network for smart roadbeds and intelligent transportation systems, MS thesis, Massachusetts Institute of Technology, Cambridge, MA, June 2000.
8. K. A. Delin and S. P. Jackson, Sensor web for in situ exploration of gaseous biosignatures, in *Proceedings of 2000 IEEE Aerospace Conference*, Big Sky, MT, March 2000.
9. X. Yang, K. G. Ong, W. R. Dreschel, K. Zeng, C. S. Mungle, and C. A. Grimes, Design of a wireless sensor network for long-term, in-situ monitoring of an aqueous environment, *Sensors*, 2(7): 455–472, 2002.
10. A. Cerpa, J. Elson, M. Hamilton, and J. Zhao, Habitat monitoring: Application driver for wireless communications technology, in *2001 ACM SIGCOMM Workshop on Data Communications in Latin America and the Caribbean*, San Jose, CA, April 2001.
11. A. J. Bell and T. J. Sejnowski, An information-maximisation approach to blind separation and blind deconvolution, *Neural Computation*, 7(6): 1129–1159, 1995.
12. J. Herault and J. Jutten, Space or time adaptive signal processing by neural network models, in *Neural Networks for Computing: AIP Conference Proceedings*, J. S. Denker, Ed., American Institute for Physics, New York, Vol. 151, 1986.
13. Y. Tan and J. Wang, Nonlinear blind source separation using higher order statistics and a genetic algorithm, *IEEE Transactions on Evolutionary Computation*, 5(6): 600–612, 2001.
14. P. Comon, Independent component analysis: A new concept, *Signal Processing*, 36(3): 287–314, April 1994.
15. A. Hyvarinen and E. Oja, Independent component analysis: A tutorial, http://www.cis.hut.fi/aapo/papers/IJCNN99_tutorialweb/, April 1999.

16. J. Karhunen, Neural approaches to independent component analysis and source separation, in *Proceedings of 4th European Symposium on Artificial Neural Networks (ESANN)*, Bruges, Belgium, 1996, pp. 249–266.

17. T. Lee, M. Girolami, A. J. Bell, and T. J. Sejnowski, A unifying information-theoretic framework for independent component analysis, *International Journal on Mathematical and Computer Modeling*, 1999.

18. S. Roberts and R. Everson, Eds., *Independent Component Analysis: Principles and Practice*, Cambridge University Press, Cambridge, MA, 2001.

19. A. Hyvarinen and E. Oja, A fast fixed-point algorithm for independent component analysis, *Neural Computation*, 9: 1483–1492, 1997.

20. R. Linsker, Local synaptic learning rules suffice to maximize mutual information in a linear network, *Neural Computation*, 4: 691–702, 1992.

21. A. Hyvarinen, Fast and robust fixed-point algorithms for independent component analysis, *IEEE Transactions on Neural Networks*, 10(3): 626–634, 1999.

22. K. H. Knuth, A Bayesian approach to source separation, in *Proceedings of First International Conference on Independent Component Analysis and Blind Source Separation: ICA'99*, Aussois, France, 1999, pp. 283–288.

23. S. Richardson and P. J. Green, On Bayesian analysis of mixtures with an unknown number of components, *Journal of the Royal Statistical Society, Series B*, 59(4): 731–758, 1997.

24. S. J. Roberts, Independent component analysis: Source assessment & separation, a Bayesian approach, *IEE Proceedings on Vision, Image, and Signal Processing*, 145(3): 149–154, 1998.

25. D. J. C. MacKay, Monte Carlo methods, in *Learning in Graphical Models*, M. I. Jordan, Ed., Kluwer, Norwell, MA, 1999, pp. 175–204.

26. H. Attias, Inferring parameters and structure of latent variable models by variational Bayes, in *Proceedings of the Fifteenth Conference on Uncertainty in Artificial Intelligence*, Stockholm, Sweden, 1999, pp. 21–30.

27. C. M. Bishop, *Neural Networks for Pattern Recognition*, Oxford University Press, Oxford, U.K., 1995.

28. R. Choudrey, W. D. Penny, and S. J. Roberts, An ensemble learning approach to independent component analysis, in *Proceedings of Neural Networks for Signal Processing*, Sydney, New South Wales, Australia, December 2000.

29. V. Raghunathan, C. Schurgers, S. Park, and M. B. Srivastava, Energy-aware wireless microsensor networks, *IEEE Signal Processing Magazine*, 19(2): 40–50, March 2002.

30. J. Chamberland and V. V. Veeravalli, Decentralized detection in sensor networks, *IEEE Transactions on Signal Processing*, 51(2): 407–416, February 2003.

31. R. Krysztofowicz and D. Long, Fusion of detection probabilities and comparison of multisensor systems, *IEEE Transactions on System, Man, and Cybernetics*, 20: 665–677, May/June 1990.

32. V. Hedges and I. Olkin, *Statistical Methods for Meta-Analysis*, Academic Press, New York, 1985.

14

Symbolic Dynamic Filtering for Pattern Recognition in Distributed Sensor Networks

14.1 Introduction 307
14.2 Symbolic Dynamics and Encoding...309
Review of Symbolic Dynamics • Transformation of Time Series
to Wavelet Domain • Symbolization of Wavelet Surface Profiles
14.3 Construction of Probabilistic Finite-State Automata
for Feature Extraction..312
Conversion from Symbol Image to State Image • Construction
of PFSA • Summary of SDF for Feature Extraction
14.4 Pattern Classification Using SDF-Based Features......................315
14.5 Validation I: Behavior Recognition of Mobile Robots
in a Laboratory Environment 316
Experimental Procedure for Behavior Identification of Mobile
Robots • Pattern Analysis for Behavior Identification of
Mobile Robots • Experimental Results for Behavior Identification
of Mobile Robots
14.6 Validation II: Target Detection and Classification
Using Seismic and PIR Sensors 323
Performance Assessment Using Seismic Data • Performance
Assessment Using PIR Data
14.7 Summary, Conclusions, and Future Work 331
Acknowledgments..332
References..332

Xin Jin
*The Pennsylvania
State University*

Shalabh Gupta
University of Connecticut

Kushal Mukherjee
*The Pennsylvania
State University*

Asok Ray
*The Pennsylvania
State University*

14.1 Introduction

Proliferation of modern sensors and sensing applications provides us with large volumes of data. By extracting useful information from these data sets in real time, we enhance the ability to better comprehend and analyze the environment around us. Tools for data-driven feature extraction and pattern classification facilitate performance monitoring of distributed dynamical systems over a sensor network, especially if the physics-based models are either inadequate or unavailable [14]. In this regard, a critical issue is real-time analysis of sensor time series for information compression into low-dimensional feature vectors that capture the relevant information of the underlying dynamics [3,9,10,22].

Time-series analysis is a challenging task if the data set is voluminous (e.g., collected at a fast sampling rate), high dimensional, and noise contaminated. Moreover, in a distributed sensor network, data collection occurs simultaneously at multiple nodes. Consequently, there is a need for low-complexity algorithms that could be executed locally at the nodes to generate compressed features and therefore reduce the communication overhead. In general, the success of data-driven pattern classification tools depends on the quality of feature extraction from the observed time series. To this end, several feature extraction tools, such as principal component analysis (PCA) [10], independent component analysis (ICA) [21], kernel PCA [31], dynamic time warping [2], derivative time-series segment approximation [12], artificial neural networks (ANN) [20], hidden Markov models (HMM) [36], and wavelet transforms [26,27,35] have been reported in technical literature. Wavelet packet decomposition (WPD) [26] and fast wavelet transform (FWT) [27] have been used for extracting rich problem-specific information from sensor signals. Feature extraction is followed by pattern classification (e.g., using support vector machines [SVM]) [3,9].

The concepts of symbolic dynamics [23] have been used for information extraction from time series in the form of symbol sequences [8,14]. Keller and Lauffer [19] used tools of symbolic dynamics for analysis of time-series data to visualize qualitative changes of electroencephalography (EEG) signals related to epileptic activity. Along this line, a real-time data-driven statistical pattern analysis tool, called *symbolic dynamic filtering* (SDF) [13,30], has been built upon the concepts of symbolic dynamics and information theory. In the SDF method, time-series data are converted to symbol sequences by appropriate partitioning [13]. Subsequently, probabilistic finite-state automata (PFSA) [30] are constructed from these symbol sequences that capture the underlying system's behavior by means of information compression into the corresponding matrices of state-transition probability. SDF-based pattern identification algorithms have been experimentally validated in the laboratory environment to yield superior performance over several existing pattern recognition tools (e.g., PCA, ANN, particle filtering, unscented Kalman filtering, and kernel regression analysis [3,9]) in terms of early detection of anomalies (i.e., deviations from the normal behavior) in the statistical characteristics of the observed time series [15,29].

Partitioning of time series is a crucial step for symbolic representation of sensor signals. To this end, several partitioning techniques have been reported in literature, such as *symbolic false nearest neighbor partitioning* (SFNNP) [4], *wavelet-transformed space partitioning* (WTSP) [28], and *analytic signal space partitioning* (ASSP) [32]. In particular, the wavelet transform-based method is well suited for time–frequency analysis of non-stationary signals, noise attenuation, and reduction of spurious disturbances from the raw time-series data without any significant loss of pertinent information [24]. In essence, WTSP is suitable for analyzing the noisy signals, while SFNNP and ASSP may require additional preprocessing of the time series for denoising. However, the wavelet transform of time series introduces two new domain parameters (i.e., scale and shift), thereby generating an image of wavelet coefficients. Thus, the (one-dimensional) time-series data is transformed into a (two-dimensional) image of wavelet coefficients. Jin et al. [17] have proposed a feature extraction algorithm from the wavelet coefficients by directly partitioning the wavelet images in the (two-dimensional) *scale-shift* space for SDF analysis.

This chapter focuses on feature extraction for pattern classification in distributed dynamical systems, possibly served by a sensor network. These features are extracted as statistical patterns using symbolic modeling of the wavelet images, generated from sensor time series. An appropriate selection of the wavelet basis function and the scale range allows the wavelet-transformed signal to be denoised relative to the original (possibly) noise-contaminated signal before the resulting wavelet image is partitioned for symbol generation. In this way, the symbolic images generated from wavelet coefficients capture the signal characteristics with larger fidelity than those obtained directly from the original signal. These symbolic images are then modeled using PFSA that, in turn, generate the low-dimensional statistical patterns, also called feature vectors. In addition, the proposed method is potentially applicable for analysis of regular images for feature extraction and pattern classification. From these perspectives, the major contributions of this chapter are as follows:

1. Development of a SDF-based feature extraction method for analysis of two-dimensional data (e.g., wavelet images of time series in the scale-shift domain)
2. Validation of the feature extraction method in two different applications:
   a. Behavior recognition in mobile robots by identification of their type and motion profiles
   b. Target detection and classification using unattended ground sensors (UGS) for border security

This chapter is organized into seven sections including the present one. Section 14.2 briefly describes the concepts of SDF and its application to wavelet-transformed data. Section 14.3 presents the procedure of feature extraction from the symbolized wavelet image by construction of a PFSA. Section 14.4 describes the pattern classification algorithms. Section 14.5 presents experimental validation for classification of mobile robot types and their motion profiles. The experimental facility incorporates distributed pressure sensors under the floor to track and classify the mobile robots. Section 14.6 validates the feature extraction and pattern classification algorithms based on the field data of UGS for target detection and classification. This chapter is concluded in Section 14.7 along with recommendations for future research.

14.2 Symbolic Dynamics and Encoding

This section presents the underlying concepts of SDF for feature extraction from sensor time-series data. Details of SDF have been reported in previous publications for analysis of (one-dimensional) time series [13,30]. A statistical mechanics-based concept of time-series analysis using symbolic dynamics has been presented in Ref. [14]. This section briefly reviews the concepts of SDF for analysis of (two-dimensional) wavelet images for feature extraction. The major steps of the SDF method for feature extraction are delineated as follows:

1. Encoding (possibly nonlinear) system dynamics from observed sensor data (e.g., time series and images) for generation of symbol sequences
2. Information compression via construction of PFSA from the symbol sequences to generate feature vectors that are representatives of the underlying dynamical system's behavior

14.2.1 Review of Symbolic Dynamics

In the symbolic dynamics literature [23], it is assumed that the observed sensor time series from a dynamical system is represented as a symbol sequence. Let Ω be a compact (i.e., closed and totally bounded) region in the phase space of the continuously varying dynamical system, within which the observed time series is confined [13,30]. The region Ω is partitioned into $|\Sigma|$ cells $\{\Phi_0, ..., \Phi_{|\Sigma|-1}\}$ that are mutually exclusive (i.e., $\Phi_j \cap \Phi_k =; \forall j \neq k$) and exhaustive (i.e., $C_{j=0}^{|\Sigma|-1}\Phi_j = \Omega$), where Σ is the *symbol alphabet* that labels the partition cells. A trajectory of the dynamical system is described by the discrete time-series data as: $\{\mathbf{x}_0, \mathbf{x}_1, \mathbf{x}_2, ...\}$, where each $\mathbf{x}_i \in \Omega$. The trajectory passes through or touches one of the cells of the partition; accordingly, the corresponding symbol is assigned to each point $\mathbf{x}_i$ of the trajectory as defined by the mapping $\mathcal{M}: \Omega \to \Sigma$. Therefore, a sequence of symbols is generated from the trajectory starting from an initial state $\mathbf{x}_0 \in \Omega$, such that

$$\mathbf{x}_0 \rightarrowtail \sigma_0\sigma_1\sigma_2 \ldots \sigma_k \ldots \tag{14.1}$$

where $\sigma_k \triangleq \mathcal{M}(\mathbf{x}_k)$ is the symbol at instant k. (Note: The mapping in Equation 14.1 is called *symbolic dynamics* if it attributes a legal [i.e., physically admissible] symbol sequence to the system dynamics starting from an initial state.) The next section describes how the time series are transformed into wavelet images in scale-shift domain for generation of symbolic dynamics.

14.2.2 Transformation of Time Series to Wavelet Domain

This section presents the procedure for generation of wavelet images from sensor time series for feature extraction. A crucial step in SDF [13,30] is partitioning of the data space for symbol sequence generation [8]. Various partitioning techniques have been suggested in literature for symbol generation, which include variance-based [34], entropy-based [6], and hierarchical clustering-based [18] methods. A survey of clustering techniques is provided in Ref. [22]. Another partitioning scheme, based on *symbolic false nearest neighbors* (SFNN), was reported by Buhl and Kennel [4]. These techniques rely on partitioning the phase space and may become cumbersome and extremely computation-intensive if the dimension of the phase space is large. Moreover, if the data set is noise-corrupted, then the symbolic false neighbors would rapidly grow in number and require a large symbol alphabet to capture the pertinent information. Therefore, symbolic sequences as representations of the system dynamics should be generated by alternative methods because phase-space partitioning might prove to be a difficult task.

Technical literature has suggested appropriate transformation of the signal before employing the partitioning method for symbol generation [30]. One such technique is the ASSP [32] that is based on the analytic signal that provides the additional phase information in the sensor data. The WTSP [28] is well suited for time–frequency analysis of non-stationary signals, noise attenuation, and reduction of spurious disturbances from the raw time-series data without any significant loss of pertinent information [13,24]. Since SFNNP and ASSP may require additional preprocessing of the time series for denoising, this chapter has used WTSP for construction of symbolic representations of sensor data as explained later.

In wavelet-based partitioning, time series are first transformed into the wavelet domain, where wavelet coefficients are generated at different shifts and scales. The choice of the wavelet basis function and wavelet scales depends on the time–frequency characteristics of individual signals [13]. The wavelet transform of a function $f(t) \in \mathbb{H}$, where $\mathbb{H}$ is a Hilbert space is given by

$$F_{s,\tau} = \int_{-\infty}^{\infty} f(t)\psi_{s,\tau}^*(t)dt, \tag{14.2}$$

where
  $s > 0$ is the scale
  τ is the time shift
  $\psi_{s,\tau}(t) = \psi((t - \tau)/s)$ is the wavelet basis function
  $\psi \in \mathbf{L}_2(\mathbb{R})$ is such that $\int_{-\infty}^{\infty} \psi(t)dt = 0$ and $\|\psi\|_2 = 1$

Wavelet preprocessing of sensor data for symbol sequence generation helps in noise mitigation. Let $\tilde{f}$ be a noise-corrupted version of the original signal f expressed as

$$\tilde{f} = f + kw, \tag{14.3}$$

where
  w is additive white Gaussian noise with zero mean and unit variance
  k is the noise level

The noise part in Equation 14.3 would be reduced if the scales over which coefficients are obtained are properly chosen.

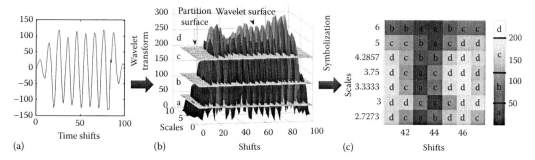

FIGURE 14.1   Symbol image generation via wavelet transform of the sensor time-series data and partition of the wavelet surface in ordinate direction. (a) Sensor time series data; (b) partition of the wavelet coefficients; and (c) symbolized wavelet image (a section).

For every wavelet, there exists a certain frequency called the center frequency F_c that has the maximum modulus in the Fourier transform of the wavelet. The pseudo-frequency f_p of the wavelet at a particular scale α is given by the following formula [1]:

$$f_p = \frac{F_c}{\alpha\,\Delta t},$$

(14.4)

where Δt is the sampling interval. Then the scales can be calculated as follows:

$$\alpha^i = \frac{F_c}{f_p^i\,\Delta t}$$

(14.5)

where

    $i = 1, 2, \ldots$

    f_p^i are the frequencies that can be obtained by choosing the locally dominant frequencies in the Fourier transform

The maximum pseudo-frequency f_p^{max} should not exceed the Nyquist frequency [1]. Therefore, the sampling frequency f_s for acquisition of time-series data should be selected at least twice the larger of the maximum pseudo-frequency f_p^{max} and the signal bandwidth B, that is, $f_s \geq 2\max(f_p^{max}, B)$.

Figure 14.1 shows an illustrative example of transformation of the (one-dimensional) time series in Figure 14.1a to a (two-dimensional) wavelet image in Figure 14.1b. The amplitudes of the wavelet coefficients over the scale-shift domain are plotted as a surface. Subsequently, symbolization of this wavelet surface leads to the formation of a symbolic image as shown in Figure 14.1c.

14.2.3 Symbolization of Wavelet Surface Profiles

This section presents partitioning of the wavelet surface profile in Figure 14.1b, which is generated by the coefficients over the two-dimensional scale-shift domain, for construction of the symbolic image in Figure 14.1c. The x–y coordinates of the wavelet surface profiles denote the shifts and the scales, respectively, and the z-coordinate (i.e., the surface height) denotes the pixel values of wavelet coefficients.

Definition 14.1 (Wavelet surface profile)

Let $\mathcal{H} \triangleq \{(i,j) : i,j \in \mathbb{N}, 1 \leq i \leq m, 1 \leq j \leq n\}$ be the set of coordinates consisting of $(m \times n)$ pixels denoting the scale-shift data points. Let $\mathcal{R}$ denote the interval that spans the range of wavelet coefficient amplitudes. Then, a wavelet surface profile is defined as

$$S : \mathcal{H} \to \mathcal{R} \tag{14.6}$$

Definition 14.2 (Symbolization)

Given the symbol alphabet Σ, let the partitioning of the interval $\mathcal{R}$ be defined by a map $P : \mathcal{R} \to \Sigma$. Then, the symbolization of a wavelet surface profile is defined by a map $S_\Sigma \equiv P \circ S$ such that

$$S_\Sigma : \mathcal{H} \to \Sigma \tag{14.7}$$

that labels each pixel of the image to a symbol in Σ.

The wavelet surface profiles are partitioned such that the ordinates between the maximum and minimum of the coefficients along the z-axis are divided into regions by different planes parallel to the $(x–y)$ plane. For example, if the alphabet is chosen as $\Sigma = \{a, b, c, d\}$, that is, $|\Sigma| = 4$, then three partitioning planes divide the ordinate (i.e., z-axis) of the surface profile into four mutually exclusive and exhaustive regions, as shown in Figure 14.1b. These disjoint regions form a partition, where each region is labeled with one symbol from the alphabet Σ. If the intensity of a pixel is located in a particular region, then it is coded with the symbol associated with that region. As such, a symbol from the alphabet Σ is assigned to each pixel corresponding to the region where its intensity falls. Thus, the two-dimensional array of symbols, called *symbol image*, is generated from the wavelet surface profile, as shown in Figure 14.1c.

The surface profiles are partitioned by using either the maximum entropy partitioning (MEP) or the uniform partitioning (UP) methods [13,28]. If the partitioning planes are separated by equal-sized intervals, then the partition is called the UP. Intuitively, it is more reasonable if the information-rich regions of a data set are partitioned finer and those with sparse information are partitioned coarser. To achieve this objective, the MEP method has been adopted in this chapter such that the entropy of the generated symbols is maximized. The procedure for selection of the alphabet size $|\Sigma|$, followed by generation of a MEP, has been reported in Ref. [13]. In general, the choice of alphabet size depends on specific data set and experiments. The partitioning of wavelet surface profiles to generate symbolic representations enables robust feature extraction, and symbolization also significantly reduces the memory requirements [13].

14.3 Construction of Probabilistic Finite-State Automata for Feature Extraction

This section presents the method for construction of a PFSA for feature extraction from the symbol image generated from the wavelet surface profile.

14.3.1 Conversion from Symbol Image to State Image

For analysis of (one-dimensional) time series, a PFSA is constructed such that its states represent different combinations of blocks of symbols on the symbol sequence. The edges connecting these states represent the transition probabilities between these blocks [13,30]. Therefore, for analysis of (one-dimensional) time series, the "states" denote all possible symbol blocks (i.e., words) within a window of certain length. Let us now extend the notion of "states" on a two-dimensional domain for analysis of wavelet surface profiles via construction of a "*state image*" from a "*symbol image.*"

Definition 14.3 (State)

Let $\mathcal{W} \subset \mathcal{H}$ be a two-dimensional window of size $(\ell \times \ell)$ that is denoted as $|\mathcal{W}| = \ell^2$. Then, the state of a symbol block formed by the window $\mathcal{W}$ is defined as the configuration $q = S_\Sigma(\mathcal{W})$.

Let the set of all possible states (i.e., two-dimensional words or blocks of symbols) in a window $\mathcal{W} \subset \mathcal{H}$ be denoted as $Q \triangleq \{q_1, q_2, \ldots, q_{|Q|}\}$, where $|Q|$ is the number of (finitely many) states. Then, $|Q|$ is bounded above as $|Q| \leq |\Sigma|^{|\mathcal{W}|}$; the inequality is due to the fact that some of the states might have zero probability of occurrence. Let us denote $\mathcal{W}_{i,j} \subset \mathcal{H}$ to be the window where (i, j) represents the coordinates of the top-left corner pixel of the window. In this notation, $q_{i,j} = S_\Sigma(\mathcal{W}_{i,j})$ denotes the state at pixel $(i, j) \in \mathcal{H}$. Thus, every pixel $(i, j) \in \mathcal{H}$ corresponds to a particular state $q_{i,j} \in Q$ on the image. Every pixel in the image $\mathcal{H}$ is mapped to a state, excluding the pixels that lie at the periphery depending on the window size. Figure 14.2 shows an illustrative example of the transformation of a *symbol image* to the *state image* based on a sliding window $\mathcal{W}$ of size (2×2). This concept of state formation facilitates capturing of long range dynamics (i.e., word to word interactions) on a symbol image.

In general, a large number of states would require a high computational capability and hence might not be feasible for real-time applications. The number of states, $|Q|$, increases with the window size $|\mathcal{W}|$ and the alphabet size $|\Sigma|$. For example, if $\ell = 2$ and $|\Sigma| = 4$, then the total number of states are $|Q| \leq |\Sigma|^{\ell^2} = 256$. Therefore, for computational efficiency, it is necessary to compress the state set Q to an effective reduced set $\mathcal{O} \triangleq \{o_1, o_2, \ldots, o_{|\mathcal{O}|}\}$ [13] that enables mapping of two or more different configurations in a window $\mathcal{W}$ to a single state. State compression must preserve sufficient information as needed for pattern classification, albeit possibly lossy coding of the wavelet surface profile.

In view of the earlier discussion, a probabilistic state compression method is employed, which chooses the m most probable symbols, from each state as a representation of that particular state. In this method, each state consisting of $\ell \times \ell$ symbols is compressed to a reduced state of length $m < \ell^2$ symbols by choosing the top m symbols that have the highest probability of occurrence arranged in descending order. If two symbols have the same probability of occurrence, then either symbol may be preferred with equal probability. This procedure reduces the state set Q to an effective set $\mathcal{O}$, where the total number of compressed states is given as $|\mathcal{O}| = |\Sigma|^m$. For example, if $|\Sigma| = 4$, $|\mathcal{W}| = 4$ and $m = 2$, then the state compression reduces the total number of states to $|\mathcal{O}| = |\Sigma|^m = 16$ instead of 256. This method of state compression is motivated from the renormalization methods in *statistical physics* that are useful in eliminating the

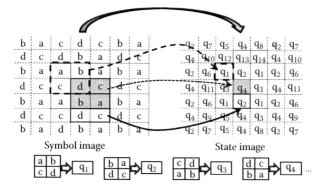

FIGURE 14.2   Conversion of the symbol image to the state image.

irrelevant local information on lattice spin systems while still capturing the long range dynamics [14]. The choice of $|\Sigma|$, ℓ, and m depends on specific applications and noise level as well as the available computational power and is made by an appropriate trade-off between robustness to noise and capability to detect small changes. For example, a large alphabet may be noise sensitive while a small alphabet could miss the information of signal dynamics [13].

14.3.2 Construction of PFSA

A PFSA is constructed such that the states of the PFSA are the elements of the compressed state set $\mathcal{O}$ and the edges are the transition probabilities between these states. Figure 14.3a shows an example of a typical PFSA with four states. The transition probabilities between states are defined as

$$P(o_k \mid o_l) = \frac{N(o_l, o_k)}{\sum_{k'=1,2,\ldots,|\mathcal{O}|} N(o_l, o_{k'})} \quad \forall\, o_l, o_k \in \mathcal{O} \tag{14.8}$$

where $N(o_l, o_k)$ is the total count of events when o_k occurs adjacent to o_l in the direction of motion. The calculation of these transition probabilities follows the principle of sliding block code [23]. A transition from the state o_l to the state o_k occurs if o_k lies adjacent to o_l in the positive direction of motion. Subsequently, the counter moves to the right and to the bottom (row-wise) to cover the entire state image, and the transition probabilities $P(o_k|o_l)$, $\forall\, o_l, o_k \in \mathcal{O}$ are computed using Equation 14.8. Therefore, for every state on the state image, all state-to-state transitions are counted, as shown in Figure 14.3b. For example, the dotted box in the bottom-right corner contains three adjacent pairs, implying the transitions $o_1 \to o_2$, $o_1 \to o_3$, and $o_1 \to o_4$ and the corresponding counter of occurrences $N(o_1, o_2)$, $N(o_1, o_3)$, and $N(o_1, o_4)$, respectively, are increased by one. This procedure generates the stochastic state-transition probability matrix of the PFSA given as

$$\Pi = \begin{bmatrix} P(o_1 \mid o_1) & \cdots & P(o_{|\mathcal{O}|} \mid o_1) \\ \vdots & \ddots & \vdots \\ P(o_1 \mid o_{|\mathcal{O}|}) & \cdots & P(o_{|\mathcal{O}|} \mid o_{|\mathcal{O}|}) \end{bmatrix} \tag{14.9}$$

where $\Pi \equiv [\pi_{j,k}]$ with $\pi_{j,k} = P(o_k|o_j)$. Note: $\pi_{j,k} \geq 0\ \forall j, k \in \{1,2,\ldots,|\mathcal{O}|\}$ and $\sum_k \pi_{j,k} = 1\ \forall j \in \{1,2,\ldots,|\mathcal{O}|\}$.

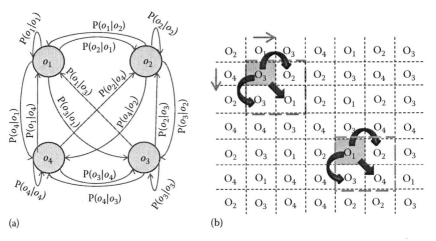

(a) (b)

FIGURE 14.3 Example of feature extraction from stage image by constructing a PFSA. (a) An example of four-state PFSA and (b) feature extraction from the state image.

In order to extract a low-dimensional feature vector, the stationary state probability vector **p** is obtained as the left eigenvector corresponding to the (unique) unity eigenvalue of the (irreducible) stochastic transition matrix Π. The state probability vectors **p** serve as the "*feature vectors*" and are generated from different data sets from the corresponding state-transition matrices. These feature vectors are also denoted as "patterns" in this chapter.

14.3.3 Summary of SDF for Feature Extraction

The major steps of SDF for feature extraction are summarized as follows:

- Acquisition of time-series data from appropriate sensor(s) and signal conditioning as necessary
- Wavelet transform of the time-series data with appropriate scales to generate the wavelet surface profile
- Partitioning of the wavelet surface profile and generation of the corresponding symbol image
- Conversion from symbol image to state image via probabilistic state compression strategy
- Construction of PFSA and computation of the state-transition matrices that in turn generate the state probability vectors as the feature vectors (i.e., patterns)

The advantages of SDF for feature extraction and subsequent pattern classification are summarized as follows:

- Robustness to measurement noise and spurious signals
- Adaptability to low-resolution sensing due to the coarse graining in space partitions [30]
- Capability for detection of small deviations because of sensitivity to signal distortion
- Real-time execution on commercially available inexpensive platforms

14.4 Pattern Classification Using SDF-Based Features

Once the feature vectors are extracted in a low-dimensional space from the observed sensor time series, the next step is to classify these patterns into different categories based on the particular application. Technical literature abounds in diverse methods of pattern classification, such as divergence measure, k-nearest neighbor (k-NN) algorithm [7], SVM [3], and ANN [16]. The main focus of this chapter is to develop and validate the tools of SDF for feature extraction from wavelet surface profiles generated from sensor time-series data. Therefore, the SDF method for feature extraction is used in conjunction with the standard pattern classification algorithms, as described in the experimental validation sections.

Pattern classification using SDF-based features is posed as a two-stage problem, that is, the training stage and the testing stage. The sensor time-series data sets are divided into three groups: (1) partition data, (2) training data, and (3) testing data. The partition data set is used to generate partition planes that are used in the training and the testing stages. The training data set is used to generate the training patterns of different classes for the pattern classifier. Multiple sets of training data are obtained from independent experiments for each class in order to provide a good statistical spread of patterns. Subsequently, the class labels of the testing patterns are generated from testing data in the testing stage. The partition data sets may be part of the training data sets, whereas the training data sets and the testing data sets must be mutually exclusive.

Figure 14.4 depicts the flow chart of the proposed algorithm that is constructed based on the theory of SDF. The partition data is wavelet-transformed with appropriate scales to convert the one-dimensional numeric time-series data into the wavelet image. The corresponding wavelet surface is analyzed using the *maximum entropy principle* [13,28] to generate the partition planes that remain invariant for both the training and the testing stages. The scales used in the wavelet transform of the partitioning data also remain invariant during the wavelet transform of the training and the testing data. In the training stage, the wavelet surfaces are generated by transformation of the training data sets corresponding

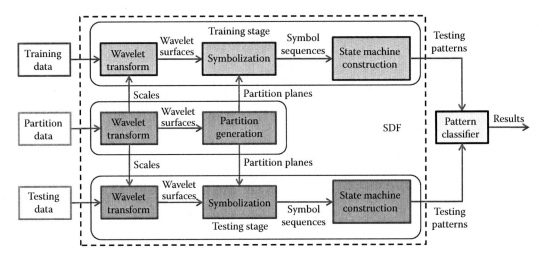

FIGURE 14.4 Flow chart of the proposed methodology.

to different classes. These surfaces are symbolized using the partition planes to generate the symbol images. Subsequently, PFSAs are constructed based on the corresponding symbol images, and the training patterns (i.e., state probability vectors **p** or state-transition matrices Π) are extracted from these PFSAs. Similar to the training stage, the PFSA and the associated pattern is generated for different data sets in the testing stage. These patterns are then classified into different classes using pattern classifier, such as SVM, k-NN, and ANN.

Consider a classification problem of $|C|$ classes, where C is the set of class labels. In the training stage, feature vectors $\mathbf{p}_j^{C_i}$, $j = 1, 2, \ldots, n_i$ are generated from the training data sets of class C_i, where n_i is the number of samples in class C_i. The same procedure is carried out for all other classes. In the testing stage, a testing feature vector $\mathbf{p}_{test}$ with unknown class labels is generated using SDF. Two examples of using the pattern classifiers with SDF are provided here. For k-NN algorithm, the estimated class label of a testing feature vector $\mathbf{p}_{test}$ is equal to the most frequent class among the k-nearest training features [7]. For SVM, a separating hyperplane/hypersurface is generated based on training feature vectors ($\mathbf{p}_j^{C_i}$, $j = 1, 2, \ldots, n_i$). The estimated class label of the testing feature vector $\mathbf{p}_{test}$ depends on which side of the hyperplane/hypersurface the testing feature vector falls [3].

14.5 Validation I: Behavior Recognition of Mobile Robots in a Laboratory Environment

This section presents experimental validation of the proposed wavelet-based feature extraction method in a laboratory environment of networked robots. The objective here is to identify the robot type and the motion profile based on the sensor time series obtained from the pressure-sensitive floor. These experiments are inspired from various real-life applications of pattern classification such as (1) classification of enemy vehicles across the battlefield through analysis of seismic and acoustic time-series data and (2) classification of human and animal movements through analysis of seismic time series.

14.5.1 Experimental Procedure for Behavior Identification of Mobile Robots

The experimental setup consists of a wireless network incorporating mobile robots, robot simulators, and distributed sensors as shown in Figures 14.5 and 14.6. A major component of the experimental setup is the pressure-sensitive floor that consists of distributed piezoelectric wires installed underneath the floor to serve as arrays of distributed pressure sensors. A coil of piezoelectric wire is placed under a 0.65 m × 0.65 m

 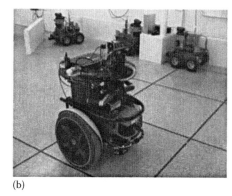

FIGURE 14.5   Robot Hardware: Pioneer 2AT (a) and Segway RMP (b).

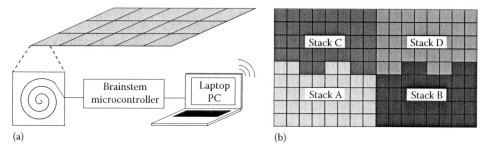

FIGURE 14.6   Layout of the distributed pressure sensors in the laboratory environment: (a) sensor and (b) distribution of sensors.

square floor tile as shown in Figure 14.6a such that the sensor generates an analog voltage due to pressure applied on it. This voltage is sensed by a *Brainstem*™ microcontroller using one of its 10-bit A/D channels thereby yielding sensor readings in the range of 0–1023. The sampling frequency of the pressure sensing device that captures the dynamics of robot motion is 10 Hz, while the maximum pseudo-frequency f_p^{max} is 4.44 Hz (see Section 14.2.2). A total of 144 sensors are placed in a 9×16 grid to cover the entire laboratory environment as shown in Figure 14.6b. The sensors are grouped into four quadrants, each being connected to a stack consisting of eight networked *Brainstem* microcontrollers for data acquisition. The microcontrollers are, in turn, connected to two laptop computers running *Player* [11] server that collects the raw sensor data and distributes to any client over the wireless network for further processing.

Figure 14.5 shows a pair of Pioneer robots and a Segway RMP that have the following features:

- Pioneer 2AT is a four-wheeled robot that is equipped with a differential drive train system and has an approximate weight of 35 kg.
- Segway RMP is a two-wheeled robot (with inverted pendulum dynamics) that has a zero turn radius and has an approximate weight of 70 kg.

Since Pioneer is lighter than Segway and Pioneer's load on the floor is more evenly distributed, their statistics are dissimilar. Furthermore, since the kinematics and dynamics of the two types of robots are different, the textures of the respective pressure sensor signals are also different.

The objective is to identify the robot type and motion type from the time-series data. The Segway RMP and Pioneer 2AT robots are commanded to execute three different motion trajectories, namely, *random motion*, *circular motion*, and *square motion*. Table 14.1 lists the parameters for the three types of robot motion. In the presence of uncertainties (e.g., sensor noise and fluctuations in robot motion), a complete solution of the robot type and motion identification problem may not be possible

TABLE 14.1 Parameters Used for Various
Types of Motion

Motion Type	Parameter	Value
Circular	Diameter	4 m
Square	Edge length	3 m
Random	Uniform distribution	x-direction 1–7 m
		y-direction 1–4 m

in a deterministic setting because the patterns would not be identical for similar robots behaving similarly. Therefore, the problem is posed in the statistical setting, where a family of patterns is generated from multiple experiments conducted under identical operating conditions. The requirement is to generate a family of patterns for each class of robot behavior that needs to be recognized. Therefore, both Segway RMP and Pioneer 2AT robots were made to execute several cycles of each of the three different types of motion trajectories on the pressure-sensitive floor of the laboratory environment. Each member of a family represents the pattern of a single experiment of one robot executing a particular motion profile. As a robot changes its type of motion from one (e.g., circular) to another (e.g., random), the pattern classification algorithm is capable of detecting this change after a (statistically quasi-stationary) steady state is reached. During the brief transient period, the analysis of pattern classification may not yield accurate results because the resulting time series may not be long enough to extract the features correctly.

Figure 14.7a shows an example of the sensor reading when the robot moves over it. The voltage generated by the piezoelectric pressure sensor gradually increases as the robot approaches the sensor, and discharge occurs in the sensor when the robot moves away from the sensor and hence the voltage resumes to be 0. The choice of mother wavelet depends on the shape of the sensor signal; the mother wavelet should match the shape of the sensor signal in order to capture the signature of the signal. Haar wavelet (*db*1), as shown in Figure 14.7b, is chosen to be the mother wavelet in this application. The sensor data collected by the 9 × 16 grid is stacked sequentially to generate a one-dimensional time series. For each motion trajectory consisting of several cycles, the time-series data collected from the pressure sensors was divided into 40–50 data sets. The length of each data set is 3.0×10^5 data points, which corresponds to about 3 min of the experiment time. The data sets are randomly divided into half training and half testing. Among the training data, 10 sets are chosen to serve as the partitioning data sets as well.

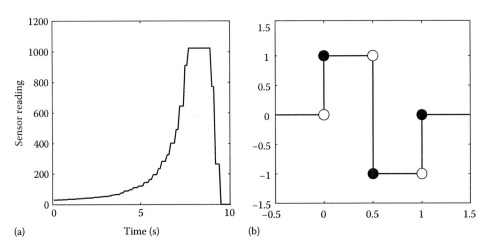

(a)   Time (s)   (b)

FIGURE 14.7 Example of sensor readings and plot of Haar wavelet: (a) sensor reading and (b) Haar wavelet.

14.5.2 Pattern Analysis for Behavior Identification of Mobile Robots

This section provides a description of the application of different pattern analysis methods to time-series data of pressure sensors for classification of the robots and their motion types.

For feature extraction using SDF, each data set of a family (or class) is analyzed to generate the corresponding state probability vectors (i.e., patterns). Thus, the patterns $\mathbf{p}_j^{C_i}$, $j = 1, 2, \ldots, n_i$ are generated for n_i samples in each class C_i corresponding to robot type and motion. Following the SDF procedure, each time-series data set is analyzed using $|\Sigma| = 8$, $\ell = 2$ and $m = 1$. Ensemble mean of pattern vectors for different motion profiles of Segway and Pioneer robots is shown in Figure 14.8. It can be observed in Figure 14.8 that the state probability vectors of Segway and Pioneer robots are quite distinct. Following Figure 14.4, for each motion type, the state probability vectors $\mathbf{p}_j^{C_i}$ were equally divided into training sets and testing sets.

In this application, the efficacy of SDF for feature extraction is evaluated by comparison with PCA. The time-series data are transformed to the frequency domain for noise mitigation and then the standard PCA method is implemented to identify the eigen directions of the transformed data and to obtain an orthogonal linear operator that projects the frequency-domain features onto a low-dimensional compressed-feature space. For the purpose of comparison, the dimension of this compressed-feature space is chosen to be the same as that of the feature vectors obtained by SDF. In this application, the SVM, k-NN algorithm, radial basis neural network (rbfNN), and multilayer perceptron neural network (mlpNN) have been used as the pattern classifiers to identify different classes of feature vectors extracted by SDF and PCA. The pattern classifiers identify the type of the robot and its motion profile, based on the acquired statistical patterns. Since, in this pattern classification problem, there are two robots and each robot has three different types of motion profiles, it is natural to formulate this problem as a two-layer classification problem, where the robot type is identified in the first layer followed by identification

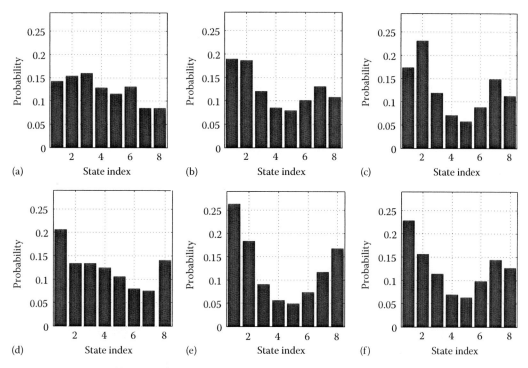

FIGURE 14.8 Ensemble mean of the state probability vectors (feature vectors): (a) segway random, (b) segway circle, (c) segway square, (d) pioneer random, (e) pioneer circle, and (f) pioneer square.

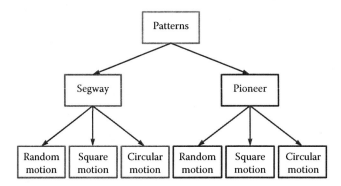

FIGURE 14.9   Tree structure for pattern classification.

of the motion type in the second layer. Thus, the earlier problem is formulated using a tree-structure classification as shown in Figure 14.9.

14.5.3 Experimental Results for Behavior Identification of Mobile Robots

The performance comparison between SDF and PCA that are used in conjunction with different classifiers is presented in Table 14.2. The left part of Table 14.2 shows the results of robot type and robot motion classification using SDF for feature extraction, and the right part shows the corresponding results using PCA for feature extraction. As stated earlier, SVM, k-NN, rbfNN, and mlpNN have been used as pattern classifiers in both cases. The *polynomial* kernel is used in SVM [5], and a neighbor size of $k = 5$ is used in the k-NN classifier. The rbfNN uses one hidden layer and one output layer with a single neuron. Optimal training is obtained with 100 neurons in the hidden layer that uses a radial basis function, while the output layer uses a linear transfer function. The mlpNN utilizes a feed-forward back-propagation network that consists of one hidden layer with 50 neurons and an output layer with a single neuron; the tangent sigmoid function has been used in the hidden layers as a transfer function, while the output layer uses a linear function.

It is noted that since the tree structure is used for pattern classification, the motion recognition results are affected by the robot recognition results. For example, the samples that are incorrectly classified in the robot recognition stage will be incorrectly classified for motion recognition also. However, this particular aspect is application dependent and the tree structure for classification can be redesigned accordingly. The classification results are presented in Table 14.2 that show the accuracy percentage equal to ((# correct classifications/# total data sets) × 100). In the left part of Table 14.2, the combination of SDF with all four classifiers yields good accuracy in recognizing the robot type, namely, 100% for Segway and more than 94% for Pioneer. Although not explicitly shown in Table 14.2, all four classifiers successfully identified the three types of motions of the Pioneer robot with 100% accuracy in the robot motion classification stage. The errors in the motion recognition, as seen in Table 14.2, originate from the robot recognition stage. The success rate in recognizing Segway motion is slightly lower due to the following possible reasons: (1) complicated kinematics of the Segway robot and (2) the nonstationarity in the samples due to uncertainties in the laboratory environment (e.g., floor friction). It is expected that this accuracy would further improve if the number of stationary samples is increased. The right part of Table 14.2 shows that PCA yields slightly worse (but still comparable) results than SDF in robot recognition. However, SDF significantly outperforms PCA in motion recognition, because the feature vectors extracted by PCA lack class separability among different types of motions, which in turn yields poor motion recognition accuracy.

The proposed SDF-based method has a computational complexity of $O(N)$ for a given algebraic structure of the PFSA, with a leading constant that is proportional to the number of scales used in the wavelet transform [25]. A comparison of the computational complexity of SDF and PCA is presented in

TABLE 14.2   Results of Robot and Motion Classification

Feature Extraction Using SDF

Pattern Classifier	Robot Recognition		Motion Recognition		
	Robot	Result	Motion	Result	Total
SVM	Segway	$100\%\left(\frac{58}{58}\right)$	Random	$92\%\left(\frac{23}{25}\right)$	$95\%\left(\frac{55}{58}\right)$
			Circular	$92\%\left(\frac{12}{13}\right)$	
			Square	$100\%\left(\frac{20}{20}\right)$	
	Pioneer	$94\%\left(\frac{61}{65}\right)$	Random	$100\%\left(\frac{20}{20}\right)$	$94\%\left(\frac{61}{65}\right)$
			Circular	$90\%\left(\frac{18}{20}\right)$	
			Square	$92\%\left(\frac{23}{25}\right)$	
k-NN	Segway	$100\%\left(\frac{58}{58}\right)$	Random	$88\%\left(\frac{22}{25}\right)$	$91\%\left(\frac{53}{58}\right)$
			Circular	$85\%\left(\frac{11}{13}\right)$	
			Square	$100\%\left(\frac{20}{20}\right)$	
	Pioneer	$94\%\left(\frac{61}{65}\right)$	Random	$95\%\left(\frac{19}{20}\right)$	$94\%\left(\frac{61}{65}\right)$
			Circular	$100\%\left(\frac{20}{20}\right)$	
			Square	$88\%\left(\frac{22}{25}\right)$	

Feature Extraction Using PCA

Pattern Classifier	Robot Recognition		Motion Recognition		
	Robot	Result	Motion	Result	Total
SVM	Segway	$91\%\left(\frac{53}{58}\right)$	Random	$84\%\left(\frac{21}{25}\right)$	$81\%\left(\frac{47}{58}\right)$
			Circular	$77\%\left(\frac{10}{13}\right)$	
			Square	$80\%\left(\frac{16}{20}\right)$	
	Pioneer	$100\%\left(\frac{65}{65}\right)$	Random	$20\%\left(\frac{4}{20}\right)$	$65\%\left(\frac{42}{65}\right)$
			Circular	$65\%\left(\frac{13}{20}\right)$	
			Square	$100\%\left(\frac{25}{25}\right)$	
k-NN	Segway	$100\%\left(\frac{58}{58}\right)$	Random	$84\%\left(\frac{21}{25}\right)$	$52\%\left(\frac{30}{58}\right)$
			Circular	$69\%\left(\frac{9}{13}\right)$	
			Square	$0\%\left(\frac{0}{20}\right)$	
	Pioneer	$88\%\left(\frac{57}{65}\right)$	Random	$0\%\left(\frac{0}{20}\right)$	$51\%\left(\frac{33}{65}\right)$
			Circular	$55\%\left(\frac{11}{20}\right)$	
			Square	$88\%\left(\frac{22}{25}\right)$	

(continued)

TABLE 14.2 (continued) Results of Robot and Motion Classification

Feature Extraction Using SDF

Pattern Classifier	Robot Recognition		Motion Recognition		
	Robot	Result	Motion	Result	Total
rbfNN	Segway	$100\% \left(\frac{58}{58}\right)$	Random	$84\% \left(\frac{21}{25}\right)$	$91\% \left(\frac{53}{58}\right)$
			Circular	$92\% \left(\frac{12}{13}\right)$	
			Square	$100\% \left(\frac{20}{20}\right)$	
	Pioneer	$97\% \left(\frac{63}{65}\right)$	Random	$95\% \left(\frac{19}{20}\right)$	$97\% \left(\frac{63}{65}\right)$
			Circular	$100\% \left(\frac{20}{20}\right)$	
			Square	$96\% \left(\frac{24}{25}\right)$	
mlpNN	Segway	$100\% \left(\frac{58}{58}\right)$	Random	$96\% \left(\frac{24}{25}\right)$	$98\% \left(\frac{57}{58}\right)$
			Circular	$100\% \left(\frac{13}{13}\right)$	
			Square	$100\% \left(\frac{20}{20}\right)$	
	Pioneer	$100\% \left(\frac{65}{65}\right)$	Random	$100\% \left(\frac{20}{20}\right)$	$100\% \left(\frac{65}{65}\right)$
			Circular	$100\% \left(\frac{20}{20}\right)$	
			Square	$100\% \left(\frac{25}{25}\right)$	

Feature Extraction Using PCA

Pattern Classifier	Robot Recognition		Motion Recognition		
	Robot	Result	Motion	Result	Total
rbfNN	Segway	$100\% \left(\frac{58}{58}\right)$	Random	$92\% \left(\frac{23}{25}\right)$	$72\% \left(\frac{42}{58}\right)$
			Circular	$31\% \left(\frac{4}{13}\right)$	
			Square	$75\% \left(\frac{15}{20}\right)$	
	Pioneer	$95\% \left(\frac{62}{65}\right)$	Random	$0\% \left(\frac{0}{20}\right)$	$66\% \left(\frac{43}{65}\right)$
			Circular	$100\% \left(\frac{20}{20}\right)$	
			Square	$92\% \left(\frac{23}{25}\right)$	
mlpNN	Segway	$100\% \left(\frac{58}{58}\right)$	Random	$96\% \left(\frac{24}{25}\right)$	$98\% \left(\frac{57}{58}\right)$
			Circular	$100\% \left(\frac{13}{13}\right)$	
			Square	$100\% \left(\frac{20}{20}\right)$	
	Pioneer	$100\% \left(\frac{65}{65}\right)$	Random	$85\% \left(\frac{17}{20}\right)$	$92\% \left(\frac{60}{65}\right)$
			Circular	$95\% \left(\frac{19}{20}\right)$	
			Square	$96\% \left(\frac{24}{25}\right)$	

TABLE 14.3   Comparison of Computational Complexity of Feature Extraction Methods

| Method | Training Stage | | Testing Stage | |
	Execution Time (s)	Memory Requirement (MB)	Execution Time (s)	Memory Requirement (MB)
SDF	5.21	65.2	5.17	64.9
PCA	6.62	233.85	0.04	37.5

Table 14.3 in terms of execution time and memory requirements for processing each data set. For the data set consisting of 3.0×10^5 data points, which is about 3.5 min of the experimentation time, it takes an average of 5.21 s for SDF and 6.62 s for PCA to process each data set in the training stage, respectively. The memory requirement is 65.2 MB for SDF and 233.9 MB for PCA. PCA takes longer execution time and consumes more memory in the training stage because it needs to calculate the covariance matrix using all training data sets. In the testing stage, the execution time and memory requirement for SDF are almost the same as those in the training stage, while the PCA requires less time and memory than those in the training stage. Both feature extraction methods have real-time implementation capability since the execution time in the testing stage is much less than the experiment time spent for collecting each data set. The rationale for SDF taking longer time than PCA in the testing stage is that the SDF-based method involves wavelet transformation and PFSA construction from the two-dimensional wavelet image in both training and testing stages, while the PCA-based method only involves Fourier transform and finding the projection of the testing data set using the projection matrix that is already constructed in the training stage; this is a price paid for the superior performance and robustness achieved in SDF-based feature extraction (see Table 14.2). It is anticipated that the PCA-based method will be relatively slower if the raw time series is (more effectively) denoised by wavelet transform instead of Fourier transform. In these experiments, the data analysis was performed on a 2.83 GHz Quad Core CPU desktop computer with 8.0 GB of RAM.

14.6   Validation II: Target Detection and Classification Using Seismic and PIR Sensors

The objective of this application is to detect and classify different targets (e.g., humans, vehicles, and animals led by human), where seismic and PIR sensors are used to capture the characteristic signatures. For example, in the movement of a human or an animal across the ground, oscillatory motions of the body appendages provide the respective characteristic signatures.

The seismic and PIR sensor data, used in this analysis, were collected on multiple days from test fields on a wash (i.e., the dry bed of an intermittent creek) and at a choke point (i.e., a place where the targets are forced to go due to terrain difficulties). During multiple field tests, sensor data were collected for several scenarios that consisted of targets walking along an approximately 150 m long trail and returning along the same trail to the starting point. Figure 14.10 illustrates a typical data collection scenario.

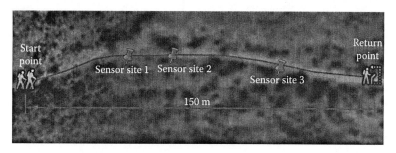

FIGURE 14.10   Illustration of the test scenario with three sensor sites.

(a)   (b)   (c)

FIGURE 14.11 Examples of test scenarios with different targets: (a) human, (b) vehicle, and (c) animal led by human.

The targets consisted of (male and female) humans, animals (e.g., donkeys, mules, and horses), and all-terrain vehicles (ATVs). The humans walked alone and in groups with and without backpacks; the animals were led by their human handlers (simply denoted as "animal" in the sequel) and they made runs with and without payloads; and ATVs moved at different speeds (e.g., 5 and 10 mph). Examples of the test scenarios with different targets are shown in Figure 14.11. There were three sensor sites, each equipped with seismic and PIR sensors. The seismic sensors (geophones) were buried approximately 15 cm deep underneath the soil surface, and the PIR sensors were collocated with the respective seismic sensors. All targets passed by the sensor sites at a distance of approximately 5 m. Signals from both sensors were acquired at a sampling frequency of 10 kHz.

The tree structure in Figure 14.12 shows how the detection and classification problem is formulated. In the detection stage, the pattern classifier detects the presence of a moving target against the null hypothesis of no target present; in the classification stage, the pattern classifiers discriminate among different targets and subsequently identify the movement type and/or payload of the targets. While the detection system should be robust to reduce the false alarm rates, the classification system must be sufficiently sensitive to discriminate among different types of targets with high fidelity. In this context, feature extraction plays an important role in target detection and classification because the performance of classifiers largely depends on the quality of the extracted features.

In the classification stage, there are multiple classes (i.e., humans, animals, and vehicles), and the signature of the vehicles is distinct from those of the other two classes. Therefore, this problem is formulated into a two-layer classification procedure. A binary classification is performed to detect the

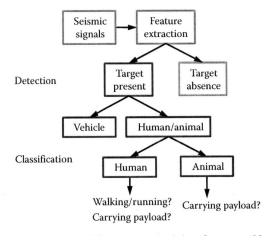

FIGURE 14.12 Tree-structure formulation of the detection and classification problem.

TABLE 14.4 Number of Feature Vectors
for Each Target Class

	Day 1	Day 2	Day 3	Total
No target	50	36	32	118
Vehicle	0	8	0	8
Human	30	22	14	66
Animal	20	6	18	44

presence of a target and then to identify whether the target is a vehicle or a human/animal. Upon recognizing the target as a human/animal, another binary classification is performed to determine its specific class. More information could be derived upon recognition of the target type. For example, if the target is recognized as a human, then further binary classifications are performed to identify if the human is running or walking, and if the human is carrying a payload or not.

Field data were collected in the scenario illustrated in Figure 14.10. Multiple experiments were made to collect data sets of all three classes, that is, human, vehicle, and animal. The data were collected over 3 days at different sites. Table 14.4 shows the number of runs of each class.

Each data set, acquired at a sampling frequency of 10 kHz, has 100,000 data points that correspond to 10 s of the experimentation time. In order to test the capability of the proposed algorithm for target detection, a similar data set was collected with no target present. The problem of target detection is then formulated as a binary pattern classification, where no target present corresponds to one class, and target present (i.e., human, vehicle, or animal) corresponds to the other class. The data sets, collected by the channel of seismic sensors that are orthogonal to the ground surface and the PIR sensors that are collocated with the seismic sensors, are used for target detection and classification. For computational efficiency, the data were downsampled by a factor of 10 with no apparent loss of information.

Figure 14.13 depicts the flow chart of the proposed detection and classification algorithm that is constructed based on the theories of SDF and SVM [3]. The proposed algorithm consists of four main steps: signal preprocessing, feature extraction, target detection, and target classification, as shown in Figure 14.13.

In the signal preprocessing step, the DC component of the seismic signal is eliminated, and the resulting zero mean signal is normalized to unit variance. The amplitude of seismic signal of an animal with a heavy payload walking far away is possibly similar to that of a pedestrian passing by at a close distance due to the fact that the signal-to-noise ratio (SNR) decreases with the distance between the sensor and the target. The normalization of all signals to unit variance makes the pattern classifier independent of the signal amplitude and any discrimination should be solely texture-dependent. For PIR signals, only the DC component is removed and the normalization is not performed because the range of PIR signals does not change during the field test.

In the feature extraction step, SDF captures the signatures of the preprocessed sensor time series for representation as low-dimensional feature vectors. Based on the spectral analysis of the ensemble of seismic data at hand, a series of pseudo-frequencies from the 1–20 Hz bands have been chosen to generate the scales for wavelet transform, because these bands contain a very large part of the footstep energy. Similarly, a series of pseudo-frequencies from the 0.2–2.0 Hz bands have been chosen for PIR signals to generate the scales. Upon generation of the scales, continuous wavelet transforms (CWT) are performed with an appropriate wavelet basis function on the seismic and PIR signals. The wavelet basis *db*7 is used for seismic signals because it matches the impulse shape of seismic signals, and *db*1 is used for the PIR case because PIR signals' shape is close to that of square waves. A maximum entropy wavelet surface partitioning is then performed. Selection of the alphabet size $|\Sigma|$ depends on the characteristics of the signal; while a small alphabet is robust against noise and environmental variations, a large alphabet has more discriminant power for identifying different objects. The same alphabet is used for both target detection and classification. The issues of optimization of the alphabet size and data set partitioning are

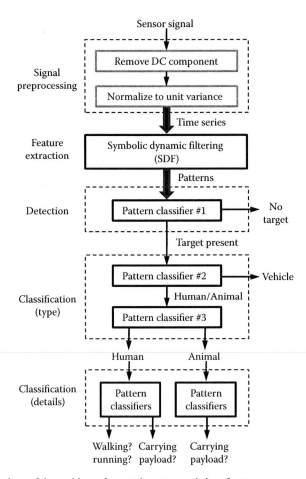

FIGURE 14.13 Flow chart of the problem of target detection and classification.

not addressed in this chapter. Subsequently, the extracted low-dimensional patterns are used for target detection and classification. One pattern is generated from each experiment, and the training patterns are used to generate the separating hyperplane in SVM.

14.6.1 Performance Assessment Using Seismic Data

This section presents the classification results using the patterns extracted from seismic signals using SDF. The leave-one-out cross-validation method [3] has been used in the performance assessment of seismic data. Since the seismic sensors are not site-independent, they require partial information of the test site, which is obtained from the training set in the cross-validation. Results of target detection and classification, movement type, and target payload identification are reported in this section.

14.6.1.1 Target Detection and Classification

Figure 14.14 shows the normalized seismic sensor signals and the corresponding feature vectors extracted by SDF of the three classes of targets and the no target case. It is observed that the feature vectors are quite different among the no target, vehicle, and human/animal cases. The feature vectors of human and animal are somewhat similar and yet still distinguishable. In the feature vector plots in Figure 14.14, the states with small index number corresponds to the wavelet coefficients with large values, and vice versa.

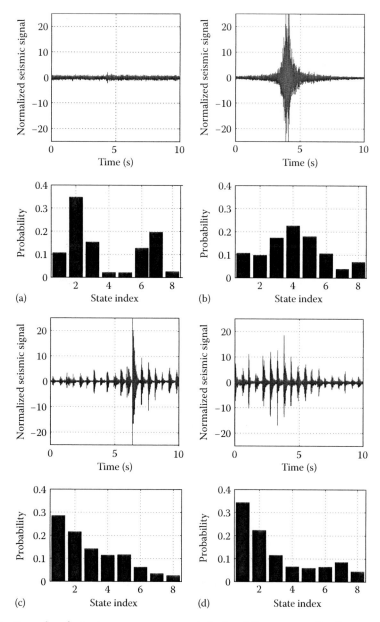

FIGURE 14.14   Examples of seismic sensor measurements (top) and the corresponding feature vectors extracted by SDF of the four classes (bottom): (a) no target, (b) vehicle, (c) human, and (d) animal.

For the purpose of comparative evaluation, kurtosis analysis [33] has been used for target detection and classification as a benchmark tool of footstep detection. Kurtosis analysis is useful for footstep detection because the kurtosis value is much higher in the presence of impulse events (i.e., target present) than in the absence of a target [33]. The results of SDF and kurtosis analysis are shown in Table 14.5. The detection and classification accuracy is summarized in Table 14.6. Although the performance of kurtosis analysis is slightly inferior, it is comparable to that of SDF for target detection and vehicle classification; however, SDF significantly outperforms kurtosis analysis for distinguishing humans from animals.

TABLE 14.5   Confusion Matrices of Leave-One-Out Cross-Validation Using SDF and Kurtosis

	No Target	Vehicle	Human	Animal
SDF				
No target	114	1	1	2
Vehicle	0	7	1	0
Human	3	0	61	2
Animal	0	0	1	43
Kurtosis				
No target	102	0	5	11
Vehicle	0	0	7	1
Human	1	0	47	18
Animal	1	0	30	13

TABLE 14.6   Comparison of Detection and Classification Accuracy Using SDF and Kurtosis

		Classification	
	Detection (%)	Vehicle versus Others (%)	Human versus Animal (%)
SDF	97.0	99.1	97.2
Kurtosis	92.4	93.1	55.6

The execution of the MATLAB® code takes 2.27 s and 43.73 MB of memory for SDF and SVM on a desktop computer to process a data set of 10,000 points and perform pattern classification with the following parameters: alphabet size $|\Sigma| = 8$, number of scales $|\alpha| = 4$, window size $\ell \times \ell = 2 \times 2$, number of most probable symbol $m = 1$, and *quadratic* kernel for SVM. Pattern classification consumes about 80% of the total execution time because by using leave-one-out cross-validation, the pattern classifier needs to be trained with all the remaining patterns (e.g., 235 in the detection stage). The choice of *quadratic* kernel in SVM improves the performance of the classifier; however, it also increases the computation time in training the classifier. It is expected that the execution time and memory requirements will be reduced significantly if fewer training patterns are used.

14.6.1.2 Movement Type Identification

Upon recognition of human, more information can be derived by performing another binary classification to identify whether the human is running or walking. The physical explanations are (1) the cadence (i.e., interval between events) of human walking is usually larger than the cadence of human running and (2) the impact of running on the ground is much stronger than that of walking, and it takes longer for the oscillation to decay. Figure 14.15 shows the seismic signal and corresponding feature vectors of human walking and running. The feature vectors of human walking and running are very different from each other, which is a clear indication that the SDF-based feature extraction method is able to capture these features (cadence and impact). It is noted that the feature vectors shown in Figure 14.15 are different from those in Figure 14.14 because different partitions are used in the target classification and movement type identification stages.

Ideally, the identification of movement type should be performed based on the results of human classification. However, in order to assess the performance of SDF in this particular application, a binary classification between human walking and human running is directly performed, and the result is shown in Table 14.7. It is seen in Table 14.7 that the proposed feature extraction algorithm and SVM are able to identify the human movement type with an accuracy of 90.9%.

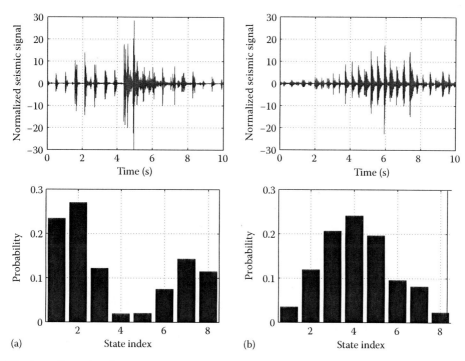

FIGURE 14.15 Examples of seismic sensor measurements (top) and the corresponding feature vectors extracted by SDF (bottom) for human walking and running: (a) walking and (b) running.

TABLE 14.7   Confusion Matrix of Leave-One-Out Cross-Validation for Movement Type Identification

	Human Walking	Human Running
Human walking	47	1
Human running	5	13

14.6.2 Performance Assessment Using PIR Data

PIR sensors are widely used for motion detection. In most applications, the signals from PIR sensors are used as discrete variables (i.e., on or off). This may work for target detection but will not work well for target classification because the time–frequency information is lost in the discretization. In this chapter, the PIR signals are considered to be continuous signals, and CWT is used to reveal the distinction among different types of targets in the time–frequency domain. Since the PIR sensor does not emit an infrared beam but merely passively accepts incoming infrared radiation, it is less sensitive to environmental variations (i.e., variation in test sites) than the seismic sensor. A three-way cross-validation [3] is used for the performance assessment of PIR data. The data are divided into three sets by date (i.e., Day 1, Day 2, and Day 3) and three different sets of experiments are performed:

1. Training: Day 1 + Day 2; Testing: Day 3
2. Training: Day 1 + Day 3; Testing: Day 2
3. Training: Day 2 + Day 3; Testing: Day 1

Training and testing on feature vectors from different days is very meaningful in practice. In each run of the cross-validation, no prior information is assumed for the testing site or the testing data. The classifiers' capability to generalize to an independent data set is thoroughly tested in the three-way cross-validation. In this section, four types of targets are considered, namely, no target, human walking, human running, animal led by human. From Figure 14.13, the following cases are tested:

1. Detection of target presence against target absence
2. Classification of target type, that is, human versus animal
3. Classification of target movement type (i.e., walking vs. running) upon recognition of the target as human

Figure 14.16 shows the PIR sensor measurements (top) and the corresponding feature vectors extracted by SDF (bottom) of the four classes. For the no target case, the PIR signal fluctuates around zero and no information is embedded in the wavelet coefficients; thus, the states in the middle (i.e., states 3–10) are occupied, whereas for the target present cases, the PIR sensors are excited

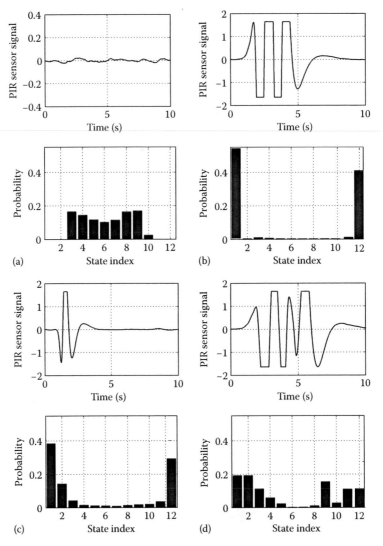

FIGURE 14.16 Examples of PIR sensor measurements (top) and the corresponding feature vectors extracted by SDF (bottom) of the four classes: (a) no target, (b) human walking, (c) human running, and (d) animal.

TABLE 14.8 Confusion Matrix of the Three-Way Cross-Validation

		Human		
	No Target	Walking	Running	Animal
No target	110	0	0	0
Human				
Walking	1	33	7	7
Running	0	5	13	0
Animal	0	2	0	42

by the presence of the targets, so states 1–2 and 11–12 that correspond to the crests and troughs in the PIR signals are more populated than other states.

The following parameters are used in SDF and SVM for processing the PIR signals: alphabet size $|\Sigma| = 12$, number of scales $|\alpha| = 3$, window size $\ell \times \ell = 2 \times 2$, number of most probable symbol $m = 1$, and *quadratic* kernel for SVM. The execution of SDF and SVM takes 1.13 s and 39.83 MB of memory on a desktop computer to process a data set of 1×10^4 points, which is a clear indication of the real-time implementation capability onboard UGS systems.

Table 14.8 shows the confusion matrix of the three-way cross-validation results using PIR sensors. It is seen in Table 14.8 that the proposed feature extraction algorithm works very well with the PIR sensor; the target detection accuracy is 99.5%; the human/animal classification accuracy is 91.7%; and the human movement type classification accuracy is 79.3%. Leave-one-out cross-validation usually underestimates the error rate in generalization because more training samples are available; it is expected that the classification accuracy will further improve for the PIR signals if leave-one-out cross-validation is used.

14.7 Summary, Conclusions, and Future Work

This chapter addresses feature extraction from sensor time-series data for situation awareness in distributed sensor networks. While wavelet transformation of time series has been widely used for feature extraction owing to their time–frequency localization properties, the work reported here presents a symbolic dynamics-based method for feature extraction from wavelet images of sensor time series in the (two-dimensional) *scale-shift* space. In this regard, symbolic dynamics-based models (e.g., PFSA) are constructed from wavelet images as information-rich representations of the underlying dynamics, embedded in the sensor time series. Subsequently, low-dimensional feature vectors are generated from the associated Perron–Frobenius operators (i.e., the state-transition probability matrices) of the PFSA. These feature vectors facilitate in situ pattern classification for decision making in diverse applications. The proposed method has been experimentally validated for two different applications: (1) identification of mobile robots and their motion profiles in a laboratory environment and (2) target detection and classification from the field data of UGS.

The proposed SDF-based feature extraction and pattern classification methodology is executable in real time on commercially available computational platforms. A distinct advantage of this method is that the low-dimensional feature vectors, generated from sensor time series in real time, can be communicated as short packets over a limited-bandwidth wireless sensor network with limited memory nodes.

Further theoretical and experimental research is recommended in the following areas:

1. Exploration of other wavelet transform techniques for wavelet image generation
2. Optimization of the partitioning scheme for symbolization of the wavelet images
3. Experimental validation in other applications

Acknowledgments

This work has been supported in part by the U.S. Army Research Laboratory and the U.S. Army Research Office under Grant No. W911NF-07-1-0376 and by the U.S. Office of Naval Research under Grant No. N00014-09-1-0688. Any opinions, findings, and conclusions or recommendations expressed in this publication are those of the authors and do not necessarily reflect the views of the sponsoring agencies.

References

1. P. Abry. *Ondelettes et Turbulence, Multirésolutions, Algorithmes Décomposition, Invariance Déchelles.* Diderot Editeur, Paris, France, 1997.
2. D. J. Berndt and J. Clifford. Using dynamic time warping to find patterns in time series. *Proceedings of the AAAI Workshop on Knowledge Discovery in Databases*, Seattle, Washington, pp. 359–370, July 31, 1994.
3. C. M. Bishop. *Pattern Recognition and Machine Learning.* Springer, New York, 2006.
4. M. Buhl and M. B. Kennel. Statistically relaxing to generating partitions for observed time-series data. *Physical Review E*, 71(4):046213, 2005.
5. S. Canu, Y. Grandvalet, V. Guigue, and A. Rakotomamonjy. *SVM and Kernel Methods MATLAB Toolbox.* Perception Systèmes et Information, INSA de Rouen, Rouen, France, 2005.
6. T. Chau and A. K. C. Wong. Pattern discovery by residual analysis and recursive partitioning. *IEEE Transactions on Knowledge and Data Engineering*, 11(6):833–852, 1999.
7. T. Cover and P. Hart. Nearest neighbor pattern classification. *IEEE Transactions on Information Theory*, 13(1):21–27, 1967.
8. C. S. Daw, C. E. A. Finney, and E. R. Tracy. A review of symbolic analysis of experimental data. *Review of Scientific Instruments*, 74(2):915–930, 2003.
9. R. O. Duda, P. E. Hart, and D. G. Stork. *Pattern Classification*, 2nd edn. Wiley Interscience, New York, 2001.
10. K. Fukunaga. *Statistical Pattern Recognition*, 2nd edn. Academic Press, Boston, MA, 1990.
11. B. Gerkey, R. Vaughan, and A. Howard. The Player/Stage project: Tools for multi-robot and distributed sensor systems. *Proceedings of the International Conference on Advanced Robotics.* Coimbra, Portugal, pp. 317–323, June 30–July 3, 2003.
12. F. Gullo, G. Ponti, A. Tagarelli, and S. Greco. A time series representation model for accurate and fast similarity detection. *Pattern Recognition*, 42(7):2998–3014, 2009.
13. S. Gupta and A. Ray. Symbolic dynamic filtering for data-driven pattern recognition. In E. A. Zoeller, ed., *Pattern Recognition: Theory and Application*, Chapter 2. Nova Science Publishers, Hauppage, NY, pp. 17–71, 2007.
14. S. Gupta and A. Ray. Statistical mechanics of complex systems for pattern identification. *Journal of Statistical Physics*, 134(2):337–364, 2009.
15. S. Gupta, A. Ray, and E. Keller. Symbolic time series analysis of ultrasonic data for early detection of fatigue damage. *Mechanical Systems and Signal Processing*, 21:866–884, 2007.
16. S. S. Haykin. *Neural Networks and Learning Machines*, 3rd edn. Prentice Hall, New York, 2009.
17. X. Jin, S. Gupta, K. Mukherjee, and A. Ray. Wavelet-based feature extraction using probabilistic finite state automata for pattern classification. *Pattern Recognition*, 44(7):1343–1356, 2011.
18. Y. Kakizawa, R. H. Shumway, and N. Taniguchi. Discrimination and clustering for multivariate time series. *Journal of the American Statistical Association*, 93(441):328–340, 1999.
19. K. Keller and H. Lauffer. Symbolic analysis of high-dimensional time series. *International Journal of Bifurcation and Chaos*, 13(9):2657–2668, 2003.
20. O. R. Lautour and P. Omenzetter. Damage classification and estimation in experimental structures using time series analysis and pattern recognition. *Mechanical Systems and Signal Processing*, 24:1556–1569, 2010.

21. T. W. Lee. *Independent Component Analysis: Theory and Applications*. Kluwer Academic Publishers, Boston, MA, 1998.

22. T. W. Liao. Clustering of time series data—A survey. *Pattern Recognition*, 38:1857–1874, 2005.

23. D. Lind and M. Marcus. *An Introduction to Symbolic Dynamics and Coding*. Cambridge University Press, Cambridge, U.K., 1995.

24. S. G. Mallat. *A Wavelet Tour of Signal Processing: The Sparse Way*, 3rd edn. Academic Press, Orlando, FL, 2009.

25. A. Muñoz, R. Ertlé, and M. Unser. Continuous wavelet transform with arbitrary scales and $o(n)$ complexity. *Signal Processing*, 82(5):749–757, 2002.

26. D. B. Percival and A. T. Walden. *Wavelet Methods for Time Series Analysis*. Cambridge University Press, Cambridge, U.K., 2000.

27. S. Pittner and S. V. Kamarthi. Feature extraction from wavelet coefficient for pattern recognition tasks. *IEEE Transactions on Pattern Analysis and Machine Intelligence*, 21(1):83–88, 1999.

28. V. Rajagopalan and A. Ray. Symbolic time series analysis via wavelet-based partitioning. *Signal Processing*, 86(11):3309–3320, 2006.

29. C. Rao, A. Ray, S. Sarkar, and M. Yasar. Review and comparative evaluation of symbolic dynamic filtering for detection of anomaly patterns. *Signal, Image, Video Processing*, 3:101–114, 2009.

30. A. Ray. Symbolic dynamic analysis of complex systems for anomaly detection. *Signal Processing*, 84(7):1115–1130, 2004.

31. R. Rosipal, M. Girolami, and L. Trejo. Kernel PCA feature extraction of event-related potentials for human signal detection performance. *Proceedings of the International Conference on Artificial Neural Networks in Medicine Biology*, Gothenburg, Sweden, pp. 321–326, May 2000.

32. A. Subbu and A. Ray. Space partitioning via Hilbert transform for symbolic time series analysis. *Applied Physics Letters*, 92(8):084107-1–084107-3, 2008.

33. G. P. Succi, D. Clapp, R. Gampert, and G. Prado. Footstep detection and tracking. *Unattended Ground Sensor Technologies and Applications III*, Orlando, FL, Vol. 4393, SPIE, pp. 22–29, 18 April 2001.

34. C. J. Veenman, M. J. T. Reinders, E. M. Bolt, and E. Baker. A maximum variance cluster algorithm. *IEEE Transactions on Pattern Analysis and Machine Intelligence*, 24(9):1273–1280, 2002.

35. K. P. Zhu, Y. S. Wong, and G. S Hong. Wavelet analysis of sensor signals for tool condition monitoring: A review and some new results. *International Journal of Machine Tools and Manufacture*, 49:537–553, 2009.

36. W. Zucchini and I. L. MacDonald. *Hidden Markov Models for Time Series: An Introduction*. CRC Press, Boca Raton, FL, p. 308, 2009.

15

Beamforming

15.1 Introduction 335
    Historical Background • Narrowband versus Wideband
    Beamforming • Beamforming for Narrowband
    Waveforms • Beamforming for Wideband Waveforms

15.2 DOA Estimation and Source Localization 343
    RF Signals • Acoustic/Seismic Signals

15.3 Array System Performance Analysis and Robust Design 353
    Computer-Simulated Results for Acoustic Sources • CRB for Source
    Localization • Robust Array Design

15.4 Implementations of Two Wideband Beamforming Systems 366
    Implementation of a Radar Wideband Beamformer Using a Subband
    Approach • iPAQS Implementation of an Acoustic Wideband
    Beamformer

References...369

J.C. Chen
*The Pennsylvania
State University*

Kung Yao
*University of California,
Los Angeles*

15.1 Introduction

15.1.1 Historical Background

Beamforming is a space–time operation in which a waveform originating from a given source but received at spatially separated sensors is coherently combined in a time-synchronous manner. If the propagation medium preserves sufficient coherency among the received waveforms, then the beamformed waveform can provide an enhanced signal-to-noise ratio (SNR) compared with a single sensor system. Beamforming can be used to determine the direction-of-arrival(s) (DOAs) and the location(s) of the source(s), as well as perform spatial filtering of two (or more) closely spaced sources. Beamforming and localization are two interlinking problems, and many algorithms have been proposed to tackle each problem individually and jointly (i.e. localization is often needed to achieve beamforming and some localization algorithms take the form of a beamformer). The earliest development of space–time processing was for enhancing SNR in communicating between the United States and the United Kingdom dating back before the World War II [1]. Phase-array antennas based upon beamforming for radar and astronomy were developed in the 1940s [2]. Since then, phase-array antennas utilizing broad ranges of radio frequencies (RFs) have been used for diverse military and civilian ground, airborne, and satellite applications. Similarly, sonar beamforming arrays have been used for more than 50 years.

Recent developments in integrated circuit technology have allowed the construction of low-cost small acoustic and seismic sensor nodes with signal processing and wireless communication capabilities that can form distributed wireless sensor network systems. These low-cost systems can be used to perform detection, source separation, localization, tracking, and identification of acoustic and seismic sources in diverse military, industrial, scientific, office, and home applications [3–7]. The design of acoustic localization algorithms mainly focuses on high performance, minimal communications load, computationally efficiency, and robust methods to reverberant and interference effects. Brandstein and Silverman [8] proposed a robust method for relative time-delay estimation by reformulating the problem as a linear regression of phase data and then estimating the time delay through minimization of a robust statistical error measure. When several signals coexist, the relative time delay of the dominant signal was shown to be effectively estimated using a second-order subspace method [9]. A recent application of particle filtering to acoustic source localization using a steered beamforming framework also promises efficient computations and robustness to reverberations [10]. Another attractive approach using the integration (or fusion) of distributed microphone arrays can yield high performance without demanding data transfer among nodes [11]. Unlike the aforementioned approaches that perform independent frame-to-frame estimation, a tracking framework has also been developed [12] to provide power-aware, low-latency location tracking that utilizes historical source information (e.g. trajectory and speed) with single-frame updates.

More recently, in cellular telephony, due to the ill-effects of multipaths and fading and the need to increase performance and data transmission rates, multiple antennas utilizing beamforming arrays have also been proposed. While several antennas at the basestations can be used, only two antennas can be utilized on hand-held mobile devices due to their physical limitation. Owing to the explosive growth of cell phones around the world, much progress is being made in both the research and technology aspects of beamforming for smart antennas.

Besides various physical phenomena, many system constraints also limit the performance of coherent array signal-processing algorithms. For instance, the system performance may suffer dramatically due to sensor location uncertainty (due to unavailable measurement in random deployment), sensor response mismatch and directivity (which may be particularly serious for some types of microphone in some geometric configurations), and loss of signal coherence across the array (i.e. widely separated microphones may not receive the same coherent signal) [13]. In a self-organized wireless sensor network, the collected signals need to be well time synchronized in order to yield good performance. These factors must be considered for practical implementation of the sensor network. In the past, most reported sensor network systems performing these processing operations usually involve custom-made hardware. However, with the advent of low-cost but quite capable processors, real-time beamforming utilizing iPAQs has been reported [14].

15.1.2 Narrowband versus Wideband Beamforming

In radar and wireless communications, the information signal is modulated on some high RF f_0 for efficient transmission purposes. In general, the bandwidth of the signal over $[0, f_s]$ is much less than the RF. Thus, the ratio of the highest to lowest transmitted frequency, $(f_0 + f_s)/(f_0 - f_s)$, is typically near unity. For example, for the 802.11b ISM wireless local-area network system, the ratio is 2.4835 GHz/2 GHz = 1.03. These waveforms are denoted as narrowband. Narrowband waveforms have a well-defined nominal wavelength, and time delays can be compensated by simple phase shifts. The conventional narrowband beamformer operating on these waveforms is merely a spatial extension of the matched filter. In the classical time-domain filtering, the time-domain signal is linearly combined with a filtering weight to achieve the desired high/low/band-pass filtering. This narrowband beamformer also combines the spatially distributed sensor collected array data linearly with the beamforming weight to achieve spatial filtering. Beamforming enhances the signal from the desired spatial direction and reduces the signal(s) from other direction(s) in addition to possible time/frequency filtering. Details on the spatial filtering aspect of this beamformer will be given in Section 15.1.3.

The movement of personnel, cars, trucks, wheeled/tracked vehicles, and vibrating machinery can all generate acoustic or seismic waveforms. The processing of seismic/vibrational sensor data is similar to that of acoustic sensors, except for the propagation medium and unknown speed of propagation. For acoustic/seismic waveforms, the ratio of the highest to lowest frequencies can be several octaves. For audio waveforms (i.e. 30 Hz–15 kHz), the ratio is about 500, and these waveforms are denoted as wideband. Dominant acoustical waveforms generated from wheeled and tracked vehicles may range from 20 Hz to 2 kHz, resulting in a ratio of about 100. Similarly, dominant seismic waveforms generated from

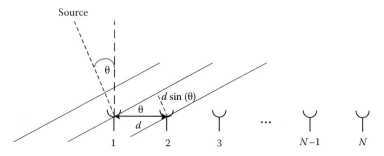

FIGURE 15.1 Uniform linear array of N sensors with inter-sensor spacing $d = \lambda/2$.

wheeled vehicles may range from 5 to 500 Hz, also resulting in a ratio of about 100. Thus, the acoustic and seismic signals of interest are generally wideband. However, even for certain RF applications, the ratio of the highest to lowest frequencies can also be considerably greater than unity. For wideband waveforms there is no characteristic wavelength, and time delays must be obtained by interpolation of the waveforms. When an acoustic or seismic source is located close to the sensors, the wavefront of the received signal is curved, and the curvature depends on the distance, then the source is in the near field. As the distances become large, all the wavefronts are planar and parallel, then the source is in the far field. For a far-field source, only the DOA angle in the coordinate system of the sensors is observable to characterize the source. A simple example is the case when the sensors are placed on a line with uniform inter-sensor spacing, as shown in Figure 15.1. Then all adjacent sensors have the same time delay and the DOA of the far-field source can be estimated readily from the time delay. For a near-field source, the collection of all relative time delays and the propagation speed of the source can be used to determine the source location. In general, wideband beamforming is considerably more complex than narrowband beamforming. Thus, the acoustic source localization and beamforming problem is challenging due to its wideband nature, near- and far-field geometry (relatively near/far distance of the source from the sensor array), and arbitrary array shape. Some basic aspects of wideband beamforming are discussed in Section 15.1.4.

15.1.3 Beamforming for Narrowband Waveforms

The advantage of beamforming for a narrowband waveform can be illustrated most simply by considering a single tone waveform

$$s(t) = a\exp(i2\pi f_0 t), \quad -\infty < t < \infty \tag{15.1}$$

where
$a = 1$ is the transmitted amplitude
The frequency f_0 is assumed to be fixed and known

Consider the waveforms at two receivers given by

$$x_1(t) = As(t - \tau_1) + v_1(t), \quad -\infty < t < \infty \tag{15.2}$$

$$x_2(t) = As(t - \tau_2) + v_2(t), \quad -\infty < t < \infty \tag{15.3}$$

where
A denotes the received amplitude, which is assumed to be the same for both channels
τ_1 and τ_2 are the propagation times from the source to the two receivers, which are allowed to be different
v_1 and v_2 are two received complex-valued uncorrelated zero-mean white noises of equal variance σ_0^2

Then the SNR of each receiver is given by

$$\text{SNR}_{\text{in}} = \text{SNR}(X_1) = \text{SNR}(X_2) = \frac{A^2}{\sigma_0^2} \tag{15.4}$$

A beamformer is a device that combines all the received waveforms in a coherent manner. Suppose we assume the propagation delays τ_1 and τ_2 are known. In practice, these delay values may be estimated or evaluated from the geometry of the problem. Then the output of a beamformer that coherently combines the two received waveforms in Equations 15.2 and 15.3 is given by

$$\begin{aligned}
y(t) &= \exp(i2\pi f_0\tau_1)x_1(t) + \exp(i2\pi f_0\tau_2)x_2(t) \\
&= \exp(i2\pi f_0\tau_1)A\exp[i2\pi f_0(t-\tau_1)] + \exp(i2\pi f_0\tau_1)v_1(t) + \exp(i2\pi f_0\tau_1)A\exp(i2\pi f_0(t-\tau_2)] \\
&\quad + \exp(i2\pi f_0\tau_2)v_2(t) \\
&= 2A\exp(i2\pi f_0 t) + \exp(i2\pi f_0\tau_1)v_1(t) + \exp(i2\pi f_0\tau_2)v_2(t)
\end{aligned} \tag{15.5}$$

The beamformer output noise variance is given by

$$E\{|\exp(i2\pi f_0\tau_1)v_1(t) + \exp(i2\pi f_0\tau_2)v_2(t)|^2\} = 2\sigma_0^2 \tag{15.6}$$

the beamformer output signal power is given by $(2A)^2/2$, and the beamformer output SNR is given by

$$\text{SNR}_{\text{output}} = \frac{(2A)^2}{2\sigma_0^2} = \frac{2A^2}{\sigma_0^2} = 2 \times \text{SNR}_{\text{in}} \tag{15.7}$$

Thus, Equation 15.7 shows that a coherent combining beamformer using two ideal receivers increases the effective SNR over a single receiver by a factor of 2 (i.e. 3 dB).

In the same manner, with N receivers, then Equations 15.2 and 15.3 become

$$x_n(t) = As(t - \tau_n) + v_n(t), \quad n = 1, \ldots, N, \, -\infty < t < \infty \tag{15.8}$$

Equation 15.5 becomes

$$y(t) = \sum_{n=1}^{N}[\exp(i2\pi f_0\tau_n)x_n(t)], \quad -\infty < t < \infty$$

$$= NA\exp(i2\pi f_0 t) + \sum_{n=1}^{N}[\exp(i2\pi f_0\tau_n)v_n(t)] \tag{15.9}$$

Equation 15.6 becomes

$$E\left\{\left|\sum_{n=1}^{N}\{\exp(i2\pi f_0\tau_n)v_n(t)\}\right|^2\right\} = N\sigma_0^2 \tag{15.10}$$

and the beamformer output SNR of Equation 15.7 now becomes

$$\text{SNR}_{\text{out}} = \frac{(NA)^2}{N\sigma_0^2} = N \times \text{SNR}_{\text{in}} \tag{15.11}$$

Thus, Equation 15.11 shows in an ideal situation, when the time delays τ_n, $n = 1, \ldots, N$, are exactly known, a beamformer that performs a coherent array processing of the N received waveforms yields an SNR improvement by a factor of N relative to a single receiver.

In general, an Nth-order narrowband beamformer has the form of

$$y(t) = \sum_{n=1}^{N} w_n^* x_n(t), \quad -\infty < t < \infty \tag{15.12}$$

where $\{w_1, \ldots, w_N\}$ is a set of complex-valued weights chosen for the beamformer to meet some desired criterion. In Equation 15.9, in order to achieve coherent combining of the received narrowband waveforms, the weights are chosen so $w_n = \exp(i2\pi\tau_n)$, $n = 1, \ldots, N$.

Consider the special case of a uniform linear array, when all the N receive sensors lie on a line with an inter-sensor spacing of d, as shown in Figure 15.1. Furthermore, we assume the source is far from the linear array (i.e. in the "far-field" scenario), so the received wavefront is planar (i.e. with no curvature) and impacts the array at an angle of θ. From Figure 15.1, the wavefront at sensor 2 has to travel an additional distance of $d \sin(\theta)$ relative to the wavefront at sensor 1. The relative time delay to travel this additional distance is then given by $d \sin(\theta)/c = d \sin(\theta)/(f_0\lambda)$, where c is the speed of propagation of the wavefront and λ is the wavelength corresponding to frequency f_0. Similarly, the relative time delay of the nth sensor relative to the first sensor becomes $(n-1)d \sin(\theta)/(f_0\lambda)$ Then the time delay expression for all the sensors is given by

$$\tau_n = \tau_1 + \frac{(n-1)d \sin(\theta)}{f_0\lambda}, \quad n = 1, \ldots, N \tag{15.13}$$

and the expression for all the received waveforms is given by

$$x_n(t) = A \exp(i2\pi f_0 t) \exp(-i2\pi f_0 \tau_1) \exp\left(\frac{-i2\pi d(n-1)\sin(\theta)}{\lambda}\right) + v_n(t), \quad n = 1, \ldots, N \tag{15.14}$$

In practice, if we use this uniform linear array, τ_1 is still known, but all the other τ_n, $n = 2, \ldots, N$, are fully known relative to τ_1 as given by Equation 15.13. Then the ideal beamformer output expression of Equation 15.9 now becomes

$$y(t) = \exp(i2\pi f_0 \tau_1) \left\{ NA \exp(i2\pi f_0 t) + \sum_{n=1}^{N} \left[\exp\left(\frac{i2\pi d \sin(\theta)}{\lambda}\right) v_n(t) \right] \right\} \tag{15.15}$$

which still achieves the desired $\text{SNR}_{\text{out}} = N \times \text{SNR}_{\text{in}}$ of Equation 15.11.

Now, suppose each sensor has a uniform response in all angular directions (i.e. isotropic over $[-\pi, \pi]$). The beamformer angular transfer function for a uniform linear array with inter-sensor spacing of $d = \lambda/2$ is given by

$$H(\theta) = \sum_{n=1}^{N} \exp[-i\pi(n-1)\sin(\theta)] = \frac{1 - \exp[-i\pi N \sin(\theta)]}{1 - \exp[-i\pi \sin(\theta)]}$$

$$= \frac{\exp\{-i[\pi/2(N-1)\sin(\theta)]\} \sin[(N\pi/2)]\sin(\theta)}{\sin[(\pi/2)\sin(\theta)]}, \quad -\pi \leq \theta < \pi \tag{15.16}$$

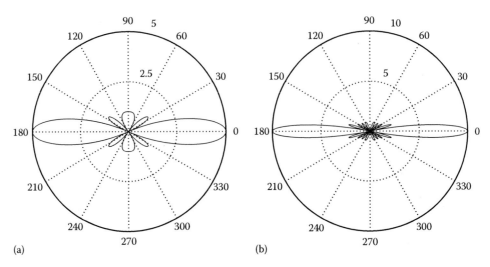

FIGURE 15.2   Polar plot of $|H(\theta)|$ versus θ for $d = \lambda/2$ and $N = 5$ (a) and $N = 10$ (b).

A polar plot of $|H(\theta)|$ displayed from 0° to 360° for $N = 5$ is shown in Figure 15.2a and for $N = 10$ is shown in Figure 15.2b. We note, in this figure, that the linear array lies on the 90° to −90° line (in these two figures 270° = −90°) and that the gain is symmetric about this line. Thus, there is a high gain at the 0° direction (the "forward broadside" of the array) as well at the 180° direction (the "backward broadside"), with various sidelobes in other directions. In some applications we may assume that all the desired and unwanted sources are known to be in the forward sector (i.e. in the −90° to 0° to −90° sector). If the desired source is in the high gain 0° direction, then other unwanted sources in the directions of the sidelobes form interferences to the reception of the desired source. Thus, an array with a mainlobe having high gain over a narrow angular sector plus sidelobes with small values is considered desirable. From Equation 15.16 and Figure 15.2, as the number of array elements N increases, the mainlobe of the beamformer angular response becomes narrower and thus able to provide a better angular resolution, while the sidelobe peak values stay the same relative to the beam peak.

On the other hand, if we set $d = \lambda$ then Equation 15.16 has the form of

$$H(\theta) = \frac{\exp\{-i[\pi(N-1)\sin(\theta)]\}\sin[(N\pi)\sin(\theta)]}{\sin[(\pi)\sin(\theta)]}, \quad -\pi \le \theta < \pi \tag{15.17}$$

A polar plot of this $|H(\theta)|$ again displayed from 0° to 360° for $N = 5$ is shown in Figure 15.3a and for $N = 10$ is shown in Figure 15.3b. We note, in these two figures, in addition to the desired high gains in the 0° and 180°, there are also two undesired equal high gains with large angular spreads at 90° and 270° These two additional large gains may cause a large interference to the desired source signal from unwanted sources in these directions. The spatial Nyquist criterion requires the inter-sensor spacing d of a uniform linear array to be less or equal to $\lambda/2$ to avoid grating lobes (also called spatial aliasing). This phenomenon is analogous to spectral aliasing due to the periodicity in the frequency domain created by sampling in time. Thus, a uniform linear array is most commonly operated at the $d = \lambda/2$ condition.

In Equation 15.5 we have assumed both time delays τ_1 and τ_2 are known. Then the optimum array weights are given by $w_1 = \exp(-i2\pi f_0 \tau_1)$ and $w_2 = \exp(-i2\pi f_0 \tau_2)$. For a single source and a uniform linear array of N sensors, as shown in Figure 15.1, as long as the DOA angle θ of the source is known (or estimated), then the remaining relative time delays relative to the first sensor can be evaluated from the geometry. However, for two or more sources, suppose the DOA θ_1 of the desired source is known (or estimated), but other unwanted interfering source DOAs are unknown. The minimum variance desired response (MVDR) method [15] provides a computationally attractive solution for constraining

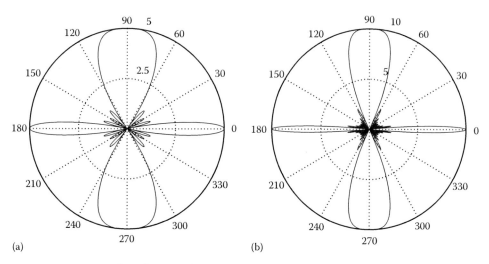

FIGURE 15.3 Polar plot of $|H(\theta)|$ versus θ for $d = \lambda$ and $N = 5$ (a) and $N = 10$ (b).

the array response of the desired source DOA angle to a fixed value (say the unity value) while minimizing the response to all other interference sources. From the output of the Nth-order beamformer in Equation 15.12, denote its $N \times N$ autocorrelation matrix as R, the array weight vector as $W = [w_1, ..., w_N]^T$, and the steering vector as $C(\theta) = [1, \exp(-i\theta), ..., \exp(-i(N-1)\theta)]^T$ Then the MVDR solution satisfies

$$\text{Min}\{W^H R W\}, \quad \text{subject to } W^C C(\theta_1) = 1 \qquad (15.18)$$

where
    Superscript C is the complex-valued operator
    T is the transpose C operator
    H is the complex conjugation operator

MVDR solution is then given by

$$W = R^{-1}C(C^H R^{-1} C)^{-1} \qquad (15.19)$$

Figure 15.4 shows the $N = 20$ uniform linear array MVDR beamformer output response (dB) versus spatial angle to a unit-amplitude single-tone source of $f = 900\,\text{Hz}$ at DOA of $\theta_1 = 30°$, subjected to a broadband white Gaussian interferer of variance $\sigma^2 = 2$ at DOA of $\theta_2 = 60°$, in the presence of additive white Gaussian noise of unity variance. The beamformer input signal-to-interference-plus-noise ratio SINR = 3.7 dB and the output SINR = 10.0 dB result in a gain of SIRN = 13.7 dB. In Figure 15.4, we note this MDVR beamformer achieved a unity gain at the known desired angle of $\theta_1 = 30°$ as constrained and placed a null of over −40 dB at the interference angle of 60°, which is not specifically specified in the algorithm. The nulling angle information of θ_2 was obtained implicitly from the autocorrelation matrix R of the array output data.

15.1.4 Beamforming for Wideband Waveforms

Consider a source waveform containing two tones at frequency f_1 and f_2 with amplitudes a_1 and a_2 respectively, as given by

$$s(t) = a_1 \exp(i2\pi f_1 t) + a_2 \exp(i2\pi f_2 t), \quad -\infty < t < \infty \qquad (15.20)$$

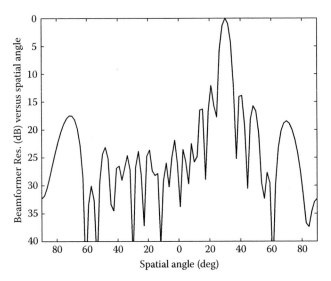

FIGURE 15.4 $N = 20$ uniform linear array MVDR beamformer response (dB) versus θ.

Let the wavefront of the above far-field source impact two sensors on a uniform linear array with a spacing of d. Then the received waveforms $x_1(t)$ and $x_2(t)$ have the form of

$$x_1(t) = A_1 \exp[i2\pi f_1(t - \tau_1)] + A_2 \exp[i2\pi f_2(t - \tau_1)] + v_1(t), \quad -\infty < t < \infty \qquad (15.21)$$

$$x_2(t) = A_1 \exp[i2\pi f_1(t - \tau_2)] + A_2 \exp[i2\pi f_2(t - \tau_2)] + v_2(t), \quad -\infty < t < \infty \qquad (15.22)$$

Suppose we want to use the narrowband beamformer as given by Equation 15.12 in the form of Figure 15.5 with $N = 2$ sensor, but only one complex weight per sensor for possible coherent combining. The issue is whether we can find two complex-valued weights $\{w_1, w_2\}$ able to achieve the desired expression on the right-hand side of Equation 15.23 given by

$$y(t) = w_1^* x_1(t) + w_2^* x_2(t) = c_1 A_1 \exp(i2\pi f_1 t) + c_2 A_2 \exp(i2\pi f_2 t) + v_1'(t) + v_2'(t), \quad -\infty < t < \infty \qquad (15.23)$$

where
    c_1 and c_2 are some arbitrary complex-valued constants
    $v_1'(t)$ and $v_2'(t)$ are two uncorrelated white noises

After some algebra, one can show that a narrowband beamformer with only one complex-valued weight per sensor channel cannot achieve the desired result in Equation 15.23.

Now, consider a wideband beamformer with N sensors and M complex-valued weights $\{w_{nm}, n = 1, \ldots, N, m = 1, \ldots, M\}$ shown in Figure 15.5. For the nth sensor channel, the M weights $\{w_{nm}, m = 1, \ldots, M\}$ with the $(M - 1)$ time delays of value T form an Mth-order tapped delay line. For the time sampled case, they form an Mth-order finite impulse response (FIR) filter with the time delay T replaced by Z^{-1}. Thus, a wideband beamformer performs spatial–time–frequency filtering operations. In the above case (considered in Equation 15.23), it can be shown (after some algebra) that a wideband beamformer with $N = 2$ and $M = 2$ (i.e. two complex-valued weights per sensor channel) can achieve the desired coherent combining. In general, it can be shown that, for N tones with distinct frequencies, a wideband beamformer using the uniform linear array of N sensors needs N complex-valued weights per sensor channel to achieve the desired coherent combining. Of course, in practice, a realistic wideband waveform is equivalent to an

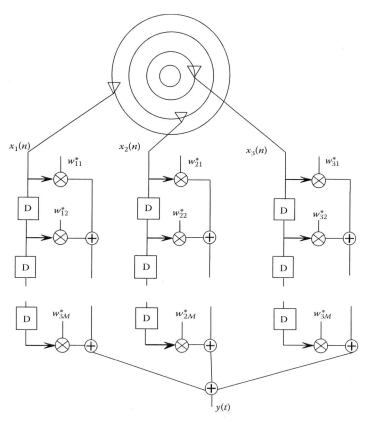

FIGURE 15.5 Wideband beamformer with $N = 3$ sensors and M taps per sensor section.

infinite number of tones in the $[f_{\text{low}}, f_{\text{high}}]$ band. Thus, a sufficiently large M number of complex-valued weights per sensor channel in the wideband beamformer may need to be used to approximate the desired coherent combining effect. The number of sensors N is mainly used to control the narrowness of the mainlobe, similar to that in the narrowband beamformer case seen in Section 15.1.3.

In Section 15.2 we deal with DOA estimation and source localization. Then, in Section 15.3 we show the array system performance analysis and robust design. Finally, in Section 15.4 we demonstrate practical implementation of two wideband beamforming systems.

15.2 DOA Estimation and Source Localization

DOA, bearing estimation, or angle of arrival (AOA), and source localization are the classical problems in array signal processing. These problems are essential in many RF and acoustic/seismic applications, and a variety of algorithms have been developed over the years. In most RF and sonar applications, only the DOA estimation is of interest due to the far-field geometry between the platform and the target, and the term "source localization" has been used interchangeably with DOA estimation. In other applications, such as acoustic and seismic sensing in a sensor network, the target may be in the near field of the sensor array, and in this case the range and the angle of the target are included as the estimation parameters. In this section we refer to near-field angle and range estimation as source localization, and far-field-angle-only estimation as DOA estimation.

The DOA estimation and source localization belong to the general class of parameter estimation problems, and in most cases closed-form solutions are not available. The algorithms can be generally categorized into two classes: parametric and time-delay methods. The parametric methods normally are based on optimizing the parameters directly from the signal model. Some examples of this type include maximum-likelihood (ML) and super-resolution subspace methods such as MUSIC [16]. Iterative algorithms are usually used to obtain the solution efficiently, or in other cases a grid-search solution can be obtained when only a crude estimation is needed. An alternative approach, which we refer to as the time-delay method, is also often used in many acoustic/seismic applications. In this case, the problem is broken down into two estimation steps. First, the relative time delays among the sensors are estimated based on some form of correlation [17]. Then, based on the estimated relative time delays, the DOA or the location of the source is estimated using some form of least-squares (LS) fit. This approach, compared with the parametric approach, has a major computational advantage, since the LS estimation can be obtained in closed form. Even though the relative time-delay estimation step is not closed form, its solution is much easier to obtain compared with the direct parametric solution. In terms of performance, the time-delay methods are suboptimal compared with the optimum parametric ML method, but in some cases they are satisfactory. However, in the case of multiple sources of comparable strength emitting simultaneously, the parametric algorithms can be expanded to estimate the DOAs or locations of the sources jointly, while the relative time-delay estimation for multiple sources is not at its mature stage [16].

The wideband extension of the ML solution has been shown to be effective [19]; however, many other suboptimal methods, such as MUSIC or its variants, do not show promising results when extended to the wideband problem. In the following, we formulate some of the DOA and source localization algorithms of the RF and acoustic/seismic problems.

15.2.1 RF Signals

For RF applications, the parametric methods are mostly used. For narrowband RF signals, the relative time delay is merely the phase difference between sensors. This narrowband property naturally lends itself to the parametric methods, and a variety of efficient algorithms have been developed. Krim and Viberg [20] have provided an excellent review and comparison of many classical and advanced parametric narrowband techniques up to 1996. Early work in DOA estimation includes the early version of the ML solution, but it did not become popular owing to its high computational cost. Concurrently, a variety of suboptimal techniques with reduced computations have dominated the field. The more well-known techniques include the minimum variance, the MUSIC, and the minimum norm. The MUSIC algorithm is perhaps one of the most popular suboptimal techniques. It provides super-resolution DOA estimation in a spatial pseudo-spectral plot by utilizing the orthogonality between the signal and noise subspaces. However, a well-known problem with some of these suboptimal techniques occurs when two or more sources are highly correlated. This may be caused by multipaths or intentional jamming, and most of the suboptimal techniques have difficulties without reverting to advanced processing or constraints. Many variants of the MUSIC algorithm have been proposed to combat signal correlation and to improve performance.

When M spatially separated RF signals impinge on an array, one snapshot of the signal received by the array can be characterized by

$$x(t) = \sum_{m=1}^{M} s_m(t) \mathbf{d}_m(\mathbf{\Theta}) + \mathbf{n}(t) \tag{15.24}$$

where
  $s_m(t)$ is the mth RF signal received by the reference sensor
  $\mathbf{d}_m(\mathbf{\Theta})$ is the steering vector that contains the relative phase of the mth signal traversing across the array
  $\mathbf{\Theta}$ is the parameter vector of interest, which in this case contains the DOAs of the sources
  $\mathbf{n}(t)$ is the noise vector of the array

For statistical estimation of the DOA, multiple independent snapshots of the array signal are collected. Denote L as the total number of snapshots. By the classical estimation theory, the estimation performance improves as L increases; however, in some systems the source could be maneuvering in time, thus L should be chosen to be just large enough before the DOA changes.

Depending on the system characteristics and problem at hand, the stochastic properties of the array signal determine which estimation method should be used. For most systems the noise can be characterized by a vector Gaussian process, and the covariance matrix of the signal $\mathbf{R}_n = E[\mathbf{nn}^H]$, usually also assumed to be white $\mathbf{R}_n = \sigma^2 \mathbf{I}$, can be estimated by its time averaged version $\tilde{\mathbf{R}}_n$. The correlation matrix of the data is then given by

$$\mathbf{R}_x = E[\mathbf{xx}^H] = \mathbf{D}(\mathbf{\Theta})\mathbf{S}(t)\mathbf{D}(\mathbf{\Theta})^H + \mathbf{R}_n \qquad (15.25)$$

where
  $\mathbf{D}(\mathbf{\Theta}) = [d_1(\mathbf{\Theta}) \ldots d_M(\mathbf{\Theta})]$ is the steering matrix
  $\mathbf{S}(t) = E[\mathbf{ss}^H]$ is the source correlation matrix
  $s = [s_1(t) \ldots s_M(t)]^T$ is the source vector

When the number of sources is less than the number of sensors, the data correlation matrix can be broken down into two orthogonal subspaces:

$$\mathbf{R}_x = E[\mathbf{xx}^H] = \mathbf{U}_s \mathbf{\Lambda}_s \mathbf{U}_s^H + \mathbf{U}_n \mathbf{\Lambda}_n \mathbf{U}_n^H \qquad (15.26)$$

where
  $\mathbf{U}_s$ and $\mathbf{\Lambda}_s$ are the matrices containing the signal subspace eigenvectors and eigenvalues, respectively
  $\mathbf{U}_n$ and $\mathbf{\Lambda}_n$ are those of the noise subspace

Define the projections into the signal and noise subspaces as

$$\mathbf{\Pi} = \mathbf{U}_s \mathbf{U}_s^H = \mathbf{D}(\mathbf{D}^H \mathbf{D})^{-1} \mathbf{D}^H \qquad (15.27)$$

$$\mathbf{\Pi}^\perp = \mathbf{U}_n \mathbf{U}_n^H = \mathbf{I} - \mathbf{D}(\mathbf{D}^H \mathbf{D})^{-1} \mathbf{D}^H \qquad (15.28)$$

respectively. The subspace estimation methods make use of the orthogonality between the signal and noise subspaces. In particular, the MUSIC spatial spectrum can be produced by generating the following plot:

$$P_{\text{MUSIC}}(\theta) = \frac{\mathbf{a}(\theta)^H \mathbf{a}(\theta)}{\mathbf{a}(\theta)^H \hat{\mathbf{\Pi}}^\perp \mathbf{a}(\theta)} \qquad (15.29)$$

where
  $\mathbf{a}(\theta)$ is the steering vector of the array pointing to a particular direction θ
  $\hat{\mathbf{\Pi}}^\perp$ is the estimated version of the noise subspace projection matrix

Owing to the orthogonality of the array steering vector and the noise subspace, the MUSIC spectrum should generate peaks in the vicinity of the true DOAs and the algorithm then becomes a peak searching one. The MUSIC algorithm is a powerful estimator, and it is computationally attractive since it reduces the search space into a one-dimensional one. However, it usually requires almost perfect environment and conditions, and it may not be able to yield good results in practical situations. A more optimum solution, despite possibly being computationally extensive, is the ML approach. The deterministic ML (DML) solution can be given when the statistics of the source are assumed to be unknown and deterministic. This involves minimizing the following metric:

$$\hat{\Theta}^{\mathrm{DML}} = \arg\min_{\Theta} \mathrm{Tr}(\mathbf{\Pi}_A^{\perp}\hat{\mathbf{R}}_x) \tag{15.30}$$

where $\mathrm{Tr}(\cdot)$ is the trace operation, and where

$$\mathbf{A} = [\mathbf{a}_1(\theta_1)\cdots\mathbf{a}_M(\theta_M)] \tag{15.31}$$

$$\mathbf{\Pi}_A = \mathbf{A}[\mathbf{A}^H\mathbf{A}]^{-1}\mathbf{A}^H \tag{15.32}$$

$$\mathbf{\Pi}_A^{\perp} = \mathbf{I} - \mathbf{\Pi}_A \tag{15.33}$$

and $\hat{\mathbf{R}}_X$ is the sample data correlation matrix obtained by time averaging. In certain cases the statistics of the source are known or can be estimated, and the statistical ML (SML) solution can be obtained instead [21]. The performance of the SML normally outperforms that of the DML.

15.2.2 Acoustic/Seismic Signals

15.2.2.1 Parametric Methods

The parametric methods developed for narrowband signals have also been extended to wideband signals such as acoustic and seismic. For wideband signals, the time delay becomes a linear phase in frequency. Thus, the narrowband signal model can be applied to each frequency snapshot of the wideband signal spectrum. Then, a composite wideband signal model that includes all relevant frequency components of this signal can be used instead. For an array of R microphones (or seismic sensors) simultaneously receiving M independent, spatially separated sound (or seismic) signals ($M < R$), the acoustic (or seismic) waveform arriving at the rth microphone is given by

$$x_r(t) = \sum_{m=1}^{M} h_r^{(m)}(t) \star s_m(t) + n_r(t) \tag{15.34}$$

for $r = 1, \ldots, R$, where
  s_m is the mth source signal
  $h_r^{(m)}$ is the impulse response from the mth source to the rth sensor (i.e. a delta function in free space corresponding to the time delay or a filtered response to include the reverberation effect)
  n_r is the additive noise
  $\star$ denotes the convolution operation

For each chosen frame time (a function of the source motion and signal bandwidth), the received signal is appropriately digitized and collected into a space–time data vector $x = [x_1(0),\ldots,x_1(L-1),\ldots,x_R(0),\ldots,x_R(L-1)]^T$ of length RL. The corresponding frequency spectrum data vector is then given by $X(\omega k) = [X_1(\omega k), \ldots, X_R(\omega k)]^T$, for $k = 0, \ldots, N-1$, where N is the number of fast Fourier transform (FFT) bins.

Denote Θ as the estimation parameter: in the near-field case this is the source location vector $[r_{s_1}^T,\ldots,r_{s_M}^T]^T$ and r_{s_m} is the mth source location; in the far-field case this is the angle vector $[\phi_s^{(1)},\theta_s^{(1)},\ldots,\phi_s^{(M)},\theta_s^{(M)}]^T$, and $\phi_s^{(m)}$ and $\theta_s^{(m)}$ are the azimuth and elevation angles of the mth source respectively.

In general, the ML estimation of parameter Θ with additive white Gaussian noise assuming idealistic non-reverberant environment is given by

$$\hat{\Theta}^{\mathrm{ML}} = \arg\max_{\Theta} J(\Theta) = \arg\max_{\Theta} \sum_{k=1}^{N/2} W(k)\,\|\,\mathbf{P}(k,\Theta)\mathrm{X}(\omega_k)\,\|^2 \qquad (15.35)$$

where [19]: $W(k)$ is the weighting function by design (e.g. lower for significant bins and stronger weighting on dominant bins and/or high frequency bins); $\mathbf{P}(k,\Theta) = \mathbf{D}(k,\Theta)\mathbf{D}^{\dagger}(k,\Theta)$ is the projection matrix that projects the data into the parameter space; $\mathbf{D}^{\dagger}(k,\Theta) = (\mathbf{D}(k,\Theta)^{\mathrm{H}}\,\mathbf{D}(k,\Theta))^{-1}\,(\mathbf{D}(k,\Theta)^{\mathrm{H}}$ is the pseudo-inverse of the steering matrix; $\mathbf{D}(k,\Theta) = [\mathbf{d}^{(1)}(k,\Theta), \ldots, \mathbf{d}^{(M)}(k,\Theta)]$ is the steering matrix; $\mathbf{d}^{(m)}(k,\Theta) = [\mathbf{d}_1^{(m)}(k,\Theta), \ldots, d_R^{(m)}(k,\Theta)]^{\mathrm{T}}$ is the mth steering vector; $d_r^{(m)}(k,\Theta) = \mathrm{e}^{(-j2\pi k t_r^{(m)}/N)}$ is one component of the mth steering vector; $t_r^{(m)}$ is the time delay for the mth signal to traverse to the rth sensor; and only the positive frequency bins are considered (negative frequencies are simply mirror images for real-valued signals).

Note, a closed-form solution is not available for Equation 15.35 and a brute-force search is prohibitive as M increases. Efficient iterative computational methods including the alternating projection [4,19], particle filtering [10], and the SAGE method [22] have been shown to be effective both in computer simulations and in real-life data for up to $M = 2$. The nice feature about the objective function in Equation 15.35 is that it is continuous and differentiable; thus, many efficient gradient methods can be applied and tailored to the specific application for fast convergence. However, the gradient approach does not guarantee convergence to the global maximum, as the objective function is multi-modal (with multiple local extrema) in general, and the selection of the initial condition may greatly impact the estimation performance. The dependency of the initial condition is de-emphasized for two sources using the alternating projection approach [19]; but, without prior information, the above algorithms may get trapped in local solutions that may be far away from the actual location(s). For robust solutions that can achieve the global maximum, a genetic algorithm and simulated annealing can be used at the cost of convergence speed.

In most practical cases of ground sensor networks, the number of sources of interest M may not be a large number since only the nearby sources will have high enough energy to contribute to the overall received power. The interfering sources that are far away and weak will only contribute to the noise floor of the system, and thus will not be of interest to the estimator. However, in the case where the number of sources M becomes large, the computational time of the estimator may be burdensome even for the aforementioned iterative methods. For M sources of roughly equal strength, the search dimension can be limited to only that of a single source, and the highest M humps can be used as the suboptimal solutions. Obviously, this approach has several drawbacks, despite its significant computational advantage as M increases. It requires the M sources to be widely separated so the interaction among them is minimal. With some difference in strength, it is also difficult to distinguish a sidelobe of a stronger source from the presence of a weaker source. The widely used narrowband super-resolution methods such as MUSIC have been extended to the wideband case to achieve high resolution with low interaction among sources. Many have proposed using focusing matrices to transform the wideband signal subspaces into a predefined narrowband subspace [23]. Then, the MUSIC algorithm can be applied afterwards as if it was the narrowband case. This class of methods is referred to as the coherent signal subspace method (CSM). However, the drawback of these methods is the preprocessing that must be performed. For instance, the CSM methods need to construct the focusing matrices that require focusing angles that are not far from the true DOAs. A simpler noncoherent wideband MUSIC algorithm without such a condition has also been compared [19] and shown to yield poor estimation results (especially in range for the near-field case) for a finite number of data samples (limited due to moving source) and low SNR.

Alternating projection: To estimate the source location or DOA of multiple sources, the alternating projection is shown to be the most effective way. The alternating projection approach breaks the multi-dimensional parameter search into a sequence of single-source parameter searches, and yields a fast convergence rate. The following describes the alternating projection algorithm for the case of two sources, and it can be easily extended to the case of M sources. Let $\boldsymbol{\Theta} = [\boldsymbol{\Theta}_1^T \boldsymbol{\Theta}_2^T]^T$ be either the source locations in the near-field case or the DOAs in the far-field case. *The alternating projection algorithm* is as follows:

Step 1. Estimate the location/DOA of the stronger source on a single source grid

$$\boldsymbol{\Theta}_1^{(0)} = \arg\max_{\boldsymbol{\Theta}_1} J(\boldsymbol{\Theta}_1) \tag{15.36}$$

Step 2. Estimate the location/DOA of the weaker source on a single source grid under the assumption of a two-source model while keeping the first source location estimate from step 1 constant

$$\boldsymbol{\Theta}_2^{(0)} = \arg\max_{\boldsymbol{\Theta}_2} J\left(\left[\boldsymbol{\Theta}_1^{(0)T}, \boldsymbol{\Theta}_2^T\right]^T\right) \tag{15.37}$$

For $i = 1, \ldots$ (repeat steps 3 and 4 until convergence).

Step 3. Iterative approximated ML (AML) parameter search (direct search or gradient) for the location/DOA of the first source while keeping the estimate of the second source location from the previous iteration constant

$$\boldsymbol{\Theta}_1^{(i)} = \arg\max_{\boldsymbol{\Theta}_1} J\left(\left[\boldsymbol{\Theta}_1^T, \boldsymbol{\Theta}_2^{(i-1)T}\right]^T\right) \tag{15.38}$$

Step 4. Iterative AML parameter search (direct search or gradient) for the location/DOA of the second source while keeping the estimate of the first source location from step 3 constant

$$\boldsymbol{\Theta}_2^{(i)} = \arg\max_{\boldsymbol{\Theta}_2} J\left(\left[\boldsymbol{\Theta}_1^{(i)T}, \boldsymbol{\Theta}_2^T\right]^T\right) \tag{15.39}$$

15.2.2.2 ML Beamforming

The main purpose of beamforming is to improve the SINR, which is often performed after a desired source location is obtained (except for the blind beamforming methods). In the most general sense of digital wideband beamforming in the time domain, the digitized received array signal is combined with appropriate delays and weighting to form the beamformer output

$$y(n) = \sum_{r=1}^{R} \sum_{\ell=0}^{L-1} w_{r\ell} x_r(n-\ell) \tag{15.40}$$

where

$w_{r\ell}$ is the chosen beamforming weight to satisfy some criterion
x_r here denotes the digitized version of the received signal

Numerous criteria exist in the design of the beamforming weight, including maximum SINR with frequency and spatial constraints. Other robust blind beamforming methods have also been proposed to enhance SINR without the knowledge of the sensor responses and locations. For instance, the blind maximum power (MP) beamformer [9] obtains array weights from the dominant eigenvector (or singular vector) associated with the largest eigenvalue (or singular value) of the space–time sample correlation (or data) matrix. This approach not only collects the maximum power of the dominant source, but also provides some rejection of other interferences and noise.

In some cases, especially for multiple sources, frequency-domain beamforming may be more attractive for acoustic signals due to their wideband nature. This is especially advantageous when an ML localization algorithm is used a priori, since the beamforming output is a direct result of the ML source signal vector estimate $\hat{\mathbf{S}}^{\mathrm{ML}}(\omega_k)$ given by

$$\mathbf{Y}(\omega_k) = \hat{\mathbf{S}}^{\mathrm{ML}}(\omega_k) = \mathbf{D}^{\dagger}(k, \hat{\mathbf{\Theta}}^{\mathrm{ML}})\mathbf{X}(\omega_k) \tag{15.41}$$

where $\mathbf{Y}(\omega_k)$ is the beamformed spectrum vector for the M sources [19]. The ML beamformer in effect performs signal separation by utilizing the physical separation of the sources, and for each source signal the SINR is maximized in the ML sense. When only a single source exists, $\mathbf{D}^{\dagger}(k, \hat{\mathbf{\Theta}}^{\mathrm{ML}})$ degenerates to a vector and only the SNR is maximized.

15.2.2.3 Time-Delay-Type Methods

The problem of determining the location of a source given a set of differential time delays between sensors has been studied for many years. These techniques include two independent steps, namely estimating relative time delays between sensor data and then source location based on the relative time-delay estimates. Closed-form solutions can be derived for the second step based on spherical interpolation [24,25], hyperbolic intersection [26], or linear intersection [27]. However, these algorithms require the knowledge of the speed of propagation (e.g. the acoustic case). Yao et al. [9,28] derived closed-form LS and constrained LS (CLS) solutions for the case of unknown speed of propagation (e.g. the seismic case). The CLS method improves the performance from that of the LS method by forcing an equality constraint on two components of the unknowns. When the speed of propagation is known, the LS method of Yao et al. [9,28], after removing such unknowns, becomes the method independently derived later by Huang et al. [29]. Huang et al. [29] show that their method is mathematically equivalent to the spherical interpolation technique [24,25], but with less computational complexity. We refer to the method of Huang et al. [29] as the LS method of known speed of propagation and use it to compare with the other methods in this section. A similar LS formulation is also available for DOA estimation [30]. In theory, these two-step methods can work for multiple sources as long as the relative time delays can be estimated from the data. However, this remains to be a challenging task in practice. Recently, the use of higher order statistics to estimate the relative time delays for multiple independent sources has been studied [8], but its practicality is yet to be shown.

Denote the source location in Cartesian by $r_s = [x_s, y_s, z_s]^{\mathrm{T}}$ and the rth sensor location by $r_r = [x_r, y_r, z_r]^{\mathrm{T}}$. Without loss of generality, we choose $r = 1$ as the reference sensor for differential time delays. Let the reference sensor be the origin of the coordinate system for simplicity. The speed of propagation v in this formulation can also be estimated from the data. In some problems, v may be considered to be partially known (e.g. acoustic applications), while in others it may be considered to be unknown (e.g. seismic applications). The differential time delays for R sensors satisfy

$$t_{r1} = t_r - t_1 = \frac{\|r_s - r_r\| - \|r_s - r_1\|}{v} \tag{15.42}$$

for $r = 2, \ldots, R$. This is a set of $R - 1$ nonlinear equations, which makes finding its solution r_s nontrivial.

To simplify the estimation to a linear problem, we make use of the following:

$$\|r_s - r_r\|^2 - \|r_s\|^2 \|r_r\|^2 - 2(x_s x_r + y_s y_r + z_s z_r) \tag{15.43}$$

The left-hand side of Equation 15.43 is equivalent to

$$\left(\left\|\mathbf{r}_s - \mathbf{r}_r\right\| - \left\|\mathbf{r}_s\right\|\right)\left(\left\|\mathbf{r}_s - \mathbf{r}_r\right\| + \left\|\mathbf{r}_s\right\|\right) = vt_{r1}\left(2\left\|\mathbf{r}_s\right\| + vt_{r1}\right) \tag{15.44}$$

in the case of $\mathbf{r}_1 = \mathbf{0}$. Upon combining both expressions, we have the following linear relation for the rth sensor:

$$\mathbf{r}_r^T \mathbf{r}_s + vt_{r1}\left\|\mathbf{r}_s\right\| + \frac{v^2 t_{r1}^2}{2} = \frac{\left\|\mathbf{r}_r\right\|^2}{2} \tag{15.45}$$

With R sensors, we formulate the LS solution by putting $R - 1$ linear equations into the following matrix form:

$$\mathbf{Ay} = \mathbf{b} \tag{15.46}$$

where

$$\mathbf{A} = \begin{bmatrix} \mathbf{r}_2^T & t_{21} & t_{21}^2/2 \\ \mathbf{r}_3^T & t_{31} & t_{31}^2/2 \\ \vdots & \vdots & \vdots \\ \mathbf{r}_R^T & t_{R1} & t_{R1}^2/2 \end{bmatrix}, \quad \mathbf{y} = \begin{bmatrix} \mathbf{r}_s \\ v\left\|\mathbf{r}_s\right\| \\ v^s \end{bmatrix}, \quad \mathbf{b} = \frac{1}{2} \begin{bmatrix} \left\|\mathbf{r}_2\right\|^2 \\ \left\|\mathbf{r}_3\right\|^2 \\ \vdots \\ \left\|\mathbf{r}_R\right\|^2 \end{bmatrix} \tag{15.47}$$

For 3-D uncertainty, the dimension of $\mathbf{A}$ is $(R - 1) \times 5$. In the case of six or more sensors, the pseudo-inverse of matrix $\mathbf{A}$ is given by

$$\mathbf{A}^\dagger = (\mathbf{A}^T \mathbf{A})^{-1} \mathbf{A}^T \tag{15.48}$$

The LS solution for the unknown vector can be given by $\mathbf{y} = \mathbf{A}^\dagger \mathbf{b}$. The source location estimate is given by the first three elements of $\mathbf{y}$ and the speed of propagation estimate is given by the square-root of the last element of $\mathbf{y}$.

In the three-dimensional case there are five unknowns in w. To obtain an overdetermined solution, we need at least five independent equations, which can be derived from the data of six sensors. However, placing sensors randomly does not provide much assurance against ill-conditioned solutions. The preferred approach would be to use seven or more sensors, yielding six or more relative delays, and to perform an LS fitting of the data. In the two-dimensions problem, the minimum number of sensors can be reduced by 1. If the propagation speed is known, then the minimum number of sensors can be further reduced by 1.

Notice in the unknown vector $\mathbf{y}$ of Equation 15.47 that the speed of propagation estimate can also be given by

$$\hat{v} = \frac{v\|\widehat{\mathbf{r}_s}\|}{\|\hat{\mathbf{r}}_s\|} \tag{15.49}$$

using the fourth and the first three elements of **y**. To exploit this relationship, we can add another non-linear constraint to ensure equivalence between the speed of propagation estimates from the fourth and the fifth elements. By moving the fifth element of **y** to the other side of the equation, we can rewrite Equation 15.46 as follows:

$$\mathbf{Ay} = \mathbf{b} + v^2\mathbf{d} \qquad (15.50)$$

where

$$\mathbf{A} = \begin{bmatrix} \mathbf{r}_2^{\mathrm{T}} & t_{21} \\ \mathbf{r}_3^{\mathrm{T}} & t_{31} \\ \vdots & \vdots \\ \mathbf{r}_R^{\mathrm{T}} & t_{R1} \end{bmatrix}, \quad \mathbf{y} = \begin{bmatrix} \mathbf{r}_s \\ v\|\mathbf{r}_s\| \end{bmatrix} \qquad (15.51)$$

$$\mathbf{b} = \frac{1}{2}\begin{bmatrix} \|\mathbf{r}_2\|^2 \\ \|\mathbf{r}_3\|^2 \\ \vdots \\ \|\mathbf{r}_R\|^2 \end{bmatrix}, \quad \mathbf{d} = -\frac{1}{2}\begin{bmatrix} t_{21}^2 \\ t_{31}^2 \\ \vdots \\ t_{R1}^2 \end{bmatrix} \qquad (15.52)$$

In this case, the dimension of **A** is $(R-1) \times 4$ for three-dimensional uncertainty. The pseudo-inverse of matrix **A** is given by

$$\mathbf{A}^\dagger = (\mathbf{A}^{\mathrm{T}}\mathbf{A})^{-1}\mathbf{A}^{\mathrm{T}} \qquad (15.53)$$

The CLS solution for the unknown vector can be given by $\mathbf{y} = \mathbf{A}^\dagger\mathbf{b} + v^2\mathbf{A}^\dagger\mathbf{d}$. Define $\mathbf{p} = \mathbf{A}^\dagger\mathbf{b}$ and $\mathbf{q} = \mathbf{A}^\dagger\mathbf{d}$. The source location and speed of propagation estimates can be given by

$$x_s = p_1 + v^2 q_1$$
$$y_s = p_2 + v^2 q_2$$
$$z_s = p_3 + v^2 q_3 \qquad (15.54)$$
$$v\|\mathbf{r}_s\| = p_4 + v^2 q_4$$

where pi and qi are the ith entries of **p** and **q** respectively. The number of unknowns appears to be 5, but the five unknowns only contribute 4 degrees of freedom due to the following nonlinear relationship:

$$\|\mathbf{r}_s\|^2 = x_s^2 + y_s^2 + z_s^2 \qquad (15.55)$$

By substituting (15.54) into (15.55), the following third order constraint equation results

$$\alpha(v^2)^3 + \beta(v^2)^2 + \gamma(v^2) + \delta = 0 \qquad (15.56)$$

where

$$\alpha = q_1^2 + q_2^2 + q_3^2$$

$$\beta = 2(p_1 q_1 + p_2 q_2 + p_3 q_3) - p_4^2$$

$$\gamma = p_1^2 + p_2^2 + p_3^2 - 2 p_4 q_4 \tag{15.57}$$

$$\delta = -p_4^2$$

At most, three solutions exist to the third-order equation. The speed of propagation estimate is given by the positive square root of the real positive solution. If there is more than one positive and real root, then the more "physical" estimate that fits the data is used. Once the speed of propagation is estimated by the constraint equation, the source location estimate can be given by $\mathbf{y} = \mathbf{P} + \hat{v}^2 \mathbf{q}$. Compared with the LS method, the minimum required number of sensors in the CLS method can be further reduced by 1.

15.2.2.4 Time-Delay Estimation Methods

Relative time-delay estimation is a classical problem in array signal processing, and plenty of algorithms have been proposed over the years. A collection of relative time-delay estimation papers has been put together by Carter [17], the pioneer in this field. The idea behind most relative time-delay estimation methods is based on maximizing the weighted cross-correlation between a pair of sensor data to extract an ML-type estimate. Denote the pair of sensor data as $x_p(t)$ and $x_q(t)$, where one is a delayed version of the other and corrupted by independent additive noise. The delay τ is estimated by the following:

$$\max_{\tau} \int_t w(t) x_p(t) x_q(t - \tau) dt \tag{15.58}$$

where $w(t)$ is the weighting function appropriately chosen for the data type. Normally, $w(t)$ is a tapering function, i.e. maximum gain at the center of data and attenuation at the edge of data, which reduces the edge effect caused by the missing data at the edges (leading and trailing data in the pair) in a finite data window. This effect is especially magnified for low-frequency data in a short data window.

Note that, when the signal is relatively narrowband, an ambiguity in the correlation peak also occurs at every period of the signal. This is directly related to the grating lobes or spatial aliasing effect observed in the beampattern of an equally spaced array. Such a flaw can be avoided by limiting the search range of the delay to within the period of the signal, or equivalently limiting the angular/range search range in the spatial domain for direct localization or DOA estimation. However, for arrays with large element spacing, the ambiguity may be difficult to resolve without any prior information of the rough direction of the source, and a leading waveform may very well be confused with a trailing waveform by the estimator, hence leading to the ambiguous solution of the relative time delay. Normally, the relative time delay is performed independently between the reference sensor and every sensor other than the reference one. An erroneous location/DOA estimate results when one of the relative time-delay estimates is ambiguous. Nonetheless, as shown by Chen et al. [19], the parametric ML location/DOA estimator equivalently maximizes the summed cross-correlation of all pairs of sensors instead of separately maximizing for each pair; thus, the ML solution is less susceptible to the ambiguity problem.

For many applications, the relative time-delay may be very small comparing to the sampling rate of the data; thus, direct time-domain cross-correlation shown in (15.58) is usually performed after interpolation of the sensor data to achieve subsample estimate of the relative time-delay. An alternative approach is to perform cross-correlation in the frequency-domain, where the following is to be maximized

$$\max_{\tau} \int_{\omega} G(\omega) X_p(\omega) X_q^*(\omega) e^{j\omega\tau} d\omega \tag{15.59}$$

where

$G(\omega)$ is the frequency weighting function
$X(\omega)$ is the Fourier transform of $x(t)$
$*$ denotes the complex conjugate operation

The classical relative time-delay estimation method [17], the generalized cross-correlation (GCC) approach, conceived by Carter, is given in the above form where the choice of $G(\omega)$ leads to specific algorithms tailored for each application. One common choice of $G(\omega)$ is

$$G(\omega) = \frac{1}{\left| X_p(\omega) X_q^*(\omega) \right|}$$

which results in the PHAT algorithm that outperforms most algorithms in a nonreverberant environment. In a reverberant (multipath) environment, where the signal traverses via multiple paths other than the direct path, the aforementioned ambiguity issue becomes more severe, since now one sensor signal can be highly correlated with any nondirect path signal of the other sensor that arrives after a tiny delay. Depending on the frequency of the signal and the delay spread, this effect may be too distressing for some systems to work properly. More recent work on time-delay estimation focuses on robust estimators in the reverberant environment [31].

15.3 Array System Performance Analysis and Robust Design

In this section we compare the performance of several representative wideband source localization algorithms. This includes the ML parametric method in Equation 15.35, which we here refer to as the AML solution due to the finite length of the data,* the LS/CLS solution described in Equations 15.46 and 15.50, and the noncoherent wideband MUSIC algorithm proposed in by Tung et al. [32]. For performance comparison purposes, well-controlled computer simulations are used. The theoretical Cramér–Rao bound (CRB), the formulation of which is shown in the next section, is also compared with the performance of each algorithm. Then, the best algorithms are applied to the experimental data collected by wirelessly connected iPAQs with built-in microphones in Section 15.4.2. The outstanding experimental results show that the sensor network technology is not just in the plain research stage, but is readily available for practical use.

15.3.1 Computer-Simulated Results for Acoustic Sources

Now, we consider some simulation examples and analysis for some acoustic sources. In principle, the methods used in the following can be applied to signals of other nature, e.g. seismic, but they will not

---

* For finite length data, the discrete Fourier transform (DFT) has a few artifacts. The circular shift property of the DFT introduces an edge effect problem for the actual linear time shift, and this edge effect is not negligible for a small block of data. To remove the edge effect, appropriate zero padding can be applied. However, it is also known that zero padding destroys the orthogonality of the DFT, which makes the noise spectrum appear correlated across frequency. As a result, there does not exist an exact ML solution for data of finite length.

be considered for the purpose of this section. In all cases, the sampling frequency is set to be 1 kHz. The speed of propagation is 345 m/s. A prerecorded tracked vehicle signal, with significant spectral content of about 50 Hz bandwidth centered about a dominant frequency at 100 Hz, is considered. For an arbitrary array of five sensors, we simulate the array data (with uniform SNR across the array) using this tracked vehicle signal with appropriate time delays. The data length $L = 200$ (which corresponds to 0.2 s), the FFT size $N = 256$ (with zero-padding), and all positive frequency bins are considered in the AML metric. To understand the fundamental properties of the AML algorithm, a normalized metric for a single source $J_N(r_s)$, which is a special case of Equation 15.35, is plotted in a range of possible locations. In this case

$$J_N(r_s) \equiv \frac{\sum_{k=1}^{N/2} \left| d(k, r_s)^H X(\omega_k) \right|^2}{RJ_{max}} \le 1 \qquad (15.60)$$

where

  $d(k, r_s)$ is the steering vector that steers to a particular position r_s

  $J_{max} = \sum_{k=1}^{N/2} \left[\sum_{p=1}^{R} \left| X_p(\omega_k) \right| \right]^2$, which is useful to verify estimated peak values

The source location can be estimated based on where $J_N(r_s)$ is maximized for a given set of locations. In Figure 15.6, $J_N(r_s)$ is evaluated at different near-field positions and plotted for a source inside the convex hull of the array (overlaid on top) under 20 dB SNR. A high peak shows up at the source location, thus indicating good estimation. When the source moves away from the array, the peak broadens and results in more range estimation error since it is more sensitive to noise. As depicted in Figure 15.7 (for the same setting), the image plot of $J_N(r_s)$ evaluated at different positions shows that the range estimation error is likely to occur in the source direction when the source moves away from the array. However, the angle estimation is not greatly impacted.

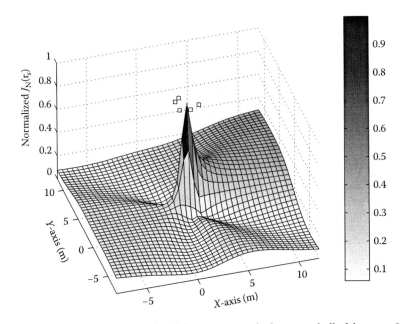

FIGURE 15.6 The three-dimensional plot of $J_N(r_s)$ for a source inside the convex hull of the array. Square: sensor locations.

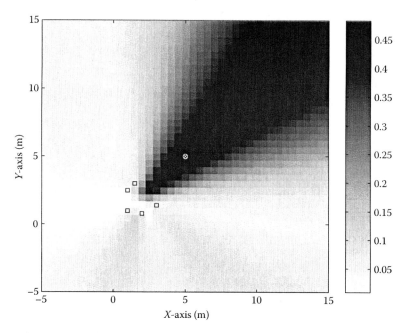

FIGURE 15.7 The image plot of $J_N(r_s)$ for a distant source. Square: sensor locations; circle: actual source location; ×: source location estimate.

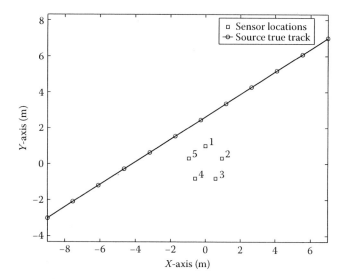

FIGURE 15.8 Single traveling source scenario.

In the following, we consider a single traveling source scenario for a circular array (uniformly spaced on the circumference), as depicted in Figure 15.8. In this case we consider the spatial loss that is a function of the distance from the source location to each sensor location; thus, the SNR is no longer uniform. For each frame of $L = 200$ samples, we simulate the array data for 1 of the 12 source locations. We apply the AML method without weighting the data according to the SNR and estimate the source location at each frame. A two-dimensional polar grid-point system is being used, where the angle is uniformly sampled and the range is uniformly sampled on a log-scale (to be physically meaningful). This two-dimensional grid-point search provides the initial estimate of the direct search simplex

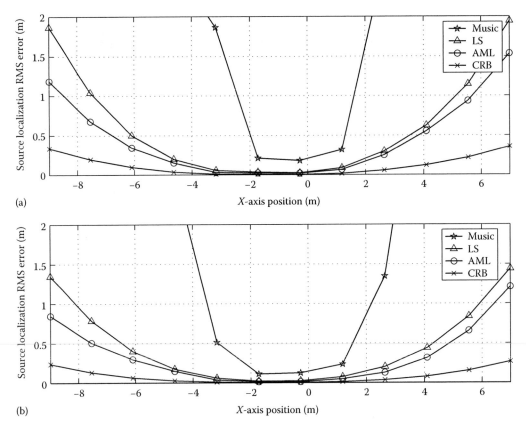

(a)

(b)

FIGURE 15.9 Source localization performance comparison for known speed of propagation case: (a) $R = 5$; (b) $R = 7$.

method in MATLAB®'s Optimization Toolbox; thus, a global maximum is guaranteed. In addition, we apply the LS method [29] and the wideband MUSIC method [32] and compute the CRB using Equation 15.75 for comparison purposes. Generally speaking, the LS is the most computationally efficient, then the AML and the wideband MUSIC, but the detailed complexities of these methods are not compared here. The wideband MUSIC divides the 200 samples into 4 subintervals of 50 samples each for the computation of the sample correlation matrix and uses the same grid-point search and direct search method as the AML case. As depicted in Figure 15.9 ((a) for $R = 5$ and (b) for $R = 7$), both the AML and LS methods approach the CRB when the source is near the array (high SNR and short range), but the AML outperforms the LS method. However, the wideband MUSIC yields much worse estimates than those of the LS and AML methods, especially when the source is far from the array. The wideband MUSIC requires a large number of samples for good performance, but in this case L is limited by the moving source. Note that when the source is far from the array (low SNR and long range) the CRB is too idealistic, since it does not take into account other factors such as the edge effect or correlated noise across frequency. Thus, the reader should not get the impression that the AML performs poorly in this region.

15.3.2 CRB for Source Localization

The CRB is most often used as a theoretical lower bound for any unbiased estimator [33]. Normally, the result of an algorithm is compared with the CRB for performance merit. However, the CRB provides not just a lower bound, it also provides very valuable information regarding the nature of the problem.

By analyzing the CRB, we can find many insights to the problem at hand and it can help the design of the array for optimum performance. In the following we provide the formulation of two CRBs based on time-delay error and SNR. The CRB based on time-delay error can be used to analyze and compare the time-delay-based algorithms, and the CRB based on SNR, which is more generalized, can be used to analyze and compare all algorithms.

15.3.2.1 CRB Based on Time-Delay Error

In this section we derive the CRB for source localization and speed of propagation estimations with respect to the relative time-delay estimation error. Denote the unknown parameter vector by $\Theta = [\mathbf{r}_s^T, v]^T$. The measured relative time-delay vector can be modeled by

$$\hat{t} = t(\Theta) + \xi \tag{15.61}$$

where
  $t = [t_{21}, \ldots, t_{R1}]^T$ is the true relative time-delay vector
  ξ is the estimation error vector, which is assumed to be zero mean white Gaussian distributed with variance σ^2

Define the gradient matrix $\mathbf{H} = \partial t / \partial \Theta^T = [\partial t / \partial \mathbf{r}_s^T, \partial t / \partial v] = (1/v)[\mathbf{B}, -t]$, where $\mathbf{B} = [u_2 - u_1, \ldots, u_R - u_1]^T$ and the unit vector $u_r = (r_s - r_r)/\|r_s - r_r\|$ indicates the direction of the source from the rth sensor. The Fisher information matrix [33] is then given by

$$\mathbf{F}_{r_s} = \mathbf{H}^T \mathbf{R}_\xi^{-1} \mathbf{H} = (1/\sigma^2) \mathbf{H}^T \mathbf{H}$$

$$= (1/\sigma^2 v^2) \begin{bmatrix} \mathbf{B}^T \mathbf{B} & -\mathbf{B}^T t \\ -t^T \mathbf{B} & \|t\|^2 \end{bmatrix} \tag{15.62}$$

The theoretical lower bound of the variance of r_s is given by the diagonal elements of the leading 3×3 submatrix of the inverse Fisher information matrix

$$\left[\mathbf{F}_{r_s}^{-1} \right]_{11:33} = \sigma^2 v^2 [\mathbf{B}^T \mathbf{P}_t^\perp \mathbf{B}]^{-1} \tag{15.63}$$

where
  $\mathbf{P}_t = tt^T / \|t\|^2$ is the orthogonal projection matrix of t
  $\mathbf{P}_t^\perp = \mathbf{I} - \mathbf{P}_t$

The variance bound of the distance d between the estimated source location and the actual location is given by trace $[\mathbf{F}_{r_s}^{-1}]_{11:33}$. The variance bound of the speed of propagation is given by

$$\sigma_v^2 \geq \left[\mathbf{F}_{r_s}^{-1} \right]_{44} = \sigma^2 v^2 \left\| \mathbf{P}_\mathbf{B}^\perp t \right\|^{-2} \tag{15.64}$$

where
  $\mathbf{P}_\mathbf{B} = \mathbf{B}(\mathbf{B}^T \mathbf{B})^{-1} \mathbf{B}^T$ is the orthogonal projection matrix of $\mathbf{B}$
  $\mathbf{P}_\mathbf{B}^\perp = \mathbf{I} - \mathbf{P}_\mathbf{B}$

The CRB analysis shows theoretically that the variance of source location estimation grows linearly with the relative time-delay estimation variance. The standard deviations of both the location and speed of propagation estimation errors grow linearly with the value of the speed of propagation. Note that, when the speed of propagation is known, the source location estimation variance bound becomes $\sigma_d^2 \geq \sigma^2 v^2$ trace $[(\mathbf{B}^\mathrm{T} \mathbf{B})^{-1}]$, which is always smaller than that of the unknown speed of propagation case. Define the *array matrix* by $\mathbf{A}_{\mathrm{r}_s} = \mathbf{B}^\mathrm{T}\mathbf{B}$. It provides a measure of geometric relations between the source and the sensor array. Poor array geometry may lead to degeneration in the rank of matrix $\mathbf{A}_{\mathrm{r}_s}$, thus resulting in a large estimation error bound.

By transforming to the polar coordinate system in the two-dimensional case, the CRB for the source range and DOA can also be given. Denote $r_s = \sqrt{x_s^2 + y_s^2}$ as the source range from a reference position such as the array centroid. The DOA can be given by $\phi_s = \tan^{-1}(x_s/y_s)$ with respect to the Y-axis. The time delay from the source to the rth sensor is given by $t_r = \sqrt{r_s^2 + r_r^2 - 2r_s r_r \cos(\phi_s - \phi_r)}/v$, where (r_r, ϕ_r) is the location of the rth sensor in the polar coordinate system. The gradient with respect to the source range is given by

$$\frac{\partial t_{r1}}{\partial r_s} = \frac{1}{v}\left[\frac{r_s - r_r \cos(\phi_s - \phi_r)}{\|\mathbf{r}_s - \mathbf{r}_r\|} - \frac{r_s - r_1 \cos(\phi_s - \phi_1)}{\|\mathbf{r}_s - \mathbf{r}_1\|} \right] \tag{15.65}$$

and the gradient with respect to the DOA is given by

$$\frac{\partial t_{r1}}{\partial \phi_s} = \frac{1}{v}\left[\frac{r_s r_r \sin(\phi_s - \phi_r)}{\|\mathbf{r}_s - \mathbf{r}_r\|} - \frac{r_s r_1 \sin(\phi_s - \phi_1)}{\|\mathbf{r}_s - \mathbf{r}_1\|} \right] \tag{15.66}$$

Define $\mathbf{B}_{\mathrm{pol}} = v[\partial t/\partial r_s, \partial t/\partial \phi_s]$. The polar array matrix is then defined by $\mathbf{A}^{\mathrm{pol}} = \mathbf{B}_{\mathrm{pol}}^\mathrm{T}\mathbf{B}_{\mathrm{pol}}$. The leading 2×2 submatrix of the inverse polar Fisher information matrix can be given by

$$\left[\mathbf{F}_{(r_s, \phi_s)}^{-1} \right]_{11.22} = \sigma^2 v^2 \left[\mathbf{B}_{\mathrm{pol}}^\mathrm{T} \mathbf{P}_\mathrm{t}^\perp \mathbf{B}_{\mathrm{pol}} \right]^{-1} \tag{15.67}$$

Then, the lower bound of range estimation variance can be given by $\sigma_{r_s}^2 \geq [\mathbf{F}_{(r_s, \phi_s)}^{-1}]_{11}$ and the lower bound of DOA estimation variance can be given by $\sigma_{\phi_s}^2 \geq [\mathbf{F}_{(r_s, \phi_s)}^{-1}]_{22}$.

To evaluate the performance of the LS and CLS source localization algorithms by comparing with the theoretical Cramér–Rao lower bound, the simulated scenario depicted in Figure 15.10 is used. A randomly distributed array of seven sensors is used to collect the signal generated by a source moving in a straight line. By perturbing the actual time delays by a white Gaussian noise with zero mean and standard deviation of $\sigma_{\mathrm{td}} = 10\,\mu\mathrm{s}$, the LS and CLS algorithms are applied to estimate the source location at each time frame. As depicted in Figure 15.11, the CLS yields much better range and DOA estimates than the LS on the average of 10,000 random realizations. The CLS estimates are also very close to the corresponding CRB.

15.3.2.2 CRB Based on SNR

Most of the derivations of the CRB for wideband source localization found in the literature are in terms of relative time-delay estimation error, as shown in the previous section. In this section we derive a more general CRB directly from the signal model. By developing a theoretical lower bound in terms of signal characteristics and array geometry, we not only bypass the involvement of the intermediate time-delay estimator, but also offer useful insights to the physics of the problem.

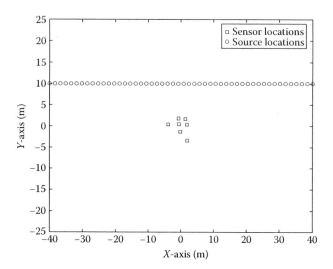

FIGURE 15.10 Traveling source scenario.

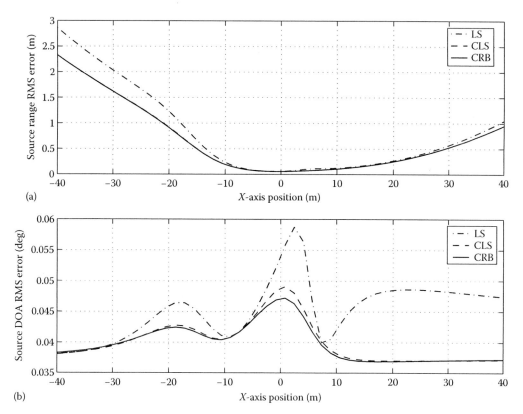

FIGURE 15.11 LS, CLS, and CRB location root-mean-square error at various source locations ($\sigma_{td} = 10\,\mu s$).

We consider the following three cases: known signal and known speed of propagation; known signal but unknown speed of propagation; and known speed of propagation but unknown signal. The comparison of the three conditions provides a sensitivity analysis that explains the fundamental differences of different problems, e.g. unknown speed of propagation for seismic sensing and known signal for radar applications. The case of unknown signal and unknown speed of propagation is not too different from the case of unknown signal and known speed of propagation; thus, it is not considered. For all cases, we can construct the Fisher information matrix [33] from the signal model by

$$\mathbf{F} = 2\,\mathrm{Re}\left[\mathbf{H}^{\mathrm{H}}\mathbf{R}_{\xi}^{-1}\mathbf{H}\right] = \left(\frac{2}{L\sigma^2}\right)\mathrm{Re}[\mathbf{H}^{\mathrm{H}}\mathbf{H}] \tag{15.68}$$

where $\mathbf{H} = \partial\mathbf{G}/\partial\mathbf{r}_s^{\mathrm{T}}$ for the first case, assuming $\mathbf{r}_s$ is the only unknown in the single source case. In this case, $\mathbf{F}_{r_s} = \zeta\mathbf{A}$, where $\zeta = (2/L\sigma^2v^2)\sum_{k=1}^{N/2}(2\pi k|s_0(k)|/N)^2$ is the scale factor that is proportional to the total power in the derivative of the source signal,

$$\mathbf{A} = \sum_{p=1}^{R} a_p^2 \mathbf{u}_p \mathbf{u}_p^{\mathrm{T}} \tag{15.69}$$

is the *array matrix*, and $\mathbf{u}_p = (\mathbf{r}_s - \mathbf{r}_p)/\|\mathbf{r}_s - \mathbf{r}_p\|$ is the unit vector indicating the direction of the source from the pth sensor. The $\mathbf{A}$ matrix provides a measure of geometric relations between the source and the sensor array. Poor array geometry may lead to degeneration in the rank of matrix $\mathbf{A}$. It is clear from the scale factor ζ, as will be shown later, that the performance does not depend solely on the SNR, but also on the signal bandwidth and spectral density. Thus, source localization performance is better for signals with more energy in the high frequencies.

For the second case, when the speed of propagation is also unknown, i.e. $\mathbf{\Theta} = [\mathbf{r}_s^{\mathrm{T}}, v]^{\mathrm{T}}$, the $\mathbf{H}$ matrix for this case is given by $\mathbf{H} = \partial\mathbf{G}/\partial\mathbf{r}_s^{\mathrm{T}}, \partial\mathbf{G}/\partial v]$. The Fisher information block matrix is given by

$$\mathbf{F}_{r_s,v} = \zeta\begin{bmatrix} \mathbf{A} & -\mathbf{U}\mathbf{A}_a\mathbf{t} \\ -\mathbf{t}^{\mathrm{T}}\mathbf{A}_a\mathbf{U}^{\mathrm{T}} & \mathbf{t}^{\mathrm{T}}\mathbf{A}_a\mathbf{t} \end{bmatrix} \tag{15.70}$$

where
$\mathbf{U} = [\mathbf{u}_1, \ldots, \mathbf{u}_R]$
$\mathbf{A}_a = \mathrm{diag}([a_1^2, \ldots, a_R^2])$
$\mathbf{t} = [t_1, \ldots, t_R]^{\mathrm{T}}$

By applying the well-known block matrix inversion lemma, the leading $D \times D$ submatrix of the inverse Fisher information block matrix can be given by

$$\left[\mathbf{F}_{(r_s,v)}^{-1}\right]_{11.DD} = \frac{1}{\zeta}\left[\mathbf{A} - \mathbf{Z}_v\right]^{-1} \tag{15.71}$$

where the *penalty matrix* due to unknown speed of propagation is defined by $\mathbf{Z}_v = (1/\mathbf{t}^{\mathrm{T}}\mathbf{A}_a\mathbf{t})\mathbf{U}\mathbf{A}_a\mathbf{t}\mathbf{t}^{\mathrm{T}}\mathbf{A}_a\mathbf{U}^{\mathrm{T}}$. The matrix $\mathbf{Z}_v$ is nonnegative definite; therefore, the source localization error of the unknown speed of propagation case is always larger than that of the known case.

For the third case, when the source signal is also unknown, i.e. $\mathbf{\Theta} = [\mathbf{r}_s^{\mathrm{T}}, |\mathbf{S}_0|^{\mathrm{T}}, \mathbf{\Phi}_0^{\mathrm{T}}]^{\mathrm{T}}$, the $\mathbf{H}$ matrix is given by

$$\mathbf{H} = \left[\frac{\partial\mathbf{G}}{\partial\mathbf{r}_s^{\mathrm{T}}}, \frac{\partial\mathbf{G}}{\partial|\mathbf{S}_0|^{\mathrm{T}}} \frac{\partial\mathbf{G}}{\partial\mathbf{\Phi}_0^{\mathrm{T}}}\right]$$

where
$\mathbf{S}_0 = [S_0(1), \ldots, S_0(N/2)^{\mathrm{T}}$
$|\mathbf{S}_0|$ and $\mathbf{\Phi}_0$ are the magnitude and phase part of $\mathbf{S}_0$ respectively

The Fisher information matrix can then be explicitly given by

$$
\mathbf{F}_{r_s,S_0} = \begin{bmatrix} \zeta\mathbf{A} & \mathbf{B} \\ \mathbf{B}^{\mathrm{T}} & \mathbf{D} \end{bmatrix}
\tag{15.72}
$$

where $\mathbf{B}$ and $\mathbf{D}$ are not explicitly given since they are not needed in the final expression. By applying the block matrix inversion lemma, the leading $D \times D$ submatrix of the inverse Fisher information block matrix can be given by

$$
\left[\mathbf{F}_{r_s,S_0}^{-1}\right]_{11:DD} = \frac{1}{\zeta}(\mathbf{A} - \mathbf{Z}_{S_0})^{-1}
\tag{15.73}
$$

where the penalty matrix due to unknown source signal is defined by

$$
\mathbf{Z}_{S_0} = \frac{1}{\displaystyle\sum_{p=1}^{R} a_p^2}\left(\sum_{p=1}^{R} a_p^2 \mathbf{u}_p\right)\left(\sum_{p=1}^{R} a_p^2 \mathbf{u}_p\right)^{\mathrm{T}}
\tag{15.74}
$$

The CRB with unknown source signal is always larger than that with known source signal, as discussed below. This can easily be shown, since the penalty matrix $\mathbf{Z}_{S_0}$ is nonnegative definite. The $\mathbf{Z}_{S_0}$ matrix acts as a penalty term, since it is the average of the square of weighted up vectors. The estimation variance is larger when the source is far away, since the $\mathbf{u}_p$ vectors are similar in direction and generate a larger penalty matrix, i.e. $\mathbf{u}_p$ vectors add up. When the source is inside the convex hull of the sensor array, the estimation variance is smaller, since $\mathbf{Z}_{S_0}$ approaches the zero matrix, i.e. $\mathbf{u}_p$ vectors cancel each other. For the two-dimensional case, the CRB for the distance error of the estimated location $[\hat{x}_s, \hat{y}_s]^{\mathrm{T}}$ from the true source location can be given by

$$
\sigma_d^2 = \sigma_{x_s}^2 + \sigma_{y_s}^2 \geq \left[\mathbf{F}_{r_s,S_0}^{-1}\right]_{11}\left[\mathbf{F}_{r_s,S_0}^{-1}\right]_{22}
\tag{15.75}
$$

where $d^2 = (\hat{x}_s - x_s)^2 + (\hat{y}_s - y_s)^2$. By further expanding the parameter space, the CRB for multiple source localizations can also be derived, but its analytical expression is much more complicated. Note that, when both the source signal and sensor gains are unknown, it is not possible to determine the values of the source signal and the sensor gains (they can only be estimated up to a scaled constant).

By transforming to the polar coordinate system in the two-dimensional case, the CRB for the source range and DOA can also be given. Denote $r_s\sqrt{x_s^2 + y_s^2}$ as the source range from a reference position such as the array centroid. The DOA can be given by $\phi_s = \tan^{-1}(x_s/y_s)$ with respect to the Y-axis. The time delay from the source to the pth sensor is then given by $t_p = \sqrt{r_s^2 + r_p^2 - 2r_s r_p \cos(\phi_s - \phi_p)}/v$, where (r_p, ϕ_p) is the polar position of the pth sensor. We can form a polar array matrix $\mathbf{A}^{\mathrm{pol}} = \displaystyle\sum_{p=1}^{R} a_p^2 v_p v_p^{\mathrm{T}}$, where

$$
v_p = \left[\frac{r_s - r_p\cos(\phi_s - \phi_p)}{\|r_s - r_p\|}, \frac{r_s r_p \sin(\phi_s - \phi_p)}{\|r_s - r_p\|}\right]^{\mathrm{T}}
\tag{15.76}
$$

Similarly, the polar penalty matrix due to the unknown source signal can be given by

$$\mathbf{Z}_{S_0}^{\text{pol}} = \frac{1}{\sum_{p=1}^{R} a_p^2} \left(\sum_{p=1}^{R} a_p^2 v_p \right) \left(\sum_{p=1}^{R} a_p^2 v_p \right)^{\mathrm{T}}$$

(15.77)

The leading 2×2 submatrix of the inverse polar Fisher information block matrix can be given by

$$\left[\mathbf{F}_{(r_s,\phi_s),S_0}^{-1} \right]_{11:22} = \frac{1}{\zeta} (\mathbf{A}^{\text{pol}} - \mathbf{Z}_{S_0}^{\text{pol}})^{-1}$$

(15.78)

Then, the lower bound of range estimation variance can be given by $\sigma_{r_s}^2 \geq [\mathbf{F}_{(r_s,\phi_s),S_0}^{-1}]_{11}$ and the lower bound of DOA estimation variance can be given by $\sigma_{\phi_s}^2 \geq [\mathbf{F}_{(r_s,\phi_s),S_0}^{-1}]_{22}$. The polar array matrix shows that a good DOA estimate can be obtained at the broadside of a linear array and a poor DOA estimate results at the endfire of a linear array. A two-dimensional array is required for better range and DOA estimations, e.g. a circular array.

To compare the theoretical performance of source localization under different conditions, we compare the CRB for the known source signal and speed of propagation, unknown speed of propagation, and unknown source signal cases using seven sensors for the same single traveling source scenario (shown in Figure 15.8). As depicted in Figure 15.12, the unknown source signal is shown to have a much

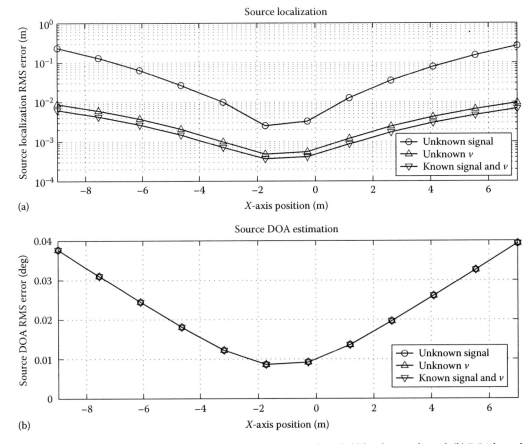

FIGURE 15.12   CRB comparison for the traveling source scenario ($R = 7$): (a) localization bound; (b) DOA bound.

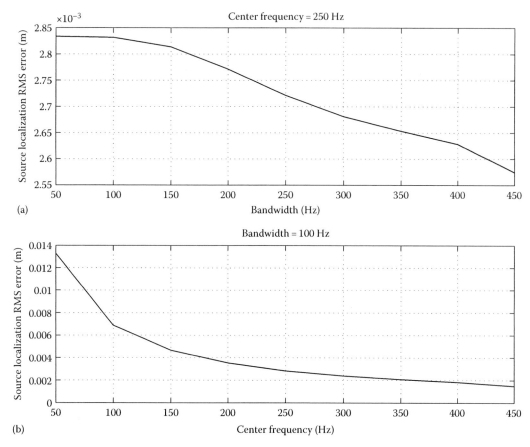

FIGURE 15.13 CRB comparison for different signal characteristics: (a) bandwidth; (b) center frequency.

more significant degrading effect than the unknown speed of propagation in source location estimation. However, these parameters are not significant in the DOA estimations. Furthermore, the theoretical performance of the estimator is analyzed for different signal characteristics. For the same setting as the one shown in Figure 15.7, we replace the tracked vehicle signal by a Gaussian signal after a band-pass filter (with fixed SNR of 20 dB). The center frequency of the source signal is first set to 250 Hz and then we let the bandwidth of this signal vary from 50 to 450 Hz. As shown in Figure 15.13a, the theoretical performance improves as the bandwidth increases. Then, for a fixed bandwidth of 100 Hz, we vary the center frequency from 50 to 450 Hz and observe the performance improvement for higher center frequency in Figure 15.13b. The theoretical performance analysis applies to the AML estimator, since the AML estimator approaches the CRB asymptotically.

15.3.3 Robust Array Design

For a coherent array beamforming operation, it is well known that the system needs to be well calibrated and that many realistic system perturbations may cause the array performance to degrade significantly [34]. Various robust array designs have been proposed [35,36]. Recently, a new class of robust array designs [37–39] based on the use of semi-definite programming (SDP) optimization methodology [40,41] and efficient MATLAB programming code [42] has been proposed.

From calculus, it is clear that, for a polynomial function of degree 2, their robust array designs have either a minimum or a maximum, and the local minimum or maximum is the global minimum or maximum, since the function is convex. Certainly, even for an arbitrary function of a single or vector variable, a local extremum solution does not guarantee a global extremum solution. However, if the optimizing function is convex and the domain is also convex, then a local extremum solution guarantees a global extremum solution. An SDP is a special class of tractable convex optimization problem in which every stationary point is also a global optimizing point, the global solution can be computed in small number of iterations, necessary and sufficient conditions for optimality can be obtained readily, and a well-developed duality theory exists that can determine whether a feasible solution exists (for the given constraints).

The general SDP problem [38] can be defined as follows. A linear matrix inequality for the vector variable $x = [x_1, ..., x_m] \in R^m$ has the form of

$$F(x) = F_0 + \sum_{i=1}^{m} x_i F_i \geq 0 \tag{15.79}$$

where $F_i = F_i^T \in R^{n \times n}, i = 0, ..., m$, are some specified matrices. $F(x)$ is an affine function of x and the feasible set is convex. Then, the general SDP problem is the convex optimization problem of

$$\text{Min}\{c^T x\}, \quad \text{subject to } F(x) \geq 0 \tag{15.80}$$

for any vector $c \in R^n$. A special case of the SDP problem called the *Second Order Cone Problem*, (SOCP) has the form of

$$\text{Min}\{c^T x\}, \quad \text{subject to } ||A_i x + b_i|| \leq c_i^T + d_i, \quad i = 1, ..., N \tag{15.81}$$

where
   $A_i \in C^n$
   $b_i \in C^N$
   $D_i \in R$

Many array-processing design problems can be formulated and solved as a second-order cone program [38].

We show a simple robust array design problem [43] based on the formulation and solution as an SOCP (but omitting details). Consider a uniformly spaced circle array of 15 elements as denoted by the open circles in Figure 15.14. The array magnitude response at the desired angle of 0° is constrained to have unity value with all the sidelobe peaks to be equal or below the 0.1 value (i.e. 20 dB below the desired response). The resulting array magnitude response using standard beamforming weight design is shown in Figure 15.15a. This response meets the desired constraints. However, for the weights designed for the ideal circular array but applied to a slightly perturbed array as denoted by the black dots in Figure 15.14, the new array response is given by Figure 15.15b, which is probably not acceptable with the large sidelobe values near 0°. Now, using an SOCP array design, we note in Figure 15.16a, the new sidelobe values for the ideal circular array are indeed slightly worse than those in Figure 15.15a. But, the array response for the perturbed array shown in Figure 15.16b is only slightly worse than that for the ideal circular array response of Figure 15.16a, and is certainly much more robust than that in Figure 15.15b. This simple example shows the importance of robust array design to meeting practical system imperfections.

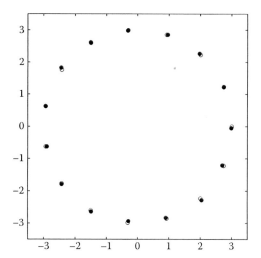

FIGURE 15.14 Placement of 15 uniformly spaced antenna elements. Ideal circular array (open circles); Perturbed array (black circles).

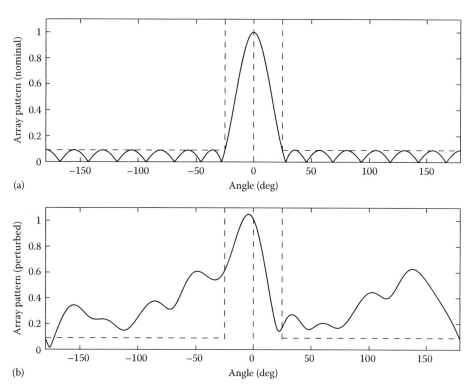

FIGURE 15.15 Nonrobust conventional array design. (a) Array response (ideal circular placements); (b) array response (perturbed placements).

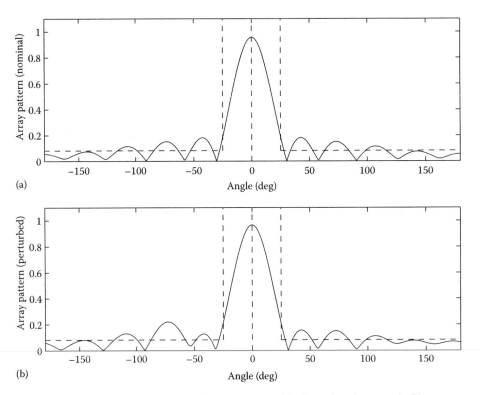

(a)

(b)

FIGURE 15.16 Robust SCOP array design: (a) array response (ideal circular placements); (b) array response (perturbed placements).

15.4 Implementations of Two Wideband Beamforming Systems

15.4.1 Implementation of a Radar Wideband Beamformer Using a Subband Approach

Modern wideband radar space–time adaptive processing problems may need to consider hundreds of sensor channels to form dozens of beams at high sampling rate [44–46]. Given the wideband nature of the radar waveform, subband domain approach is a known attractive method to decompose the wideband signals into sets of narrowband signals. The signals in each subband can then be decimated and processed at lower sampling rate using an analysis filter and reconstructed in a synthesis filer in conjunction with efficient polyphase filtering [47]. As an example, consider two tone signals of length 4096 at the input of a wideband beamformer. The desired signal has unit amplitude and a frequency of $f_1 = 1300\,\text{Hz}$ at a DOA angle of 30°, while the interference has an amplitude of 2 with a frequency of $f_2 = 2002\,\text{Hz}$ at a DOA angle of 50°. The sampling rate is take at $f_s = 10,000\,\text{Hz}$. The number of channels, FFT size, the decimation, and interpolation factors are all taken to be 8 and a polyphase FIR filter of length 120 is used. Figure 15.17a shows a section of the desired signal in the time domain, Figure 15.17b shows the magnitude of the two signals at the input to the wideband beamformer, Figure 15.17c shows a section of the time-domain beamformer output with a slightly distorted desired signal, and Figure 15.17d shows the magnitude of the beamformer output with the interference signal greatly suppressed with respect to the desired signal. In a practical system [44,45] the array weights can be implemented using a field programmable gate array architecture and various adaptive schemes need to be used to update the weights.

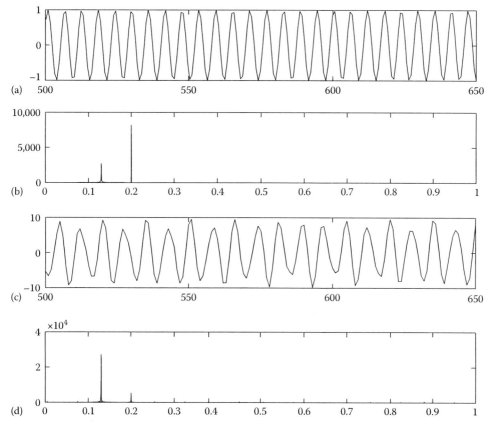

FIGURE 15.17 Wideband beamformer with numbers of channels, FFT size, decimation and interpolation factors set to 8: (a) section of desired signal in time domain; (b) magnitude of beamformer input; (c) section of beamformer output in time domain; (d) magnitude of beamformer output.

15.4.2 iPAQS Implementation of an Acoustic Wideband Beamformer

In this section we consider the implementation of a Linux-based wireless networked acoustic sensor array testbed, utilizing commercially available iPAQs with built-in microphones, codecs, and microprocessors, plus wireless Ethernet cards, to perform acoustic source localization. Owing to space limitation, we only consider the cases where several subarrays are available to obtain independent DOA estimation of the same source(s), and the bearing crossings from the subarrays are used to obtain the location estimate. Direct localization for sources in the near field is also possible and has been demonstrated successfully [14]. We consider the AML and the time-delay-based CLS DOA estimator [30], where the CLS method uses the LS solution with a constraint to improve the accuracy of estimation. In the case of multiple sources, only the AML method can perform estimation and the alternating projection procedure is applied.

The first experimental setting is depicted in Figure 15.18, where three linear subarrays, each with three iPAQs, form the sensor network. In this far-field case (relative to each subarray), the DOA of the source is independently estimated in each subarray and the bearing crossing is used to obtain the location estimate. The speaker is placed at four distinct source locations S1, …, S6, simulating source movement, and the same vehicle sound is played each time. Figure 15.19 depicts one snapshot (for clear illustration) of the AML and CLS results at six distinct source locations. We note that better results are clearly obtained when the source is inside the convex hull of the overall array. The second experimental setting is depicted in Figure 15.20, where four square subarrays each with four iPAQs form a

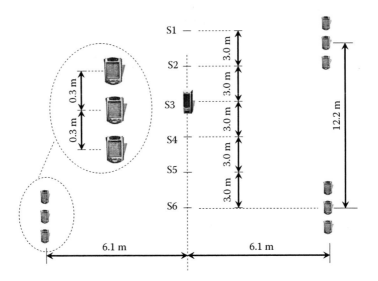

FIGURE 15.18   Linear subarray configuration.

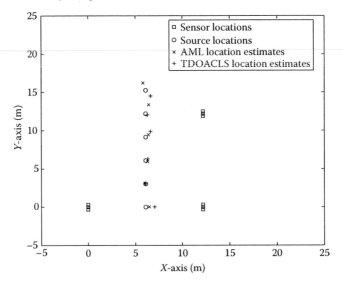

FIGURE 15.19   Cross-bearing localization of one source at different locations.

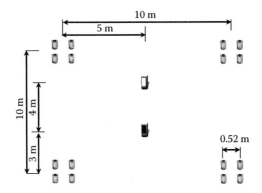

FIGURE 15.20   Square subarray configuration.

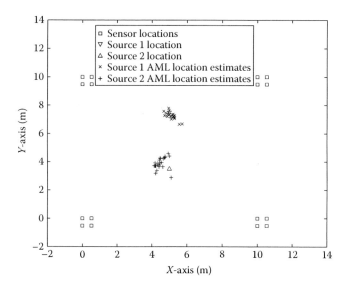

FIGURE 15.21 AML cross-bearing localization of two sources using alternating projection.

single network. Two speakers, one playing the vehicle sound and the other one playing the music sound simultaneously, are placed inside the convex hull of the overall array. When both sources are playing simultaneously, as shown in Figure 15.21, the AML method is able to estimate both source locations simultaneously with alternating projection. Note that, when the number of subarray element increases, the localization accuracy of the results reported above improves, which agrees with the CRB analysis reported by Stoica and Nehorai [48] and in the previous section.

References

1. Friis, H.T. and Feldman, C.B.A., A multiple unit steerable antenna for shear-wave reception, *Proceedings of the IRE*, 25, 841, 1937.
2. Fourikis, N., *Advanced Array Systems, Applications, and RF Technologies*, Academic Press, London, U.K., 2000.
3. Agre, J. and Clare, L., An integrated architecture for cooperative sensing and networks, *Computer*, 33, 106, 2000.
4. Pottie, G.J. and Kaiser, W.J., Wireless integrated network sensors, *Communications of the ACM*, 43, 51, 2000.
5. Kumar, S. et al. (eds.), Special issue on Collaborative signal and information processing in microsensor networks, *IEEE Signal Processing Magazine*, 19, 13, 2002.
6. Iyengar, S.S. and Kumar, S. (eds.), Special issue on Advances in information technology for high performance and computational intensive distributed sensor networks, *International Journal of High Performance Computing Applications*, 16, 203, 2002.
7. Yao, K. et al. (eds.), Special issue on sensor networks, *European Journal on Applied Signal Processing*, 2003(4), 319, 2003.
8. Brandstein, M.S. and Silverman, H.F., A robust method for speech signal time-delay estimation in reverberant rooms, in *Proceedings of IEEE ICASSP*, Vol. 1, 1997, Munich, Germany, p. 375.
9. Yao, K. et al., Blind beamforming on a randomly distributed sensor array system, *IEEE Journal on Selected Areas in Communication*, 16(8), 1555, 1998.
10. Ward, D.B. and Williamson, R.C., Particle filter beamforming for acoustic source localization in a reverberant environment, in *Proceedings of IEEE ICASSP*, Vol. 2, May 2002, Orlando, FL, p. 1777.

11. Aarabi, P., The fusion of distributed microphone arrays for sound localization, *European Journal on Applied Signal Processing*, 2003(4), 338, 2003.
12. Zhao, F. et al., Information-driven dynamic sensor collaboration, *IEEE Signal Processing Magazine*, 19, 61, 2002.
13. Chen, J.C. et al., Source localization and beamforming, *IEEE Signal Processing Magazine*, 19, 30, 2002.
14. Chen, J.C. et al., Coherent acoustic array processing and localization on wireless sensor networks, *Proceedings of the IEEE*, 91, 1154, 2003.
15. Capon, J., High-resolution frequency–wavenumber spectrum analysis, *IEEE Proceedings*, 57, 1408, 1969.
16. Schmidt, R.O., Multiple emitter location and signal parameter estimation, *IEEE Transactions on Antennas and Propagation*, AP-34(3), 276, 1986.
17. Carter, G.C. (ed.), *Coherence and Time Delay Estimation: An Applied Tutorial for Research, Development, Test, and Evaluation Engineers*, IEEE Press, New York, 1993.
18. Emile, B. et al., Estimation of time-delays with fewer sensors than sources, *IEEE Transactions on Signal Processing*, 46(7), 2012, 1998.
19. Chen, J.C. et al., Maximum-likelihood source localization and unknown sensor location estimation for wideband signals in the near-field, *IEEE Transactions on Signal Processing*, 50(8), 1843, 2002.
20. Krim, J. and Viberg, M., Two decades of array signal processing research: The parametric approach, *IEEE Signal Processing Magazine*, 13, 67, 1996.
21. Jaffer, A.G., Maximum likelihood direction finding of stochastic sources: A separable solution, in *Proceedings of IEEE ICASSP*, Vol. 5, 1988, New York, p. 2893.
22. Chung, P.J. and Böhme, J.F., Comparative convergence analysis of EM and SAGE algorithms in DOA estimation, *IEEE Transactions on Signal Processing*, 49(12), 2940, 2001.
23. Wang, H. and Kaveh, M., Coherent signal-subspace processing for the detection and estimation of angles of arrival of multiple wideband sources, *IEEE Transactions on Acoustics, Speech and Signal Processing*, ASSP-33, 823, 1985.
24. Schau, H.C. and Robinson, A.Z., Passive source localization employing intersecting spherical surfaces from time-of-arrival differences, *IEEE Transactions on Acoustics, Speech, and Signal Processing*, ASSP-35(8), 1223, 1987.
25. Smith, J.O. and Abel, J.S., Closed-form least-squares source location estimation from range-difference measurements, *IEEE Transactions on Acoustics, Speech, and Signal Processing*, ASSP-35(12), 1661, 1987.
26. Chan, Y.T. and Ho, K.C., A simple and efficient estimator for hyperbolic location, *IEEE Transactions on Signal Processing*, 42(8), 1905, 1994.
27. Brandstein, M.S. et al., A closed-form location estimator for use with room environment microphone arrays, *IEEE Transactions on Speech and Audio Processing*, 5(1), 45, 1997.
28. Chen, J.C. et al., Source localization and tracking of a wideband source using a randomly distributed beamforming sensor array, *International Journal of High Performance Computing Applications*, 16(3), 259, 2002.
29. Huang, Y. et al., Passive acoustic source localization for video camera steering, in *Proceedings of IEEE ICASSP*, Vol. 2, 2000, Istanbul, Turkey, p. 909.
30. Wang, H. et al., A wireless time-synchronized COTS sensor platform part II: Applications to beamforming, in *Proceedings of IEEE CAS Workshop on Wireless Communications and Networking*, September 2002, Pasadena, CA.
31. Brandstein, M.S. and Ward, D.B., *Microphone Arrays: Techniques and Applications*, Springer-Verlag, New York, 2001.
32. Tung, T.L. et al., Source localization and spatial filtering using wideband MUSIC and maximum power beamforming for multimedia applications, in *Proceedings of IEEE SiPS*, October 1999, Taipei, Taiwan, p. 625.
33. Kay, S.M., *Fundamentals of Statistical Signal Processing: Estimation Theory*, Prentice-Hall, Englewood Cliffs, NJ, 1993.

34. Weiss, A.J. et al., Analysis of signal estimation using uncalibrated arrays, in *Proceedings of IEEE ICASSP*, Vol. 3, 1995, Detroit, MI, p. 1888.
35. Cox, H. et al., Robust adaptive beamforming, *IEEE Transactions Signal Processing*, 35, 1335, 1987.
36. Er, M.H. and Cantoni, T., An alternative formulation for an optimum beamformer with robust capacity, *Proceedings of IEE Radar, Sonar, and Navigation*, 32, 447, 1985.
37. Lebret, S. and Boyd, S., Antenna array pattern synthesis via convex optimization, *IEEE Transactions on Signal Processing*, 45, 5256, 1997.
38. Wang, F. et al., Optimal array synthesis using semidefinite programming, *IEEE Transactions on Signal Processing*, 51, 1172, 2003.
39. Vorobyov, S.A. et al., Robust adaptive beamforming using worst-case performance optimization: A solution to the signal mismatch problem, *IEEE Transactions on Signal Processing*, 51, 313, 2003.
40. Vandenberghe, L. and Boyd, S., Semidefinite programming, *SIAM Review*, 38, 49095, 1996.
41. Boyd, S. and Vandenberghe, L., Convex optimization, Class lecture notes, 2003.
42. Sturm, J.F., CeDuMi: A Matlab toolbox for optimization over symmetric cones, http:fewcal.kub.nl/sturm/software/sedumi.html, 2001 (last accessed in February 2004).
43. Vandenberghe, L., Personal communication, September 2003.
44. Moeller, T.J., Field programmable gate arrays for radar front-end digital signal processing, MS dissertation, Massachusetts Institute of Technology, Cambridge, MA, 1999.
45. Martinez, D.R. et al., Application of reconfigurable computing to a high performance front-end radar signal processor, *Journal of VLSI Signal Processing*, 28(1–2), 65, 2001.
46. Rabankin, D.V. and Pulsone, N.B., Subband-domain signal processing for radar array systems, *Proceedings of SPIE*, 3807, 174, 1999.
47. Mitra, S.K., *Digital Signal Processing: A Computer-Based Approach*, 2nd edn., McGraw-Hill, New York, 2001.
48. Stoica, P. and Nehorai, A., MUSIC, maximum likelihood, and Cramer–Rao bound, *IEEE Transactions on Acoustics, Speech, and Signal Processing*, 37, 720, 1989.

III

Information Fusion

16 Foundations of Data Fusion for Automation *S. Sitharama Iyengar, Shivakumar Sastry, and N. Balakrishnan* 377
Introduction • Automation Systems • Data Fusion Foundations • Security Management for Discrete Automation • Conclusions • Acknowledgments • References

17 Measurement-Based Statistical Fusion Methods for Distributed Sensor Networks *Nageswara S.V. Rao* 387
Introduction • Classical Fusion Problems • Generic Sensor Fusion Problem • Empirical Risk Minimization • Statistical Estimators • Applications • Performance of Fused System • Metafusers • Conclusions • Acknowledgment • References

18 Soft Computing Techniques *Richard R. Brooks* 415
Problem Statement • Genetic Algorithms • Simulated Annealing • TRUST • Tabu Search • Artificial Neural Networks • Fuzzy Logic • Linear Programming • Summary • References

19 Estimation and Kalman Filters *David L. Hall* 429
Introduction • Overview of Estimation Techniques • Batch Estimation • Sequential Estimation and Kalman Filtering • Sequential Processing Implementation Issues • References

20 Data Registration *Richard R. Brooks, Jacob Lamb, Lynne L. Grewe, and Juan Deng* 455
Problem Statement • Coordinate Transformations • Survey of Registration Techniques • Objective Functions • Results from Meta-Heuristic Approaches • Feature Selection • Real-Time Registration of Video Streams with Different Geometries • Beyond Registration • Closely Mounted Sensors and Temporal Changes • Ensemble Registration • Non-Rigid Body Registration • Summary • References

21 Signal Calibration, Estimation for Real-Time Monitoring and Control *Asok Ray and Shashi Phoha* 495
Introduction • Signal Calibration and Measurement Estimation • Sensor Calibration in a Commercial-Scale Fossil Power Plant • Summary and Conclusions • Appendix • Acknowledgment • References

22 Semantic Information Extraction *David S. Friedlander*..................................513
Introduction • Symbolic Dynamics • Formal Language Measures • Behavior
Recognition • Experimental Verification • Conclusions and Future
Work • Acknowledgments and Disclaimer • References

23 Fusion in the Context of Information Theory *Mohiuddin Ahmed
and Gregory Pottie*..523
Information Processing in Distributed Networks • Evolution toward Information
Theoretic Methods for Data Fusion • Probabilistic Framework for Distributed
Processing • Bayesian Framework for Distributed Multi-Sensor Systems • Concluding
Remarks • References

24 Multispectral Sensing *N.K. Bose*...541
Motivation • Introduction to Multispectral Sensing • Mathematical Model for
Multisensor Array Based Superresolution • Color Images • Conclusions • References

**25 Chebyshev's Inequality-Based Multisensor Data Fusion in Self-Organizing
Sensor Networks** *Mengxia Zhu, Richard R. Brooks, Song Ding, Qishi Wu,
Nageswara S.V. Rao, and S. Sitharama Iyengar* 553
Introduction • Problem Formulation • Local Threshold Adjustment
Method • Threshold-OR Fusion Method • Simulation Results • Conclusions and Future
Work • References

26 Markov Model Inferencing in Distributed Systems *Chen Lu, Jason M. Schwier,
Richard R. Brooks, Christopher Griffin, and Satish Bukkapatnam* 569
Introduction • Hidden Markov Models • Inferring
HMMs • Applications • Conclusion • Acknowledgments • References

27 Emergence of Human-Centric Information Fusion *David L. Hall* 581
Introduction • Participatory Observing: Humans as Sensors • Hybrid Human/Computer
Analysis: Humans as Analysts • Collaborative Decision Making: Crowd-Sourcing of
Analysis • Summary • Acknowledgments • References

Once signals and images have been locally processed, additional work is performed to make global decisions from the local information. This part considers issues concerning information and data fusion. Fusion can occur at many different levels and in many different ways. The chapters in this part give an overview of the most important technologies.

Iyengar et al. provide a conceptual framework for data fusion systems in Chapter 16. This approach is built upon two primary concepts:

1. A model describes the conceptual framework of the system. This is the structure of the global system.
2. A goal-seeking paradigm is used to guide the system in combining information.

Rao's revision for the second edition (Chapter 17) discusses the statistical concepts that underlie information fusion. Data are retrieved from noise-corrupted signals and features inferred. The task is to extract information from sets of features that follow unknown statistical distributions. Rao provides a statistical discussion that unifies neural networks, vector space, and Nadaraya–Watson methods.

Brooks reviews soft computing methodologies that have been applied to information fusion in Chapter 18. The methods discussed include the following families of meta-heuristics: genetic algorithms, simulated annealing, tabu search, artificial neural networks, TRUST, fuzzy logic, and linear programming. From this viewpoint, information fusion is phrased as an optimization problem. A solution is sought that minimizes a given objective function. Among other things, this function could be the amount of ambiguity in the system.

Hall provides an overview of estimation and Kalman filters in Chapter 19. This approach uses control theoretical techniques. A system model is derived and an optimization approach is used to fit the data to the model. These approaches can provide optimal solutions to a large class of data fusion problems. The recursive nature of the Kalman filter algorithm has made it attractive for many real-time applications.

Brooks et al. tackle the problem of data registration in Chapter 20. Readings from different sources must be mapped to a common coordinate system. This chapter has been revised for the second edition. Registration is a difficult problem, which is highly dependent on sensor geometry. The chapter provides extensive mathematical background and a survey of the best-known techniques. An example application is given using soft computing techniques. Other problems addressed include selecting the proper features for registering images and registering images from sensors with different geometries.

Ray and Phoha provide a distributed sensor calibration approach in Chapter 21. A set of sensors monitors an ongoing process. The sensor hardware will degrade over time, causing the readings to drift from the correct value. By comparing the readings over time and calculating the variance of the agreement, it is possible to assign trust values to the sensors. These values are then used to estimate the correct reading, allowing the sensors to be recalibrated online.

Friedlander uses an innovative approach to extract symbolic data from streams of sensor data in Chapter 22. The symbols are then used to derive automata that describe the underlying process. In this chapter, he uses the derived automata to recognize complex behaviors of targets under surveillance. Of particular interest is the fact that the abstraction process can be used to combine data of many different modes.

Ahmed and Pottie use information theory to analyze the information fusion problem in Chapter 23. Distributed estimation and signal detection applications are considered. A Bayesian approach is provided and justified using information measures.

Bose considers the use of multispectral sensors in Chapter 24. This class of sensors simultaneously collects image data using many different wavelengths. He describes the hardware used in multispectral sensing and how it is possible to achieve subpixel accuracy from the data.

Zhu et al.'s Chapter 25 is a new chapter for the second edition that uses the Chebyshev inequality to determine how to best fuse distributed sensor readings. This approach is used to simultaneously increase the target detection rates while lowering the number of false alarms.

Schwier et al.'s Chapter 26 is another new chapter for the second edition. This chapter explains zero-knowledge techniques for inferring Markov models that can be used for detecting complex behaviors in distributed sensing environments. This approach is related to the symbolic dynamics techniques mentioned in earlier chapters. These tools can be used for either monitoring sensor feeds or network traffic patterns.

Hall provides a new chapter (Chapter 27) that discusses issues with integrating humans into the sensor fusion process. This is an important issue, since, in almost all applications, the human is the ultimate decision maker. Fusion needs to be done in the manner most appropriate for exploitation in this context.

This part provides a broad overview of information fusion technology. The problem is viewed from many different perspectives. Many different data modalities are considered, as are the most common applications of information fusion.

16

Foundations of Data Fusion for Automation*

16.1 Introduction 377
16.2 Automation Systems...378
    System Characteristics • Operational Problems • Benefits of Data Fusion
16.3 Data Fusion Foundations 381
    Representing System Structure • Representing System Behavior
16.4 Security Management for Discrete Automation..........................383
    Goal-Seeking Formulation • Applying Data Fusion
16.5 Conclusions..385
Acknowledgments..385
References..385

S. Sitharama Iyengar
Florida International University

Shivakumar Sastry
The University of Akron

N. Balakrishnan
Carnegie Mellon University
Indian Institute of Science

16.1 Introduction

Data fusion is a paradigm for integrating data from multiple sources to synthesize new information such that the whole is greater than the sum of its parts. This is a critical task in contemporary and future systems that are distributed networks of low-cost, resource-constrained sensors [1,2]. Current techniques for data fusion are based on general principles of distributed systems and rely on cohesive data representations to integrate multiple sources of data. Such methods do not extend easily to systems in which real-time data must be gathered periodically, by cooperative sensors, where some decisions become more critical than other decisions episodically.

There has been an extensive study in the areas of multi-sensor fusion and real-time sensor integration for time-critical sensor readings [3]. A distributed sensor data network is a set of spatially scattered sensors designed to derive appropriate inferences from the information gathered. The development of such networks for information gathering in unstructured environments is receiving a lot of interest, partly because of the availability of new sensor technology that is economically feasible to implement [4]. Sensor data networks represent a class of distributed systems that are used for sensing and in situ processing of spatially and temporally dense data from limited resources and harsh environments, by routing and cooperatively processing the information gathered. In all these systems, the critical step is the fusion of data gathered by sensors to synthesize new information.

Our interest is to develop data fusion paradigms for sensor–actuator networks that perform engineering tasks and we use Automation Systems as an illustrative example. Automation systems represent an important, highly engineered, domain that has over a trillion dollars of installed base in the United States. The real-time and distributed nature of these systems, with the attendant demands for safety,

---

* First published in *IEEE Instrumentation and Measurement Magazine*, 6(4), 35–41, 2003, and used with permission.

determinism, and predictability, represent significant challenges, and hence these systems are a good example. An Automation system is a collection of devices, equipment, and networks that regulate operations in a variety of manufacturing, material and people moving, monitoring, and safety applications.

Automation systems evolved from early centralized systems to large distributed systems that are difficult to design, operate, and maintain [5]. Current hierarchical architectures, the nature and use of human–computer-interaction (HCI) devices, and the current methods for addressing and configuration increase system life-cycle costs. Current methods to integrate system-wide data are hardcoded into control programs and not based on an integrating framework. Legacy architectures of existing automation systems are unable to support future trends in distributed automation systems [6]. Current methods for data fusion are also unlikely to extend to future systems because of system scale and simplicity of nodes.

We present a new integrating framework for data fusion that is based on two systems concepts: a conceptual framework and the goal-seeking paradigm [7]. The conceptual framework represents the structure of the system and the goal-seeking paradigm represents the behavior of the system. Such a systematic approach to data fusion is essential for proper functioning of future sensor–actuator networks [8] and SmartSpace [9]. In the short term, such techniques help to infuse emerging paradigms in to existing automation architectures. We must bring together knowledge in the fields of sensor fusion, data and query processing, automation systems design, and communication networks to develop the foundations. While extensive research is being conducted in these areas, as evidenced by the chapters compiled in this book, we hope that this special issue will open a window of opportunity for researchers in related areas to venture in to this emerging and important area of research [2].

16.2 Automation Systems

An automation system is a unique distributed real-time system that comprises a collection of sensors, actuators, controllers, communication networks, and user-interface devices. Such systems regulate the coordinated operation of physical machines and humans to perform periodic and precise tasks that may sometimes be dangerous for humans to perform. Examples of automation systems are: a factory manufacturing cars, a baggage handling system in an airport, and an amusement park ride. *Part*, *process*, and *plant* are three entities of interest. A plant comprises a collection of stations, mechanical fixtures, energy resources, and control equipment that regulate operations using a combination of mechanical, pneumatic, hydraulic, electric, and electronic components or subsystems. A process specifies a sequence of stations that a part must traverse through and operations that must be performed on the part at each station.

Figure 16.1 shows the major aspects of an automation system. All five aspects, namely input, output, logic processing, behavior specification, and HCI must be designed, implemented, and commissioned to operate an automation system successfully. Sensors and actuators are transducers that are used to acquire inputs and set outputs respectively. The controller periodically executes logic to determine new output values for actuators. HCI devices are used to specify logic and facilitate operator interaction at runtime.

The architecture of existing automation systems is hierarchical and the communication infrastructure is based on proprietary technologies that do not scale well. Ethernet is emerging as the principal control and data-exchange network. The transition from rigid, proprietary networks to flexible, open networks introduces new problems into the automation systems domain and security is a critical problem that demands attention.

16.2.1 System Characteristics

Automation systems operate in different modes. For example, k-hour-run is a mode that is used to exercise the system without affecting any parts. Other examples of modes are automatic, manual, and semiautomatic. In all modes, the overriding concern is to achieve deterministic, reliable, and safe operations.

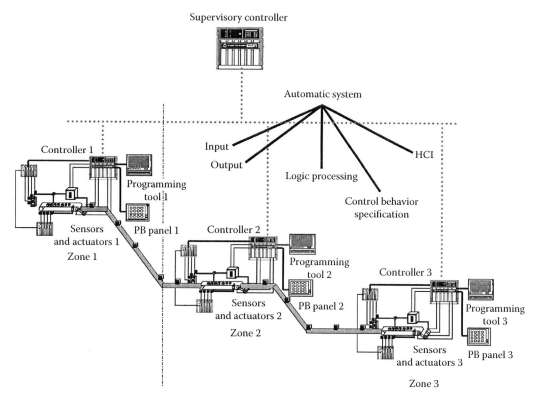

FIGURE 16.1   Major aspects of an automation system.

The mode of the system dictates the degree to which humans interact with the system. Safety checks performed in each of these modes usually decrease inversely with the degree of user interaction.

Communication traffic changes with operating mode. For example, when the system is in automatic mode, limited amounts of data (a few bytes) are exchanged; such data flows occur in localized areas of the system. In this mode, small disruptions and delays in message delivery could be tolerated. However, when there is a disruption (in part, process, or plant), large bursts of data will be exchanged between a large number of nodes across the system. Under such conditions, disruptions and delays significantly impair system capabilities.

Because of both the large capital investments required and the changing market place, these systems are designed to be flexible with respect to part, process, and plant. Automation systems operate in a periodic manner. The lack of tools to evaluate performance makes it difficult to evaluate the system's performance and vulnerability. These systems are highly engineered and well documented in the design and implementation stages. Demands for reconfigurable architectures and new properties, such as self-organization, require a migration away from current hierarchical structures to loosely coupled networks of devices and subsystems. Control behavior is specified using a special graphical language called *Ladder*, and typical systems offer support for both online and offline program editing.

16.2.2 Operational Problems

Hierarchical architecture and demands for backward compatibility create a plethora of addressing and configuration problems. Cumbersome, expensive, implementations and high commissioning costs are a consequence of such configuration problems. Figure 16.2 shows a typical configuration of input and output (IO) points connected to controllers.

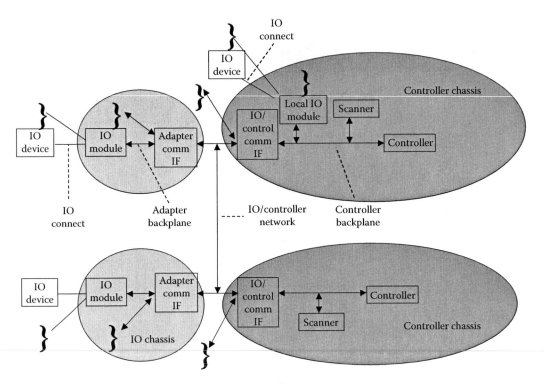

FIGURE 16.2   Connecting input–output points to controllers.

Unique addresses must be assigned to every IO point and single structural component in Figure 16.2. These addresses are typically related to the jumper settings on the device or chassis; thus, a large number addresses are related and structured by manual naming conventions. In addition, depending on the settings on the device or chassis, several items in software packages must also be configured manually. Naturally, such a landscape of addresses leads to configuration problems. State-of-the-art implementations permit specification of a global set of tags and maintaining multiple links behind the scenes—while this simplifies the user's chores, the underlying problems remain. The current methods of addressing and configuration will not extend to future systems that are characterized by large-scale, complex interaction patterns, and emergent behaviors.

Current methods for commissioning and fault recovery are guided by experience and based on trial-and-error. User-interface devices display localized, controller-centric data and do not support holistic, system-wide decision-making. Predictive nondeterministic models do not accurately represent system dynamics, and hence approaches based on such models have met with limited success. The state-of-practice is a template-based approach that is encoded into Ladder programs. These templates recognize a few commonly occurring errors.

Despite these operational problems, automation systems are a benchmark for safe, predictable, and maintainable systems. HCI devices are robust and reliable. Methods for safe interaction, such as the use of dual palm switches, operator-level de-bouncing, safety mats, etc., are important elements of any safe system. Mechanisms that are used to monitor, track, and study trends in automation systems are good models for such tasks in general distributed systems.

16.2.3 Benefits of Data Fusion

Data fusion can alleviate current operational problems and support development of new architectures that preserve the system characteristics. For example, data fusion techniques based on the foundations discussed in this chapter can provide a more systematic approach to commissioning, fault management (detection, isolation, reporting, and recovery), programming, and security management. Data fusion techniques can be embodied in distributed services that are appropriately located in the system, and such services support a SmartSpace in decision making [9].

16.3 Data Fusion Foundations

The principle issue for data fusion is to manage uncertainty. We accomplish this task through a goal-seeking paradigm. The application of the goal-seeking paradigm in the context of a multi-level system, such as an automation system, is simplified by a conceptual framework.

16.3.1 Representing System Structure

A conceptual framework is an experience-based stratification of the system, as shown in Figure 16.3. We see a natural organization of the system into levels of a node, station, line, and plant. At each level, the dominant considerations are depicted on the right. There are certain crosscutting abstractions that do not fit well into such a hierarchical organization. For example, the neighborhood of a node admits nodes that are not necessarily in the same station and are accessible. For example, there may be a communications node that has low load that could perform data fusion tasks for a duration when another node in the neighborhood is managing a disruptive fault. Similarly, energy resources in the environment affect all levels of the system.

A conceptual framework goes beyond a simple layering approach or a hierarchical organization. The strata of a conceptual framework are not necessarily organized in a hierarchy. The strata do not provide a *service abstraction* to other strata, like a layer in a software system. Instead, each stratum imposes performance requirements for other strata that it is related to. At runtime, each stratum is responsible for monitoring its own performance based on sensed data. As long as the monitored performance is within tolerance limits specified in the goal-seeking paradigm, the system continues to perform as expected.

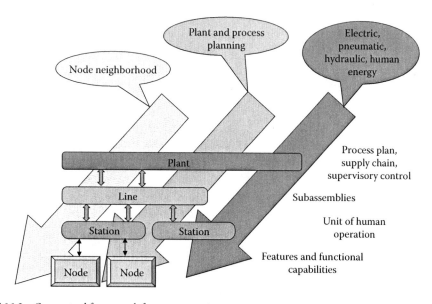

FIGURE 16.3 Conceptual framework for an automation system.

16.3.2 Representing System Behavior

We represent the behavior of the system using the goal-seeking paradigm. We briefly review the funda-mental state transition paradigm and certain problems associated with this paradigm before discussing the goal-seeking paradigm.

16.3.2.1 State Transition Paradigm

The state-transition paradigm is an approach to modeling and describing systems that is based on two key assumptions: first, that states of a system are precisely describable; and second, that the dynamics of the system are also fully describable in terms of states, transitions, inputs that initiate transitions, and outputs that are produced in states.

A state-transition function $S_1: Z_{t1} \otimes X_{t1,t2} \rightarrow Z_{t2}$ defines the behavior of the system by mapping inputs to a new state. For each state, an output function $S_2: Z_{ti} \rightarrow \psi_{ti}$, where $X_{t1,t2}$ is the set of inputs presented to the system in the time interval between $t1$ and $t2$; ψ_i is the set of outputs produced at time ti; Z_{t1} and Z_{t2} are states of the automation system at times $t1$ and $t2$, respectively; the symbol $\otimes$ the Cartesian product, indicates that the variables before the arrow (i.e. inputs $X_{t1,t2}$ and Z_{t1}) are causes for change in the vari-ables after the arrow (i.e. outputs Z_{t2}).

In order to understand such a system, one needs to have complete data on Z_{t1} and $X_{t1,t2}$ and knowl-edge about S_1 and S_2. This paradigm assumes that only lack of data and knowledge prevent us from completely predicting the future behavior of a system. There is no room for uncertainty or indetermin-ism. Such a paradigm (sometimes called an IO or stimulus–response paradigm) can be useful in certain limited circumstances for representing the interaction between two systems; and it can be erroneous if it is overextended. There is no consistent or uniform definition of a state, and contextual information that is based on collections of states is ignored. In such circumstances, the specifications in a state-transition paradigm are limited by what is expressed and depend heavily on environmental influences that are received as inputs. While it appears that the state-transition paradigm is simple, natural, and easy to describe, such a formulation can be misleading, especially if the true nature of the system is goal seeking.

16.3.2.2 Goal-Seeking Paradigm

A goal-seeking paradigm is an approach to modeling and describing systems that explicitly supports uncertainty management by using additional artifacts and transformations discussed in the following paragraphs.

The system can choose actions from a range of *alternate actions*, Π, in response to events occurring or expected to occur. These actions represent a choice of decisions that can be made in response to a given or emerging situation.

There is a range of *uncertainties*, Δ, that impact on the success of selected decisions. Uncertainties arise from two sources: first, from an inability to anticipate inputs correctly, either from the automa-tion system or from users; and second, from an incomplete or inaccurate view of the outcome of a decision, even if the input is correctly anticipated. For example, an operator may switch the mode of the system from automatic to manual by mistake, because of malicious intent, or because of poor training. Assuming that the user made appropriate choices, the outcome of a decision can still be uncertain because a component or subsystem of the automation system may fail just before executing the decision.

The range of *consequences*, ψ, represent outcomes that result from an implementation of system deci-sions. Consequences are usually outputs that are produced by the system. Some of these outputs may be consumed by users to resolve uncertainties further, and other outputs may actuate devices in the automation system.

The system maintains a *reflection*, $\Xi: \Pi \times \Delta \rightarrow \psi$ which is its view of the environment. Suppose that the system makes a decision $\pi \in \Pi$, the system benefits from an understanding of what consequence,

$\pi \in \Pi$, ψ produces. The consequence ψ is presented as an output, either to humans within a SmartSpace or to the automation system. ψ does not obviously follow ψ as specified by S_2 because of uncertainties in the system.

The *evaluation set*, Λ, represents a *performance scale* that is used to compare the results of alternate actions. That is, suppose the system could make two decisions $\pi_1 \in \Pi$ or $\pi_2 \in \Pi$, and these decisions have consequences $\psi_1, \psi_2 \in \psi$ respectively; Λ helps to determine which of the two decisions is preferable.

An *evaluation mapping*, Ω: $\Psi \times \Pi \rightarrow \Lambda$ is used to compare outcomes of decisions using the performance scale. Λ is specified by taking into account the *extent* or *cost of the effort* associated with a decision, i.e. $\pi \in \Pi$. For any $\pi \in \Pi$ and $\pi \in \Pi$, Ω assigns a value $\lambda \in \Lambda$ and helps to determine the system's preference for a decision–consequence pair (π, ψ).

A *tolerance function*, Γ: $\Pi \times \Delta \rightarrow \Lambda$, indicates the degree of satisfaction with the outcome if a given uncertainty $\delta \in \Delta$ comes to pass. For example, if the conditions are full of certainty, then the best (i.e. optimal) decision can be identified. If, however, there are several events that are anticipated (i.e. $|\Delta| \gg 1$), then the performance of the system, as evaluated by Ω, can be allowed to deteriorate for some $\delta \in \Delta$, but this performance must stay within a tolerance limit that will ensure *survival* of the system.

Based on the above artifacts and transformations, the functioning of a system, in a goal-seeking paradigm, is defined as: *Find a decision $\pi \in \Pi$ so that the outcome is acceptable (e.g. within tolerance limits) for any possible occurrence of uncertainty $\delta \in \Delta$, i.e. $\Omega\ (\Xi(\pi, \delta), \pi) \leq \Gamma\ (\delta, \pi), \forall \delta \in \Delta$.*

16.4 Security Management for Discrete Automation

Security management in a discrete automation system is critical because of the current trends towards using open communication infrastructures. Automation systems present challenges to traditional distributed systems, to emerging sensor networks, and for the need to protect the high investments in contemporary assembly lines and factories. Formulating the security management task in the state-transition paradigm is a formidable task and perhaps may never be accomplished because of the scale and the uncertainties that are present. We demonstrate how the goal-seeking formulation helps and the specific data fusion tasks that facilitate security management.

A future automation system is likely to comprise large sensor–actuator networks as important subsystems [8]. Nodes in such a system are resource constrained. Determinism and low jitter are extremely important design considerations. Since a node typically contains a simple processor without any controller hierarchy or real-time operating system, it is infeasible to implement a fully secure communications channel for all links without compromising determinism and performance. Further, because of the large installed base of such systems, security mechanisms must be designed to mask the uneven conditioning of the environment [10]. The goal-seeking formulation presented here, and the cooperative data fusion tasks associated, supports such an implementation.

16.4.1 Goal-Seeking Formulation

This section details the artifacts and transformations necessary for security management.

16.4.1.1 Alternate Actions

The set of actions includes options for security mechanisms that are available to a future automation system. Asymmetric cryptography may be applied to either a specific link, a broadcast, a connection, or a session. Digital signatures or certificates may be required or suggested. Frequency hopping may be used to make it difficult for malicious intruders to masquerade or eavesdrop messages. Block ciphering, digest function, or μ-tesla can be applied. These possible actions comprise the set of alternate actions. At each time step, the automation system selects one or more of these alternate actions to maintain security of the system. Because the nodes are resource constrained, it is not possible to implement full encryption for each of the links. Thus, to make the system secure, one or more mechanisms must be selected in a systematic manner depending on the current conditions in the system.

16.4.1.2 Uncertainties

As already discussed in Section 16.3.2.2, there are two sources of uncertainty. First, when a specific user choice is expected as input (a subjective decision), the system cannot guess the choice. Second, component or subsystem malfunctions cannot be predicted. For example, an established channel, connection, packet, or session may be lost. A node receiving a request may be unable to respond to a query without compromising local determinism or jitter. The node, channel, or subsystem may be under a denial-of-service attack. There may be a malicious eavesdropper listening to the messages or someone may be masquerading production data to mislead management.

16.4.1.3 Consequences

It is not possible to predict either the occurrence or the time of occurrence of an uncertainty (if it occurs). However, the actions selected may not lead to the consequences intended if one or more uncertainties come to pass. Some example consequences are the communications channel is highly secure at the expected speeds, the channel may be secure at a lower speed, or the channel may be able to deliver data at desired speed without any security. The authentication supplied, digital signature or certificate, is either verified or not verified.

16.4.1.4 Reflection

This transformation maps every decision–uncertainty pair in to a consequence that is presented as an output. The reflection includes all possible choices, without any judgment about either the cost of effort or the feasibility of the consequence in the current system environment.

16.4.1.5 Evaluation Set

The three parameters of interest are security of channel, freshness of data, and authenticity of data. Assuming a binary range for each of these parameters, we get the following scale to evaluate consequences:

1. Highly secure channel with strongly fresh, truly authenticated data.
2. Highly secure channel with strongly fresh, weakly authenticated data.
3. Highly secure channel with weakly fresh, truly authenticated data.
4. Highly secure channel with weakly fresh, weakly authenticated data.
5. Weakly secure channel with strongly fresh, truly authenticated data.
6. Weakly secure channel with strongly fresh, weakly authenticated data.
7. Weakly secure channel with weakly fresh, truly authenticated data.
8. Weakly secure channel with weakly fresh, weakly authenticated data.
9. Total communication failure, no data sent.

16.4.1.6 Evaluation Mapping

Ideally, consequences follow from decisions directly. Because of uncertainties, the consequence obtained may be more desirable or less desirable, depending on the circumstances. The evaluation mapping produces such an assessment for every consequence–decision pair.

16.4.1.7 Tolerance Function

The tolerance function establishes a minimum performance level on the evaluation set that can be used to decide whether or not a decision is acceptable. In an automation system, the tolerance limit typically changes when the system conditions are different. For example, a highly secure channel with strongly fresh, truly authenticated data may be desirable when a user is trying to reprogram a controller, and a weakly secure channel with weakly fresh, weakly authenticated data may be adequate during commissioning phases. The tolerance function can be defined to include such considerations as operating mode and system conditions, with the view of evaluating a decision in the current context of a system.

16.4.2 Applying Data Fusion

Only a few uncertainties may come to pass when a particular alternate action is selected by the system. Hence, we first build a list of uncertainties that apply to every alternate action $\{\Delta_1, \Delta_2, ..., \Delta_{|\Pi|}\}$.

For an action π_k, if the size of the corresponding set of uncertainties $|\Delta_k| > 1$, then data fusion must be applied.

To manage security in an automation system, we need to work with two kinds of sensor. The first kind of sensor is ones that is used to support the automation system, such as to sense the presence of a part, completion of a traversal, or failure of an operation. In addition, there are sensors that help resolve uncertainties. Some of these redundant sensors could either be additional sensors that sense a different modality or another sensor whose value is used in another context to synthesize new information. Some uncertainties can only be inferred as the absence of certain values. For every uncertainty, the set of sensor values (or lack thereof) and inferences are recorded a priori. Sensors that must be queried for fresh data are also recorded a priori.

For each uncertainty, we also record the stratum in the conceptual framework that dominates the value of an uncertainty. For example, suppose there is an uncertainty regarding the mode of a station. The resolution of this uncertainty is based on whether there is a fault at the station or not. Suppose there is a fault at the station, then the mode at the station must dominate the mode at the line to which the station belongs. Similarly, when the fault has been cleared, the mode of the line must dominate, if the necessary safety conditions are met. Thus, the conceptual framework is useful for resolving uncertainties. The specific implementation of data fusion mechanisms will depend on the capabilities of the nodes. In a resource-constrained environment, we expect such techniques to be integrated with the communications protocols.

16.5 Conclusions

In this chapter we provide a new foundation for data fusion based on two concepts: a conceptual framework and the goal-seeking paradigm. The conceptual framework emphasizes the dominant structures in the system. The goal-seeking paradigm is a mechanism to represent system evolution that explicitly manages uncertainty. The goal-seeking formulation for data fusion helps to distinguish between subjective decisions that resolve uncertainty by involving humans and objective decisions that can be executed by computers. These notions are useful for critical tasks, such as security management in large-scale distributed systems. Investigations in this area, and further refinement of the goal-seeking formulation for instrumentation and measurement applications, are likely to lead to future systems that facilitate holistic user decision-making.

Acknowledgments

This work is supported in part by a University of Akron, College of Engineering Research Startup Grant, 2002–2004, to Dr. Sastry and an NSF Grant # IIS-0239914 to Professor Iyengar.

References

1. Brooks, R.R. and Iyengar, S.S., *Multi-Sensor Fusion: Fundamentals and Applications with Software*, Prentice Hall, Upper Saddle River, NJ, 1997.
2. Iyengar, S.S. et al., Distributed sensor networks for real-time systems with adaptive configuration, *Journal of the Franklin Institute*, 338, 571, 2001.

3. Kannan, R. et al., Sensor-centric quality of routing in sensor networks, in *Proceedings of IEEE Infocom*, San Francisco, CA, April 2003.
4. Akyildiz, I.F. et al., Wireless sensor networks: A survey, *Computer Networks*, 38, 393, 2002.
5. Agre, J. et al., A taxonomy for distributed real-time control systems, *Advances in Computers*, 49, 303, 1999.
6. Slansky, D., Collaborative discrete automation systems define the factory of the future, ARC Strategy Report, May 2003.
7. Mesarovic, M.D. and Takahara, Y., *Mathematical Theory of General Systems*, Academic Press, New York, 1974.
8. Sastry, S., and Iyengar, S.S., Sensor technologies for future automation systems, *Sensor Letters*, 2(1), 9–17, 2004.
9. Sastry, S., A SmartSpace for automation, *Assembly Automation*, 24(2), 201–209, 2004.
10. Sathyanarayanan, M., Pervasive computing: Vision and challenges, *Pervasive Computing*, 8(4), 10, August 2001.

17

Measurement-Based Statistical Fusion Methods for Distributed Sensor Networks

17.1 Introduction 387
17.2 Classical Fusion Problems...388
17.3 Generic Sensor Fusion Problem 389
     Related Formulations
17.4 Empirical Risk Minimization..392
     Feedforward Sigmoidal Networks • Vector Space Methods
17.5 Statistical Estimators..394
17.6 Applications...396
17.7 Performance of Fused System...398
     Isolation Fusers • Projective Fusers • Localization-Based Fusers
17.8 Metafusers..409
17.9 Conclusions..410
Acknowledgment...411
References...411

Nageswara S.V. Rao
Oak Ridge National
Laboratory

17.1 Introduction

In distributed sensor networks (DSNs), the fusion problems naturally arise when overlapping regions are covered by a set of sensor nodes. The sensor nodes typically consist of specialized sensor hardware and/or software, and consequently their outputs are related to the actual object features in a complicated manner, which is often modeled by probability distributions. While the fusion problems have been solved for centuries in various disciplines, such as political economy, the specific nature of fusion problems of DSNs require nonclassical approaches. Early information fusion methods required statistical independence of sensor errors, which greatly simplified the fuser design; for example, a weighted majority rule suffices in detection problems. Such solution is not applicable to DSNs since the sensors could be highly correlated while sensing common regions or objects, and thereby violate the statistical independence property. Another classical approach to fuser design relies on the Bayesian method that minimizes a suitable expected risk. A practical implementation of this method requires closed-form analytical expressions for sensor distributions to generate efficiently computable fusers. Several popular distributed decision fusion methods belong to this class [58]. In DSNs, the sensor distributions can be arbitrarily complicated. In addition, deriving closed form expressions for sensor distributions is a very difficult and expensive task since it requires the knowledge of a variety of areas such as device physics, electrical engineering, and statistical modeling. Furthermore, the problem of selecting a fuser from a carefully chosen function class is easier in an information-theoretic sense than inferring a completely unknown distribution [57].

In operational DSNs, it is quite practical to collect "data" by sensing objects and environments with known parameters. Thus, fusion methods that utilize empirical data available from the observation and/or experimentation will be of practical value. In this chapter, we present an overview of rigorous methods for fusion rule estimation from empirical data, based on empirical process theory and computational learning theory. Our main focus is on methods that provide *performance guarantees based on finite samples* from a statistical perspective. We do not cover ad hoc fusion rules with no performance bounds or results based on asymptotic guarantees valid *only* as the sample size approaches infinity. This approach is based on a statistical formulation of the fusion problem and may not fully capture the nonstatistical aspects such as calibration and registration. These results, however, provide an analytical justification for sample-based approaches to a very general formulation of the sensor fusion problem.

The organization of this chapter is as follows. We briefly describe the classical sensor fusion methods from a number of disciplines in Section 17.2. We present the formulation of a generic sensor fusion problem in Section 17.3. In Section 17.4, we present two solutions based on the empirical risk minimization methods using neural networks and vector space methods. In Section 17.5, we present solutions based on Nadaraya–Watson statistical estimator. We describe applications of these methods in Section 17.6. In Section 17.7, we address the issues of relative performance of the fused system compared to the component sensors. We briefly discuss metafusers that combine individual fusers in Section 17.8.

17.2 Classical Fusion Problems

Historically, the information fusion problems predate DSNs by a few centuries. Fusion of information from multiple sources to achieve performances exceeding those of individual sources has been recognized in diverse areas such as political economy models [13] and composite methods [8]; a brief overview of these works can be found in Ref. [20]. The fusion methods continued to be applied in a wide spectrum of areas such as reliability [60], forecasting [12], pattern recognition [4], neural networks [15], decision fusion [7,58], and statistical estimation [3].

If the sensor error distributions are known, several fusion rule estimation problems have been solved typically by methods that do not require samples. Earlier work in pattern recognition is due to Chow [4] who showed that a weighted majority fuser is optimal in combining outputs from pattern recognizers under statistical independence conditions. Furthermore, the weights of the majority fuser can be derived in closed form in terms of the individual detection probabilities of pattern recognizers. A simpler version of this problem has been studied extensively in political economy models (e.g., see Ref. [13] for an overview). Under the Condorcet jury model of 1786, the simple majority rule has been studied in combining the 1–0 probabilistic decisions of a group of N statistically independent members. If each member has probability p of making a correct decision, the probability that the majority makes the correct decision is $P_N = \sum_{i=N/2}^{N} \binom{N}{i} p^i (1-p)^{N-i}$. Then, we have an interesting dichotomy: (1) if $p > 0.5$, then $P_N > p$ and $P_N \to 1$ as $N \to \infty$ and (2) if $p < 0.5$, then $P_N < p$ and $P_N \to 0$ as $N \to \infty$. For the boundary case $p = 0.5$ we have $P_N = 0.5$. Interestingly, this result was rediscovered by von Neumann in 1959 in building reliable computing devices using unreliable components by taking a majority vote of duplicated components.

The distributed detection problem [58], studied extensively in the target tracking area, can be viewed as a generalization of the aforementioned two problems. The Boolean decisions from a system of detectors are combined by minimizing a suitably formulated Bayesian risk function. The risk function is derived from the densities of detectors, and the minimization is typically carried out using analytical or deterministic optimization methods. A special case of this problem is very similar to Ref. [4] where the risk function corresponds to the probability of misclassification and its minima is achieved by a weighted majority rule. Another important special case is the correlation coefficient method [10] that explicitly accounts for the correlations between the subsets of detectors in designing the fuser. In these works, the sensors distributions are assumed to be known, which is quite reasonable in the areas these methods are applied. While several of these solutions can be converted into sample-based ones [29,39], these are not designed with measurements as primary focus; furthermore, they address only special cases of the generic sensor fusion problem.

17.3 Generic Sensor Fusion Problem

We consider a *generic sensor system* of N sensors, where the sensor S_i, $i = 1, 2, \ldots, N$, outputs $Y^{(i)} \in \mathfrak{R}^d$ corresponding to input $X \in \mathfrak{R}^d$ according to the distribution $P_{Y^{(i)}|X}$. Informally, the input X is the quantity that needs to be "estimated" or "measured" by the sensors, such as the presence of a target or a value of the feature vector. We define the *expected error* of the sensor S_i as

$$I(S_i) = \int C(X, Y^{(i)}) dP_{Y^{(i)}, X}$$

where $C: \mathfrak{R}^d \times \mathfrak{R}^d \mapsto \mathfrak{R}$ is the cost function. Intuitively, $I(S_i)$ is a measure of how good the sensor S_i is in "sensing" the input feature X. If S_i is a Boolean detector [58], we have $X \in \{0, 1\}$ and $Y^{(i)} \in \{0, 1\}$, where $X = 1$ (0) corresponds to the presence (absence) of a target. Then,

$$I(S_i) \int [X \oplus Y^{(i)}] dP_{Y^{(i)}, X}$$

is the probability of misclassification (false alarm and missed detection) of S_i, where $\oplus$ is the exclusive-OR operation. If S_i is a feature-based detector, $X > 0$, for $d = 1$, indicates the presence of a target, and $Y^{(i)} \in \{0, 1\}$ indicates sensor output. Then, we have the probability of misclassification given by

$$I(S_i) \int [1_{X>0} \oplus Y^{(i)}] dP_{Y^{(i)}, X}$$

where 1_C is the indicator function that takes value 1 if condition C is true and 0 otherwise. In some detection networks, S_i does not perform detection and $Y^{(i)}$ corresponds to measurement of X. In radiation detection networks [50], $X \in \mathfrak{R}^3$ corresponds to the location $(x_S, y_S) \in \mathfrak{R}^2$ and intensity $A_S \in \mathfrak{R}$ of a radiation source, and $Y^{(i)} \in \mathfrak{R}$ is Poisson distributed with the parameter given by the propagated source intensity value at sensor S_i.

The *measurement error* corresponds to the randomness involved in measuring a particular value of the feature X, which is distributed according to $P_{Y^{(i)}|X}$. The *systematic error* at X corresponds to $E[C(X, Y^{(i)})|X]$, which must be 0 in the case of a perfect sensor. This error is often referred to as the bias error.

We consider an example of a sensor system of two sensors such that for the first sensor, we have $Y^{(1)} = a_1 X + Z$, where Z is normally distributed with zero mean and is independent of X. Thus, the first sensor has a constant scaling error and a random additive error. For the second sensor, we have $Y^{(2)} = a_2 X + b_2$, which has a scaling and bias error. Let X be uniformly distributed over $[0, 1]$, and $C[X, Y] = (X - Y)^2$. Then, we have $I(S_1) = (1 - a_1)^2$ and $I(S_2) = (1 - a_2 - b_2)^2$, which are nonzero in general.

We consider a fuser $f: \mathfrak{R}^{Nd} \mapsto \mathfrak{R}^d$ that combines the outputs of sensors $Y = (Y^{(1)}, Y^{(2)}, \ldots, Y^{(N)})$ to produce the fused output $f(Y)$. We define the *expected error* of the fuser f to be

$$I_F(f) = \int C(X, f(Y)) dP_{Y,X}$$

where $Y = (Y^{(1)}, Y^{(2)}, \ldots, Y^{(N)})$. The objective of fusion is to achieve low values of $I_F(f)$ for which both systematic and measurement errors must be accounted for. The fuser is typically chosen from a family of fusion rules $\mathcal{F} = \{f: \mathfrak{R}^N \mapsto \mathfrak{R}\}$ that could be either explicitly or implicitly identified. The expected best fusion rule f^* minimizes $I_F(\cdot)$ over $\mathcal{F}$, that is,

$$I_F(f^*) = \min_{f \in \mathcal{F}} I_F(f)$$

For example, $\mathcal{F}$ could be the set of sigmoidal neural networks obtained by varying the weight vector for a fixed architecture. In this case, $f^* = f_{w^*}$ corresponding to the weight vector w^* that minimizes $I_F(\cdot)$ over all the weight vectors. For the detection problems, where each sensor performs detection, that is, $Y^{(i)} \in \{0, 1\}$, the fuser class $\mathcal{F}$ is a subset of all Boolean functions of N variables.

Continuing our example, we consider the fuser

$$f(Y^{(1)}, Y^{(2)}) = \frac{Y^{(1)}}{2a_1} + \frac{1}{2a_2}(Y^{(2)} - b)$$

For this fuser, we have $I_F(f) = 0$, since the bias b is subtracted from $Y^{(2)}$ and the multipliers cancel the scaling error. In practice, however, such fuser can be designed only with a significant insight into sensors, in particular with a detailed knowledge of the distributions.

In this formulation, $I_F(\cdot)$ depends on the error distribution $P_{Y,X}$, and hence f^* cannot be computed even in principle if the former is not known. We consider that only an independently and identically distributed (iid) l-sample

$$(X_1, Y_1), (X_2, Y_2), \ldots, (X_l, Y_l)$$

is given, where $Y_i = \left(Y_i^{(1)}, Y_i^{(2)}, \ldots, Y_i^{(N)}\right)$ and $Y_i^{(j)}$ is the output of S_j in response to input X_i. We consider an estimator $\hat{f}$, based *only* on a sufficiently large sample, such that

$$P_{Y,X}^l[I_F(\hat{f}) - I_F(f^*) > \in] < \delta \tag{17.1}$$

where

$\in > 0$ and $0 < \delta < 1$

$P_{Y,X}^l$ is the distribution of iid l-samples

For simplicity, we subsequently denote $P_{Y,X}^l$ by P. This condition states that the "error" of $\hat{f}$ is within $\in$ of optimal error (of f^*) with an arbitrary high probability $1 - \delta$, *irrespective* of the underlying sensor distributions. It is a reasonable criterion since $\hat{f}$ is to be "chosen" from an infinite set, namely, $\mathcal{F}$, based only on a finite sample. The conditions that are strictly stronger than Equation 17.1 are generally not possible. To illustrate this, consider the condition $P_{Y,X}^l[I_F(\hat{f}) > \in] < \delta$ for the case $\mathcal{F} = \{f: [0,1]^N \mapsto \{0,1\}\}$. This condition cannot be satisfied, since for any $f \in \mathcal{F}$, there exists a distribution for which $I_F(f) > 1/2 - \rho$ for any $\rho \in [0, 1]$; see Theorem 7.1 of Ref. [9] for details.

To illustrate the effects of finite samples in the previous example, consider that we generate three values for X given by $\{0.1, 0.5, 0.9\}$ with corresponding Z values given by $\{0.1, -0.1, -0.3\}$. The corresponding values for $Y^{(1)}$ and $Y^{(2)}$ are given by $\{0.1a_1 + 0.1, 0.5a_1 - 0.1, 0.9a_1 - 0.3\}$ and $\{0.1a_2 + b_2, 0.5a_2 + b_2, 0.9a_2 + b_2\}$, respectively. Consider the class of linear fusers such as $f(Y^{(1)}, Y^{(2)}) = w_1 Y^{(1)} + w_2 Y^{(2)} + w_3$. Based on the measurements, the following weights enable the fuser outputs to exactly match X values for each of the measurements:

$$w_1 = \frac{1}{0.2 - 0.4a_1}, \quad w_2 = \frac{1}{0.4a_2}, \quad \text{and} \quad w_3 = \frac{0.1a_1 + 0.1}{0.4a_1 + 0.1} - \frac{0.1a_2 + b_2}{0.4a_2}$$

While the fuser with these weights achieves zero error on the measurements it does not achieve zero value for I_F. Note that a fuser with zero expected error exists and can be computed if the sensor distributions are given. The idea behind the criterion in Equation 17.1 is to achieve performances close to optimal fuser using only a sample. To meet this criterion one needs to select a suitable $\mathcal{F}$, and then achieve small error on a sufficiently large sample, as will be illustrated subsequently.

The generic sensor fusion problem formulated here is fairly general and requires very little information about the sensors. In the context of DSN, each sensor could correspond to a node consisting of a hardware device, a software module, or a combination. We describe some concrete examples in Section 17.6.

17.3.1 Related Formulations

Due to the generic nature of the sensor fusion problem described here, it is related to a number of similar problems in a wide variety of areas. Here, we briefly show its relationship to some of the well-known methods in engineering areas. If the sensor error distributions are known, several fusion rule estimation problems have been solved by methods not requiring the samples. The distributed detection problem based on probabilistic formulations has been extensively studied [58]. These problems are special cases of the generic fusion problem such that $X \in \{0, 1\}$ and $Y \in \{0, 1\}^N$, but the difference is that they assume that various probabilities are available. If only measurements are available, these methods are not applicable. While solutions to the generic sensor fusion problem are applicable here, much tighter performance bounds are possible since distribution detection is a special (namely, Boolean) case of the generic sensor fusion problem [45,46]. Also, in many cases, the solutions based on known distributions case can be converted to sample-based ones [29].

Many of the existing information integration techniques are based on maximizing a posteriori probabilities of hypotheses under a suitable probabilistic model. However, in situations where the probability densities are unknown (or difficult to estimate) such methods are ineffective. One alternative is to estimate the density based on a sample. But, as illustrated, in general, by Ref. [57], the density estimation is more difficult than the subsequent problem of estimating a function chosen from a family with bounded capacity. This property holds for several pattern recognition and regression estimation problems [57]. In the context of feedforward neural networks that "learn" a function based on sample, the problem is to identify the weights of a network of chosen architecture. The choice of weights corresponds to a particular network $\hat{f}$ from a family $\mathcal{F}$ of neural networks of a particular architecture. This family $\mathcal{F}$ satisfies bounded capacity [2] and Lipschitz property [55]. Both these properties are conducive for the statistical estimation of $\hat{f}$ as explained in the next section. On the other hand, no such information is available about the class from which the unknown density is chosen, which makes it difficult to estimate the density.

It is not necessary that all sensor distributions be unknown to apply our formulation. Consider that the joint conditional distribution $P_{Y^{(1)},...,Y^{(M)}|Y^{(M+1)},...,Y^{(N)}}$ of the sensors $S_1, ..., S_M$, for $M < N$, is known. Then, we can rewrite

$$I_F(f) = \int \Phi(X, f(Y)) dP_{Y^{(M+1)},...,Y^{(N)},X}$$

where $\Phi(\cdot)$ is suitably derived from the original cost function $C(\cdot)$ and the known part of the conditional distribution. Then, the solutions to the generic sensor fusion problem can be applied to this new cost function with only a minor modification. Since the number of variables with unknown distribution is reduced now, the statistical estimation process is easier. It is important to note that it is not sufficient to know the individual distributions of the sensors, but the *joint* conditional distributions are required to apply this decomposition. In the special case of statistical independence of sensors, the joint distribution is just the product, which makes the transformation easier. In general, for sensor fusion problems, however, the interdependence between the sensors is a main feature to be exploited to overcome the limitations of single sensors.

17.4 Empirical Risk Minimization

In this section, we present fusion solutions based on the empirical risk minimization methods [57]. Consider that the empirical estimate

$$I_{emp}(f) = \frac{1}{l}\sum_{i=1}^{l}\left[X_i - f\left(Y_i^{(1)}, Y_i^{(2)}, ..., Y^{(N)}\right)\right]^2$$

is minimized by $\hat{f} \in \mathcal{F}$. Using Vapnik's empirical risk minimization method [57], for example, we can show [28] that if $\mathcal{F}$ has finite capacity, then under bounded error or bounded relative error for sufficiently large sample

$$P_{Y,X}^l[I_F(\hat{f}) - I_F(f^*) > \in] < \delta$$

for arbitrarily specified $\in > 0$ and δ $0 < \delta < 1$. Typically, the required sample size is expressed in terms of $\in$ and δ and the parameters of $\mathcal{F}$. The most general result that ensures this condition is based on the scale-sensitive dimension [35]. This result establishes the basic tractability of the sensor fusion problem but often results in very loose bounds for the sample size. By utilizing specific properties of $\mathcal{F}$, tighter sample size estimates are possible. In this section, we describe two such classes of $\mathcal{F}$ and their sample size estimators.

17.4.1 Feedforward Sigmoidal Networks

We consider a feedforward network with a single hidden layer of k nodes and a single output node. The output of the jth hidden node is $\sigma(b_j^T y + t_j)$, where $y \in [-B, B]^N$, $b_j \in \mathfrak{R}^N, t_j \in \mathfrak{R}$, and the nondecreasing $\sigma: \mathfrak{R} \to [-1, +1]$ is called the *activation function*. The output of the network corresponding to input y is given by

$$f_w(y) = \sum_{j=1}^{k} a_j\sigma\left(b_j^T y + t_j\right)$$

where $w = (w_1, w_2, ..., w_{k(d+2)})$ is the *weight vector* of the network consisting of $a_1, a_2, ..., a_k, b_{11}, b_{12}, ..., b_{1d}, ..., b_{k1}, ... b_{kd}$, and $t_1, t_2, ..., t_k$. Let the set of *sigmoidal feedforward networks* with *bounded weights* be denoted by

$$\mathcal{F}_W^\gamma = \{f_w: w \in [-W, W]^{k(d+2)}\}$$

where $0 < \gamma < \infty$, and $\sigma(z) = \tanh^{-1}(\gamma z)$, $0 < W < \infty$.

The following theorem provides several sample size estimates for the fusion rule estimation problem based on the different properties of neural networks.

Theorem 17.1

Consider the class of feedforward neural networks $\mathcal{F}_W^\gamma$. Let $X \in [-A, A]$ and $R = 8(A + kW)^2$. Given a sample of size at least [34]

$$\frac{16R}{\epsilon^2}\left(\ln\left(\frac{18}{\delta}\right) + 2\ln\left(\frac{8R}{\epsilon^2}\right) + \ln\left(\frac{2\gamma^2 W^2 kR}{\epsilon}\right)\right) + \frac{\gamma W^2 kR}{\epsilon}\left[\left(\frac{\gamma W^2 kR}{\epsilon} - 1\right)^{N-1} + 1\right]\right)$$

the empirical best neural network $\hat{f}_w$ in $\mathcal{F}_W^\gamma$ approximates the expected best $\hat{f}_w$ in $\mathcal{F}_W^\gamma$ such that

$$P[I_F(\hat{f}_w) - I_F(f_w^*) > \epsilon] < \delta$$

The same condition can also be ensured under the sample size

$$\frac{16R}{\epsilon^2}\left(\ln\left(\frac{18}{\delta}\right) + 2\ln\left(\frac{8R}{\epsilon^2}\right) + k(d+2)\ln\left(\frac{L_w R}{\epsilon}\right)\right)$$

where $L_w = \max(1, WB\gamma^2/4, \gamma^2/4)$ or, for $\gamma = 1$,

$$\frac{128R}{\epsilon^2}\max\left\{\ln\left(\frac{8}{\delta}\right), \quad \ln\left(\frac{16e(k+1)R}{\epsilon}\right)\right\}$$

These sample sizes are based on three qualitatively different parameters of $\mathcal{F}$, namely, (1) Lipschitz property of $f(y) \in \mathcal{F}$ with respect to input y [27], (2) compactness of weight set and smoothness of $f \in \mathcal{F}$ with respect to weights, and (3) VC-dimension of translates of sigmoid units [19]. The three sample estimates provide three different means for controlling the sample size depending on the available information and intrinsic characteristics of the neural network class $\mathcal{F}_W^\gamma$. The sample sizes in the first and second bounds can be modified by changing the parameter γ. For example, by choosing $\gamma = \epsilon/(W^2 kR)$ the first sample size can be reduced to a simpler form

$$\frac{16R}{\epsilon^2}\left[\ln\left(\frac{18}{\delta}\right) + \ln\left(\frac{128R}{W^2 k \epsilon^2}\right)\right]$$

Also, by choosing $\gamma^2 = 4/(W_{\max(1,B)})$, we have a simpler form of the second sample size estimate

$$\frac{16R}{\epsilon^2}\left[\ln\left(\frac{1152}{\delta}\right) + k(d+2)\ln\left(\frac{R}{\epsilon}\right)\right]$$

for $R \geq 1$. In practice, it could be useful to compute all three bounds and choose the smallest one.

The problem of computing the empirical best neural network $\hat{f}_w$ is NP-complete for very general subclass of $\mathcal{F}_W^\gamma$ [54]. In several practical cases, very good results have been obtained using the backpropagation algorithm that provides an approximation to $\hat{f}_w$. For the vector space method in the next section, the computation problem is polynomial-time solvable.

17.4.2 Vector Space Methods

We now consider that $\mathcal{F}$ forms a finite dimensional vector space, and, as a result, (1) the sample size is a simple function of the dimensionality of $\mathcal{F}$, (2) $\hat{f}$ can be easily computed by well-known least square methods in polynomial time, and (3) no smoothness conditions are required on the functions or distributions.

Theorem 17.2

Let f^* and $\hat{f}$ denote the expected best and empirical best fusion functions chosen from a vector space $\mathcal{F}$ of dimension d_V and range $[0, 1]$. Given an iid sample of size [33]

$$\frac{512}{\epsilon^2}\left[d_V\ln\left(\frac{64e}{\epsilon}+\ln\frac{64e}{\epsilon}\right)+\ln\left(\frac{8}{\delta}\right)\right]$$

we have $P[I_F(\hat{f}) - I_F(f^*) > \epsilon] < \delta$.

If $\{f_1, f_2, ..., f_{dv}\}$ is a basis of $\mathcal{F}, f \in \mathcal{F}$ can be written as $f(y) = \sum_{i=1}^{dv} a_i f_i(y)$ for $a_i \in \mathfrak{R}$. Then, consider $\hat{f} = \sum_{i=1}^{dv} \hat{a}_i f_i(y)$ such that $\hat{a} = (\hat{a}_1, \hat{a}_2, ..., \hat{a}_{dv})$ minimizes the cost expressed as

$$I_{emp}(a) = \frac{1}{l}\sum_{i=1}^{l}\left(X_k - \sum_{i=1}^{dv} a_i f_i(Y_k)\right)^2$$

where $a = (a_1, a_2, ..., a_{dv})$. Then, I_{emp} (a) can be written in the quadratic form $a^T C a + a^T D$, where $C = [c_{ij}]$ is a positive definite symmetric matrix, and D is a vector. This form can be minimized in polynomial-time using quadratic programming methods [59].

This method subsumes two very important cases:

1. *Potential functions:* The potential functions of Aizerman et al. [1], where $f_i(y)$ is of the form $\exp((y - \alpha)^2/\beta)$ for suitably chosen constants α and β, constitute an example of the vector space methods. An incremental algorithm was originally proposed for the computation of the coefficient vector a, for which finite sample results have been derived recently [49] under certain conditions.
2. *Special neural networks:* In two-layer sigmoidal networks of Ref. [18], the unknown weights are only in the output layer, which enables us to express each network in the form $\sum_{k=1}^{dv} a_i \eta_i(y)$ universal $\eta_i(\cdot)$s. These networks have been shown to approximate classes of the continuous functions with arbitrarily specified precision in a manner similar to the general single-layer sigmoidal networks as shown in Ref. [6].

17.5 Statistical Estimators

The fusion rule estimation problem is very similar to the regression estimation problem. In this section, we present a polynomial-time (in sample size l) computable Nadaraya–Watson estimator that guarantees the criterion in Equation 17.1 under additional smoothness conditions. We first present some preliminaries needed for the main result. Let Q denote the unit cube $[0, 1]^N$ and $C(Q)$ denote the set of all continuous functions defined on Q. The modulus of smoothness of $f \in C(Q)$ is defined as

$$\omega_\infty(f;r) = \sup_{\|y-z\|_\infty < r, y, z \in Q} |f(y) - f(z)|$$

where $\|\|y - z\|\|_\infty = \max\limits_{i=1}^{M} |y_i - z_i|$. For $m = 0, 1, \ldots$, let Q_m denote a family of dyadic cubes (Haar system) such that $Q = \bigcup\limits_{J \in Q_m} J, J \cap J' = \emptyset$ for $J \neq J'$, and the N-dimensional volume of J, denoted by $|J|$, is 2^{-Nm}. Let $1_J(y)$ denote the indicator function of $J \in Q_m$: $1_J(y) = 1$, if $y \in J$, and $1_J(y) = 0$ otherwise. For given m, we define the map P_m on $C(Q)$ as follows: for $f \in C(Q)$, we have $P_m(f) = P_m f$ defined by

$$P_m f(y) = \frac{1}{|J|} \int_J f(z) dz$$

for $y \in J$ and $J \in Q_m$ [5]. Note that $P_m f : Q \to [0, 1]$ is a discontinuous (in general) function that takes constant values on each $J \in Q_m$. The *Haar kernel* is given by

$$P_m(y, z) = \frac{1}{|J|} \sum_{J \in Q_m} 1_J(y) 1_J(z)$$

for $y, z \in Q$.

Given l-sample, the Nadaraya–Watson estimator based on Haar kernels is defined by

$$\hat{f}_{m,l}(y) = \frac{\sum_{j=1}^{l} X_j P_m(y, Y_j)}{\sum_{j=1}^{l} P_m(y, Y_j)} = \frac{\sum_{Y_j \in J} X_j}{\sum_{Y_j \in J} 1_J(Y_j)}$$

for $y \in J$ [11, 26]. This expression indicates that $\hat{f}_{m,l}(y)$ is the mean of the function values corresponding to Y_js in J that contains y. The Nadaraya–Watson estimator based on more general kernels is well known in statistics literature [23]. Typical performance results of this estimator are in terms of asymptotic results and are not particularly targeted toward fast computation. The aforementioned computationally efficient version based on Haar kernels is due to Ref. [11], which was subsequently shown to yield finite sample guarantees in Ref. [48] under the finiteness of capacity in addition to smoothness. This estimator can be used to solve the generic sensor fusion problem.

Theorem 17.3

Consider a family of functions $\mathcal{F} \subseteq C(Q)$ with range $[0, 1]$ such that $\omega_\infty(f; r) \leq k_r$ for some $0 < k < \infty$. We assume that (1) there exists a family of densities $\mathcal{P} \subseteq C(Q)$; (2) for each $p \in \mathcal{P}, \omega_\infty(p; r) \leq kr$; and (3) there exists $\mu > 0$ such that for each $p \in \mathcal{P}, p(y) > \mu$, for all $y \in [0, 1]^N$. Suppose that the sample size, l, is larger than [30]

$$\frac{2^{2m+4}}{\epsilon_1^2} \left[\left(\frac{k2^m}{\epsilon_1} \left[\left(\frac{k2^m}{\epsilon_1} - 1 \right)^{N-1} + 1 \right] + m \right) \ln \frac{2^{m+1} k}{\epsilon_1} + \ln \left(\frac{2^{2m+6}}{(\delta - \lambda) \epsilon_1^4} \right) \right]$$

where

$$\epsilon_1 = \frac{\epsilon\,(\mu - \epsilon)}{4,0} < \beta < \frac{N}{2(N+1)}, \quad m = \left\lceil \frac{\log l\beta}{N} \right\rceil$$

and

$$\lambda = b\left(\frac{2}{\epsilon}\right)^{1/N+1-1/2\beta} + b\left(\frac{2}{\epsilon_1}\right)^{1/N+1-1/2\beta}$$

Then, for any $f \in \mathcal{F}$, we have $P\left[I_F(\hat{f}_{m,l}) - I_F(f^*)| > \epsilon\right] < \delta$.

We note that the value of $\hat{f}_{m,l}(y)$ at a given y is the ratio of local sum of X_is to the number of Y_is in J that contains y. The range tree [25] can be constructed to store the cells J that contain at least one Y_i; with each such cell, we store the number of the Y_is that are contained in J and the sum of the corresponding X_is. The time complexity of this construction is $O(1(\log 1)^{N-1})$ [25]. Using the range tree, the values of J containing y can be retrieved in $O((\log l)^N)$ time [48].

The smoothness conditions required in Theorem 17.3 are not very easy to verify in practice. However, this estimator is found to perform well in a number of applications including those that do not have smoothness properties (see next section). Several other statistical estimators can also be used for fusion rule estimation, but finite sample results must be derived to ensure the condition in Equation 17.1. Such finite sample results are available for adapted nearest neighbor rules and regressograms [48] that can also be applied for the fuser estimation problem.

17.6 Applications

We describe three concrete applications to illustrate the performance of methods described in the previous sections—the first two are simulation examples and the third one is an experimental system. In addition, the first two examples also provide results obtained with the nearest neighbor rule, which is analyzed in Ref. [27]. In the second example, we also consider another estimate, namely, the empirical decision rule described in Ref. [44]. Pseudorandom number generators are used in both the simulation examples.

Example 17.1

Fusion of noisy function estimators: Consider five estimators of a function g: $[0,1] \rightarrow [0,1]$ such that ith estimator outputs a corrupted value $Y^{(i)} = g_i(X)$ of $g(X)$ when presented with input $X \in [0,1]^d$ [30]. The fused estimate $f(g_1(X), \ldots, g_5(X))$ must closely approximate $g(X)$. Here, g is realized by a feedforward neural network, and, for $i = 1, 2, \ldots, 5$, $g_i(X) = g(X)(1/2 + iZ/10)$, where Z is uniformly distributed over $[-1, 1]$. Thus, we have $1/2 - i/10 < g_i(X)/g(X) \leq 1/2 + i/10$. Table 17.1 corresponds to the mean square error in the estimation of f for $d = 3$ and $d = 5$, respectively, using the Nadaraya–Watson estimator, nearest neighbor rule, and a feedforward neural network with backpropagation learning algorithm. Note the superior performance of the Nadaraya–Watson estimator.

TABLE 17.1   Fusion of Function Estimators: Mean Square Error over Test Set

Training Set	Testing Set	Nadaraya–Watson	Nearest Neighbor	Neural Network
(a) $a = 3$				
100	10	0.000902	0.002430	0.048654
1,000	100	0.001955	0.003538	0.049281
10,000	1000	0.001948	0.003743	0.050942
(b) $d = 5$				
100	10	0.004421	0.014400	0.018042
1,000	100	0.002944	0.003737	0.021447
10,000	1000	0.001949	0.003490	0.023953

TABLE 17.2   Performance of Nadaraya–Watson Estimator for Decision Fusion

Sample Size	Test Set	S_1	S_2	S_3	S_4	S_5	Nadaraya–Watson
100	100	7.0	20.0	33.0	35.0	55.0	12.0
1,000	1,000	11.3	18.5	29.8	38.7	51.6	10.6
10,000	10,000	9.5	20.1	30.3	39.8	49.6	8.58
50,000	50,000	10.0	20.1	29.8	39.9	50.1	8.860

TABLE 17.3   Comparative Performance

Sample Size	Test Size	Bayesian Fuser	Empirical Decision	Nearest Neighbor	Nadaraya–Watson
100	100	91.91	23.00	82.83	88.00
1,000	1,000	91.99	82.58	90.39	89.40
10,000	10,000	91.11	90.15	90.81	91.42
50,000	50,000	91.19	90.99	91.13	91.14

Example 17.2

Distributed detection: We consider five sensors such that $Y \in \{H_0, H_1\}^5$ such that $X \in \{H_0, H_1\}$ corresponds to "correct" decision, which is generated with equal probabilities, that is, $P(X = H_0) = P(X = H_1) = 1/2$ [34,44]. The error of sensor S_i, $i = 1, 2, \ldots, 5$, is described as follows: the output $Y^{(i)}$ is correct decision with probability of $1 - i/10$, and is the opposite with probability $i/10$. The task is to combine the outputs of the sensors to predict the correct decision. The percentage error of the individual detectors and the fused system based on the Nadaraya–Watson estimator is presented in Table 17.2. Note that the fuser is consistently better than the best sensor S1 beyond the sample sizes of the order of 1000. The performance results of the Nadaraya–Watson estimator, empirical decision rule, nearest neighbor rule, and the Bayesian rule based on the analytical formulas are presented in Table 17.3. The Bayesian rule is computed based on the formulas used in the data generation and is provided for comparison only.

Example 17.3

Door detection using ultrasonic and infrared sensors: Consider the problem of recognizing a door (an opening) wide enough for a mobile robot to move through. The mobile robot (TRC Labmate) is equipped with an array of four ultrasonic and four infrared Boolean sensors on each of four sides as shown in Figure 17.1. We address only the problem of detecting a wide enough door when the sensor array of any side is facing it. The ultrasonic sensors return a measurement corresponding to distance to an object within a certain cone as illustrated in

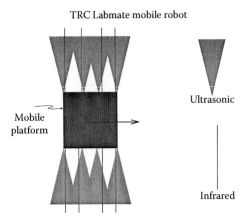

TRC Labmate mobile robot

Mobile
platform

Ultrasonic

Infrared

FIGURE 17.1   Schematic of sensory system (only the side sensor arrays are shown for simplicity).

Figure 17.1. The infrared sensors return Boolean values based on the light reflected by an object in the line-of-sight of the sensor; white smooth objects are detected due to high reflectivity, while objects with black or rough surface are generally not detected. In practice, both ultrasonic and infrared sensors are unreliable, and it is very difficult to obtain accurate error distributions of these sensors. The ultrasonic sensors are susceptible to multiple reflections and the profiles of the edges of the door. The infrared sensors are susceptible to surface texture and color of the wall and edges of the door. Accurate derivation of probabilistic models for these sensors requires a detailed knowledge of the physics and engineering of the device as well as a priori statistical information. Consequently, a Bayesian solution to this problem is very hard to implement. On the other hand, it is relatively easy to collect experimental data by presenting to the robot doors that are wide enough as well as those that are narrower than the robot. We employ the Nadaraya–Watson estimator to derive a nonlinear relationship between the width of the door and the sensor readings. Here, the training sample is generated by actually recording the measurements, while the sensor system is facing the door. Positive examples are generated if the door is wide enough for the robot, and the sensory system is facing the door. Negative examples are generated when the door is not wide enough or the sensory system is not correctly facing a door (wide enough or not). The robot is manually located in various positions to generate the data. Consider the sensor array of a particular side of the mobile robot. Here, $Y^{(1)}$, $Y^{(2)}$, $Y^{(3)}$, $Y^{(4)}$ correspond to the normalized distance measurements from the four ultrasonic sensors, and $Y^{(5)}$, $Y^{(6)}$, $Y^{(7)}$, $Y^{(8)}$ correspond to the Boolean measurements of the infrared sensors. X is 1 if the sensor system is correctly facing a wide enough door and is 0 otherwise. The training data included 6 positive examples and 12 negative examples. The test data included three positive examples and seven negative examples. The Nadaraya–Watson estimator predicted the correct output in all examples of test data.

17.7 Performance of Fused System

We now address the issue of the relative performance of the composite system, composed of the fuser and $S_1, S_2, \ldots, S_N$, and the individual sensors or sensor subsets. We describe sufficiency conditions under which the composite system can be shown to be at least as good as the best sensor or best subset of sensors. In the empirical risk minimization methods, $I_F(\hat{f})$ is shown to be close to $I_F(f^*)$, which depends on $\mathcal{F}$. In general, $I_F(f^*)$ could be very large for particular fuser classes. Note that one cannot simply choose an arbitrary large $\mathcal{F}$: if so, the performance guarantees of the type in Equation 17.1 cannot be guaranteed. If $I_F(f^*) > I(S_i)$, then fusion is not useful, since one is better off just using S_i. In practice, however, such condition cannot be verified if the distributions are not known.

For simplicity, we consider a system of N sensors such that $X \in [0,1]$, $Y^{(i)} \in [0,1]$ and the *expected square error is given by*

$$I_S(S_i) = \int [X - Y^{(i)}]^2 dP_{Y^{(i)},X}$$

The *expected square error* of the fuser f is given by

$$I_F(f) = \int [X - fY]^2 dP_{Y,X}$$

respectively, where $Y = (Y^{(1)}, Y^{(2)}, ..., Y^{(N)})$.

17.7.1 Isolation Fusers

If the distributions are known, one can derive the best sensor S_{i^*} such that $I_S(S_{i^*}) = \min_{i=1}^{N} I_S(S_i)$. In the present formulation, the availability of *only* a sample makes the selection (with probability 1) of the best sensor infeasible, even in the special case of the target detection problem [9]. In this section, we present a method that circumvents this difficulty by fusing the sensors such that the performance of best sensor is achieved as a minimum. The method is fully sample based; in that, no comparative performance of the sensors is needed—in particular, the best sensor may be unknown.

A function class $\mathcal{F} = \{f : [0,1]^k \to [0,1]\}$ has the *isolation property* if it contains the functions $f^i(y_1, y_2, ..., y_k) = y_i$, for all $i = 1, 2, ..., k$. If $\mathcal{F}$ has the isolation property, we have

$$I_F(f^*) = \min_{f \in \mathcal{F}} \int X - f(Y))^2 dP_{Y,X} \leq \int (X - f^i(Y))^2 dP_{Y,X}$$

$$= \int \left(X - Y^{(i)} \right)^2 dP_{Y,X} = I_S(S_i)$$

which implies $I_F(f^*) = \min_{i=1}^{N} I_S(S_i) - \Delta$, for some $\Delta \in [0, \infty)$. Due to the isolation property, we have $\Delta \geq 0$, which implies that the error of f^* is no higher than $I(S_{i^*})$, but can be significantly smaller. The precise value of Δ depends on $\mathcal{F}$, but the isolation property guarantees that $I_F(f^*) \leq \min_{i=1}^{N} I_S(S_i)$ as a minimum.

Let the set S be equipped with a pseudometric ρ. The *covering number* $N_C(\epsilon, \rho, S)$ under metric ρ is defined as the smallest number of closed balls of radius ϵ, and centers in S, whose union covers S. For a set of functions $\mathbf{G} = \{g : \mathfrak{R}^M \mapsto [0,1]\}$, we consider two metrics defined as follows: for $g_1, g_2 \in G$ we have

$$d_P(g_1, g_2) = \int_{z \in \mathfrak{R}^M} |g_1(z) - g_2(z)| dP$$

for the probability distribution P defined on $\mathfrak{R}^M$, and

$$d_\infty(g_1, g_2) = \sup_{z \in \mathfrak{R}^M} |g_1(z) - g_2(z)|$$

This definition is applied to functions defined on $A \subseteq \mathfrak{R}^M$ by extending them to take value 0 on $\mathfrak{R}^M \backslash A$.

Theorem 17.4

Consider a fuser class $\mathcal{F} = \{f : [0,1]^N \to [0,1]\}$, such that $I_F(f^*) \min_{f \in \mathcal{F}} I_F(f)$ and $\hat{I}_F(\hat{f}) = \min_{f \in \mathcal{F}} \hat{I}_F(f)$. If $\mathcal{F}$ has the isolation property, we have [31]

$$I_F(f^*) = \min_{i=1}^{N} I_S(S_i) - \Delta$$

for $\Delta \in [0, \infty)$, and

$$P_{Y,X}^l \left[I_F(\hat{f}) - \min_{i=}^{N} I_S(S_i) + \Delta > \epsilon \right] < \delta$$

given the sample size l of at least

$$\frac{2048}{\epsilon^2} \left[\ln N_C \left(\frac{\epsilon}{64, \mathcal{F}} \right) + \ln \left(\frac{4}{\delta} \right) \right]$$

for cases: (1) $N_C(\epsilon, \mathcal{F}) = N_C(\epsilon, d_\infty, \mathcal{F})$ and (2) $N_C(\epsilon, \mathcal{F}) = N_C(\epsilon, d_P, \mathcal{F})$ for all distributions P.

If $\mathcal{F}$ has the isolation property, the fuser is guaranteed to perform at least as good as the best sensor in PAC sense. No information other than the iid sample is needed to ensure this result. Since $\Delta \geq 0$, under the sample size of Theorem 17.4, we trivially have

$$P \left[I_F(\hat{f}) - \min_{i=}^{N} I_S(S_i) > \epsilon \right] < \delta$$

The sample size needed is expressed in terms of d_∞ or distribution-free covers for $\mathcal{F}$. For smooth fusers such as sigmoid neural networks, we have simple d_∞ cover bounds. In other cases, the pseudodimension and scale-sensitive dimension of $\mathcal{F}$ provide the distribution-free cover bounds needed in Theorem 17.4. The isolation property was first proposed in Ref. [27,47] for concept and sensor fusion problems. For linear combinations, that is, $f(y_1, y_2, \ldots, y_k) = w_1 y_1 + w_2 y_2 + \cdots + w_k y_k$, for $w_i \in \mathfrak{R}$, this property is trivially satisfied. For potential functions [1] and feedforward sigmoid networks [53], this property is not satisfied in general; see Ref. [38] for a more detailed discussion on the isolation property and various function classes that have this property.

Consider the special case where S_is are classifiers obtained using different methods as in Ref. [22]. For Boolean functions, the isolation property is satisfied if $\mathcal{F}$ contains all Boolean functions on k variables. S_i computed based an iid l-sample is *consistent* if $I_S(S_i) \to I_S(S^*)$, where S^* is the Bayes classifier. By the isolation property, if one of the classifiers is consistent, the fused classifier system (trained by l-sample independent from n-sample used by the classifiers) can be seen to be consistent. Such result was obtained in Ref. [22] for linear combinations (for which $N_C(\epsilon, \mathcal{F})$ is finite). The aforementioned result does not require the linearity, but pinpoints the essential property, namely, the isolation. Linear fusers have been extensively used as fusers in combining neural network estimators [14], regression estimators [56,3], and classifiers [22]. Since the linear combinations possess the isolation property, Theorem 17.4 provides some analytical justification for these methods.

17.7.2 Projective Fusers

A *projective fuser* [37], f_P, corresponding to a *partition* $P = \{\pi_1, \pi_2, \ldots, \pi_k\}$, $k \leq N$, of input space $\mathfrak{R}^d (\pi_i \subseteq [0, 1]^d, \bigcup_{i=1}^{k} \pi_i = \mathfrak{R}^d$, and $\pi_i \cap \pi_j = \emptyset$ for $i \neq j)$ assigns each block π_i to a sensor S_j that

$$f_P(Y) = Y^{(j)}$$

for all $X \in \pi_i$, that is, the fuser simply transfers the output of the sensor S_j for every point in π_i. An *optimal projective fuser*, denoted by f_{P^*}, minimizes $I(\cdot)$ over all projective fusers corresponding to all partitions of $\mathfrak{R}^d$ and assignments of blocks to sensors $S_1, S_2, \ldots, S_N$.

We define the *error regression* of the sensor S_i and fuser f_F as

$$\varepsilon(X, S_i) = \int C(X, Y^{(i)} dP_{Y|X}$$

and

$$\varepsilon(X, f_P) = \int C(X, f_P Y)) dP_{Y|X}$$

respectively. The *projective fuser* based on the lower envelope of error regressions of sensors is defined by

$$f_{LE}(Y) = Y^{(i)LE(X))}$$

where

$$i_{LE}(X) = \arg\min_{i=1,2,\ldots,N} \varepsilon(X_i, S_i)$$

We have $\varepsilon(X, f_{LE}) = \min_{i=1,\ldots,N} \varepsilon(X, S_i)$, or equivalently the error regression of f_{LE} is the lower envelope with respect to X of the set of error regressions of sensors given by $\{\varepsilon(X, S_1), \ldots, \varepsilon(X, S_N)\}$.

Example 17.4

Consider that X is uniformly distributed over $[0, 1]$, which is measured by two sensors S_1 and S_2. Let $C(X, Y^{(i)}) = (X - Y^{(i)})^2$. Consider $Y^{(1)} = X + |X - 1/2| + U$ and $Y^{(2)} = X + 1/[4(1 + |X - 1/2|)] + U$, where U is an independent random variable with zero mean [37].

Thus, for both sensors the measurement error at any X is represented by U. Note that

$$E[Y^{(1)} - X] = \left| X - \frac{1}{2} \right|$$

$$E[Y^{(2)} - X] = \frac{1}{[4(1 + |X - 1/2|)]}$$

Thus, S_1 achieves a low error in the middle of the range $[0, 1]$, and S_2 achieves a low error toward the end point of the range $[0, 1]$. The error regressions of the sensors are given by

$$\varepsilon(X, S_1) = \left(X - \frac{1}{2} \right)^2 + E[U^2]$$

$$\varepsilon(X, S_2) = \frac{1}{[16(1 + |X - 1/2|)^2] + E[U^2]}$$

We have

$$I(S_1) = 0.0833 + E[U_2] \quad \text{and} \quad I(S_2) = 0.125 + E[U_2]$$

which indicates that S_1 is the better of the two sensors. Now consider the projective fuser f_{LE} specified as follows, which corresponds to the lower envelope of $\varepsilon(X, S_1)$ and $\varepsilon(X, S_2)$.

Range for X	Sensor to be Projected
$[0, 0.134]$	S_2
$[0.134, 0.866]$	S_1
$[0.866, 1]$	S_2

Then, we have $I(f_{LE}) = 0.0828 + E[U^2]$, which is lower than that of the best sensor.

Example 17.5

We consider a classification example such that $X \in [0,1] \times \{0,1\}$ is specified by a function $f_X = [1/4, 3/4]$, where $1a\,(z)$ is the *indicator function* (which has a value 1 if and only if $z \in A$ and has value 0 otherwise) [37]. The value of X is generated as follows: a random variable Z is generated uniformly in the interval $[0,1]$ as the first component, and then $f_X(Z)$ forms the second component, that is, $X = (Z, f_X(Z))$. In the context of the detection problem, the second component of X corresponds to the presence $(f_X(Z) = 1)$ or absence $(f_X(Z) = 0)$ of a target, which is represented by a feature Z taking a value in the interval $[1/4, 3/4]$. Each sensor consists of a device to measure the first component of X and an algorithm to compute the second component. We consider that S_1 and S_2 have ideal devices that measure Z without an error but make errors in utilizing the measured features. Consider that $Y^{(1)} = (Z, 1_{[1/4-\in_1 3,4]}(Z))$ and $Y^{(2)} = (Z, 1_{[1/4,3/4-\in_2]}(Z))$ for some $0 < \in_1, \in_2 < 1/4$ (see Figure 17.2). In other words, there is no measurement noise in the sensors but just a systematic error due to how the feature value is utilized; addition of independent measurement noise as in Example 17.1 does not change the basic conclusions of the example. Now consider the quadratic cost function $C(X, Y^{(i)}) = (X - Y^{(i)})^T(X - Y^{(i)})$ The error regressions are given by $\varepsilon(X, S_1) = 1_{[1/4-\in_1, 1/4]}(Z)$ and $\varepsilon(X, S_2) = 1_{[3/4-\in_2, 3/4]}(Z)$, which corresponds to disjoint intervals of Z as shown in Figure 17.3. The lower envelope of the two regressions is the zero function hence $I(f_{LE}) = 0$, where as both $I(S_1)$ and $I(S_2)$ are positive. The profile of f_{LE} is shown at the bottom of Figure 17.2, wherein S_1 and S_2 are projected based on the first component of X in the intervals $[3/4 - \in_2, 3/4]$ and $[1/4 - \in_1, 1/4]$, respectively, and in other regions either sensor can be projected.

The projective fuser based on error regressions is optimal as in the following theorem.

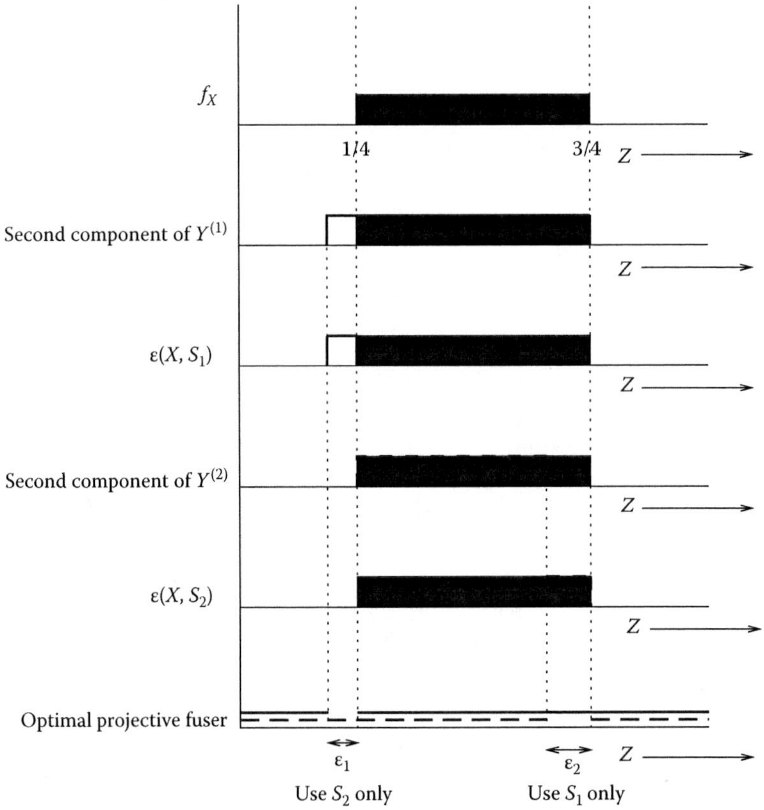

FIGURE 17.2   Processing introduces errors into the idealized sensor readings.

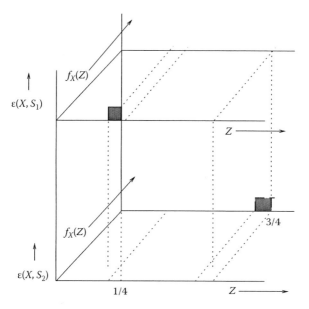

FIGURE 17.3   Fusion makes the systematic errors disjoint.

Theorem 17.5

The projective fuser based on the lower envelope of error regressions is optimal among all projective fusers [37].

A special case of this theorem for function estimation can be found in Ref. [36], and for classifiers can be found in Ref. [32]. A sample-based version of the projective fuser is presented for function estimation using the nearest neighbor concept in Ref. [40], where finite-sample performance bounds are derived using the total variances of various individual estimators. Furthermore, f_{LE} may not be optimal in a larger class of fusers where some function of the sensor output (as opposed to just the output) can be projected [37].

Example 17.6

In Example 17.5, consider $f_X = 1_{[1/4)3/4]}$ [37],

$$Y^{(1)}(X) = (Z, 1_{[1/4-\epsilon_1 3, 4-\epsilon_1]}(Z))$$

$$Y^{(2)}(X) = (Z, 1_{[1/4, 3/4-\epsilon_2]}(Z))$$

for some $0 < \epsilon_1, \epsilon_2 < 1/8$, and $\epsilon_1 < \epsilon_2$. Thus, we have $\epsilon(X, S_1) = 1_{[1/4-\epsilon_1, 1/4]}(Z)$ and $\epsilon(X, S_2) = 1_{[3/4-\epsilon_2, 3/4]}(Z)$, whose lower envelope is not the zero function. Thus, we have $\epsilon(X, f_{LE}) = 1_{[3/4-\epsilon_2, 3/4-\epsilon_1]}(Z)$ and $I(f_{LE}) = \int_{[3/4-\epsilon_2, 3/4-\epsilon_1]} dP_Z$. By changing the assignment $Y^{(1)}$ of f_{LE} to $1 - Y^{(1)}$ for $Z \in [3/4 - \epsilon_2, 3/4 - \epsilon_1]$, one can easily achieve zero error.

17.7.3 Localization-Based Fusers

We consider a class of detection networks that are motivated by the detection of low-level radiation sources [50], where the source parameter estimates can be used to achieve performances superior to projective fusers under certain conditions. In this section, we describe only a case of Lipschitz smoothness conditions [43]; similar results are presented under more restrictive smoothness conditions in Ref. [52], general non-smoothness conditions in Ref. [42], and network losses in Ref. [41].

17.7.3.1 Detection Problem

We consider a two-dimensional monitoring area $\mathcal{M} \subseteq \Re^2$, such as $[0, D] \times [0, D]$-grid, for detecting the presence of a source S with unknown intensity $A_S \in A$, $A = (0, A]$, $A < \infty$ located at an unknown location $(x_S, y_S) \in \mathcal{M}$. The source parameters $(A_S, x_S, y_S) \in \Re^+ \times \mathcal{M}$ constitute the parameter-space $\mathcal{Z} = A \times \mathcal{M}$ and are distributed according to $P_{(AS),zS,yS)}$. The source appears inside $\mathcal{M}$ with a priori probability

$$\mathcal{P}_{\mathcal{M}} = \int\limits_{A_S \in \mathcal{A};(x_S,y_S)\in\mathcal{M}} dP_{(A_S,x_S,y_S)}$$

Both distributions $P_{\mathcal{M}}$ and $P(A_S, x_S, y_S)$ are unknown. There is a background noise process characterized by the intensity parameter $B_{(x,y)} \in B$, $B = [0, B]$, $< \infty$ that depends on the location $(x, y) \in \Re^2$, and thus background noise process is parameterized by $P_{(B_{(x,y)},x,y)}$.

Sensors are located at $M_i = (x_i, y_i) \in \Re^2$, $i = 1, 2, \ldots, N$ to monitor the area $\mathcal{M}$; the sensors may not necessarily be located inside $\mathcal{M}$. For any point $P = (x, y) \in R^2$, we have the distance $d(P, M_i) = \sqrt{(x - x_i)^2 + (y - y_i)^2}$, for $1 \le i \le N$. For two points in state-space $z_1 = \{a_1, x_1, y_1\}$, $z_2 = (a_2, x_2, y_2) \in \mathcal{Z}$, we define $d(z_1, z_2) = \sqrt{(a_1 - a_2)^2 + (x_1 - x_2)^2 + (y_1 - y_2)^2}$. The sensor measurements are characterized as follows:

1. *Background measurements:* When there is no source present, the "background" measurements of M_i are distributed according to P_{B_i}, $B_i = B_{(x_i, y_i)}$.
2. *Source measurements:* When the source is present in $\mathcal{M}$, the intensity at sensor location (x_i, y_i) is A_i, which is a function of A_S and $d(S, M_i) = d((x_S, y_S), M_i))$. We represent this dependence explicitly as a function $A_i = F_S(A_S, x_S, y_S, x_i, y_i)$. The measurements of A_i collected at M_i are distributed according to $P_{A_i+B_i}$.

It is assumed that the functional form of the underlying measurement distributions P_{B_i} and $P_{A_i+B_i}$ is known; for example, for detecting point radiation sources, these distributions are approximated by Poisson process with parameters B_i and $A_i + B_i$, respectively [17,24]. For the Gaussian source detection problem, these distributions are given by Gaussian distribution with mean parameters B_i and $A_i + B_i$, respectively, and standard deviation σ [61,62].

Let $m_{i,1}, m_{i,2}, \ldots, m_{i,n}$ be the sequence of measurements collected by sensor M_i over an observation time window W, such that $m_{i,t}$, $i = 1, 2, \ldots, N$, are collected at the same time t at all sensors.

We consider the *detection problem* that deals with inferring the presence of a source inside $\mathcal{M}$ based on measurements collected at $M_1, M_2, \ldots, M_N$. We characterize the solution of the detection problem by the (1) *false-alarm probability* $P_{0,1}$, corresponding to the probability of declaring the presence of a source when none exists, and (2) *missed detection probability* $P_{1,0}$, corresponding to the probability of declaring the presence of only the background radiation when a source is present in the monitoring area. The detection probability is given by $P_{1,1} = 1 - P_{1,0}$.

17.7.3.2 SPRT Detection

Consider the measurements $m_{i,1}, m_{i2}, \ldots, m_{i,n}$ collected by sensor M_i within a given time window and the background noise level $B_i = B_{(xi, yi)}$ at this sensor location. Let H_C, for $C \in \{A_i + B_i, B_i\}$, denote the hypothesis that the measurements correspond to intensity level C at the sensor M_i. Now consider the likelihood function $L(m_{i,1}, m_{i,2}, \ldots, m_{i,n}|H_C)$, which represents the probability that the measurements were produced by the source if $C = A_i + B_i$ and just the background if $C = B_i$. The ratio of these likelihood functions can be utilized to decide between these hypotheses. We now consider the following SPRT based on sensor measurements at M_i

$$\mathcal{L}_{A_i, B_i, n} = \frac{L(m_{i,1}, m_{i,2}, \ldots, m_{i,n} \mid H_{A_i + B_i})}{L(m_{i,1}, m_{i,2}, \ldots, m_{i,n} \mid H_{B_i})}$$

which can be used for detecting the source with false-positive and missed detection probability parameters $P_{0,1}$ and $P_{1,0}$, respectively, as follows [16]:

1. If $\mathcal{L}_{A_i, B_i; n} < \dfrac{P_{0,1}}{1 - P_{1,0}}$ then declare the background, namely, H_{B_i}
2. Else if $\mathcal{L}_{A_i, B_i; n} > (1 - P_{0,1})/P_{1,0}$ then declare that a source is present, that is, $H_{A_i + B_i}$
3. Otherwise, declare that the measurements are not sufficient to make a decision and continue collecting additional measurements.

This test can be compactly expressed as

$$\frac{P_{0,1}}{1 - P_{1,0}} \leq \mathcal{L}_{A_i, B_i; n} \leq \frac{1 - P_{0,1}}{P_{1,0}}$$

Typically, $\mathcal{L}_{A_i, B_i; n}$ cannot be directly applied for our detection problem since it depends on A_i, which, in turn, depends on source location and intensity both of which are unknown. By utilizing the domain knowledge, this test is often expressed in terms of measurements, and we consider such a generic case.

Definition 17.1

We define a likelihood ratio test to be separable if it can be expressed as

$$F_L(P_{0,1}, P_{1,0} A_i, B_i, n) < \sum_{j=1}^{n} m_{ij}$$

$$< F_U(P_{0,1}, P_{1,0}, A_i, B_i, n)$$

for suitable lower and upper threshold function $F_L(\times) \in \mathcal{F}_L$ and $F_U(\cdot) \in \mathcal{F}_U$, respectively.

In practice, suitable scalar values τ_L and τ_H are chosen for the upper and lower thresholds, respectively, based on domain-specific considerations, Bayesian inference, or other method (in addition to choosing appropriate values for $P_{1,0}$ and $P_{0,1}$). We denote the SPRT with such selected threshold by $\mathcal{L}_{\tau L, \tau H}$, which will be called *fixed-threshold* SPRT.

Definition 17.2

We define a separable SPRT to be Lipschitz-separable if the threshold functions are Lipschitz in the following sense: for any $P_{0,1}, P_{1,0}, B_i$, there exists scalars K_L and K_U such that

$$|F_L(.,A_i,.) - F_L(.,A_i + \gamma,.)| \leq K_{L\gamma}$$

and

$$|F_U(.,A_i,.) - F_U(.,A_i + \gamma,.)| \leq K_{U\gamma}$$

The Lipschitz parameters K_L and K_U denote the sensitivity of the threshold functions to intensity value at sensor M_i, which, in turn, depends both on source location and intensity through the function $A_i = F_S(A_S, x_S, y_S x_i, y_i)$. We show in the next section that this property is satisfied for both Gaussian and point radiation sources.

17.7.3.3 Detection Using Localization

The *localization* corresponds to estimating the location and strength of the source using measurements $m_{i,j}, i = 1,2, ..., N, j = 1,2, ..., T$. The estimates of A_S and (x_S, y_S) are denoted by $\hat{A}_S$ and $(\hat{x}_S, \hat{y}_S)$, respectively. The estimated source parameters will be substituted into the SPRT as follows:

$$F_L(P_{0,1}, P_{1,0}, \hat{A}_i, B_i, n) < \sum_{j=1}^{n} m_{i,j} < F_U(P_{0,1}, P_{1,0}, \hat{A}_i, B_i, n)$$

such that $\hat{A}_i = F(\hat{A}_S, \hat{x}_S, \hat{y}_S, x_i, y_i)$. We denote this SPRT as $\mathcal{L}_{\hat{S}}$, and refer to as the *localization-based* SPRT. When a localization algorithm is executed using "background only" measurements, the estimated parameters correspond to ghost sources. By generalizing the approach in Ref. [52], we now show that the absence of ghost source and hence presence of real source (namely, detection) can be asserted more effectively than $\mathcal{L}_{\tau L, \tau H}$, thereby improving the detection rate.

Definition 17.3

A localization method is 6-robust if the following condition can be ensured: there exits $\delta(\in, n, N))$, which is a nonincreasing function of number of measurements n and number of sensors N and nondecreasing function of precision e such that

$$P\{(\hat{x}_S, \hat{y}_S, \hat{A}_S) \in \Re_{S,\in}\} > \delta(\in, n, N)$$

where $\Re_{S,\in}\{z \in \Re^3 | d(z, z_S) \leq \in; z_S = (A_S, x_S, y_S)\}$ is called $\in$-precision region.

This condition ensures that the estimate is within $\in$ of source parameter z_S with probability δ, which improves as more measurements are collected and more sensors are deployed. This condition is a reasonable requirement and is satisfied by algorithms used for localizing point radiation sources under isotropic shielding conditions [51]. However, it is much harder to satisfy under arbitrary shielding environments. A general approach in such cases is to execute the localization algorithm on three-sensor combinations, wherein the sensors are "close-by." Then results from such localized three-sensor combinations can be combined to derive the source parameters [51].

For a given SPRT $\mathcal{L}$, we denote the detection and false-alarm probabilities by $\varepsilon_D(\mathcal{L})$ and $\varepsilon_F(\mathcal{L})$, respectively. Let $\mathcal{F} = \{f \colon D \mapsto D\}$. A *spherical cell* with center $(z_k, f_k) \in \mathfrak{R}^3 \times \mathcal{F}$ and radius ρ is defined as

$$C(z_k, f_k) = \{(z, f) \,|\, d_\infty(z, z_k) < \rho \quad \text{and} \quad \| f - f_k \|_\infty < p\}$$

where

$$d_\infty((a_1, a_2, a_3), (b_1, b_2, b_3)) = \max_{i=1,2,3} |a_i - b_i|$$

and

$$\| f - f_k \|_\infty = \max_{x \in D} |f(x) - f_K(x)|$$

Definition 17.4

A ρ-packing of state-space $\mathcal{Z} = \mathcal{Z} \times \mathcal{F} = \mathcal{A} \times \mathcal{M} \times \mathcal{F}$ corresponds to disjoint spherical cells with cell centers at (z_k, f_k), $k = 1, 2, \ldots, K$ of radius p all contained inside state-space $\mathcal{Z}$. We define such a packing to be translation invariant if all cells are still inside $\mathcal{Z}$ when centers are translated as $z + z_k$, for all $z \in \mathcal{Z}$. The *state packing number* $\mathcal{N}_\infty(\mathcal{Z}, \rho)$ denotes the maximum size of translation invariant p-packing of state-space $\mathcal{Z}$.

We characterize the state-space by the following two parts:

$$\mathcal{Z}_L = \mathcal{Z} \times \mathcal{F}_L = \mathcal{A} \times \mathcal{M} \times \mathcal{F}_L$$

and

$$\mathcal{Z}_H = \mathcal{Z} \times \mathcal{F}_H = \mathcal{A} \times \mathcal{M} \times \mathcal{F}_H$$

We define two sets that represent all possible source parameters and SPRT-bound functions that correspond to the thresholds of $\mathcal{L}_{\tau L, \tau H}$ as follows:

$$\mathcal{S}_{\tau L} = \{z \in \mathcal{Z}_L \,|\, \tau_L = F_L(P_{0,1}, P_{1,0}, A_i, B_i, n);$$
$$A_i = F_S(A_S, x_S, y_S, x_i, y_i)\}$$

and

$$\mathcal{S}_{\tau H} = \{z \in \mathcal{Z}_H \,|\, \tau_H = F_H(P_{0,1}, P_{1,0}, A_i, B_i);$$
$$A_i = F_S(A_S, x_S, y_S, x_i, y_i)\}$$

The following theorem presents a general result on the relative performance of the threshold-based SPRT LTL)TH and localization-based SPRT $\mathcal{L}_{\hat{S}}$ In particular, this theorem specifies a sufficiency condition under which the latter performs better than SPRT implemented at any single sensor. Note that the threshold-based SPRT can be computed at any sensor, whereas the localization-based SPRT requires a network of at least three sensors. In this sense, this result captures the conditions under which a network of sensors can be shown to outperform any single sensor within SPRT framework.

Theorem 17.6

Consider the detection of a source under the conditions of Lipschitz-separable SPRT. Then for SPRT $\mathcal{L}_{\hat{S}}$ based on δ-robust localization method and any threshold-based SPRT, $\mathcal{L}_{\tau L, \tau H}$, for sufficiently large n and N [43]:

1. Detection rates satisfy

$$\varepsilon_D(\mathcal{L}_{\hat{S}}) > [\varepsilon_D(\mathcal{L}_{\tau L, \tau H}) + (\mathcal{N}_\infty(\mathcal{Z}_L, \in_D) - 1)]\delta(\in_D, n, N)$$

2. False-alarm rates satisfy

$$\varepsilon_F(\mathcal{L}_S) < [\varepsilon_F(\mathcal{L}_{\tau L, \tau H}) - (\mathcal{N}_\infty(\mathcal{Z}_H, \in_F) - 1)]\delta(\in_F, n, N)$$

where

$$\in_D = \max_{z_1, z_2 \in S_\tau H} d_\infty(z_1, z_2)$$

$$\in_F = \max_{z_1, z_2 \in S_\tau L} d_\infty(z_1, z_2)$$

The performance, in terms of both ε_D and ε_F, of $\mathcal{L}_{\hat{S}}$ is better than $\mathcal{L}_{\tau L, \tau H}$ by the factor proportional to the packing number $\mathcal{N}(\mathcal{Z}, \in_D)$ and $\delta(\cdot)$. Informally speaking, "larger" state-space will have larger packing number, and hence $\mathcal{L}_{\hat{S}}$ will lead to a more effective detection. In particular, performance of $\mathcal{L}_{\hat{S}}$ will be increasingly better as one considers larger parameter space $\mathcal{Z}$, larger functional spaces of SPRT-bound functions $\mathcal{F}_L$ and $\mathcal{F}_U$, more sensors N, and more measurements n. Under these conditions, localization-based fuser can perform better than projective fuser, since the latter achieves the best performance of individual sensor within each region, but the former can perform better than any single sensor in every region. The performance inequalities of Theorem 17.6 are valid no matter how thresholds are chosen for $\mathcal{L}_{\tau L, \tau H}$; for example, they can be based on domain-specific knowledge as in radiation source detection, Bayesian inference, or Dempster–Shaefer theory.

Theorem 17.6 imposes no smoothness conditions on the source function $A_i = F_S(A_S, x_S, y_S, x_i, y_i)$, and indeed $F_S(\cdot)$ can have discrete drops. Under certain shielding conditions such discrete drops occur for the case of radiation sources [17,21].

A more restricted Lipschitz condition was considered in Ref. [52] for the source profile function $A_i = F_S(A_S, x_S, y_S, x_i, y_i)$ that requires the intensity to smoothly vary as one moves away from the source. This restriction enables us to specify the equivalent conditions of Theorem 17.6 entirely in terms of the packing number for the parameter-space $\mathcal{Z} \subset \mathfrak{R}^3$, as opposed to more complex state-space $\mathcal{Z}_L$ or $\mathcal{Z}_U$ This, in turn, leads to an intuitive interpretation of the packing number as the number of cells within the parameter space: the precision of source parameter estimates $\in$ is suitably multiplied by the Lipschitz constant of $F_S(\cdot)$ to obtain the precision parameter for the packing number. In contrast, the more complex state-space $\mathcal{Z}_L$ or $\mathcal{Z}_U$ of Theorem 17.6 has the additional functional component from $\mathcal{F}_L$ or $\mathcal{F}_U$, respectively. Consequently, it leads to a less intuitive interpretation of the packing number in terms of precision of the source parameter estimation, although same $\in$ is used in the packing number.

The approach of Theorem 17.6 can be adapted to derive more general results, for example, the source intensity could be a finite dimensional vector and the source location can be specified in a higher dimensional space. A generalization that does not require SPRT to be Lipschitz is presented in Ref. [42], which requires a more complex packing number specified by two parameters. These results are generalized to account for network losses in Ref. [41], and have been applied to the detection of low-level radiation sources in Refs. [42,52].

17.8 Metafusers

In this section, we first show that the projective and linear fusers offer complementary performances, which leads us to the idea of combining them exploit their relative merits. Such approach lead to the concept of *metafusers*, and here we describe how the isolation property can be utilized to design them.

The output of linear fuser corresponding to input X and sensor output $Y = (Y^{(1)}, \dots, Y^{(N)})$ is defined as

$$f_L(Y) = \sum_{i=1}^{N} \alpha_i Y^{(i)}$$

where α_i is a $d \times d$ matrix. For simplicity, we consider the case $d = 1$ such that $(\alpha_1, \dots, \alpha_N) \in \mathfrak{R}^N$.

An *optimal linear combination fuser*, denoted by f_{L^*}, minimizes $I(\cdot)$ over all linear combinations. In general, f_{LE} is better than f_{L^*} if the individual sensors perform better in certain localized regions of $\mathfrak{R}^d$. On the flip side, if the sensors are equally distributed around certain values in global sense, f_L performs better as illustrated in the following.

In Example 17.5, for $f_L = \alpha_1 Y^{(1)} + \alpha_2 Y^{(N)}$, we have

$$I(f_L) = \alpha_1^2 \int\limits_{[1/4-\in_1, 1/4)} dP_Z$$

$$+ (1 - \alpha_1 - \alpha_2)^2 \int\limits_{[1/4, 3/4-\in_2)} dP_Z$$

$$+ (1 - \alpha_1)^2 \int\limits_{[3/4-\in_2, 3/4)} dP_Z$$

which is nonzero no matter what the coefficient are. The error regressions of S_1 and S_2 take nonzero values in the intervals $[1/4 - \in_1, 1/4]$ and $[3/4 - \in_2, 3/4]$ of Z, respectively. Since, these intervals are disjoint, there is no possibility of the error of one sensor being canceled by a scalar multiplier of the other. This argument is true in general: if the error regressions of the sensors take nonzero values on disjoint intervals, then any linear fuser will have nonzero residual error. On the other hand, the disjointness yields $\varepsilon(X, f_{LE}) = 0$, for all X, and hence $I(f_{LE}) = 0$. Consider that in Example 17.5, $f_X = 1$ for $Z \in [0, 1]$,

$$Y^{(1)}(X) = (Z, \in Z + 1 - \in)$$

$$Y^{(2)}(X) = (Z, - \in Z + 1 + \in)$$

for $0 < \in < 1$. The optimal linear fuser is given by $f_{L^*}(Y) = 1/2(Y^{(1)} + Y^{(2)}) = 1$, and $I(f_{L^*}) = 0$.

At every $X \in [0, 1]$, we have

$$\varepsilon(X, S_1) = \varepsilon(X, S_2) = \epsilon^2 (1 - Z)^2 = \varepsilon(X, f_{LE})$$

Thus, $I(f_{LE}) = \epsilon^2 \int_{[0,1]} (1 - Z)^2 dP_Z > 0$, whereas $I(f_{L^*}) = 0$.

Thus, the performance of the optimal linear and projective fusers are complementary in general. We now combine linear and projective fusers to realize various metafusers that are guaranteed to be at least as good as the best sensor as well as best sensor. By including the optimal linear combination as S_{N+1}, we can guarantee that $I(f_{LE}) \leq I(f_{L^*})$ by the isolation property of projective fusers [37].

Since linear combinations also satisfy the isolation property, we, in turn, have $I(f_{L^*}) \leq \min_{i=1,\ldots,N} I(S_i)$.

The roles of f_{L^*} and f_{LE} can be switched—by including f_{LE} as one of the components of f_{L^*}—to show that

$$I(f_{L^*}) \leq (f_{LE}) \leq \min_{i=1,\ldots,N} I(S_i)$$

One can design a metafuser by utilizing the available sensors that are combined using a number of fusers including a fuser based on isolation property (e.g., a linear combination). Consider that we employ a metafuser based on a linear combination of the fusers. Then, the fused system is guaranteed to be at least as good as the best of the fusers as well as the best sensor. If at a latter point, a new sensor or a fuser is developed, it can be easily integrated into the system by retraining the fuser and/or metafuser as needed. As a result, we have a system guaranteed (in PAC sense) to perform at least as good as the best available sensor and fuser at all times. Also, the computational problem of updating the fuser and/or metafuser is a simple least squares estimation that can be solved using a number of available methods.

17.9 Conclusions

In a DSN, we considered that the sensor outputs are related to the actual feature values according to a probability distribution. For such system, we presented an overview of informational and computational aspects of a fuser that combines the sensor outputs to more accurately predict the feature, when the sensor distributions are unknown but iid measurements are given. Our performance criterion is the probabilistic guarantee in terms of distribution-free sample bounds based entirely on a finite sample. We first described two methods based on the empirical risk minimization approach, which yield a fuser that is guaranteed, with a high probability, to be close to an optimal fuser. Note that the optimal fuser is computable only under a complete knowledge of sensor distributions. Then, we described the isolation fusers that are guaranteed to perform at least as good as the best sensor. We then described the projective fusers that are guaranteed to perform at least as good as the best subset of sensors. For a class of detection problems, we showed that localization-based fusers can be employed to achieve performances superior to projective fusers under certain conditions. We briefly discussed the notion of metafusers that can combine fusers of different types.

The overall focus of this chapter is very limited: we only considered sample-based fuser methods that provide finite sample performance guarantees. Even then, there are a number of important issues for the fuser rule computation that have not been covered here. An important aspect is the utilization of fusers that have been designed for known distribution cases for the sample-based case. In many important cases, the fuser formulas expressed in terms of probabilities can be converted into sample-based ones by utilizing suitable estimators [29]. It would be interesting to see an application of this approach to the generic sensor fusion problem. For the most part, we only considered stationary systems, and it would be of future interest to study sample-based fusers for time-varying systems.

Acknowledgment

This work was funded by the Mathematics of Complex, Distributed, Interconnected Systems Program, Office of Advanced Computing Research, U.S. Department of Energy; SensorNet Program, Office of Naval Research; Defense Advanced Projects Research Agency under MIPR No. K153; and National Science Foundation and was performed at Oak Ridge National Laboratory managed by UT-Battelle, LLC for U.S. Department of Energy under Contract No. DE-AC05–00OR22725.

References

1. M. A. Aizerman, E. M. Braverman, and L. I. Rozonoer. Extrapolative problems in automatic control and method of potential functions. *American Mathematical Society Translations*, 87:281–303, 1970.

2. M. Anthony. Probabilistic analysis of learning in artificial neural networks: The PAC model and its variants. NeuroCOLT Technical Report Series NC-TR-94-3, Royal Holloway, University of London, London, U.K., 1994.

3. L. Breiman. Stacked regressions. *Machine Learning*, 24(1):49–64, 1996.

4. C. K. Chow. Statistical independence and threshold functions. *IEEE Transactions on Electronic Computers*, EC-16:66–68, 1965.

5. Z. Ciesielski. Haar system and nonparametric density estimation in several variables. *Probability and Mathematical Statistics*, 9:1–11, 1988.

6. G. Cybenko. Approximation by superpositions of a sigmoidal function. *Mathematics of Controls, Signals, and Systems*, 2:303–314, 1989.

7. B. V. Dasarathy. *Decision Fusion*. IEEE Computer Society Press, Los Alamitos, CA, 1994.

8. P. S. de Laplace. Deuxième supplément à la théorie analytique des probabilités. 1818. Reprinted (1847) in *Oeuvres Complétes de Laplace*, Vol. 7. Paris, France, Gauthier-Villars, pp. 531–580.

9. L. Devroye, L. Gyorfi, and G. Lugosi. *A Probabilistic Theory of Pattern Recognition*. Springer-Verlag, New York, 1996.

10. E. Drakopoulos and C. C. Lee. Optimal multisensor fusion of correlated local decision. *IEEE Transactions on Aerospace Electronics Systems*, 27(4):593–605, 1991.

11. J. Engel. A simple wavelet approach to nonparametric regression from recursive partitioning schemes. *Journal of Multivariate Analysis*, 49:242–254, 1994.

12. C. W. J. Granger. Combining forecasts—Twenty years later. *Journal of Forecasting*, 8:167–173, 1989.

13. B. Grofman and G. Owen (Eds.). *Information Pooling and Group Decision Making*. Jai Press Inc., Greenwich, CT, 1986.

14. S. Hashem. Optimal linear combinations of neural networks. *Neural Networks*, 10(4):599–614, 1997.

15. S. Hashem, B. Schmeiser, and Y. Yih. Optimal linear combinations of neural networks: An overview. In *Proceedings of 1994 IEEE Conference on Neural Networks*, pp. 1507–1512, 1994.

16. N. L. Johnson. Sequential analysis: A survey. *Journal of Royal Statistical Society, Series A*, 124(3):372–411, 1961.

17. G. F. Knoll. *Radiation Detection and Measurement*. John Wiley, New York, 2000.

18. V. Kurkova. Kolmogorov's theorem and multilayer neural networks. *Neural Networks*, 5:501–506, 1992.

19. G. Lugosi and K. Zeger. Nonparametric estimation via empirical risk minimization. *IEEE Transactions on Information Theory*, 41(3):677–687, 1995.

20. R. N. Madan and N. S. V. Rao. Guest editorial on information/decision fusion with engineering applications. *Journal of Franklin Institute*, 336B(2):199–204, 1999.

21. D. Mihalas and B. W. Mihalas. *Foundations of Radiation Hydrodynamics*. Courier Dover Publications, Mineola, NY, 2000.

22. M. Mojirsheibani. A consistent combined classification rule. *Statistics and Probability Letters*, 36:43–47, 1997.

23. E. A. Nadaraya. *Nonparametric Estimation of Probability Densities and Regression Curves*. Kluwer Academic Publishers, Dordrecht, the Netherlands, 1989.

24. K. E. Nelson, J. D. Valentine, and B. R. Beauchamp. Radiation detection method and system using the sequential probability ratio test, 2007. U.S. Patent 7244930 B2.

25. F. P. Preparata and I. A. Shamos. *Computational Geometry: An Introduction*. Springer-Verlag, New York, 1985.

26. B. L. S. Prakasa Rao. *Nonparametric Functional Estimation*. Academic Press, New York, 1983.

27. N. S. V. Rao. Fusion methods for multiple sensor systems with unknown error densities. *Journal of Franklin Institute*, 331B(5):509–530, 1994.

28. N. S. V. Rao. Fusion rule estimation in multiple sensor systems using training. In H. Bunke, T. Kanade, and H. Noltemeier (Eds.), *Modelling and Planning for Sensor Based Intelligent Robot Systems*, World Scientific Publications, Singapore, pp. 179–190, 1995.

29. N. S. V. Rao. Distributed decision fusion using empirical estimation. *IEEE Transactions on Aerospace and Electronic Systems*, 33(4):1106–1114, 1996.

30. N. S. V. Rao. Nadaraya-Watson estimator for sensor fusion. *Optical Engineering*, 36(3):642–647, 1997.

31. N. S. V. Rao. A fusion method that performs better than best sensor. In *First International Conference on Multisource-Multisensor Data Fusion*, Las Vegas, NV, pp. 19–26, 1998.

32. N. S. V. Rao. To fuse or not to fuse: Fuser versus best classifier. In *SPIE Conference on Sensor Fusion: Architectures, Algorithms, and Applications II*, pp. 25–34. 1998.

33. N. S. V. Rao. Vector space methods for sensor fusion problems. *Optical Engineering*, 37(2):499–504, 1998.

34. N. S. V. Rao. Fusion methods in multiple sensor systems using feedforward neural networks. *Intelligent Automation and Soft Computing*, 5(1):21–30, 1999.

35. N. S. V. Rao. Multiple sensor fusion under unknown distributions. *Journal of Franklin Institute*, 336(2):285–299, 1999.

36. N. S. V. Rao. On optimal projective fusers for function estimators. In *Second International Conference on Information Fusion*, Sunnyvale, CA, pp. 296–301, 1999.

37. N. S. V. Rao. Projective method for generic sensor fusion problem. In *Proceedings of IEEE/SICE/RSJ International Conference on Multisensor Fusion and Integration for Intelligent Systems*, pp. 1–6, Taipei, Taiwan, 1999.

38. N. S. V. Rao. Finite sample performance guarantees of fusers for function estimators. *Information Fusion*, 1(1):35–44, 2000.

39. N. S. V. Rao. On sample-based implementation of non-smooth decision fusion functions. In *Proceedings of IEEE/SICE/RSJ International Conference on Multisensor Fusion and Integration for Intelligent Systems*, 2001.

40. N. S. V. Rao. Nearest neighbor projective fuser for function estimation. In *Proceedings of International Conference on Information Fusion*, 2002.

41. N. S. V. Rao. Localization-based detection under network losses. In *International Conference on Information Fusion*, 2011.

42. N. S. V. Rao, J. C. Chin, D. K. Y. Yau, and C. Y. T. Ma. Localization leads to improved distribution detection under non-smooth distributions. In *International Conference on Information Fusion*, 2010.

43. N. S. V. Rao, J. C. Chin, D. K. Y. Yau, C. Y. T. Ma, and R. N. Madan. Cyber-physical trade-offs in distributed detection networks. In *International Conference on Multisensor Fusion and Integration*, 2010.

44. N. S. V. Rao and S. S. Iyengar. Distributed decision fusion under unknown distributions. *Optical Engineering*, 35(3):617–624, 1996.

45. N. S. V. Rao and E. M. Oblow. Majority and location-based fusers for PAC concept learners. *IEEE Transactions on Systems, Man and Cybernetics*, 24(5):713–727, 1994.

46. N. S. V. Rao and E. M. Oblow. N-learners problem: System of PAC learners. In *Computational Learning Theory and Natural Learning Systems, Vol. IV: Making Learning Practical*, MIT Press, Cambridge, MA, pp. 189–210, 1997.

47. N. S. V. Rao, E. M. Oblow, C. W. Glover, and G. E. Liepins. N-learners problem: Fusion of concepts. *IEEE Transactions on Systems, Man and Cybernetics*, 24(2):319–327, 1994.

48. N. S. V. Rao and V. Protopopescu. On PAC learning of functions with smoothness properties using feedforward sigmoidal networks. *Proceedings of the IEEE*, 84(10):1562–1569, 1996.

49. N. S. V. Rao, V. Protopopescu, R. C. Mann, E. M. Oblow, and S. S. Iyengar. Learning algorithms for feedforward networks based on finite samples. *IEEE Transactions on Neural Networks*, 7(4): 926–940, 1996.

50. N. S. V. Rao, M. Shankar, J. C. Chin, D. K. Y. Yau, S. Srivathsan, S. S. Iyengar, Y. Yang, and J. C. Hou. Identification of low-level point radiation sources using a sensor network. In *International Conference on Information Processing in Sensor Networks*, 2008.

51. N. S. V. Rao, M. Shankar, J. C. Chin, D. K. Y. Yau, Y. Yang, J. C. Hou, X. Xu, and S. Sahni. Localization under random measurements with applications to radiation sources. In *International Conference on Information Fusion*, 2008.

52. N. S. V. Rao, M. Shankar, J. C. Chin, D. K. Y. Yau, Y. Yang, X. Xu, and S. Sahni. Improved SPRT detection using localization with application to radiation sources. In *International Conference on Information Fusion*, 2009.

53. V. Roychowdhury, K. Siu, and A. Orlitsky (Eds.). *Theoretical Advances in Neural Computation and Learning*. Kluwer Academic Publishers, Boston, MA, 1994.

54. J. Sima. Back-propagation is not efficient. *Neural Networks*, 9(6):1017–1023, 1996.

55. Z. Tang and G. J. Koehler. Lipschitz properties of feedforward neural networks. Technical report, Department of Decision and Information Science, University of Florida, Gainesville, FL, 1994.

56. M. Taniguchi and V. Tresp. Averaging regularized estimators. *Neural Computation*, 9:1163–1178, 1997.

57. V. Vapnik. *Estimation of Dependences Based on Empirical Data*. Springer-Verlag, New York, 1982.

58. P. K. Varshney. *Distributed Detection and Data Fusion*. Springer-Verlag, New York, 1997.

59. S. A. Vavasis. *Nonlinear Optimization*. Oxford University Press, New York, 1991.

60. J. von Neumann. Probabilistic logics and the synthesis of reliable organisms from unreliable components. In C. E. Shannon and J. McCarthy (Eds.), *Automata Studies*, Princeton University Press, Princeton, NJ, pp. 43–98, 1956.

61. G. Xing, R. Tan, B. Liu, J. Wang, X. Jia, and C. Yi. Data fusion improves the coverage of wireless sensor networks. In *Proceedings of Mobicom*, Beijing, China, 2009.

62. M. Zhu, S. Ding, R. R. Brooks, Q. Wu, S. S. Iyengar, and N. S. V. Rao. Fusion of threshold rules for target detection in wireless sensor networks. *ACM Transactions on Sensor Networks*, 2010.

18

Soft Computing
Techniques

18.1 Problem Statement...415
18.2 Genetic Algorithms 415
18.3 Simulated Annealing..416
18.4 TRUST..417
18.5 Tabu Search...420
18.6 Artificial Neural Networks 421
18.7 Fuzzy Logic...422
18.8 Linear Programming 423
18.9 Summary..425
         References..425

Richard R. Brooks
Clemson University

18.1 Problem Statement

When trade-offs in an implementation have been identified, many design issues become optimization problems. A very large class of combinatorial optimization problems can be shown to be NP-hard. So that large instances of the problems cannot be solved given current technologies. There is a good chance that it may never be tractable to find the optimal solutions to any of these problems within a reasonable period of time. In spite of this, some techniques have been shown capable of tractably finding near optimal or reasonable answers to these problems. These techniques are usually referred to as heuristics. The general classes of techniques given here are thus known as meta-heuristics. Meta-heuristics, fuzzy sets and artificial neural networks also make up the set of what has come to be known as soft computing technologies. These are advanced computational techniques for problem solving.

This chapter presents a brief introduction to: genetic algorithms, linear programming, simulated annealing, tabu search, TRUST, artificial neural networks and fuzzy sets. An example that uses some of these techniques can be found in Chapter 20. More in depth treatments of this work can be found in [Brooks 96, Brooks 98a, Brooks 98b, Chen 02].

18.2 Genetic Algorithms

Genetic algorithms, first developed in the mid-1960s, apply a Darwinian concept of survival of the fittest to optimization problems. Refer to [Holland 75] for details. An in depth comparison of genetic algorithms versus exhaustive search for a reliability design problem can be found in [Kumar 95]. Several different reproduction strategies are attempted in the literature. The metaphorical approach of genetic search has been shown experimentally to be useful for solving many difficult optimization problems.

Possible solutions to a problem are called chromosomes and a diverse set of chromosomes is grouped into a gene pool. The relative quality of these answers are determined using a fitness function, and this quality is used to determine whether or not the chromosomes will be used in producing the next generation of chromosomes. The contents of high quality solutions are more likely to continue into the next generation. The next generation is generally formed via the processes of crossover: combining elements of two chromosomes from the gene pool, and mutation: randomly altering elements of a chromosome.

A large number of strategies exist for determining the contents of a new generation. Two different genetic algorithms are discussed here. They differ only in their reproduction strategies. The first strategy has been described in [Holland 75]. Each string in the gene pool is evaluated by the fitness function. Based on the quality of the answer represented by the string it is assigned a probability of being chosen for the pool of strings used to produce the next generation. Those with better answers are more likely to be chosen. A mating pool is then constructed by choosing strings at random from the gene pool following the probability distribution derived.

The new generation is formed by mixing the elements of two strings in the mating pool chosen at random. This is generally called crossover. Different crossover probabilities may be used. Where the string is split can be chosen at random, or deterministic schemes are possible as well. A certain amount of mutation usually exists in the system, where one or more elements at random are replaced by random values.

A second strategy that may be applied has been described in [Bean 94]. This strategy is described as elitist since some percent of the strings with the best fitness function values are copied directly into the next generation. In addition to this in our work [Brooks 96, Brooks 98a,b], some of the strings for the next generation are the result of random mutations. Random mutations may be strings where all elements are chosen at random. Performing crossover between random strings in the current generation forms the rest of the new generation. The choice is done entirely at random, no weighting based on the quality of the string is performed. Bean reports that this strategy has been found to be stable experimentally. Its implementation is straightforward.

Genetic algorithms are not sensitive to the presence of local minima since they work on a large number of points in the problem space simultaneously. By comparing many possible solutions they achieve what Holland has termed implicit parallelism, which increases the speed of their search for an optimal solution [Holland 75]. Discussion of the advantages gained by using genetic algorithms instead of exhaustive search for a different optimization problem based on system reliability can be found in [Painton 95].

18.3 Simulated Annealing

Simulated annealing attempts to find optimal answers to a problem in a manner analogous to the formation of crystals in cooling solids. A material heated beyond a certain point will become fluid, if the fluid is cooled slowly the material will form crystals and revert to a minimal energy state. Refer to [Laarhoven 87] for a full description of simulated annealing and a discussion of its scientific basis.

The strategy of the algorithm is again based on a *fitness function* comparing the relative merit of various points in the problem space. As before, vectors corresponding to possible system configurations describe the problem space. The fitness function is the value of the configuration described by the vector. The algorithm starts at a point in the search space. From the algorithm's current position a neighboring point is chosen at random. The cost difference between the new point and the current point is calculated. This difference is used together with the current system temperature to calculate the probability of the new position being accepted. This probability is given by a Boltzmann distribution $e^{-\Delta C/\tau}$. The process continues with the same temperature τ for either a given number of iterations, or until a given number of positions have been occupied, at which time the value τ is decreased. The temperature decreases until no transitions are possible, so the system remains frozen in one position. This occurs only when ΔC is positive for all neighboring points, therefore the position must be a local minimum and may be the global minimum [Press 86].

The simulated annealing method used in our research is based on the algorithm given in [Laarhoven 87]. The algorithm was always modified so that the parameters being optimized and the fitness function are appropriate for our application, and a cooling schedule was found that allows the algorithm to converge to a reasonable solution.

Just as many different reproduction schemes exist for genetic algorithms, several possible *cooling schedules* exist for simulated annealing. A cooling schedule is defined by the initial temperature, the number of iterations performed at each temperature, the number of position modifications allowed at a given temperature and the rate of decrease of the temperature. The answers found by the algorithm are directly dependent on the cooling schedule and no definite rules exist for defining the schedule [Laarhoven 87]. The cooling schedule is important in that it determines the rate of convergence of the algorithm as well as the quality of the results obtained. The complexity of this approach could potentially increase in the order of J^2. This increase is significantly less than the exponential growth of the exhaustive search, but greater than the increase for the genetic search.

18.4 TRUST

The global minimization problem can be stated as follows: Given a function f over some (possibly vector valued) domain D, compute $\bar{x}_{GM} \in D$ such that $f(\bar{x}_{GM}) \leq f(\bar{x})$, $\forall \bar{x} \in D$. Usually, but not necessarily, f is assumed to be continuous and differentiable.

One strategy for global minimum determination is shown in Figure 18.1. Given an initial starting position in the search space, find the local minimum closest to this value. This can be done using gradient descent or a probabilistic search [Bonet 97], such as genetic algorithms and simulated annealing. Once a local minimum is found, it is reasonable to assume that it is not globally optimal. Therefore attempt to escape the local basin of values.

Global optimization research concentrates on finding methods for escaping the basin of attraction of a local minimum. Often, the halting condition is difficult to determine. This means there is a tradeoff between accepting a local minimum as the solution and performing an exhaustive search of the state space. In certain cases, global optimization can become prohibitively expensive when the search space has a large number of dimensions. Thus, there is a natural trade-off between solution accuracy and the time spent searching. Additionally, if an analytic form of the cost function is unknown, as is typically the case, then this problem is more pronounced.

The following characteristics are desirable in a global optimization method:

- Avoid entrapment in local minimum basins
- Avoid performing an exhaustive search on the state space
- Minimize the number of object function evaluations
- Have a clearly defined stopping criteria

In practice, a good global optimization method judiciously balances these conflicting goals.

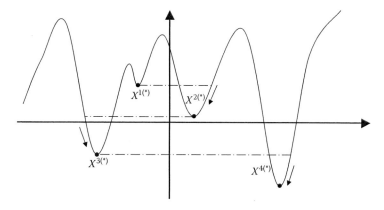

FIGURE 18.1   Tunneling method for determining global minimum.

A number of approaches have been proposed for solving global optimization problems (see references in [Barhen 97]). TRUST [Barhen 97, Cetin 93] is a deterministic global optimization method, which avoids local minima entrapment and exhaustive search. This method defines a dynamic system using two concepts:

1. Subenergy tunneling
2. Non-Lipschitz terminal repellers

to avoid being stuck in local minima. TRUST makes the following assumptions about the cost function *f* and its domain *D* [Barhen 96]:

1. $f: D \rightarrow R$ is a lower semicontinuous function with a finite number of discontinuities.
2. $D \subseteq R^n$ is compact and connected.
3. Every local minima of *f* in *D* is twice differentiable. Furthermore, for any local minima $\boldsymbol{x}_{LM}$, we have

$$\left. \frac{\partial f}{\partial x_i} \right|_{x_{LM}} = 0, \quad \forall i = 1, 2, \ldots, d$$

and

$$\boldsymbol{y}^{\mathrm{T}} \frac{\partial^2 f(x_{LM})}{\partial x^2} \boldsymbol{y} \geq 0, \quad \forall y \in D \tag{18.1}$$

where $\partial^2 f(x_{LM})/\partial x^2$ is the Jacobian matrix given by $\left[\left. \dfrac{\partial^2 f}{\partial x_i \partial x_j} \right|_{x_{LM}} \right]$.

TRUST uses tunneling [Levy 85] to avoid being trapped in local minima. Tunneling is performed by transforming the function *f* into a new function $E(\boldsymbol{x}, \boldsymbol{x}^*)$ with similar extrema properties such that the current local minimum of *f* at some value $\boldsymbol{x}^*$ is a global maximum of $E(\boldsymbol{x}, \boldsymbol{x}^*)$. A value of *f* strictly less than $f(\boldsymbol{x}^*)$ is then found by applying gradient descent. The general algorithm is as follows:

1. Use a gradient descent method to find a local minimum at $\boldsymbol{x}^*$
2. Transform *f* into the following *virtual objective function*

$$E(\boldsymbol{x}, \boldsymbol{x}^*) = E_{sub}(\boldsymbol{x}, \boldsymbol{x}^*) + E_{rep}(\boldsymbol{x}, \boldsymbol{x}^*) \tag{18.2}$$

where

$$E_{sub}(x, x^*) = \log\left(\frac{1}{1 + e^{-(f(x) - f(x^*)) + a}} \right) \tag{18.3}$$

and

$$E_{rep}(x, x^*) = -\frac{3}{4} \rho (x - x^*)^{4/3} u\left(f(x) - f(x^*) \right) \tag{18.4}$$

E_{sub} is the *subenergy tunneling term* and is used to isolate the function range of *f* less than the functional value $f(x^*)$. The difference $f(x) - f(x^*)$ offsets the *f* such that $f(x^*)$ tangentially intersects the x-axis. E_{rep} is the *terminal repeller term* and is used to guide the gradient descent search in the next step. In this term, $u(y)$ is the Heaviside function. Note that $E(x,x^*)$ is a well defined function with a global maximum at x^*.

3. Apply gradient descent to $E(x,x^*)$. This yields the dynamical system

$$\dot{x} = -\frac{\partial E}{\partial x} = -\frac{\partial f}{\partial x}\frac{1}{1+e^{(f(x)-f(x^*))+a}} + \rho(x-x^*)^{1/3}u(f(x)-f(x^*)) \qquad (18.5)$$

An equilibrium state of $\dot{x}$ is a local minimum of $E(x,x^*)$ which in turn is a local or global minimum of the original function f.

4. Until the search boundaries are reached, repeat Step 2 with the new local minima found in Step 3 as the initial extrema.

Complete theoretical development of E_{sub} and E_{rep}, including a discussion of terminal repeller dynamics, can be found in [Barhen 96].

In the one-dimensional case, TRUST guarantees convergence to the globally optimal value. The reason for this is: the transformation of f into $E(x,x^*)$ and subsequent gradient descent search generates monotonically decreasing minimal values of f (see Figures 18.2 and 18.3). This behavior is a result of the compactness of the domain D since the function is not allowed to diverge to infinity. When the last local minimum is found and the method attempts to search $E(x,x^*)$, the difference $f(x) - f(x^*)$ becomes equivalent to the x-axis. Thus, $E(x,x^*) = E_{rep}(x,x^*)$, and the subsequent search on this curve will proceed until the endpoints of the domain D are reached.

In the more difficult multi-dimensional case, there is no theoretical guarantee TRUST will indeed converge to the globally optimal value. To address this problem, [Barhen 96] presents two strategies. The first involves augmenting the repeller term E_{rep} with a weight based on the gradient behavior of f (basically, this is the same concept as momentum in conjugate gradient descent). The effect is to guide the gradient descent search in Step 3 to the closest highest ridge value on the surface of $E(x,x^*)$. The second strategy is to reduce the multidimensional problem into a one-dimensional problem for which TRUST is guaranteed to converge by using hyperspiral embedding from differential geometry.

TRUST is computationally efficient in the number of function evaluations made during the convergence process. Usually, the most computationally demanding aspect of global optimization is evaluating the cost function [Srinivas 94]. In [Barhen 97], TRUST was compared to several other global optimization methods and was found to need far fewer function evaluations to converge. Popular methods of eluding local minima basins other than conjugate gradient descent include simulated

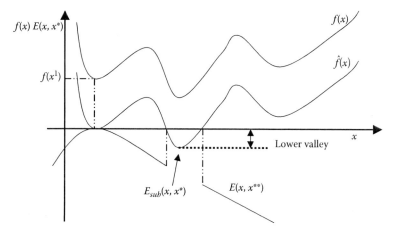

FIGURE 18.2 TRUST in one dimensional case.

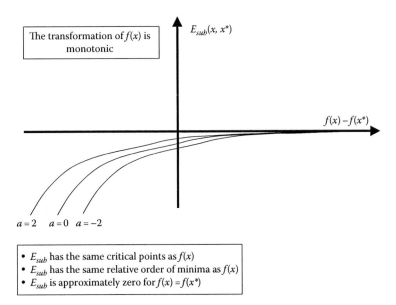

FIGURE 18.3   TRUST: subenergy tunneling function.

annealing and genetic algorithms. Both of these are probabilistic since their entrapment avoidance mechanism requires randomness. TRUST, on the other hand, is deterministic and has a well-defined stopping criteria.

18.5 Tabu Search

Tabu search is one of a class of non-monotonic search methods, which have been proposed for solving optimization problems. All methods discussed have been constructed to find globally optimal solutions in problems containing local minima. Local minima in optimization problems can be either the result of the feasible set being defined by a non-convex surface, or the function $f(x)$ being non-linear. Optimization problems containing local minima are more difficult than the linear problems, which can be solved by linear programming.

Monotonic methods include simulated annealing and threshold acceptance. The majority of search heuristics start at a point in the parameter space and move to neighboring points whose value of $f(x)$ is inferior to the current value. Monotonic methods avoid local minima by occasionally moving to a neighboring point whose value of $f(x)$ is superior to the current value. The amount of increase that will be accepted by the method can be either probabilistic (simulated annealing) or deterministic (threshold acceptance) in nature and is dependent on a parameter whose value decreases as the search progresses.

Non-monotonic methods also occasionally move to points in the search space where the value of $f(x)$ is superior to the current value, but the algorithm's ability to make this move is not dependent on a strictly decreasing parameter. A number of methods have been proposed including Old Bachelor Acceptance [Hu 95] and many variations of Tabu Search [Glover 89]. Old Bachelor Acceptance is non-monotone in that it uses a parameter which is dependent on whether or not previous moves have been accepted, and on the number of moves remaining before the final answer is due [Hu 95]. Tabu search is non-monotone in that it disqualifies a number of moves due to the recent history of moves made by the algorithm [Glover 95]. We limit our discussion to Tabu search since it is the most widely implemented heuristic of this class.

Tabu search involves modifying an existing heuristic search by keeping a list of the nodes in the search space which were visited most recently by the search algorithm. These points then become "tabu" for the algorithm, where "tabu" means that these points are not revisited as long as they are on the list.

This simple modification will allow a search algorithm to eventually climb out of shallow local minima in the search space. It requires less computation than simulated annealing, while providing roughly equivalent or superior results [Glover 95]. Several questions are being studied as to how to optimize Tabu searches, such as the optimal length for the tabu list [Battiti 94], and methods for implementing parallel searches [Taillard 94].

The implementation of the Tabu search we have used relies on a "greedy" heuristic. The search can move one position in each direction. The algorithm evaluates the fitness function for each of these possibilities. Naturally, the search chooses to visit the next node in the direction with the minimum value for the fitness function. When one of these parameter sets is visited it is placed on the tabu list.

Values on the tabu list are disqualified for consideration in the future. Should the search arrive at a neighboring point later, the fitness function value given to parameter sets on the tabu list is set to a very large value essentially disqualifying it from consideration.

As each parameter set is visited by the search, the value attributed to it by the fitness function is compared to the parameter set already visited with the smallest value for the fitness function up to this point. If the value is smaller the parameter set now becomes the best fit found.

It is impossible to find a clear stopping criteria for this algorithm since the only way to be sure that the global minimum for the fitness function has been found is through an exhaustive search of the search space.

Tabu search has been successfully implemented in a number of systems solving problems related to intelligent sensor processing. Chiang and Kouvelis have used a variation of tabu search in order to find paths for navigation by autonomous guided vehicles (AGVs) [Chiang 94]. AGVs are of increasing importance in material handling and manufacturing applications. Another area of direct relevance for the sensor architectures topic is the use of distributed systems. Research has been done using tabu search to map tasks onto processors in distributed systems in a way that minimizes communication time requirements for the system [Chakrapani 95]. The design of electro-magnetic sensor devices has also been optimized through research using tabu search methods. The variables involved in designing a magnetic that produces a homogeneous field can be reduced to a finite alphabet that the tabu search strategy uses to design near-optimal components for Magnetic Resonance Imaging systems [Fanni 96].

18.6 Artificial Neural Networks

Artificial neural networks, like simulated annealing and genetic algorithms, originated as a simulation of a natural phenomena. The first artificial neural network application was developed in the 1940s by McCulloch and Pitts as a system designed to mimic a theory of how biological nervous systems work [Davalo 89]. Since then, a number of variations of artificial neural networks have been developed, some of which bear little resemblance to functioning biological systems [Davalo 89, Gulati 91, Hinton 92, Kak 94].

Neural networks are also called *connectionist learning schemes*, since they store information in weights given to connections between the artificial neurons [Hinton 92]. In the vast majority of implementations, the artificial neurons are a very primitive computational device. The individual neuron takes weighted inputs from a number of sources, performs a simple function, and then produces a single output. The thresholding function is commonly a step function, a sigmoid function, or another similar function [Kak 94]. The computational power of a neural network comes from the fact that a large number of these simple devices are interconnected in parallel. The overall system may be implemented either in hardware or software.

Probably the most common architecture for neural networks is termed a *multi-layer feed-forward* network. These networks can consist of any number of neurons arranged in a series of layers. Each layer i of the network takes input from the previous layer $i - 1$, and its output is used as input by layer $i + 1$. Layer 0 of the network is the input data, and the outputs of the last layer n of artificial neurons is the output of the network. The layers between 0 and n are termed *hidden layers*. Since no loops exist within the network, the system can be treated as a simple black box.

The number of neurons at each layer in multi-layer feed-forward networks can be any integer greater than zero, and the number of inputs and outputs are not necessarily equal. A neural network of this type, with n inputs and m outputs, can approximate any function that maps binary vectors of length n onto binary vectors of length m. In doing this, each neuron defines a hyperplane in the solution space. The neuron fires if the region defined by the inputs is on one side of the hyperplane, and does not fire if the region defined by the inputs is on the other side [Rojas 96].

The architecture needed for a neural network depends on the number of hyperplanes needed to define the desired mapping, and whether the mapping is defined by a convex region, a non-convex region without holes, or a non-convex region with holes. A convex region can be defined by a network with only one layer. Non-convex regions require two layers, and regions containing holes require three layers. Although no limit exists to the number of layers that could be used, more than three layers are not required for this type of application [Kak 94]. No clear guidelines exist for determining the number of neurons required at each level for a given application.

Each neuron at layer i outputs the value of a function of the linearly weighted outputs of all the neurons at layer $i - 1$. As mentioned earlier, this function is normally a step function, a sigmoid function, or a similar function such as the hyperbolic tangent. A method is needed for determining the weights to be applied to each input. The most widely used method for doing this is *back propagation* [Rojas 96]. For back propagation to be used, the function computed must be differentiable. A more computationally efficient way for computing the desired weight tables can be found in [Kak 94].

Neural networks can also be implemented to serve as self-associative memories. The network used for this type of application generally consists of a single layer of neurons, all of which are fully connected. These types of networks are called *Hopfield Networks*. The weights can then be defined so that the network has a number of stable states. The input to the network is used as initial values for the neurons. Each neuron then automatically readjusts its output to move towards one of the stable states. This type of learning is called *Hebbian learning* [Kak 94].

In addition to the above applications, neural networks can automatically classify inputs into equiprobable sets. This approach was first pioneered by Kohonen. These networks compute their own weights to determine rational groupings of the input vectors [Davalo 89].

Probably the most appealing aspect of neural networks is that feed forward networks can infer functions that differentiate between groups of objects based on raw data in the form of training sets. By using the network responses to data to readjust the weighting parameters networks can be designed to infer any m to n mapping function. Unfortunately, knowing what sets of data will work adequately as training data is often difficult. Training sets that do not accurately represent the overall population of data will result in a network that is faulty. Training sets that are too large can result in *over-training* the network so that the network does not correctly generalize the problem and only recognizes the instances used in training the network. Recently Rao has derived bounds on the size of the training set necessary to be certain of correctly learning a concept [Rao 95, Rao 96].

One drawback to the use of neural networks is the potential size of the system. For systems involving $n \times m$ two-dimensional images, each neuron at the lowest level would require an $n \times m$ matrix of weights to be able to treat the raw data. This is generally unfeasible. For this reason, most neural networks that treat visual images require a preprocessing step to extract features of interest. This step amounts to encoding the image into an alphabet that a neural network can easily manipulate. Often this step is the most difficult one in developing these types of pattern recognition systems.

18.7 Fuzzy Logic

The last soft computing method we discuss is fuzzy logic and fuzzy sets. Fuzzy logic is a type of multivalued logic that has been identified as a new technology with special promise for use with sensor fusion problems [Luo 92] as are fuzzy sets [Waltz 91]. Fuzzy sets and fuzzy logic can be presented as a general theory that contains probability [Klir 95] and Dempster-Shafer evidential reasoning [Dubois 92] as special-cases. Fuzzy sets are similar to traditional set theory, except that every member of a fuzzy set has an associated value assigned by a *membership function* that quantifies the amount of membership the element has in the function [Waltz 91].

In addition to membership, a number of functions exist for fuzzy sets that are analogous to the functions performed on traditional, or *crisp*, sets. In fact, a number of possible functions exist for the operations of complementation, union, and intersection with fuzzy sets. Similarly, fuzzy logic presents a simple framework for dealing with uncertainty [Klir 95]. The number of possible implementations of fuzzy sets and fuzzy logic make difficult saying exactly which implementation is best suited for use in sensor fusion methodology.

Sensor fusion is most closely associated with two areas in which fuzzy sets and logic have been used with success: measurement theory and the development of fuzzy controllers [Klir 95]. Both areas are closely related to the sensor fusion problems and possibly a number of existing methods could be phrased in terms of fuzzy methodology. Other related areas that fuzzy techniques have successfully been applied to are pattern recognition and inference of statistical parameters [Klir 95]. These point to the potential for successful research applying fuzzy methods directly to sensor fusion questions. Of particular interest is the growing synergism between the use of fuzzy sets, artificial neural networks, and genetic algorithms [Klir 95].

18.8 Linear Programming

Although it is not a soft computing method, linear programming is discussed here for completeness, since it is probably the most widely used optimization methodology. Given a cost vector $c = (c_1, c_2, \ldots c_n)$, linear programming searches for the solution vector $x = (x_1, x_2, \ldots x_n)$, which optimizes $f(x) = \Sigma c_i x_i$ and also satisfies a number of given constraints $\Sigma a_{ij} x_j \leq b_i$ (or $\Sigma a_{ij} x_j \geq b_i$). Note that the cost function optimization can solve either maximization or minimization problems, depending on the signs of the coefficients c_j. The set of problems that can be put into this form is large and it includes a number of practical applications. Note that the linear nature of this problem formulation excludes the existence of local minima of $f(x)$ in the parameter space.

Two major classes of methods exist for solving linear programming problems: the simplex method and interior methods. The simplex method was developed by a research team led by Dantzig in 1947 [Alj 90]. Conceptually it considers the problem as an N-dimensional space where each vector x is a point in the space, and each constraint defines a hyper-plane dividing the space into two half-spaces. The constraints therefore define a convex region containing the set of points which are possible answers to the problem, called the feasible set [Strang 76]. The set of optimal answers must contain a vertex. The simplex algorithm starts from a vertex on the N-dimensional surface and moves in the direction of greatest improvement along the surface until the optimal point is found [Alj 90, Strang 76].

The first interior method was found by Karmarkar [Alj 90]. Interior methods start with a point in the feasible set and move through the interior of the feasible set towards the optimal answer without being constrained to remain on the defining surface. Research is active in finding less computationally expensive implementations of both interior and simplex methods.

To represent how linear programming works, we solve an example problem using a variation of the simplex method. We attempt to maximize the equation $5x_1 + 3x_2 + 2x_3$ within the constraints:

$$
\begin{aligned}
x_1 - x_2 &\leq 6 &\quad \text{inequality 1} \\
x_1 + 3x_2 - x_3 &\leq 14 &\quad \text{inequality 2} \\
3x_1 + x_2 + 2x_3 &\leq 22 &\quad \text{inequality 3}
\end{aligned}
\tag{18.6}
$$

To make the three constraints equations instead of inequalities, we create fictitious variables: $x_{1'}$ is the difference between the left-hand side of inequality 1 (i.e., $x_1 - x_2$) and 6; similarly, $x_{2'}$ ($x_{3'}$) is the difference between the left-hand side of inequality 2 (3) and 14 (22). This definition also provides us with a starting point on the hyper-surface that defines the feasible set. The starting point is $x_1 = 0$, $x_2 = 0$, $x_3 = 0$, $x_{1'} = 6$, $x_{2'} = 14$, and $x_{3'} = 22$, or in vector form (0, 0, 0, 6, 14, 22).

We can now phrase the problem at this starting point in tabular form:

c_{ij}	j	x_1	x_2	x_3	x_1'	x_2'	x_3'	u
0	x_1'	1	−1	0	1	0	0	6
0	x_2'	1	3	−1	0	1	0	14
0	x_3'	3	1	2	0	0	1	22
								F
Δ		5	3	2	0	0	0	0

Conceptually, each constraint defines a hyper-plane that separates the feasible set from the rest of the answer space. If an optimal answer exists, it will be located on at least one of the hyperplanes. It may exist at an intersection of all the hyperplanes. The left-most box in the tabular representation above contains two columns: j says which of the variables considered is the limiting factor for the hyperplane defined by the corresponding constraint, c_{ij} is the value of one unit of j. The rightmost box u shows the number of units of j in the current answer. Beneath the right-most box is F, the value of the current answer, which is the summation of c_{ij} times the corresponding element of the right-most box, in this case zero.

The box marked Δ is the coefficients of the corresponding variables in the equation to be optimized. The middle box is simply a translation of the constraint equations into matrix form. Verify that you understand the derivation of these tables before continuing.

The algorithm follows Dantzig's simplex rules by moving the solution in the search space in the direction that produces the greatest rate of improvement. This variable has the largest coefficient in Δ. For the table above, this is variable x_1 since it has the coefficient 5. To determine which constraint will limit the growth of x_1, the value of the coefficient of x_1 for each constraint is used to divide the corresponding value in u. The smallest resulting value determines the limiting constraint. In this case the values are 6/1, 14/1, and 22/3. Since 6/1 is the smallest value, x_1 replaces $x_{1'}$ as the limiting variable in the first constraint. The new table is now

c_{ij}	j	x_1	x_2	x_3	x_1'	x_2'	x_3'	u
5	x_1	1	−1	0	1	0	0	6
0	x_2'	0	4	−1	−1	1	0	8
0	x_3'	0	4	2	−3	0	1	4
								F
Δ		0	8	2	-5	0	0	30

When we replace $x_{1'}$ with x_1 in constraint 1, the following steps are performed:

- The row corresponding to constraint 1 in the leftmost box is changed to contain the coefficient of the limiting variable in the column c_{ij} and to contain the limiting variable in column j.
- The row corresponding to constraint 1 is set to *canonical form* by dividing all elements of the middle box and the corresponding element of u by the coefficient of x_1 in that row. In this case the value is 1 so the row remains unchanged.
- The value of the coefficient of x_1 is removed from all the other rows by subtracting a multiple of row 1 from the other rows. Note that the corresponding modifications are made to u as well.
- Similarly, the value of the coefficient of x_1 is set to 0 in Δ by subtracting a multiple of row 1 from Δ.

The value of F is recomputed using the new values in c_{ij} and u.

The variable with the largest coefficient in Δ is now x_2. Since the coefficient of x_2 is negative in the row corresponding to constraint 1, we do not consider that constraint. The two remaining constraints are 2 with a value of 8/4, and constraint 3 with a value of 4/4. The value corresponding to 3 is the smallest, so we replace variable x_3, with variable x_2 as the limiting factor in constraint 3. Using the same process as above, this gives us the following table:

c_{ij}	j	x_1	x_2	x_3	x_1'	x_2'	x_3'	u
5	x_1	1	0	1/2	1/4	0	1/4	7
0	x_2'	0	0	−3	2	1	−1	4
3	x_2	0	1	1/2	−3/4	0	1/4	1
								F
Δ		0	0	−2	1	0	−2	38

The only positive coefficient in Δ is now that of x_1', so it will be used to replace one of the existing coefficients. Since its coefficient is negative in constraint 3, we do not consider that constraint. The value corresponding to constraint 2 (4/2) is less than the value corresponding to constraint 3 (4 * 7). The same process as before is used to replace x_2' with x_1'. This final table results:

c_{ij}	j	x_1	x_2	x_3	x_1'	x_2'	x_3'	u
5	x_1	1	0	7/8	0	−1/8	3/8	61/2
0	x_1'	0	0	−3/2	1	1/2	−1/2	2
3	x_2	0	1	−5/8	0	3/8	−1/8	21/2
								F
Δ		0	0	−1/2	0	−1/2	−3	40

Since all elements of Δ are now negative or zero, no improvements can be made to the current answer. The process stops and 40 is the optimal answer to our equation. The optimal answer consists of 6.5 units of x_1, 2.5 units of x_2, and no units of x_3.

18.9 Summary

This chapter provided a brief description of optimization and soft computing techniques. These techniques are widely used decision making tools for difficult optimization problems. Each method presented can be considered a class of heuristic. [Brooks 98a,b] compares the use of many of these techniques for many sensor network design issues. The results in [Brooks 98a,b] can be used to help decide which technique is best suited for a given application.

References

[Alj 90] A. Alj and R. Faure, *Guide de la Recherche Operationelle*, Vols. 1–2, Masson, Paris, France, 1990.

[Barhen 96] J. Barhen and V. Protopopescu, Generalized TRUST algorithms for global optimization, *State of the Art in Global Optimization*, C. A. Floudas and P. M. Pardalos, Eds., Kluwer Academic Publishers, Dordrecht, the Netherlands, 1996, pp. 163–180.

[Barhen 97] J. Barhen, V. Protopopescu, and D. Reister, TRUST: A deterministic algorithm for global optimization, *Science*, 276, 1094–1097, May 16, 1997.

[Battiti 94] R. Battiti and G. Tecchioli, The reactive tabu search, *ORSA Journal on Computing*, 6(2), 126–140, Spring 1994.

[Bean 94] J. C. Bean, Genetic algorithms and random keys for sequencing and optimization, *ORSA Journal on Computing*, 6(2), 154–160, Spring 1994.

[Bonet 97] J. S. De Bonet, C. L. Isbell, Jr., and Paul Viola, MIMIC: Finding optima by estimating probability densities, *Advances in Neural Information Processing Systems*, MIT Press, Cambridge, MA, 1997.

[Brooks 96] R. R. Brooks, S. S. Iyengar, and J. Chen, Automatic correlation and calibration of noisy sensor readings using elite genetic algorithms, *Artificial Intelligence*, 84(1–2), 339–354, July 1996.

[Brooks 98a] R. R. Brooks, S. S. Iyengar, and S. Rai, A comparison of GAs and simulated annealing for cost minimization in a multi-sensor system, *Optical Engineering*, 37(2), 505–516, February 1998.

[Brooks 98b] R. R. Brooks and S. S. Iyengar, *Multi-Sensor Fusion: Fundamentals and Applications with Software*, Prentice Hall PTR, Saddle River, NJ, 1998.

[Cetin 93] B. C. Cetin, J. Barhen, and J. W. Burdick, Terminal repeller unconstrained subenergy tunneling (TRUST) for fast global optimization, *Journal of Optimization Theory and Applications*, 77(l), 97–126, 1993.

[Chakrapani 95] J. Chakrapani and J. Skorin-Kapov, Mapping tasks to processors to minimize communication time in a multiprocessor system, *The Impact of Emerging Technologies on Computer Science and Operations Research*, Kluwer Academic Publishers, Norwell, MA, 1995.

[Chen 02] Y. Chen, R. R. Brooks, S. S. Iyengar, S. V. N. Rao, and J. Barhen, Efficient global optimization for image registration, *IEEE Transactions on Knowledge and Data Engineering*, 14(1), 79–92, January 2002, http://www.computer.org/tkde/Image-processing.html

[Chiang 94] W. C. Chiang and P. Kouvelis, Simulated annealing and tabu search approaches for unidirectional flowpath design for automated guided vehicle systems, *Annals of Operation Research*, 50, 115–142, 1994.

[Davalo 89] E. Davalo and P. Daïm, *Des Reseaux de Neurones*, Editions Eyrolles, Paris, France, 1989.

[Dubois 92] D. Dubois and H. Prade, Combination of fuzzy information in the framework of possibility theory, *Data Fusion in Robotics and Machine Intelligence*, M. A. Abidi and R. C. Gonzales, Eds., Academic Press, Boston, MA, 1992, pp. 481–503.

[Fanni 96] A. Fanni, G. Giacinto, and M. Marchesi, Tabu search for continuous optimization of electromagnetic structures, submitted to *International Workshop on Optimization and Inverse Problems in Electromagnetism*, June 1996.

[Glover 89] F. Glover, Tabu search—Part 1, *ORSA Journal on Computing*, 1, 190–206, 1989.

[Glover 95] F. Glover, Tabu thresholding: Improved search by nonmonotonic techniques, *ORSA Journal on Computing*, 7(4), 426–442, Fall 1995.

[Gulati 91] S. Gulati, J. Barhen, and S. S. Iyengar, Neurocomputing formalisms for computational learning and machine intelligence, *Advances in Computers*, Academic Press, Boston, MA, 1991.

[Hinton 92] G. E. Hinton, Connectionist learning procedures, *Machine Learning*, J. Carbonnell, Ed., MIT/Elsevier, Cambridge, MA, 1992, pp. 185–234.

[Holland 75] J. H. Holland, *Adaptation in Natural and Artificial Systems*, University of Michigan Press, Ann Arbor, MI, 1975.

[Hu 95] T. C. Hu, A. B. Kahng, and C. A. Tsao, Old bachelor acceptance: A new class of non-monotone threshold accepting methods, *ORSA Journal on Computing*, 7(4), 417–425, Fall 1995.

[Kak 94] S. Kak, *Neural Networks, Iterative Maps, and Chaos*, Course Notes, Louisiana State University, Baton Rouge, LA, 1994.

[Klir 95] G. J. Klir and B. Yuan, *Fuzzy Sets and Fuzzy Logic: Theory and Applications*, Prentice Hall, Englewood Cliffs, NJ, 1995.

[Kumar 95] A. Kumar, R. Pathak, and Y. Gupta, Genetic-algorithm-based optimization for computer network expansion, *IEEE Transactions on Reliability*, 44(1), 63–72, March 1995.

[Laarhoven 87] P. J. M. van Laarhoven and E. H. L. Aarts, *Simulated Annealing: Theory and Applications*, D. Reidel Publishing Co., Dordrecht, the Netherlands, 1987.

[Levy 85] A. V. Levy and A. Montalvo, The tunnelling algorithm for the global minimization of functions, *SIAM Journal on Scientific and Statistical Computing*, 6, 15–29, 1985.

[Luo 92] R. Luo and M. Kay, Data fusion and sensor integration: State-of-the-art 1990s, *Data Fusion in Robotics and Machine Intelligence*, M. A. Abidi and R. C. Gonzales, Eds., Academic Press, Boston, MA, 1992, pp. 7–136.

[Painton 95] L. Painton and J. Campbell, Genetic algorithms in optimization of system reliability, *IEEE Transactions on Reliability*, 14(2), 172–178, June 1995.

[Press 86] W. Press, S. Teukolsky, W. Vetterling, and B. Flannery, *Numerical Recipes in Fortran*, Cambridge University Press, Cambridge, U.K., 1986, pp. 436–448.

[Rao 95] N. S. V. Rao, Fusion rule estimation in multiple sensor systems with unknown noise distributions, *Parallel and Distributed Signal and Image Integration Problems*, R. N. Madan, N. S. V. Rao, V. P. Bhatkar and L. M. Patnaik, Eds., World Scientific, Singapore, 1995, pp. 263–279.

[Rao 96] N. S. V. Rao, Multiple sensor fusion under unknown distributions, *Proceedings of Workshop on Foundations of Information/Decision Fusion with Applications to Engineering Problems*, N. S. V. Rao, V. Protopepscu, J. Barhen, and G. Seetharaman, Eds., Washington, DC, 1996, pp. 174–183.

[Rojas 96] R. Rojas, *Neural Networks: A Systematic Introduction*, Springer Verlag, Berlin, Germany, 1996.

[Srinivas 94] M. Srinivas and L. M. Patnaik, Genetic algorithms: A survey, *IEEE Computer*, 27(6), 17–26, June 1994.

[Strang 76] G. Strang, *Linear Algebra and Its Applications*, Academic Press, New York, 1976.

[Taillard 94] E. Taillard, Parallel taboo search techniques for the job shop scheduling problem, *ORSA Journal on Computing*, 6(2), 108–117, Spring 1994.

[Waltz 91] E. L. Waltz and J. Llinas, *Sensor Fusion*, Artech House, Norwood, MA, 1991.

19

Estimation and Kalman Filters

19.1 Introduction 429
19.2 Overview of Estimation Techniques..432
         System Models • Optimization Criteria • Optimization
         Approach • Processing Approach
19.3 Batch Estimation.. 440
         Derivation of Weighted Least Squares Solution • Processing
         Flow • Batch Processing Implementation Issues
19.4 Sequential Estimation and Kalman Filtering............................. 445
         Deviation of Sequential Weighted Least Squares
         Solution • Sequential Estimation Processing Flow
19.5 Sequential Processing Implementation Issues 449
         Filter Divergence and Process Noise • Filter
         Formulation • Maneuvering Targets • Software Tools
References...452

David L. Hall
*The Pennsylvania
State University*

19.1 Introduction

Within the overall context of multi-sensor data fusion, Kalman filters provide a classic sequential estimation approach for fusion of kinematic and attribute parameters to characterize the location, velocity and attributes of individual entities (e.g., targets, platforms, events, or activities). This chapter (based on material originally presented in Chapter 4 of Hall [1992]) provides an introduction to estimation, sequential processing, and Kalman filters. The problem of fusing multi-sensor parametric data (from one or more sensors) to yield an improved estimate of the state of the entity is a classic problem. Examples of estimation problems include

1. Using positional data such as line-of-bearing (angles), range or range-rate observations to determine the location of a stationary entity (e.g., determining the location and velocity of a ground-based target using a distributed network of ground sensors)
2. Combining positional data from multiple sensors to determine the position and velocity of a moving object as a function of time (the tracking problem)
3. Estimating attributes of an entity, such as size or shape, based on observational data
4. Estimating the parameters of a model (e.g., the coefficients of a polynomial), which represents or describes observational data

The estimation problem involves finding the value of a state vector (e.g., position, velocity, polynomial coefficients) that best fits, in a defined mathematical sense, the observational data. From a mathematical viewpoint, we have a redundant set of observations (viz., more than the minimum number of observations for a minimum data-set solution), and we seek to find the value of set of parameters that provides

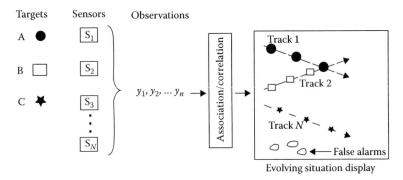

FIGURE 19.1 Conceptual multi-target, multi-sensor target-tracking problem. Data from multiple sensors, observing multiple targets are fused to obtain estimates of the position, velocity, attributes, and identity of the targets. (From Hall, D., *Mathematical Techniques in Multisensor Data Fusion*, Artech House, Inc., Boston, MA, 2003.)

a "best fit" to the observational data. In general, the observational data are corrupted by measurement errors, signal propagation noise, and other factors, and we may or may not know a priori the statistical distribution of these sources of error. Figure 19.1 shows a sample target-tracking problem in which multiple targets are tracked by multiple sensors.

In this example, it is easy to determine which observations "belong" to which target tracks (because of our artificial representation of data using geometrical icons). However, in general, we do not know a priori how many targets exist, which observations belong to which targets, or whether observations are evidence of a moving target or are simply false alarms.

Estimation problems may be dynamic, in which the state vector changes as a function of time, or static in which the state vector is constant in time. This chapter introduces the estimation problem and develops specific solution strategies. For simplicity, it is assumed that the related problems of data registration and data correlation have been (or can be) solved. That is, we assume that the observational data are allocated to distinct groups or sets, each group belonging to a unique entity or object. In practice, data association and estimation techniques must be interleaved to develop an overall solution, especially for multi-target tracking problems. A detailed discussion of techniques for data correlation and association is provided by Hall (2003).

The history of estimation techniques has been summarized by Sorenson (1970). The first significant effort to address estimation was Karl Friedrick Gauss invention of the method of least squares to determine the orbits of planets, asteroids, and comets from redundant data sets. In celestial mechanics, techniques for determining orbital elements from minimum data sets are termed initial orbit methods or minimum data set techniques. Gauss utilized the method of least squares in 1795 and published a description of the technique in 1809 (Gauss 1963). Independently, Legendre invented the least-squares method and published his results in 1806 (Legendre 1806). The resulting controversy of intellectual propriety prompted Legendre to write to Gauss complaining, "Gauss, who was already so rich in discoveries, might have had the decency not to appropriate the method of least-squares" (quoted in Sorenson [1970] and Bell [1961]). Gauss' contribution included not only the invention of the least-squares method, but also the introduction of such modern notions as

1. *Observability*—the issue of how many and what types of observations are necessary to develop an estimate of the state vector
2. *Dynamic modeling*—the need for accurate equations of motion to describe the evaluation of a state vector in time
3. *A priori estimate*—the role of an initial (or starting) value of the state vector in order to obtain a solution
4. *Observation noise*—set the stage for a probabilistic interpretation of observational noise

Subsequent historical developments of estimation techniques include Fisher's probabilistic interpretation of the least squares method (Fischer 1912) and definition of the maximum likelihood method, Wiener (1949) and Kolmogorov's (1941) development of the linear minimum mean square error method, and Kalman's formulation of a discrete-time, recursive, minimum mean square filtering technique (viz the Kalman filter [Kalman 1963]). The Kalman filter (also independently described by Bucy and Sterling) was motivated by the need for rapid prediction of the position of early spacecraft using very limited computational capability.

Numerous texts and papers have been published on the topic of estimation (and in particular on sequential estimation). Blackman (1986) provides a detailed description of sequential estimation for target tracking, Gelb (1974) describes sequential estimation from the viewpoint of control theory and Ramachandra (2000) describes the application of Kalman filtering for radar tracking. In addition, Grewal and Andrews (2001) provide practical advice and MATLAB® code for implementing Kalman filters.

An conceptual view of the general estimation processing flow is illustrated in Figure 19.2.

The situation is illustrated for a positional fusion (e.g., target tracking and identification) problem. A number of sensors observe location parameters such as azimuth, elevation, range, or range rate and attribute parameters such as radar cross section. The location parameters may be related to the dynamic position and velocity of an entity via observation equations. For each sensor, a data alignment function transforms the "raw" sensor observations into a standard set of units and coordinate reference frame. An association/correlation process groups observations into meaningful groups— each group representing observations of a single physical entity or event. The associated observations represent collections of observation-to-observation pairs, or observation-to-track pairs, which "belong" together. An estimation process combines the observations to obtain a new or improved estimate of a state vector, $\mathbf{x}(t)$, which best fits the observed data. The estimation problem illustrated in Figure 19.2 is the level 1 process within the Joint Directors of Laboratories data fusion process model (Hall 1992; Kessler 1992; Steinberg 2001). It also assumes a centralized architecture in which

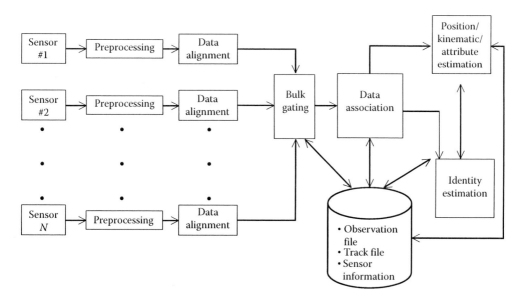

FIGURE 19.2 A conceptual view of the estimation process. Data from each sensor or source must be preprocessed to align the data with respect to a common coordinate frame and grouped (associated/correlated) such that each data set represents data belonging to an individual target. Positional, kinematic, and attribute estimation can be performed by various techniques including Kalman filtering. (From Hall, D., *Mathematical Techniques in Multisensor Data Fusion*, Artech House, Inc., Boston, MA, 2003.)

observations are input to an estimation process for combination. In practice, other architectures such as distributed processing or hybrid-processing architectures could be used for the estimation fusion process.

19.2 Overview of Estimation Techniques

Estimation techniques have a rich and extensive history. An enormous amount has been written, and numerous methods have been devised for estimation. This section provides a brief overview of the choices and common techniques available for estimation. Figure 19.3 summarizes the alternatives and issues related to estimation. These include the following:

- *System models*: What models will be selected to define the problem under consideration? What is to be estimated (viz. what is the state vector sought)? That is, what set of parameters are sufficient and necessary to provide a description of the system "state"? How do we predict the state vector in time? How are the observations related to the state vector? What assumptions (if any) can we make about the observation process (e.g., noise, biases, etc.)?
- *Optimization criteria*: How will we define a criteria to specify *best fit*? That is, what equation will be used to specify that a state vector best fits a set of observations?
- *Optimization approach*: Having defined a criteria for best fit, what method will be used to find the unknown value of the state vector which satisfies the criterion?
- *Processing approach*: Fundamentally, how will the observations be processed, e.g., on a batch mode in which all observations are utilized after they have been received, or sequentially, in which observations are processed one at a time as they are received?

19.2.1 System Models

An estimation problem is defined by specifying the state vector, observation equations, equations of motion (for dynamic problems), and other choices such as data editing, convergence criteria, and coordinate systems necessary to specify the estimation problem. We will address each of these in turn.

A fundamental choice in estimation is to specify what parameters are to be estimated, i.e., what is the independent variable or state vector $\mathbf{x}(t)$ whose value is sought? For positional estimation, a typical choice for $\mathbf{x}$ is the coordinates necessary to locate a target or entity. Examples include the geodetic

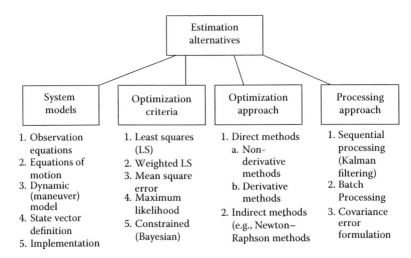

FIGURE 19.3 Overview of estimation alternatives. Design of a process for state vector estimation requires selection of system models, optimization criteria, an optimization approach, and a basic data processing approach.

latitude and longitude (ϕ, λ) of a object on the surface of the earth, the three dimensional Cartesian coordinates (x, y, z) of an object with respect to Earth-centered inertial coordinates, or the range and angular direction (r, azimuth, elevation) of an object with respect to a sensor. For non-positional estimation, the state vector may be selected as model coefficients (e.g., polynomial coefficients) to represent or characterize data. State vectors may also include coefficients that model sensor biases and basic system parameters. For example, at the National Aeronautics and Space Administration (NASA) Goddard Space Flight Center, large-scale computations are performed to estimate a state vector having several hundred components including; position and velocity of a spacecraft, spherical harmonic coefficients of the earth's geo-potential, sensor biases, coefficients of atmospheric drag, precise location of sensors, and many other parameters (Zavaleta and Smith 1975).

The choice of what to estimate depends upon defining what parameters are necessary to characterize a system and to determine the future state of a system, and what parameters may be estimated based on the observed data. The latter issue is termed *observability*. The issue concerns the extent to which it is feasible to determine components of a state vector based on observed data. A weak relationship may exist between a state vector element and the data. Alternatively, two or more components of a state vector may be highly correlated with the result that variation of one component to fit observed data may be indistinguishable from variation in the related state vector component. Hall and Waligora (1978) provide an example of the inability to distinguish between camera biases and orientation (attitude) of Landsat satellite data. Deutsch (1963) also provides a discussion of observability. The general rule of thumb in selecting a state vector is to choose the minimum set of components necessary to characterize a system under consideration. It is often tempting to choose more, rather than fewer, components of a state vector. An example of this occurs when researchers attempt to represent observational data using a high-order polynomial. Hence, a cautious, *less is better* approach is recommended for selecting the components of a state vector.

A second choice required to define the estimation problem is the specification of the observation equations. These equations relate the unknown state vector to predicted observations. Thus, if $\mathbf{x}(t)$ is a state vector, and $y_i(t_i)$ is an observation, then

$$z_i = g(x_j(t_i)) + \eta \tag{19.1}$$

predicts an observation, $z_i(t_i)$, which would match $y_i(t_i)$ *exactly if we* knew the value of $\mathbf{x}$ and we also knew the value of the observational noise, η. The function $g(x_j(t_i))$ represents the coordinate transformations necessary to predict an observation based on an assumed value of a state vector. If the state vector $\mathbf{x}$ varies in time, then the estimation problem is dynamic and requires further specification of an equation of motion which propagates the state vector at time, t_0, to the time of an observation, t_i, i.e.,

$$x(t_i) = \Phi(t_i, t_0)x(t_0) \tag{19.2}$$

The propagation of the state vector in time (Equation 19.2) may involve a simple truncated Taylor series expansion

$$x(t_0) = x(t_0) + \dot{x}(t_0)\Delta t + \frac{1}{2}\ddot{x}(t_0)\Delta t^2 \tag{19.3}$$

where
  $\Delta t = t_i - t_0$
  $\dot{x}$ represents the velocity at time t_0
  $\ddot{x}$ represents the acceleration at time, t_0

In other situations, more complex equations of motion may be required. For example, in astrodynamical problems, the equation of motion may involve second-order, nonlinear, simultaneous differential equations in which the acceleration depends upon the position, velocity, and orientation of a body and the positions of third bodies such as the sun and moon. In that case, the solution of the differential equations of motion requires significant computational effort utilizing numerical integration techniques. An example of such a problem may be found in Zavaleta and Smith (1975) and Deutsch (1963). The selection of the equations of motion (19.2) depends on the physics underlying the dynamic problem. Tracking problems may require models to predict propulsion, target maneuvering, motion over terrain or through a surrounding media, or even move-stop-move motions. Selection of an appropriate equation of motion must trade off physical realism and accuracy versus computational resources and the required prediction interval. For observations closely spaced in time (e.g., for a radar tracking an object for a brief interval) a linear model may be sufficient. Otherwise, more complex (and computationally expensive) models must be used. A special difficulty for positional estimators involves maneuvering targets. Both Blackman (1986) and Waltz and Llinas (1990) discuss these issues.

Closely coupled with the selection of the equations of motion is the choice of coordinate systems in which the prediction is performed. Some coordinate reference frames may be *natural* for defining an equation of motion. For example, earth centered (geocentric) inertial Cartesian coordinates provide an especially simple formulation of the equations which describe the motion of a satellite about the earth (see Deutsch [1963] and Escobol [1976]). By contrast, for the same case, topocentric (earth surface) non-inertial coordinates provide an equation of motion that must introduce artificial acceleration components (viz., due to corriolis *forces*) to explain the same motion. However, despite this, the use of a topocentric non-inertial coordinate frame may be advisable from a system viewpoint. Kamen and Sastry (1990) provide an example of such a tradeoff. Sanza et al. (1994) present the equations of motion and sequential estimation equations for tracking an object in a spherical (r, θ, φ) coordinate reference frame.

19.2.2 Optimization Criteria

19.2.2.1 *t* Criteria

Having established the observation equations that relate a state vector to predicted observations, and equations of motion (for dynamic problems), a key issue involves the definition of *best fit*. We seek to determine a value of a state vector, $\mathbf{x}(t)$, which best fits the observed data. There are several ways to define *best fit*. Each of these formulations involves a function of the residuals:

$$v_i = [y_i(t_i) - z_i(t_i)] \tag{19.4}$$

Here v_i is the vector difference between the ith observation, $y_i(t_i)$, at time t_i and the predicted observation, $z_i(t_i)$. The predicted observation, z_i is a function of the state vector $\mathbf{x}(t_o)$ via Equation 19.1, and hence v_i, is also a function of $\mathbf{x}(t_o)$.

Various functions of v_i have been defined and used for estimation. A summary of some of these criteria is provided in Table 19.1. A function is chosen which provides a measure of the extent to which the predicted observations match the actual observations. This function of the unknown state vector, $\mathbf{x}$, is sometimes termed a loss function because it provides a measure of the penalty (i.e., poor data fit) for an incorrect estimate of $\mathbf{x}$. The state vector, $\mathbf{x}$, is varied until the loss function is either a minimum or maximum, as appropriate. The solution of the estimation problem then becomes an optimization problem. Such optimization problems have been treated by many texts (see for example Wilde and Beightler [1967]).

Perhaps the most familiar definition of best fit are the least squares and weighted least squares formulations. The weighted least squares expression in vector form may be written as

$$L(x) = vWv^{\mathrm{T}} \tag{19.5}$$

TABLE 19.1 Examples of Optimization Criteria

Criteria	Description	Mathematical Formulation	Comments
Least squares (LS)	Minimize the sum of the squares of the residuals	$L(x) = vv^{\mathrm{T}}$	Earliest formulation proved by Gauss—no a priori knowledge assumed
Weighted least squares (WLS)	Minimize the sum of the weighted squares of the residuals	$L(x) = vwv^{\mathrm{T}}$	Yields identical results to MLE when noise is Gaussian and weight matrix equals inverse covariance matrix
Mean square error (MSE)	Minimize the expected value of the squared error	$L(x) = \int (x - \bar{X})w(x - \bar{X})P(x\|y)(x - \bar{X})^{\mathrm{T}} dx$	Minimum covariance solution
Bayesian weighted least squares (BWLS)	Minimize the sum of the weighted squares of the residuals constrained by a priori knowledge of x	$L(x) = vwv(x - x_0)P_{\Delta x_0}(x - x)^{\mathrm{T}}$	Constrains the solution of x to a reasonable value close to the a priori estimate of x
Maximum likelihood estimate (MLE)	Maximize the multi-variate probability distribution function	$L(x) = \prod_{i-1}^{n} t_1(n_1/x)t_2(n_2/x_2)...t_n(n_n/x)$	Allows specification of the probability distribution for the noise process

Equivalently,

$$L(x) = [(y_1 - z_1),...,(y_n - z_n)]\begin{pmatrix} w_{11} & \cdots & 0 \\ \vdots & \ddots & \vdots \\ 0 & \cdots & w_{nn} \end{pmatrix}\begin{bmatrix} (y_1 - z_1) \\ \vdots \\ (y_n - z_n) \end{bmatrix} \tag{19.6}$$

or,

$$L(x) = \sum_{i=1}^{n}(y_i - z_i)w_{ij}(y_i - z_i) \tag{19.7}$$

The loss function $L(x)$ is a scalar function of x which is the sum of the squares of the observation residuals weighted by W. The least squares or weighted least squares criterion is used when there is no basis to assign probabilities to x and y, and there is limited information about the measurement errors. A special case involves linear least squares, in which the predicted observations are a linear function of the state vector. In that case, $L(x)$ may be solved explicitly with a closed form solution (see Press et al. (1986) for the formulation).

A variation of the weighted least squares objective function is the constrained loss function

$$L(x) = vWv^{\mathrm{T}} + (x - x_0)P_{\Delta x_0}(x - x_0)^{\mathrm{T}} \tag{19.8}$$

This expression, sometimes termed the Bayesian weighted least squares criterion, constrains the weighted least squares solution for x to be *close to* an a priori value of x (i.e., x_0). In Equation 19.8, the quantity $P_{\Delta x_0}$ represents an estimate of the covariance of x given by the symmetric matrix:

$$P_{\Delta x_0} = \begin{pmatrix} \sigma_{x1}^2 & \cdots & \sigma_{x1}\sigma_{xn} \\ & \ddots & \\ \sigma_{x1}\sigma_{xn} & & \sigma_{xn}^2 \end{pmatrix} \tag{19.9}$$

If the components of x are statistically independent, then $P_{\Delta x_0}$ is a diagonal matrix. The Bayesian criterion is used when there is prior knowledge about the value of x and a priori knowledge of associated uncertainty via $P_{\Delta x_0}$. The resulting optimal solution for x lies nearby the a priori value x_0.

The mean square error formulation minimizes the expected (mean) value of the squared error, i.e., minimize

$$L(x) = \int (x - \hat{x})^{\mathrm{T}} W (x - \hat{x}) P(x|y) dx \tag{19.10}$$

$P(x|y)$ is the conditional probability of state vector **x** given the observations y. This formulation assumes that x and y are jointly distributed random variables. The quantity x is the conditional expectation of x

$$\hat{x} = \int x P(x|y) dx \tag{19.11}$$

The solution for x yields the minimum covariance of x.

The final optimization criterion shown in Table 19.1 to define *best fit* is the maximum likelihood criterion

$$L(x) = \prod_{i=1}^{n} l_i(n_i/x) \tag{19.12}$$

$L(x)$ is the multivariate probability distribution to model the observational noise n_i. The function, $L(x)$ is the conditional probability that the observational noise at times $t_0, t_1, \ldots, t_i$ will have the values $n_0, n_1, \ldots, n_i$, if x is the actual or true value of the state vector. The maximum likelihood criteria selects the value of x which maximizes the multivariate probability of the observational noise.

If the measurement errors, n_i are normally distributed about a zero mean, then $l_i(n_i/x)$ is given by

$$l_i(n_i/x) = \frac{1}{(2\pi)^{m/2} \left| M_i^{1/2} \right|} \exp\left(-\frac{1}{2} n_i^{\mathrm{T}} M_i^{-1} n_i \right) \tag{19.13}$$

The quantity, m, refers to the number of components, at each time, t_i, of the observation vector and M_i is the variance of the observation at time t_i. The maximum likelihood criterion allows us to postulate non-Gaussian distributions for the noise statistics.

Selection of an optimization criterion from among the choices shown in Table 19.1 depends upon the a priori knowledge about the observational process. Clearly, selection of the MLE criterion presumes that the probability distributions of the observational noise are known. Similarly, the MSE criterion presumes knowledge of a conditional probability function, while the Bayesian weighted least squares assumes a priori knowledge of the variance of the state vector. Under the following restricted conditions, the use of these criterion result in an identical solution for x. The conditions are as follows:

1. The measurement (observational) errors are Gaussian distributed about a zero mean.
2. The errors n_i at time t_i are stochastically independent of the errors n_j, at time t_j.
3. The weight for the weighted least squares criterion is the inverse covariance of x.

Under these conditions, the WLS solution is identical to the MLE, MSE, and BWLS solutions.

19.2.3 Optimization Approach

Solution of the optimization criterion to determine an estimate of the state vector **x** may be performed by one of several techniques. Several texts present detailed algorithms for optimization, e.g., Wilde and Beightler (1967), Press et al. (1986), Brent (1973), and Shoup (1987). Press et al. (1986) and Shoup (1987) provide computer codes to solve the optimization problem. In this section, we will provide an overview of optimization approaches and give additional detail in Sections 19.3 and 19.4 for batch and sequential estimation, respectively.

Optimization techniques may be categorized into two broad classes as illustrated in Table 19.2.

Direct methods treat the optimization criteria, without modification, seeking to determine the value of x which finds an extremum (i.e., minimum or maximum) of the optimization criterion. Geometrically, direct methods are hill climbing (or valley seeking) techniques which seek to find the value of x for which $L(x)$ is a maxima or minima. By contrast, indirect methods seek to solve the simultaneous non-linear equation given by

$$\frac{\partial L(x)}{\partial x} = \begin{bmatrix} \dfrac{\partial L}{\partial x_1} \\ \vdots \\ \dfrac{\partial L}{\partial x_m} \end{bmatrix} \tag{19.14}$$

where m is the number of components of the state vector. Indirect methods require that the optimization criterion be explicitly differentiated with respect to x. The problem is transformed from finding the maximum (or minimum) of a nonlinear equation, to one of finding the roots of m simultaneously non-linear equations given by (19.14).

A summary of techniques for a direct solution of the optimization problem is given in Table 19.3. An excellent reference with detailed algorithms and computer program listings is Press et al. (1986). The direct techniques may be subdivided into non-derivative techniques (i.e., those methods that require on the ability to compute $L(x)$, and derivative techniques which rely on the ability to compute $L(x)$ as well as derivatives of $L(x)$. Non-derivative techniques include simplex methods such as that described by Nelder and Mead (1965), and direction set methods which seek successive minimization along preferred coordinate directions. Specific techniques include conjugate direction methods and Powell's methods (see Press et al. [1986] and Brent [1973]).

TABLE 19.2 Categories of Optimization Techniques

Category	Optimization Technique	Description
Direct methods	Non-derivative methods • Downhill simplex • Direction set	Direct methods find the value of x that satisfies the optimization criteria (i.e., find x such that the loss function is either a minimum or maximum)
	Derivative methods • Conjugate gradient • Variable metric (quasi-Newton)	Techniques fall into two classes: Derivative methods require knowledge of derivative of loss function with respect to x, while non-derivative methods require only the ability to compute the loss function
Indirect methods	Newton–Raphson methods	Indirect methods find the roots of a system of equations involving partial derivatives of the loss function with respect to the state vector, **x**; i.e., the partial derivative of $L(x)$ with respect to x set equal to zero. The only successful techniques are multi-dimensional Newton–Raphson methods

TABLE 19.3   Summary of Direct Methods for Optimization

Type of Method	Class of Techniques	Algorithm/Strategy	References
Non-derivative methods (do not require derivative of $L(x)$)	Downhill simplex techniques	Utilize a geometric simplex (polygonal figure) to map out and bracket extrema of $L(x)$	Press et al. (1986) Nelder and Mead (1965)
	Direction set methods	Successive minimization along preferred coordinate directions; examples include conjunctive direction techniques and Powell's method	Press et al. (1986) Brent (1973)
Derivative methods (requires derivative of $L(x)$)	Conjunctive gradient methods	Utilize multi-dimensional derivative (gradient) information to seek an extremum. Techniques include Fletcher–Reeves and Polak–Ribiere methods	Press et al. (1986) Polak (1971)
	Variable metric (quasi-Newton techniques)	Variations of multi-dimensional Newton's method. Techniques include Davidson–Fletcher–Powell and Broyden–Fletcher–Goldfarb–Shannon methods	Press et al. (1986)

Derivative methods for direct optimization utilize first or higher order derivatives of $L(x)$ to seek an optimum. Specific methods include conjugate gradient methods such as Fletcher–Reeves method and the Polak–Ribiere method (see Press et al. [1986] and Polak [1971]). Variable metric methods utilize a generalization of the one-dimensional Newton's approach. Effective techniques include the Davidon–Fletcher–Powell method and the Broyden–Fletcher–Goldfarb–Shanns method. A well-known, but relatively ineffective gradient technique is the method of steepest descent. Press et al. provides a discussion of tradeoffs and relative performance of these methods.

Press et al. (1986) points out that there is only one effective technique for finding the roots of Equation 19.14, namely the multiple dimensional Newton–Raphson method. For a one-dimensional state vector, the technique may be summarized as follows. We seek to find x such that

$$\frac{\partial L(x)}{\partial x} = f(x) = 0 \tag{19.15}$$

Expand $f(x)$ in a Taylor series

$$f(x + \delta x) = f(x) + \frac{\partial f}{\partial x} \delta x + \left(\frac{\partial^2 f}{\partial x^2}\right) \delta x^2 + \cdots \tag{19.16}$$

Neglecting second and higher order terms yield a linear equation

$$f(x + \delta x) = f(x) + \frac{\partial f}{\partial x} \delta x \tag{19.17}$$

To find the root of (19.17), set $f(x + \delta x)$ equal to zero and solve for δx

$$\delta x = f(x) - \left(\frac{f(x)}{(df/dx)} \right) \tag{19.18}$$

In order to find the root of Equation 19.15, we begin with an initial value of x, say x_i. An improved value for x_i is given by

$$x_i = x_i + \delta x_i \tag{19.19}$$

Equation 19.19 is solved iteratively until $\delta x_i < \varepsilon$, where epsilon is an arbitrarily small convergence criterion. A multi-dimensional description of the Newton–Raphson technique is described by Press et al. (1986).

19.2.4 Processing Approach

In introducing the estimation problem and alternative design choices, we have implicitly ignored the fundamental question of how data will be processed. As illustrated in Figure 19.3, there are two basic alternatives; (1) batch processing, and (2) sequential processing. Batch processing assumes that all data are available to be considered simultaneously. That is, we assume that all n observations are available, select an optimization criterion, and proceed to find the value of x which best fits the n observations (via one of the techniques described in Section 19.3). The batch approach is commonly used in modeling or curve fitting problems. Batch estimation is often used in situations in which there is no time critical element involved. For example, estimating the heliocentric orbit of a new comet or asteroid involves observations over a period of several days with subsequent analysis of the data to establish an ephemeris. Another example would entail modeling in which various functions (e.g., polynomials, log-linear functions, etc.) are used to describe data. Batch approaches have a number of advantages, particularly for situations in which there may be difficulty in association. At least in principle, one approach to find an optimal association of observations-to-tracks or observations-to-observations is simply to exhaustively try all $\{n(n - 1)/2\}$ combinations. While in practice such exhaustive techniques are not used, batch estimation techniques have more flexibility in such approaches than do sequential techniques. Section 19.3 of this chapter provides more detail on batch estimation including a processing flow and discussion of implementation issues.

An alternative approach to batch estimation is the sequential estimation approach. This approach incrementally updates the estimate of the state vector as each new observation is received. Hence if $x(t_0)$ is the estimate of a state vector at time t_0 based on n previous observations, then sequential estimation provides the means of obtaining a new estimate for x, (i.e., $x_{n+1}(t_0)$), based on $n + 1$ observations by modifying the estimate $x_n(t_0)$. This new estimate is obtained without revisiting all previous n observations. By contrast in batch estimation, if a value of $x_n(t_0)$ had been obtained utilizing n observations, then to determine $x_{n+1}(t_0)$, all $n + 1$ observations would have to be processed. The Kalman filter is a commonly used approach for sequential estimation.

Sequential estimation techniques provide a number of advantages including

1. Determination of an estimate of the state vector with each new observation
2. Computationally efficient scalar formulations
3. Ability to adapt to changing observational conditions (e.g., noise, etc.)

Disadvantages of sequential estimation techniques involve potential problems in data association, divergence (in which the sequential estimator ignores new data), and problems in initiating the process. Nevertheless, sequential estimators are commonly used for tracking and positional estimation. Section 19.4 of this chapter provides more detail on sequential estimation including a process flow and discussion of implementation issues.

19.3 Batch Estimation

19.3.1 Derivation of Weighted Least Squares Solution

In order to illustrate the formulation and processing flow for batch estimation, consider the weighted least squares solution to a dynamic tracking problem. One or more sensors observe a moving object, reporting a total of n observations, $y_M(t_i)$, related to target position. Note: henceforth in this chapter the subscript M denotes a measured value of a quantity. The observations are assumed to be unbiased with observational noise whose standard deviation is σ_i. The (unknown) target position and velocity at time t_i is represented by an one dimensional vector, $\mathbf{x}(t_i)$. Since the target is moving, $\mathbf{x}$ is a function of time given by the equations of motion

$$\mathbf{x}(t) = f(x, \dot{x}, t) \qquad (19.20)$$

Equation 19.20 represents one simultaneous non-linear differential equations. This is an initial value problem in differential equations. Given a set of initial conditions (viz. specification of x_o, and x_o at time t_o) Equation 19.20 can be solved to predict $\mathbf{x}(t)$ at an arbitrary time, t. Numerical approaches to solve (19.20) are described in a number of texts such as Brent (1973) and Henrici (1962). Specific techniques include numerical integration methods such as Runge–Kutta methods, predictor–corrector methods, perturbation methods, or in simple cases, analytical solutions.

An observational model allows the observations to be predicted as a function of the unknown state vector

$$y(t) = g(x, t) \qquad (19.21)$$

The residuals are the differences between the sensor data and the computed measurement at time t_i, given by

$$v_i = \left[y_M(t_i) - g(x, t_i) \right] \qquad (19.22)$$

The weighted least squares criteria or loss function for best fit is specified as

$$L(x) = \sum_{i=1}^{n} \left(\frac{v_i}{\sigma_i} \right)^2 = \sum_{i=1}^{n} \left(\frac{1}{\sigma_i^2} \right) \left[y_M(t_i) - g(x, t_i) \right]^2 \qquad (19.23)$$

$L(x)$ is a measure of the closeness of fit between the observed data and the predicted data as a function of x. Note that if different sensor types are utilized (e.g., radar, optical tracker, etc.), then the form of the function, $g(x, t_i)$ changes for each sensor type. Hence, for example, $g(x, t_i)$ may represent one set of equations for a radar to predict range, azimuth, elevation, and range-rate, and another set of equations for an optical tracker to predict right ascension and declination angles. Further, the observation represented by $y_M(t_i)$ may be a vector quantity, e.g.,

$$y_M(t_i) = \begin{bmatrix} \text{range}(t_i) \\ \text{range} - \text{rate}(t_i) \\ \text{azimuth}(t_i) \\ \text{elevation}(t_i) \end{bmatrix}$$

for a radar observation, or

$$y_M(t_i) = \begin{bmatrix} \text{Right ascension}(t_i) \\ \text{Declination}(t_i) \end{bmatrix}$$

for an optical tracker. In that case, $g(x, t_i)$ and v_i would correspondingly be vector quantities.

In matrix notation, Equation 19.23 becomes

$$L(x) = [y_M - g(x)]^T W [y_M - g(x)] \tag{19.24}$$

where

$$y_M = \begin{bmatrix} y_M(t_1) \\ y_M(t_2) \\ \vdots \\ y_M(t_n) \end{bmatrix}, \quad \text{and} \quad g(x) = \begin{bmatrix} g(x, t_1) \\ g(x, t_2) \\ \vdots \\ g(x, t_n) \end{bmatrix} \tag{19.25}$$

and

$$W = \begin{bmatrix} \dfrac{1}{\sigma_1^2} & 0 & \cdots & 0 \\ 0 & \dfrac{1}{\sigma_2^2} & \cdots & 0 \\ 0 & 0 & \cdots & \dfrac{1}{\sigma_n^2} \end{bmatrix} \tag{19.26}$$

The weighted least squares estimate is the value of x_o, denoted $\hat{x}_o$, that minimizes the function $L(x)$.

Using an indirect approach, we seek $\hat{x}_o$, such that

$$\frac{\partial L(x)}{\partial x} = 2[y_M - g(x)]^T W \frac{\partial g}{\partial x} = 0 \tag{19.27}$$

Equation 19.27 represents a set of nonlinear equations in one unknowns (the one components of x_o).

The solution to Equations 19.27 may be obtained by a multi-dimensional Newton–Raphson approach as indicated in the previous section. The following derivation explicitly shows the linearized, iterative solution. First, expand the measurement prediction, $g(x, t)$, in a Taylor series about a reference solution $x_{REF}(t)$, namely

$$g(x, t_i) = g(x_{REF}, t) + H_i \Delta x_i \tag{19.28}$$

where

$$\Delta x_i = x(t_i) - x_{REF}(t_i) \tag{19.29}$$

and

$$H_i = \frac{\partial g(x, t_i)}{\partial x} \tag{19.30}$$

The value x_{REF} is a particular solution to the dynamical equation of motion (Equation 19.20).

A further simplification may be obtained if the equation of motion is linearized, i.e.,

$$\Delta x_i = \Phi(t_i, t_j)\Delta x_j \tag{19.31}$$

where, $\Phi(t_i, t_j)$ represents a state transition matrix which relates variations about $x_{REF}(t)$ at times t_i, and t_j,

$$\Phi(t_i, t_j) = \frac{\partial x(t_i)}{\partial x(t_j)} \tag{19.32}$$

Substituting (19.32) into (19.28) yields

$$g(x, t_i) = g(x_{REF}, t_i) + H_i\Phi(t_i, t_j)\Delta x_j \tag{19.33}$$

Using this expression for $g(x, t_i)$ in Equation 19.27 yields

$$\left[\Delta y - H(x_j)\Delta x_j\right]^{\mathrm{T}} WH(x_j) = 0 \tag{19.34}$$

where

$$\Delta y = y_M - g(x_{REF}) = \begin{bmatrix} y_M(t_1) - g(x_{REF}, t_1) \\ y_M(t_2) - g(x_{REF}, t_2) \\ \vdots \\ y_M(t_n) - g(x_{REF}, t_n) \end{bmatrix} \tag{19.35}$$

and

$$H(x_j) = \begin{bmatrix} H_1\Phi(t_1, t_j) \\ H_2\Phi(t_2, t_j) \\ \vdots \\ H_n\Phi(t_n, t_j) \end{bmatrix} \tag{19.36}$$

Solving Equation 19.34 for Δx_j yields

$$\Delta \hat{x}_j = [H(x_j)^{\mathrm{T}} WH(x_j)]^{-1}[H(x_j)^{\mathrm{T}} W\Delta y] \tag{19.37}$$

The increment Δx_j is added to $x_{REF}(t_j)$

$$\hat{x}(t_j) = x_{REF}(t_i) + \Delta \hat{x}_j \tag{19.38}$$

to produce an updated estimate of the state vector. The improved value of $x(t_j)$ is used as a new reference value of $x(t_j)$. Equations 19.37 and 19.38 are applied iteratively until Δx_j becomes arbitrarily small.

Table 19.4 illustrates the linearized iterative solution for several optimization criteria including; least squares, weighted least squares, Bayesian weighted least squares, and the maximum likelihood optimization criteria. In Equation 19.37, Δy represents the difference between predicted and actual observations, while $H(x_j)$ expresses the relationship between changes in predicted observations and changes in

TABLE 19.4   Summary of Batch Estimation Solutions

Optimization Criterion	Mathematical Formulation	Linearized Iterative Solution
Least squares	$L(x) = vv^T$	$\Delta \hat{x}_j = [H^T H]^{-1}[H^T v]$
Weighted least squares	$L(x) = vwv^T$	$\Delta \hat{x}_j = [H^T W H]^{-1}[H^T W v]$
Bayesian weighted least squares	$L(x) = vwv + (x - x_0)P_{\Delta x_0}(x - x_0)^T$	$\Delta \hat{x}_j = [H^T W H + P_{\Delta x}^{-1}][H^T W v + P_{\Delta x}^{-1}\Delta \hat{x}_{j-1}]$
Maximum likelihood estimate	$L(x) = \prod_{i=1}^{n} l_i(n_i/x)$	$\Delta \hat{x}_j = [H^T M^{-1} H]^T [H^T M^{-1} v]$

components of the state vector. For static (non-dynamic) problems, the state vector $\mathbf{x}(t)$ is constant in time, and the state transition matrix reduces to the identity matrix,

$$\Phi(t_i, t_j) = I = \begin{bmatrix} 1 & 0 & \cdots & 0 \\ 0 & 1 & \cdots & 0 \\ & & \ddots & \\ 0 & 0 & \cdots & 1 \end{bmatrix} \qquad (19.39)$$

19.3.2 Processing Flow

The processing flow to solve the batch estimation problem is illustrated in Figure 19.4 for a weighted least squares formulation. The indirect approach discussed in the previous section is used. Inputs to the process include an initial estimate of the state vector, $\hat{x}_0$, at an epoch time, t_0, and n observations, y_{Mk} at times, t_k, with associated uncertainties σ_k. The output of the process is an improved estimate of the state vector, $x_{i+1}(t_0)$ at the epoch time, t_0.

The processing flow shown in Figure 19.4 utilizes two nested iterations. An inner iteration (or processing loop letting $k = 1, 2, ..., n$), cycles through each observation, y_{Mk}, performing a series of calculations:

1. Retrieve the kth observation, y_{Mk}, and its associated time of observation, t_k, and observational uncertainty, σ_k.
2. Solve the differential equation of motion (Equation 19.20) to obtain $x_{ki}(t_k)$. That is, using the current estimate of the state vector, $x_i(t_0)$, update to time t_k.

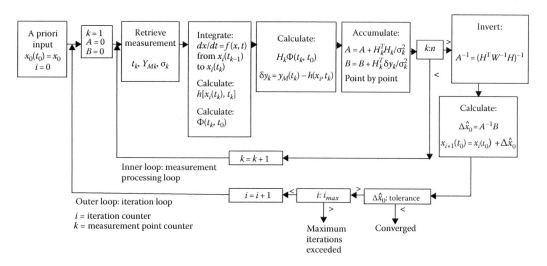

FIGURE 19.4   Weighted least squares batch process computational flow sequence.

3. Compute a predicted observation, $g[x_i(t_k), t_k]$ based on sensor models. Note that the predicted observation utilizes a model appropriate to the sensor type (e.g., radar, optical, etc.).
4. Compute the transition matrix $\Phi(t_k, t_o)$.
5. Calculate the quantity $H_k \Phi(t_k, t_o)$, and the observation residual $\delta y_k = [y_M(t_k) - g(x_i, t_k)]$.
6. Accumulate the matrices

$$A = A + \frac{H_k^T H_k}{\sigma_k^2} \tag{19.40}$$

and

$$B = B + \frac{H_k^T \delta y_k}{\sigma_k^2} \tag{19.41}$$

Steps 1–6 are performed for each observation $k = 1, 2, \ldots, n$.

An outer processing loop ($i = 0, 1, \ldots$) iteratively computes and applies corrections to the state vector until convergence is achieved. For each iteration, we compute

$$A^{-1} = (H^T W^{-1} H)^{-1} \tag{19.42}$$

and

$$\Delta \hat{x}_0 = A^{-1} B \tag{19.43}$$

with

$$x_{i+1}(t_o) = x_i(t_o) + \Delta \hat{x}_0 \tag{19.44}$$

Hence, the state vector is successively improved until $\Delta \hat{x}_0$, becomes arbitrarily small.

The process described here and illustrated in Figure 19.4 is meant to convey the essence of the batch solution via a linearized iterative approach. It can be seen that the solution may become computationally demanding. For each observation (which may number in the hundreds to thousands), we must solve a non-linear set of differential equations, perform coordinate transformations to predict an observation, compute the transition matrix, and perform several matrix multiplications. These calculations are performed for all n observations. Moreover, the complete set (for all n observations) of computations is iteratively performed to achieve an estimate of $x(t_o)$. Upward of 10–30 iterations of Δx_j may be required to achieve convergence, depending upon the initial value chosen for $x(t_o)$.

19.3.3 Batch Processing Implementation Issues

The processing flow for batch estimation shown in Figure 19.4 is meant to be illustrative rather than a prescription or flowchart suitable for software implementation. Several implementation issues often arise. We will discuss a few of these issues, including convergence, data editing, the initial estimate of x, and observability.

The processing flow in Figure 19.4 shows an outer processing loop in which successive improvements are made to the state vector estimate. The convergence criteria tests the magnitude of Δx, and declares convergence when

$$|\Delta x_0| \leq \varepsilon \tag{19.45}$$

The iterations are terminated when the incremental changes to the state vector fall within an arbitrarily small increment (for each component of the state vector). This is a logical criterion since we can use physical arguments to establish the values for ε. For example, we might declare that distances within 1 m, velocities within 1 cm/s, frequencies within 1 Hz, etc. are arbitrarily small. Other convergence criteria might equally well be used. An example is the ratio criterion

$$\left| \frac{\Delta x_0}{x_0} \right| \le \varepsilon \tag{19.46}$$

A number of convergence criteria have been used including multiple logical conditions (e.g., Equation 19.45 or 19.46, etc.). There is no guarantee that the iterative solution will converge in a finite number of iterations. Thus logical checks should be made to determine how many iterations have been performed with some upper bound (e.g., $k \le 50$) established to terminate the iterations.

In batch estimation it is tempting to perform data editing within the outer processing loop. A common practice is to reject all observations for which the residual ($|v_k|$) exceeds either an a priori limit, or a standard deviation (e.g., 3σ) test. Such a practice is fraught with potential pitfalls. Deutsch (1963) discusses some of these pitfalls. Two problems are most notable. Iterative data editing can prolong the outer-loop iteration. It is possible in the ith estimate of $x(t_o)$ to reject one or more observations only to find in the $(i + 1)$st estimate that these observations are acceptable. Hence the iteration for $x(t_o)$ can sometimes oscillate, alternatively rejecting and accepting observations. A second problem involves the case in which all observations are valid and highly accurate. Statistically, there will be some valid observations whose residuals exceed 3σ. Rejecting or editing out such observations corrupts the solution for $x(t_o)$ by rejecting perfectly good data. A rule of thumb is not to reject any data unless these are valid physical reasons for such editing.

Another implementation issue for batch processing involves how to obtain an initial estimate of the state vector, $x_0(t_o)$. Generally, several observations may be used in a minimum data solution to obtain a value of $x(t_o)$. Alternatively, some sensors may provide estimates of the state vector, or an estimate of x may be available from other a priori information. The generation of a starting value for x is very much dependent upon the particular sensors and observing geometry. Sometimes several minimum data sets are used (i.e., observations y_M, y_{M3}, and y_{M5}; observations y_{M2}, y_{M4}, and y_{M6}) with an initial estimate of $x(t_p)$ developed from each data set. Subsequently, the initial estimates are averaged to produce a starting value for the estimation process.

A final implementation issue is the question of observability. We briefly introduced this issue in Section 19.2. Observability is the question of whether improvements in components of $\mathbf{x}(t)$ can be obtained based on the observational data. Mathematically this problem is exhibited via an ill-conditioned matrix

$$[H^T W^{-1} H] \tag{19.47}$$

Several methods have been introduced to address such an ill-conditioned system. For example, we may require that the determinant of matrix $H^T H$ be non-zero, at each iterative step. Alternatively, nonlinear terms may be introduced to treat the ill-conditioned linear system. While these techniques may be useful, there is no substitute for care in selecting and analyzing the choice of state vector components.

19.4 Sequential Estimation and Kalman Filtering

During the early 1960s, a number of technical papers were published describing a sequential approach to estimation. Swerling (1959), Kalman (1960), and Kalman and Bucy (1961) published papers describing a linearized sequential technique to update an estimate of a state vector. The work was a discrete implementation of earlier continuous formulations by Wiener (1949) and Kolmogorov (1941) in the late 1940s. An informal history of the development of the Kalman filter, and its application to space flight is provided by McGee and Schmidt (1985). Since then, much work has been performed on discrete estimation. A number of techniques exist. A extensive review of recursive filtering techniques is provided by Sayed and Kailath (1994) who compare variants of the Kalman filter including the covariance Kalman filter, the information filter, the square-root covariance filter, the extended square-root information filter, the square-root Chandrasekhar filter, and the explicit Chandrassekhar filter. We will describe one such approach utilizing the Kalman filter.

Several approaches may be used to derive the sequential estimation equations. In particular, the reader is referred to the feedback-control system approach described in Gelb (1974). In this section, we will derive the equations for a weighted least squares optimization criterion, for a dynamic tracking problem, identical to that in the previous section. Following a derivation, we will present a processing flow, and discussion of implementation issues.

19.4.1 Deviation of Sequential Weighted Least Squares Solution

Assume that n observations are available from multiple sensors, and that a weighted least squares solution has been obtained, i.e.,

$$\Delta \hat{x}_n = \left[H_n W_n^{-1} H_n \right]^{-1} H_n^{\mathrm{T}} W_n^{-1} \Delta y_n \tag{19.48}$$

H_n denotes $H(x_n)$ the partial derivatives of the observation components with respect to the state vector (Equation 19.36), and Δy_n denotes the difference between the measured and predicted observation (19.22). Suppose that one additional observation, $y_M(t_{n+1})$ is received. How does the $(n + 1)$st observation affect the estimate of x? Let us utilize the weighted least squares formulation separating the $(n + 1)$st data from the previous n observations. Thus,

$$\Delta \hat{x}_{n+1} = [H_{n+!} W_{n+1}^{-1} H_{n+1}]^{-1} H_{n+!}^{\mathrm{T}} W_{n+1}^{-1} \Delta y_{n+1} \tag{19.49}$$

where

$$H_{n+1} = \left| \frac{H_n \phi(t_n, t_{n+1})}{H_{n+1}} \right|; \quad W_{n+1} = \begin{pmatrix} W_n & 0 \\ 0 & \sigma_{n+1}^2 \end{pmatrix}; \quad \Delta y_{n+1} = \left| \begin{matrix} \Delta y_n \\ \delta y_{n+1} \end{matrix} \right| \tag{19.50}$$

and the $(n + 1)$st residual is

$$\delta y_{n+1} = y_M(t_{n+1}) - g(x_{REF}, t_{n+1}) \tag{19.51}$$

Substituting Equations 19.50 into 19.49 yields

$$\Delta \hat{x}_{n+1} = \left[[H_n \phi, H_{n+1}] \begin{pmatrix} w_n^{-1} & 0 \\ 0 & \sigma_{n+1}^{-2} \end{pmatrix} \begin{bmatrix} H_n \phi \\ H_{n+1} \end{bmatrix} \right]^{-1} [H_n \phi, H_{n+1}] \begin{pmatrix} w_n^{-1} & 0 \\ 0 & \sigma_{n+1}^{-2} \end{pmatrix} \begin{bmatrix} \Delta y_m \\ \delta y_{n+1} \end{bmatrix} \tag{19.52}$$

which can be manipulated to obtain

$$\Delta \hat{x}\left(\frac{t_{n+1}}{t_n} \right) = \Delta \hat{x}\left(\frac{t_{n+1}}{t_n} \right) - K \left[H_{n+1} \Delta \hat{x}\left(\frac{t_{n+1}}{t_n} \right) - \delta y_{n+1} \right] \tag{19.53}$$

where

$$K = P_n(t_{n+1})H_{n+1}^{\mathrm{T}}\left[H_{n+1}P_n(t_{n+1})H_{n+1}^{\mathrm{T}} + \sigma_{n+1}^2 \right]^{-1} \tag{19.54}$$

$$P_{n+1}(t_{n+1}) = P_n(t_{n+1}) - KH_{n+1}P_n(t_{n+1}) \tag{19.55}$$

$$\Delta\hat{x}\left(\frac{t_{n+1}}{t_n}\right) = \Phi(t_{n+1}, t_n)\Delta\hat{x}\left(\frac{t_n}{t_n}\right) \tag{19.56}$$

$$P_n(t_{n+1}) = \Phi(t_{n+1}, t_n)P_n(t_n)\Phi \tag{19.57}$$

Equations 19.53 through 19.57 constitute a set of equations for recursive update of a state vector. Thus, given n observations $y_M(t_n)$ ($t = 1, 2, \ldots, n$), and an associated estimate for $\mathbf{x}(t)$, these equations prescribe an update of $y_M(t_n)$ based on a new observation $y_M(t_{n+1})$. Clearly the equations can be applied recursively, replacing the solution for $x_n(t_n)$ by $x_{n+1}(t_{n+1})$ and processing yet another observation, $y_M(t_{n+2})$, etc.

Equation 19.53 is a linear expression that expresses the updated value of $x(t_{n+1})$ as a function of the previous value, $x(t_n)$, a constant K, and the observation residual δy_{n+1}. The constant K is called the Kalman gain, which in turn is a function of the uncertainty in the state vector (viz., the covariance of x_n given by $P_n(t_{n+1})$, and the uncertainty in the observation, σ_{n+1}). Equation 19.55 updates the uncertainty in the state vector, while Equation 19.56 uses the transition matrix to update the value of x from time, t_n, to time, t_{n+1}. Similarly Equation 19.57 updates the covariance matrix at time t_n to time t_{n+1}. In these equations, the parenthetical expression (t_{n+1}/t_n) denotes that the associated quantity is based on the previous n observations, but is extrapolated to time t_{n+1}. Correspondingly, (t_{n+1}/t_{n+1}) indicates that the associated quantity is valid at time t_{n+1} and also has been updated to include the effect of the $(n + 1)$st observation.

The Kalman gain, K, directly scales the magnitude of the correction to the state vector, $\Delta\mathbf{x}_n$. The gain, K, will be relatively large (and hence will cause a large change in the state vector estimate) under two conditions:

1. When the uncertainty in the state vector is large (i.e., when $P_n(t_{n+1})$ is large)
2. When the uncertainty in the $(n + 1)$st observation is small (i.e., when σ_{n+1} is small)

Conversely, the Kalman gain will be small when the state vector is well-known (i.e., when $P_n(t_{n+1})$ is small) and/or when the $(n + 1)$st observation is very uncertain (i.e., when σ_{n+1} is large). This result is conceptually pleasing, since we want to significantly improve inaccurate state vector estimates with accurate new observations but do not want to corrupt accurate state vectors with inaccurate data.

There are two main advantages of sequential estimation over batch processing. First, Equations 19.53 through 19.57 can be formulated entirely as scalar equations requiring no matrix inversions. Even when each observation, $y_M(t_n)$ is a vector quantity (e.g., a radar observation comprising range, range-rate, azimuth, and elevation), we can treat each observation component as a separate observation occurring at the same observation time, t_{n+i}. This scalar formulation allows very computationally efficient formulations. The second advantage is that the sequential process allows the option of updating the reference solution ($x_{REF}(t_o)$) after each observation is processed. This option is sometimes referred to as the Extended Kalman Filter. This option provides an operational advantage for dynamic problems such as target tracking because a current estimate of $\mathbf{x}(t)$ is available for targeting purposes, or to guide the sensors.

19.4.2 Sequential Estimation Processing Flow

A processing flow for sequential estimation is shown in Figure 19.4. The processing flow is shown for an extended Kalman filter with dynamic noise. Input to the process are an initial estimate of the state vector, $x_0(t_0)$, and associated covariance matrix, $P_0(t_0)$. For each measurement, $k = 1, 2, \ldots, n$ the following steps are performed:

1. Retrieve the observation $y_M(t_k)$ and its associated uncertainty, σ_k.
2. Solve the differential equations of motion (Equation 19.20) to propagate the state vector from time t_{k-1} to t_k.
3. Compute the transition matrix $\Phi(t_k/t_{k-1})$.
4. Propagate Δx and P_{k-1} from time t_{k-1} to t_k using the transition matrix (i.e., Equations 19.56 and 19.57 respectively).
5. Compute a predicted observation, $g(x(t_k), t_k)$ (Equation 19.21), the observation residual, δ_{vk}, and H_k, the partial derivative of the predicted observation with respect to the state vector.
6. Compute the Kalman gain via Equation 19.54.
7. Update the state vector correction, $\Delta x(t_k, t_k)$ (Equation 19.56) and the covariance matrix, $P_k(t_k)$ (Equation 19.55).
8. Update the reference state vector

$$\hat{x}(t_k) = \hat{x}(t_k) + \Delta \hat{x}(t_k/t_k) \tag{19.58}$$

Steps 1–8 are repeated until all observations have been processed.

Output from the sequential estimation is an updated state vector, $\mathbf{x}(t_k)$ and associated covariance matrix, $P_n(t_n)$ based on all n observations (Figure 19.5).

The concept of how the Kalman filter updates the state vector and associated quantities from one observation to the next is illustrated in Figure 19.6.

Figure 19.6 illustrated a timeline with two observations at time t_{k-1} and t_k. At time, t_{k-1}, we have an estimate of the state vector, $x_{k-1}(-)$ and its associated covariance, $P_{k-1}(-)$. The parenthetical use of the minus sign indicates that the values of the state vector and covariance matrix are the estimates at time, t_{k-1}, but prior to incorporating the effect of the new observation. When the observation at time, t_k, is received, the Kalman equations are used to updated to estimate a new value of the state vector. The value of the Kalman Gain, K, and the observation uncertainty, R, are used to develop updates to the state vector, $\mathbf{x}$, and to the covariance matrix. At a subsequent time, t_k, we receive a new observation. It is first necessary to propagate the state vector, x_{k-1}, and the covariance matrix, P_{k-1} forward in time from time, t_{k-1}, to time, t_k. Using the equations of motion for the state vector and the covariance matrix does this. As a result, we have an estimate of the state vector, $\mathbf{x}$, at time t_k based on the $k - 1$ processed observations.

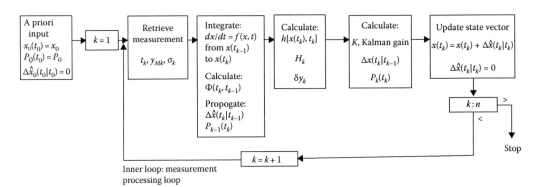

FIGURE 19.5   Recursive filer process with dynamic noise (extended Kalman filter).

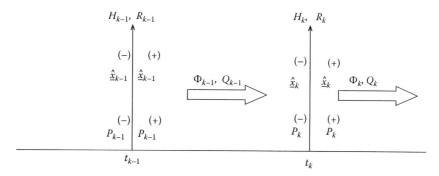

FIGURE 19.6 Kalman filter update process. The Kalman filter equations are used as each new observation is received to update the estimates of the state vector and associated covariance matrix. (Adapted from Gelb, A., *Applied Optimal Estimation*, MIT Press, Cambridge, MA, 1974.)

At time, t_k, we again apply the Kalman update equations to result in improved estimates of the state vector and covariance matrix due to the new information (via the observation at time, t_k). This process continues until all observations are processed.

19.5 Sequential Processing Implementation Issues

While implementation of a sequential estimation process can be performed in a computationally efficient, scalar formulation, there are a number of other implementation issues. Some of these issues are described below.

19.5.1 Filter Divergence and Process Noise

One potential problem of a sequential estimation is the termed divergence. The problem occurs when the magnitude of the state vector covariance matrix becomes relatively small. This decrease in $P_{\Delta x}$ occurs naturally as more observations are processed, since the knowledge of the state vector increases with the number of observations processed. When $P_{\Delta x}$ becomes relatively small, the Kalman gain becomes correspondingly small (since $P_{\Delta x}$ is a multiplicative factor in Equation 19.43). The result is that the estimator *ignores* new data and does not make significant improvements to Δx. While this would seem to be a desirable result of the estimation process, sometimes $P_{\Delta x}$ becomes artificially small, resulting in the filter disregarding valid observations. In order to correct this divergence problem, two techniques are often used: (1) introduction of process noise, and (2) use of a fading memory factor.

Process or dynamic noise is white noise added to the linear state perturbation model, i.e.,

$$\Delta \hat{x}(t_{n+1}) = \Phi(t_{n+1}, t_n)\Delta \hat{x}(t_n) + \eta \tag{19.59}$$

The vector, η, is assumed to have zero mean and covariance Q

$$\text{cov}(\eta) = Q \tag{19.60}$$

This dynamic noise represents imperfections or random errors in the dynamic state model. The noise vector in turn affects the propagation of the covariance propagation equation, yielding

$$P_n(t_{n+1}) = \Phi(t_{n+1}, t_n)P_n(t_n)\Phi(t_{n+1}, t_n)^{\mathrm{T}} + Q(t_n) \tag{19.61}$$

instead of Equation 19.57. Hence the covariance matrix has a minimum magnitude of $Q(t_n)$. This reduces the divergence problem.

Another technique used to avoid divergence is the use of a fading memory. The concept here is to weight recent data more than older data. This can be accomplished by multiplying the covariance matrix by a memory factor

$$s = e^{\Delta t / \tau}$$

such that

$$P_n(t_{n+1}) = s\Phi(t_{n+1}, t_n)P_n(t_n)\Phi(t_{n+1}, t_n)^{\mathrm{T}} \tag{19.62}$$

where
  Δt is the interval $(t_{n+1}, -t_n)$
  τ is an a priori specified memory constant

This is chosen so that $s \geq 1$. The fading memory factor, s, amplifies or scales the entire covariance matrix. By contrast, process noise is additive and establishes a minimum value for P_n. Either of these techniques can effectively combat the divergence problem. Their use should be based on physical or statistical insight into the estimation problem.

19.5.2 Filter Formulation

A second implementation issue for sequential estimation involves the formulation of the filter equations. The previous section derived a linearized set of equations. Non-linear formulations can also be developed. Gelb (1974) describes a second-order approximation to sequential estimation. Even if a linear approximation is used, there are variations possible in the formulation of the equations. A number of techniques have been developed to ensure numerical stability of the estimation process. Gelb (1974) describes a so-called square root formulation and a $\mathrm{UDU^T}$ formulation both aimed at increasing the stability of the estimators. It is beyond the scope of this book to describe these formulations in detail. Nevertheless, the reader is referred to work by Tapley (see for example Tapley and Peters [1980] for a detailed discussion and comparison of results).

The issue of observability is just as relevant to sequential estimation as it is to batch estimation. The state vector may be only weakly related to the observational data, or there may exist a high degree of correlation between the components of the state vector. In either case, the state vector will be indeterminate based on the observational data. As a result, the filter will fail to obtain an accurate estimate of the state parameters.

A final implementation addressed here is that of data editing. As in the batch estimation process, indiscriminate editing (rejection) of data is not recommended without a sound physical basis. One technique for editing residuals is to compare the magnitude of the observation residual, $v(t_k)$, with the value $[HPH^\mathrm{T} + \sigma_{OBS}]$

$$\frac{|v(t_k)|}{\left[HPH^\mathrm{T} + \sigma_{OBS}^2\right]} > O_{MAX} \tag{19.63}$$

If the ratio specified by Equation 19.63 exceeds O_{MAX}, then the observation is rejected and the state vector and covariance matrix are not updated.

19.5.3 Maneuvering Targets

One of the special difficulties with estimation techniques for target tracking involves maneuvering targets. The basic problem is that acceleration cannot be observed directly. Instead, we can only observe the after the fact results of an acceleration maneuver (e.g., turn, increase in speed). So in one sense, the estimate of a target's state vector "lags behind" what the target is actually doing. For benign maneuvers, this is not a large problem since, given new observations, the estimated state vector "catches up" with the actual state of the observed target. A challenge occurs however, if one or more targets are deliberately maneuvering to avoid detection or accurate tracking (e.g., terrain following tactical aircraft or aircraft involved in a "dog fight"). Because a sequential estimator such as a Kalman filter processes data on an observation-by-observation basis, it is possible for a sequential estimator to be very challenged by such maneuvers. Examples of work in this area include Schultz et al. (1999), McIntyre and Hintz (1998), and Lee and Tahk (1999). A series of surveys of maneuvering target methods has been conducted by Li and Jilkov; see Li and Jilkov (2000), Li and Jilkov (2002), and Li and Jilkov (2001a,b).

Several methods have been used to try to address such maneuvering targets. These include the following:

1. *Estimate the acceleration*—One way to address the problem is to augment the state vector to include acceleration terms (e.g., add components of acceleration to the state vector, **x**). This poses a problem in the estimation process because components of acceleration are highly correlated with components of velocity. Hence, the filter can become numerically unstable.

2. *Model the acceleration and estimate the model parameters*—Various researchers have attempted to develop models for acceleration using assumptions such as; the target only performs horizontal maneuvers at fixed turn rates, etc. Under these assumptions, the state vector can be augmented to include parameters in the acceleration model.

3. *Detect maneuvers and use decision-based methods to select an acceleration model*—Another technique used to address maneuvers is to monitor the size of the observation residuals to try to detect when a maneuver has occurred (based on anomalies in the predicted versus actual observations). Subsequently, automated decision methods can be used to select the appropriate acceleration model based on the observed observation residual anomalies.

4. *Use multiple Kalman filters in parallel (each having a different model of acceleration)*—Finally, some researchers have used an approach in which multiple estimators are used in parallel, each using a different model of acceleration. In this case, each estimator provides a different estimate (as a function of time) of the target's position. An expert system "overseer" can be used to monitor the observation residuals to determine which model is "correct" at a given time.

Fundamentally this remains a challenge for any estimation process because the sought for state (viz., knowledge of the acceleration) cannot be directly observed.

19.5.4 Software Tools

There are a wide variety of numerical methods libraries, mathematical software, simulation tools, and special toolkits that support the development of Kalman filters and related estimation algorithms. A sample of software tools is provided in Table 19.5. A survey of various estimation software tools is available at the web site: http://www.lionhrtpub.com/orms/surveys/FSS/fss9.html

TABLE 19.5 Summary of Kalman Filter Software Tools

Tool Name	Company	Web Site Reference
KFTool 2.5.1	Navtech	http://www.navtechgps.com/supply/kftool.asp
M-KFTOOL	MATLAB®	http://www.navtechgps.com/pdf/mkftoolcontents.pdf
ReBEL	Machine Learning and Signal Processing Group	http://choosh.ece.ogi.edu/rebel/index.html
MathLibX	Newcastle Scientific	http://www.virtualsoftware.com/ProdPage.cfm?ProdID=1405
Bayesian Filtering	Australian Center for Field Robotics	http://www.acfr.usyd.edu.au/technology/bayesianfilters/Bayes++.htm
Flight Dynamics S/S	Telesat	http://www.telsat.ca/eng/international_software1.htm
IMSL	Visual Numerics Inc.	http://www.vni.com/press/cn150.html

References

Bell, E. T. (1961). *Men of Mathematics*. New York: Simon & Schuster.

Blackman, S. S. (1986). *Multiple Target Tracking with Radar Applications*. Norwood, MA: Artech House, Inc.

Brent, R. P. (1973). *Algorithms for Minimization without Derivatives*. Englewood Cliffs, NJ: Prentice-Hall.

Deutsch, R. (1963). *Orbital Dynamics of Space Vehicles*. Englewood Cliffs, NJ: Prentice-Hall.

Escobol, R. R. (1976). *Methods of Orbit Determination*. Melbourne, Victoria, Australia: Krieger Publishing Co.

Fischer, R. A. (1912). On the absolute criteria for fitting frequency curves. *Messenger of Mathematics* **41**: 155.

Gauss, K. G. (1963). *Theory of Motion of the Heavenly Bodies* (reprinted from the 1809 original entitled, Theoria Motus Corporum Coelestium). New York: Dover.

Gelb, A. (1974). *Applied Optimal Estimation*. Cambridge, MA: MIT Press.

Grewal, M. S. and A. P. Andrews (2001). *Kalman Filtering Theory and Practice using MATLAB*. New York: John Wiley & Sons, Inc.

Hall, D. (1992). *Mathematical Techniques in Multisensor Data Fusion*. Norwood, MA: Artech House, Inc.

Hall, D. (2003). *Mathematical Techniques in Multisensor Data Fusion*. Boston, MA: Artech House, Inc.

Hall, D. and S. R. Waligora (1978). Orbit/attitude estimation using Landsat-1 and Landsat-2 landmark data. *NASA Goddard Space Flight Center Flight Mechanics/Estimation Theory Symposium*, NASA Goddard Space Flight Center, MD: NASA.

Henrici, P. (1962). *Discrete Variable Methods in Ordinary Differential Equations*. New York: John Wiley & Sons.

Kalman, R. E. (1960). A new approach in linear filtering and prediction problems. *Journal of Basic Engineering* **82**: 34–45.

Kalman, R. E. (1963). New methods in Wiener filtering theory. *Proceedings of the First Symposium of Engineering Application of Random Function Theory and Probability*. New York: John Wiley & Sons, pp. 270–388.

Kalman, R. E. and R. S. Bucy (1961). New results in linear filtering and prediction theory. *Journal of Basic Engineering* **83**: 95–108.

Kamen, E. W. and C. R. Sastry (1990). Multiple target tracking using an extended Kalman filter. *SPIE Signal and Data Processing of Small Targets*, Orlando, FL: SPIE.

Kessler, O. (1992). Functional description of the data fusion process. Warminster, PA: Office of Naval Technology, Naval Air Development Center.

Kolmogorov, A. N. (1941). Interpolation and extrapolation von stationaren zufalliegen Folgen. *Bulletin of the Academy of Sciences, USSR, Ser. Math. S.*: 3–14.

Lee, H. and M.-J. Tahk (1999). Generalized input estimation techniques for tracking maneuvering targets. *IEEE Transactions on Aerospace Electronic Systems* **35**(4): 1388–1402.

Legendre, A. M. (1806). *Nouvelles Methods pour la Determination des Orbits des Commetes*. Paris, France.

Li, X. R. and V. P. Jilkov (2000). A survey of maneuvering target tracking: Dynamic models. *SPIE Conference on Signal and Data Processing of Small Targets*, Orlando, FL, SPIE.

Li, X. R. and V. P. Jilkov (2001a). A survey of maneuvering target tracking—Part II: Ballistic target models. *SPIE Conference on Signal and Data Processing of Small Targets*, Orlando, FL, SPIE.

Li, X. R. and V. P. Jilkov (2001b). A survey of maneuvering target tracking—Part III: Measurement models. *SPIE Conference on Signal and Data Processing of Small Targets*, Orlando, FL, SPIE.

Li, X. R. and V. P. Jilkov (2002). A survey of maneuvering target tracking—Part IV: decision-based methods. *SPIE Conference on Signal and Data Processing of Small Targets*, Orlando, FL, SPIE.

McGee, A. and S. F. Schmidt (1985). *Discovery of the Kalman Filter as a Practical Tool for Aerospace and Industry*. Ames Research Center, Mountain View, CA: NASA.

McIntyre, G. A. and K. J. Hintz (1998). A comparison of several maneuvering target tracking models. *SPIE Conference on Signal Processing, Sensor Fusion and Target Recognition*, Orlando, FL, SPIE.

Nelder, J. A. and R. Mead (1965). A simplex method for function minimization. *Computer Journal* **7**: 308.

Polak, E. (1971). *Computational Methods in Optimization*. New York: Academic Press.

Press, W. H., B. P. Flannery et al. (1986). *Numerical Recipes: The Art of Scientific Computing*. New York: Cambridge University Press.

Ramachandra, K. V. (2000). *Kalman Filtering Techniques for Radar Tracking*. New York: Marcel Dekker, Inc.

Sanza, N. D., M. A. McClure et al. (1994). Spherical target state estimators. *American Control Conference*, Baltimore, MD.

Sayed, A. H. and T. Kailath (1994). A state-space approach to adaptive RLS filtering. *IEEE Signal Processing Magazine* 11(3): 18–70.

Schultz, R., B. Engleberg et al. (1999). Maneuver tracking algorithms for AEW target tracking applications. *SPIE Conference on Signal Processing and Data Processing of Small Targets*, Denver, CO: SPIE.

Shoup, T. E. (1987). *Optimization Methods with Applications for Personal Computers*. Englewood Cliffs, NJ: Prentice-Hall.

Sorenson, H. W. (1970). Least squares estimation: From Gauss to Kalman. *IEEE Spectrum* 7: 63–68.

Steinberg, A. (2001). Revisions to the JDL data fusion process model. *Handbook of Multisensor Data Fusion*. D. Hall and J. Llinas (eds.). Boca Raton, FL: CRC Press, pp. 2-1–2-19.

Swerling, P. (1959). First order error propagation in a stagewise smoothing procedure for satellite observations. *Journal of Astronautical Science* 6: 46–52.

Tapley, D. B. and J. G. Peters (1980). Sequential estimation using a continuous UDU^T covariance factorization. *Journal of Guidance and Control* 3(4): 326–331.

Waltz, E. and J. Llinas (1990). *Multi-sensor Data Fusion*. Norwood, MA: Artech House, Inc.

Wiener, N. (1949). *The Extrapolation, Interpolation and Smoothing of Stationary Time Series*. New York: John Wiley & Sons.

Wilde, D. J. and C. S. Beightler (1967). *Foundations of Optimization*. Englewood Cliffs, NJ: Prentice Hall.

Zavaleta, E. L. and E. J. Smith (1975). *Goddard Trajectory Determination System User's Guide*. Silver Spring, MD: Computer Sciences Corporation.

20

Data Registration

20.1 Problem Statement...455
20.2 Coordinate Transformations 456
20.3 Survey of Registration Techniques 459
20.4 Objective Functions..462
20.5 Results from Meta-Heuristic Approaches 464
20.6 Feature Selection...469
20.7 Real-Time Registration of Video Streams
     with Different Geometries...475
20.8 Beyond Registration 485
20.9 Closely Mounted Sensors and Temporal Changes.....................487
20.10 Ensemble Registration 488
20.11 Non-Rigid Body Registration 489
20.12 Summary..490
References...490

Richard R. Brooks
Clemson University

Jacob Lamb
*The Pennsylvania
State University*

Lynne L. Grewe
California State University

Juan Deng
Clemson University

20.1 Problem Statement

To fuse two sensor readings, they must be in a common coordinate system. The assumption that the mapping is known a priori is unwarranted in many dynamic systems. Finding the correct mapping of one image onto another is known as registration. An image can be thought of as a two-dimensional (2D) sensor reading. In this chapter we will provide examples using 2- and 2½D data. The same approaches can be trivially applied to 1D readings. Their application to data of higher dimensions is limited by the problem of occlusion, where data in the environment is obscured by the relative position of objects in the environment. The first step in fusing multiple sensor readings is registering the images to find the correspondence between them [Gonzalez 92].

Existing methods are primarily based on methods used by cartographers. These methods often make assumptions concerning the input data that may not be true. As shown in Figure 20.1, the general problem is: given two N-dimensional sensor readings, find the function F which best maps the reading from sensor 2 $S_2(x_1, ..., x_n)$ onto the reading from sensor 1 $S_1(x_1, ..., x_n)$ so that ideally $F(S_2(x_1, ..., x_n)) = S_1(x_1, ..., x_n)$. In practice, all sensor readings contain some amount of measurement error or noise so that the ideal case occurs rarely, if ever.

We will initially approach the registration problem as an attempt to automatically find a gruence (translation and rotation) correctly calibrating two 2D sensor readings with identical geometries. We can make these assumptions without a loss of generality:

- A method for two readings can be sequentially extended to any number of images.
- Most sensors currently work in one or two dimensions.
- We presuppose known sensor geometries. If they are known, a function can be derived to map readings as if they were identical.

Given two images:

Observed S_2 Reference S_1

Find the function that best maps the observed image
to the reference image:

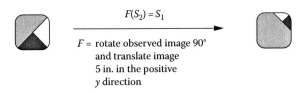

$$F(S_2) = S_1$$

F = rotate observed image 90°
and translate image
5 in. in the positive
y direction

FIGURE 20.1 Registration is finding the mapping function $F(S_2)$. (Adapted from Brooks, R.R. and Iyengar, S.S., *Multi-Sensor Fusion: Fundamentals and Applications with Software*, Prentice Hall, Upper Saddle River, NJ, 1998.)

Most of the work given here finds gruences (translations and rotations) since these functions are representative of the most common problems. Extending these approaches to include the class of affine transformations by adding scaling transformations [Hill 90] is straightforward. "Rubber sheet" transformations also exist that warp the contents of the image [Wolberg 90]. The final case study in this chapter transforms images of radically different geometries to a common mapping.

20.2 Coordinate Transformations

The math we present for coordinate transformations is based on the concept of rigid body motion. Rigid body motion generally refers to the set of *affine* transformations, which are used to describe either object motion or image transformations. We will refer to the entity transformed as an *image* throughout this section.

Affine transformations are combinations of four elementary operations: *rotation, scaling, shearing,* and *translation*. Rotation refers to the angular movement of an image. Scaling refers to changes in the size of the image. Shearing is movement of an image proportional to the distance along an axis. To visualize shearing, imagine a letter printed on a sheet of rubber. If the bottom edge of the sheet of rubber is held in place while the top edge is moved 6 in. to the right, the stretching of the sheet of rubber *shears* the image. Translation is simple movement of the image in a given direction (ex. moving an object 4 in. to the left).

These transformations can exist in any number of dimensions. Most commonly they are defined in two or three dimensions, describing either an image or an object in space. Often 3D data will be written as *quaternions*, which are 4D vectors. An object with x coordinate a, y coordinate b, and z coordinate c would be represented by the quaternion $(a\ b\ c\ 1)^T$. This formalism simplifies many image operations on images. Quaternions are also useful for the geometry of projections [Faugeras 93, Hill 90]. The geometry of projections is outside of the area treated in this book, but is extremely useful for interpreting sensor data.

To illustrate the use of quaternions, we calculate the position $(x\ y\ z)^T$ of point $(a\ b\ c)$ after it has been rotated by θ degrees around the z-axis and then translated by 14 units in the y-axis:

$$\begin{bmatrix} x \\ y \\ z \end{bmatrix} = \begin{bmatrix} \cos\theta & -\sin\theta & 0 \\ \sin\theta & \cos\theta & 0 \\ 0 & 0 & 1 \end{bmatrix} \begin{bmatrix} a \\ b \\ c \end{bmatrix} + \begin{bmatrix} 0 \\ 14 \\ 0 \end{bmatrix}$$

Using quaternions, this can be represented as a simple matrix multiplication:

$$\begin{bmatrix} x \\ y \\ z \\ 1 \end{bmatrix} = \begin{bmatrix} \cos\theta & -\sin\theta & 0 & 0 \\ \sin\theta & \cos\theta & 0 & 14 \\ 0 & 0 & 1 & 0 \\ 0 & 0 & 0 & 1 \end{bmatrix} \begin{bmatrix} a \\ b \\ c \\ 1 \end{bmatrix}$$

You are advised to calculate the values of x, y, and z for both expressions and verify that they are in fact identical.

We now present the matrices for individual transformations:

$R_X(\theta)$ Rotation about the x-axis by θ degrees:

$$\begin{bmatrix} 1 & 0 & 0 & 0 \\ 0 & \cos\theta & -\sin\theta & 0 \\ 0 & \sin\theta & \cos\theta & 0 \\ 0 & 0 & 0 & 1 \end{bmatrix}$$

$R_Y(\theta)$ Rotation about the y-axis by θ degrees:

$$\begin{bmatrix} \cos\theta & 0 & -\sin\theta & 0 \\ 0 & 1 & 0 & 0 \\ \sin\theta & 0 & \cos\theta & 0 \\ 0 & 0 & 0 & 1 \end{bmatrix}$$

$R_Z(\theta)$ Rotation about the z-axis by θ degrees:

$$\begin{bmatrix} \cos\theta & -\sin\theta & 0 & 0 \\ \sin\theta & \cos\theta & 0 & 0 \\ 0 & 0 & 1 & 0 \\ 0 & 0 & 0 & 1 \end{bmatrix}$$

$T(t_x, t_y, t_z)$ Translation by t_x in the x-axis, t_y in the y-axis, and t_z in the z-axis:

$$\begin{bmatrix} 1 & 0 & 0 & t_x \\ 0 & 1 & 0 & t_y \\ 0 & 0 & 1 & t_z \\ 0 & 0 & 0 & 1 \end{bmatrix}$$

$S(s_x, s_y, s_z)$ Scaling by s_x in the x-axis, s_y in the y-axis, and s_z in the z-axis:

$$\begin{bmatrix} s_x & 0 & 0 & 0 \\ 0 & s_y & 0 & 0 \\ 0 & 0 & s_z & 0 \\ 0 & 0 & 0 & 1 \end{bmatrix}$$

A total of six shearing transformations exist, one for each possible combination of the three axes. Since a shear is not technically a rigid body transform, they are omitted from our list; interested readers are referred to [Hill 90].

Note that affine transforms preserve linearity, parallelism of lines, and proportional distances [Hill 90]. The volume of an object going through an affine transform represented by the matrix **M** is increased by a factor equal to the determinant of **M** [Hill 90]:

$$\frac{\text{Volume after transform}}{\text{Volume before transform}} = \text{Determinant}(\mathbf{M})$$

You can also use matrix multiplication to construct a matrix that will perform any combination of affine transformations. You do so by simply constructing an equation that would perform the transformations sequentially and then calculate the matrix product. Following this logic, translating the vector **v** by 4 in the z-axis and then rotating it by θ around the y-axis is: $R_Y(\theta)T(0,0,4)\mathbf{v}$.

Consider the transformation that takes place when moving a camera C relative to a 3D box B. The camera initially looks at the middle of one side of the box from a distance of 3 m. The boxes vertices are at positions (2,–1,3), (2,1,3), (–2,–1,3), (–2,1,3), (2,–1,6), (2,1,6), (–2,–1,6), and (–2,1,6) relative to the camera. The camera makes the following sequence of motions: the camera moves 10 m further away from the box, pans 20° to the right, and rotates 60° counterclockwise about its axis. We will calculate the new positions of the box's vertices relative to the camera.

Care must be taken that the coordinate system used is consistent throughout the process. In this case the observer is in motion, and all measurements are made in the frame of reference of the observer. For problems of this type, the natural choice of coordinates is called a *viewer-centered* coordinate system. The origin is the center of mass of the observer, and every motion made by the observer is considered as a diametrically opposed motion made by the observed environment. For example, a video camera on a train moving east at 40 miles/h shows the world moving west at 40 miles/h.

The transformation matrix for this problem is easily derived. Moving C 10 m away from B is a simple translation. Since C moves –10 m along the z-axis, the matrix **m1** moves B + 10 m along the same axis:

$$\begin{bmatrix} 1 & 0 & 0 & 0 \\ 0 & 1 & 0 & 0 \\ 0 & 0 & 1 & 10 \\ 0 & 0 & 0 & 1 \end{bmatrix}$$

Matrix **m2** describes the 20° pan taken by C around B. In essence, this is a 3D rotation of B about the y-axis. The rotation, however, is about the mid-point of B and not the mid-point of C. In order to perform this transformation, the midpoint of C must first be translated to the midpoint of B, the rotation performed, and C must then be translated along the z-axis. Care must be taken to note that rotating C by 20° is seen as rotating B by –20°. The matrix **m2** is the result of this matrix multiplication:

$$\begin{bmatrix} 1 & 0 & 0 & 0 \\ 0 & 1 & 0 & 0 \\ 0 & 0 & 1 & 14.5 \\ 0 & 0 & 0 & 1 \end{bmatrix} \begin{bmatrix} 0.94 & 0 & -0.34 & 0 \\ 0 & 1 & 0 & 0 \\ 0.34 & 0 & 0.94 & 0 \\ 0 & 0 & 0 & 1 \end{bmatrix} \begin{bmatrix} 1 & 0 & 0 & 0 \\ 0 & 1 & 0 & 0 \\ 0 & 0 & 1 & -14.5 \\ 0 & 0 & 0 & 1 \end{bmatrix}$$

The final transformation is done by rotating C about its own z-axis. The matrix **m3** is therefore a simple rotation about the **z**-axis of $-60°$:

$$\begin{bmatrix} 0.5 & -0.87 & 0 & 0 \\ 0.87 & 0.5 & 0 & 0 \\ 0 & 0 & 1 & 0 \\ 0 & 0 & 0 & 1 \end{bmatrix}$$

Again, using these three affine transformations, we can find one single transformation by performing the matrix multiplication $\mathbf{M} = \mathbf{m3\ m2\ m1}$. This gives the following equation, which maps any point $(x\ y\ z)$ in the environment to a point $(x'\ y'\ z')$ as now seen by C:

$$\begin{bmatrix} 0.47 & -0.87 & -0.17 & 0.7650 \\ 0.8178 & 0.5 & -0.2958 & 1.3311 \\ 0.34 & 0 & 0.94 & 10.27 \\ 0 & 0 & 0 & 1 \end{bmatrix}$$

This means that the vertices of the cube are: (2.065,1.5793,13.77), (0.325,2.5793,13.77), (0.185,−1.6919,12.41), (−1.555,−0.6919,12.41), (1.555,0.6919,16.59), (−0.185,1.6919,16.59), (−0.325,−2.5793,15.23), and (−2.065,−1.5793,15.23) after the transformations. C programs are available in [Brooks 98] that perform these transformations for camera images.

20.3 Survey of Registration Techniques

Many processes require that data from one image called the observed image be compared with or mapped to another image called the reference image. Perhaps the largest amount of current image registration research is in the field of medical imaging. Outputs from different medical imaging technologies, such as CT (Computed Tomography) and MRI (Magnetic Resonance Imaging), are combined to obtain more complete information for illness diagnosis, monitoring illness evolution, radiotherapy, etc. Image registration is also required in remote sensing (e.g., environment monitoring, weather forecast, information integration in geographic information system), robotics (e.g., location tracking, motion detection), computer vision (e.g., target localization, automatic quality control), and a few others. Many methods exist for registering images. A comprehensive survey of traditional methods of image registration that were published before 1992 can be found in [Brown 92]. The representative traditional methods became either classic, which are still in use, or key ideas for registration methods developed later. Table 20.1 summarizes the features of the representative traditional methods. Surveys of later image registration methods can be found in [Wyawahare 09, Zitova 03]. The techniques that are mainly used in remote sensing are described and evaluated in [Fonseca 96, Gulch 91, Moigne 98].

Image registration usually consists of four steps [Zitova 03]:

- Feature detection: Salient and distinctive objects (e.g., close-boundary regions, edges, contours, line intersections, corners) in the observed image and the reference image are detected. These features are represented by the points, which are called control points.
- Feature matching: Match the control points in the observed image to the control points in the reference image.
- Transform model estimation: Use the control points matching to estimate the mapping functions, which align the observed image with the reference image.
- Image resampling and transformation: The observed image is transformed by means of the mapping functions.

TABLE 20.1 Image Registration Methods

Algorithm	Image Type	Matching Method	Interpolation Function	Transforms Supported	Comments
Andrus	Boundary maps	Correlation	None	Gruence	Noise intolerant
					Small rotations
Barnea	No restriction	Improved correlation	None	Translation	No rotation, scaling, noise, rubber sheet
Barrow	No restriction	Hill climbing	Parametric chamfer	Gruence	Noise intolerant
					Small displacement
Brooks Iyengar	No restriction	Elitist Gen. Alg.	None	Gruence	Noise tolerant
					Tolerates periodicity
Cox	Line segments	Hill climbing	None	Gruence	Matches using small number of features
Davis	Specific shapes	Relaxation	None	Affine	Matches shapes
Goshtasny 1986	Control points	Various	Piecewise linear	Rubber sheet	Fits images using mapped points
Goshtasby 1987	Control points	Various	Piecewise cubic	Rubber sheet	Fits images using mapped points
Goshtasby 1988	Control points	Various	Least squares	Rubber sheet	Fits images using mapped points
Jain	Sub-images	Hill climbing	None	Translation	Small translations
					No rotation, no noise
Mandara	Control points	Classic G.A S.A.	Bi-linear	Rubber sheet	Fits four fixed points using error fitness
Mitiche	Control points	Least squares	None	Affine	Uses control points
Oghabian	Control points	Sequential search	Least squares	Rubber sheet	Assumes small displacement
Pinz	Control points	Tree search	None	Affine	Difficulty with local minima
Stockman	Control points	Cluser	None	Affine	Assumes landmarks, periodicity problem
Wong	Intensity differences	Exhaustive search	None	Affine	Uses edges, intense computation

Several algorithms exist for each step. In feature detection, significant regions (lakes, fields, buildings) [Roux 96, Wong 79], lines (region boundaries, coastlines, roads, rivers) [Moss 97], points (corners, line intersections, points of high curvature) [Goshtasby 85, Vasileisky 98] are treated as features. Table 20.2 summarizes the features and methods to detect them. A large number of methods have been proposed for matching control points in the observed image to the control points in the reference image. The obvious method is to correlate a template of the observed image against the reference image. The classic correlation-based method is the sequential similarity detection algorithm [Barnea 72]. Although it is less accurate than the later developed cross-correlation algorithms [Berthilsson 98, Kaneko 02], it is faster.

TABLE 20.2 Features and Detection Methods

Features	Type	Detection Methods
Region features	Projections of high contrast closed-boundary region, water reservoirs and lakes, buildings, forests, urban areas, shadows	Segmentation methods [Pal 93]
Line features	line segments, object contours, coastal lines, roads	Edge detection methods [Canny 86, Marr 80, Ziou 97]
Point features	Line intersections, road crossings, centroids of water regions, oil and gas pads, high variance points, local curvature discontinuities, inflection points of curves, local extrema of wavelet transform, corners	Conner detector [Rohr 01, Smith 97, Zheng 99]

Relaxation-based image registration methods can be found in [Cheng 96, Ton 89, Wang 83]. Another widely used approach is to calculate the spatial relationships (e.g., transformation matrix) between the control points in the observed and reference images [Barrow 77, Goshtasby 85, Stockman 82]. [Goshtasby 85] evaluated the number of features in the observed image that, after the particular transformation, fall within a given range next to the features in the reference image. The transformation parameters with the highest score were set as a valid estimate. Clustering technique was presented in [Stockman 82], which tires to match control pointed connected by abstract edges or line segments. [Barrow 77] introduced the chamfer matching for image registration. Line features detected in the image are matched by means of the minimization of the generalized distance between them. An improved chamfer matching [Borgefors 88] was proposed with better distance measurement and faster speed.

As an alternative to registering images by calculating spatial relationships, the matching of features can be estimated using their description. Features from the observed image and the reference image with the most similar descriptions are matched. The choice of description depends mainly on the feature characteristics. Forests are described in [Sester 98] by elongation parameter, compactness, number of holes, and several characteristics of the minimum bounding rectangle. [Zana 99] described each feature point by means of angles between relevant intersecting lines. [Montesinos 00] proposed using differential descriptors of the image function in the neighborhood of the control points. [Yang 99] used border triangles generated by object convex hull and computed affine geometric invariants on them.

A number of researchers have used multiresolution methods to prune the search space considered by their algorithms. [Mandara 89] uses a multiresolution approach to reduce the size of their initial search space for registering medical images using simulated annealing and genetic algorithms. This work influenced [Oghabian 92], who similarly reduced their search space when registering brain images with small displacements. [Pinz 95] adjusted both multiresolution scale space and step size in order to reduce the computational complexity of a hill-climbing registration method. By starting with low-resolution images, these researchers believe that rejecting large numbers of possible matches is possible, and that the correct match can be found by progressively increasing the resolution. Note that, in images with a strong periodic component, a number of low-resolution matches may be feasible. In which case, the multiresolution approach will be unable to prune the search space and instead will increase the computational load.

The approach in [Grewe 99], which we discuss in more detail in Section 20.6, uses a multi-resolution technique, the wavelet transform, to extract features used to register images. Others have also applied wavelets to this problem, including using locally maximum wavelet coefficient values as features from two images [Sharman 97]. The centroids of these features are used to compute the translation offset between the two images. This use of a simple centroid difference is subject to difficulties when the scenes only partially overlap and hence contain many other different features. A principle components analysis is then performed and the eigenvectors of the covariance matrix provide an orthogonal reference system for computing the rotation between the two images. In another example [DeVore 97], the wavelet transform is used to obtain a complexity index for two images. They use the complexity measure to determine the amount of compression appropriate for the image. Compression is then performed, giving a small number of control points. They then test images made up of control points for rotations to determine the best fit.

The recent *Mutual Information* (MI) methods originated from the information theory represent the leading technique in multimodal registration. MI is a statistical dependency between two data sets. [Viola 97] proposed using MI for image registration for the first time. The method is based on the maximization of MI. Various algorithms to maximize MI are studied. [Viola 97] maximized MI by using gradient descent optimization method. [Ritter 99] used hierarchical search strategy together with simulated annealing to find the maximum of the MI, while in [Studholme 99] maximization was achieved by using a multiresolution hill-climbing algorithm. Other MI maximization algorithms include Parzen window [Thevenaz 97], the Brent's method and the Powell's multi-dimensional direction set method [Maes 97]. A review of MI based image registration methods can be found in [Josien 03].

In the next section, we examine the registration problem generically from the point of view of an optimization problem. Beyond what data is to be matched between data sets, there is also the task of how to go about proposing matches without exhaustively trying all of them. We mentioned work above that uses hierarchical spaces or genetic algorithms to prune the search space. Another technique is discussed in [Yin 03] which mimics aspects of medical immunology to the task of registration with good convergence to globally optimum registration result.

Most systems, and especially true for real-time system, view registration as the problem of finding a global transformation to perform registration. However, there are situations where a single global transformation is not appropriate and in this case different techniques can be applied like piecewise transformations, elastic membrane or optical flow techniques.

20.4 Objective Functions

In optimization literature, the function to be optimized is often called the *objective function* [Barhen 96, Barhen 97, Cetin 93]. The term fitness function is used in genetic algorithms literature [Brooks 96, Brooks 96a, Brooks 98] in a manner similar to the use of objective function in optimization literature. Mappings can be defined by affine transformations. The terms objective function and fitness function will be used interchangeably in this chapter. Here we present a few functions that are appropriate for data registration. The proper function to use is dependent on the amount and type of noise present in the data.

If the noise is approximately Gaussian (i.e., white noise), it follows a normal distribution and has an expected value of zero. A fitness function can be derived by first computing the intersection between the two images, sensor 1 and sensor 2, using the proposed mapping function. The gray level of every pixel from sensor 1 in the intersection are compared with the gray level of the corresponding sensor 2 pixel. We define $read_1(x,y)$ as the value returned by sensor 1 at point (x,y), and $read_2(x',y')$ as the reading returned by sensor 2 at point (x',y'). Point (x',y') is found by reversing the translation and rotation defined by the parameters being tested. You can present the difference of $read_1(x,y)$ and $read_2(x',y')$ as

$$read_1(x,y) - read_2(x',y') = (v_1(x,y) + noise_1(x,y)) - (v_2(x',y') + noise_2(x',y')) \qquad (20.1)$$

where

$v_1(x,y)$ and $v_2(x',y')$ are the actual gray scale values
$noise_1(x,y)$ and $noise_2(x',y')$ are the noise in the sensor 1 and sensor 2 readings, respectively

This expression can be rewritten as

$$read_1(x,y) - read_2(x',y') = (v_1(x,y) - v_2(x',y')) + (noise_1(x,y) - noise_2(x',y')) \qquad (20.2)$$

If we square this value and sum it over the entire intersection, we get

$$\sum (read_1(x,y) - read_2(x',y'))^2 = \sum ((v_1(x,y) - v_2(x',y')) + (noise_1(x,y) - noise_2(x',y')))^2 \qquad (20.3)$$

Note that when the parameters are correct, the gray scale values $v_1(x,y)$ and $v_2(x',y')$ will be identical, and expression (20.3) becomes

$$\sum (read_1(x,y) - read_2(x',y'))^2 = \sum ((noise_1(x,y) - noise_2(x',y'))^2 \qquad (20.4)$$

Since all noise follows the same distribution with the same variance, the expected value of this is identical for all intersections of the same area and, as such, the minimum value for the function over all intersections of a given area. Variation in this value thus consists of two parts: the difference in the gray scale values of the noise-free image, and a random factor that is distributed according to a Chi-square distribution of unknown variance. The number of degrees-of-freedom for the Chi-square distribution is the number of pixels in the intersection.

We can have small intersections that match coincidentally. In order to favor intersections of larger area, we divide by the number of pixels, denoted by N, in the intersection squared. The fitness function thus becomes

$$\frac{\sum (read_1(x,y) - read_2(x',y'))^2}{N^2} \tag{20.5}$$

The expected value of a Chi-square function is the number of degrees of freedom, and the number of degrees of freedom in this case is equal to the number of pixels in the intersection. In the case of a perfect fit (i.e., $v_1(x,y) = v_2(x',y')$), the expected value of this function is therefore a constant factor of

$$\frac{1}{N} \tag{20.6}$$

This function is the summation of the error per pixel squared over the intersection of the sensor 1 and sensor 2 readings. As shown above, the unique global minimum of this function is found when using the parameters that define the largest intersection where the gray scale values of sensor 1 are the same as the gray scale values of the translated and rotated sensor 2 reading. The fitness function is thus

$$\frac{\sum (read_1(x,y) - read_2(x',y'))^2}{K(w)^2} = \frac{\sum ((gray_1(x,y) - gray_2(x',y')) + (noise_1(x,y) - noise_2(x',y')))^2}{K(w)^2} \tag{20.7}$$

where
- w is a point in the search space
- $K(w)$ is the number of pixels in the overlap
- $read_1(x,y)$ ($read_2(x',y')$) is the pixel value from sensor 1 (2) at point (x,y) $((x',y'))$
- $gray_1(x,y)$ ($gray_2(x',y')$) is the noiseless value for sensor 1 (2) at (x,y) $((x',y'))$
- $noise_1(x,y)$ ($noise_2(x',y')$) is the noise in the sensor 1 (2) reading at (x,y) (x',y')

This function has been shown to reflect the problem adequately when the noise at each pixel follows a Gaussian distribution of uniform variance. In our experiments this is not strictly true, when the gray scale is limited to 256 discrete values. Because of this, when the gray scale value is 0 (255) the noise is limited to positive (negative) values. For large intersections, however, this factor is not significant.

Other noise models can be accounted for by simply modifying the fitness function. Another common noise model is the salt-and-pepper noise. Either malfunctioning pixels in electronic cameras or dust in optical systems commonly causes this type of noise. In this model the correct gray scale value in a picture is replaced by a value of 0 (255) with an unknown probability $p(q)$. An appropriate fitness function for this type of noise is Equation 20.8:

$$\sum_{\substack{read_1(x,y) \neq 0,255 \\ read_2(x,y) \neq 0,255}} \frac{(read_1(x,y) - read_2(x',y'))^2}{K(w)^2} \tag{20.8}$$

A similar function can be derived for uniform noise by using the expected value $E[(U_1 - U_2)^2]$ of the squared difference of two uniform variables U_1 and U_2. An appropriate fitness function is then given by Equation 20.9:

$$\sum \frac{(read_1(x, y) - read_2(x', y'))^2}{E[(U_1 - U_2)^2]K(w)^2} \tag{20.9}$$

20.5 Results from Meta-Heuristic Approaches

In this section, we consider 2D images, without occlusion and projection, although the results in this section could easily be extended to account for projection. We consider two images, which can be matched using gruence (translation and rotation) transformations. Addition of scaling would make the class of transformations include all affine transformations. Occlusion and projection require knowledge of the scene structure and interpretation of the sensor data before registration. It is doubtful that is feasible in real-time.

Given two noisy overlapping sensor readings, compute the optimal gruence (translation and rotation) mapping one to the other. The sensors return 2D gray level data from the same environment. Both sensors have identical geometric characteristics. They cover circular regions such that the two readings overlap. Since the size, position, and orientation of the overlaps are unknown, traditional image processing techniques are unsuited to solving the problem. For example, the method of using moments of a region is useless in this context [Russ 95].

Readings from both sensors are corrupted with noise. In Section 20.4, we derive appropriate fitness (objective) functions for different noise models. The Gaussian noise model is applicable to a large number of real-world problems. It is also the limiting case when a large number of independent sources of error exist.

This section reviews results from [Brooks 98, Chen 02] where we attempted to find the optimal parameters (x_T, y_T, θ) defining the relative position and orientation of the two sensor readings. The search space is a 3D vector space defined by these parameters, where a point is denoted by the vector $w = [x_T, y_T, \theta]^T$. Figure 20.2 shows the transformation between the two images given Equation 20.10:

$$\begin{bmatrix} x' \\ y' \\ 1 \end{bmatrix} = \begin{bmatrix} \cos\theta & -\sin\theta & x_T \\ \sin\theta & \cos\theta & y_T \\ 0 & 0 & 1 \end{bmatrix} \begin{bmatrix} x \\ y \\ 1 \end{bmatrix} \tag{20.10}$$

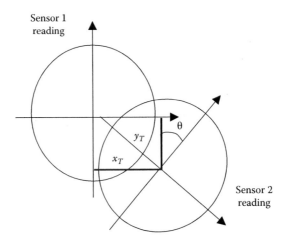

FIGURE 20.2 Geometric relationship of two sensor readings. (From Chen, Y. et al., *IEEE Trans. Knowl. Data Eng.*, 14(1), 79, January 2002)

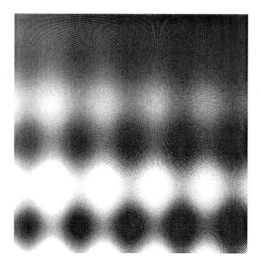

FIGURE 20.3   Terrain model.

Figure 20.3 shows the artificial terrain used in the experiments described in [Brooks 96, Chen 02]. It contains significant periodic and non-periodic components. Two overlapping circular regions are taken from the terrain. One of them is rotated. These are virtual sensor readings. The readings are also corrupted with noise. The experiments use optimization methods to search for the proper mapping of sensor 1 to sensor 2. Figures 20.4 and 20.5 give examples of the two sensor readings used with different noise levels.

Figure 20.6 shows paths taken by the tabu search algorithm when searching for an optimal match between the two sensor readings with increasing noise. The search starts in the middle of the sensor 1 reading. Note that the correct answer would have been at the bottom right-hand edge of the sensor 2 reading. The left-hand image shows the path taken during 75 iterations when the error variance was increased to 90. Note that even in the presence of noise that is strong enough to obscure most of the information contained in the picture, the search took approximately the same path as with very little noise.

Figure 20.7 shows gene pool values after a number of iterations of a genetic algorithm using a classic reproduction scheme with variance values of 1 (right) and 90 (left). This figure shows that even in the presence of noise, the values contained in the gene pool tend to converge. Unfortunately convergence is not to globally optimal values.

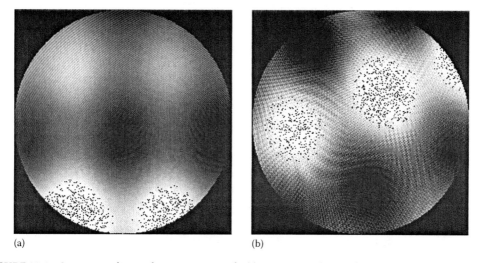

(a)   (b)

FIGURE 20.4   Sensor 1 reading with noise variance of 1 (a). Sensor 2 reading with noise variance of 1 (b).

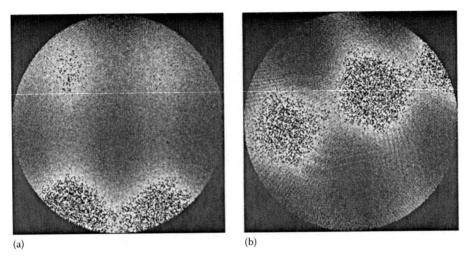

FIGURE 20.5   Sensor 1 reading with noise variance of 33 (a). Sensor 2 reading with noise variance of (b).

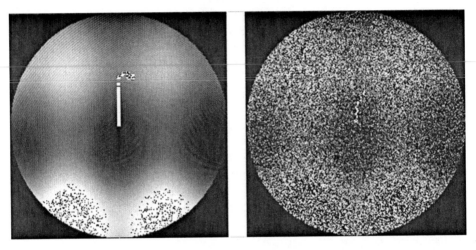

FIGURE 20.6   Search path taken by tabu search registration method. Noise increases on the left.

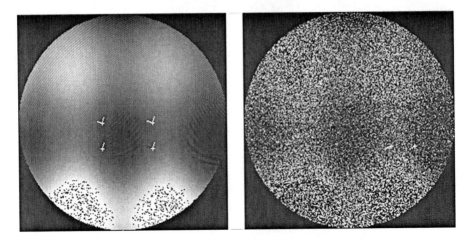

FIGURE 20.7   Gene pools from classic genetic algorithm seeking image mapping parameters.

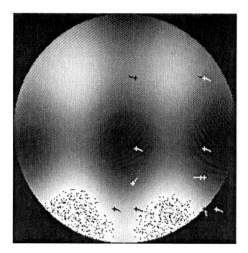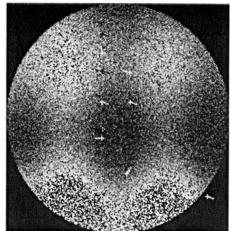

FIGURE 20.8   Gene pools from elite genetic algorithm.

Figure 20.8 shows gene pools found by the elite reproduction scheme after several iterations. Notice that the images contain values very close to the globally optimal value (the values near the lower left edge of the sensor reading). The genetic algorithm with the elite reproduction scheme tended to converge towards the globally optimal value even in the presence of moderate noise.

The elitist approach converged rapidly towards good solutions to the problem. In fact, the shapes of the graphs are surprisingly similar considering the differences in the images they are treating. In spite of the fact that the algorithm converged toward good solutions even in the presence of overwhelming amounts of noise, a limit existed to its ability to find the globally optimal solution. Note that the globally optimal parameter values are: x displacement = 91, y displacement = 91, rotation = 2.74889 rad. The algorithm does not always find the globally optimal values, but tends to do a good job even in the presence of moderate amounts of noise. However, once the noise reaches a point where it obscures too much of the information present in the image, it no longer locates the optimal values.

Simulated annealing and TRUST are the methods used with clear termination criteria. This makes a direct comparison with tabu search and genetic algorithms difficult. The final answers found by simulated annealing were roughly comparable to the answers found by tabu search. The number of iterations used to find these answers was much larger than the number of iterations taken by tabu search. Figure 20.9 displays

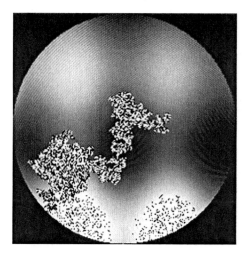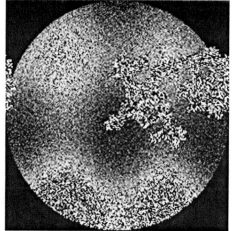

FIGURE 20.9   Search paths taken by simulated annealing.

the paths taken by simulated annealing searching for the correct registration in the presence of varying noise levels. Since the correct answer is in the lower right-hand corner of the sensor reading, obviously simulated annealing did not converge to the global optimum. The simulated annealing approach searched within any of a number of local optima. It did not remain trapped in the region defined by the first local optima, but did not find the global optimum, either.

Figure 20.10 shows paths taken by TRUST when the global optimum is in the upper right hand corner with a rotation of 2.49 rad. The line from the center $(0, 0)$ goes to bound $(255, 255)$ and the search stops. This illustrates TRUST's ability to find local minima and quickly climb out of their basins of attraction.

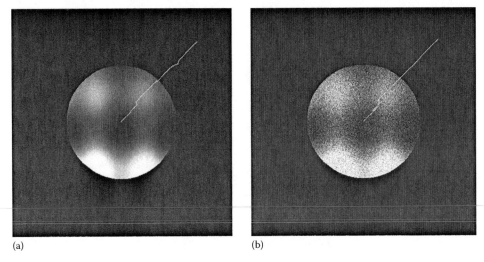

(a)   (b)

FIGURE 20.10   Paths taken by TRUST with noise variance 0.0 (a) and 30.0 (b).

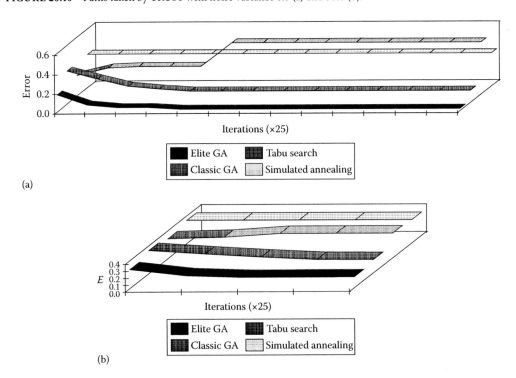

FIGURE 20.11   Fitness function results variance 1 (a) and 50 (b).

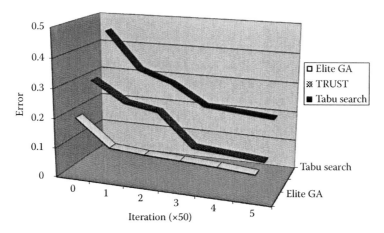

FIGURE 20.12   Average fitness function value for Elite GA, TRUST, and tabu search.

After 300 iterations, the results of tabu search are not even close to the global minimum. The results from TRUST and Elite Genetic Algorithms are almost identical, except TRUST has concrete stopping criteria. It has finished examining the search space. We can be more certain that a global minimum has been located. We tested TRUST using noise with seven different variances under different conditions. We compare the results with the elitist genetic algorithm. Both the elitist genetic algorithm and TRUST can handle noise with a variance of up to 30. The algorithms do not always find the global minima. But the TRUST results show the optimal value can be found even in the presence of large amounts of noise. When the noise reaches levels such as 70, or 90, it obscures the images and it becomes impossible to find the correct answer.

Figures 20.11 and 20.12 show the value of the best parameter set found by the approaches for two separate sets of experiments. Tabu search, elite genetic algorithms and TRUST tend to move towards locally optimal values and are stable manner. They show the relationship of the fitness function value of the best parameter set to the number of iterations used by the algorithm. In Figure 20.11, the best answer found by simulated annealing is represented by a straight line. This is not an entirely fair representation. The first several iterations of the algorithm are when the temperature parameter is at its highest point, in which case the algorithm amounts to a random walk. This is intentional; convergence is delayed until a later point in the algorithm when the system starts to cool.

20.6   Feature Selection

It is possible, though with high computational complexity to try to match all possible combinations of raw sensor data points to product potential registrations. Many systems to reduce the complexity extract a smaller set of features that are used to create hypothetical registrations that are then verified. In this section, we will discuss a few feature-based registration systems. Later in this chapter, we discuss the specific topic of beyond data level registration where the goal may not be to register the entire data sets.

It is possible to extract features in the spatial domain, but, many have found it useful to perform registration in other domains like the wavelet domain as discussed in [Grewe 99]. Figure 20.13 shows a block diagram of the WaveReg system from [Grewe 99]. The registration process begins with the transformation of range image data to the wavelet domain. Registration may be done using one decomposition level of this space to greatly reduce the computational complexity of registration. Alternatively, hierarchical registration across multiple levels may be performed. Features are extracted from the user selected wavelet decomposition level. Users also determine the amount of compression desired in this level. Matches between features from the two range images are used to hypothesize transformations between the two images, which are then evaluated. The "best" transformations are retained.

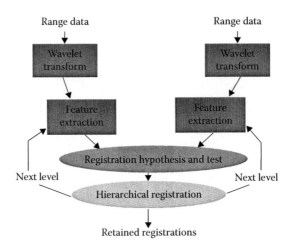

FIGURE 20.13   Block diagram of WaveReg system.

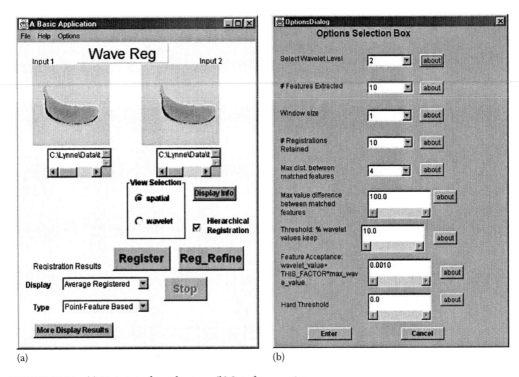

(a)   (b)

FIGURE 20.14   (a) Main interface of system. (b) Set of user options.

Figure 20.14a shows the main user interface. The user can select to perform registration from an initial estimate if one is known. Other options can be altered from defaults in the Options Selection Box shown in Figure 20.14b. Features include data compression thresholds, the wavelet level for use in registration, and the number of features to use.

The registration process starts by applying a Daubechies-4 wavelet transform to each range image. The Daubechies-4 wavelet was chosen for compactness. Wavelet data is compressed by thresholding to eliminate low magnitude wavelet coefficients. The wavelet transform produces a series of 3D edge maps at different resolutions. Maximal wavelet values indicate a sharp change in depth. Figure 20.15 illustrates this. Figure 20.15a is the original range map. Figure 20.15b is the corresponding intensity image

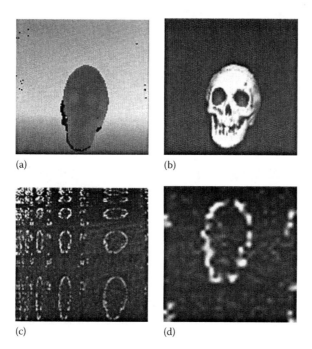

FIGURE 20.15 (a) Range image, (b) corresponding intensity image, (c) wavelet space, and (d) blow-up of level 2 of wavelet decomposition space (originally 32 × 32 pixels).

of the human skull. Figure 20.15c is the resulting wavelet transform. Figure 20.15d is a blow-up of one decomposition level (one level of resolution). Note how maximum values in Figure 20.15d correspond to regions where large depth changes occur. The edges may be due to object boundaries, often called jump edges, or may be due to physical edges/transitions on the object, called roof edges.

Features, "special points of interest" in the wavelet domain, are simply points of maximum value in the wavelet decomposition level under examination. They are selected so that no two points are close to each other. Users can specify the distance, or a default value is used. The distance is scaled consistent with the wavelet decomposition level under examination. Users may specify the number of features to extract at the first decomposition level. For hierarchical registration, this parameter is appropriately scaled for each level. Thresholds can also be changed from their defaults to influence the number of features extracted at each wavelet level. We found that the empirical defaults work well for a range of scenes. Figure 20.16 shows features detected for different range scenes at different wavelet levels. Notice how they correspond to sharp changes in depth.

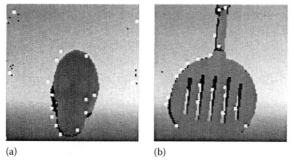

FIGURE 20.16 Features detected, approximate location indicated by white squares. (a) For wavelet level 2. (b) For wavelet level 1.

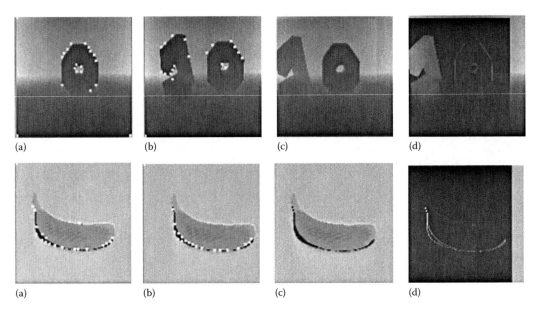

(a)   (b)   (c)   (d)

(a)   (b)   (c)   (d)

FIGURE 20.17 (a) Features extracted level 1, image 1 of scene 1. (b) Features extracted level 1, image 2 of scene 1. (c) Merged result via averaging registered images of scene 1. (d) Merged via subtraction of registered images of scene 1. (e) Features extracted level 1, image 1 of scene 2. (f) Features extracted level 1, image 1 of scene 2. (g) Merged result via averaging registered images of scene 2. (h) Merged via subtraction of registered images of scene 2.

The next step is verifying possible correspondences between features extracted from the unregistered range images. Each such hypothesis is a possible registration and is evaluated for goodness of fit. The best fits are retained. Hypothesis formation begins at the user selected wavelet decomposition level L. The default value for L is 2. If hierarchical registration is performed, registrations retained at level L are refined at level $L - 1$. The process continues until we reach the lowest level in the wavelet space.

For each hypothesis, the transformation for matching the features is calculated and the features from one range image transformed to the other's coordinate space. This reduces the number of computations needed for hypothesis evaluation compared to non-feature-based registration like Section 20.5. Features are compared and they match if they are close in value and location. Hypotheses are ranked by the number of features matched and how closely their values match. Figure 20.17 illustrates registration results on a range of scenes. These results are restricted to translations.

For an $N \times N$ image, the number of features extracted is typically $O(c)$. The search algorithm involves $O(c^2) = O(c)$ iterations. Compare this to a data-based approach which must compare $O(N^2)$ points for $O(N^2)$ iterations. Compared to other feature-based approaches, the compactness of the wavelet-domain and its multi-resolution nature reduce the number of features extracted. Hierarchical registration also allows efficient refinement of registration at higher-resolution yielding progressively more accurate results.

We tested this approach on several scenes. The system successfully found the correct registration in a retained set of 10 registrations for all scenes. Usually, only correct registrations are retained. Figure 20.17 shows images being registered and the results. Both averages and subtractions of registered images are shown.

One interesting point is that the location of an object in a scene can significantly change the underlying wavelet values [Grewe 97]. Values will still be of the same order of magnitude but direct comparison of values can be problematic, especially at higher levels of decomposition. It may be necessary to perform registration only at low decomposition levels. One way to resolve this is by using hierarchical registration. It usually eliminates incorrect registrations through further refinement at lower wavelet levels. Figure 20.18 illustrates this situation where in (a) an incorrect registration is retained at wavelet level 2 that is rejected in (b) during refined registration at wavelet level 1 and replaced by the correct registration.

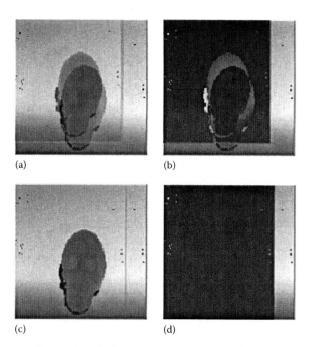

FIGURE 20.18 Correction of registration by hierarchical registration (a) incorrect registration, level 2, merged via averaging. (b) Same incorrect registration merged via subtraction. (c) Correct registrations retained after refined registration at level 1, merged via averaging. (d) same as (a), merged via subtraction.

When an image contains strong periodic components, our feature points may not define a unique mapping of the observed image to the reference image. This can result in incorrect registrations. Note that this is a problem with any feature-based registration approach and also, can cause problems for simple correlation-type registration systems [Brooks 98].

Another system that does not work in the spatial domain is described in [Hassen 09] which uses an old idea from the field of object recognition in using the phase information for representing the data in this case of registration. In particular, the phase information of a wavelet transform is used. Raw data in this phase space is used in a weighted formula which also uses phase histogram information. As with the older object recognition researches use of phase, the idea here is to reduce the dependency of exact spatial data values or changes in contrast.

Related in some sense to the work discussed above, is the concept of SIFT, Scale Invariant Feature Transform, discussed in [Lowe 99] which was developed for the task of Object Recognition. The concept presented by [Lowe 99] is that through SIFT a large number of ideally scale invariant features are detected so that recognition (matching to a model) for even articulated objects whose features change may be successful. SIFT points are defined as maxima and minima of the result of difference of Gaussians function applied in scale-space to a series of smoothed and resampled images. This alone yields too many feature points (called key point) and reduction is done first by discarding low contrast candidate points. Also, points which are more sensitive to noise in changing of their location are discarded. These points have poorly defined peaks in the DoG function and this is used to further eliminate feature points that are more sensitive to noise in location (possibly in one direction like along an edge) but, are at values of higher contrast.

An example of using SIFT in registration is given in [Zhang 09] where the authors discuss a system that uses both SIFT and maximally stable external region (MSER) features for registration that first analyzes the quality of features before selecting them for use in multi-sensor registration. In [Zhang 09], they create metrics to measure the "quality" of SIFT feature points using in part the idea of Information Entropy. Another metric involves the idea that greater spatial distribution, and in fact uniform distribution, is more idea. The suggest partitioning images into non-overlapping regions and removing features with low information entropy. This work is at beginning stages and it is unclear as to how many SIFT points should be retained in each region, how to make the regions, etc. But, the concept of looking at features and selecting them based on their goodness—even in the case of simple features like points is an important one. Another system using SIFT for the case of image registration is given in [Yang 07].

There are a number of systems that uses edge information as features for matching for the registration task. In [Li-pi 10] a simple Canny Edge detector is applied to the images and only longer edges are retained. The edges are described by a chain code and this is used to match the edges together accounting for shape. This particular version of edge feature based registration will work best in situations where there are longer semantically meaningful edges between the multiple sensor data. Figure 20.19 shows some results on aerial imagery where such edges (i.e., road boundaries and land mass boundaries) exist. Another edge feature based registration system again applied to the task of aerial image registration is found in [Zhu 10].

A system that uses higher-level features for registration is described in [Coiras 00] that performs registration of segments found in visual and infrared images. The system in [Weijie 09] uses multiple features at different levels to perform visual to SAR image registration including segments (regions), edges and special tie points with matching performed in that order. We further discuss higher level feature based registration in a later part of this chapter.

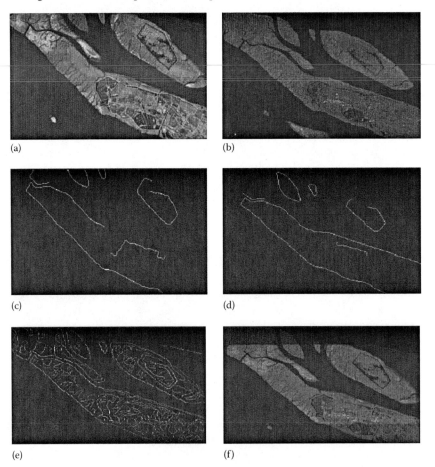

FIGURE 20.19 Results from Li-pi et al.: (a) IR image, (b) SAR image, (c) longer edges from image (a), (d) longer edges from image (b), (e) matched edges, and (f) resulting registration. (From Li-pi, N. et al., A multi-sensor image registration approach based on edge-correlation, *International Conference on Signal Processing Systems*, Dalian, China, pp. 36–40, 2010.)

20.7 Real-Time Registration of Video Streams with Different Geometries

This section discusses a method for registering planar video images to images from a camera with a 360° field of view in real-time. First we discuss three primary concepts vital to the implementation of the fast alignment algorithm. Covered first is catadioptric sensor geometry, which is used to obtain omni images. The second section covers featureless alignment of planar images and provides an algorithm proposed by Picard and Mann to accomplish such a goal. The final section describes the CAMSHIFT algorithm as it is used in a face tracking application.

The limited fields of view inherent in conventional cameras restrict their application. Researchers have investigated methods for increasing field of view for years [Nayar 97], [Rees 70]. Recently, there have been a number of implementations of cameras the use conic mirrors to address this issue [Nayar 97]. Catadioptric systems combine refractive and reflective elements in a vision system [Danilidis 00]. The properties of these systems have been well studied in literature on telescopes. The first panoramic field camera was proposed by [Rees 70]. Recently, Nayar proposed a compact catadioptric imaging system using folded geometry that provides an effective reflecting surface geometrically equivalent to a paraboloid [Nayar 99]. This system uses orthographic projection rather than perspective projection.

Cameras providing a field of view greater than conventional cameras are useful in many applications. A simple but effective mechanism for providing this functionality is the placement of a paraboloid mirror immediately in front of the camera lens. Though this solution would provide information about an expanded portion of the environment, constraints need to be applied for the information to be easily used. If an omnidirectional imaging system could view the world in 360° × 180° from a single effective pinhole, then planar projection images can be generated from the omni image by projecting the omni image onto a plane [Nayar 97]. Similarly, panoramic images can be constructed from an omni image by projecting the omni image onto the inside surface of a cylinder. Figure 20.20 shows how a paraboloid mirror can be used to expand field of view.

The cross section of the reflecting surface is given by $z(r)$. The cross section is then rotated about the z-axis to give a solid of revolution, the mirror surface. The viewpoint, v, is at the focus of the paraboloid. Light rays from the scene headed in the direction of v are reflected by the mirror in the direction of the orthographic projection. The angle θ gives the relation between the incoming ray and $z(r)$, then

$$\tan\theta = \frac{r}{z} \tag{20.11}$$

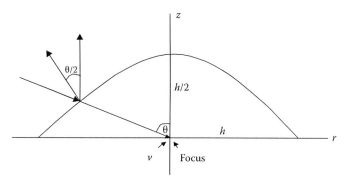

FIGURE 20.20 Mirror cross-section.

The surface of the mirror is specular, making the angles of incidence and reflectance $\theta/2$. The slope of the mirror surface at the point of reflectance is

$$\frac{dz}{dr} = \tan\frac{\theta}{2} \tag{20.12}$$

Substituting the trigonometric identity

$$\tan\theta = \frac{2\tan(\theta/2)}{1-\tan^2(\theta/2)} \tag{20.13}$$

yields

$$\frac{-2(dz/dr)}{1-(dz/dr)^2} = \frac{r}{z} \tag{20.14}$$

which indicates that the reflecting surface must satisfy a quadratic first-order differential equation [Nayar 97]. This can be solved to give the following equation for a reflecting surface that guarantees a single effective viewpoint:

$$z = \frac{h^2+r^2}{2h} \tag{20.15}$$

This is a paraboloid with h being the radius of the paraboloid at $z = 0$. The distance between the vertex and its focus is $h/2$. If the paraboloid if terminated at $z = 0$, the field of view equals exactly one hemisphere. A virtual planar image may be created by placing a plane at a tilt angle θ and pan angle ϕ and projecting the omni image onto the plane. Figure 20.21 shows the geometry of this projection.

The world coordinate system is represented by **X**-, **Y**-, and **Z**-axes vectors, with the origin located at *O*. The equation for the mirror is related to the world coordinates as

$$Z = \frac{h^2+r^2}{2h} \tag{20.16}$$

where
$r = X^2 + Y^2$
$h/2$ is the focal distance of the mirror

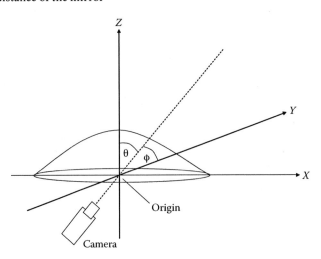

FIGURE 20.21 Projection geometry.

FIGURE 20.22 Omni and virtual planar images.

The tilt angle is measured clockwise from the z-axis and the pan angle is measured counter-clockwise from the x-axis. Given a line of sight from the origin to a point in the scene (x_p, y_p, z_p) at a distance ρ from its focus where

$$\rho = \frac{h}{(1 + \cos\theta)} \qquad (20.17)$$

The projection of the omni-directional image pixel at (θ, ϕ) onto the perspective plane is then given by

$$x = \rho\sin\theta\cos\phi, \quad y = \rho\sin\theta\sin\phi \qquad (20.18)$$

Figure 20.22 shows an example of an omni image and a virtual perspective image derived from the omni image.

[Mann 97] presents a featureless alignment of images method. They prove that an eight parameter projective transformation is capable of exactly describing the motion of a camera relative to a scene. Their algorithm is robust under noisy conditions, and applicable in scenes with varied depth. It is suited for the alignment needed to provide the initial bounding area for tracking.

Small changes from one image to another can be measured with optical flow. The optical flow equation assumes that every point $\mathbf{x}$, where $\mathbf{x} = (x, y)^T$, in a frame t is a translated version of some point $\mathbf{x} + \Delta\mathbf{x}$ in frame $t + \Delta t$. The optical flow velocity $\mu_f = (u, v)$ at point $\mathbf{x}$ is $\Delta\mathbf{x}/\Delta t$. If $\mathbf{E}_t$ and $\mathbf{E}_\mathbf{x} = (E_x, E_y)$ are the temporal and spatial derivatives of the frame then

$$\mu_f^T \mathbf{E}_\mathbf{x} + \mathbf{E}_t \approx 0 \qquad (20.19)$$

gives the traditional optical flow equation.

Given the optical flow between two frames, g and h, a projective transformation that aligns g with h can be found in the following manner. Represent the coordinates of a pixel in g as $\mathbf{x}$ and the coordinates of the corresponding pixel in h as $\mathbf{x}'$. The coordinate transformation from $\mathbf{x}$ to $\mathbf{x}'$ is given as

$$\mathbf{x}' = [x', y']^T = \frac{A[x, y]^T + b}{c^T [x, y]^T + 1} \tag{20.20}$$

where

the eight parameters of the transformation are given by $\mathbf{p} = [A, b, c, 1]$
$A \in \Re^{2\times 2}$
$b \in \Re^{2\times 1}$
$c \in \Re^{2\times 1}$

The model velocity μ_m is then given as $\mu_m = \mathbf{x}' - \mathbf{x}$. There will be some discrepancy between the model and flow velocities due to errors in flow calculation and other errors in the model assumption. The error between the two can be defined as

$$\varepsilon_{fit} = \sum_x (\mu_m - \mu_f)^2 = \sum_x \left(\mu_m + \frac{E_t}{E_{x,y}} \right)^2 \tag{20.21}$$

Differentiating with respect to the eight free parameters and setting the result to zero yields the linear solution

$$\left(\sum \phi\phi^T \right) [a_{11}, a_{12}, b_1, a_{21}, a_{22}, b_2, c_1, c_2]^T = \sum \left(x^T E_{x,y} - E_t \right) \phi \tag{20.22}$$

where $\phi^T = [E_x(x,y,1), E_y(x,y,1), xE_t - x^2 E_x - xyE_y, xE_y - xyE_x - y^2 E_y]$. This provides the approximate model of the transformation of the image h to the image g.

In order to compute the effectiveness of the transformation, the image h can be transformed to the image g using the parameters yielded by the above process. Then, some method of determining the difference between the two must be used. In order to yield the exact parameters of the transformation given the approximate model, [Mann 97] utilizes a "Four point method:"

1. Select the four corners of the bounding box containing the region to be aligned. When the entire image is to be aligned, the four corners are the corners of the image. These points are denoted as $\mathbf{s} = [s_1, s_2, s_3, s_4]$.
2. Transform the region using the approximate model described above to yield the set of transformed corners $\mathbf{t} = \mathbf{u}_m(\mathbf{s})$.
3. The correspondences between $\mathbf{t}$ and $\mathbf{s}$ are given by solving the four linear equations

$$\begin{bmatrix} x'_k \\ y'_k \end{bmatrix} = \begin{bmatrix} x_k, y_k, 1, 0, 0, 0, -x_k x'_k, -y_k x'_k \\ 0, 0, 0, x_k, y_k, 1, -x_k y'_k, -y_k y'_k \end{bmatrix} \cdot [a_{x'x}, a_{x'y}, b_{x'}, a_{y'x}, a_{y'y}, b_{y'}, c_x, c_y]^T \tag{20.23}$$

for $1 < k < 4$ to give the parameters of the transformation $\mathbf{p}$.

The "Four point method" is applied repetitively to the image h until $\mathbf{p}$ represents the exact transformation parameters using the law of composition.

1. Set $h_o = h$ and $\mathbf{p}_{0,0}$ to be the identity operator.
2. Repeat until the error between h_k and g falls below a certain threshold or a maximum number of iterations is reached:
   a. Estimate the eight parameters, $\mathbf{q}_k$ of the approximate model between g and h_{k-1}.
   b. Relate $\mathbf{q}_k$ to the exact parameters using the "Four point method" to yield the new set of parameters $\mathbf{p}_k$.
   c. Apply the law of composition to accumulate the effect of the $\mathbf{p}_k$'s. The composite parameters are denoted $\mathbf{p}_{0,k}$ where $\mathbf{p}_{0,k} = \mathbf{p}_k \circ \mathbf{p}_{0,k-1}$. Finally set $h_k = \mathbf{p}_{0,k} \circ h$.

The CAMSHIFT (Continuously Adaptive Mean Shift) algorithm is designed to track human faces based on the mean shift algorithm [Bradski 98]. In standard form, the mean shift algorithm attempts to find an object by finding the peak projection probability distribution image. The mean shift algorithm assumes that this probability distribution stays relatively constant as the object moves through an image [Cheng 98]. CAMSHIFT modifies the mean shift algorithm by dropping this constancy assumption. CAMSHIFT is able to track dynamically changing probability distributions by using color histograms that are relatively insensitive to rotation and translation of an object.

Given a set of objects o and a local measure set M, for a single measurement vector m_k, at a point in the image the probability of one of the objects o_n containing the point is

$$p(o_n \mid m_k) = \frac{p(m_k \mid o_n)p(o_n)}{\sum_i p(m_k \mid o_i)p(o_i)} \tag{20.24}$$

Equation 20.24 gives the probability of an object at a point in the image given a region surrounding that point, where a probability of 0.5 indicates complete uncertainty. The probability at each point in the vicinity of the last known position of the object is calculated, forming an object probability image. CAMSHIFT is then applied to this object probability image to track in which direction the object moved by climbing the gradient of the probability distribution to find the nearest peak.

CAMSHIFT involves five primary steps.

1. Select a search window size and shape for the probability distribution.
2. Place the search window.
3. Compute the probability distribution image of a region centered at the search window center but slightly larger than the window itself.
   a. Find the zeroth moment

$$M_{00} = \sum_x \sum_y I(x, y) \tag{20.25}$$

   b. Find the first moment for x and y

$$M_{10} = \sum_x \sum_y xI(x, y); \quad M_{01} = \sum_x \sum_y yI(x, y) \tag{20.26}$$

   c. Find the mean search window location

$$x_c = \frac{M_{10}}{M_{00}}; \quad y_c = \frac{M_{01}}{M_{00}} \tag{20.27}$$

   where $I(x,y)$ is the pixel probability at position (x,y) in the image.
4. Shift the window to (x_c, y_c) and repeat step 3 until convergence.
5. For the next frame in the video sequence, center the window at the location obtained in step 4 and set the size of the window based upon the zeroth moment found there.

The mean shift calculation in step 3 will tend to converge to the mode of the distribution. This causes the CAMSHIFT algorithm to track the mode of color objects moving in the video scene. The second moments of the distribution are used to determine its orientation. The second moments are

$$M_{20} = \sum_x \sum_y x^2 I(x,y); \quad M_{02} = \sum_x \sum_y y^2 I(x,y) \tag{20.28}$$

and the major axis of the object being tracked is

$$\theta = \frac{\arctan\left(2\dfrac{((M_{11}/M_{10}) - x_c y_c)}{((M_{20}/M_{00}) - x_c^2) - ((M_{02}/M_{00}) - y_c^2)}\right)}{2} \tag{20.29}$$

The length and width of the probability distribution region are found by solving the following equations:

$$a = \frac{M_{20}}{M_{00}} - x_c^2 \tag{20.30}$$

$$b = 2\left(\frac{M_{11}}{M_{00}} - x_c y_c\right) \tag{20.31}$$

and

$$c = \frac{M_{02}}{M_{00}} - y_c^2 \tag{20.32}$$

Then the length I and width w are

$$I = \sqrt{\frac{(a+c) + \sqrt{b^2 + (a-c)^2}}{2}} \tag{20.33}$$

$$w = \sqrt{\frac{(a+c) - \sqrt{b^2 + (a-c)^2}}{2}} \tag{20.34}$$

The size of the window used in the search is adapted according to the first zeroth moment of the probability distribution

$$s = 2 * \sqrt{\frac{M_{00}}{256}} \tag{20.35}$$

This equation converts the area found to a number of pixels. 256 is the number of color intensity values possible in an 8 bit color scheme. These three concepts provide the basis for the fast alignment algorithm.

Omni perspective cameras are well suited to navigation and surveillance. They are relatively sensitive to the difference between small rotation and small translation, and provide information about a wide field of view. However, current implementations of omni cameras invariably suffer from poor resolution in some regions of their images [Nayar 97]. In nearly direct opposition, planar perspective cameras offer high resolution throughout their images, but suffer from a limited field of view and have difficulty distinguishing small rotation from small translation. Some hunting spiders use the combination of a pair of high resolution forward looking eyes with an array of lower resolution eyes that provide view in a wide field. The advantages of this type of system are obvious. Navigation and gross surveillance can be accomplished through the omni capabilities of the system, while tasks such as range-finding, object recognition and object manipulation can be addressed by the planar perspective capabilities. Inherent in this type of system is the ability of the control system to integrate the information gathered by all the vision sensors at its disposal.

In order to integrate the information gathered by an omni camera and a planar camera, we need to determine how the two sets of information overlap spatially. When time is not a constraining factor, there are a number of methods available for alignment of omni images and planar images. The geometry used to unwrap omni images into sets of planar images or a panoramic is relatively straightforward. With a set of planar images or panoramic derived from an omni image, there are a number of alignment algorithms that can be used to transform one image to the coordinate system of another. However, these operations are computationally expensive and are not suitable for video processing.

We present a system that provides on-demand alignment of any pair of image frames while still processing the frames at video rate. The method takes advantage of the fact that a histogram of a planar image projected into an omni image should be equivalent to the histogram of a subset of the omni image itself. The proposed method modifies CAMSHIFT to track the projection of a set of planar images through a set of omni images. CAMSHIFT tracks a dynamic probability distribution through a video sequence. The probability distribution may be dynamic due to translation and rotation of an object moving through the scene. We provide CAMSHIFT with a dynamically changing probability distribution to track. The current probability distribution of the planar image is supplied to CAMSHIFT frame by frame. CAMSHIFT tracks the dynamic probability distribution, changing because of a changing planar view instead of a rotating and translating object. CAMSHIFT tracks the object at video rate.

CAMSHIFT requires an initial search window. For the first pair of frames, the first omni-image will be unwrapped into a panoramic image. The first planar camera image will then be aligned with this panoramic image, and the four corners of the best-fit transformation provide the initial search window. With a search window in place, the task becomes one of tracking the probability distribution of the moving planar image through the omni image. The probability distribution of each planar image is provided to CAMSHIFT for processing each omni image. The geometry of the search window is altered to reflect the shape of the planar image projection into the omni-image. Pixels closer to the center of the omni image are weighted more heavily when generating probability distributions within the omni image. The modified CAMSHIFT algorithm tracks the probability distribution of the planar image through the omni image, giving an estimate of the projection location. When an exact alignment is required between any set of frames, only a subset of the panoramic image need be considered as dictated by the projection estimate.

We assume that the focal point of the planar camera is aligned with the center of projection of the omni camera. If the planar camera can only pan and tilt, then the vertical edges of any image taken from the planar camera will be aligned with radial lines of the omni image. All planar projections into the omni image will occupy a sector. With this in mind, the CAMSHIFT algorithm is modified to use a search window that is always in the shape of a sector. Traditionally the CAMSHIFT algorithm search window varies in terms of its size, shape, location and orientation. In this application, shape and orientation are constrained to be those of a sector of the omni-image. The projection of the color histogram of a planar image to the color histogram of a sector of unknown size and location is a critical component of the CAMSHIFT adaptation. Pixels at the top of the planar image must be weighted more heavily in order to account for compression of the planar image at the bottom of its sector projection.

CAMSHIFT takes the last known location of the object and follows a probability gradient to the most likely location of the object in the next frame. When tracking the projection of one image into another, there is the advantage of knowing the color histogram being sought at each frame. By deriving a method for transforming a planar color histogram into a sector color histogram, CAMSHIFT can be provided with an accurate description of what it is looking for frame-to-frame.

The function for the projection of the planar image into the omni image is dependent on the mirror in the catadioptric system. The omni camera used utilizes a paraboloid single mirror. Ideally, any

FIGURE 20.23   Projection of planar-image into omni image.

algorithm for use with an omni camera will not require calibration for the shape of mirror used in any particular system. The nature of CAMSHIFT provides an opportunity to generalize to this effect. CAMSHIFT is robust to dynamically changing probability distributions, which in effect means that it tolerates shifts in the probability distribution it is tracking from one frame to the next. As CAMSHIFT will be provided with a true calculation of the probability distribution at each frame, it may be possible to generalize the plane-to-sector transformation in order to eliminate specific mirror equations from calculations.

Given a sector representing the projection of a planar image into the omni image, see Figure 20.23, it is possible to assume that the aspect ratio, R, of the planar image is reflected in the ratio of the sector height, h, to the length of the arc passing through x, y and bounded by the radial sides of the sector, w' (Figure 20.24).

If the aspect ratio of the sector is constrained by the aspect ratio of the planar image, then it is possible to determine the exact dimensions of a sector given the coordinates of the sector center and area of the sector. If this assumption is not made, an interesting problem arises. During fast alignment, the histogram of the planar image must be transformed to approximate the histogram of a corresponding sector in the omni image. For this to be done, either the exact parameters of the planar image projection into the omni image must be known or an assumption must be made about these parameters. Exact parameters can be obtained if the mirror equation is known and the exact location and area of the projection are known.

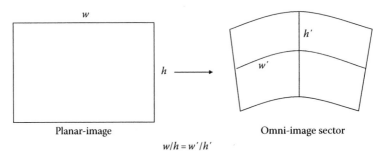

FIGURE 20.24   Projection ratio.

The mirror equation is available a priori, but the exact location and area of the projection are certainly not known, and are exactly what the algorithm attempts to estimate. Each estimation of position and area in the fast alignment algorithm depends on a past position and area of projection that is itself an estimate. Thus, the exact parameters are unavailable and an assumption must be made.

We expect the length of the median arc of the projection sector to be the most stable with respect to the height of the sector. As the sector area changes and the distance between the projection center and image center changes the outer and inner sector arc lengths will fluctuate relative to the sector height. For this reason the median arc is chosen for the needed assumption. Measurements of the median arc length and the sector height show that the ratio between the two remains highly stable, fluctuating between 76% and 79%. Thus, the assumption that the aspect ratio of the virtual planar image is similar to the ratio of sector median arc width to sector height will be used in this thesis (Figure 20.25).

The distance from the center of the omni-image to the center of the section is $r = \sqrt{(x^2 + y^2)}$ (Figure 20.25). If the inner arc-edge of the sector is w_1 and the outer arc-edge is w_2, then the area of the ring bounded by circles centered at the image center and of radius r_1 and r_2, where $r_1 = r - h/2$ is the distance from image center to w_1 and $r_2 = r + h/2$ is the distance from image center to w_2 is given as

$$\pi\left(\frac{r+h}{2}\right)^2 - \pi\left(\frac{r-h}{2}\right)^2 = 2\pi rh \tag{20.36}$$

The area of the sector bounded by ϕ, w_1 and w_2 is given by

$$\frac{AREA}{2\pi rh} = \frac{\phi}{2\pi} \quad AREA = \phi rh \tag{20.37}$$

ϕ can be described in terms of r and w:

$$\frac{\phi}{2\pi} = \frac{w}{2\pi r} \tag{20.38}$$

Substituting $h = w/R$, where R is the aspect ratio of the planar image, into (20.27) gives

$$\phi = \frac{Rh}{r} \tag{20.39}$$

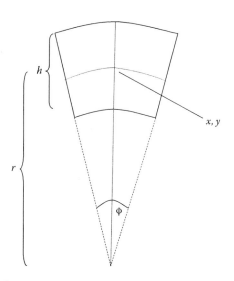

FIGURE 20.25 Sector centered at x, y.

Substituting (20.39) into (20.37) gives

$$AREA = Rh^2 \quad h = \sqrt{\frac{AREA}{R}} \quad (20.29) \quad \text{and} \quad w = R\sqrt{\frac{AREA}{R}} \qquad (20.40)$$

This result is important for defining how the modified CAMSHIFT filter establishes a probability gradient and how the size of the actual search window itself is defined. When the initial alignment has been completed, CAMSHIFT establishes a probability gradient by moving the search window through the region directly surrounding the initial location. In a planar image application of CAMSHIFT, the search window is moved by translation in the x-axis and y-axis. In an omni-image, the search window sector must be translated along the radius of the circle and the arcs of concentric circles within the omni-image. With the results shown in (20.39) and (20.40), the sector dimensions can be easily calculated as the center of the sector translates while searching, given the projection area. The area of the search window for the nth frame is a parameter returned by CAMSHIFT after the alignment of frame $n - 1$. Given the dimensions of the search sector, it is possible to transform the color histogram of the planar image to fit the projection into the omni image.

If the arc-length of w_0 is l_0, weighting the pixels on the planar image can be done as follows. If the weight of a pixel in row $y = 0$ (the bottom row) in the planar image is $weight = l_0/w'$, then the weight of a pixel at row $y = n$ in the planar image is given as

$$\frac{r + n * h/h'}{r} * weight \qquad (20.41)$$

This equation can easily transform the color histogram calculation of the planar image to fit the color histogram found in a corresponding sector. Figure 20.26 shows the histogram of a projection section as compared to transformed and untransformed planar image histograms. Clearly the transformed planar histogram much more closely approximates the histogram of the original sector than does the untransformed planar histogram. The MSE between the sector histogram and warped planar histogram was less than 15% of the MSE between the sector histogram and the unwrapped planar histogram.

There are other techniques that are applied to the problem of video registration [Krotosky 07]. Some technique ignore the temporal component of video signals such as in [Lee 05] where edge matching of individual image frames between two video sequences is performed in the DCT domain.

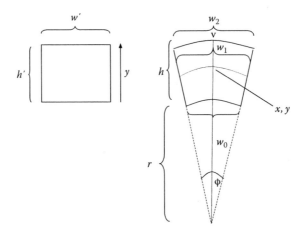

FIGURE 20.26 Diagram of histogram transformation geometry.

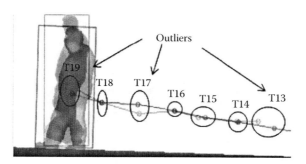

FIGURE 20.27 Matching of object and trajectory points from thermal and visible video. (From Torabi, A. et al., Feedback scheme for thermal-visible video registration, sensor fusion, and people tracking, *IEEE Conference on Computer Vision and Pattern Recognition Workshops*, San Francisco, CA, pp. 15–22, 2010.)

Other techniques expose the time element in videos when the system objectives include object tracking and registration is done using trajectories found in the separate video streams [Han 03], [Caspi 06], [Morin 08]. The successfulness of the registration is directly dependent on the ability to accurately calculate the object trajectories independently in each video sequence.

A recent example in [Torabi 10] discusses registration of simultaneous thermal and visible video sequences by augmenting the concept of trajectory matching with data-level registration simultaneously. The idea being that video registration solely using trajectories will be impaired when there are few objects that have trajectories and the addition of data-level matching can assist. In addition, this system employs a feedback loop that allows more accurate trajectory refinement using fused data from the registered data that then can be reused in a registration process. Figure 20.27 shows some interim results when trajectory points matching is applied. A comparison of results with and without the feedback loop show improved performance with the feedback loop.

20.8 Beyond Registration

This chapter has been primarily concerned with data (level) registration of entire images or data sets. In the case of images, this has focused at the pixel-level or fairly lower-level features. However, there are some systems that only register parts of the images or register higher level image data. The latter certainly leads us to the main process of multi-sensor fusion—but, as some methods pertain to unique spaces of registering information we mention it here.

When performing registration as in the case of fusion, and feature-based registration is chosen there is the selection of the "level" of features that are to be used. The answer to this question can be dependent on the system objectives. In [Apatean 09], a comparison of data-level to different feature level and finally decision level fusion is compared for the task of classifying pedestrian/vehicle obstacles using both IR and visible spectrum cameras. In this case, the data fusion gave the best results followed by feature fusion and then decision level fusion. While feature level fusion is different than feature-level registration we think the concept of choice of registration "level" (data, low feature level to higher feature level) being a function of system objectives is applicable for the registration task.

An example of a system that performs registration over only portions of the images is given in [Blum 04] which uses IR, millimeter wave and visual images for concealed weapon detection. First, either the IR or millimeter wave image is used to detect areas that contain potential weapons. It is only these regions that are registered and merged. A multi-resolution image mosaic technique is applied for the registration and merging of these areas to the visual image. The multi-resolution image mosaic algorithm involves applying weighted averaging operations in the predefined area at different resolutions. Figure 20.28 shows the operations involved in this system and Figure 20.29 the results.

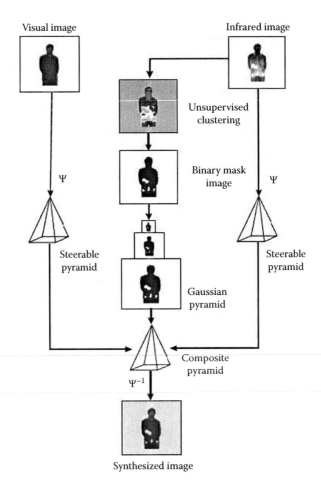

FIGURE 20.28 Shows the stages involved in the system that only aligns and merges IR and/or millimeter wave image regions (binary mask image) highlighting potential weapons. (From Blum, R. et al., Multisensor concealed weapon detection by using a multiresolution Mosaic approach, *IEEE Vehicular Technology Conference*, Los Angeles, CA, pp. 4597–4601, 2004.)

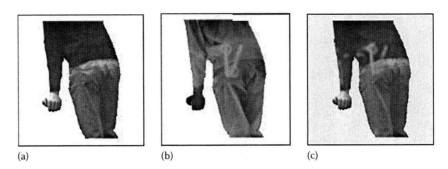

FIGURE 20.29 (a) Original visual image, (b) IR image, and (c) images aligned and merged only in potential weapons area via multi-resolution mosaic algorithm. (From Blum, R. et al., Multisensor concealed weapon detection by using a multiresolution Mosaic approach, *IEEE Vehicular Technology Conference*, Los Angeles, CA, pp. 4597–4601, 2004.)

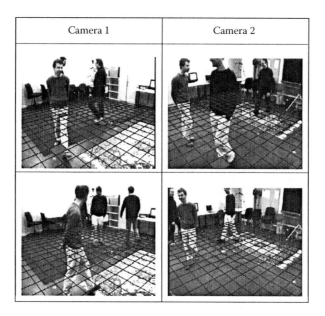

FIGURE 20.30   System for person detection shows beyond data-level fusion where use occupancy grid of Bayesian occupancy filter for two cameras registering (fusing) points of high occupancy. (From Ros, J. and Mekhnacha, K., Multi-sensor human tracking with the Bayesian occupancy filter, *International Conference on Digital Signal Processing*, Santorini, Greece, 2009.)

As discussed previously, in the application of registration with video streams you see more often the application of object-level matching for registration as discussed in [Torabi 10], [Morin 08], [Caspi 06], [Han 03] which perform object trajectory tracking and matching. The differences between these focus on topics such as what objects to track (i.e., humans), how to track, how to match tracks and then from this how to calculate a registration transformation.

In [Ros 09] the multi-sensor system first attempt's to do a higher-level processing of the different data and then perform registration or if you like almost fusion of this higher level information. In [Ros 09] both visual spectrum and IR images are captured for the purpose of person detection. Specifically, they introduce a sensor model within the Bayesian Occupancy Filter framework allowing to perform efficiently the fusion of infrared and CCD cameras without the prior registration step. First, the visual and IR images are processed separately for the detection of people. Now, a fusion of information takes place not in the original image space but, in a new space called the Bayesian Occupancy Filter space which is a 2D grid decomposition of the environment where each cell of the grid contains two probability distributions: one the probability of occupancy of the cell and two the probability of its velocity (useful for tracking a person through a scene). Figure 20.30 shows two image sequences from two cameras each one with the superimposed BOF output. The red squares indicating high occupancy.

20.9 Closely Mounted Sensors and Temporal Changes

In this section we discuss the work of [Heather 05] that shows an example treatment of the case when we have the special sub-problem of general registration where the multiple sensors are closely mounted and in addition discusses the issue of temporal changes in multi-sensor registration.

In closely mounted sensors there are two generic cases, one where they are mounted side by side, usually will parallel optical axes and the second where the sensors share the same optical aperture. Some work has been done to preconfigured registration information using a "calibration" stage done prior to the system running. As we discuss below it may be necessary or desirable to occasionally re-register the sensors.

In [Heather 05], they discuss a system to perform registration between closely aligned visible image and IR image sensors that has a component for temporal re-registration. In their algorithm, an extension of [Irani 98], they look for matching data in the different image modalities by representing the image in a multi-scale space that captures spatial structure. The importance of this is to reduce the effects of different meaning in the raw data values in this case between visible and IR images. The goal is to in essence suppress low frequency information that can vary more greatly between the visible and IR. A set of derivative functions is used to create this "invariant" space. The multi-scale is achieve through the creation of a set of different resolution images in a Gaussian pyramid for both the visible and IR data. It is to these two pyramids the derivative functions are applied. The result of this is very much like the wavelet transforms discussed above in [Grewe 97, Grewe 99].

One thing that is different than many systems that use a multi-scale approach, rather than first finding an initial registration in the coarser/lower resolution scale and then refining up into higher resolutions (going down the pyramid), in [Heather 05] they instead perform "simultaneous" processing at all levels. While this does not capture the performance improvement available in registration at different levels [Grewe 99], the authors suggest they may achieve the optimal solution more often. The [Heather 05] system formulates registration as the minimization of a global objective function comprised of local correlation measurements each local area originating from a non-overlapping partition in the multi-scale image representation space. However, their global optimization problem treats each level independently to create a simpler optimization function and to improve performance. In doing this, they may sacrifice the likelihood of obtaining the optimal solution. Given this, it is unclear whether a level changing approach or their "simultaneous" level processing approach will be superior.

The effects of sensors moving over time and in some cases large changes in scene distance that create differences due to parallax effects can be negated by the re-running of the registration process. To keep within real-time constraints of the [Heather 05] system, they perform re-registration at different temporal points using the past registration information as the starting point for the optimization search and allow for only one iteration. The limitation of one iteration is to meet with real-time constraints but, allows the system to recover to some extent temporal registration changes. This will probably not work for radically altered sensor alignments and in these cases any system will need to go through a complete re-registration phase.

20.10 Ensemble Registration

Most of the work reviewed so far has concentrated on registration for a few sensors, often two. The term Ensemble Registration indicates the registration of a larger number of sensors. In [Orchard 10], a system for performing ensemble registration is discussed for different multiple sensor images. The algorithm behind this system has two steps, one called density estimation and another called motion adjustment. To understand each, we must first define a "joint intensity scatter plot" (JISP). Let's suppose we had two perfectly registered images and we created a map, distribution for each pixel that showed the intensity value in image 1 versus the intensity value in image 2. In a perfectly registered image, it would mean that each pixel would be represented by a single point I this 2D mapping which is called the Joint Intensity Scatter Plot or JISP. As the two images become un-registered, what was once a single point blooms out. Thus, what once was a set of tight clusters now has spread and overlap. The goal of registration then is to minimize and bring these clusters into coherence. This can be further extrapolated for N sensors to a ND JSIP space.

Each JSIP can be assumed to have several clusters each with its own distribution. During the determination of the registration transformation, the scatter points move in the JISP as the images themselves have interim spatial transformations. The process of moving these scatter points is done iteratively with two steps: density estimation and motion adjustment. Density estimation involves the estimation of the

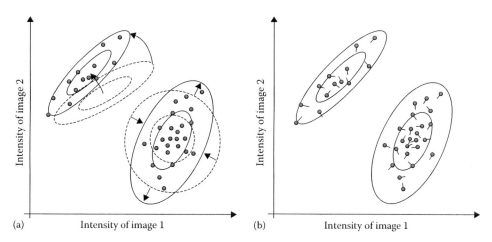

(a)   Intensity of image 1   (b)   Intensity of image 1

FIGURE 20.31 (a) Density estimation: the registration parameters are held fixed while a better density estimate is computing by altering cluster location and shape. (b) Motion adjustment: density estimate is held fixed and registration parameters are altered to overall minimize scattering. (From Orchard, J. and Mann, R., *IEEE Trans. Image Process.*, 19(5), 1236, 2010.)

clusters in the JISP space using Gaussian Mixture Models with the current set of registration (called in [Orchard 10] motion parameters) parameters. The difficulty here is that given an initial starting registration transformation a "natural" cluster in the current JISP space can have pixel alignment from very different parts of the scene (they are mis-registered). The motion adjustment phase is the alteration of the registration parameters with the goal of minimizing the inter-cluster dispersion. Figure 20.31 illustrates this two step process. The two step process is iterated consecutively until an optimization function is minimized (convergence).

20.11 Non-Rigid Body Registration

Most of the systems we have discussed implicitly are working on registration for rigid-bodied object scenes or at least not explicitly discussing issues related to non-rigid deformations. However, there are some applications where the main subjects are of non-rigid bodied objects. An example of this discussed in [Kim 10] is the area of multi-sensor registration of brain scan data from different sensors including MR-T1, MR-T2, MR-PD, CT, and PET scan sensors.

In general, when trying to match between different data sources with non-rigid bodies the registration modeling will involve some component of a deformable registration scheme [Rueckert 99]. This leads to techniques and research found most often in the field of computer graphics and there can be many techniques applied to the deformable modeling.

For example, in [Kim 10] they use the technique of free-form deformation modeling using B-splines. The main processes involve the two step algorithm discussed previously in ensemble image registration [Orchard 10] but, added to it is the deformation component for non-rigid objects which is performed through a manipulation of a mesh of control points using B-splines. Now, the optimization function includes a cost metric associated with the mean elastic energy of B-spline deformed images.

Figure 20.32 shows the results of the system including original images on the top row, followed by the simulation of deformed images created by applying a controlled deformation randomly generated within a range of values to the original data. It is these simulated deformed images that are used as input into the [Kim 10] system for testing. The last row in Figure 20.32 shows the results of this non-rigid registration scheme. The authors expect that similar results would be achieved on real, not simulated non-rigid multi-sensor data.

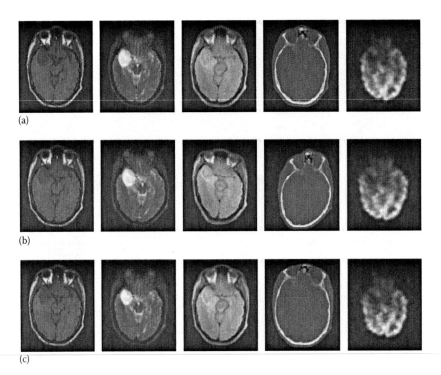

FIGURE 20.32 (a) Top row contains original images, (b) simulated deformed images input into system, and (c) results of non-rigid body registration process. (From Kim, H. and Orchard, J., Registering a non-rigid multi-sensor ensemble of images, *International Conference of the IEEE Engineering in Medicine and Biology Society*, Buenos Aires, Argentina, pp. 5935–5938, 2010.)

20.12 Summary

This chapter provides an overview of data registration. Examples given here concentrated on 2D images, since they are most intuitively understood. The information given here can easily be generalized to deal with data of an arbitrary number of dimensions. Registration is the process of finding the set of parameters that describe an accurate mapping of one set of data onto another set of data. The basic mathematical tools for affine mappings were described, and a survey of existing work was presented.

Of special interest is the ability to register data sets containing noise. To that end, we derived a number of objective functions that can be used to compare data registrations when the data has been corrupted. A number of different types of noise were considered. These objective functions are appropriate for registering data using optimization techniques [Brooks 98, Chen 02].

We then presented two case studies for data registration using different sensing modalities. One approach used wavelet filters to hierarchically register range images. The other approach considered the geometries of different camera lenses, and explained how to maintain registration for a specific camera configuration in real-time.

References

[Apatean 09] A. Apatean, A. Rogozan, and A. Bensrhair, Information fusion for obstacle recognition in visible and infrared images, *International Symposium on Signals, Circuits and Systems*, Iasi, Romania, 2009.

[Barhen 96] J. Barhen and V. Protopopescu, Generalized TRUST algorithms for global optimization, *State of the Art in Global Optimization*, C. A. Floudas and P. M. Pardalos, Eds., Kluwer Academic Publishers, Dordrecht, the Netherlands, 1996, pp. 163–180.

[Barhen 97] J. Barhen, V. Protopopescu, and D. Reister, TRUST: A deterministic algorithm for global optimization, *Science*, 276, 1094–1097, May 16, 1997.

[Barnea 72] D. Barnea and H. Silverman, A class of algorithms for fast digital image registration, *IEEE Transactions on Computers*, C-21(2), 179–186, 1972.

[Barrow 77] H. G. Barrow, J. M. Tennenbaum, R. C. Bolles, and H. C. Wolf, Parametric correspondence and chamfer matching: Two new techniques for image matching, *Proceedings of the International Joint Conference on Artificial Intelligence*, San Francisco, CA, 1977, pp. 659–663.

[Berthilsson 98] R. Berthilsson, Affine correlation, *Proceedings of the International Conference on Pattern Recognition ICPR'98*, Brisbane, Queensland, Australia, 1998, pp. 1458–1461.

[Blum 04] R. Blum, X. Zhiyun, Z. Liu, and D. Forsyth, Multisensor concealed weapon detection by using a multiresolution Mosaic approach, *IEEE Vehicular Technology Conference*, Los Angeles, CA, 2004, pp. 4597–4601.

[Borgefors 88] G. Borgefors, Hierarchical chamfer matching: A parametric edge matching algorithm, *IEEE Transactions on Pattern Analysis and Machine Intelligence*, 10, 849–865, 1988.

[Bradski 98] G. Bradski, Real time face and object tracking as a component of a perceptual user interface, *Proceedings of the Fourth IEEE Workshop on Applications of Computer Vision, WACV '98*, Princeton, NJ, 1998, pp. 214–219.

[Brooks 96] R. R. Brooks, S. S. Iyengar, and J. Chen, Automatic correlation and calibration of noisy sensor readings using elite genetic algorithms, *Artificial Intelligence*, 84, 339–354, 1996.

[Brooks 96a] R. R. Brooks, *Robust Sensor Fusion Algorithms: Calibration and Cost Minimization*, PhD dissertation, Louisiana State University, Baton Rouge, LA, August 1996.

[Brooks 98] R. R. Brooks and S. S. Iyengar, *Multi-Sensor Fusion: Fundamentals and Applications with Software*, Prentice Hall, Upper Saddle River, NJ, 1998.

[Brown 92] L. G. Brown, A survey of image registration techniques, *ACM Computing Surveys*, 24(4), 325–376, December 1992.

[Canny 86] J. Canny, A computational approach to edge detection, *IEEE Transactions on Pattern Analysis and Machine Intelligence*, 8, 679–698, 1986.

[Caspi 06] Y. Caspi, D. Simakov, and M. Irani, Feature-based sequence-to-sequence matching, *International Journal of Computer Vision*, 68, 53–64, 2006.

[Cetin 93] B. C. Cetin, J. Barhen, and J. W. Burdick, Terminal repeller unconstrained subenergy tunneling (TRUST) for fast global optimization, *Journal of Optimization Theory and Applications*, 77(l), 97–126, 1993.

[Charoentam 06] O. Charoentam, V. Patanavijit, and S. Jitapunkul, A stable region-based multiscale image fusion scheme with thermal and visible image application for mis-registration problem, *IEEE North-East Workshop on Circuits and Systems*, Gatineau, Quebec, Canada, 2006.

[Chen 02] Y. Chen, R. R. Brooks, S. S. Iyengar, S. V. N. Rao, and J. Barhen, Efficient global optimization for image registration, *IEEE Transactions on Knowledge and Data Engineering*, 14(1), 79–92, January 2002.

[Cheng 96] J. K. Cheng and T. S. Huang, Image registration by matching relational structures, *Pattern Recognition*, 17, 149–159, 1984.

[Cheng 98] Y. Cheng, Mean shift, mode seeking, and clustering, *IEEE Transactions on Pattern Analysis and Machine Intelligence*, 17, 790–799, 1998.

[Coiras 00] E. Coiras, J. Santamaria, and C. Miravet, Segment-based registration technique for visual-infrared images, *Optical Engineering*, 39, 282–289, 2000.

[Daniilidis 00] K. Daniilidis and C. Geyer, Omnidirectional vision: Theory and algorithms, *Proceedings of the 15th International Conference on Pattern Recognition*, Vol. 1, Barcelona, Spain, 2000, pp. 89–96.

[De Vore 97] R. A. De Vore, W. Shao, J. F. Pierce, E. Kaymaz, B. T. Lerner, and W. J. Campbell, Using non-linear wavelet compression to enhance image registration, *SPIE Proceedings*, 3078, 539–551, 1997.

[Faugeras 93] O. Faugeras, *Three-Dimensional Computer Vision: A Geometric Viewpoint*, MIT Press, Cambridge, MA, 1993.

[Fonseca 96] L. M. G. Fonseca and B. S. Manjunath, Registration techniques for multisensor remotely sensed imagery, *Photogrammetric Engineering and Remote Sensing*, 62, 1049–1056, 1996.

[Gonzalez 92] R. C. Gonzalez and R. E. Woods, *Digital Image Processing*, Addison-Wesley, Menlo Park, CA, 1992.

[Goshtasby 85] A. Goshtasby and G. C. Stockman, Point pattern matching using convex hull edges, *IEEE Transaction on Systems, Man and Cybernetics*, 15, 631–637, 1985.

[Grewe 97] L. Grewe and R. Brooks, On localization of objects in the wavelet domain, *1997 IEEE Symposium on Computational Intelligence in Robotics and Automation*, Monterey, CA, July 1997, pp. 412–418.

[Grewe 99] L. Grewe and R. R. Brooks, Efficient registration in the compressed domain, *Wavelet Applications VI, SPIE Proceedings*, Vol. 3723, H. Szu, Ed., Aerosense, Orlando, FL, 1999.

[Gulch 91] E. Gulch, Results of test on image matching of ISPRS WG, *ISPRS Journal of Photogrammetry and Remote Sensing*, 6, 1–18, 1991.

[Han 03] J. Han and B. Bhanu, Detecting moving humans using color and infrared video, *Proceedings of the IEEE International Conference on Multisensor Fusion and Integration for Intelligent Systems*, Tokyo, Japan, 2003, pp. 228–233.

[Hassen 09] R. Hassen, W. Zhou, and M. Salama, Multi-sensor image registration based-on local phase coherence, *Proceedings of 16th International Conference on Image Processing*, Nov. 7–10, 181–184, 2009.

[Heather 05] J. Heather and M. Smith, Multimodal image registration with applications to image fusion, *8th International Conference on Image Fusion*, Philadelphia, PA, 2005, pp. 372–379.

[Hill 90] F. S. Hill, *Computer Graphics*, Prentice Hall, Englewood, NJ, 1990.

[Irani 98] M. Irani and P. Anandan, Robust multi-sensor image alignment, *Sixth International Conference on Computer Vision*, Bombay, India, 1998, pp. 959–966.

[Josien 03] P. W. Josien, A. Maintz, and M. A. Viergever, Mutual information based registration of medical images: A survey, *IEEE Transactions on Medical Imaging*, 22(8), 986–1004, 2003.

[Kaneko 02] S. Kaneko, I. Murase, and S. Igarashi, Robust image registration by increasing sign correlation, *Pattern Recognition*, 35, 2223–2234, 2002.

[Kim 10] H. Kim and J. Orchard, Registering a non-rigid multi-sensor ensemble of images, *International Conference of the IEEE Engineering in Medicine and Biology Society*, Buenos Aires, Argentina, 2010, pp. 5935–5938.

[Krotosky 07] A. Kroto, S. J. Krotosky, and M. M. Trevidi, Mutual information based registration of multimodal stereo videos for person tracing, *Computer Vision Image Understanding*, 106, 270–287, 2007.

[Lee 05] M. Lee, M. Shen, A. Yoneyama, and C. Kuo, DCT-domain image registration techniques for compressed video, *IEEE International Symposium on Circuits and Systems*, Kobe, Japan, 2005, pp. 4562–4565.

[Li-pi 10] N. Li-pi, Y. Ying-yun, and Z. Wen-hui, A multi-sensor image registration approach based on edge-correlation, *International Conference on Signal Processing Systems*, Dalian, China, 2010, pp. 36–40.

[Lowe 99] D. Lowe, Object recognition from local scale-invariant features, *Proceedings of the International Conference on Computer Vision 2*, Corfu, Greece, 1999, pp. 1150–1157.

[Lester 99] H. Lester and S. R. Arridge, A survey of hierarchical non-linear medical image registration, *Pattern Recognition*, 32, 129–149, 1999.

[Maes 97] F. Maes, A. Collignon, D. Vandermeulen, G. Marchal, and P. Suetens, Multimodality image registration by maximization of mutual information, *IEEE Transactions on Medical Imaging*, 16, 187–198, 1997.

[Mandara 89] V. R. Mandara and J. M. Fitzpatrick, Adaptive search space scaling in digital image registration, *IEEE Transactions on Medical Imaging*, 8(3), 251–262, September 1989.

[Mann 97] S. Mann and R. Picard, Video orbits of the projective group a simple approach to featureless estimation of parameters, *IEEE Transactions on Image Processing*, 6(9), 1281–1295, September 1997.

[Marr 80] D. Marr and E. Hildreth, Theory of edge detection, *Proceedings of the Royal Society of London*, B 207, 187–217, 1980.

[Moigne 98] J. Moigne, First evaluation of automatic image registration methods, *Proceedings of the International Geoscience and Remote Sensing Symposium, IGARSS'98*, Seattle, WA, 1998, pp. 315–317.

[Montesinos 00] P. Montesions, V. Gouet, R. Deriche, and D. Pele, Matching color uncalibrated images using differential invariants, *Image and Vision Computing*, 18, 659–671, 2000.

[Morin 08] F. Morin, A. Torabi, and G. Bilodeau, Automatic registration of color and infrared videos using trajectories obtained from a multiple object tracking Algorithm, *Proceedings of Canadian Conference on Computer and Robot Vision*, Canada, 311–318, 2008.

[Moss 97] S. Moss and E. R. Hancock, Multiple-line-template matching with EM algorithm, *Pattern Recognition Letters*, 18, 1283–1292, 1997.

[Nayar 97] S. Nayar, Catadioptric omnidirectional camera, *Proceedings of the IEEE Computer Society Conference on Computer Vision and Pattern Recognition*, San Juan, Puerto Rico, 1997, pp. 482–488.

[Nayar 99] S. Nayar and V. Peri, Folded catadioptric cameras, *Proceedings of the IEEE Computer Society Conference on Computer Vision and Pattern Recognition*, Vol. 2, Ft. Collins, CO, 1999, p. 223.

[Oghabian 92] M. A. Oghabian and A. Todd-Pokropek, Registration of brain images by a multi-resolution sequential method, *Information Processing in Medical Imaging*, Springer, New York, 1991, pp. 165–174.

[Orchard 10] J. Orchard and R. Mann, Registering a multi-sensor ensemble of images, *IEEE Transactions on Image Processing*, 19(5), 1236–1247, 2010.

[Pal 93] N. R. Pal and S. K. Pal, A review on image segmentation techniques, *Pattern Recognition*, 26, 1277–1294, 1993.

[Pinz 95] A. Pinz, M. Prontl, and H. Ganster, Affine matching of intermediate symbolic presentations, *CAIP'95 Proceedings LNCS 970*, V. Hlavac and R. Sara, Eds., Springer Verlag, Berlin, Germany, 1995, pp. 359–367.

[Rees 70] D. Rees, Panoramic television viewing system, U.S. Patent (3505465), April 1970.

[Ritter 99] N. Ritter, R. Owens, J. Cooper, R. H. Eikelboom, and P. P. Van Saarloos, Registration of stereo and temporal images and the retina, *IEEE Transactions on Medical Imaging*, 18, 404–418, 1999.

[Rohr 01] K. Rohr, Landmark-based image analysis: Using geometric and intensity models, *Computational Imaging and Vision Series*, Vol. 21, Kluwer Academic Publishers, Dordrecht, the Netherlands, 2001.

[Ros 09] J. Ros and K. Mekhnacha, Multi-sensor human tracking with the Bayesian occupancy filter, *International Conference on Digital Signal Processing*, Santorini, Greece, 2009.

[Roux 96] M. Roux, Automatic registration of SPOT images and digitized maps, *Proceedings of the IEEE International Conference on Image Processing ICIP'96*, Lausanne, Switzerland, 1996, pp. 625–628.

[Rueckert 99] D. Rueckert, L. I. Sonoda, C. Hayes, D. L. G. Hill, M. O. Leach, and D. J. Hawkes, Nonrigid registration using free-form deformations: Application to breast MR images, *IEEE Transactions on Medical Imaging*, 18, 712–721, 1999.

[Russ 95] J. C. Russ, *The Image Processing Handbook*, CRC Press, Boca Raton, FL, 1995.

[Sester 98] M. Sester, H. Hild, and D. Fritsch, Definition of ground control features for image registration using GIS data, *Proceedings of the Symposium on Object Recognition and Scene Classification from Multispectral and Multisensor Pixes*, Columbus, OH, 1998.

[Sharman 97] R. Sharman, J. M. Tyler, and O. S. Pianykh, Wavelet based registration and compression of sets of images, *SPIE Proceedings*, 3078, 497–505, 1997.

[Smith 97] S. M. Smith and J. M. Brady, SUSAN—A new approach to low level image processing, *International Journal of Computer Vision*, 23, 45–78, 1997.

[Stockman 82] G. Stockman, S. Kopstein, and S. Benett, Matching images to models for registration and object detection via clustering, *IEEE Transactions on Pattern Analysis and Machine Intelligence*, 4(3), 229–241, 1982.

[Studholme 99] C. Studholme, D. L. G. Hill, and D. J. Hawkes, An overlap invariant entropy measure of 3D medical image alignment, *Pattern Recognition*, 32, 71–86, 1999.

[Thevenaz 97] P. Thevenaz and M. Unser, Spline pyramids for inter-modal image registration using mutual information, *Proceedings of the SPIE: Avalet Applications in Signal and Image Processing*, San Diego, CA, 1997, pp. 236–247.

[Ton 89] J. Ton and A. K. Jain, Registering landsat images by point matching, *IEEE Transactions on Geoscience and Remote Sensing*, 27, 642–651, 1998.

[Torabi 10] A. Torabi, G. Masse, and G. Bilodeau, Feedback scheme for thermal-visible video registration, sensor fusion, and people tracking, *IEEE Conference on Computer Vision and Pattern Recognition Workshops*, San Francisco, CA, 2010, pp. 15–22.

[Vasileisky 98] A. S. Vasileisky, B. Zhukov, and M. Berger, Automated image co-registration based on linear feature recognition, *Proceedings of the 2nd Conference Fusion of Earth Data*, Sophia Antipolis, France, 1998, pp. 59–66.

[Viola 97] P. Voila and W. M. Wells, Alignment by maximization of mutual information, *International Journal of Computer Vision*, 24, 137–154, 1997.

[Wang 83] C. Y. Wang, H. Sun, S. Yadas, and A. Rosenfeld, Some experiments in relaxation image matching using corner features, *Pattern Recognition*, 16, 167–182, 1983.

[Weijie 09] J. Weijie, Z. Jixian, and Y. Jinghui, Automatic registration of SAR and optics image based on multi-features on suburban areas, *IEEE Urban Remote Sensing Joint Event*, Shanghai, China, 2009.

[Wolberg 90] G. Wolberg, *Digital Image Warping*, IEEE Computer Society Press, Los Alamitos, CA, 1990.

[Wong 79] R. Y. Wong and E. L. Hall, Performance comparison of scene matching techniques, *IEEE Transaction on Pattern Analysis and Machine Intelligence*, PAMI-1(3), 325–330, July 1979.

[Wyawahare 2009] M. V. Wyawahare, P. M. Patil, and H. K. Abhyankar, Image registration techniques: An overview, *International Journal of Signal Processing, Image Processing and Pattern Recognition*, 2(3), 11–18, 2009.

[Yang 99] Z. Yang and F. S. Cohen, Image registration and object recognition using affine invariants and convex hulls, *IEEE Transactions on Image Proceeding*, 8, 934–946, 1999.

[Yang 07] G. Yang, C. Stwart, M. Sofka, and C. Tsai, Registration of challenging image pairs: Initialization, estimation, and decision, *IEEE Transaction on Pattern Analysis and Machine Intelligence*, 29(11), 1973–1989, 2007.

[Yin 03] G. Yin and Q. Wu, The multi-sensor fusion: Image registration using artificial immune algorithm, *International Workshop on Soft Computing Techniques in Instrumentation, Measurement and Related Applications*, Provo, UT, 2003, pp. 32–36.

[Zana 99] F. Zana and J. C. Klein, A multimodal registration algorithm of eye fundus images using vessels detection and Hough transform, *IEEE Transactions on Medical Imaging*, 18, 419–428, 1999.

[Zhang 09] Y. Zhang, Y. Guo, and Y. Gu, Robust feature matching and selection methods for multisensor image registration, *IEEE Geoscience and Remote Sensing Symposium*, Cape Town, South Africa, 2009, pp. 255–258.

[Zheng 99] Z. Zheng, H. Wang, and E. K. Teoh, Analysis of gray level corner detection, *Pattern Recognition Letters*, 20, 149–162, 1999.

[Zhou 97] Y. Zhou, H. Leung, and E. Bosse, Registration of mobile sensors using the parallelized extended Kalman filter, *Optical Engineering*, 36(3), 780–788, March 1997.

[Zhu 10] Z. Zhu and J. Luo, Automatic multisensor image registration based on global and local geometric consistent edge segments, *IEEE International Geoscience and Remote Sensing Symposium (IGARSS)*, Honolulu, HI, 2010, pp. 1621–1624.

[Ziou 97] D. Ziou and S. Tabbone, Edge detection techniques—An overview, *International Journal of Pattern Recognition and Image Analysis*, 8(4), 537–559, 1998.

[Zitova 03] B. Zitova and J. Flusser, Image registration methods: A survey, *Image and Vision Computing*, 21, 977–1000, 2003.

21

Signal Calibration, Estimation for Real-Time Monitoring and Control

21.1 Introduction 495
21.2 Signal Calibration and Measurement Estimation 496
Degradation Monitoring • Possible Modifications of the Calibration Filter
21.3 Sensor Calibration in a Commercial-Scale Fossil Power Plant.. 500
Filter Parameters and Functions • Filter Matrices • Filter Performance Based on Experimental Data
21.4 Summary and Conclusions...508
21.A Appendix: Multiple Hypotheses Testing Based on Observations of a Single Variable 508
Acknowledgment...511
References..511

Asok Ray
The Pennsylvania State University

Shashi Phoha
The Pennsylvania State University

21.1 Introduction

Performance, reliability, and safety of complex dynamical processes such as aircraft and power plants depend upon validity and accuracy of sensor signals that measure plant conditions for information display, health monitoring, and control (Dickson et al., 1996). Redundant sensors are often installed to generate spatially averaged time-dependent estimates of critical variables so that reliable monitoring and control of the plant are assured. Examples of redundant sensor installations in complex engineering applications are

- Inertial navigational sensors in both tactical and transport aircraft for guidance and control (Potter and Suman, 1977; Daly et al., 1979)
- Neutron flux detectors in the core of a nuclear reactor for fuel management, health monitoring, and power control (Ray et al., 1983)
- Temperature, pressure, and flow sensors in both fossil and nuclear steam power plants for health monitoring and feedforward-feedback control (Deckert et al., 1983)

Sensor redundancy is often augmented with analytical measurements that are obtained from physical characteristics and/or model of the plant dynamics in combination with other available sensor data (Desai et al., 1979; Ray et al., 1983). The redundant sensors and analytical measurements are referred to as redundant measurements in the sequel.

Individual measurements in a redundant set may often exhibit deviations from each other after a length of time. These differences could be caused by slowly time-varying sensor parameters (e.g., amplifier gain), plant parameters (e.g., structural stiffness, and heat transfer coefficient), transport delays, etc. Consequently, some of the redundant measurements could be deleted by a fault detection and isolation (FDI) algorithm (Ray and Desai, 1986) if they are not periodically calibrated. On the other hand, failure to isolate a degraded measurement could cause an inaccurate estimate of the measured variable by, for example, increasing the threshold bound in the FDI algorithm. In this case, the plant performance may be adversely affected if that estimate is used as an input to the decision and control system. This problem can be resolved by adaptively filtering the set of redundant measurements as follows:

- All measurements, which are consistent relative to the threshold of the FDI algorithm, are simultaneously calibrated on-line to compensate for their relative errors.
- The weights of individual measurements for computation of the estimate are adaptively updated on-line based on their respective a posteriori probabilities of failure instead of being fixed a priori.

In the event of an abrupt disruption of a redundant measurement in excess of its allowable bound, the respective measurement is isolated by the FDI logic, and only the remaining measurements are calibrated to provide an unbiased estimate of the measured variable. On the other hand, if a gradual degradation (e.g., a sensor drift) occurs, the faulty measurement is not immediately isolated by the FDI logic. But its influence on the estimate and calibration of the remaining measurements is diminished as a function of the magnitude of its residual (i.e., deviation from the estimate) that is an indicator of its degradation. This is achieved by decreasing the relative weight of the degraded measurement as a monotonic function of its deviation from the remaining measurements. Thus, if the error bounds of the FDI algorithm are appropriately increased to reduce the probability of false alarms, the resulting delay in detecting a gradual degradation could be tolerated. The rationale is that an undetected fault, as a result of the adaptively reduced weight, would have smaller bearing on the accuracy of measurement calibration and estimation. Furthermore, since the weight of a gradually degrading measurement is smoothly reduced, the eventual isolation of the fault would not cause any abrupt change in the estimate. This feature, known as *bumpless* transfer in the process control literature, is very desirable for plant operation.

This chapter presents a calibration and estimation filter for redundancy management of sensor data and analytical measurements. The filter is validated based on redundant sensor data of throttle steam temperature collected from an operating power plant. Development and validation of the filter algorithm are presented in the main body of the chapter along with concluding remarks. Appendix 21.A presents the theory of multiple hypotheses based on a posteriori probability of failure of a single measurement (Ray and Phoha, 2002).

21.2 Signal Calibration and Measurement Estimation

A redundant set of ℓ sensors and/or analytical measurements of a n-dimensional plant variable are modeled at the kth sample as

$$m_k = (H + \Delta H_k)x_k + b_k + e_k \qquad (21.1)$$

where
m_k is the $(\ell \times 1)$ vector of (uncalibrated) redundant measurements
H is the $(\ell \times n)$ a priori determined matrix of scale factor having rank n, with $l > n \geq 1$
ΔH_k is the $(\ell \times n)$ matrix of scale factor errors
x_k is the $(n \times 1)$ vector of true (unknown) value of the measured variable
b_k is the $(\ell \times 1)$ vector of bias errors
e_k is the $(\ell \times 1)$ vector of measurement noise, such that $E[e_k] = 0$ and $[e_k e_l^T] = R_k \delta_{kl}$

The noise covariance matrix R_k of uncalibrated measurements plays an important role in the adaptive filter for both signal calibration and measurement estimation. It is shown in the sequel how R_k is recursively tuned based on the history of calibrated measurements.

Equation 21.1 is rewritten in a more compact form as

$$m_k = Hx_k + c_k + e_k \qquad (21.2)$$

where the correction c_k due to the combined effect of bias and scale factor errors is defined as

$$c_k \equiv \Delta H_k x_k + b_k \qquad (21.3)$$

The objective is to obtain an unbiased predictor estimate $\hat{c}_k$ of the correction c_k so that the sensor output m_k can be calibrated at each sample. A recursive relation of the correction c_k is modeled similar to a random walk process as

$$c_{k+1} = c_k + v_k$$

$$E[v_k] = 0; \quad E[v_K v_j] = Q\delta_{kj} \quad \text{and} \quad E[v_k e_j] = 0 \quad \forall k, j \qquad (21.4)$$

where the stationary noise v_k represents uncertainties of the model in Equation 21.4.

We construct a filter to calibrate each measurement with respect to the remaining redundant measurements. The filter input is the parity vector p_k of the uncalibrated measurement vector m_k, which is defined (Potter and Suman, 1977; Ray and Luck, 1991) as

$$p_k = Vm_k \qquad (21.5)$$

where the rows of the projection matrix $V \in \Re^{(\ell-n)\times\ell}$ form an orthonormal basis of the left null space of the measurement matrix $H \in \Re^{\ell\times n}$ in Equation 21.1, i.e.,

$$VH = 0_{(\ell-n)\times n}$$

$$VV^{\mathrm{T}} = I_{(\ell-n)\times(\ell-n)} \qquad (21.6)$$

and the columns of V span the parity space that contains the parity vector. A combination of Equations 21.2, 21.4 through 21.6 yields

$$p_k = Vc_k + \varepsilon_k \qquad (21.7)$$

where the noise $\varepsilon_k \equiv Ve_k$ having $E[\varepsilon_k] = 0$ and $E[\varepsilon_k \varepsilon_j^{\mathrm{T}}] \equiv VR_k V^{\mathrm{T}} \delta_{kj}$. If the scale factor error matrix ΔH_k belongs to the column space of H, then the parity vector p_k is independent of the true value x_k of the measured variable. Therefore, for $\|V\Delta H_k x_k\| \ll \|Vb_k\|$ that includes relatively small scale factor errors, the calibration filter operates approximately independent of x_k.

Now we proceed to construct a recursive algorithm to predict the estimated correction $\hat{c}_k$ based on the principle of best linear least square estimation that has the structure of an optimal minimum-variance filter (Jazwinski, 1970; Gelb, 1974) and uses Equations 21.4 and 21.7:

$$\left. \begin{aligned} \hat{c}_{k+1} &= \hat{c}_k + K_k \gamma_k \quad \text{given } \hat{c}_0 \\ P_{k+1} &= (I - K_k V)P_k + Q \quad \text{given } P_0 \text{ and } Q \\ K_k &= P_k V^{\mathrm{T}}(V[R_k + P_k]V^{\mathrm{T}})^{-1} \quad \text{given } R_k \\ \gamma_k &= p_k - V\hat{c}_k \quad \text{innovation} \end{aligned} \right\} \qquad (21.8)$$

Upon evaluation of the unbiased estimated correction $\hat{c}_k$, the uncalibrated measurement m_k is compensated to yield the calibrated measurement y_k as

$$y_k = m_k - \hat{c}_k \qquad (21.9)$$

Using Equations 21.5 and 21.9, the innovation γ_k in Equation 21.8 can be expressed as the projection of the calibrated γ_k measurement y_k onto the parity space, i.e.,

$$\gamma_k = Vy_k \qquad (21.10)$$

By setting $\Gamma_k \equiv K_k V$, we obtain an alternative form of the recursive relations in Equation 21.8 as

$$\hat{c}_{k+1} = \hat{c}_k + \Gamma_k y_k \quad \text{given } \hat{c}_0$$

$$P_{k+1} = (I - \Gamma_k)P_k + Q \quad \text{given } P_0 \text{ and } Q \qquad (21.11)$$

$$\Gamma_k = P_k V^{\mathrm{T}}(V[R_k + P_k]V^{\mathrm{T}})^{-1}V \quad \text{given } R_k$$

Note that inverse of the matrix $(V[R_k + P_k]V^{\mathrm{T}})$ in Equations 21.8 and 21.11 exists because the rows of V are linearly independent, $R_k > 0$, and $P_k \geq 0$.

Next we obtain an unbiased weighted least squares estimate $\hat{x}_k$ of the measured variable x_k based on the calibrated measurement y_k as

$$\hat{x}_k = (H^{\mathrm{T}}R_k^{-1}H)^{-1}H^{\mathrm{T}}R_k^{-1}y_K \qquad (21.12)$$

The inverse of the (symmetric positive-definite) measurement covariance matrix R_k serves as the weighting matrix for generating the estimate $\hat{x}_k$, and is used as a filter matrix. Compensation of a (slowly varying) undetected error in the jth measurement out of l redundant measurements causes the largest jth element $|j\hat{c}_k|$ in the correction vector $\hat{c}_k$. Therefore, a limit check on the magnitude of each element of $\hat{c}_k$ will allow detection and isolation of the degraded measurement. The bounds of limit check, which could be different for the individual elements of $\hat{c}_k$, are selected by trade-off between the probability of false alarms and the allowable error in the estimate $\hat{x}_k$ of the measured variable (Ray, 1989).

21.2.1 Degradation Monitoring

Following Equation 21.12, we define the residual η_k of the calibrated measurement y_k as

$$\eta = y_k - H\hat{x}_k \qquad (21.13)$$

The residuals represent a measure of relative degradation of individual measurements. For example, under the normal condition, all calibrated measurements are clustered together, i.e., $\|\eta_k\| \approx 0$, although this may not be true for the residual $(m_k - H\hat{x}_k)$ of uncalibrated measurements.

While large abrupt changes in excess of the error threshold are easily detected and isolated by a standard diagnostics procedure [e.g., Ray and Desai (1986)], small errors (e.g., slow drift) can be identified from the a posteriori probability of failure that is recursively computed from the history of residuals based on the following trinary hypotheses:

H^0: Normal behavior with a priori conditional density function $_jf^0(\bullet) \equiv_j f\left(\bullet \middle| H^0\right)$

H^1: High (positive) failure with a priori conditional density function $_jf^1(\bullet) \equiv_j f\left(\bullet \middle| H^1\right)$ $\qquad (21.14)$

H^2: Low (negative) failure with a priori conditional density function $_jf^2(\bullet) \equiv_j f\left(\bullet \middle| H^2\right)$

where
   The left subscript refers to of the jth measurement for $j = 1, 2, \ldots, \ell$
   The right superscript indicates the normal behavior or failure mode

The density function for each residual is determined a priori from experimental data and/or instrument manufacturers' specifications. Only one test is needed here to accommodate both positive and negative failures in contrast to the binary hypotheses that require two tests. We now apply the recursive relations for multi-level hypotheses testing of single variables, derived in Appendix 21.A, to each residual of the redundant measurements. Then, for the jth measurement at the kth sampling instant, a posteriori probability of failure $_j\Pi_k$ is obtained following Equation 21.A.17 as

$$\left.\begin{aligned}
_j\Psi_k &= \left(\frac{_jp + _j\Psi_{k-1}}{2(1 - _jp)}\right)\left(\frac{_jf^{1'}(_j\eta_k) + _jf^2(_j\eta_k)}{_jf^0(_j\eta_k)}\right) \\
_j\Pi_k &= \frac{_j\Psi_k}{1 + _j\Psi_k}
\end{aligned}\right\} \tag{21.15}$$

where $_jp$ is the a priori probability of failure of the jth sensor during one sampling period, and the initial condition of each state, $_j\Psi_0, j = 1, 2, \dots, \ell$, needs to be specified.

Based on the a posteriori probability of failure, we now proceed to formulate a recursive relation for the measurement noise covariance matrix R_k that influences both calibration and estimation as seen in Equations 21.8 through 21.12. Its initial value R_0, which is determined a priori from experimental data and/or instrument manufacturers' specifications, provides the a priori information on individual measurement channels and conforms to the normal operating conditions when all measurements are clustered together, i.e., $\|\eta_k\| \approx 0$. In the absence of any measurement degradation, R_k remains close to its initial value R_0. Significant changes in R_k may take place if one or more sensors start degrading. This phenomenon is captured by the following model:

$$R_k = \sqrt{R_k^{rel}} R_0 \sqrt{R_k^{rel}} \quad \text{with } R_0^{rel} = I \tag{21.16}$$

where R_k^{rel} is a positive-definite diagonal matrix representing relative performance of the individual calibrated measurements and is recursively generated as follows:

$$R_{k+1}^{rel} = \text{diag}[h(_j\Pi_k)], \quad \text{i.e., } jr_{k+1}^{rel} = h(_j\Pi_k) \tag{21.17}$$

where

jr_k^{rel} and $_j\Pi_k$ are respectively the relative variance and a posteriori probability of failure of the jth measurement at the kth instant

$h: [0, 1) \rightarrow [1, \infty)$ is a continuous monotonically increasing function with boundary conditions $h(0) = 1\, h(\varphi) \rightarrow \infty$ as $\varphi \rightarrow 1$

The implication of Equation 21.17 is that credibility of a sensor monotonically decreases with increase in its variance that tends to infinity as its a posteriori probability of failure approaches 1. The magnitude of the relative variance $_jr_k^{rel}$ is set to the minimum value of 1 for zero a posteriori probability of failure. In other words, the jth diagonal element $_jw_k^{rel} \equiv 1/_jr_k^{rel}$ of the weighting matrix $W_k^{rel}(R_k^{rel})^{-1}$ tends to zero as $_j\Pi_k$ approaches 1. Similarly, the relative weight $_jw_k^{rel}$ is set to the maximum value of 1 for $_j\Pi_k = 0$. Consequently, a gradually degrading sensor carries monotonically decreasing weight in the computation of the estimate $\hat{x}_k$ in Equation 21.12.

Next we set the bounds on the states $_j\Psi_k$ of the recursive relation in Equation 21.15. The lower limit of $_j\Pi_k$ (which is an algebraic function of $_j\Psi_k$) is set to the probability $_jp$ of intra-sample failure. On the other extreme, if $_j\Pi_k$ approaches 1, the weight $_jw_k^{rel}$ (that approaches zero) may prevent fast restoration of a degraded sensor following its recovery. Therefore, the upper limit of $_j\Pi_k$ is set to $(1 - {_j}\alpha)$ where $_j\alpha$ is the allowable probability of false alarms of the jth measurement. Consequently, the function $h(\cdot)$ in Equation 21.17 is restricted to the domain $[_jp, (1 - {_j}\alpha)]$ to account for probabilities of intra-sampling failures and false alarms. Following Equation 21.15, the lower and upper limits of the states $_j\Psi_k$ are thus become $_jp/(1 - {_j}p)$ and $(1 - {_j}\alpha)/j^\alpha$, respectively. Consequently, the initial state in Equation 21.15 is set as: $_j\Psi_0 = {_j}p/(1 - {_j}p)$ for $j = 1, 2, \ldots, \ell$.

21.2.2 Possible Modifications of the Calibration Filter

The calibration filter is designed to operate in conjunction with a FDI system that is capable of detecting and isolating abrupt disruptions (in excess of specified bounds) in one or more of the redundant measurements (Ray and Desai, 1986). The consistent measurements, identified by the FDI system, are simultaneously calibrated at each sample. Therefore, if a continuous degradation, such as a gradual monotonic drift of a sensor amplifier, occurs sufficiently slowly relative to the filter dynamics, then the remaining (healthy) measurements might be affected, albeit by a small amount, due to simultaneous calibration of all measurements including the degraded measurement. Thus, the fault may be disguised in the sense that a very gradual degradation over a long period may potentially cause the estimate $\hat{x}_k$ to drift. This problem could be resolved by modifying the calibration filter with one or both of the following procedures:

- *Adjustments via limit check on the correction vector $\hat{c}_k$*: Compensation of a (slowly varying) undetected error in the jth measurement out of l redundant measurements will cause the largest jth element $|{_j}\hat{c}_k|$ in the correction vector $\hat{c}_k$. Therefore, a limit check on the magnitude of each element of $\hat{c}_k$ will allow detection and isolation of the degraded measurement. The bounds of limit check, which could be different for the individual elements of $\hat{c}_k$, are selected by trade-off between the probability of false alarms and the allowable error in the estimate $\hat{x}_k$ (1989).
- *Usage of additional analytical measurements*: If the estimate $\hat{x}_k$ is used to generate an analytic measurement of another plant variable that is directly measured by its own sensor(s), then a possible drift of the calibration filter can be detected whenever this analytical measurement disagrees with the sensor data in excess of a specified bound. The implication is that either the analytical measurement or the sensor is faulty. Upon detecting such a fault, the actual cause needs to be identified based on additional information including reasonability check. This procedure not only checks the calibration filter but also guards against simultaneous and identical failure of several sensors in the redundant set possibly due to a common cause, known as the common-mode fault.

21.3 Sensor Calibration in a Commercial-Scale Fossil Power Plant

The calibration filter, derived above, has been validated in a 320 MWe coal-fired supercritical power plant for on-line sensor calibration and measurement estimation at the throttle steam condition of ~1040°F (560°C) and ~3625 psia (25.0 MPa). The set of redundant measurements is generated by four temperature sensors installed at different spatial locations of the main steam header that carries superheated steam from the steam generator into the high-pressure turbine via the throttle valves and governor valves (Stultz and Kitto, 1992). Since these sensors are not spatially collocated, they can be asynchronous under transient conditions due to the transport lag. The filter simultaneously calibrates the sensors to generate a time-dependent estimate of the throttle steam temperature that is spatially averaged over the main steam header. This information on the estimated average temperature is used for health monitoring and damage prediction in the main steam header as well as for coordinated feedforward-feedback control of the power plant under both steady-state and transient operations (Kallappa et al., 1997; Kallappa and Ray, 2000). The filter software is hosted in a Pentium platform.

The readings of all four temperature sensors have been collected over a period of 100 h at the sampling frequency of once every 1 min. The collected data, after bad data suppression (e.g., elimination of obvious outliers following built-in tests such as limit check and rate check), shows that each sensor exhibits temperature fluctuations resulting from the inherent thermal-hydraulic noise and process transients as well as the instrumentation noise. For this specific application, the parameters, functions, and matrices of the calibration filter are selected as described below.

21.3.1 Filter Parameters and Functions

We start with the filter parameters and functions that are necessary for degradation monitoring. In this application, each element of the residual vector η_k of the calibrated measurement vector y_k is assumed to be Gaussian distributed that assures existence of the likelihood ratios in Equation 21.15. The structures of the a priori conditional density functions are chosen as follows:

$$_jf^0(\varphi) = \frac{1}{\sqrt{2\pi}\,j\sigma}\exp\left(-\frac{1}{2}\left(\frac{\varphi}{j\sigma}\right)^2\right)$$

$$_jf^1(\varphi) = \frac{1}{\sqrt{2\pi}\,j\sigma}\exp\left(-\frac{1}{2}\left(\frac{\varphi - j\theta}{j\sigma}\right)^2\right) \qquad (21.18)$$

$$_jf^2(\varphi) = \frac{1}{\sqrt{2\pi}\,j\sigma}\exp\left(-\frac{1}{2}\left(\frac{\varphi + j\theta}{j\sigma}\right)^2\right)$$

where
$_j\sigma$ is the standard deviation
$_j\theta$ and $-_j\theta$ are the thresholds for positive and negative failures, respectively, of the *j*th residual

Since it is more convenient to work in the natural-log scale for Gaussian distribution than for the linear scale, an alternative to Equation 21.17 is to construct a monotonically decreasing continuous function $g: (-\infty, 0) \to (0, 1]$ in lieu of the monotonically increasing continuous function $h: [0, 1) \to [1, \infty)$ so that

$$W^{rel}_{k+1} \equiv \left(R^{rel}_{k+1}\right)^{-1} = \text{diag}\left[g\left(\ell n_j\Pi_k\right)\right], \quad \text{i.e., the weight } _jw^{rel}_{k+1} \equiv \left(_jr^{rel}_{k+1}\right)^{-1} = g\left(\ell n_j\Pi_k\right) \qquad (21.19)$$

The linear structure of the continuous function $g(\cdot)$ is chosen to be piecewise linear as given below:

$$g(\varphi) = \begin{cases} w^{max} & \text{for } \varphi \le \varphi^{min} \\ \dfrac{(\varphi^{max} - \varphi)w^{max} + (\varphi - \varphi^{min})w^{min}}{\varphi^{max} - \varphi^{min}} & \text{for } -\infty \le \varphi^{min} \le \varphi \le \varphi^{max} < 0 \\ w^{min} & \text{for } \varphi \ge \varphi^{max} \end{cases} \qquad (21.20)$$

The function $g(\cdot)$ maps the space of $_j\Pi_k$ in the log scale into the space of the relative weight $_jw_{k+1}^{rel}$ of individual sensor data. The domain of $g(\cdot)$ is restricted to $[\ell n(_jp), \ell n(1 - _j\alpha)]$ to account for probability $_jp$ of intra-sampling failure and probability $_j\alpha$ of false alarms for each of the four sensors. The range of $g(\cdot)$ is selected to be $[_jw^{min}, 1]$ where a positive minimum weight (i.e., $_jw^{min} > 0$) allows the filter to restore a degraded sensor following its recovery. Numerical values of the filter parameters, j^σ, j^θ, j^p, j^α, and $_jw^{min}$ are presented below:

- The standard deviations of the a priori Gaussian density functions of the four temperature sensors are

$$_1\sigma = 4.1°F(2.28°C); \quad _2\sigma = 3.0°F(1.67°C); \quad _3\sigma = 2.4°F(1.33°C); \quad _4\sigma = 2.8°F(1.56°C)$$

  The initial condition for the measurement noise covariance matrix is set as: $R_0 = \text{diag}\left[_j\sigma\right]$. The failure threshold parameters are selected as: $_j\theta = _j\sigma/2$ for $j = 1,2,3,4$.
- The probability of intra-sampling failure is assumed to be identical for all four sensors as they are similar in construction and operate under identical environment. Operation experience at the power plant shows that the mean life of a resistance thermometer sensor, installed on the mean steam header, is about 700 days (i.e., about 2 years) of continuous operation. For a sampling interval of 1 min, this information leads to

$$_jp \approx 10^{-6} \quad \text{for } j = 1,2,3,4$$

- The probability of false alarms is selected in consultation with the plant operating personnel. On the average, each sensor is expected to generate a false alarm after approximately 700 days of continuous operation (i.e., once in 2 years). For a sampling interval of 1 min, this information leads to

$$_j\alpha \approx 10^{-6} \quad \text{for } j = 1,2,3,4.$$

- To allow restoration of a degraded sensor following its recovery, the minimum weight is set as

$$_jw_{min} \approx 10^{-3} \quad \text{for } j = 1,2,3,4.$$

21.3.2 Filter Matrices

After conversion of the four temperature sensor data into engineering units, the scale factor matrix in Equation 21.1 becomes: $H = [1\ 1\ 1\ 1]^T$. Consequently, following Potter and Suman (1977) and Ray and Luck (1991), the parity space projection matrix in Equation 21.6 becomes

$$V = \begin{bmatrix} \sqrt{\dfrac{3}{4}} & -\sqrt{\dfrac{1}{12}} & -\sqrt{\dfrac{1}{12}} & -\sqrt{\dfrac{1}{12}} \\ 0 & \sqrt{\dfrac{2}{3}} & -\sqrt{\dfrac{1}{6}} & -\sqrt{\dfrac{1}{6}} \\ 0 & 0 & \sqrt{\dfrac{1}{2}} & -\sqrt{\dfrac{1}{2}} \end{bmatrix}$$

In the event of a sensor being isolated as faulty, sensor redundancy reduces to 3, for which

$$H = [1\ 1\ 1]^{\mathrm{T}} \quad \text{and} \quad V = \begin{bmatrix} \sqrt{\dfrac{2}{3}} & -\sqrt{\dfrac{1}{6}} & -\sqrt{\dfrac{1}{6}} \\[2ex] 0 & \sqrt{\dfrac{1}{2}} & -\sqrt{\dfrac{1}{2}} \end{bmatrix}$$

The ratio, $R_k^{-1/2} Q\, R_k^{-1/2}$, of covariance matrices Q and R_k in Equations 21.4 and 21.1 largely determines the characteristics of the minimum variance filter in Equation 21.8 or Equation 21.11. The filter gain Γ_k increases with a larger ratio $R_k^{-1/2} Q R_k^{-1/2}$ and vice versa. Since the initial steady-state value R_0 is specified and R_k^{rel} is recursively generated thereon to calculate R_k via Equation 21.16, the choice is left only for selection of Q. As a priori information on Q may not be available, its choice relative to R_0 is a design feature. In this application, we have set $Q = R_0$.

21.3.3 Filter Performance Based on Experimental Data

The filter was tested on-line in the power plant over a continuous period of 9 months except for two short breaks during plant shutdown. The test results showed that the filter was able to calibrate each sensor under both pseudo-steady-state and transient conditions under closed loop control of throttle steam temperature. The calibrated estimate of the throttle steam temperature was used for plant control under steady-state, load following, start-up, and scheduled shutdown conditions. No natural failure of the sensors occurred during the test period and there was no evidence of any drift of the estimated temperature. As such the modifications (e.g., adjustments via limit check on $\hat{c}_k$, and additional analytical measurements) of the calibration filter, described earlier in this chapter, were not implemented. In addition to testing under on-line plant operation, simulated faults have been injected into the plant data to evaluate efficacy of the calibration filter under sensor failure conditions. Based on the data of four temperature sensors that were collected at an interval of 1 min over a period of 0–100 h, the following three cases of simulated sensor degradation are presented below:

21.3.3.1 Case 1 (Drift Error and Recovery in a Single Sensor)

Starting at 12.5 h, a drift error was injected into the data stream of Sensor#1 in the form of an additive ramp at the rate of 1.167°F (0.648°C) per hour. The injected fault was brought to zero at 75 h signifying that the faulty amplifier in the sensor hardware was corrected and reset.

Simulation results in the six plates of Figure 21.1 exhibit how the calibration filter responds to a gradual drift in one of the four sensors while the remaining three are normally functioning. Plate (a) in Figure 21.1 shows the response of the four uncalibrated sensors as well as the estimate generated by simple averaging (i.e., fixed identical weights) of these four sensor readings at each sample. The sensor data profile includes transients lasting from ~63 to ~68 h. From time 0 to 12.5 h when no fault is injected, all sensor readings are clustered together. Therefore, the uncalibrated estimate, shown by a thick solid line, is in close agreement with all four sensors during the period 0–12.5 h. Sensor#1, shown by the dotted line, starts drifting at 12.5 h while the remaining sensors stay healthy. Consequently, the uncalibrated estimate starts drifting at one quarter of the drift rate of Sensor#1 because of equal weighting of all sensors in the absence of the calibration filter. Upon termination of the drift fault at 75 h, when Sensor#1 is brought back to the normal state, the uncalibrated estimate resumes its normal state close to all four sensors for the remaining period from 75 to 100 h.

Plate (b) in Figure 21.1 shows the response of the four calibrated sensors as well as the estimate generated by weighted averaging (i.e., varying non-identical weights) of these four sensor readings at each sample. The calibrated estimate in Plate (b) stays with the remaining three healthy sensors even though Sensor#1 is gradually drifting. Plate (f) shows that, after the fault injection, Sensor#1 is weighted less

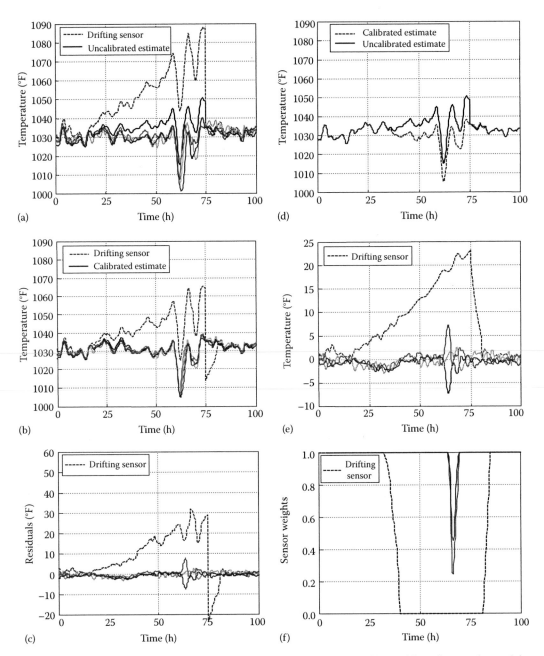

FIGURE 21.1 Performance of the calibration filter for drift error in a sensor: (a) Uncalibrated sensor data and the estimate; (b) calibrated sensor data and the estimate; (c) residuals of the calibrated sensor data; (d) calibrated and the uncalibrated estimates; (e) correction for sensor data calibration; (f) weights for sensor calibration.

than the remaining sensors. This is due to the fact that the residual η_k^1 (see Equation 21.13) of Sensor#1 in Plate (c) increases in magnitude with the drift error. The profile of $_1 w^{rel}$ in Plate (f) is governed by its non-linear relationship with η_k^1 given by Equations 21.15, 21.19, and 21.20. As seen in Plate (f), $_1 w^{rel}$ initially changes very slowly to ensure that it is not sensitive to small fluctuations in sensor data due to spurious noise such as those resulting from thermal-hydraulic turbulence. The significant reduction in $_1 w^{rel}$ takes place after about 32 h and eventually reaches the minimum value of 10^{-3} when η_k^1 is sufficiently large.

Therefore, the calibrated estimate $\hat{x}_k$ is practically unaffected by the drifting sensor and stays close to the remaining three healthy sensors. In essence, $\hat{x}_k$ is the average of the three healthy sensors. Upon restoration of Sensor#1 to the normal state, the calibrated signal $_1y_k$ temporarily goes down because of the large value of correction $_1\hat{c}_k$ at that instant as seen in Plate (e). However, the adaptive filter quickly brings back $_1\hat{c}_k$ to a small value and thereby the residual $_1\eta_k$ is reduced and the original weight (i.e., ~1) is regained. Calibrated and uncalibrated estimates are compared in Plate (d) that shows a peak difference of about 12°F (6.67°C) over a prolonged period.

In addition to the accuracy of the calibrated estimate, the filter provides fast and smooth recovery from abnormal conditions under both steady-state and transient operations of the power plant. For example, during the transient disturbance after about 65 h, the steam temperature undergoes a relatively large swing. Since the sensors are not spatially collocated, their readings are different during plant transients as a result of transport lag in the steam header. Plate (f) shows that the weights of two sensors out of the three healthy sensors are temporarily reduced while the remaining healthy sensor enjoys the full weight and the drifting Sensor#1 has practically no weight. As the transients are over, three healthy sensors resume the full weight. The cause of weight reduction is the relatively large residuals of these two sensors as seen in Plate (c). During this period, the two affected sensors undergo modest corrections: one is positive and the other negative as seen in Plate (e) so that the calibrated values of the three healthy sensors are clustered together. The health monitoring system and the plant control system rely on the spatially averaged throttle steam temperature (Kallappa et al., 1997; Holmes and Ray, 1998, 2001; Kallappa and Ray, 2000).

Another important feature of the calibration filter is that it reduces the deviation of the drifting Sensor#1 from the remaining sensors as seen from a comparison of its responses in Plates (a) and (b). This is very important from the perspectives of FDI for the following reason. In an uncalibrated system, Sensor#1 might have been isolated as faulty due to accumulation of the drift error. In contrast, the calibrated system makes Sensor#1 temporarily ineffective without eliminating it as faulty. A warning signal can be easily generated when the weight of Sensor#1 diminishes to a small value. This action will draw the attention of maintenance personnel for possible repair or adjustment. Since the estimate $\hat{x}_k$ is not poisoned by the degraded sensor, a larger detection delay can be tolerated. Consequently, the allowable threshold for fault detection can be safely increased to reduce the probability of false alarms.

21.3.3.2 Case 2 (Zero-Mean Fluctuating Error and Recovery in a Single Sensor)

We examine the filter performance by injecting a zero-mean fluctuating error to Sensor#3 starting at 12.5 h and ending at 75 h. The injected error is an additive sine wave of period ~36 h and amplitude 25°F (13.9°C). Simulation results in the six plates of Figure 21.2 exhibit how the calibration filter responds to the fluctuating error in Sensor#3 while the remaining three sensors (i.e., Sensor#1, Sensor#2, and Sensor#4) are normally functioning. To some extent, the filter response is similar to that of the drift error in Case 1. The major difference is the oscillatory nature of the weights and corrections of Sensor#3 as seen in Plates (f) and (e) in Figure 21.2, respectively. Note that this simulated fault makes the filter autonomously switch to the normal state from either one of the two abnormal states as the sensor error fluctuates between positive and negative limits. Since this is a violation of the assumption A-3 in Appendix 21.A, the recursive relation in Equation 21.A.17 represents an approximation of the actual situation. The results in Plates (b) to (f) in Figure 21.2 show that the filter is sufficiently robust to be able to execute the tasks of sensor calibration and measurement estimation in spite of this approximation. The filter not only exhibits fast response but also its recovery is rapid regardless of whether the fault is naturally mitigated or corrected by an external agent.

21.3.3.3 Case 3 (Drift Error in One Sensor and Zero-Mean Fluctuating Error in Another Sensor)

This case investigates the filter performance in the presence of simultaneous faults in two out of four sensors. Note that if the two affected sensors have similar types of faults (e.g., common mode faults),

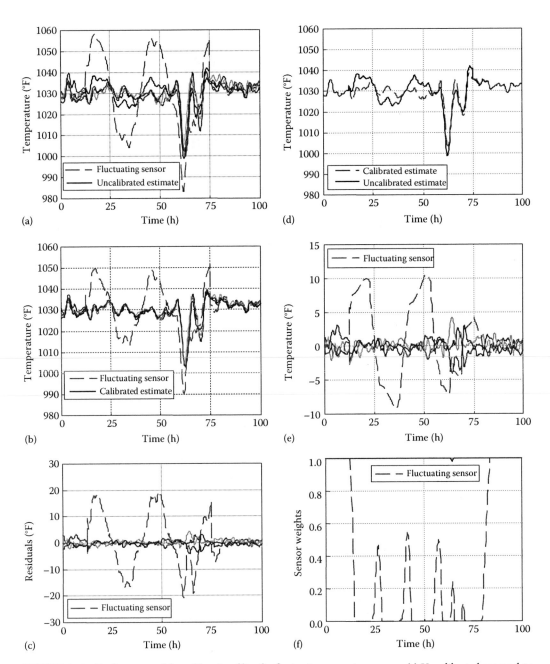

FIGURE 21.2 Performance of the calibration filter for fluctuation error in a sensor: (a) Uncalibrated sensor data and the estimate; (b) calibrated sensor data and the estimate; (c) residuals of the calibrated sensor data; (d) calibrated and uncalibrated estimates; (e) correction for sensor data calibration; (f) weights for sensor calibration.

the filter will require additional redundancy to augment the information base generated by the remaining healthy sensors. Therefore, we simulate simultaneous dissimilar faults by injecting a drift error in Sensor#1 and a fluctuating error in Sensor#3 exactly identical to those in Case 1 and Case 2, respectively. A comparison of the simulation results in the six plates of Figure 21.3 with those in Figures 21.1 and 21.2 reveals that the estimate $\hat{x}_k$ is essentially similar in all three cases except for small differences during the transients at ~65 h. It should be noted that, during the fault injection period from 12.5 to 75 h, $\hat{x}_k$ is

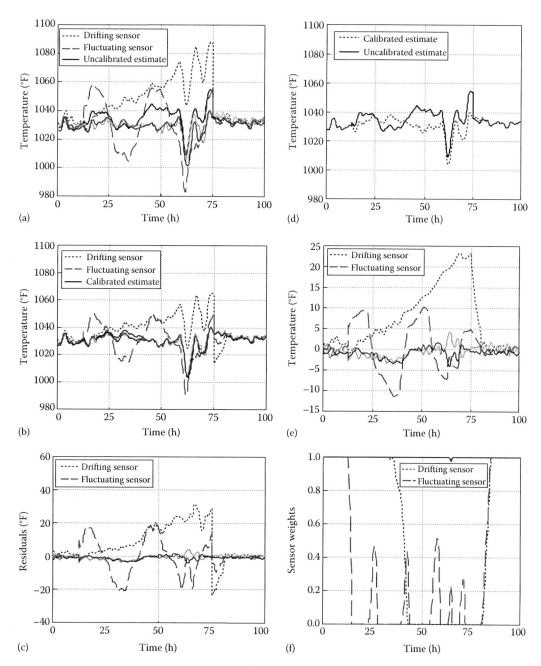

FIGURE 21.3 Performance of the calibration filter for drift error and fluctuation error in two sensor: (a) Uncalibrated sensor data and the estimate; (b) calibrated sensor data and the estimate; (c) residuals of the calibrated sensor data; (d) calibrated and uncalibrated estimates; (e) correction for sensor data calibration; (f) weights for sensor calibration.

strongly dependent on: Sensors #2, #3, and #4 in Case 1; Sensors #1, #2, and #4 in Case 2; and Sensors #2 and #4 in Case 3. Therefore, the estimate $\hat{x}_k$ cannot be exactly identical for these three cases. The important observation in this case study is that the filter can handle simultaneous faults in two out of four sensors provided that these faults are not strongly correlated; otherwise, additional redundancy or equivalent information would be necessary.

21.4 Summary and Conclusions

This chapter presents formulation and validation of an adaptive filter for real-time calibration of redundant signals consisting of sensor data and/or analytically derived measurements. Individual signals are calibrated on-line by an additive correction that is generated by a recursive filter. The covariance matrix of the measurement noise is adjusted as a function of the a posteriori probabilities of failure of the individual measurements. An estimate of the measured variable is also obtained in real time as a weighted average of the calibrated measurements. These weights are recursively updated in real time instead of being fixed a priori. The effects of intra-sample failure and probability of false alarms are taken into account in the recursive filter. The important features of this real-time adaptive filter are summarized below:

- A model of the physical process is not necessary for calibration and estimation if sufficient redundancy of sensor data and/or analytical measurements is available.
- The calibration algorithm can be executed in conjunction with a FDI system.
- The filter smoothly calibrates each measurement as a function of its a posteriori probability of failure that is recursively generated based on the current and past observations.

The calibration and estimation filter has been tested by injecting faults in the data set collected from an operating power plant. The filter exhibits speed and accuracy during steady-state and transient operations of the power plant. It also shows fast recovery when the fault is corrected or naturally mitigated. The filter software is portable to any commercial platform and can be potentially used to enhance the Instrumentation & Control System Software in tactical and transport aircraft, and nuclear and fossil power plants.

21.A Appendix: Multiple Hypotheses Testing Based on Observations of a Single Variable

Let $\{\eta_k, k = 1,2,3,...\}$ be (conditionally) independent values of a single variable (e.g., residual of a measurement) at consecutive sampling instants. We assume M distinct possible modes of failure in addition to the normal mode of operation that is designated as the mode 0. Thus, there are $(M + 1)$ mutually exclusive and exhaustive hypotheses defined at the kth sample as

$$H_k^0: \text{Normal behavior with a priori density function } f^0(\bullet) \equiv f\left(\bullet \middle| H^0\right)$$

$$H_k^i: \text{Abnormal behavior with a priori density function } f^i(\bullet) \equiv f\left(\bullet \middle| H^1\right), \quad i = 1,2,...,M$$

(21.A.1)

where each hypothesis H_k^j, $j = 0,1,2, ..., M$ is treated as a Markov state.

We define the a posteriori probability π_k^j of the jth hypothesis at the kth sample as

$$\pi_k^j \equiv P[H_k^j | Z_k], \quad j = 0,1,2, ..., M$$

(21.A.2)

based on the history $Z_k \equiv \bigcap_{i=1}^{k} z_i$ where $z_i \equiv \{\eta_i \in B_i\}$ B_i is the region of interest at the sample.

The problem is to derive a recursive relation for a posteriori probability of failure Π_k at the kth sample:

$$\Pi_k \equiv P\left[\bigcup_{j=1}^{M} H_k^j | Z_k\right] = \sum_{j=1}^{M} P\left[H_k^j | Z_k\right] \Rightarrow \Pi_k = \sum_{j=1}^{M} \pi_k^j$$

(21.A.3)

because of the exhaustive and mutually exclusive properties of the Markov states, $H_k^j, j = 1, 2, ..., M$. To construct a recursive relation for, Π_k we introduce the following three definitions:

$$\text{Joint probability: } \xi_k^j \equiv P\left[H_k^j, Z_k\right] \tag{21.A.4}$$

$$\text{A priori probability: } \lambda_k^j \equiv P\left[z_k \big| H_k^j\right] \tag{21.A.5}$$

$$\text{Transition probability: } a_k^{i,j} \equiv P\left[H_k^j \big| H_{k-1}^i\right] \tag{21.A.6}$$

Then, because of conditional independence of z_k and Z_{k-1}, Equation 21.A.4 takes the following form:

$$\xi_k^j = P\left[H_k^j, z_k, Z_{k-1}\right]$$

$$= P\left[z_k \big| H_k^j\right] P[H_k^j, Z_{k-1}] \tag{21.A.7}$$

Furthermore, the exhaustive and mutually exclusive properties of the Markov states $H_k^j, j = 0, 1, 2, ..., M$ and independence of Z_{k-1} and H_k^j lead to

$$P\left[H_k^j, Z_{k-1}\right] = \sum_{i=0}^{M} P\left[H_k^j, H_{k-1}^i, Z_{k-1}\right]$$

$$= \sum_{i=0}^{M} P\left[Z_{k-1} \big| H_{k-1}^i\right] P\left[H_k^j \big| H_{k-1}^i\right] P\left[H_{k-1}^i\right]$$

$$= \sum_{i=0}^{M} P\left[H_k^j \big| H_{k-1}^i\right] P\left[H_{k-1}^i \big| Z_{k-1}\right] \tag{21.A.8}$$

The following recursive relation is obtained from a combination of Equations 21.A.4 through 21.A.8 as

$$\xi_k^j = \lambda_k^j \sum_{i=0}^{M} a_k^{i,j} \xi_{k-1}^i \tag{21.A.9}$$

We introduce a new term

$$\psi_k^j \equiv \frac{\xi_k^j}{\xi_k^0} \tag{21.A.10}$$

that reduces to the following form by use of Equation 21.A.9:

$$\psi_k^j = \left(\frac{\lambda_k^j}{\lambda_k^0}\right) \left(\frac{a_k^{0,j} + \sum_{i=1}^{M} a_k^{i,j} \psi_{k-1}^i}{a_k^{0,0} + \sum_{i=1}^{M} a_k^{i,0} \psi_{k-1}^i}\right) \tag{21.A.11}$$

to obtain the a posteriori probability π_k^j in Equation 21.A.2 in terms of ξ_k^j and ψ_k^j as

$$\pi_k^j = \frac{P[H_k^j, Z_k]}{P[Z_k]} = \frac{P[H_k^j, Z_k]}{\displaystyle\sum_{i=0}^{M} P[H_k^i, Z_k]}$$

$$= \frac{\xi_k^j}{\xi_k^0 + \displaystyle\sum_{i=1}^{M} \xi_k^i} = \frac{\psi_k^j}{1 + \displaystyle\sum_{i=1}^{M} \psi_k^i} \tag{21.A.12}$$

A combination of Equations 21.A.3 and 21.A.12, leads to the a posteriori probability Π_k of failure as

$$\Pi_k = \frac{\Psi_k}{1 + \Psi_k} \quad \text{with} \quad \Psi_k \equiv \sum_{j=1}^{M} \psi_k^j \tag{21.A.13}$$

The above expressions can be realized by a simple recurrence relation under the following four assumptions:

- *Assumption 21.A-1*: At the starting point (i.e., $k = 0$), all measurements operate in the normal mode, i.e.,

$$p[H_0^0] = 1 \quad \text{and} \quad P[H_0^j] = 0 \quad \text{for } j = 1, 2, ..., M.$$

Therefore,

$$\xi_0^0 = 1 \quad \text{and} \quad \xi_0^j = 0 \quad \text{for } j = 1, 2, ..., M$$

- *Assumption 21.A-2*: Transition from the normal mode to any abnormal mode is equally likely. That is, if p is the a priori probability of failure during one sampling interval, then $a_k^{0,0} = 1 - p$ and $a_k^{0,i} = p/M$ for $i = 1, 2, ..., M$, and all k.
- *Assumption 21.A-3*: No transition takes place from an abnormal mode to the normal mode implying $a_k^{i,0} = 0$ for $i = 1, 2, ..., M$, and all k. The implication is that a failed sensor does not return to the normal mode (unless replaced or repaired).
- *Assumption 21.A-4*: Transition from an abnormal mode to any abnormal mode including itself is equally likely. That is, $a_k^{i,j} = 1/M$ for $i, j = 1, 2 ..., M$, and all k.

A recursive relation for Ψ_k is generated based on the above assumptions and using the expression in Equation 21.A.11 as

$$\psi_k^j = \frac{p + \displaystyle\sum_{i=1}^{M} \psi_{k-1}^i}{(1-p)M} \left(\frac{\lambda_k^j}{\lambda_k^0} \right) \quad \text{given } \psi_0^j = 0 \quad \text{for } j = 1, 2, ..., M \tag{21.A.14}$$

which is simplified by use of the relation $\psi_k \equiv \displaystyle\sum_{i=1}^{M} \psi_k^i$ in Equation 21.A.13 as

$$\Psi_k = \left(\frac{p + \Psi_{k-1}}{(1-p)M} \right) \sum_{j=1}^{M} \frac{\lambda_k^j}{\lambda_k^0} \quad \text{given } \Psi_0 = 0 \tag{21.A.15}$$

If the probability measure associated with each abnormal mode is absolutely continuous relative to that associated with the normal mode, then the ratio λ_k^j/λ_k^0 of a priori probabilities converges to a Radon–Nikodym derivative as the region B_k in the expression $z_k \equiv \{\eta_k \in B_k\}$ approaches zero measure (Wong and Hajek, 1985). This Radon–Nikodym derivative is simply the likelihood ratio $f^j(\eta_k)/f^0(\eta_k)$, $j = 1, 2, \ldots, M$, where $f^i(\cdot)$ is the a priori density function conditioned on the hypothesis H^i, $i = 0, 1, 2, \ldots, M$. Accordingly, Equation 21.A.15 becomes

$$\Psi_k = \left(\frac{p + \Psi_{k-1}}{(1-p)M}\right) \sum_{j=1}^{M} \frac{f^j(\eta_k)}{f^0(\eta_k)} \quad \text{given } \Psi_0 = 0 \tag{21.A.16}$$

For the specific case of two abnormal hypotheses (i.e., $M = 2$) representing positive and negative failures, the recursive relations for Ψ_k and Π_k in Equations 21.A.16 and 21.A.13 become

$$\left.\begin{array}{l} \Psi_k = \left(\dfrac{p + \Psi_{k+1}}{2(1-p)}\right)\left(\dfrac{f^1(\eta_k) + f^2(\eta_k)}{f^0(\eta_k)}\right) \\[1.5em] \Pi_k = \dfrac{\Psi_k}{1 + \Psi_k} \end{array}\right\} \quad \text{given } \Psi_0 = 0 \tag{21.A.17}$$

Acknowledgment

The research work reported in this chapter has been supported in part by the Army Research Office under Grant No. DAAD19-01-1-0646.

References

Daly, K.C., E. Gai, and J.V. Harrison (1979), Generalized likelihood test for FDI in redundant sensor configurations, *Journal of Guidance and Control*, 2(1), 9–17.

Deckert, J.C., J.L. Fisher, D.B. Laning, and A. Ray (1983), Signal validation for nuclear power plants, *ASME Journal of Dynamic Systems, Measurement and Control*, 105(1), 24–29.

Desai, M.N., J.C. Deckert, and J.J. Deyst (1979), Dual sensor identification using analytic redundancy, *Journal of Guidance and Control*, 2(3), 213–220.

Dickson, B., J.D. Cronkhite, and H. Summers (1996), Usage and structural life monitoring with HUMS, *American Helicopter Society 52nd Annual Forum*, Washington, DC, June 4–6, pp. 1377–1393.

Gelb, A., ed. (1974), *Applied Optimal Estimation*, MIT Press, Cambridge, MA.

Holmes, M. and A. Ray (1998), Fuzzy damage mitigating control of mechanical structures, *ASME Journal of Dynamic Systems, Measurement and Control*, 120(2), 249–256.

Holmes, M. and A. Ray (2001), Fuzzy damage mitigating control of a fossil power plant, *IEEE Transactions on Control Systems Technology*, 9(1), 140–147.

Jazwinski, A.H. (1970), *Stochastic Processes and Filtering Theory*, Academic Press, New York.

Kallappa, P.T., M. Holmes, and A. Ray (1997), Life extending control of fossil power plants for structural durability and high performance, *Automatica*, 33(6), 1101–1118.

Kallappa, P.T. and A. Ray (2000), Fuzzy wide-range control of fossil power plants for life extension and robust performance, *Automatica*, 36(1), 69–82.

Potter, J.E. and M.C. Suman (1977), Thresholdless redundancy management with arrays of skewed instruments, Integrity in electronic flight control systems, NATO AGARDOGRAPH-224, pp. 15-1–15-15.

Ray, A. (1989), Sequential testing for fault detection in multiply-redundant systems, *ASME Journal of Dynamic Systems, Measurement and Control*, 111(2), 329–332.

Ray, A. and M. Desai (1986), A redundancy management procedure for fault detection and isolation, *ASME Journal of Dynamic Systems, Measurement and Control*, 108(3), 248–254.

Ray, A., R. Geiger, M. Desai, and J. Deyst (1983), Analytic redundancy for on-line fault diagnosis in a nuclear reactor, *AIAA Journal of Energy*, 7(4), 367–373.

Ray, A. and R. Luck (1991), Signal validation in multiply-redundant systems, *IEEE Control Systems Magazine*, 11(2), 44–49.

Stultz, S.C. and J.B. Kitto, eds. (1992), *STEAM: Its Generation and Use*, 40th edn., Babcock & Wilcox Co., Baberton, OH.

Wong, E. and B. Hajek (1985), *Stochastic Processes in Engineering Systems*, Springer-Verlag, New York.

22

Semantic Information Extraction

22.1 Introduction 513
22.2 Symbolic Dynamics..514
    The Conversion of System Dynamics into Formal
    Languages • Determination of ε-Machines
22.3 Formal Language Measures...516
22.4 Behavior Recognition...517
22.5 Experimental Verification 517
22.6 Conclusions and Future Work..520
Acknowledgments and Disclaimer 521
References..521

David S. Friedlander
*Defense Advanced
Research Projects Agency*

22.1 Introduction

This chapter describes techniques for extracting *semantic* information from sensor networks and applying them to recognizing the behaviors autonomous vehicles based on their trajectories, and predicting anomalies in mechanical systems based on a network of embedded sensors.

Sensor networks generally observe systems that are too complex to be simulated by computer models based directly on their Physics. We therefore use a semi-empirical model based on time series measurements. These systems can be stochastic but not necessarily stationary. In this case, we make a simplifying assumption, that the time-scale for changes in the equations of motion is much longer than the time-scale for changes in the dynamical variables. The system is then in semi-equilibrium so its dynamics can be determined from a data sample whose duration is long compared to changes in the dynamical variables but short compared to changes in the dynamical equations.

The techniques are based on integrating and converting sensor measurements into formal languages and using a *formal language measure* to compare the language of the observations to the languages associated with known behaviors stored in a database. Based on the hypothesis that behaviors represented by similar formal languages are semantically similar, this method provides a form of computer perception for physical behaviors through the extension of traditional pattern-matching techniques.

One intriguing aspect of this approach is that people represent their perception of the environment with *natural language*. Statistical approaches to analyzing formal languages have been successfully applied to *natural language processing* (NLP) [Charniak 93]. This suggests that formal languages may be a promising approach for representing sensor network data.

22.2 Symbolic Dynamics

In symbolic dynamics, the numeric time series associated with a system's dynamics are converted into streams of symbols. The streams define a formal language where any substring in the stream belongs to the language. Conversions of physical measurements to symbolic dynamics and the analysis of the resulting strings of symbols have been used for characterizing nonlinear dynamical systems as they simplify data handling while retaining important qualitative phenomena. This also allows usage of complexity measures defined on formal languages made of symbol strings to characterize the system dynamics [Kurths 95]. The distance between individual symbols is not defined so there is no notion of linearity.

22.2.1 The Conversion of System Dynamics into Formal Languages

One method for generating a stream of symbols from the resampled sensor network data divides the phase-space volume of the network into hyper-cube shaped regions and assigns a symbol to each region. When the phase-space trajectory enters a region, its symbol in added to the symbol stream as shown in Figure 22.1.

Any set containing strings of symbols defines a formal language. If the language contains an infinite number of strings, it cannot be fully represented in this way. The specification of a formal language can be compressed allowing finite representations of infinite languages. Usually, greater compression provides greater insight into the language. Two equivalent representations are generally used: finite state machines and formal grammars.

[Chomsky 57] developed a classification of formal languages based on their complexity. From least to most complex, they are: regular, context free, context sensitive and recursively enumerable. The simplest is regular languages, which can be represented by finite-state automata. Since the dynamics of complex systems are generally stochastic, we use probabilistic finite-state automata (PFSA).

Sometimes PFSA for very simple systems can be specified intuitively. There is also a method to determine them analytically [Shalizi 02]. Finite state machines determined this way are called ε-machines. Unfortunately, the method is currently limited to regular languages.

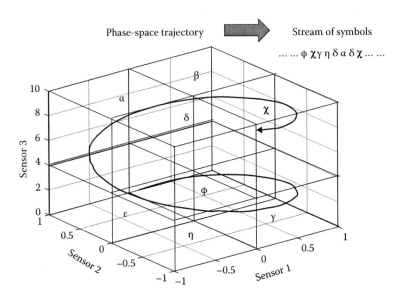

FIGURE 22.1   Continuous to symbolic dynamics.

22.2.2 Determination of ε-Machines

The symbol stream is converted into a PFSA or, equivalently, a probabilistic regular language. A sample of the symbol stream of some length, L, is used to determine the model. Shalizi's method generates a PFSA from the sample. Since the symbol stream is unbounded, each state is considered an accepting state. Each state, s, in the PFSA is assigned a set of substrings, U, such that the path of any string that is accepted by the automaton and ends in $u \in U$, will be accepted at state s.

Each state also contains a *morph*, which is a list of the probabilities of each symbol being emitted from that state, i.e. $\bar{M}(S_i) : M_j(S_i) \equiv P(e_j \mid S_i)$, where $\bar{M}(S_i)$ is the morph of state S_i and $P(e_j \mid S_i)$ is the probability of emitting the jth symbol, e_j, when the system is in state S_i. The probabilities are approximated by the statistics of the sample. Let $U^i \equiv \{s_k^i\}$ be the set of substrings assigned to state S_i. The quantity $M_j(S_i) \approx \sum_k |e_j s_k^i| / |s_k^i|$, where $|s_i|$ is the count of the substring s_i in the sample and new symbols are appended on the left hand side of strings.

The PFSA is initialized to a single state containing the empty string. Its morph is $\bar{M}(S_0) = \{P(e_i)\}$, where S_0 is the initial state and the ith component of $\bar{M}(S_0)$ is the unconditional probability of symbol e_i. Then add transitions $S_0 \xrightarrow{e_i, P(e_i)} S_0$ for each symbol e_i. In other words, the initial morph contains the probability of each individual symbol. The initial automaton is shown in Figure 22.2.

The initial automaton is expanded using [Shalizi 02]. A simplified version is given in Figure 22.3. Strings of length one through some maximum length are added to the *PFSA*. Given a state, containing string s, the string $s' = e_k \| s$, where e_k is the kth symbol and "$\|$" is the concatenation operator, will have the morph $\bar{M}(s') = \{P(e_k \mid s')\}$. If an existing state, $\hat{S}$, has a morph close to $\bar{M}(S)$, the transition $S \xrightarrow{e_i} \hat{S}$ is added to the PFSA and the string s' is added to $\hat{S}$. Otherwise, a new state, $\tilde{S}$, and the transition $S \xrightarrow{e_i} \tilde{S}$ are added to the *PFSA*. The string s' and the morph $\bar{M}(s')$ are assigned to $\tilde{S}$.

The next stage is to *determinize* [Shalizi 02] the PFSA by systematically adding states whenever a given state has two or more transitions leaving it with the same symbol. Finally the *transient states* are eliminated, i.e. a state is transient if it cannot be reached from any other state.

Most complexity measures are based on entropy and therefore are a minimum for constant data streams and a maximum for random data streams. This, however, contradicts the intuitive notion of complexity, which is low for both constant and random behavior of dynamical systems (see Figure 22.2). [Crutchfield 94] introduced a measure called ε-complexity that is defined based on the construction of a *PFSA* for the symbol stream. The ε-complexity is defined as the Shannon entropy of the state probabilities of the automaton: $C_\varepsilon \equiv \sum_i P(S_i) \log P(S_i)$. It is minimal for both constant and random behavior and diverges when chaotic behavior is exhibited, i.e. the number of states in the *PSFA* goes to infinity as some system parameter goes to its critical value for chaotic behavior.

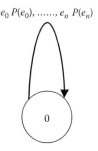

$$e_0\, P(e_0),\, \ldots\ldots,\, e_n\, P(e_n)$$

FIGURE 22.2   Initial automaton.

```
Function Automaton (string-sample, S₀, {transitions})
    for each L = 2, Lₘₐₓ
        for each state, Sᵢ
            for each string sᵢᵏ ∈ Sᵢ
                if |sᵢᵏ|+1 = L
                    for each symbol eⱼ
                        M̄(sᵢᵏeⱼ) = morph(sᵢᵏeⱼ)
                        if M̄(sᵢᵏeⱼ) ≈ M̄(Sᵢ)
                            add-string-to-state (sᵢᵏeⱼ, Sᵢ, Sᵢ, {transition})
                        elseif M̄(sᵢᵏeⱼ) ≈ M̄(Sₘ) where m ≠ i
                            add-string-to-state (sᵢᵏeⱼ, Sᵢ, Sᵢ, {transitions})
                        else
                            create new state Sₙ
                            add-string-to-state (sᵢᵏeⱼ, Sᵢ, Sᵢ, {transitions})
                        end
                    end
                end
            end
        end
    end
    return {Sᵢ}, {transitions}
```

Function *morph (string)*
  for each symbol e_j

$$M_j(string) = \frac{|string|}{|string||e_j|}$$

  end
  return $\bar{M}(string)$

Function *add-string-to-state (string, S_0, S_f {transitions})*
  add *string* to S_f
  $e_k = last\text{-}symbol(string)$
  add $S_0 \xrightarrow{e_k} S_f$ to {*transitions*}

FIGURE 22.3 Algorithm for building ε-machine.

22.3 Formal Language Measures

One shortcoming of complexity measures for detecting, predicting or classifying anomalous behaviors is that they are scalars. That is, two different behaviors of a complex system may have the same complexity measure.

[Ray 02, Wang 02] have addressed this problem by representing each possible formal language with a given alphabet Σ. The language of all possible strings is denoted as Σ^*. It is represented as the unit vector in an infinite-dimensional vector space $(2^{\Sigma^*}, \oplus)$ over the finite field $GF(2)$ where $\oplus$ is the exclusive-OR operator for vector addition and the zero vector in this space is the null language $\varnothing$.

There are at least two methods to determine $\mu(L)$, the measure of language L. If a *PFSA* can be derived for the language, an exact measure developed [Ray 02, Wang 02] can be used. If the language is not regular and cannot be well approximated by a PFSA, an approximate measure developed by the author can be used instead. In either case, the distance between two formal languages, L_1 and L_2, is defined as $d(L_1, L_2) \equiv |\mu(L_1 \cup L_2 - L_1 \cap L_2)|$, i.e. the measure of the "exclusive or" of the strings in the languages. The only restriction on the two languages is that they come from the same alphabet of symbols. In other words, the two languages must represent dynamical processes defined on the same phase-space.

The measures can be applied to a single language or to the vector difference between any two languages where the vector difference corresponds to the exclusive-OR operation of the strings belonging to the languages. Since the exclusive-OR of the language vectors maps back to the *symmetric set differ-ence* of the languages, this vector addition operation can be considered as taking the *difference* between two languages.

[Friedlander 03a] has proposed another measure that is a real positive measure $\mu: 2^{\Sigma^*} \to [0, \infty)$, called the *weighted counting measure*, defined as

$$\mu(L) \equiv \sum_{i=1}^{\infty} w_i n_i(L)$$

where $n_\ell(L)$ is the number of strings of length ℓ in the language L, and $w_\ell = 1/(2k)^\ell$ where the positive integer $k = |\Sigma|$ is the alphabet length. The weighting factor w_ℓ was designed so that $\mu(\Sigma^*) = 1$. The weighting factor, w_ℓ decays exponentially with the string length, ℓ. This feature allows good approximations to the language measure from a relatively small sample of a language with a large number of strings.

22.4 Behavior Recognition

If we define a *behavior* as a pattern of activity in the system dynamics, and represent it as a formal language, we can compare an observed behavior with a database of known behaviors and determine the closest match using a distance based on a formal language measure [Friedlander 03b]. We can also discover new behaviors that are based on clusters of formal language vectors. When the behavior is based on an object's trajectory, the techniques can be applied to surveillance and defense.

The concepts of ε-machines, language measures and distance functions can be used to apply traditional pattern matching techniques to behavior recognition. For example, we can store a set of languages $\{L_i\}$ corresponding to known behaviors and use them as exemplars. When the sensor network records some unknown target behavior with language L_u, it can be compared to the database to find the best matching language of some known behavior using

$$\text{Behavior } (L_k): k = \text{index } \max_i d(L_u, L_i)$$

Target behaviors will change over time and it is desirable to track those changes as they occur. This can be done with the method presented here as long as the time scale for detecting behaviors, i.e. the length of the language sample, is shorter that the time scale for behavior changes. The sensor data are sampled at regular intervals, the behavior for each interval is determined and changes in the corresponding languages can be analyzed.

If we define an *anomaly* as an abrupt and significant change in system dynamics, it will include faults (recoverable errors) and failures (unrecoverable errors). When the behaviors are based on anomaly precursors, the technique can be applied to *condition-based maintenance* providing early prediction of failures in mechanical systems. Taking corrective action in advance could increase safety, reliability and performance.

22.5 Experimental Verification

This section contains the results of early experiments that test our method for extracting semantic information from sensor network data. We attempted to distinguish between two type of robot behavior: following a perimeter and a random search. This system was able to consistently recognize the correct behavior and detect changes from one behavior to the other. Although preliminary, these results suggest the new methods are promising.

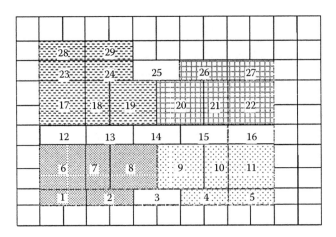

FIGURE 22.4   Pressure sensitive floor.

The experiments use a pressure sensitive floor measuring simple, single robot behaviors [Friedlander 03b]. Due to the noisiness and unreliability of the pressure sensors, they were used only to determine the quadrant of the floor where the robot was located. The results show the robustness of our technique. Pressure sensitive wire was placed under "panels" of either 2 × 2 or 2 × 1 square floor tiles that were 584 mm on a side.

Each panel is numbered in the diagram. The floor is divided into four quadrants (as shown in Figure 22.4). The panels 12, 13, 14, 15, 16, 25, and 3 are between quadrants and their data were not used in the experiment. Each panel is a sensor and provides time series data that were analyzed in real-time. The upper-left quadrant had seven panels and the others had five each. This redundancy provided experimental robustness while using unreliable and noisy sensing devices. One or more of the panels did not work, or worked incorrectly during most of the experimental runs.

The experiments involved dynamically differentiating between *wall-following* and *random search* behaviors of a single robot. The sensors were built by coiling pressure sensitive wire under each panel as shown in Figure 22.5. The robot had four wheels that passed over multiple points in the coiled wire as it ran over the panel. Each panel provided a series of data peaks as the robot crossed over it.

The first step in processing the 29 channels of time series data was to localize the robot in terms of which panel it was crossing when the data sample was taken. Two unsynchronized servers, one for panels 1–16 and the other for panels 17–29 provided the data. The data were *pushed* in the sense that a server provided a sample whenever one or more of the panels has an absolute value over an adjustable cutoff.

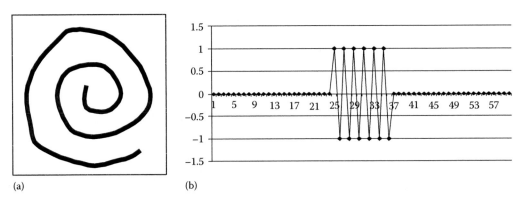

(a)   (b)

FIGURE 22.5   Pressure sensor data: (a) pressure panel; (b) data sample.

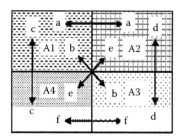

Symbol	Event description
a	A1 to A2 or A2 to A1
b	A1 to A3 or A3 to A1
c	A1 to A4 or A4 to A1
d	A2 to A3 or A3 to A2
e	A4 to A2 or A2 to A4
f	A4 to A3 or A3 to A4

FIGURE 22.6 Event definitions for circling behavior.

When the real-time behavior recognition software receives a data packet, it preprocesses the data. The first stage is to remove the data from the panels between quadrants. If there is no large value for any of the remaining panels, the packet is ignored; otherwise, the panel with the highest absolute value is considered the location of the robot. This transforms the time series data into a symbol stream of panel id numbers. The next stage filters the panel number stream to reduce noise. The stream values flow into a buffer of length 5. Whenever the buffer is full of identical panel numbers this number is emitted to the filtered stream, if an inconsistent number enters the buffer, the buffer is flushed. This eliminates false positives by requiring five peaks per panel.

The panel id stream is then converted to a stream of quadrants as shown in Figures 22.4 and 22.6. The stream of quadrants is then converted into a stream of events. An event occurs when the robot changes quadrants as shown in Figure 22.6. The event depends only on the two quadrants involved, not the order in which they were crossed. This was done to lower the number of symbols in the alphabet from 12 to 6.

The next stage is to use the event stream to recognize behaviors. Because the language of wall following is a subset of the language of random searching, the event stream can prove the random search and disprove the wall following hypotheses, but not prove the wall following and disprove the random search hypotheses. The longer the event stream is recognized by the wall following automaton, however, the more evidence there is that the robot is wall following rather than performing a random search. The finite state automaton in Figure 22.7 recognizes the stream of symbols from wall following behavior, starting in any quadrant and going in either direction.

The initial behavior is given as *unknown* and goes to *wall following* or *random walk*. It then goes between these two behaviors during the course of the experiment depending on the frequency of string rejections in the wall-following automaton. If there is less than one string rejection for every six events, the behavior is estimated to be *wall following*, otherwise it is estimated to be *random walk*.

The displays associated with the behavior recognition software demonstration are shown in Figure 22.8. There are four displays. The behavior recognition software shows the current estimate of the robot's

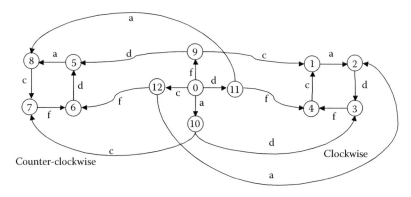

FIGURE 22.7 Automaton to recognize circling behavior.

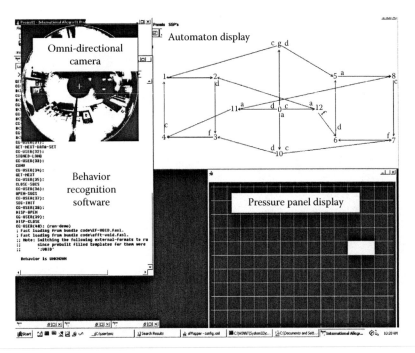

FIGURE 22.8 Behavior recognition demonstration displays.

behavior, the symbol and time of each event and the time of each string rejection. The omni-directional camera shows the physical location of the robot and the floor panel display shows the panel being excited by the robot. The automaton display shows the current state of the wall following model based on the event stream.

22.6 Conclusions and Future Work

Traditional pattern matching techniques measure the distances between the feature vectors of an observed object and a set of stored exemplars. Our research extends these techniques to dynamical systems using symbolic dynamics and a recent advance in formal language theory defining a formal language measure. It is based on a combination of Nonlinear Systems theory and Language theory. It is assumed that mechanical systems under consideration exhibit nonlinear dynamical behavior on two time scales. Anomalies occur on a *slow time scale* that is at least two or more orders of magnitude larger than the *fast time scale* of the system dynamics. It is also assumed that the dynamical system is stationary at the fast time scale and that any nonstationarity is observable only on the slow time scale.

Finite state machine representations of complex, nonlinear systems have had success in capturing essential features of a process while leaving out irrelevant details [Friedlander 03a, Shalizi 01]. These results suggest that the behavior recognition mechanism would be effective in artificial perception of scenes and actions from sensor data. Applications could include image understanding, voice recognition, fault prediction and intelligent control. We have experimentally verified the method for two simple behaviors.

Future research may include collection and analysis of additional data on gearboxes and other mechanical systems of practical significance. Another area of future research is integration of formal language measures into damage mitigating control systems. The methods should be tested on behaviors that are more complex. We are in the process of analyzing data from observations of such behaviors using LADAR. They include coordinated actions of multiple robots. One planned experiment contains behavior specifically designed to create by a context free, but not regular, language.

Another application is data compression. The definition of the formal language describing the observation is transmitted, rather than the sensor data itself. At the receiving end, the language can be used to classify the behavior or to regenerate sensor data that are statistically equivalent to the original observations. We have begun research to develop this technique in the context of video data from a wireless network of distributed cameras.

Acknowledgments and Disclaimer

This material is based upon work supported in part by the ESP MURI Grant No. DAAD19-01-1-0504; and by the NASA Glenn Research Center under Grant No. NAG3-2448. Any opinions, findings, and conclusions or recommendations expressed in this publication are those of the authors and do not necessarily reflect the views of the Defense Advanced Research Projects Agency (DARPA), the Army Research Office or the NASA Glenn Research Center.

References

[Abarbanel 95] Abarbanel, H.D.I., *Analysis of Observed Chaotic Data*, Springer-Verlag, New York, November 1995.

[Begg 99] Begg, C., T. Merdes, C.S. Byington, and K.P. Maynard, *Mechanical System Modeling for Failure Diagnosis and Prognosis, Maintenance and Reliability Conference (MARCON 99)*, Gatlinburg, TN, May 10–12, 1999.

[Charniak 93] Charniak, E., *Statistical Language Learning*, MIT Press, Cambridge, MA, 1993.

[Chomsky 57] Chomsky, N., *Syntactic Structures*, Mouton, Gravenhag, the Netherlands, 1957.

[Crutchfield 94] Crutchfield, J.P., The calculi of emergence, dynamics, and induction, *Physica D*, 75, 11–54, 1994.

[Friedlander 03a] Friedlander, D.S., I. Chattopadhayay, A. Ray, S. Phoha, and N. Jacobson, Anomaly prediction in mechanical system using symbolic dynamics, *Proceedings of the American Control Conference*, Boulder, CO, 2003a.

[Friedlander 03b] Friedlander, D.S., S. Phoha, and R. Brooks, Determination of vehicle behavior based on distributed sensor network data, to be published in the *Proceedings of SPIE's 48th Annual Meeting*, San Diego, CA, August 3–8, 2003b.

[Hanson 97] Hanson, J.E. and J.P. Crutchfield, Computational mechanics of cellular automata: An example, *Physica D*, 1997.

[Kurths 95] Kurths, J., U. Schwarz, A. Witt, R.Th. Krampe, and M. Abel, Measures of complexity in signal analysis, *3rd Technical Conference on Nonlinear Dynamics (Chaos) and Full Spectrum Processing*, Mystic, CT, July 10–13, 1995.

[Ray 02] Ray, A. and S. Phoha, A language measure for discrete-event automata, *Proceedings of the International Federation of Automatic Control (IFAC) World Congress b'02*, Barcelona, Spain, July 2002.

[Shalizi 01] Shalizi, C.R., *Causal Architecture, Complexity, and Self-Organization in Time Series and Cellular Automata*, PhD dissertation, University of Wisconsin-Madison, Madison, WI, 2001.

[Shalizi 02] Shalizi, C.R., K.L. Shalizi, and J.P. Crutchfield, An algorithm for pattern discovery in time series, Santa Fe Institute Working Paper 02-10-060, 2002, available at http://arxiv.org/abs/cs.LG/0210025

[Wang 02] Wang, X. and A. Ray, Signed real measure of regular languages, *Proceedings of the American Control Conference*, Anchorage, AK, May 2002.

23

Fusion in the Context of Information Theory

23.1 Information Processing in Distributed Networks.......................523
23.2 Evolution toward Information Theoretic Methods
for Data Fusion...525
Sensor Fusion Research • Sensor Administration Research
23.3 Probabilistic Framework for Distributed Processing.................526
Sensor Data Model for Single Sensors • A Bayesian Scheme for
Decentralized Data Fusion • Distributed Detection Theory and
Information Theory
23.4 Bayesian Framework for Distributed Multi-Sensor Systems.....534
Information Theoretic Justification of the Bayesian Method
23.5 Concluding Remarks..538
References...538

Mohiuddin Ahmed
University of California

Gregory Pottie
University of California

In this section, we selectively explore some aspects of the theoretical framework that has been developed to analyze the nature, performance, and fundamental limits for information processing in the context of data fusion. In particular, we discuss how Bayesian methods for distributed data fusion can be interpreted from the point of view information theory. Consequently, information theory can provide a common framework for distributed detection and communication tasks in sensor networks.

Initially, the context is established for considering distributed networks as efficient information processing entities (Section 23.1). Next, in Section 23.2, the approaches taken toward analyzing such systems and the path leading toward the modern information theoretic framework for information processing are discussed. The details of the mathematical method are highlighted in Section 23.3, and applied specifically for the case of multi-sensor systems in Section 23.4. Finally our conclusions are presented in Section 23.5.

23.1 Information Processing in Distributed Networks

Distributed networks of sensors and communication devices provide the ability to electronically network together what were previously isolated islands of information sources and sinks, or more generally, *states of nature*. The states can be measurements of physical parameters (e.g., temperature, humidity, etc.) or estimates of operational conditions (network loads, throughput, etc.), among other things, distributed over a region in time and/or space. Previously, the aggregation, fusion and interpretation of this mass of data representing some phenomena of interest were performed by isolated *sensors*, requiring human supervision and control. However, with the advent of powerful hardware platforms and networking technologies, the possibility and advantages of *distributed sensing* information processing has been recognized [32].

A sensor can be defined to be any device that provides a quantifiable set of outputs in response to a specific set of inputs. These outputs are useful if they can be mapped to a state of nature that is under consideration.

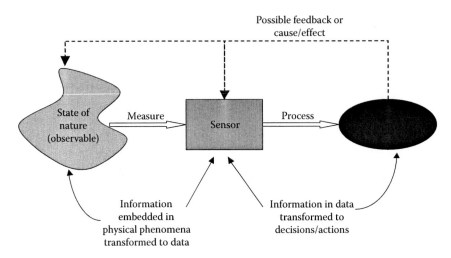

FIGURE 23.1 Information processing in sensors.

The end goal of the sensing task is to acquire a description of the external world, predicated upon which can be a series of *actions*. In this context, sensors can be thought of as *information gathering, processing and dissemination* entities, as diagrammed in Figure 23.1. The data pathways in the figure illustrate an abstraction of the flow of information in the system. In a distributed network of sensors, the sensing system may be comprised of multiple sensors that are physically disjoint or distributed in time or space, and that work cooperatively. Compared to a single sensor platform, a network has the advantages of *diversity* (different sensors offer complementary viewpoints), and *redundancy* (reliability and increased resolution of the measured quantity) [24]. In fact, it has been rigorously established from the theory of distributed detection that higher reliability and lower probability of detection error can be achieved when observation data from multiple, distributed sources is intelligently fused in a decision making algorithm, rather than using a single observation data set [44]. Intuitively, any practical sensing device has limitations on its sensing capabilities (e.g., resolution, bandwidth, efficiency, etc.). Thus, descriptions built on the data sensed by a single device are only approximations of the true state of nature. Such approximations are often made worse by incomplete knowledge and understanding of the environment that is being sensed and its interaction with the sensor. These uncertainties, coupled with the practical reality of occasional sensor failure greatly compromises reliability and reduces confidence in sensor measurements. Also, the spatial and physical limitations of sensor devices often means that only partial information can be provided by a single sensor.

A network of sensors overcomes many of the shortcomings of a single sensor. However new problems in efficient information management arise. These may be categorized into two broad areas [30]:

1. *Data Fusion*: This is the problem of combining diverse and sometimes conflicting information provided by sensors in a multi-sensor system, in a consistent and coherent manner. The objective is to infer the relevant states of the system that is being observed or activity being performed.
2. *Resource Administration*: This relates to the task of optimally configuring, coordinating and utilizing the available sensor resources, often in a dynamic, adaptive environment. The objective is to ensure efficient* use of the sensor platform for the task at hand.

In comparison to lumped-parameter sensor systems (Figure 23.1), the issues mentioned above for multi-sensor systems can be diagrammed as shown in Figure 23.2 [24].

---

* *Efficiency*, in this context, is very general and can refer to power, bandwidth, overhead, throughput, or a variety of other performance metrics, depending upon the particular application.

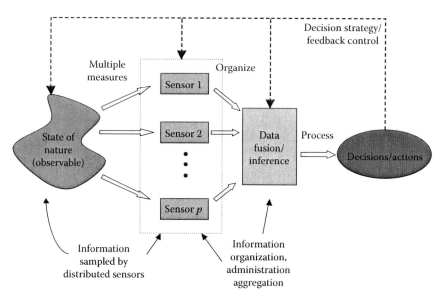

FIGURE 23.2 Information processing in distributed sensors.

23.2 Evolution toward Information Theoretic Methods for Data Fusion

Most of the early research effort in probabilistic and information theoretic methods for data fusion focused on techniques motivated by specific applications, such as in vision systems, sonar, robotics platforms, etc. [19,22,23,26]. As the inherent advantages of using multi-sensor systems were recognized [2,46], a need for a comprehensive theory of the associated problems of distributed, decentralized data fusion and multi-user information theory became apparent [7,16,43]. Advances in integrated circuit technology have enabled mass production of sensors, signal processing elements and radios [12,34], spurring new research in wireless communications [20], and in *ad hoc networking* [36,40]. Subsequently, it was only natural to combine these two disciplines—sensors and networking—to develop a new generation of distributed sensing devices that can work cooperatively to exploit diversity [31,32]. An abridged overview of the research in sensor fusion and management is now summarized [24].

23.2.1 Sensor Fusion Research

Data fusion is the process by which data from a multitude of sensors is used to yield an optimal estimate of a specified state vector pertaining to the observed system [44], whereas sensor administration is the design of communication and control mechanisms for the efficient use of distributed sensors, with regards to power, performance, reliability, etc. Data fusion and sensor administration have mostly been addressed separately. Sensor administration has been addressed in the context of wireless networking and not necessarily in conjunction with the unique constraints imposed by data fusion methodologies.

To begin with, sensor models have been aimed at interpretation of measurements. This approach to modeling can be seen in the sensor models used by Kuc and Siegel [19], among others. Probability theory, and in particular, a Bayesian treatment of data fusion [9] is arguably the most widely used method for describing uncertainty in a way that abstracts from a sensor's physical and operational details. Qualitative methods have also been used to describe sensors, for example by Flynn [12] for sonar and infra-red applications. Much work has also been done in developing methods for intelligently combining information from different sensors. The basic approach has been to pool the information using what are essentially "weighted averaging" techniques of varying degrees of complexity. For example Berger [2] discusses a majority voting technique based on a probabilistic representation of information. Non-probabilistic methods [17] used inferential techniques, for example for multi-sensor target identification. Inferring the state of nature given a probabilistic representation is, in general, a well understood problem in classical estimation. Representative methods are Bayesian estimation, Least Squares estimation, Kalman Filtering, and its various derivatives. However, the question of how to use these techniques in a distributed fashion has not been addressed to date in a systematic fashion except for some specific physical layer cases [45].

23.2.2 Sensor Administration Research

In the area of sensor network administration, protocol development and management have mostly been addressed using application specific descriptive techniques for specialized systems [46]. Tracking radar systems provided the impetus for much of the early work. Later, robotic applications led to the development of models for sensor behavior and performance that could then be used to analyze and manage the transfer of sensor data. The centralized or hierarchical nature of such systems enabled this approach to succeed. Other schemes that found widespread use were based on determining cost functions and performance trade-offs a priori [1], e.g., cost-benefit assignment matrices allocating sensors to targets, or Boolean matrices characterizing sensor-target assignments based on sensor availability and capacity. Expert system approaches have also been used, as well as decision-theoretic (*normative*) techniques. However, optimal sensor administration in this way has been shown by Tsitsiklis [43] to be very hard in the general framework of distributed sensors, and practical schemes use a mixture of heuristic techniques (for example in data fusion systems involving wired sensors in combat aircrafts). Only recently have the general networking issues for wireless ad hoc networks been addressed (Sohrabi et al. [41], Singh et al. [39]), where the main problems of self-organization, bootstrap, route discovery etc., have been identified. Application specific studies, e.g., in the context of antenna arrays (Yao et al. [47]) have also discussed these issued. However, few general fusion rules or data aggregation models for networked sensors have been proposed, with little analytical or quantitative emphasis. Most of these studies do not analyze in detail the issues regarding the network-global impact of administration decisions, such as choice of fusion nodes, path/tree selections, data fusion methodology, or physical layer signaling details.

23.3 Probabilistic Framework for Distributed Processing

The information being handled in multi-sensor systems almost always relates to a state of nature, and consequently, it is assumed to be unknown prior to observation or estimation. Thus, the model of the information flow shown in Figure 23.2 is probabilistic, and hence can be quantified using the principles of information theory [6,15]. Furthermore, the process of data detection and processing that occurs within the sensors and fusion node(s) can be considered as elements of classical statistical decision theory [29]. Using the mature techniques that these disciplines offer, a probabilistic information processing relation can be quantified for sensor networks, and analyzed within the framework of the well-known Bayesian paradigm [35]. The basic tasks in this approach are the following:

1. Determination of appropriate information processing techniques, models and metrics for fusion and sensor administration
2. Representation of the sensors process, data fusion, and administration methodologies using the appropriate probabilistic models
3. Analysis of the measurable aspects of the information flow in the sensor architecture using the defined models and metrics

4. Design of optimal data fusion algorithms and architectures for optimal inference in multi-sensor systems

5. Design, implementation and test of associated networking and physical layer algorithms and architectures for the models determined in (4)

We now consider two issues in information combining in multi-sensor systems: (1) the nature of the information being generated by the sensors, and (2) the method of combining the information from disparate sources.

23.3.1 Sensor Data Model for Single Sensors

Any observation or measurement by any sensor is always uncertain to a degree determined by the *precision* of the sensor. This uncertainty, or measurement *noise*, requires us to treat the data generated by a sensor probabilistically. We therefore adopt the notation and definitions of probability theory to determine an appropriate model for sensor data [14,24].

Definition 23.1

A *state vector* at time instant *t*, is a representation of the *state of nature* of a process of interest, and can be expressed as a vector $x(t)$ in a measurable, finite-dimensional vector space, Ω, over a discrete or continuous field, $\mathcal{F}$:

$$x(t) \in \Omega \subseteq \mathbb{R}^n \tag{23.1}$$

The state vector is arbitrarily assumed to n-dimensional and can represent a particular state of nature of interest, e.g., it can be the three dimensional position vectors of an airplane. The state space may be either continuous or discrete (e.g., the on or off states of a switch).

Definition 23.2

A *measurement vector* at time instant *t* is the information generated by a single sensor (in response to an observation of nature), and can be represented by an m-dimensional vector, $z(t)$ from a measurement vector space Ψ:

$$z(t) = \begin{pmatrix} z_1 \\ z_2 \\ \vdots \\ z_m \end{pmatrix} \in \Psi \subseteq \mathbb{R}^m \tag{23.2}$$

Intuitively, the measurement vector may be thought of as m pieces of data that a single sensor generates from a single observation at a single instant of time. Because of measurement error, the sensor output $z(t)$ is an approximation of $x(t)$—the true state of nature. It is important to note that $z(t)$ may itself not be directly visible to the user of the sensor platform. A noise corrupted version $\Gamma\{z(t), v(t)\}$, as defined below, may be all that is available for processing. Furthermore, the dimensionality of the sensor data may not be the same as the dimension of the observed parameter that is being measured. For example, continuing with the airplane example, a sensor may display the longitude and latitude of the airplane at a particular instant of time via GPS (a two-dimensional observation vector), but may not be able to measure the altitude of the airplane (which completes the three-dimensional specification of the actual location of the airplane in space).

The *measurement error* itself can be considered as another vector, $v(t)$, or a *noise process* vector, of the same dimensionality as the observation vector $z(t)$. As the name suggests, noise vectors are inherently stochastic in nature, and serve to render all sensor measurements uncertain, to a specific degree.

Definition 23.3

An *observation model*, Γ, for a sensor is a mapping from state space Ω to observation space Ψ, and is parameterized by the statistics of the noise process:

$$\Gamma_v : \Omega \mapsto \Psi \tag{23.3}$$

Functionally, the relationship between the state, observation and noise vectors can be expressed as

$$z(t) = \Gamma\{x(t), v(t)\} \tag{23.4}$$

Objective: The objective in sensing applications is to infer the unknown state vector $x(t)$ from the error corrupted and (possibly lower dimensional) observation vector $z(t), v(t)$. If the functional specification of the mapping in Equation 23.3, and the noise vector $v(t)$, were known for all times t, then finding the inverse mapping for one-to-one cases would be trivial, and the objective would be easily achieved. It is precisely because either or both parameters may be random that gives rise to various estimation architectures for inferring the state vector from the imperfect observations. A geometric interpretation of the objective can be presented as shown in Figure 23.3a. The simplest mapping relationship Γ that can be used as a sensor data model is the *additive* model of noise corruption, as shown in Figure 23.3b, which can be expressed as

$$x = \Gamma(z + v) \tag{23.5}$$

Typically, for well designed and matched sensor platforms, the noise vector is small compared to the measurement vector, in which case a Taylor approximation can be made

$$x = \Gamma(z) + (\nabla_z \Gamma)z + \text{(higher order terms)} \tag{23.6}$$

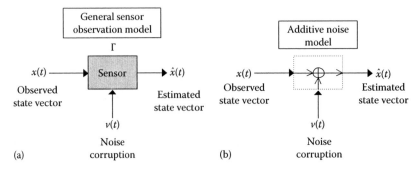

FIGURE 23.3 Sensor data models: (a) General case. (b) Noise additive case.

where ∇_z is the Jacobian matrix of the mapping Γ with respect to the state measurement vector z. Since the measurement error is random, the state vector observed is also random, and we are in essence dealing with random variables. Thus, we can use well established statistical methods to quantify the uncertainty in the random variables [35]. For example, the statistics of the noise process $v(t)$ can be often be known a priori. Moments are the most commonly used measures for this purpose, and in particular, if the covariance of the noise process is known, $E\{vv^T\}$, then the covariance of the state vector is [24]

$$E\{xx^T\} = (\nabla_z\Gamma)E\{vv^T\}(\nabla_z\Gamma)^T \qquad (23.7)$$

For uncorrelated noise v, the matrix $(\nabla_z\Gamma)E\{vv^T\}(\nabla_z\Gamma)^T$ is symmetric and can be decomposed using singular value decomposition [37]:

$$(\nabla_z\Gamma)E\{vv^T\}(\nabla_z\Gamma)^T = (SDS^T) \qquad (23.8)$$

where
    S is an $(n \times n)$ matrix of orthogonal vectors e_j
    D are the eigenvalues of the decomposition

$$S = (e_1, e_2, \ldots, e_n), \quad e_i e_j = \begin{pmatrix} 1 & \text{for } i = j \\ 0 & \text{for } i \neq j \end{pmatrix} \qquad (23.9)$$

$$D = \text{diag}(d_1, d_2, \ldots, d_n) \qquad (23.10)$$

The components of D correspond to the scalar variance in each of direction. Geometrically, all the directions for a given state x can be visualized as an ellipsoid in n-dimensional space, with the principal axes in the directions of the vectors e_k and $2\sqrt{d_j}$ as the corresponding magnitudes. The volume of the ellipsoid is the uncertainty in x. The 2-dimensional case is shown in Figure 23.4. From this perspective, the basic objective in the data fusion problem is then to reduce the volume of the uncertainty ellipsoid. All the techniques for data estimation, fusion, and inference are designed toward this goal [27].

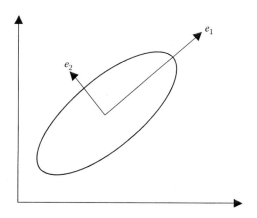

FIGURE 23.4   Ellipsoid of state vector uncertainty.

23.3.2 A Bayesian Scheme for Decentralized Data Fusion

Given the inherent uncertainty in measurements of states of nature, the end goal in using sensors, as mentioned in the previous section, is to obtain the best possible estimates of the states of interest for a particular application. The Bayesian approach to solving this problem is concerned with quantifying *likelihoods* of events, given various types of partial knowledge or observations, and subsequently determining the state of nature that is most probably responsible for the observations as the "best" estimate.

The issue of whether the Bayesian approach is intrinsically the "best" approach for a particular problem* is a philosophical debate that is not discussed here further. It may be mentioned, however, that arguably, the Bayesian paradigm is most *objective* because it is based only on observations and "impartial" models for sensors and systems.

The information contained in the (noise corrupted) measured state vector *z* is first described by means of probability distribution functions (PDF). Since all observations of states of nature are causal manifestations of the underlying processes governing the state of nature,[†] the PDF of *z* is conditioned by the state of nature at which time the observation/measurement was made. Thus, the PDF of *z* conditioned by *x* is what is usually measurable and is represented by

$$F_Z(z \mid x) \tag{23.11}$$

This is known as the *Likelihood Function* for the observation vector. Next, if information about the possible states under observation is available (e.g., a priori knowledge of the range of possible states), or more precisely the probability distribution of the possible states $F_X(x)$, then the prior information and the likelihood function (23.11) can be combined to provide the a posteriori conditional distribution of *x*, given *z*, by Bayes' Theorem [14]:

Theorem 23.1

$$F_X(x \mid z) = \frac{F_Z(x \mid z)F_X(x)}{\displaystyle\int_\Omega F_z(z \mid x)F_x(x)dF(x)} = \frac{F_Z(z \mid x)F_X(x)}{F_Z(z)} \tag{23.12}$$

Usually, some function of the actual likelihood function, $g(T(z)|x)$, is commonly available as the processable information from sensors. $T(z)$ is known as the *sufficient statistic* for *x* and Equation 23.12 can be reformulated as

$$F_X(x \mid z) = F_X(x \mid T(z)) = \frac{g(T(z) \mid x)F_X(x)}{\displaystyle\int_\Omega g(T(z) \mid x)F_X(x)dF(x)} \tag{23.13}$$

When observations are carried out in discrete time steps according to a desired resolution, then a vector formulation is possible. Borrowing notation from [24], all observations up to time index *r* can be defined as

$$Z^r \triangleq \{z(1), z(2), \ldots, z(r)\} \tag{23.14}$$

---

* In contrast with various other types of inferential and subjective approaches [35].
† Ignoring the observer-state interaction difficulties posed by Heisenberg Uncertainty considerations.

from where the posterior distribution of x given the set of observations Z^r becomes

$$F_X(x \mid Z^r) = \frac{F_{z^r}(Z^r \mid x)F_x(x)}{F_{z^r}(Z^r)} \qquad (23.15)$$

Using the same approach, a recursive version of Equation 23.15 can also be formulated:

$$F_X(x \mid Z^r) = \frac{F_z(z(r) \mid x)F_x(x \mid Z^{r-1})}{F_z(z(r) \mid Z^{r-1})} \qquad (23.16)$$

in which case all the r observations do not need to be stored, and instead only the current observation $z(r)$ can be considered at the rth step. This version of the Bayes' Law is most prevalent in practice since it offers a directly implementable technique for fusing observed information with *prior beliefs*.

23.3.2.1 Classical Estimation Techniques

A variety of inference techniques can now be applied to estimate the state vector x (from the time series observations from a single sensor). The estimate, denoted by $\hat{x}$, is derived from the posterior distribution $F_{vecx}(x|Z^r)$ and is a point in the uncertainty ellipsoid of Figure 23.4. The basic objective is to reduce the volume of the ellipsoid, which is equivalent to minimizing the probability of error based on some criterion. Three classical techniques are now briefly reviewed: *Maximum Likelihood, Maximum A Posteriori* and *Minimum Mean Square Error* estimation.

Maximum Likelihood (ML) estimation involves maximizing the likelihood function (Equation 23.11) by some form of search over the state space Ω:

$$\hat{x}_{ML} = \arg\max_{x \in \Omega} F_{z^r}(Z^r \mid x) \qquad (23.17)$$

This is intuitive since the PDF is greatest when the correct state has been guessed for the conditioning variable. However, a major drawback is that for state vectors from large state spaces, the search may be computationally expensive, or infeasible. Nonetheless, this method is widely used in many disciplines, e.g., digital communication reception [33].

Maximum a posteriori (MAP) estimation technique involves maximizing the posterior distribution from observed data as well as from prior knowledge of the state space:

$$\hat{x}_{MAP} = \arg\max_{x \in \Omega} F_x(x \mid Z^r) \qquad (23.18)$$

Since prior information may be subjective, objectivity for an estimate (or the inferred state) is maintained by considering only the likelihood function (i.e. only the observed information). In the instance of no prior knowledge, and the state space vectors are all considered to be equally likely, the MAP and ML criterion can be shown to be identical.

Minimum Mean Square Error (MMSE) techniques attempt to minimize the estimation error by searching over the state space, albeit in an organized fashion. This is the most popular technique in a wide variety of information processing applications, since the variable can often be found analytically, or the search space can be reduced considerably or investigated systematically. The key notion is to reduce the covariance of the estimate. Defining the mean and variance of the posterior observation variable as

$$\bar{x} \triangleq \mathrm{E}_{F(x|Z^r)}\{x\} \qquad (23.19)$$

$$\text{Var}(\boldsymbol{x}) \triangleq \mathrm{E}_{F(\boldsymbol{x}|\boldsymbol{Z}^r)}\{(\boldsymbol{x} - \overline{x})(\boldsymbol{x} - \overline{x})^{\mathrm{T}}\} \tag{23.20}$$

it can be shown that the least squares estimator is one that minimizes the Euclidean distance between the true state $\boldsymbol{x}$ and the estimate $\hat{\boldsymbol{x}}$, given the set of observations $\boldsymbol{Z}^r$. In the context of random variables, this estimator is referred to as the MMSE estimate and can be expressed as

$$\hat{\boldsymbol{x}}_{MMSE} = \arg\min_{\boldsymbol{x} \in \Omega} \mathrm{E}_{F(\boldsymbol{x}|\boldsymbol{Z}^r)}\{(\boldsymbol{x} - \overline{x})(\boldsymbol{x} - \overline{x})^{\mathrm{T}}\} \tag{23.21}$$

To obtain the minimizing estimate, Equation 23.21 can be differentiated with respect to $\hat{\boldsymbol{x}}$ and set equal to zero, which yields $\hat{\boldsymbol{x}} = E\{\boldsymbol{x} \mid \boldsymbol{Z}^r\}$. Thus the MMSE estimate is the conditional mean. It also can be shown that the MMSE estimate is the minimum variance estimate, and when the conditional density coincides with the mode, the MAP and MMSE estimators are equivalent.

These estimation techniques and their derivatives such as the Wiener and Kalman filters [18] all serve to reduce the uncertainty ellipsoid associated with state $\boldsymbol{x}$ [27]. In fact, direct applications of these mathematical principles formed the field of radio frequency *signal detection* in noise, and shaped the course of developments in digital communication technologies.

23.3.3 Distributed Detection Theory and Information Theory

Information theory was developed to determine the fundamental limits on the performance of communication systems [38]. Detection theory on the other hand, involves the application of statistical decision theory to estimate states of nature, as discussed in the previous section. Both these disciplines can be used to treat problems in the transmission and reception of information, as well as the more general problem of data fusion in distributed systems. The synergy was first explored by researchers in the 1950s and 1960s [25], and the well established source and channel coding theories were spawned as a result. With respect to data fusion, the early research in the fields of information theory and fusion proceeded somewhat independently. Whereas information theory continued exploring the limits of digital signaling, data fusion, on the other hand, and its myriad ad hoc techniques were developed by the practical concerns of signal detection, aggregation and interpretation for decision making. Gradually, however, it was recognized that both issues, at their abstract levels, dealt fundamentally with problems of information processing.

Subsequently, attempts were made to unify distributed detection and fusion theory, as it applied e.g., in sensor fusion, with the broader field of information theory. Some pioneering work involved the analysis of the hypothesis testing problem using discrimination (Kullback, [21]), employing cost functions based on information theory for optimizing signal detection (Middleton [25]) and formulating the detection problem as a coding problem for asymptotic analysis using error exponent functions (Csiszar and Longo [8], Blahut [3]). More recently, research in these areas have been voluminous, with various theoretical studies exploring the performance limits and asymptotic analysis of fusion and detection schemes [4,43].

In particular, some recent results [44] are relevant to the case of a distributed system of sensor nodes. As has been noted earlier, the optimal engineering trade-offs for the efficient design for such a system is not always clear cut. However, if the detection/fusion problem can be recast in terms of information theoretic cost functions, then it has been shown that system optimization techniques provide useful design paradigms.

For example consider the block diagrams of a conventional binary detection system and a binary communication channel as shown in Figure 23.5. The source in the detection problem can be viewed as the information source in the information transmission problem. The decisions in the detection model can be mapped as the channel outputs in the channel model. Borrowing the notation from [44], if the input is considered a random variable $H = i$, $i = 0, 1$ where probability $P(H = 0) = P_0$, the output $u = i$, $i = 0, 1$ is then a decision random variable, whose probabilities of detection (P_D), miss (P_M),

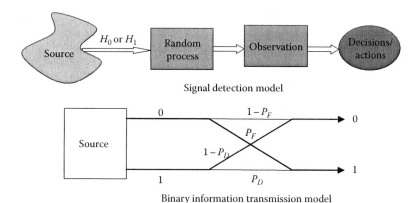

FIGURE 23.5 Signal detection vs. information transmission.

false alarm (P_F), etc. can be interpreted in terms of the transition probabilities of the information transmission problem. This is the classic example of the binary channel [33].

If the objective of the decision problem is the minimization of the information loss between the input and output, then it can be shown that the objective is equivalent to the maximization of the *mutual information, $I(H;u)$* (see Section 23.4 for formal definitions of entropy and information measures). This provides a mechanism for computing practical likelihood test ratios as a technique for *information-optimal* data fusion. Thus, the case of the binary detection problem, the a posteriori probabilities are

$$P(u = 0) = P_0(1 - P_F) + (1 - P_0)(1 + P_D) \triangleq \alpha_0 \tag{23.22}$$

$$P(u = 1) = P_0 P_F + (1 - P_0)P_D \triangleq \alpha_1 \tag{23.23}$$

whereupon it can be shown that the optimal decision threshold for the received signal is

$$threshold = \frac{-P_0\{\log(\alpha_0/\alpha_1) - \log[(1 - P_F)/P_F]\}}{(1 - P_0)\{\log(\alpha_0/\alpha_1) - \log[(1 - P_D)/P_D]\}} \tag{23.24}$$

This approach can be extended to the case of distributed detection. For example, for a detection system in a parallel topology without a fusion center, and assuming the observations at the local detectors are conditionally independent, the goal is then to maximize the mutual information $I(H;\boldsymbol{u})$ where the vector $\boldsymbol{u}$ contains the local decisions. Once again, it can be shown that the optimal detectors are threshold detectors, and likelihood ratio tests can then be derived for each detector. Using the second subscript in the variables below to refer to the detector number, the thresholds are

$$threshold_1 = -\frac{P_0\left[\log\left(\dfrac{\alpha_{00}}{\alpha_1 0}\right) + P_{F2}\log\left(\dfrac{\alpha_{01}\alpha_1 0}{\alpha_{00}\alpha_1 1}\right) - \log\left(\dfrac{1 - P_{F1}}{P_{F1}}\right)\right]}{(1 - P_0)\left[\log\left(\dfrac{\alpha_{00}}{\alpha_1 0}\right) + P_{D2}\log\left(\dfrac{\alpha_{01}\alpha_1 0}{\alpha_{00}\alpha_1 1}\right) - \log\left(\dfrac{1 - P_{D1}}{P_{D1}}\right)\right]} \tag{23.25}$$

with a similar expression for *threshold$_2$*. In a similar manner, other entropy-based information theoretic criteria (e.g., logarithmic cost functions) can be successfully used to design the detection and distributed fusion rules in an integrated manner for various types of fusion architectures, (e.g., serial, parallel with fusion center, etc.) This methodology provides an attractive, unified approach for system design, and has the intuitive appeal of treating the distributed detection problem as an information transmission problem.

23.4 Bayesian Framework for Distributed Multi-Sensor Systems

When a number of spatially and functionally different sensor systems are used to observe the same (or similar) state of nature, then the data fusion problem is no longer simply a state space uncertainty minimization issue. The distributed and multi-dimensional nature of the problem requires a technique for checking the usefulness and validity of the data from each of the not necessarily independent sensors. The data fusion problem is more complex, and general solutions are not readily evident. This section explores some of the commonly studied techniques and proposes a novel, simplified methodology that achieves some measure of generality.

The first issue is the proper modeling of the data sources. If there are p sensors observing the same state vector, but from different vantage points, and each one generates its own observations, then we have a collection of observation vectors $z_1(t)$, $z_2(t)$, ..., $z_p(t)$, which can be represented as a combined matrix of all the observations from all sensors (at any particular time t):

$$\mathbf{Z}(t) = (z_1(t) \quad z_2(t) \cdots z_p(t)) = \begin{bmatrix} z_{11} & z_{21} & \cdots & z_{p1} \\ z_{12} & z_{22} & \cdots & z_{p2} \\ & & \ddots & \\ z_{1m} & z_{2m} & \cdots & z_{pm} \end{bmatrix} \tag{23.26}$$

Furthermore, if each sensor makes observations up to time step r for a discretized (sampled) observation scheme, then the matrix of observations $\mathbf{Z}(r)$ can be used to represent the observations of all the p sensors at time-step r (a discrete variable, rather than the continuous $\mathbf{Z}(t)$). With adequate memory allocation for signal processing of the data, we can consider the super-matrix $\{\mathbf{Z}^r\}$ of all the observations of all the p sensors from time step 0 to r:

$$\{\mathbf{Z}^r\} = \bigcup_{i=1}^{p} \mathbf{Z}_i^r \tag{23.27}$$

$$\text{where } \mathbf{Z}_i^r = \{z_i(1), z_i(2), \ldots z_i(r)\} \tag{23.28}$$

This suggests that to use all the available information for effectively fusing the data from multiple sensors, what is required is the global posterior distribution $F_x(x|\{\mathbf{Z}^r\})$, given the time-series information from each source. This can be accomplished in a variety of ways, the most common of which are summarized below [24].

The *Linear Opinion Pool* [42] aggregates probability distributions by linear combinations of the local posterior PDF information $F_x(x \mid \mathbf{Z}_i^r)$ (or appropriate likelihood functions, as per Equation 23.11):

$$F(x \mid \{\mathbf{Z}^r\}) = \sum_j w_j F(x \mid \mathbf{Z}_j^r) \tag{23.29}$$

where
   the weights w_j sum to unity
   each weight w_j represents a subjective measure of the reliability of the information from sensor j

The process can be illustrated as shown in Figure 23.6. Bayes' theorem can now be applied to Equation 23.29 to obtain a recursive form, which is omitted here for brevity. One of the shortcomings of the linear opinion pool method is its inability to reinforce opinion because the weights are usually unknown except in very specific applications.

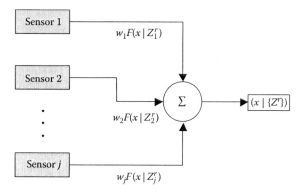

FIGURE 23.6 Multi-sensor data fusion by linear opinion pool.

The *Independent Opinion Pool* is a product form modification of the linear opinion pool and is defined by the product

$$F(\boldsymbol{x} \,|\, \{\boldsymbol{Z}^r\}) = \alpha \prod_j F(\boldsymbol{x} \,|\, \boldsymbol{Z}_j^r) \tag{23.30}$$

where α is a normalizing constant. The fusion process in this instance can be illustrated as shown in Figure 23.7.

This model is widely used since it represents the case when the observations from the individual sensors are essentially independent. However, this is also its weakness, since if the data is correlated at a group of nodes, their opinion is multiplicatively reinforced, which can lead to error propagation in faulty sensor networks. Nevertheless, this technique is appropriate when the prior state space distributions are truly independent and equally likely (as is common in digital communication applications).

To counter the weaknesses of the two common approaches summarized above, a third fusion rule is the *Likelihood Opinion Pool*, defined by the following recursive rule:

$$F(\boldsymbol{x} \,|\, \{\boldsymbol{Z}^r\}) = \alpha F(\boldsymbol{x} \,|\, \{\boldsymbol{Z}^{r-1}\}) \left[\prod_j \underbrace{F(\boldsymbol{z}_j(r) \,|\, \boldsymbol{x})}_{\text{likelihood}} \right] \tag{23.31}$$

The Likelihood Opinion Pool method of data fusion can be illustrated as shown in Figure 23.8. The likelihood opinion pool technique is essentially a Bayesian update process and is consistent with the

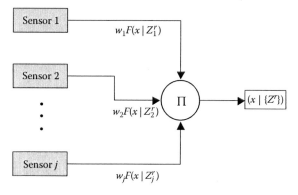

FIGURE 23.7 Multi-sensor data fusion by independent opinion pool.

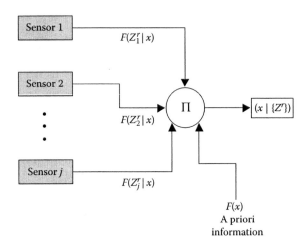

FIGURE 23.8 Multi-sensor data fusion by likelihood opinion pool.

recursive process derived in general in Equation 23.16. It is interesting to note that a simplified, specific form of this type of information processing occurs in the so called *belief propagation* [28] type of algorithms that are widespread in artificial intelligence and the decoding theory for channel codes. In the exposition above, however, the assumptions and derivations are and explicitly identified and derived, and is thus in a general form that is suitable for application to heterogeneous multi-sensor systems. This provides intuitive insight as to how the probabilistic updates help to reinforce "opinions" when performing a distributed state space search.

23.4.1 Information Theoretic Justification of the Bayesian Method

Probability distributions allow a quantitative description of the observables, the observer, and associated errors. As such, the likelihood functions and distributions contain information about the underlying states that they describe. This approach can be extended further to actually incorporate measures for the information contained in these random variables. In this manner, an information theoretic justification can be obtained for the Likelihood Opinion Pool for multi-sensor data fusion, as discussed in the previous section. Some key concepts from Information Theory [6] are required first.

23.4.1.1 Information Measures

The connections between information theory and distributed detection [44] were briefly surveyed in Section 23.2. In this section, some formal information measures are defined to enable an intuitive information theoretic justification of the utility of the Bayesian update method. This approach also provides an insight toward the practical design of algorithms based on the likelihood opinion pool fusion rules that has been discussed earlier.

To build an information theoretic foundation for data fusion, the most useful fundamental metric is the Shannon definition of *Entropy*.

Definition 23.4

Entropy is the uncertainty associated with a probability distribution, and is a measure of the descriptive complexity of a PDF [5]. Mathematically:

$$h\{F(\boldsymbol{x})\} \triangleq \mathrm{E}\{-\ln F(\boldsymbol{x})\} \tag{23.32}$$

Note that alternative definitions of the concept of information which predate Shannon's formulation, e.g., the *Fisher Information Matrix* [10], are also relevant and useful, but not discussed here further.

Using this definition, an expression for the entropy of the posterior distribution of x given Z^r at time r (which is the case of multiple observations from a single sensor) can be expressed as

$$h(r) \triangleq h\{F(x \mid Z^r)\} = -\sum F(x \mid Z^r) \ln F(x \mid Z^r) \tag{23.33}$$

Now, the entropy relationship for Bayes Theorem can be developed as follows:

$$E\{-\ln[F(x \mid Z^r)]\} = E\{-\ln[F(x \mid Z^{r-1})]\} - E\left\{\ln\left[\frac{F(z(r) \mid x)}{F(z(r) \mid Z^{r-1})}\right]\right\} \tag{23.34}$$

$$\Rightarrow h(r) = h(r-1) - E\left\{\ln\left[\frac{F(z(r) \mid x)}{F(z(r) \mid Z^{r-1})}\right]\right\} \tag{23.35}$$

This is an alternative form of the result that conditioning with respect to observations reduces entropy (cf. [6]). Using the definition of mutual information, Equation 23.34 can be written in an alternative form as shown below.

Definition 23.5

For an observation process, *mutual information* at time r is the information about x contained in the observation $z(r)$:

$$I(x, z(r)) \triangleq E\left\{\ln\left[\frac{F(z(r) \mid x)}{F(z(r))}\right]\right\} \tag{23.36}$$

from where

$$h(r) = h(r-1) - I(r) \tag{23.37}$$

which means that the entropy following an observation is reduced by an amount equal to the information inherent in the observation.

The insight to be gained here is that by using the definitions of entropy and mutual information, the recursive Bayes update procedure derived in Equation 23.16 can now be seen as an information update procedure:

$$E\{\ln[F(x \mid Z^r)]\} = E\{\ln[F(x \mid Z^{r-1})]\} + E\left\{\ln\left[\frac{F(z(r) \mid x)}{F(z(r) \mid Z^{r-1})}\right]\right\} \tag{23.38}$$

which can be interpreted as [24]

posterior information = prior information + observation information.

The information update equations for the Likelihood Opinion Pool fusion rule thus becomes

$$E\{\ln[F(\boldsymbol{x} \,|\, \boldsymbol{Z}^r)]\} = E\{\ln[F(\boldsymbol{x} \,|\, \boldsymbol{Z}^{r-1})]\} + \sum_{j} E\left\{\ln\left[\frac{F(z_j(r) \,|\, \boldsymbol{x})}{F(z_j(r) \,|\, \boldsymbol{Z}^{r-1})}\right]\right\} \tag{23.39}$$

The utility of the log-likelihood definition is that the information update steps reduce to simple additions, and are thus amenable to hardware implementation without such problems as overflow and dynamic range scaling.

Thus the Bayesian probabilistic approach is theoretically self-sufficient for providing a unified framework for data fusion in multi-sensor platforms. The information theoretic connection to the Bayesian update makes the approach intuitive, and shows rigorously how the Likelihood Opinion Pool method serves to reduce the ellipsoid uncertainty. This framework answers the question of how to weight or process outputs of diverse sensors, whether they have different sensing modes of signal to noise ratios, without resort to ad hoc criteria. Acoustic, visual, magnetic and other signals can all be combined [11]. Further, since tradeoffs in information rate and distortion can be treated using entropies (rate distortion theory [15]) as of course can communication, questions about fundamental limits in sensor networks can now perhaps be systematically explored.

Of course, obvious practical difficulties remain, such as how to determine the uncertainty in measurements, the entropy of sources, and in general how to efficiently convert sensor measurements into entropies.

23.5 Concluding Remarks

In this section, the approach of using a probabilistic, information processing approach to data fusion in multi-sensor networks was discussed. The Bayesian approach was seen to be the central unifying tool in formulating the key concepts and techniques for decentralized organization of information. Thus, it offers an attractive paradigm for implementation in a wide variety of systems and applications. Further, it allows one to use information theoretic justifications of the fusion algorithms, and also offers preliminary asymptotic analysis of large scale system performance.

The information theoretic formulation makes clear how to combine the outputs of possibly entirely different sensors. Moreover, it allows sensing, signal processing and communication to be viewed in one mathematical framework. This may allow systematic study of many problems involving the cooperative interplay of these elements. This further can lead to the computation of fundamental limits on performance against with practical reduced complexity techniques can be compared.

References

1. J. Balchen et al., Structural solution of highly redundant sensing in robotic systems, in *Highly Redundant Sensing in Robotic Systems*, Vol. 58, NATO Advanced Science Institutes Series, Berlin, Germany: Springer-Verlag, 1991.
2. T. Berger et al., Model distribution in decentralized multi-sensor fusion, in *Proceedings of the American Control Conference (ACC)*, 1991, Boston, MA, pp. 2291–2294.
3. R. E. Blahut, Hypothesis testing and information theory, *IEEE Transactions on Information Theory*, 20(4), 405–417, July 1974.
4. R. S. Blum and S. A. Kassam, On the asymptotic relative efficiency of distributed detection schemes, *IEEE Transactions on Information Theory*, 41(2), 523–527, March 1995.
5. D. Catlin, *Estimation, Control and the Discrete Kalman Filter*. New York: Springer-Verlag, 1989.

6. T. M. Cover and J. A. Thomas, *Elements of Information Theory*. Hoboken, NJ: Wiley-Interscience, 1991.

7. I. Csiszar and J. Korner, Towards a general theory of source networks, *IEEE Transactions on Information Theory*, IT-26, 155–165, 1980.

8. I. Csiszar and G. Longo, On the error exponent for source coding and for testing simple statistical hypotheses, *Studia Scientiarum Mathematicarum Hungarica*, 6, 181–191, 1971.

9. H. Durrant-Whyte, Sensor models and multi-sensor integration, *International Journal of Robotics*, 7(6), 97–113, 1988.

10. R. A. Fisher, On the mathematical foundations of theoretical statistics, *Philosophical Transactions of the Royal Society of London*, Sec A(222), 309–368, 1922.

11. J. W. Fisher III, M. Wainwright, E. Sudderth, and A. S. Willsky, Statistical and information-theoretic methods for self-organization and fusion of multimodal, networked sensors, *International Journal of High Performance Computing Applications*, 2001.

12. A. M. Flynn, Combining ultra-sonic and infra-red sensors for mobile robot navigation, *International Journal of Robotics Research*, 7(5), 5–14, 1988.

13. R. Frank, *Understanding Smart Sensors*. Norwood, MA: Artech House, 2000.

14. B. Fristedt and L. Gray, *A Modern Approach to Probability Theory*. Probability and Its Applications Series. Boston, MA: Birkhäuser, 1997.

15. R. G. Gallager, *Information Theory and Reliable Communications*. New York: John Wiley & Sons, 1968.

16. E. Gamal and T. M. Cover, Multiple user information theory, *Proceedings of the IEEE*, 68, 1466–1483, 1980.

17. T. Garvey et al., Model distribution in decentralized multi-sensor fusion, *Proceedings of the American Control Conference*, 1991, Boston, MA, pp. 2291–2294.

18. R. E. Kalman, A new approach to linear filtering and prediction problems, *Transactions of the ASME Journal of Basic Engineering*, 82(D), 34–35, 1969.

19. R. Kuc and M. Siegel, Physically based simulation model for acoustic sensor robot navigation, *IEEE Transactions on Pattern Analysis and Machine Intelligence*, 9(6), 766–778, 1987.

20. A. D. Kucar, Mobile radio—An overview, *IEEE Personal Communications Magazine*, 72–85, November 1991.

21. S. Kullback, *Information Theory and Statistics*. New York: John Wiley & Sons, 1959.

22. J. J. Leonard, Directed sonar sensing for mobile robot navigation, PhD dissertation, University of Oxford, Oxford, U.K., 1991.

23. R. Luo and M. Kay, Multi-sensor integration and fusion in intelligent systems, *IEEE Transactions on Systems Man and Cybernetics*, 19(5), 901–931, 1989.

24. J. Manyika and H. Durrant-Whyte, *Data Fusion and Sensor Management*, Ellis Horwood Series in Electrical and Electronic Engineering. West Sussex, England: Ellis Horwood, 1994.

25. D. Middleton, *Statistical Communication Theory*. New York: McGraw-Hill, 1960.

26. A. Mitchie and J. K. Aggarwal, Multiple sensor integration through image processing: A review, *Optical Engineering*, 23(2), 380–386, 1986.

27. Y. Nakamura, Geometric fusion: Minimizing uncertainty ellipsoid volumes, in *Data Fusion, Robotics and Machine Intelligence*. Boston, MA: Academic Press, 1992.

28. J. Pearl, *Probabilistic Reasoning in Intelligent Systems: Networks of Plausible Inference*. San Francisco, CA: Morgan Kaufmann, 1997.

29. H. V. Poor, *An Introduction to Signal Detection and Estimation*. New York: Springer-Verlag, 1988.

30. R. Popoli, The sensor management imperative, in *Multi-Target Multi-Sensor Tracking*, Y. Bar-Shalom, Ed., Norwood, MA: Artech House, 1992, pp. 325–392.

31. G. Pottie, Hierarchical information processing in distributed sensor networks, *IEEE International Symposium on Information Theory*, August 16–21, 1998, Cambridge, MA.

32. G. Pottie et al., Wireless sensor networks, *Information Theory Workshop Proceedings*, Killamey, Ireland, June 22–26, 1998.

33. J. G. Proakis, *Digital Communications*. New York: McGraw-Hill, 2000.

34. P. Rai-Choudhury, Ed., *MEMS and MOEMS Technology and Applications.* Bellingham, WA: Society of Photo-Optical Instrumentation Engineers, 2000.

35. G. G. Roussas, *A Course in Mathematical Statistics,* 2nd edn. Burlington, MA: Harcourt/Academic Press, 1997.

36. E. Royer and C.-K. Toh, A review of current routing protocols for ad hoc wireless networks, *IEEE Personal Communications Magazine,* 6(2), 46–55, April 1999.

37. J. T. Scheick, *Linear Algebra with Applications.* New York: McGraw-Hill, 1996.

38. C. E. Shannon, A mathematical theory of communication, *Bell Systems Technical Journal,* 27, 279–423, 1948.

39. S. Singh et al., Power-aware routing in mobile ad hoc networks, *Proceedings of the 4th Annual IEEE/ACM International Conference on Mobile Computing and Networking (MOBICOM),* 1998, Dallas, TX, pp. 181–190.

40. K. Sohrabi and G. Pottie, Performance of a self-organizing algorithm for wireless ad-hoc sensor networks, *IEEE Vehicular Technology Conference,* Fall 1999.

41. K. Sohrabi et al., Protocols for self-organization for a wireless sensor network, *IEEE Personal Communications Magazine,* 6–17, October 2000.

42. M. Stone, The opinion pool, *The Annals of Statistics,* 32, 1339–1342, 1961.

43. J. N. Tsitsiklis, On the complexity of decentralized decision-making and detection problems, *IEEE Transactions on Automatic Control,* 30(5), 440–446, 1985.

44. P. K. Varshney, *Distributed Detection and Data Fusion.* New York: Springer-Verlag, 1997.

45. S. Verdu, *Multiuser Detection.* New York: Cambridge University Press, 1998.

46. E. Waltz and J. Llinas, *Multi-Sensor Data Fusion.* Norwood, MA: Artech House, 1991.

47. K. Yao et al., Blind beamforming on a randomly distributed sensor array, *IEEE Journal on Selected Areas in Communications,* 16(8), 1555–1567, October 1998.

24
Multispectral Sensing

24.1 Motivation..541
24.2 Introduction to Multispectral Sensing.......................................542
Instruments for Multispectral Data Acquisition • Array and Super
Array Sensing • Multisensor Array Technology for Superresolution
24.3 Mathematical Model for Multisensor Array Based
Superresolution 544
Image Reconstruction Formulation • Other Approaches
to Superresolution
24.4 Color Images..548
24.5 Conclusions...549
References..550

N.K. Bose
*The Pennsylvania
State University*

24.1 Motivation

Sensing is ubiquitous in multitudinous applications that include biosensing, chemical sensing, surface acoustic wave sensing, sensing coupled with actuation in control, and imaging sensors. This article will be concerned primarily with multispectral sensing that spans acquisition, processing, and classification of data from multiple images of the same scene at several spectral regions. The topic of concern here finds usage in surveillance, health care, urban planning, ecological monitoring, geophysical exploration and agricultural assessment. The field of sensing has experienced a remarkable period of progress that necessitated the launching of an *IEEE journal* devoted exclusively to that topic in June 2001.

Sensors, photographic or nonphotographic, are required in data acquisition prior to processing and transmission. Light and other forms of electromagnetic (EM) radiation are commonly described in terms of their wavelengths (or frequencies) and the sensed data may be in different portions of the EM spectrum. Spectroscopy is the study of EM radiation as a function of wavelength that has been emitted, reflected or scattered from a solid, liquid, or gas. The complex interaction of light with matter involves reflection and refraction at boundaries of materials, a process called scattering, and absorption by the medium as light passes through the medium. Scattering makes reflectance spectroscopy possible. The amount of light scattered and absorbed by a grain is dependent on grain size. Reflectance spectroscopy can be used to map exposed minerals from aircrafts including detailed clay mineralogy. Visual and near-infrared spectroscopy, on the other hand, is insensitive to some minerals which do not have absorptions in this wavelength region. For a comprehensive survey, the reader is referred to Ref. [1].

Photographic film is limited for use in the region from near ultraviolet (wavelength range: 0.315–0.380 μm) to near infrared (IR; wavelength range: 0.780–3 μm). Electronic sensors like radar, scanners, photoconductive or tube sensors, and solid-state sensors like charge coupled device (CCD) arrays, though more complicated and bulkier than comparable photographic sensors, are usable over a wider frequency range, in diurnal as well as nocturnal conditions and are more impervious to fog, clouds, pollution and bad weather. IR sensors are indispensable under nocturnal and limited visibility conditions while electro-optic (EO) sensing systems using both absorption lidar and IR spectroscopy have been widely used in both active and passive sensing of industrial and atmospheric pollutants, detection of concealed explosives for airport security applications, detection of land mines, and weather monitoring through the sensing and tracking of vapor clouds. A 40 year review of the IR imaging system modeling activities of the U.S. Army Night Vision and Electronic Sensor Directorate (NVESD) is available in the inaugural issue of the *IEEE Sensors Journal* [2].

A vast majority of image sensors today are equipped with wavelength-sensitive optical filters that produce multispectral images which are characterized as locally-correlated but globally-independent random processes. For monochrome as well as color television, solid-state sensors are being increasingly preferred over photoconductive sensors because of greater compactness and well-defined structure in spite of the fact that solid-state sensors have lower signal-to-noise ratio (SNR) and lower spatial resolution. The disadvantages, due to physical constraints like the number of sensor elements that can be integrated on a chip, are presently being overcome through the development of an innovative technical device that gave birth to the superresolution imaging technology.

24.2 Introduction to Multispectral Sensing

A significant advance in sensor technology stemmed from the subdividing of spectral ranges of radiation into bands. This allowed sensors in several bands to form multispectral images [3]. From the time that Landsat 1 was launched in 1972, multispectral sensing has found diverse uses in terrain mapping, agriculture, material identification and surveillance. Typically, multispectral sensors collect several separate spectral bands with spectral regions selected to highlight particular spectral characteristics. The number of bands range from one (panchromatic sensors) to, progressively, tens, hundreds and thousands of narrow adjacent bands in the case of multispectral, hyperspectral and ultraspectral sensing. The *spectral resolution* of a system for remote sensing (RS) depends on the number and widths (spectral bandwidths) of the collected spectral bands. The next subsection differentiates between different types of resolution.

Reflectance is the percentage of incident light that is reflected by a material and the reflectance spectrum shows the reflectance of a material across a range of wavelengths that can often permit unique identification of the material. Many terrestrial minerals have very unique spectral signatures, like human fingerprints, due to the uniqueness of their crystal geometries. Multispectral sensing is in multiple, separated and narrow wavelength bands. Hyperspectral sensors, on the other hand, operate over wider contiguous bands. Multispectral sensors can usually be of help in detecting, classifying and, possibly, distinguishing between materials but hyperspectral sensors may be required to actually characterize and identify the materials. Ultraspectral is beyond hyperspectral with a goal of accommodating, ultimately, millions of very narrow bands for a truly high resolution spectrometer that may be capable of quantifying and predicting. The need for ultraspectral sensing and imaging is because of the current thrust in chemical, biological and nuclear warfare monitoring, quantification of ecological pollutants, gaseous emission and nuclear storage monitoring and improved crop assessments though weed identification and prevention.

24.2.1 Instruments for Multispectral Data Acquisition

In RS, often the input imagery is obtained from satellite sensors like Landsat multispectral scanners (MSS) or airborne scanners, synthetic aperture radars (for obtaining high-resolution imagery at microwave frequencies), IR photographic film, image tubes, and optical scanners (for IR images) and electro-optical line scanners in addition to the commonly used photographic and television devices (for capturing the visible spectrum) [4]. Multispectral instruments image the earth in a few strategic areas of the EM spectrum, omitting entire wavelength sections. *Spatial resolution* is the smallest ground area that can be discerned in an image. In Landsat images (non-thermal bands), the spatial resolution is about 28.5 m × 28.5 m. The smallest discernible area on the ground is called the *resolution cell* and determines the sensor's maximum resolution. For a homogeneous feature to be detected, its size, generally, has to be equal to or larger than the resolution cell. *Spectral resolution* is the smallest band or portion of the EM spectrum in which objects are discernible. This resolution defines the ability of a sensor to define wavelength intervals. *Temporal resolution* is the shortest period of time in which a satellite will revisit a spot on the earth's surface. Landsat 5, for example, has a temporal resolution of 16 days. *Radiometric resolution* is the smallest size of a band or portion of the EM spectrum in which the reflectance of a feature may be assigned a digital number, i.e. the finest distinction that can be made between objects in the same part of the EM spectrum. It describes the imaging system's ability to discriminate between very slight differences in energy. Greater than 12 bits and less than 6 bits correspond, respectively, to very high and low radiometric resolutions. Continuous improvements in spatial, spectral, radiometric, and temporal resolution coupled with decreasing cost are making RS techniques very popular.

A scanning system used to collect data over a variety of different wavelengths is called a MSS. MSS systems have several advantages over conventional aerial photographic systems including the following:

- The ability to capture data from a wider portion of the EM spectrum (about $0.3–14\,\mu m$)
- The ability to collect data from multiple spectral bands simultaneously
- The data collected by MSS systems can be transmitted to earth to avoid storage problems
- The data collected by MSS systems are easier to calibrate and rectify

24.2.2 Array and Super Array Sensing

Acquisition of multivariate information from the environment often require extensive use of sensing arrays. A localized miniature sensing array can significantly improve the sensing performance and deployment of a large number of such arrays as a distributed sensing network (super array) will be required to obtain high quality information from the environment. Development of super arrays is a trend in many fields like the chemical field, where the environment could be gaseous. Such arrays could provide higher selectivity, lower thresholds of detection, broader dynamic range and long term baseline stability. Array sensors have been used in a variety of applications that are not of direct interest here. It suffices to single out two such potential areas. An array of plasma-deposited organic film-coated quartz crystal resonators has been studied for use in indoor air-monitoring in cabin of aircraft, automobile, train or a clean room [5]. Multiple sensors are also capable of carrying out remote sewer inspection tasks, where close circuit television based platforms are less effective for detecting a large proportion of all possible damages because of the low quality of the acquired images [6]. Multisensors, further discussed below, and the superresolution technology, discussed in subsequent sections, are therefore very powerful tools for solving challenging problems in military, civil and health care problems.

24.2.3 Multisensor Array Technology for Superresolution

Multiple undersampled images of a scene are often obtained by using a CCD detector array of sensors which are shifted relative to each other by subpixel displacements. This geometry of sensors, where each sensor has a subarray of sensing elements of suitable size has recently been popular in the task of attaining spatial resolution enhancement from the acquired low-resolution degraded images that comprise the set of observations. The multisensor array technology is particularly suited to microelectromechanical systems (MEMS) applications where accuracy, reliability, and low transducer failure rates are essential in applications spanning chronic implantable sensors, monitoring of semiconductor processes, mass-flow sensors, optical cross-connect switches, pressure and temperature sensors. The benefits include application to any sensor array or cluster, reduced calibration and periodic maintenance costs, higher confidence in sensor measurements based on statistical average on multiple sensors, extended life of the array compared to a single-sensor system, improved fault tolerance, lower failure rates, and low measurement drift. Due to hardware cost, size, and fabrication complexity limitations, imaging systems like CCD detector arrays often provide only multiple low-resolution degraded images. However, a high-resolution image is indispensable in applications including health diagnosis and monitoring, military surveillance, and terrain mapping by RS. Other intriguing possibilities include substituting expensive high resolution instruments like scanning electron microscopes by their cruder, cheaper counterparts and then applying technical methods for increasing the resolution to that derivable with much more costly equipment. Small perturbations around the ideal subpixel locations of the sensing elements (responsible for capturing the sequence of undersampled degraded frames), because of imperfections in fabrication, limit the performance of the signal-processing algorithms for processing and integrating the acquired images for the desired enhanced resolution and quality. Resolution improvement by applying tools from digital signal processing technique has, therefore, been a topic of very great interest.

24.3 Mathematical Model for Multisensor Array Based Superresolution

A very fertile arena for applications of some of the developed theory of multidimensional systems has been spatio-temporal processing following image acquisition by, say a single camera, multiple cameras or an array of sensors. An image acquisition system composed of an array of sensors, where each sensor has a subarray of sensing elements of suitable size, has recently been popular for increasing the spatial resolution with high SNR beyond the performance bound of technologies that constrain the manufacture of imaging devices.

A brief introduction to a mathematical model in high-resolution image reconstruction is provided first. Details can be found in Ref. [7]. Consider a sensor array with $L_1 \times L_2$ sensors in which each sensor has $N_1 \times N_2$ sensing elements (pixels) and the size of each sensing element is $T_1 \times T_2$. The goal is to reconstruct an image of resolution $M_1 \times M_2$, where $M_1 = L_1 N_1$ and $M_2 = L_2 N_2$. To maintain the aspect ratio of the reconstructed image the case where $L_1 = L_2 = L$ is considered. For simplicity, L is assumed to be an even positive integer in the following discussion.

To generate enough information to resolve the high-resolution image, subpixel displacements between sensors are necessary. In the ideal case, the sensors are shifted from each other by a value proportional to $T_1/L \times T_2/L$. However, in practice there can be small perturbations around these ideal subpixel locations due to imperfections of the mechanical imaging system during fabrication. Thus, for $l_1, l_2 = 0, 1, \ldots, L-1$ with $(l_1, l_2) \neq (0, 0)$, the horizontal and vertical displacements $d_{l_1 l_2}^x$ and $d_{l_1 l_2}^y$, respectively, of the $[l_1, l_2]$-th sensor with respect to the $[0, 0]$-th reference sensor are given by

$$d_{l_1 l_2}^x = \frac{T_1}{L}(l_1 + \overline{\epsilon}_{l_1 l_2}^x) \quad \text{and} \quad d_{l_1 l_2}^y = \frac{T_2}{L}(l_2 + \overline{\epsilon}_{l_1 l_2}^y),$$

where $\overline{\epsilon}_{l_1 l_2}^x$ and $\overline{\epsilon}_{l_1 l_2}^y$ denote, respectively, the actual normalized horizontal and vertical displacement errors. The estimates, $\epsilon_{l_1 l_2}^x$ and $\epsilon_{l_1 l_2}^y$, of these parameters, $\overline{\epsilon}_{l_1 l_2}^x$ and $\overline{\epsilon}_{l_1 l_2}^y$ can be obtained by manufacturers during camera calibration. It is reasonable to assume that

$$|\overline{\epsilon}_{l_1 l_2}^x| < \frac{1}{2} \quad \text{and} \quad |\overline{\epsilon}_{l_1 l_2}^y| < \frac{1}{2},$$

because if that is not the case, then the low-resolution images acquired from two different sensors may have more than the desirable overlapping information for reconstructing satisfactorily the high-resolution image [7].

Let $f(x_1, x_2)$ denote the original bandlimited high-resolution scene, as a function of the continuous spatial variables, x_1, x_2. Then the observed low-resolution digital image $\overline{g}_{l_1 l_2}$ acquired from the (l_1, l_2)-th sensor, characterized by a point-spread function, is modeled by

$$\overline{g}l_1 l_2[n_1, n_2] = \int_{T_2\left(n_2 - \frac{1}{2}\right) + d^y_{l_1 l_2}}^{T_2\left(n_2 + \frac{1}{2}\right) + d^y_{l_1 l_2}} \int_{T_1\left(n_1 - \frac{1}{2}\right) + d^x_{l_1 l_2}}^{T_1\left(n_1 + \frac{1}{2}\right) + d^x_{l_1 l_2}} f(x_1, x_2) dx_1 dx_2, \tag{24.1}$$

for $n_1 = 1, \ldots, N_1$ and $n_2 = 1, \ldots, N_2$. These low-resolution images are combined to yield the $M_1 \times M_2$ high-resolution image $\overline{g}$ by assigning its pixel values according to

$$\overline{g}[L(n_1 - 1) + l_1, L(n_2 - 1) + l_2] = \overline{g}_{l_1 l_2}[n_1, n_2], \tag{24.2}$$

for $l_1, l_2 = 0, 1, \ldots, (L-1)$, $n_1 = 1, \ldots, N_1$, and $n_2 = 1, \ldots, N_2$.

The continuous image model $f(x_1, x_2)$ in (24.1) can be discretized by the rectangular rule and approximated by a discrete image model. Let **g** and **f** be, respectively, the vectors formed from discretization of $g(x_1, x_2)$ and $f(x_1, x_2)$ using a column ordering. The Neumann boundary condition [8] is applied on the images. This assumes that the scene immediately outside is a reflection of the original scene at the boundary, i.e.,

$$f(i, j) = f(k, l) \quad \text{where} \quad \begin{cases} k = 1 - i, & i < 1, \\ k = 2M_1 + 1 - i, & i > M_1, \\ l = 1 - j, & j < 1, \\ l = 2M_2 + 1 - j, & j > M_2. \end{cases}$$

Under the Neumann boundary condition, the blurring matrices are banded matrices with bandwidth $L + 1$, but the entries at the upper left part and the lower right part of the matrices are changed. The resulting matrices, denoted by $\mathbf{H}^x_{l_1 l_2}(\overline{\epsilon}^x_{l_1, l_2})$ and $\mathbf{H}^y_{l_1 l_2}(\overline{\epsilon}^y_{l_1, l_2})$, each has a Toeplitz-plus-Hankel structure as shown in Equation 24.4. The blurring matrix corresponding to the (l_1, l_2)-th sensor under the Neumann boundary condition is given by the Kronecker product,

$$\mathbf{H}_{l_1 l_2}(\overline{\epsilon}_{l_1, l_2}) = \mathbf{H}^x_{l_1 l_2}(\overline{\epsilon}^x_{l_1, l_2}) \otimes \mathbf{H}^y_{l_1 l_2}(\overline{\epsilon}^y_{l_1, l_2}),$$

where the 2×1 vector $(\overline{\epsilon}_{l_1, l_2})$ is denoted by $(\overline{\epsilon}^x_{l_1, l_2}, \overline{\epsilon}^y_{l_1, l_2})^t$. The blurring matrix for the whole sensor array is made up of blurring matrices from each sensor:

$$\mathbf{H}_L(\overline{\epsilon}) = \sum_{l_1 = 0}^{L-1} \sum_{l_2 = 0}^{L-1} \mathbf{D}_{l_1 l_2} \mathbf{H}_{l_1 l_2}(\overline{\epsilon}_{l_1, l_2}), \tag{24.3}$$

where the $2L^2 \times 1$ vector $\overline{\epsilon}$ is defined as $\overline{\epsilon} = [\overline{\epsilon}^x_{00} \overline{\epsilon}^y_{00} \overline{\epsilon}^x_{01} \overline{\epsilon}^y_{01} \cdots \overline{\epsilon}^x_{L-1L-2} \overline{\epsilon}^y_{L-1L-2} \overline{\epsilon}^x_{L-1L-1} \overline{\epsilon}^y_{L-1L-1}]^t$. Here $\mathbf{D}_{l_1 l_2}$ are diagonal matrices with diagonal elements equal to 1 if the corresponding component of **g** comes from the (l_1, l_2)-th sensor and zero otherwise. The Toeplitz-plus-Hankel matrix $\mathbf{H}^x_{l_1 l_2}(\overline{\epsilon}^x_{l_1, l_2})$, referred to above is explicitly written next.

$$
\mathbf{H}^x_{l_1 l_2}(\overline{\in}^x_{l_1,l_2}) = \frac{1}{L}
\begin{pmatrix}
1 & \cdots & 1 & \frac{1}{2}-\overline{\in}^x_{l_1,l_2} & & 0 \\
\vdots & \ddots & \ddots & & \ddots & \\
1 & & \ddots & \ddots & \ddots & \frac{1}{2}-\overline{\in}^x_{l_1,l_2} \\
\frac{1}{2}+\overline{\in}^x_{l_1,l_2} & \ddots & \ddots & \ddots & \ddots & 1 \\
& \ddots & \ddots & \ddots & \ddots & \vdots \\
0 & & \frac{1}{2}+\overline{\in}^x_{l_1,l_2} & 1 & \cdots & 1
\end{pmatrix}
$$

$$
+ \frac{1}{L}
\begin{pmatrix}
1 & \cdots & 1 & \frac{1}{2}+\overline{\in}^x_{l_1,l_2} & & 0 \\
\vdots & \ddots & \ddots & & & \\
1 & & \ddots & & & \frac{1}{2}-\overline{\in}^x_{l_1,l_2} \\
\frac{1}{2}+\overline{\in}^x_{l_1,l_2} & & & \ddots & & 1 \\
& & & \ddots & \ddots & \vdots \\
0 & & \frac{1}{2}-\overline{\in}^x_{l_1,l_2} & 1 & \cdots & 1
\end{pmatrix}
\tag{24.4}
$$

The matrix $\mathbf{H}^y_{l_1 l_2}(\overline{\in}^y_{l_1,l_2})$ is defined similarly. See [7] for more details.

24.3.1 Image Reconstruction Formulation

CCD image sensor arrays, where each sensor consists of a rectangular subarray of sensing elements, produce discrete images whose sampling rate and resolution are determined by the physical size of the sensing elements. If multiple CCD image sensor arrays are shifted relative to each other by exact sub-pixel values, the reconstruction of high-resolution images can be modeled by

$$
\bar{g} = Hf \quad \text{and} \quad g = \bar{g} + \eta,
\tag{24.5}
$$

where
f is the desired high-resolution image
H is the blur operator
$\bar{g}$ is the output high-resolution image formed from low-resolution frames
η is the additive Gaussian noise

However, as perfect subpixel displacements are practically impossible to realize, blur operators in multisensor high-resolution image reconstruction are space-variant.

Since the system described in (24.5) is ill-conditioned, solution for f is constructed by applying the maximum a posteriori (MAP) regularization technique that involves a functional $R(f)$, which measures the regularity of f, and a regularization parameter α that controls the degree of regularity of the solution to the minimization problem

$$
\min_f \{\| Hf - g \|^2_2 + \alpha R(f)\}.
\tag{24.6}
$$

The boundary values of g are not completely determined by the original image f inside the scene, because of the blurring process. They are also affected by the values of f outside the scene. Therefore, when solving for *f* from (24.5), one needs some assumptions on the values of *f* outside the scene, referred to as boundary conditions. In Ref. [7], Bose and Boo imposed zero boundary condition outside the scene. Ng and Yip [8] recently showed that the model with the Neumann boundary condition gives better reconstructed high-resolution image than obtainable with the zero boundary condition. In the case of Neumann boundary conditions, discrete cosine transform (DCT) based preconditioners have been effective in the high-resolution reconstruction problem [8]. In Ref. [9], an analysis and proof has been given of convergence of the iterative method deployed to solve the transform based preconditioned system. The proof offered of linear convergence of the conjugate gradient method on the displacement errors caused from the imperfect locations of subpixels in the sensor array fabrication process has been also substantiated by results from simulation.

The observed signal vector $\bar{g}$ is, as seen above, subject to errors. It is assumed that the actual signal $g = [g_1, ..., g_{M_1 M_1}]^t$ can be represented by

$$\bar{g} = g + \delta g, \tag{24.7}$$

where

$$\delta_g = [\delta g_1 + \delta g_2 ..., \delta g_{M_1 M_2}]^t$$

and δg_i's are independent identically distributed noise with zero-mean and variance σ_g^2.

Thus the image reconstruction problem is to recover the vector *g* from the given inexact point spread function $h_{l_1 l_2}$ ($l_1 = 0,1, ..., L_1 - 1, l_2 = 0,1, ..., L_2 - 1$) and a observed and noisy signal $\bar{g}$. A constrained total least squares (TLS) approach to solving the image reconstruction problem has been advanced in Ref. [10].

24.3.2 Other Approaches to Superresolution

Multiple undersampled images of a scene are often obtained by using multiple identical image sensors which are shifted relative to each other by subpixel displacements [11,12]. The resulting high-resolution image reconstruction problem using a set of currently-available image sensors is interesting because it is closely related to the design of high-definition television (HDTV) and very high-definition (VHD) image sensors.

Limitations of image sensing lead to the formation of sequences of undersampled, blurred, and noisy images. High-resolution image reconstruction algorithms, which increase the effective sampling rate and bandwidth of observed low-resolution degraded images usually accompany a series of processing tasks such as subpixel motion estimation, interpolation and image restoration for tasks in surveillance, medical and commercial applications. Since 1990, when a method based on the recursive least squares estimation algorithm in the wavenumber domain was proposed to implement simultaneously the tasks of interpolation and filtering of a still image sequence [13], considerable progress has been made. First, the TLS recursive algorithm was developed to generalize the results in Ref. [13] to the situation often encountered in practice, when not only the observation is noise-corrupted but also the data [14]. The latter scenario originates from the inaccuracies in estimation of the displacements between frames. Second, it was shown that four image sensors are often sufficient from the standpoint of the human visual system to satisfactorily deconvolve moderately-degraded multispectral images [15]. Third, it was shown how a three-dimensional (3-D) linear minimum mean squares error (LMMSE) estimator for a sequence of time-varying images can be decorrelated into a set of 2-D LMMSE equations, which can subsequently be solved by approximating the Karhunen Loeve transform (KLT) by other transforms like the Hadamard transform or the DCT [16]. Fourth, as already discussed, the mathematical model of shifted undersampled images with subpixel displacement errors was derived in the presence of blur and noise, and the MAP formulation was adapted for fast high resolution reconstruction in the presence of subpixel displacement errors [7]. For a discursive documentation of the image acquisition system composed of the array of sensors followed by the iterative methods for high resolution reconstruction and scopes for further research, see Refs. [9,17].

A different approach towards superresolution from that in Ref. [13] was suggested in 1991 by Irani and Peleg [18], who used a rigid model instead of a translational model in the image registration process and then applied the iterative back-projection technique from computer-aided tomography. A summary of these and other research during the last decade is contained in a recent paper [19]. Mann and Picard [20] proposed the projective model in image registration because their images were acquired with a video camera. The projective model was subsequently used by Lertrattanapanich and Bose [21] for videomosaicing and high-resolution. Very recently, an approach towards superresolution using spatial tessellations has been presented in Ref. [22]. Analysis from the wavelet point of view of the construction of a high resolution (HR) image from low resolution (LR) images acquired through a multi-sensor array in the approach of Bose and Boo [7] was recently conducted in Ref. [23]. Absence of displacement errors in the LR samples was assumed and this resulted in a spatially invariant blurring operator. The algorithms developed decomposed the function from the previous iteration into different wavenumber components in the wavelet transform domain and, subsequently, added them into the new iterate to improve the approximation. Extension of the approach when some of the LR images are missing, possibly due to sensor failure was also implemented in Ref. [23]. The wavelet approach towards HR image formation was generalized to the case of spatially varying blur associated with the presence of subpixel displacement errors due to improper alignment of the sensors [24].

24.4 Color Images

Multispectral restoration of a single image is a 3-D reconstruction problem, where the third axis incorporates different wavelengths. We are interested in color images because there are many applications. Color plays an important role in pattern recognition and digital multimedia, where color based-features and color segmentation has proven pertinent in detecting and classifying objects in satellite and general purpose imagery. In particular, the fusion of color and edge based features has improved the performance of image segmentation and object recognition.

Color image can be regarded as a set of three images in their primary color channels (red, green, and blue). Monochrome processing algorithms applied to each channel independently are not optimal because they fail to incorporate the spectral correlation between the channels. Under the assumption that the spatial intrachannel and spectral interchannel correlation functions are product-separable, Hunt and Kübler [25] showed that a multispectral (e.g. color) image can be decorrelated by the Karhunen-Loeve transform (KLT). After decorrelating multispectral images, the Wiener filter can be applied independently to each channel, and the inverse KLT gives the restored color image. In the literature, Galatsanos and Chin [26] proposed and developed the 3-D Wiener filter for processing multispectral image. The 3-D Wiener filter is implemented by using the two-dimensional (2-D) block–circulant–circulant–block (BCCB) approximation to a block–Toeplitz–Toeplitz–block (BTTB) matrix [27]. Moreover, Tekalp and Pavlovic [28] considered to use the 3-D Kalman filtering to the multispectral image restoration problem.

The visual quality can always be improved by using multiple sensors with distinct transfer characteristics [29]. Boo and Bose [15] developed a procedure to restore a single color image, which has been degraded by a linear shift-invariant blur in the presence of additive noise. Only four sensors, namely, red (R), green (G), blur (B), and luminance (Y), are used in the color image restoration problem. In the NTSC YIQ representation, the restoration of the Y component is critical because this component contains 85%–95% of the total energy and has a large bandwidth. Two observed luminance images are used to restore the Y component. In their method, the 3-D Wiener filter on a sequence of these two luminance component images and two 2-D Wiener filters on each of the chrominance component images are considered. In Ref. [15], circulant approximations are used in the 2-D or 3-D Wiener filters and therefore the computational cost can be reduced significantly. The resulting well-conditioned problem is shown to provide improved restoration over the decorrelated component and the independent channel restoration methods, each of which uses one sensor for each of the three primary color components.

In Ref. [30], Ng and Boo formulate the color image restoration problem with only four sensors (R, G, B, Y) by using NTSC YIQ decorrelated component method and use the Neumann boundary condition, i.e., the data outside the domain of consideration is a reflection of the data inside, in the color image restoration process. Boo and Bose [15] used the traditional choice of imposing the periodic boundary condition outside the scene, i.e. data outside the domain of consideration are exact copies of data inside. The most important advantage of using periodic boundary condition is that circulant approximations can be used and therefore the fast Fourier transforms can be employed in the computations. Note that when this assumption is not satisfied by the images, ringing effects will occur at the boundary of the restored images, see for instance [15, Figure 6]. The Neumann image model gives better restored color images than that under the periodic boundary condition.

Besides the issue of boundary conditions, it is well-known that the color image restoration problem is very ill-conditioned, and restoration algorithms will be extremely sensitive to noise. In Ref. [30], the regularized least squares filters [31,32] are used to alleviate the restoration problem. It is shown that the resulting regularized least squares problem can be solved efficiently by using DCTs. In the regularized least squares formulation, regularization parameters are introduced to control degree of bias of the solution. The generalized cross-validation function is also used to obtain estimates of these regularization parameters, and then to restore high-quality color images. Numerical examples are given in Ref. [30] to illustrate the effectiveness of the proposed methods over other restoration methods.

In Ref. [33], Ng et al. extended the high-resolution image reconstruction method to multiple undersampled color images. The key issue is to employ the cross-channel regularization matrix to capture the changes of reflectivity across the channels.

24.5 Conclusions

With the increasing need for higher resolution multispectral imagery (MSI) in military as well as civilian applications, it is felt that the needs of the future can be best addressed by combining the task of deployment of systems with larger collection capability with the technical developments during the last decade in the area of superresolution, briefly summarized in this chapter. The need for model accuracy is undeniable in the attainment of superresolution along with the design of the algorithm whose robust implementation will produce the desired quality in the presence of model parameter uncertainty. Since the large volume of collected multispectral data might have to be transmitted prior to the deployment of superresolution algorithms it is important to attend to schemes for multispectral data compression. Fortunately, the need for attention to MSI compression technology was anticipated about a decade back [34] and the suitability of second generation wavelets (over and above the standard first generation wavelets) for multispectral (and, possibly hyperspectral and ultraspectral) image coding remains to be investigated.

In Ref. [30], the optimization problem from regularized least squares is formulated from the red, green, and blue components (from the RGB sensor). Space-invariant regularization based on GCV was used. There is considerable scope for incorporating recent regularization methods (like space-variant regularization [35]) in this approach for further improvement in quality of restoration both in the case of a single image as well as multispectral video sequences.

In Ref. [36], a *RSTLS* (Regularized Structured Total Least Squares) algorithm was proposed to perform the image restoration and estimations of the subpixel displacement errors at the same time unlike the alternate minimization algorithm in Ref. [10]. The work in Ref. [36] is based on the use of space-invariant regularization. With the oversimplification of the nature of real image, however, space-invariant algorithms produce unwanted effects such as smoothing of sharp edges, "ringing" in the vicinity of edges and noise enhancement in smooth areas of the image [37]. To overcome these unwanted artifacts, many type of space-variant image restoration algorithms have been proposed. In Ref. [35], an iterative image restoration technique using space-variant regularization was reported. In Ref. [37], multichannel restoration of single channel images using wavelet-based subband decomposition was proposed. Also, image restoration using subband or wavelet-based approach have been reported [38], and more recently in Refs. [23,24].

In surveillance systems of tomorrow, signals generated by multiple sensors need to be processed, transmitted and presented at multiple levels in order to capture the different aspects of the monitored environment. Multiple level representations to exploit perception augmentation for humans interacting with such systems are also needed in civilian applications to facilitate infrastructure development and urban planning, especially because of the perceived gap in the knowledge base that planners representing different modes of transportation, vegetation, etc. have of each other's constraints or requirements. If both panchromatic and multispectral (or hyperspectral, ultraspectral) images are available, improved automated fusion strategies are needed for the fused image to display sharp features from the panchromatic image while preserving the spectral attributes like color from the multispectral, hyperspectral and ultraspectral image. Situation awareness techniques that utilize multisensor inputs can provide enhanced indexing capabilities needed for focusing human or robot, fixed or mobile, attention on information of interest. Multiterminal mobile and cooperative alarm detection in surveillance, industrial pollution monitoring, chemical and biological weapon sensing is another emerging problem where multisensor signal/video data acquisition, compression, transmission and processing approaches become more and more relevant.

References

1. Clark, R.N., Spectroscopy of rocks and minerals, and principles of spectroscopy, in *Manual of Remote Sensing, Vol. 3, Remote Sensing for the Earth Sciences*, Rencz, A.N., Ed., John Wiley & Sons, New York, 1999, Chapter 1.
2. Ratches, J.A., Vollmerhausen, R.H., and Diggers, R.G., Target acquisition performance modeling of infrared imaging systems, *IEEE Sensors Journal*, 1, 31, June 2001.
3. Landgrebe, D.A., *Signal Theory Methods in Multispectral Remote Sensing*, John Wiley, Hoboken, NJ, 2003.
4. Hord, R.M., *Digital Image Processing of Remotely Sensed Data*, Academic Press, New York, 1982.
5. Seyama, M., Sugimoto, I., and Miyagi, T., Application of an array sensor based on plasma-deposited organic film coated quartz crystal resonators to monitoring indoor volatile compounds, *IEEE Sensors Journal*, 1, 422, October 2001.
6. Duran, O., Althoefer, K., and Seneviratne, L.D., State of the art in sensor technologies for sewer inspection, *IEEE Sensors Journal*, 2, 73, April 2002.
7. Bose, N.K. and Boo, K.J., High-resolution image-reconstruction with multisensors, *International Journal of Imaging Systems and Technology*, 9, 294, 1998.
8. Ng, M. and Andy, A., A fast MAP algorithm for high-resolution image reconstruction with multisensors, *Multidimensional Systems and Signal Processing*, 12(2), 143, 2001.
9. Ng, M. and Bose, N.K., Analysis of displacement errors in high-resolution image reconstruction with multisensors, *IEEE Transactions on Circuits and Systems, Part I*, 49, 806, 2002.
10. Ng, M., Bose, N.K., and Koo, J., Constrained total least squares computations for high resolution image reconstruction with multisensors, *International Journal on Imaging Systems and Technology*, 12, 35–42, 2002.
11. Komatsu, T. et al., Signal-processing based method for acquiring very high resolution images with multiple cameras and its theoretical analysis, *Proceedings of the IEEE, Part I*, 140(3), 19, 1993.
12. Jacquemod, G., Odet, C., and Goutte, R., Image resolution enhancement using subpixel camera displacement, *Signal Processing*, 26, 139, 1992.
13. Kim, S.P., Bose, N.K., and Valenzuela, H.M., Recursive reconstruction of high-resolution image from noisy undersampled multiframes, *IEEE Transactions on Acoustics, Speech and Signal Processing*, 38(6), 1013, June 1990.

14. Bose, N.K., Kim, H.C., and Valenzuela, H.M., Recursive total least squares algorithm for image reconstruction from noisy undersampled frames, *Multidimensional Systems and Signal Processing*, 4(3), 253, July 1993.

15. Boo, K.J. and Bose, N.K., Multispectral image restoration with multisensors, *IEEE Transactions on Geoscience and Remote Sensing*, 35(5), 1160, September 1997.

16. Boo, K.J. and Bose, N.K., A motion-compensated spatio-temporal filter for image sequences with signal-dependent noise, *IEEE Transactions on Circuits and Systems for Video Technology*, 8(3), 287, June 1998. (Reduced ver. appeared in *Proceedings of Workshop on Foundations of Information/ Decision Fusion with Applications to Engineering Problems*, Acadiana Printing Inc., Washington DC, August 8, 1996.)

17. Ng, M. and Bose, N.K., Mathematical analysis of super-resolution methodology, *IEEE Signal Processing Magazine*, 20(3), 62, May 2003.

18. Irani, M. and Peleg, S., Improving resolution by image registration, *CVGIP: Graphical Models and Image Processing*, 53, 231, 1991.

19. Elad, M. and Hel-Or, Y., A fast superresolution reconstruction algorithm for pure translational motion and common space-invariant blur, *IEEE Transactions on Image Processing*, 10, 1187, August 2001.

20. Mann, S. and Picard, R.W., Video orbits of the projective group: A simple approach to featureless estimation of parameters, *IEEE Transactions on Image Processing*, 6, 1281, September 1997.

21. Lertrattanapanich, S. and Bose, N.K., Latest results on high-resolution reconstruction from video sequences, Technical Report of IEICE, DSP 99-140, The Institute of Electronic, Information and Communication Engineers, Japan, p. 59, December 1999.

22. Lertrattanapanich, S. and Bose, N.K., High resolution image formation from low resolution frames using Delaunay triangulation, *IEEE Transactions on Image Processing*, 17, 1427, December 2002.

23. Chan, R.F. et al., Wavelet algorithms for high resolution image reconstruction, *SIAM Journal on Scientific Computing*, 24, 1408, 2003.

24. Chan, R.F. et al., Wavelet deblurring algorithms for spatially varying blur from high resolution image reconstruction, *Linear Algebra and its Applications*, 366, 139, 2003.

25. Hunt, B. and Kübler, O., Karhunen-Loeve multispectral image restoration, part I: Theory, *IEEE Transactions on Acoustics, Speech, and Signal Processing*, 32, 592, 1984.

26. Galatsanos, N. and Chin, R., Digital restoration of multichannel images, *IEEE Transactions on Acoustics, Speech, and Signal Processing*, 37, 415, 1989.

27. Bose, N.K. and Boo, K.J., Asymptotic eigenvalue distribution of block-Toeplitz matrices, *IEEE Transactions on Information Theory*, 44(2), 858, March 1998.

28. Tekalp, A. and Pavlovic, C., Multichannel image modeling and Kalman filtering for multispectral image restoration, *Signal Processing*, 19, 221, 1990.

29. Berenstein, C. and Patrick, E., Exact deconvolution for multiple convolution operators—An overview, plus performance characterization for imaging sensors, *Proceedings of IEEE*, 78, 723, 1990.

30. Ng, M. and Bose, N.K., Fast color image restoration with multisensors, *International Journal on Imaging Systems and Technology*, 12(5), 189, 2003.

31. Galatsanos, N. et al., Least squares restoration of multichannel images, *IEEE Transactions on Signal Processing*, 39, 2222, 1991.

32. Ng, M. and Kwan, W., Comments on least squares restoration of multichannel images, *IEEE Transactions on Signal Processing*, 49, 2885, 2001.

33. Ng, M., Bose, N.K., and Koo, J., Constrained total least squares for color image reconstruction, in *Total Least Squares and Errors-in-Variables Modelling III: Analysis, Algorithms and Applications*, Huffel, S., and Lemmerling, P., Eds., Kluwer Academic Publishers, Dordrecht, the Netherlands, 2002, p. 365.

34. Vaughan, V.D. and Atkinson, T.S., System considerations for multispectral image compression designs, *IEEE Signal Processing Magazine*, 12(1), 19, January 1995.

35. Reeves, S.J. and Mersereau, R.M., Optimal estimation of the regularization parameter and stabilizing functional for regularized image restoration, *Optical Engineering*, 29(5), 446, 1990.
36. Fu, H. and Barlow, J., A regularized structured total least squares algorithm for high resolution image reconstruction, *Linear Algebra and its Applications*, to appear.
37. Banham, M.R. and Katsaggelos, A.K., Digital image restoration, *IEEE Signal Processing Magazine*, 14(2), 24, March 1997.
38. Charbonnier, P., Blanc-Feraud, L., and Barlaud, M., Noisy image restoration using multiresolution Markov random fields, *Journal of Visual Communications and Image Representation*, 3, 338, December 1992.

25

Chebyshev's Inequality-Based Multisensor Data Fusion in Self-Organizing Sensor Networks

Mengxia Zhu
Southern Illinois University

Richard R. Brooks
Clemson University

Song Ding
Louisiana State University

Qishi Wu
University of Memphis

Nageswara S.V. Rao
Oak Ridge National Laboratory

S. Sitharama Iyengar
Florida International University

25.1 Introduction 553
25.2 Problem Formulation..555
    Signal Attenuation Model • Covariance among Binary Sensor Decisions
25.3 Local Threshold Adjustment Method..559
    Detection Matrix Construction • False-Alarm and Hit Rate Recalculation • Critical Values and Phase Changes
25.4 Threshold-OR Fusion Method..561
25.5 Simulation Results..564
25.6 Conclusions and Future Work..566
References 567

25.1 Introduction

Wireless sensor networks are being increasingly deployed in a number of applications such as detection of missiles; identification of chemical, biological, or nuclear plumes; monitoring of rain forests; and command and control operations in battlefield environments. In general, such sensor networks could be quite varied, ranging from thousands or millions of unattended tiny sensing devices to a small number of sensor nodes equipped with large instruments and workstations. Since sensor failures and topological changes are common, sensor network should self-organize itself to maintain viability, which assures that an object that traverses the terrain is detected and the user is alerted. In addition, redundant nodes are usually necessary to ensure uninterrupted and reliable operations. Measurements from multiple sensors are sent to a fusion center over wireless channels to achieve a more accurate situation assessment.

We consider a network of sensors that are distributed in a three-dimensional environment to monitor a region of interest (ROI) for possible intrusion by a target. The sensors collect environmental measurements that are subject to independent additive noise. Each sensor node employs a local threshold rule on the measurements to detect a target in the presence of random background noise. The local threshold value employed by sensor nodes is dynamically chosen based on random graph analysis. Lower local threshold value is accompanied by undesirable higher false-alarm rate and higher detection range to compensate for gap created by battery-drained nodes. When a target enters the monitoring region, it could be detected by multiple sensor nodes depending on their distances from the target. Typically, nearby sensors produce larger measurements while distant sensors receive less quantity of the measured signal. Our objective is to combine the decisions made by individual sensor nodes to achieve system detection performance beyond a weighted average of individual sensor nodes. The increased false-alarm rates of individual sensor resulting from local threshold adjustment can be effectively refrained under reasonable bounds using our fusion method.

This chapter has two main contributions. The first is presenting a local threshold rule based on random graph and percolation theory as opposed to traditional methods, which are not appropriate due to the nonfixed infrastructure of an ad hoc wireless sensor network [1]. The second and with the most description is proposing a binary hard fusion rule to determine the presence/absence of the target by deriving a bounds range.

At the core, the fusion problem is a specific instance of the classical distributed detection problem that has been studied extensively over past several decades [12]. The problem of fusing binary decisions taken at various sensor nodes has been solved using various schemes such as logical AND, OR, voting, Neyman–Pearson (NP), and Bayes rule [6–8,10,11]. Logical AND, OR, and voting techniques are examples of simple fusion schemes that do not require sensor probability distribution functions. Voting schemes can be further classified into threshold schemes and plurality schemes. Unanimity voting, majority voting, and m-out-of-n voting are threshold schemes, which try to minimize the probability of wrong decisions. A simple voting scheme generally lacks necessary performance guarantees in terms of error rates.

Chair and Varshney [3] proposed an optimal fusion rule based on Bayes rule, requiring a priori probabilities of $P(H_0)$ and $P(H_1)$. However, in practice, these a priori probabilities are often unavailable. Niu et al. [4] presented a fusion method that chooses an optimal sensor threshold to achieve the maximal system hit rate given a certain system false-alarm rate. This method demands a very large number of sensor nodes to apply the central limit theorem. Unfortunately, this criterion cannot be always satisfied in real scenarios. NP decision rule provides an optimal solution minimizing Type II error when subject to a chosen upper bound on Type I error probability or vice versa based on the likelihood ratio test. Although no a priori probability is required, NP fusion rule requires soft sensor readings and continuous sensor probability distribution functions to apply the notion of confidence level, and these information may incur high computational cost or even not be available. Rao [5] developed fusion methods that do not require the knowledge of sensor probability distribution functions but need to be trained using measurements. Such training measurements may not be available in some cases, and, in others, too many measurements may be needed to ensure reasonable levels of performance guarantees.

We propose a centralized threshold-OR fusion rule for combining the individual sensor node decisions in the sensor network described earlier. We derive threshold bounds for accumulated decisions using Chebyshev's inequality based on individual hit and false-alarm probabilities calculated from probability density function (pdf) and local threshold but without requiring an a priori knowledge of the underlying probabilities. We analytically show that the fused method achieves a higher hit rate and lower false-alarm rate compared to the weighted averages of individual sensor nodes. Restraining conditions are also derived. Simulation results using Monte Carlo method show that the error probabilities in the fused system are significantly reduced to near zero.

The rest of the chapter is organized as follows: an analytical model of the proposed sensor network is presented in Section 25.2 with a detailed discussion on covariance between sensor decisions. Section 25.3 gives the local threshold rule based on random graph theory for the system to remain in viable state despite of topological dynamics. In Section 25.4, we present a technical solution that derives the proper system threshold bounds for hard fusion systems. Simulation results are given in Section 25.5. We finally conclude our work in Section 25.6.

25.2 Problem Formulation

25.2.1 Signal Attenuation Model

We consider N sensor nodes deployed in a three-dimensional ROI, centering around a target within radius R, as shown in Figure 25.1. At sensor i, the noise n_i in a sensor measurement is independently and identically distributed (iid) according to the normal distribution:

$$n_i \sim \aleph(0,1) \tag{25.1}$$

The sensor measurements are subjective to an additive term due to such noise. Each sensor i makes a binary local decision as

$$
\begin{aligned}
H_1: & \quad k_i = w_i + n_i \\
H_0: & \quad k_i = n_i
\end{aligned}
\tag{25.2}
$$

where

k_i is the actual sensor reading
w_i is the ideal sensor measurement
n_i is the noise term

The region between target and sensor receiver is treated as free space, which is free of objects that might absorb or reflect radio frequency (RF) energy. We consider an isotropic signal attenuation power model based on path loss factor defined by

$$w_i = \frac{w_0}{\sqrt{1 + \beta d_i^n}} \tag{25.3}$$

where

w_0 is the original signal power emitted from the target located at point (x_0, y_0, z_0)
β is a system constant
d_i represents the Cartesian distance between the target and the sensor node, which is defined in Equation 25.4

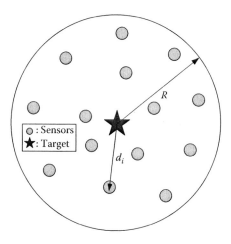

FIGURE 25.1 Sensor deployment for target detection.

Parameter n is the signal attenuation exponent typically ranging from 2 to 3, and the distance is defined as

$$d_i = \sqrt{(x_i - x_0)^2 + (y_i - y_0)^2 + (z_i - z_0)^2} \qquad (25.4)$$

This signal propagation model describes a three-dimensional unobstructed free space region monitored by a set of sensors that detect the signal emitted from a target. However, for many practical scenario, this idealized uniform free space model is not satisfactory to predict system performance under complex environment where obstacles absorb and reflect RF energy.

If there are sufficiently number of objects that scatter the RF energy such as in an urban terrain that leads to multiple reflective paths, the envelope of the channel response will therefore follow Rayleigh statistical distribution as described in Equation 25.5. Such multipath fading effect is superimposed on mean-path loss, which is proportional to an nth power of distance d between target and sensor. Variation of the mean-path loss is log-normally distributed and is usually site and distance dependent [9]:

$$p(r) = \frac{r}{\sigma^2} \exp\left(-\frac{r^2}{2\sigma^2}\right) \quad r \geq 0 \qquad (25.5)$$

where
  r is the power amplitude of the received signal
  $2\sigma^2$ is the predetection mean power of the multipath signal

If there is a dominate nonfading signal component such as line-of-sight propagation path, the power of the received signal can be described by a Rician fading. Our proposed approach can easily adapt to three basic mechanisms including reflection, diffraction (shadowing), and scattering that affect signal propagation in a complex environment using appropriate statistical modeling and measurements from real experiments. Yao et al. presented a unified theory on the modeling of fading channel statistics using spherically invariant random processes and Fox H-function [13]. Here, we use a simplified free space model to illustrate our fusion method.

Suppose that every sensor node employs the same threshold τ for decision making regardless of its distance to the target. The expected signal strength w_i for sensor i can be computed from Equation 25.3 according to its distance to the target. Thus, the hit rate p_{h_i} and false-alarm rate p_{f_i} for sensor i can be derived as follows [4]:

$$p_{h_i} = \int_{\tau}^{\infty} \frac{1}{\sqrt{2\pi}} e^{-(x-w_i)^2/2} dx \qquad (25.6)$$

$$p_{f_i} = \int_{\tau}^{\infty} \frac{1}{\sqrt{2\pi}} e^{-x^2/2} dx \qquad (25.7)$$

Figure 25.2 illustrates the use of a simple normal distribution function to model signal channel statistics for computing individual sensor hit rate and false-alarm rate in a heterogeneous system. A simplified homogeneous system ignores the impact of various distances to the target on sensor detection capabilities so that every sensor has the same hit rate and false-alarm rate. The heterogeneous

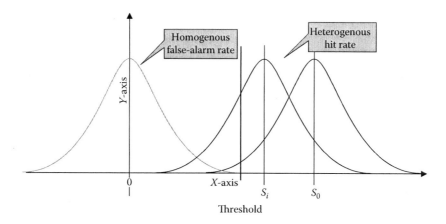

FIGURE 25.2 Normal distribution-based hit rate and false-alarm rate calculation for heterogeneous system.

sensor network system considers different hit rates and the same false-alarm rates as computed by Equations 25.6 and 25.7, respectively.

25.2.2 Covariance among Binary Sensor Decisions

Each sensor node makes its own decision regarding to presence and absence of the target. Let us consider covariance of binary decision $S_i(0/1)$ and $S_j(0/1)$ from a pair of local sensors i and j under the condition of target presence. The covariance between S_i and S_j can be expressed by Equation 25.8 (E is used to denote expected value). $E(S_iS_j|H_1)$ can be computed in Equation 25.9 by enumerating all possible combinations of S_i and S_j:

$$Cov(S_i \mid H_1, S_j \mid H_1) = E(S_iS_j \mid H_1) - E(S_i \mid H_1)E(S_j \mid H_1)$$

$$= E(S_iS_j \mid H_1) - p_{h_i}p_{h_j} \qquad (25.8)$$

$$E(S_iS_j \mid H_1) = 1 \times P(S_i = 1, S_j = 1 \mid H_1) + 0 \times P(S_i = 1, S_j = 0 \mid H_1)$$

$$+ 0 \times P(S_i = 0, S_j = 1 \mid H_1) + 0 \times P(S_i = 0, S_j = 0 \mid H_1)$$

$$= P(S_i = 1, S_j = 1 \mid H_1) \qquad (25.9)$$

We need to calculate the value of $P(S_i = 1, S_j = 1|H_1)$, which is the probability that target is detected by both sensors, implying that target falls within the overlapping detection area as shown by the shadow area S in Figure 25.3. R_i and R_j are the detection range for node i and j, respectively, and d here is the distance between two nodes. In fact, the analytical way to compute the probability is to integrate two pdfs in terms of detection range over area S:

$$P(S_i = 1, S_j = 1 \mid H_1) = \int_s p_{R_i}(r)p_{R_j}(r)ds \qquad (25.10)$$

Given the complexities involved in the pdfs, we propose an geometric approximation method to compute $P(S_i = 1, S_j = 1|H_1)$, the joint probability under H_1. The basic idea behind our approach is to first calculate the area of S, then use the ratio of the area S over entire detection circle to approximate the

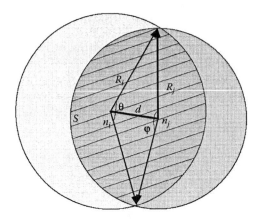

FIGURE 25.3 Calculation of covariance between two sensor measurements.

hit probability. Here, we assume that target has equal chance of appearing at any location within the detection circle:

$$Area(S) = \frac{\theta}{2\pi}\pi R_i^2 + \frac{\varphi}{2\pi}\pi R_j^2 - \frac{1}{2}R_i^2 \sin\theta - \frac{1}{2}R_j^2 \sin\varphi$$

$$\approx \frac{\theta}{2\pi}\pi R_i^2 + \frac{\varphi}{2\pi}\pi R_j^2 - \frac{1}{2}R_i^2(\pi-\theta) - \frac{1}{2}R_j^2(\pi-\varphi)$$

$$= \left(\theta - \frac{\pi}{2}\right)R_i^2 + \left(\varphi - \frac{\pi}{2}\right)R_j^2 \tag{25.11}$$

The area of *S* can be calculated by Equation 25.11 using simple geometry. We also derive the relations between angles θ and φ and distance d, as derived by Equations 25.12 through 25.14. The approximations in these equations are based on the assumption that detection range *R* is usually much larger than sensor distance *d*:

$$\begin{cases} d = R_i \cos\frac{\theta}{2} + R_j \cos\frac{\varphi}{2} \\ R_i \sin\frac{\theta}{2} = R_j \sin\frac{\varphi}{2} \end{cases} \tag{25.12}$$

$$\begin{cases} \cos\frac{\theta}{2} = \frac{R_i^2 + d^2 - R_j^2}{2dR_i} \\ \cos\frac{\varphi}{2} = \frac{R_j^2 + d^2 - R_i^2}{2dR_j} \end{cases} \tag{25.13}$$

$$\theta \approx \pi - \frac{R_i^2 + d^2 - R_j^2}{dR_i}$$

$$\varphi \approx \pi - \frac{R_j^2 + d^2 - R_i^2}{dR_j} \tag{25.14}$$

After we acquire the area of overlapping region *S*, the joint hit probability of two sensors can be written as

$$P(S_i = 1, S_j = 1 \mid H_1) = \frac{Area(S)}{\pi R_i^2}p_{h_i}\frac{Area(S)}{\pi R_j^2}p_{h_j} \tag{25.15}$$

The covariance can be expressed as

$$Cov(S_i \mid H_1, S_j \mid H_1) = P(S_i = 1, S_j = 1 \mid H_1) - p_{h_i} p_{h_j}$$

$$= \left[\frac{\left[(\theta - (\pi/2))R_i^2 + (\varphi - (\pi/2))R_j^2 \right]^2}{\pi^2 R_i^2 R_j^2} - 1 \right] p_{h_i} p_{h_j}$$

$$\approx \left[\frac{(\pi R_i^2)^2}{\pi^2 R_i^4} - 1 \right] p_{h_i} p_{h_j}$$

$$= 0 \tag{25.16}$$

if the following assumptions hold, $R_i = R_j$ and $d \upsilon R$, then $\theta = \varphi \approx \pi$. That implies that maximal detection range R is set to be the same for all sensors in a fusion system and is much larger than the distance d between two sensors. Based on these assumptions, we can declare that the covariance $Cov(S_i|H_1,S_j|H_1)$ between any two sensors is close to zero, and the local sensor decisions are assumed to be statistically independent of each other. Such independent assumption is used in the following section to simplify fusion rule derivation. If the covariance is too large to be neglected, a covariance matrix is used to calculate the variance of our fusion system instead of using a binomial distribution model in Section 25.4.

25.3 Local Threshold Adjustment Method

First, we construct the detection matrix for the sensor network based on a range-limited graph model. The critical point of maximal detection range is then derived based on manipulation of detection matrix such that the system starts to enter viable state at transitional detection range value and above. The detection range is inversely proportional to the local threshold of individual sensor nodes.

25.3.1 Detection Matrix Construction

If the distance between two nodes is no larger than twice of the maximal detection range τ_r, an edge exists with a probability of p. In practical, we substitute sensor hit rate p_h for p. We use r to denote $2\tau_r$. For such range-limited graph model with n randomly distributed sensor nodes in surveillance region, we build the detection matrix M, and the entry (i, j) represents the probability for a target to be detected while traversing between nodes n_i and n_j. Element (i, j) of detection matrix M has value of $p(2c - c^2)$, where c is a constant, which equals to

$$c = r^2 - \left(\frac{i}{n+1} - \frac{j}{n+1} \right)^2 \tag{25.17}$$

when $r^2 \geq (i/(n+1) - j/(n+1))^2$ and otherwise zero.

Similar procedure can be adopted to build a connectivity matrix, the smaller element value from connectivity and detection matrices will be saved to ensure both detection and communication capabilities. The derivation of Equation 25.17 consists of two steps: (1) sort nodes by the x- or y-coordinate value and use order statistics to find the expected value of the x- or y-coordinate for each node, and x- and y-coordinates and r are normalized to the range [0,1] and (2) determine the probabilities that an edge existing between two nodes using the expected values from step 1. More detailed steps can be found in Ref. [1] and was omitted due to limited space.

25.3.2 False-Alarm and Hit Rate Recalculation

We convert local threshold in terms of signal strength to detection range according to the attenuation model by Equation 25.3. We can restate hit rate and false-alarm rate for individual sensor in Equations 25.18 and 25.19 as a function of detection range τ_r as opposed to signal strength as in Equations 25.7 and 25.6:

$$p_{h_i} = \frac{n\beta}{2w_0^2} \int_0^{\tau_r} \frac{w_r^3 r^{n-1}}{\sqrt{2\pi}} e^{-(w_r - w_i)^2} dr \tag{25.18}$$

$$p_{f_i} = \frac{n\beta}{2w_0^2} \int_0^{\tau_r} \frac{w_r^3 r^{n-1}}{\sqrt{2\pi}} e^{-w_r^2} dr \tag{25.19}$$

where

$$w_r = \frac{w_0}{\sqrt{1 + \beta r^n}}$$

w_i is the ideal sensor measurement
n is the power index in attenuation model as defined in Section 25.2

We can also obtain the relationship between detection range and local threshold as

$$\ln K_r \approx -\frac{2}{n} \ln K_w$$

where
K_w is the ratio of local threshold (signal strength) before change to local threshold (signal strength) after change
K_r is the corresponding ratio of adjusted detection ranges

25.3.3 Critical Values and Phase Changes

For range-limited graphs, first-order monotone increasing graph properties follow 0–1 laws. These properties appear with probability asymptotically approaching either 0 or 1, as the parameters defining the random graph decrease or increase. A plot of property of probability versus parameter value forms an S-shaped curve with an abrupt phase transition between the 0 and 1 phases. In the first phase, the detection range is small and the network has a large number of isolated components. As the detection range grows, the expected size of the largest component grows logarithmically. The network will be dominated by a unique giant component that contains most of the sensor nodes. The parameter value where the phase transition occurs is referred to as the critical point. The detection matrices M can be used to identify phase transition point in terms of maximal detection range τ_r.

Brooks in Ref. [1] states that phase change point can be estimated when $M = M^2$ is satisfied. We can calculate the transition value of τ_r since M is represented as a function of n and τ_r. n is the number of sensor nodes and is usually known. According to percolation theory, for a graph with a giant component, a path can be found that connects the terrain's external boundaries. Thus, for a sensor network

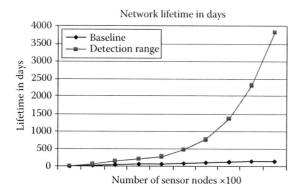

FIGURE 25.4 Comparison of network lifetime between baseline and dynamic detection range systems.

with a giant component, targets traversing the terrain will, with a high probability, be detected by at least one node that can report the detection to the user community. Therefore, the network fulfills our viability criterion.

Random graph model enables us to evaluate the viability of an ad hoc sensor network and adjust the detection range dynamically for uninterrupted surveillance task. The local threshold decreases as the detection range increases to extend the coverage scope. Figure 25.4 compares viable network lifetime in days between baseline system without any self-organizing features and dynamic detection range-based system from MATLAB® simulation results. The simulations were started with 100 nodes and increased till 1000 in steps of 100 nodes deployed within a 50 m × 50 m area. The initial detection range was set to be 5.5 m. Each node was given an initial energy of 2000 J. The simulation environment resembles the field test we performed at Twentynine Palms Marine Base [2]. We observe significant lifetime extension for system with adjustable detection range capability. However, both hit rate and false-alarm rate get higher when local threshold is reduced. Our binary fusion rule discussed in Section 25.4 can be efficiently used to restrain inflated system false-alarm rate.

25.4 Threshold-OR Fusion Method

Each sensor i makes an independent binary decision S_i as either 0 or 1. The fusion center collects local decisions and computes S as

$$S = \sum_{i=1}^{N} S_i \tag{25.20}$$

which is then compared with a threshold T to make a final decision. Under the assumption that sensor measurements are statistically independent under H_1 as we neglect covariance in Section 25.2.2, the mean and variance of S are given as follows when a target is present:

$$E(S \mid H_1) = \sum_{i=1}^{N} p_{h_i}$$

$$Var(S \mid H_1) = \sum_{i=1}^{N} p_{h_i}(1 - p_{h_i}) \tag{25.21}$$

Similarly, under the assumption that sensor measurements are statistically independent under H_0, the mean and variance of S when a target is absent are defined as

$$E(S \mid H_0) = \sum_{i=1}^{N} p_{f_i}$$

$$Var(S \mid H_0) = \sum_{i=1}^{N} p_{f_i}(1 - p_{f_i})$$

(25.22)

The threshold value T is critical to the system performance. Let P_h and P_f denote the hit rate and false-alarm rate of the fused system. If we let $P_h > 0.5$ and $P_f < 0.5$, T should be bounded by $\sum_{i=1}^{N} p_{f_i} < T < \sum_{i=1}^{N} p_{h_i}$.

The weighted averages of p_{hi} and p_{fi}, $i = 1, 2, \ldots, N$ are defined as follows:

$$\sum_{i=1}^{N} \frac{p_{h_i}}{\sum_{j=1}^{N} p_{h_j}} p_{h_i} = \frac{\sum_{i=1}^{N} p_{h_i}^2}{\sum_{i=1}^{N} p_{h_i}}$$

(25.23)

$$\sum_{i=1}^{N} \frac{1 - p_{f_i}}{\sum_{j=1}^{N} (1 - p_{f_j})} p_{f_i} = \frac{\sum_{i=1}^{N} (1 - p_{f_i}) p_{f_i}}{\sum_{i=1}^{N} (1 - p_{f_i})}$$

(25.24)

We desire better detection performance of the fused system than the corresponding weighted averages in terms of higher hit rate and lower false-alarm rate such that

$$P_h > \frac{\sum_{i=1}^{N} p_{h_i}^2}{\sum_{i=1}^{N} p_{h_i}}$$

(25.25)

$$P_f < \frac{\sum_{i=1}^{N} (1 - p_{f_i}) p_{f_i}}{\sum_{i=1}^{N} (1 - p_{f_i})}$$

(25.26)

We first consider a lower bound of the hit rate of the fused system:

$$P_h = P\{S \geq T \mid H_1\}$$

$$\geq P\left\{ \left| S - \sum_{i=1}^{N} p_{h_i} \right| \leq \left(\sum_{1}^{N} p_{h_i} - T \right) \mid H_1 \right\}$$

$$\geq 1 - \frac{\sigma^2}{k^2}$$

$$= 1 - \frac{\sum_{i=1}^{N} p_{h_i}(1 - p_{h_i})}{\left(\sum_{i=1}^{N} p_{h_i} - T \right)^2}$$

(25.27)

Probability density $p(S)$

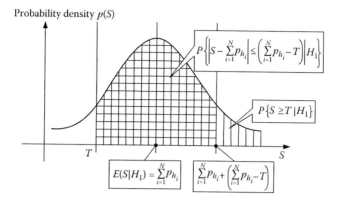

FIGURE 25.5 Application of Chebyshev's inequality in calculating the lower bound of the system hit rate.

where we applied Chebyshev's inequality in the third step as shown in Figure 25.5 and k equals $\left(\sum_{1}^{N} p_{h_i} - T\right)$. Now, the condition in Equation 25.25 can be ensured by the following sufficient condition:

$$1 - \frac{\sum_{i=1}^{N} p_{h_i}(1 - p_{h_i})}{\left(\sum_{i=1}^{N} p_{h_i} - T\right)^2} \geq \frac{\sum_{i=1}^{N} p_{h_i}^2}{\sum_{i=1}^{N} p_{h_i}} \qquad (25.28)$$

Following that, an upper bound of T can be derived from Equation 25.28 as follows:

$$T \leq \sum_{i=1}^{N} p_{h_i} - \sqrt{\sum_{i=1}^{N} p_{h_i}} \qquad (25.29)$$

Similarly, for the false-alarm rate, we carry out a similar procedure to compute the lower bound from Equations 25.30 through 25.34. Again, Chebyshev's inequality is applied in the second step in Equation 25.32:

$$P_f = P\{S \geq T \mid H_0\} = 1 - P\{S < T \mid H_0\} \qquad (25.30)$$

$$P\{S < T \mid H_0\} \geq P\left\{\left|S - \sum_{i=1}^{N} p_{f_i}\right| \leq \left(T - \sum_{i=1}^{N} p_{f_i}\right)\middle| H_0\right\} \qquad (25.31)$$

$$P_f \leq 1 - P\left\{\left|S - \sum_{i=1}^{N} p_{f_i}\right| \leq \left(T - \sum_{i=1}^{N} p_{f_i}\right)\middle| H_0\right\} \leq \frac{\sum_{i=1}^{N} p_{f_i}(1 - p_{f_i})}{\left(T - \sum_{i=1}^{N} p_{f_i}\right)^2} \qquad (25.32)$$

Now, we consider the condition that ensures false probability of fuser is smaller than that of weighted average given by

$$\frac{\sum_{i=1}^{N} p_{f_i}(1-p_{f_i})}{\left(T-\sum_{i=1}^{N} p_{f_i}\right)^2} \leq \frac{\sum_{i=1}^{N}(1-p_{f_i})p_{f_i}}{\sum_{i=1}^{N}(1-p_{f_i})} \qquad (25.33)$$

$$T \geq \sum_{i=1}^{N} p_{f_i} + \sqrt{\sum_{i=1}^{N}(1-p_{f_i})} \qquad (25.34)$$

Therefore, we define the range of T using the upper bound in Equation 25.29 and lower bound in Equation 25.34 as follows:

$$\left[\sum_{i=1}^{N} p_{f_i} + \sqrt{\sum_{i=1}^{N}(1-p_{f_i})}, \ \sum_{i=1}^{N} p_{h_i} - \sqrt{\sum_{i=1}^{N} p_{h_i}}\right] \qquad (25.35)$$

For the case of a homogeneous system, the bounds of threshold T are simplified as follows:

$$\left[N p_f + \sqrt{N(1-p_f)}, \ N p_h - \sqrt{N p_h}\right] \qquad (25.36)$$

To ensure that the upper bound is larger than the lower bound, we have the following restrictions on individual hit rates, individual false-alarm rates, and the number of sensor nodes, for the heterogeneous and homogeneous systems in Equations 25.37 and 25.38, respectively:

$$\sum_{i=1}^{N} p_{f_i} + \sqrt{\sum_{i=1}^{N}(1-p_{f_i})} - \sum_{i=1}^{N} p_{h_i} + \sqrt{\sum_{i=1}^{N} p_{h_i}} \leq 0 \qquad (25.37)$$

$$p_h - p_f \geq \frac{\sqrt{p_h} + \sqrt{1-p_f}}{\sqrt{N}} \qquad (25.38)$$

Note that for the homogeneous system, the number of sensor nodes N can be set large enough to satisfy Equation 25.38.

25.5 Simulation Results

Our simulation based on 100,000 runs* produced the receiver operative characteristic (ROC) curve, a plot of the system hit rate against the false-alarm rate for different possible thresholds. There is a trade-off between sensitivity and specificity, namely, any increase in the sensitivity of hit rate will be accompanied by an increase in nonspecificity of false-alarm rate. The curvature of the ROC curve determines the detection accuracy of the system, namely, the closer does the curve follow the left-hand border and then the top border of the ROC space, the more accurate is the system. The closer does the curve approach the 45° diagonal of the ROC space, the less accurate is the test.

---

* To simplify calculation, we restrict the sensor deployment in a 2D plane by setting coordinate z to be zero. The simulation results for a 3D region are qualitatively similar to the 2D case.

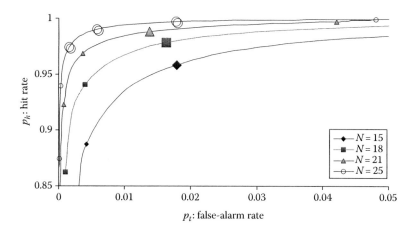

FIGURE 25.6 Homogeneous system ROC curve with different sensor node number.

For the homogeneous case, we set sensor hit rate to be 0.65, and sensor false-alarm rate to be 0.2. Please note that our sensing specifications are chosen subjectively, and we intend to test our fusion method on sensors with ordinary or less than ordinary performance. In Figure 25.6, four system ROC curves with sensor node number going from 15 to 25 are plotted. The desirable segment on the ROC curve is located by restricting threshold selection. Due to the discrete nature of our binary decision system, we identify the selected ROC segment as individual enlarged markers. From Figure 25.6, we observe that all selected points fall at the top left corner of the ROC space and bear hit rate and false-alarm rate that is by far superior to that of a single sensor node. When the number of sensor nodes increases, our system performances are further enhanced due to ample available resources.

For the heterogeneous case, from Equations 25.6 and 25.7, we conclude that sensors have the same false-alarm rate and different hit rates due to various distances to the target. In our simulation program, the given hit rate and false-alarm rate for sensors in close proximity to the target are represented as p_h and p_f, respectively. Sensor threshold τ can be calculated from Equation 25.7 given a known false-alarm rate p_f. Consequently, original signal power w_0 can be computed from Equation 25.6. w_i can be derived from Equation 25.3, and, therefore, p_{hi} can be computed according to Equation 25.6. We recalculate the detection matrix M upon node failure and evaluate the viability of the system. Local sensor threshold will be dynamically adjusted to cover gap area resulting from dead nodes.

We set the total number of sensor nodes to be 25. The p_h is set to be 0.75 and p_f is set to be 0.2 in simulation tests. Sensor deployment radius ranges from 1 to 3 to compare different deployment strategies. Results are tabulated in Table 25.1 as system threshold bounds (T_l, T_u), weighted average hit rate and false-alarm rate as W_{ph} and W_{pf}, respectively, system hit rate bounds (P_{hl}, P_{hu}), and system false-alarm rate bounds (P_{fl}, P_{fu}) under different radius R. Dispersed sensor deployment negatively affects the system performance as a result of depressed detection performance and truncated scope of threshold bounds. It further validates the rule that sensors should be deployed as close to potential targets as possible. From Figure 25.7 with radius 3, our fusion system achieves excellent hit rate close to 0.97 and low false-alarm rate below 0.02 as compared to weighted average of 0.5612 and 0.2, respectively, before fusion.

TABLE 25.1 Numeric Results with Different Deployment Radius for Heterogeneous System

$N = 25$, $W_{pf} = 0.2$	T_l	T_u	W_{ph}	P_{hl}	P_{hu}	P_{fl}	P_{fu}
$R = 1$	10	12	0.6788	0.9821	0.9982	0.0015	0.0164
$R = 2$	10	11	0.628	0.9836	0.9947	0.0054	0.017
$R = 3$	10	10	0.5612	0.966	0.966	0.0175	0.0175

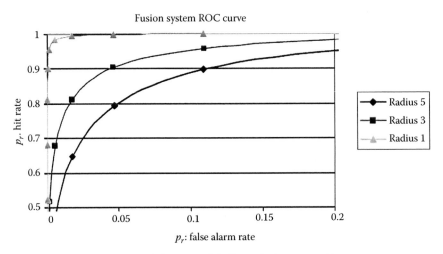

FIGURE 25.7 Heterogeneous system ROC curve with different deployment radius.

25.6 Conclusions and Future Work

We proposed a threshold-OR fusion method for a sensor network wherein each node employs a dynamic local threshold to the measurement to decide the presence of a target in the field of view. The local threshold is chosen based on random graph theory such that the system is always in viable state despite of node failure and topological changes. Sensor nodes make their binary decisions with the new local threshold. Current non-model- and model-based fusion methodologies are derived from some variants of decision rules such as voting, Bayes criterion, maximum a posterior (MAP) criterion, and NP. A comparison of fusion types in terms of input sensor data, hypothesis, sensor pdfs, a priori probability, and performance is provided in Table 25.2. Despite the simplicity, non-model-based voting rule gives no system performance guarantee in terms of probability of errors. Bayes criterion minimizes the cost function based on the knowledge of the a priori probability, which is not always available. MAP, a similar decision rule to Bayes criterion, simply minimizes the overall probability of errors since it seeks to maximize the posterior probability. However, a priori probability is still desired. NP is attractive since it does not require knowledge of priors and a cost function as opposed to Bayes criterion [12]. Nevertheless, NP-based fusion rule is essentially a type of soft fusion, which demands continuous local sensor readings and pdfs to control one error probability under certain confidence level.

Our method falls into the category of hard fusion accepting discrete sensor decisions and does not require the knowledge of the a priori probability of H_0 and H_1. Since our conditions are sufficient and not necessary in deriving the threshold bounds, simulation results under such restrictive conditions yield significantly reduced probability of errors compared to that of single sensor or weighted average with both Type I and Type II errors close to zero.

TABLE 25.2 Comparison of Different Decision-Rule-Based Fusions

Rules	Fusion Type	Hypothesis	Pdf	A Priori Probability	Performance
Voting	Hard	Multiple	No	No	No guarantee
Bayes criterion	Soft/hard	Multiple	Yes	Yes	Minimize cost function
MAP	Soft/hard	Multiple	Yes	Yes	Minimize error probability
NP	Soft	Binary	Yes	No	Fix one error probability; minimize the other
Proposed method	Hard	Binary	Yes	No	Significantly reduce error probability

The given threshold bounds allow users certain freedom in shifting between sensitivity and specificity instead of a single cutting point. For example, if Type I error is not tolerable, user can choose a threshold close to the upper bound. If Type II error needs to be minimized, user may tend to pick a threshold close to the lower bound. In addition, our approach has a low computational cost, making practical deployment feasible with limited resource. For future work, we plan to apply weighted factors to local decisions based on various distances and signal noise ratios when calculating binomial distribution in Equation 25.20. Deployment of the proposed fusion rule in practical sensor network application applying some complicated fading channel modeling and covariance calculation is also of our future interest.

References

1. R. R. Brooks, B. Pillai, S. Racunas, and S. Rai. Mobile network analysis using probabilistic connectivity matrices. *IEEE Transactions on Systems, Man, and Cybernetics Part C*, 37(4):694–702, 2006.
2. R. R. Brooks, P. Ramanathan, and A. Sayeed. Distributed target tracking and classification in sensor networks. *Proceedings of the IEEE*, 91(8):1163–1171, August 2003.
3. Z. Chair and P. K. Varshney. Optimal data fusion in multiple sensor detection systems. *IEEE Transactions on Aerospace and Electronic Systems*, AES(22):98–101, January 1986.
4. R. Niu, P. Varshney, M. Moore, and D. Klamer. Decision fusion in a wireless sensor network with a large number of sensors. In *The 7th International Conference on Information Fusion*, Stockholm, Sweden, pp. 21–27, June 2004.
5. N. S. V. Rao. Distributed decision fusion using empirical estimation. *IEEE Transactions on Aerospace and Electronic Systems*, 33(4):1106–1114, 1996.
6. A. R. Reibman and L. W. Nolte. Design and performance comparison of distributed detection networks. *IEEE Transactions on Aerospace and Electronic Systems*, AES(23):789–797, November 1987.
7. A. R. Reibman and L. W. Nolte. Optimal detection and performance of distributed sensor systems. *IEEE Transactions on Aerospace and Electronic Systems*, AES(23):24–30, January 1987.
8. F. A. Sadjadi. Hypothesis testing in a distributed environment. *IEEE Transactions on Aerospace and Electronic Systems*, AES(22):134–137, March 1986.
9. B. Sklar. Rayleigh fading channels in mobile digital communication systems. Part I: Characterization. *IEEE Communication Magazine*, 90–100, July 1997.
10. R. R. Tenney and N. R. Sandell. Detection with distributed sensors. *IEEE Transactions on Aerospace and Electronic Systems*, AES(17):501–510, July 1981.
11. F. A. Thomopoulos, R. Viswanathan, and D. C. Bougoulias. Optimal decision fusion in multiple sensor systems. *IEEE Transactions on Aerospace and Electronic Systems*, AES(23):644–653, September 1987.
12. P. K. Varshney. *Distributed Detection and Data Fusion*. Springer-Verlag, New York, 1997.
13. K. Yao, M. Simon, and E. Biglieri. *Theory on the Modeling of Wireless Communication Fading Channel Statistics*. Princeton, NJ, March 2004.

26

Markov Model Inferencing in Distributed Systems

Chen Lu
Clemson University

Jason M. Schwier
Clemson University

Richard R. Brooks
Clemson University

Christopher Griffin
The Pennsylvania State University

Satish Bukkapatnam
Oklahoma State University

26.1 Introduction 569
26.2 Hidden Markov Models...570
26.3 Inferring HMMs..571
    Baum–Welch Algorithm • CSSR Algorithm • Zero-Knowledge HMM Identification
26.4 Applications...574
26.5 Conclusion 578
Acknowledgments..578
References..578

26.1 Introduction

Hidden Markov Models (HMMs) are a common tool in pattern recognition. Applications of HMMs include voice recognition [1,2], texture recognition [3], handwriting recognition [4,5], gait recognition [6], tracking [7], and human behavior recognition [8,9]. Variations of these applications can also be used in distributed sensor networks. A lot of work has been done using HMMs to identify target behaviors from sensor data. For example, in Refs. [10,11], HMMs are used for activity monitoring and action recognition in body sensor networks. In Ref. [12], Amutha et al. used HMMs for target tracking in wireless sensor networks. HMMs can also be used to solve sensor network localization problem, which estimates the geographical location of sensors in wireless sensor networks [13]. Using HMMs, it is possible to design data transmission protocols in sensor networks for energy efficiency [14]. Other possible application for HMMs in sensor networks can be side channel analysis, such as power monitoring and network traffic timing analysis. This kind of analysis can extract extra source of information that can be exploited to analyze the sensor networks.

In the applications, HMMs are inferred from data streams in sensor network using different approaches. In this chapter, three different HMM inferencing algorithms are discussed.

Traditionally, the Baum–Welch algorithm is used to infer the state transition matrix of a Markov chain and symbol output probabilities associated to the states of the chain, given an initial Markov model and a sequence of symbolic output values (see Ref. [1]). The Baum–Welch algorithm uses dynamic programming to solve a nonlinear optimization problem. The fundamental approach of constructing Markov models from data streams has been heavily researched for specific applications. Methods in Refs. [10,11], for example, illustrate model construction and training in target action monitoring and recognition.

However, in sensor networks, it is sometimes difficult to find an appropriate initial model structure. To construct a Markov model without a priori structural information, Crutchfield and Shalizi [15–17] developed an approach that derives the HMM state structure and transition matrix from available data samples. In this chapter, we focus on the approach from Shalizi et al. Other approaches may also be used to construct models from data streams, such as Ref. [18], for specific areas, such as speech recognition.

Shalizi's approach finds statistically significant groupings of the training data that correspond to HMM states. This is done by analyzing the conditional next symbol probabilities for a data window that slides over the training data. These data window increase gradually from a size of two to an a priori known maximum window size L. Except for the training data, the only initial information required to construct the HMM model using their approach is the parameter L. The parameter L expresses the maximum number of symbols that are statistically relevant to the next symbol in the sequence. The state structure of the Markov model is inferred from the symbol groupings of length $\leq L$ by adding those states to the model that lower system entropy [19,20].

Schwier et al. [21] extended the work of Crutchfield and Shalizi so that they determined parameter L with no prior knowledge and therefore derived minimum entropy HMMs with no a priori information.

In this chapter, we introduced these methods of inferring HMMs from data streams. The remainder of this chapter is organized as follows. Section 26.2 provides background on HMMs. Section 26.3 explains several HMM inference algorithms. Applications of HMM inference are given in Section 26.4. Conclusions are in Section 26.5.

26.2 Hidden Markov Models

A Markov model is a tuple $I = (V, E, L, \phi)$, where V is a set of vertices of a graph, E is a set of directed edges between the vertices, L is a set of labels, and $\phi: L \to E$ is a labeling of the edges. A *path* through I with label $y = y_1 y_2 \dots y_n$ is an ordered set of vertices $(v_1, v_2, \dots, v_{n+1})$ such that for each pair of vertices (v_i, v_j)

1. $(v_i, v_j) \in E$
2. $\phi(v_i, v_j) = y_i$

In Markov models, the vertices of I are referred to as states and the edges are referred to as transitions, where V is the state space of size n and P is the $n \times n$ transition matrix. Each element $p_{i,j} \in P$ expresses the probability that the process transitions to state j once it is in state i. If $(v_i, v_j) \in E$, then we assume $p_{i,j} \neq 0$ and for any i, $\sum_j p_{i,j} = 1$. The fundamental property of Markov models is that they are "memoryless." The conditional probability of a transition to a new state depends only on the current state, not on the path taken to reach the current state.

We require that Markov models be deterministic in transition label; that is, if there is a pair (v_i, v_j) with $\phi(v_i, v_j) = y_i$, then $\phi(v_i, v_k) \neq y_i$, for $\forall k \neq j$. Therefore, the probability that a particular symbol will occur next is the probability $p_{i,j}$ of moving to the next state using a transition associated with that symbol. If no transition exists between state i and state j, then $p_{i,j} = 0$.

A HMM is a statistical Markov model in which the system being modeled is assumed to be a Markov process with unobserved (hidden) states. The standard HMM in Refs. [1] has two sets of random processes: states and outputs. In Ref. [16], an alternate model is used, also mentioned in Refs. [1,22], with only a single set of random processes. In this model, the path labels are associated with an output alphabet [1,23]. A series of successive y_i symbols representing a path of transitions is called an observed output sequence and is generated by the Markov model I_k, where k is a unique identification number for the model. Given a Markov model, the following procedure may be used to generate a sequence of length l:

1. Choose an initial state $v_i = v_0$.
2. Using the probabilities of the outgoing transitions, select a transition $p_{i,j}$ to move to state v_j from state v_i.
3. Record the label $y_i = \phi(v_i, v_j)$ associated with chosen transition $p_{i,j}$.
4. Repeat steps 2 and 3 until l labels have been recorded.
5. Record the sequence $\mathbf{y} = y_i y_{i+1} y_{i+2} \dots y_{i+l-1}$.

Observation sequence **y** is generated by the HMM. Baum-Welch algorithm estimates the model parameters from the observation sequence **y**. And CSSR algorithm infers HMMs from the sequence **y**. Note that if the model contains at least one absorbing state (i.e., has no outgoing transitions), it is possible for the model to be unable to generate a sequence of length l.

The process developed by Shalizi [16] produces state machines, deterministic in transition symbol, that is, where each symbol is mapped to at most one transition leaving a state. And the model is ergodic with all transient states removed from the model. This property does result in some loss of prediction power of transient effects. Like Ref. [16], Schwier [21] assume that the HMMs contain cycles. His extension would not be appropriate for HMMs without cycles.

26.3 Inferring HMMs

In this section, we first briefly review the Baum–Welch algorithm to infer HMMs. Then, we discuss the CSSR algorithm defined by Shalizi [16]. At last, zero-knowledge HMM identification algorithm developed by Schwier [21] is described.

26.3.1 Baum–Welch Algorithm

The Baum–Welch algorithm makes use of the forward–backward algorithm to estimate the transition matrix P of a HMM [1,24]. In addition to the observed sequence of symbols **y**, the algorithm requires an initial estimate of the transition structure (in the form of a guess for P and the initial probability distribution across the states π_0) be available. As a result, this algorithm requires initial knowledge of the structure of the Markov process governing the dynamics of the system producing the output. The algorithm is iterative and the stopping criteria may be given in terms of the convergence of the solution in any number of metrics.

Given a model I_k, the maximum likelihood (ML) of a sequence **y** can be calculated by iteratively exploring all possible paths and finding the probability that sequence $\hat{y}$ occurs for each path [25,26]. The forward–backward procedure [27–30] applies dynamic programming and induction to calculate the probability, $P(\hat{y}|\lambda_k)$, reducing the algorithm complexity from exponential time to polynomial time. This algorithm typically associates symbols with states. Since it is possible for any given state machine to construct an equivalent state machine that switches the role of states and transitions [31], without loss of generality we can associate symbols with transitions.

The forward–backward procedure is as follows:

1. Initialization: $\alpha_1(i) = \pi_i$, $1 \le i \le n$*
2. Induction: $\alpha_{t+1}(j) = \sum_{i=1}^{n} \alpha_t(i) p_{i,j}$, $1 \le t \le M-1$
3. Termination: $P(\hat{y}|\lambda_k) = \sum_{i=1}^{n} \alpha_M(i)$

The α-values are called "forward" variables and the aforementioned three steps are considered the "forward" or ML part of the algorithm. The "backward" or expectation maximization (EM) portion of the forward–backward procedure calculates the "backward" variables and is used to find the following:

1. The path most likely taken to produce a sequence $\hat{y}$
2. How the parameters of I_k can maximize $P(\hat{y}|\lambda_k)$ [1]

The Baum–Welch algorithm uses EM part to estimate the transition matrix of the model to best fit the observed data. Maximum mutual information (MMI) [32,33] and minimum classification error (MCE) [34,35] are alternatives to EM.

---

* π_i can refer to either an initial estimate probability distribution across states or to the stable distribution where $\pi = P^l \pi$.

26.3.2 CSSR Algorithm

The drawback of Baum–Welch algorithm is that it requires a priori knowledge of the structure of the HMM. Shalizi developed the causal state splitting and reconstruction (CSSR) algorithm [16] to infer HMMs without this prior information. Given a sequence **y** produced by a stationary process, the CSSR algorithm infers a set of causal states and a transition structure for a HMM that provides a minimum entropy estimation of the true underlying process dynamics. The states are defined as conditional probability distributions over the next symbol that can be generated by the process. The set of states found in this manner accounts for the deterministic behavior of the process while tolerating random noise that may be caused by either measurement error or play in the system under observation. The CSSR algorithm has useful information-theoretic properties in that it maximizes the mutual information among state structure and next output symbol and minimizes the remaining uncertainty (entropy).

The CSSR algorithm is straightforward. A sequence $\mathbf{y} \in \mathcal{A}^*$ and a priori value $L \in \mathbb{N}$ are given. The parameter L defines the number of symbols in the past that are necessary to find the proper state structure of the model. For values of i increasing from 0 to L, the algorithm identifies the set of sequences W that are subsequences of **y** and have length i. (When $i = 0$, the empty string is considered to be a subsequence of **y**.) It then computes the conditional distribution of the next symbol following each $\mathbf{x} \in W$ using the data provided in **y** and partition the subsequences according to these distributions. These partitions become states in the inferred HMM. CSSR algorithm groups **x** with consistent conditional probability over the next symbol into states. Therefore, subsequences **x** are used to form states in the inferred HMM. If states already exist, it compares the conditional distribution of subsequence (computed using Equations 26.1 and 26.2) **x** to the conditional distribution of the existing states and add **x** to this state if the conditional distributions are ruled statistically identical. If sequence **x** may be added to many states, the state with the best fit is chosen. Distribution comparison can be carried out using either a Kolmogorov–Smirnov test or a χ^2 test with a specified level of confidence. The level of confidence chosen affects the type I error rate. Once state generation is complete, the states are further split to ensure that the inferred model has a deterministic transition relation δ. Reconstruction merges states when possible in order to avoid creating an unnecessary large number of states. Algorithm 26.2 shows the CSSR algorithm.

The following formulas are used to compute $f_{q_i|\mathbf{y}}$ and $f_{a\mathbf{x}|\mathbf{y}}$ in Algorithm 26.2. Let $\#(\mathbf{x}, \mathbf{y})$ be the number of times the sequence **x** is observed as a subsequence of **y**:

$$f_{\mathbf{x}|\mathbf{y}}(a) = \Pr(a \mid \mathbf{x}, \mathbf{y}) = \frac{\#(\mathbf{x}a, \mathbf{y})}{\#(\mathbf{x}, \mathbf{y})} \tag{26.1}$$

$$f_{q_i|\mathbf{y}}(a) = \Pr(a \mid q_i, \mathbf{y}) = \frac{\displaystyle\sum_{\mathbf{x} \in q_i} \neq (\mathbf{x}a, \mathbf{y})}{\displaystyle\sum_{\mathbf{x} \in q_i} \neq (\mathbf{x}, \mathbf{y})} \tag{26.2}$$

The complexity of CSSR is $O(k^{L+1}) + O(N)$, where k is the size of the alphabet, L is the maximum subsequence length considered, and N is the size of the input symbol sequence. Given a stream of symbols **y**, of fixed length N, from alphabet $\mathcal{A}$, the algorithm is linear in the length of the input data set but exponential in the size of the alphabet.

Note that CSSR [16] estimates conditional probabilities by analyzing grouped sets of outputs from a stochastic process. As long as Shalizi's assumption [16] that the volume of training data is sufficient, the law of large numbers dictates that this is almost surely true. This would not be the case if one were considering only one instance of an output string. As noted in Ref. [36], the dependencies in any single trace of a Markov process go back further than L states. In the same work, it is made clear that this is not true for the dependencies of the states themselves and therefore not true for the estimated distributions.

26.3.3 Zero-Knowledge HMM Identification

The CSSR algorithm depends on the parameter L, which defines the maximum subsequence length considered when inferring the model. The model assumes that at time t, symbol y_{t+1} is a random function of symbols $y_t, y_{t-1}, \ldots, y_{t-L}$. When inferring models with the CSSR algorithm, if L is too small, the state structure of the inferred machine is incorrect because it does not capture all the statistical dependencies in the data. The number of states is incorrect and symbols are incorrectly assigned to states. Since the parameter L defines the exponent of the algorithm complexity, it is imperative that L not be larger than absolutely necessary. Therefore, Schwier et al. extended the CSSR algorithm to find the correct value of L.

Algorithm Description 26.1—CSSR Algorithm from Ref. [16]

Input: Observed sequence $\mathbf{y}$; Alphabet $\mathcal{A}$, Integer L

Initialization:

1. Define state q_0 and add λ (the empty string) to state q_0. Set $Q = \{q_0\}$.
2. Set $N: = 1$.

Splitting (repeat for each $i \leq L$)

1. Set $W = \{\mathbf{x} \mid \exists q \in Q(\mathbf{x} \in q \wedge |\mathbf{x}| = i - 1)\}$. The set of strings in states of the current model with length equal to $i - 1$.
2. Let N be the number of states.
3. For each $\mathbf{x} \in W$, for each $a \in \mathcal{A}$, if $a\mathbf{x}$ is a subsequence of $\mathbf{y}$, then
   a. Estimate $f_{ax|y} : \mathcal{A} \to [0,1]$, the probability distribution over the next input symbol.
   b. Let $f_{qj|y} : \mathcal{A} \to [0,1]$ be the joint state conditional probability distributions, that is, the probability, given the system is in state q_j, that the next symbol observed will be a. For each j, compare $f_{qj|y}$ with $f_{ax|y}$ using an appropriate statistical test with confidence level α. Add $a\mathbf{x}$ to the state that has the most similar probability distribution as measured by the p-value of the test. If all tests reject the null hypothesis that $f_{qj|y}$ and $f_{ax|y}$ are the same, then create a new state q_{N+1} and add $a\mathbf{x}$ to it. Set $N: = N + 1$.

Reconstruction

1. Let $N_0 = 0$.
2. Let N be the number of states.
3. Repeat while $N_0 \neq N$:
   a. For each $i \in 1, \ldots, N$: Set $k: = 0$. Let M be the number of sequences in state q_i. Choose a sequence $\mathbf{x}_0$ from state q_i. Create state p_{ik} and add $\mathbf{x}_0$ to it. For all sequences $\mathbf{x}_j$ $(j > 0)$ in state q_i:
      i. For each $a \in \mathcal{A}$, $\mathbf{x}_j a$ produces a sequence that is resident in another state q_k. Let $(\mathbf{x}_j, a, q_k) \in \delta$.
      ii. For $l = 0, \ldots, k$, choose $\mathbf{x}$ from sequences within p_{ik}. If $\delta(\mathbf{x}_j, a) = \delta(\mathbf{x}, a)$, for all $a \in \mathcal{A}$, then add $\mathbf{x}_j$ to p_{ik}. Otherwise, create a new state p_{ik+1} and add $\mathbf{x}_j$ to it. Set $k: = k + 1$.
   b. Reset $Q = \{p_{ik}\}$; recompute the state conditional probabilities $f_{q|y}$ for $q \in Q$ and assign transitions using the δ functions defined earlier.
   c. Let $N_0 = N$.
   d. Let N be the number of states.
4. The model of the system has state set Q and transition probability function computed from the δ relations and state conditional probabilities.

To identify incorrect state structures, a symbol-to-state mapping is defined. As L increases, this symbol-to-state mapping is observed. This mapping stabilizes once the correct L is found. This occurs because the construction method has access to all the statistically relevant information. At which point, adding additional information by using a larger L simply reproduces the same correct model.

This problem is quite different from the order problem discussed in Ref. [36]. Order estimation is a difficult problem that occurs when using the standard HMM. Since that model has two different sets of probability distributions (states and outputs), it is desirable to find the smallest number of states that can be used to express properly the set of output distributions. The smallest number of states is called the model order. Since the model used in this chapter has one set of probability distributions, where state transitions output symbols, the question of order does not arise. Reference [16] proves that CSSR finds the most compact representation of the statistical dependencies in the data analyzed.

To find the correct window size L, the HMM inferred using CSSR with window size L is checked with the model structure inferred using window size $L + 1$. If their interpretation of the symbolic dataset **y** output by the process being analyzed is consistent, the two HMMs are consistent.

Algorithm 26.2 works iteratively. It starts by inferring HMM G_2 by using CSSR with parameter $L = 2$. The training data are then input to G_2. For each symbol, it records the state in G_2 that is associated with the symbol. Since the HMM is deterministic, this mapping is unique. Since there is no specified start state, it performs this process for each state in G_2. If this process fails at some point for a specific start state, it does not store that information.

It then compute HMM G_3 using CSSR with $L = 3$ and calculate the mappings of states to symbols in exactly the same way it was done for G_2. For each pair of start states (i.e., each state in G_2 paired with each state in G_3), it determines how many times they agreed on the mapping of symbols to states. Since there is no clear start state, it keeps the largest value m_3.

This value m_3 measures how similar machines G_2 and G_3 are. (It is technically similar to the concept of *bisimulation* used in model checking literature [37].) Determining if two graphs are isomorphic is computationally challenging. Determining the equivalence of two Markov chains would be more difficult, since that would also require comparing probability distributions. Algorithm 26.2 sidesteps that issue with this step, what is important in this process is the mapping of symbols to states. If two machines assign the symbols in the training data to the same states, then their interpretations are identical.

The process repeats with increasing values of L producing a new machine G_L and value m_L with each iteration. Informally, as L increases, CSSR has more information regarding the history of the system. CSSR monotonically improves the ability of the HMM G_L to explain the training data. As the HMMs asymptotically approach the true structure of the process that produced the data, the amount of agreement between G_L and G_{L-1} increases. When the correct value of L is found, there is no new information to be gained by using $L + 1$. At which point, the mapping of symbols to states will remain stable (i.e., $m_L = = m_{L+1} = = m_{L+2} ...$) and the process can terminate. These steps are presented more formally in Algorithm 26.2.

For a model with correct string length L_{max}, the runtime operation of Algorithm 26.2 is $O(k^{L_{max}+1}) + O((L_{max} - 1)N)$. This algorithm provides a method to infer HMM from data without any prior information.

26.4 Applications

In this section, we discuss the applications of the three HMM inferencing algorithms.

A lot of applications have been done using the Baum–Welch algorithm in sensor networks. For example, in Ref. [11], they presented an action recognition approach in a body sensor network based on an HMM, which is capable of both segmenting and classifying continuous actions. Sensor data samples are labeled based on clusters generated from the training data. The initial structure they use is in Figure 26.1. The Baum–Welch algorithm is used to train and improve the initial model for different actions. The models are then adapted for distributed processing on the sensor network and for action recognition.

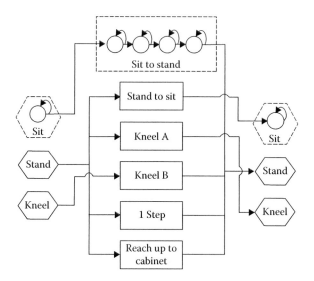

FIGURE 26.1 HMM for continuous action recognition.

Algorithm Description 26.2—Zero-Knowledge HMM Identification Algorithm

Input: Observed sequence **y**; Alphabet $\mathcal{A}$.

Initialization:

1. Set $L = 1$.
2. The set $Q_{L-1} = \{q_0\}$ and $G_{L-1} = \langle Q_{L-1}, \mathcal{A}, \delta_{L-1}, p_{L-1} \rangle$, where $(q_0, y, q_0) \in \delta_{L-1}$ for all $y \in \mathcal{A}$ and $p_{L-1}(q_0, y, q_0)$ is the proportion of times symbol y occurs in history **y**. (This is the ε-machine that results when Algorithm 26.1 is run with $L = 0$.)
3. Let the length of $N = |\mathbf{y}|$.

Main loop:

1. Let $G_L = \langle Q_L, \mathcal{A}, \delta_L, p_L \rangle$ be the ε-machine output of Algorithm 26.1 with **y**, $\mathcal{A}$ and L.
2. For every state $q_0 \in Q_L$, record the sequence of states $\mathbf{q}_L^{q_0} = \{q_1, q_2, \ldots, q_N\}$ that occurs when $\hat{\delta}_L$ is recursively applied with input **y** starting at state q. That is, $q_1 = \hat{\delta}_L(q_0, y_1)$, $q_2 = \hat{\delta}_L(q_1, y_2)$, etc. If there is some $i \leq N$ for which $q_i = \uparrow$, then we discard sequence $\mathbf{q}_L^{q_0}$ as undefined.
3. Each sequence $\mathbf{q}_L^{q_0}$ defines a partial function $f_L^{q_0} : [N] \times \mathcal{A} \to Q_L$. If q_k is the kth element of sequence $\mathbf{q}_L^{q_0}$, then $f_L^{q_0}(k, y_k) = q_k$. That is, position k with symbol y_k is associated to state q_k. Let $\mathcal{F}_L$ be the set of functions $f_L^{q_0}$ defined in this way.
4. Compare the functions in $\mathcal{F}_L$ to the elements of $\mathcal{F}_{L-1}$: We will use these sets to define a matching problem whose optimal solution will be used to define a stopping criterion.
   a. Let $\mathcal{I}$ be a set of indices corresponding to elements of Q_{L-1} and $\mathcal{J}$ be a set of indices corresponding to elements of Q_L.
   b. Define binary variables x_{ij} ($i \in \mathcal{I}, j \in \mathcal{J}$). We will declare $x_{ij} = 1$ if and only if state q_i of Q_{L-1} is matched with state q_j of Q_L.
   c. Define the following coefficients:

$$r_{ij} = \sum_{q_{0_L} \in Q_L, q_{0_{L-1}} \in Q_{L-1}} \left| (f_L^{q_{0_L}})^{-1}(q_i) \cap (f_{L-1}^{q_{0_{L-1}}})^{-1}(q_j) \right|$$

d. Solve the matching problem

$$\max_{x_{ij}} \quad m_L = \sum_{ij} r_{ij} x_{ij}$$

$$\text{s.t.} \quad \sum_{j} x_{ij} = 1$$

$$x_{ij} \in \{0,1\}$$

to obtain a matching between states in G_{L-1} and states in G_L.
5. If $|m_L - m_{L-1}| = 0$, then stop. The current value of L is the correct value.

In Ref. [21], CSSR algorithm is used for modeling behaviors of ships. For symbols, they use the Military Grid Reference System (MGRS) [38]. As a ship traverses the globe, it produces symbols corresponding to the grid regions in MGRS. They can model the behavior of the ship using CSSR. The result is a Markov model describing the probabilistic motion of a vessel as it traverses the planet. Within the regions, they also use linear models of a ship's behavior to further refine the predictions. The CSSR model and the linear models work together to produce predictions of ship location and to identify anomalies in ship behavior. Figure 26.2 shows an example of a ship prediction using models derived from CSSR for the *Disney Magic* available from http://sailwx.org

The zero-knowledge HMM identification algorithm is useful for HMM inferencing, since it does not require any a priori information about the model. In Ref. [39], it is used for inferring protocols tunneled through a Tor network. Markov models are used to represent network protocols. Each symbol represents an inter-packet delay. Each state represents a protocol state. Protocols tunneled through Tor are inferred from observed inter-packet delays. A server and a client are set up to communicate through a Tor network using a protocol. Traffic timing data are collected in the client side and using these data Markov model is inferred by zero-knowledge HMM identification algorithm. Figure 26.3 shows the model inferred from traffic timing data of a private Tor network. This shows an example of using HMMs for side channel analysis in the network.

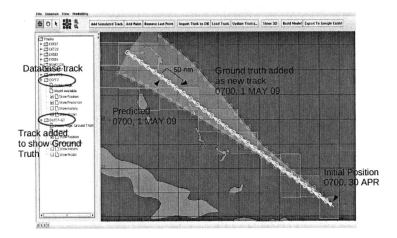

FIGURE 26.2 Two day sample of the *Disney Magic* with predictions provided in part by CSSR.

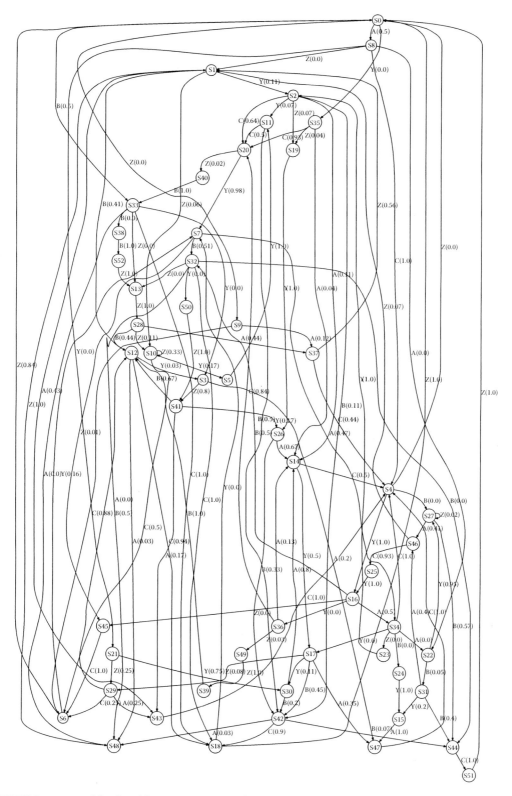

FIGURE 26.3 Model inferred from Tor network traffic.

26.5 Conclusion

HMMs are important tools for pattern recognition. HMMs have been used in sensor networks as well for a wide range of applications, that is, target action recognition, target tracking, sensor localization, and side channel analysis. The Baum–Welch algorithm uses EM methods to find the optimal parameters for a HMM. This traditional method requires the initial state structure to identify a HMM, which has been a weakness of this approach. The CSSR algorithm has alleviated much of this problem. It maximizes mutual information among state structure and next output symbol and minimizes the remaining uncertainty. The CSSR algorithm still depends on the a priori knowledge of the parameter L. The zero-knowledge HMM identification algorithm extends the CSSR algorithm to find the parameter L and, consequently, a Markov model of the dynamics of a process that generated a sequence **y**. With this extension, it is possible to generate a Markov model of a stochastic process from a sequence of symbolic observations with *zero knowledge*.

Acknowledgments

This material is based upon work supported in part by the Air Force Office of Scientific Research contract/grant number FA9550-09-1-0173, NSF grant EAGER-GENI Experiments on Network Security and Traffic Analysis contract/grant number CNS-1049765, and NSF-OCI 1064230 EAGER: Collaborative Research: A Peer-to-Peer based Storage System for High-End Computing. Opinions expressed are those of the author and neither the National Science Foundation nor the U.S. Department of Defense.

References

1. Rabiner, L.R., A tutorial on hidden Markov models and selected applications in speech recognition, *Proceedings of the IEEE*, 77(2), 257–286, February 1989.
2. Damper, R.I., Higgins, J.E., Improving speaker identification in noise by subband processing and decision fusion, *Pattern Recognition Letters*, 24(13), 2167–2173, 2003.
3. Chen, J.-L., Kundu, A., Rotation and gray scale transform invariant texture identification using wavelet decomposition and hidden Markov model, *IEEE Transactions on Pattern Analysis and Machine Intelligence*, 16(2), 208–214, February 1994.
4. Xue, H., Govindaraju, V., Hidden Markov models combining discrete symbols and continuous attributes in handwriting recognition, *IEEE Transactions on Pattern Analysis and Machine Intelligence*, 28(3), 458–462, March 2006.
5. Mozaffari, S., Faez, K., Märgner, V., El-Abed, H., Lexicon reduction using dots for off-line Farsi/Arabic handwritten word recognition, *Pattern Recognition Letters*, 29(6), 724–734, 2008.
6. Liu, Z., Sarkar, S., Improved gait recognition by gait dynamics normalization, *IEEE Transactions on Pattern Analysis and Machine Intelligence*, 28(6), 863–876, June 2006.
7. Chen, Y., Rui, Y., Huang, T.S., Muticue HMM-UKF for real-time contour tracking, *IEEE Transactions on Pattern Analysis and Machine Intelligence*, 28(9), 1525–1529, September 2006.
8. Yang, J., Xu, Y., Chen, C., Human action learning via hidden Markov model, *IEEE Transactions on Systems, Man and Cybernetics A*, 27(1), 34–44, 1997.
9. Hu, T., De Silva, L.C., Sengupta, K., A hybrid approach of NN and HMM for facial emotion classification, *Pattern Recognition Letters*, 23(11), 1303–1310, 2002.
10. Dong, L., Wu, J., Chen, X., Real-time physical activity monitoring by data fusion in body sensor networks, *10th International Conference on Information Fusion*, Québec, Canada, pp. 1–7, July 9–12, 2007.
11. Guenterberg, E., Ghasemzadeh, H., Loseu, V., Jafari, R., *Distributed Continuous Action Recognition Using a Hidden Markov Model in Body Sensor Networks*, Springer-Verlag, Berlin, Germany, DCOSS 2009, LNCS 5516, pp. 145–158, 2009.

12. Amutha, B., Ponnavaikko, M., Energy efficient hidden Markov model based target tracking mechanism in wireless sensor networks, *Journal of Computer Science*, 5(12): 1085–1093, 2009.

13. Arthi, R., Murugan, K., Localization in wireless sensor networks by hidden Markov model, *Second International Conference on Advanced Computing (ICoAC)*, Chennai, India, pp. 14–18, December 14–16, 2010.

14. Liu, S., Srivastava, R., Koksal, C.E., Sinha, P., Pushback: A hidden Markov model based scheme for energy efficient data transmission in sensor networks, *Ad Hoc Networks*, 7(5), July 2009.

15. Shalizi, C.R., Crutchfield, J.P., Computational mechanics: Patterns and prediction, structure and simplicity, *Santa Fe Institute Working Paper 99-07-044*, 2001.

16. Shalizi, C.R., Causal architecture, complexity and self-organization in time series and cellular automata, PhD dissertation, Physics Department, University of Wisconsin-Madison, Madison, WI, May 2001.

17. Shalizi, C.R., Shalizi, K.L., Crutchfield, J.P., Pattern discover in time series, Part I: Theory, algorithm, analysis, and convergence, *Santa Fe Institute Working Paper 02-10-060*, 2002.

18. Ostendorf, M., Singer, H., HMM topology design using maximum likelihood successive state splitting, *Computer Speech and Language*, 11(1), 17–42, 1997.

19. Shalizi, C.R., Shalizi, K.L., Crutchfield, J.P., An algorithm for pattern discovery in time series, SFI Working Paper 02-10-060, November 2002. Available: arXiv:cs.LG/0210025v3.

20. Shalizi, C.R., Shalizi, K.L., Blind construction of optimal nonlinear recursive predictors for discrete sequences, pp. 504–511, *UAI'04 Proceedings of the 20th Conference on Uncertainty in Artificial Intelligence*, June 2004.

21. Schwier, J.M., Brooks, R.R., Griffin, C., Bukkapatnam, S., Zero knowledge hidden Markov model inference, *UAI'04 Proceeding of the 20th Conference on Uncertainty in Artificial Intelligence, Pattern Recognition Letters*, 30(14), 1273–1280, 2009.

22. Bahl, L.R., Jelinek, F., Mercer, R.L., A maximum likelihood approach to continuous speech recognition, *IEEE Transactions on Pattern Analysis and Machine Intelligence*, 5, 179–190, 1983.

23. Upper, D.R., Theory and algorithms for hidden Markov models and generalized hidden Markov models, PhD dissertation, Mathematics Department, University of California at Berkeley, Berkeley, CA, 1989.

24. Baum, L.E., Petrie, T., Soules, G., Weiss, N., A maximization technique occurring in the statistical analysis of probabilistic functions of Markov chains, *Annals of Mathematical Statistics*, 41(1), 164–171, 1970.

25. Zaki, M.J., Jin, S., Bystroff, C. Mining residue contacts in proteins using local structure predictions, *IEEE Transactions on Systems, Man, and Cybernetics, Part B*, 33(5), 789–801, October 2003.

26. Yang, J., Xu, Y., Chen, C.S., Human action learning via hidden Markov model, *IEEE Transactions on Systems, Man and Cybernetics, Part A*, 27(1), 34–44, January 1997.

27. Baum, L.E., Egon, J.A., An inequality with applications to statistical estimation for probabilistic functions of a Markov process and to a model for ecology, *Bulletin of the American Meteorological Society*, 73, 360–363, 1967.

28. Baum, L.E., Sell, G.R., Growth functions for transformations on manifolds, *Pacific Journal of Mathematics*, 27(2), 211–227, 1968.

29. Min, L., Shun-Zheng, Y., A network-wide traffic anomaly detection method based on HSMM, *2006 International Conference on Communications, Circuits and Systems Proceedings*, Guilin, China, Vol. 3, pp. 1636–1640, June 25–28, 2006.

30. Shun-Zheng, Y., Kobayashi, H., An efficient forward-backward algorithm for an explicit-duration hidden Markov model, *IEEE Signal Processing Letters*, 10, 11–14, 2003.

31. Aho, A.V., Hopcroft, J.E., Ullman, J.D., *The Design and Analysis of Computer Algorithms*, Addison-Wesley, Reading, MA, 1974.

32. Dan, Q., Bingxi, W., Honggang, Y., Guannan, D., Discriminative training of GMM based on maximum mutual information for language identification, *Proceedings of 6th World Congress on Intelligent Control and Automation*, Dalian, China, June 21–23, 2006.

33. Kim, D., Yook, D., Spectral transformation for robust speech recognition using maximum mutual information, *IEEE Signal Processing Letters*, 14(7), 496–499, July 2007.

34. Biem, A., Minimum classification error training for online handwriting recognition, *IEEE Transactions on Pattern Analysis and Machine Intelligence*, 28(7), 1041–1051, July 2006.

35. Juang, B.-H., Discriminative learning for minimum error classification, *IEEE Transactions on Signal Processing*, 40(12), 3043–3054, December 1992.

36. Cappe, O., Moulines, E., Ryden, T. *Inference in Hidden Markov Models*, Springer Verlag, New York, 2005.

37. Milner, R., *Communications and Concurrency*, Prentice Hall, Upper Saddle River, NJ, 1989.

38. National Geospatial-Intelligence Agency, *Technical Manual 8358.1: Datums, Ellipsoids, Grids and Grid Reference Systems*, Chapter 3, 1990. Available: http://www.cartome.org/nimagrids.htm.

39. Craven, R., Traffic analysis of anonymity systems, Masters thesis, Clemson University, Clemson, SC, May 2010.

27

Emergence of Human-Centric Information Fusion

27.1 Introduction 581
27.2 Participatory Observing: Humans as Sensors 583
27.3 Hybrid Human/Computer Analysis: Humans as Analysts 586
27.4 Collaborative Decision Making: Crowd-Sourcing of Analysis....587
27.5 Summary..589
Acknowledgments..590
References..590

David L. Hall
The Pennsylvania
State University

27.1 Introduction

The conventional focus of the data fusion community has involved using physical sensor sources such as visual and infrared imagery, radar, satellite, and acoustic sensor data to observe physical entities like troops, vehicles, weapon systems, or other objects. This type of sensor activity has been useful for performing tracking, situation assessment, and threat assessment for military operations [1,2]. Two recent factors have caused a major reassessment of this paradigm. First, military emphasis has largely shifted from conventional warfare to the challenges of counterinsurgency and counterterrorism [3]. Second, the emerging concept of human-centered information fusion [4] involves exploring new ways in which humans and computer systems can work together to address challenges to optimally utilize the capabilities of physical sensors, computer hardware and software, supporting cyber-infrastructure, and human beings [5–7].

Although counterinsurgency tasks such as defeating improvised explosive devices (IEDs) still rely on observation of physical entities (such as explosives, vehicles, and communication devices), it also requires an in-depth understanding of the social networks, intent, belief systems, connectivity, policies, and procedures that drive the process [8]. While physical sensors can be useful in detecting some of the physical manifestations of these abstract concepts, they are largely unable to effectively classify their most important aspects without human intervention. Historically, the human role in the fusion process has largely been limited to analyzing the completed output of the fusion system. Unfortunately, when the human ability to observe and ascertain intent, beliefs, and cultural influences is critical to the sense-making process, any fusion system that withholds the role of the human until the end of the process will necessarily be suboptimal.

The data fusion process is evolving to effectively integrate humans in a variety of roles [4]. The concept of human-centric information fusion is shown in Figure 27.1. First, there is the role of humans as observers (i.e., "soft sensors" or participatory sensing [5–7]). Although not able to compete with conventional hard sensors at many tasks, humans have an unparalleled ability to instinctively and intuitively make

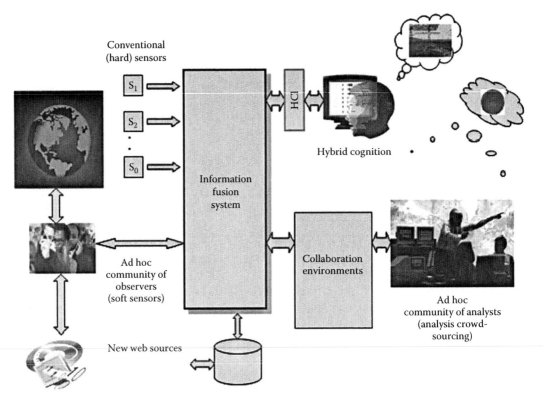

FIGURE 27.1 Human-centric information fusion. (Adapted from Liggins, M.E. et al., *Handbook of Multisensor Data Fusion: Theory and Practice*, 2nd edn., CRC Press, Boca Raton, FL, 2008.)

sense of complex situations and interactions that would leave the most advanced state-of-the-art computerized sense-making systems at a loss. With the assistance of ubiquitous smart phones and robust data networks, human observers have the capability to transmit annotated, geospatially and temporally stamped high-resolution imagery and video nearly instantly. Additionally, social networking tools such as Twitter facilitate crowd-sourced sensing that can be either tasked or opportunistic [9]. When tasked sensing is performed, the collection of desired data is requested of the human sensor. Opportunistic sensing uses data that were gathered and annotated for other purposes and published to the open source community.

A second role of humans in the data fusion process is to augment the computational capability of a computer system by performing tasks such as advanced pattern recognition, context-based analysis, and disambiguation for certain knowledge-based tasks that computers are poorly suited to perform. For example, a modern computer vision system might be very effective at detecting the presence of a backpack within a video feed of a crowded area. However, computer vision systems perform very poorly at determining if the person carrying that backpack has malicious intents. Humans, on the other hand, have an innate ability (which can be further improved by training) to almost instantly process the appearance, gestures, movements, and other factors necessary to make rapid and accurate threat assessments.

A third role of humans in data fusion systems is that of collaborative decision maker or member of a social cognition team. Advances in collaborative technology are enabling richer and more seamless exchange of information over geographically distributed groups. Web-based versions of these tools facilitate the rapid formation and reformation of ad hoc cognition teams in response to natural disasters, terrorist events, or other threats to the public health and well-being. Moreover, digital natives [10] tend to access their peers (the "hive mind") to assist in solving problems and seeking advice. The use of a group or crowd to solve problems has been termed "crowd-sourcing" [11], and online collaborations are routine using web sites such as the *Mechanical Turk* (https://www.mturk.com/mturk/welcome).

Although the promise of human-centric data fusion is great, so are the difficulties. While physical sensors are designed to perform a specific task, human observers are complex entities that require a new model of tasking and knowledge elicitation to be effective. While physical sensors are calibrated to detect information over a very specific domain, human "sensors" have a strong ability to detect intricacies of human behavior, mood, and motivation. Although this type of information is invaluable, it is also exceedingly difficult to classify and represent in a format that is machine readable and facilitates fusion between hard and soft sensors. Semantic data representation formats such as resource description framework (RDF) and the web ontology language (OWL) offer a good starting point, but many challenges remain in effective representation of this type of knowledge.

An additional challenge is in determining the quality and reliability of human observations. The test and evaluation process of data fusion systems have typically relied on measures of performance (MOPs) and measures of effectiveness (MOEs) [12]. MOPs evaluate the accuracy and reliability with which a certain task can be preformed. Examples include detection probability, false-alarm rate, location estimation accuracy, identification probability, and time from transmission to detection. MOEs evaluate the system at a higher level of abstraction. They evaluate the capability of a data fusion system to contribute to the overall success of a mission or operation. They include target nomination, timeliness of information, warning time, target leakage, and countermeasure immunity.

Calculating these metrics for conventional hard sensor systems is relatively straightforward; human-based soft sensor systems are nearly impossible to evaluate using these existing methods [13]. While there have been excellent attempts at alignment of human information perception through "cultural lenses" [14] and discussion of team-based versus individual cognition [15], there has not yet been an effective and comprehensive model developed for human tasking and evaluating the role of humans as participatory sensors in a distributed hard and soft information fusion system.

This chapter provides a discussion of the evolving roles of humans as part of information fusion systems, acting as sensors, analysts, and collaborative decision makers. The chapter introduces each role and describes challenges and issues for each area.

27.2 Participatory Observing: Humans as Sensors

The basic concept of humans acting as observers or sensors is certainly not new. In the history of science, human vision, for example, was augmented by telescopes, microscopes, cameras, and multispectral devices. However, prior to the advent of digital instruments, such as photomultipliers, observations by a trained observer were more valued than instruments. For example, in astronomical observations, a trained observer can view an object through a telescope, and the eye–brain combination intelligently integrates the view so that the observer can "ignore" the effects of variable atmospheric seeing and mentally connect the periods of good seeing. By contrast, celestial photography integrates both the "good" and the "bad" seeing (variations in the atmosphere), resulting in a photograph that is blurred compared to the best glimpse by a human observer. At one time, the declarations (and drawings) of a trained observer were more valued than that of photographic evidence. Human observers also have the ability to react to the unexpected. Meadows [16], for example, cites the discovery of orange-colored glass during the Apollo 17 lunar mission—a surprising observation that would likely have been missed by an unmanned probe.

In addition to the impact of individual observers, groups of ad hoc observers can be tasked. Charles Darwin is sometimes described as the original crowd-sourced scientist because of his habit of writing up 1500 letters per year to gather biological evidence to support his theory of evolution (see http://www.cbsnews.com/stories/2011/04/29/scitech/main20058629.shtml). Individual and groups of amateur observers have made discoveries that simply would not or could not be made by utilization of collected sensor data. A recent example is the discovery by Ruth Brooks, a 69 year old gardener, that snails have a sophisticated homing ability—capable of returning to a preestablished "home" from up to 300 m [17]. In other cases, citizen scientists or participatory observers are tasked with collecting data to augment more traditional physical sensors. Examples include the following: (1) Japanese citizens mapping radiation leaking from the damaged Fukushima nuclear plant (http://worldblog.msnbc.msn.com/_news/2011/07/12/7036501-japans-citizen-scientists-map-radiation-diy-style), (2) the United States Geological Service (USGS) earthquake hazards program that invites amateurs to report on how strong an earthquake felt (http://earthquake.usgs.gov/dyfi) using the concept of "did you feel it" to augment seismic data, (3) twitter reports of crime and information from first responders (http://blog.crimereports.com/tag/twitter/), and (4) the use of twitter reports for responding to emergency events [18].

A review of concepts and applications of participatory sensing is provided by Ref. [19]. Hall et al. [20] summarize the challenges involving how to characterize human observers. Issues in the utilization of humans as soft sensors involve knowledge elicitation and focus of attention. Numerous examples are available of the failure of human observers to notice critical information or, conversely, to report mistaken observations. An example of the former is the failure of students at the Virginia Tech University to observe the precursors to the shooting spree in which a 23 year old student, Seung-Hui Cho, killed 27 students, 5 teachers, and himself. Observable precursors included Cho practicing locking a building with chains (to prevent police from entering and stopping his shootings), 2 days prior to the actual event. Kaplan and Kaplan [21] cite the following example (p. 82): "In 1978 the Rotterdam zoo reported the escape of one of its red pandas; hundreds of helpful people called in, having spotted it in places all over the Netherlands—when in fact it had been run over by a train just a few yards from the zoo fence." Other examples of challenges in human observing and focus of attention include experiments involving the use of distractions. In an experiment conducted by Simons and Chabris [22] at the University of Illinois, for example, subjects were shown a video clip of two teams of people in white and black uniforms passing a basketball back and forth. The subjects were asked to count the number of passes by one team during a 60 s period. During the video, a person in a gorilla suit walks between the players, stops and waves, and continues on. A surprisingly high number of subjects fail to notice the gorilla (http://viscog.beckman.illinois.edu/flashmovie/15.php). Thus, part of the effort associated with evaluating the use of humans as observers is to understand issues such as focus of attention, knowledge elicitation, observer bias and training, and other effects.

Despite these issues related to humans acting as observers, there are also significant benefits from using human inputs. For example, human observers can provide contextual information that significantly assists in understanding an evolving situation. Moreover, humans can infer concepts such as intent and human interrelationships that cannot be observed by physical sensors. Examples of the effective use of human observers (viz., participatory sensing) include the following:

- *Law enforcement*: for example, monitoring police reports and events and posting these for public use [23–25] or tasking non-police workers such as sanitation workers to identify anomalies [26]
- *Social movements or cultural identity support*: for example, the Photovoice Movement project in which rural Chinese women documented their daily lives via 35 mm cameras, which raised awareness of government officials regarding childcare and midwife needs [9], use of web mashup mechanisms such as Ushahidi [27] to report political oppression
- *Environmental monitoring*: for example, use of cell phones and specialized sensors to report air quality [28]
- *Crisis management and reporting*: sharing of information by volunteers to support disaster relief [29,30]
- *Citizen scientists*: use of civilians as observers for scientific phenomena such as earthquakes [31]

It can be argued that the combination of proliferation of mobile computing/communications devices, the increasing WiFi and cellular communications networks, and the evolution of the digital natives [10] combine to ensure that soft sensing will increasingly become an important information source.

Examples of types of known challenges in observing/reporting are provided in Table 27.1 (adapted from Ref. [32]). The table summarizes four observer "functions/actions" (detection, meaning extraction, remembering information, and reporting) and indicates the types of cognitive processes, examples of deficiencies and inhibitors, and known biases. It would be conceptually possible to explore these functions in detail, developing a sort of soft sensor version of a hard sensor functional processing flow. In turn, one might attempt to model these functions and the associated biases and error types. However, this remains a challenging and difficult area. By contrast, this conceptual functional "model" could be useful in guiding methods to explore the impact of observer limitations and even provide methods for bias detection.

TABLE 27.1 Examples of Challenges in Human Observing/Reporting

Observer Action	Cognitive Processes	Example Deficiencies/Inhibitors	Known Biases
Detection	• Selective attention	• Low salience of event • High effort required to observe event • Low expectancy of event • Low value/cost of missing event	• Vigilance failure • Stereotype fixation • Signal unreliability • Signal absent • Signal-discrimination failure
	• Divided attention	• High resource demand • High structural similarity in tasks • Task management	• Cognitive/stimulus overload • Cognitive tunneling
Meaning extraction	• Perception	• Perceptual processing • Degradation of stimuli • Low frequency and poor context for event	• Attention failure • Multiple signal coding • Thematic vagabonding • Encystment • Crew-function problem • Cultural lens effects
Remembering information	• Working memory	• Capacity • Time • Confusability and similarity • Attention and similarity	• Information failure after magic number 7
	• Long-term memory (storage)	• Weak strength (low frequency of recency) • Weak/few associations • Interfering associations • Incongruent semantic networks • Incomplete or inaccurate mental models	• Schema modification
Reporting	• Long-term memory (retrieval) and articulation	• Incomplete or inaccurate information • Inability to formulate textual information related to perception/memory	• Confidence bias • Hindsight bias • Judgment bias • Political bias • Sponsor bias • Professional bias • Recognition bias • Confirmation bias • Frequency bias • Recency bias • Guilty bias

Source: Adapted from Hall, D. et al., Perspectives on the human side of data fusion: Prospects for improved effectiveness using advanced human-computer interfaces, In *Handbook of Multisensor Data Fusion*, 2nd edn., M. Liggins, D. Hall, and J. Llinas (Eds.), Chapter 20, CRC Press, Boca Raton, FL, 2008.

27.3 Hybrid Human/Computer Analysis: Humans as Analysts

A second role of humans in human-centric information fusion is acting as a collaborator with automated computer-based pattern recognition or inference processes. Traditional data-driven fusion systems often provided a situation display or common operational picture for the analyst/decision maker; the role of the human was relatively passive. That is, the information fusion "engine" conducts automated pattern recognition (e.g., for functions such as target identification or target aggregation) and automated reasoning via level-2 or level-3 fusion, and the results of this processing are provided for the analyst to peruse. This is an oversimplification. However, the human interaction generally occurs after the automated processing was performed. In part, this is due to the time frame of processing (viz., the difference between the speed of data ingestion and computational processing versus the human cognition rate) and also, in part, due to a design consideration related to how human cognition is viewed.

Hall et al. [33] have discussed the use of new types of visual displays (e.g., three-dimensional [3D] full immersion interactions) to support collaboration between the automated computing and human cognition. New display technologies such as 3D, full immersion displays allow a user to be surrounded by a 3D display (see Figure 27.2). This allows new concepts such as using a third dimension (e.g., height above the "floor") to represent concepts such as the passage of time (e.g., showing activity over the period of an hour, day, or year), level of abstraction (using increasing height to represented levels of abstraction with the bottom layer representing "raw" data and the higher levels representing abstractions of the data such as semantic labels or icons), etc.

A significant amount of literature exists on the use of graphic displays to support interpretation of data (see, e.g., Refs. [34–36]), and visualization scientists/artists such as Edward Tufte [37,38] provide wonderful examples of how creative displays of information can lead to significant insights.

In addition to visualization, other human senses can be used to assist in interpreting data including hearing (sonification) and the sense of touch (haptic interfaces). Rosenblum [39] provides interesting examples of the power of our human senses for understanding the environment and assisting in making inferences. Regarding sonification, Ballora et al. [40,41] have demonstrated the utility of using sound to represent information about the state and health of a complex cyber network, showing how, with proper design, the human ear can readily detect and identify different types of cyber attacks. Similarly, haptic interfaces are emerging to emulate the sense of touch for data in a computer. Research at the University of Colorado [42] has focused on the combination of visual and haptic interfaces for understanding data. Menelas et al. [43] provide a survey of haptic techniques for understanding large

FIGURE 27.2   Example of immersive data fusion display.

data sets represented by scalar and vector fields. Finally, one might use haptic interfaces to assist in understanding of second-order uncertainty (viz., determining how reliable an uncertainty estimate is). For example, an error ellipsoid, representing the uncertainty of a target location could be represented as a hard surface if the uncertainty of the covariance is well known and a soft or "squishy" surface if the uncertainty is less well known.

While the technologies for computer interfaces in the visual and haptic arenas are rapidly evolving, the use of these technologies to improve data and situation understanding remains challenging. Both for visual and sound interfaces, a certain amount of art or design must be used to fully evoke the power of human senses to understand the represented data. Nevertheless, this is a promising area for research and continued development.

In addition to using our senses for pattern recognition, another aspect of human support of situation awareness is the idea of use of our semantic ability to support contextual information. Pinker [44–46] provides elegant discussions of the nature of language and human cognition. The human role to augment automated computer processing can use the power of language and context to improve the interpretation of an evolving assessment of a situation, activities, or events. Hall and Jordan [4] have suggested that information fusion systems might seek to translate data into semantic terms to allow a "storification" process in which data and information are transformed into semantics so that the human in the loop can interpret information via evolving explanations or "stories" of the evolving scene. This would emulate the natural self-talk process we use for understanding the dynamic world.

A dramatic example of how even a single word can change our interpretation of a situation is provided the famous paragraph presented by Burton [47]. To understand this, the reader is asked to read and reflect upon the following paragraph provided by Burton, without reading ahead:

A newspaper is better than a magazine. A seashore is a better place than the street. At first it is better to run than to walk. You may have to try several times. It takes some skill, but it is easy to learn. Even young children can enjoy it. Once successful, complications are minimal. Birds seldom get too close. Rain, however, soaks in very fast. Too many people doing the same thing can also cause problems. One needs lots of room. If there are no complications, it can be very peaceful. A rock will serve as an anchor. If things break loose from it, however, you will not get a second chance.

Upon first reading, this paragraph provides a sense of uncertainty. The paragraph is in clear English, but what does it mean? However, as soon as you are provided with a single word, *kite*, the feeling of uncertainty disappears and you immediately see that the paragraph makes sense. This contextual information provided by the word kite allows you to understand the relationships among the places, actions, environment, and activities. While such a description as presented in Burton's paragraph is unusual (and deliberately made to be ambiguous), ordinary collections of data, tweets, news reports, and human observations can also be ambiguous. The power of a human in the loop can be the contextual interpretation of such evolving descriptions to achieve improved situation awareness.

27.4 Collaborative Decision Making: Crowd-Sourcing of Analysis

A final role of humans interacting with information fusion systems involves the role of the human decision maker and collaborative decision maker/analyst. The combination of rapidly emerging information technologies, such as virtual world technologies and social networking web tools, along with the propensity for digital natives to seek group collaboration on a routine basis provides both the opportunity and motivation for groups of people to be involved in data analysis and decision making related to information fusion. This type of collaboration can be among an established but distributed team, a small group of friends, or an ad hoc collaboration among a large, distributed group. The collaboration can take place in semi-real time to address a particular problem or over a longer period of time to address multiple problems. Examples of large-scale collaborations over time include the creation of Wikipedia [48] containing over 3.6 million articles, Amazon's online Mechanical Turk service that provides opportunities for posting of problems for both paid and unpaid crowd-based solutions, and sites such as http://www.domywork.net/ that provides a place for students and workers to post problems and projects to be solved for a specified fee.

The emerging area of virtual world technologies enables multiperson distributed collaboration in a common virtual environment. Pursel [49] provides an overview of these technologies and describes their use for collaborative problem solving and analysis. These environments typically enable the creation of 3D artifacts, such as buildings, landscapes, and meeting rooms, and all the creation of avatars to represent the collaborating participants. Examples of tools include the following: *Second Life* [50], *OLIVE* [51], and *Protosphere* [52]. A sample screen shot from OLIVE is shown in Figure 27.3.

It remains to be seen how effective and widespread the use of virtual world technologies will be in practice. On the one hand, the gaming community routinely involves huge collaborations for dynamic problem solving (and competition) in a gaming environment. For example, the popular World of Warcraft [53] engages over 12 million players worldwide. Teams of 40 players self-organize to address the most challenging elements of the game (e.g., fierce dragons, mystical gods of the elements, etc.) and interact to collectively analyze data, develop team strategies, and execute very detailed and intricate plans of action. The U.S. military has taken advantage of virtual worlds for simulated training using the tool OLIVE, by Forterra Systems, Inc. Interestingly, a new concept of "gamification" is evolving [54], in which the concept of games is deliberately used to engage groups of people to address a challenging problem.

The power of a challenge is illustrated by the Defense Advanced Research Project Agency's (DARPA's) red balloon contest [54]. Upon the 40th anniversary of the creation of DARPA, the federal agency issued a national contest in which teams were challenged to find and verify the location of 10 red weather balloons tethered in 10 different locations around the continental United States on a Saturday, December 5, 2009. A prize of $40,000 was promised as an award to the winning team who could first locate and verify the position of all 10 balloons. Approximately 3800 teams registered to compete. Many of these teams used a social networking strategy in which they created web sites and sought to motivate participatory observers to report their observations. Not surprisingly, there were a very large number of false reports and deliberate deceptions involving doctored photographs and misleading information. A team of the Pennsylvania State University undergraduate students participated and used a combined strategy using two subteams. One subteam used the standard social networking approach, while the other team

FIGURE 27.3   Screenshot from OLIVE. (Taken from http://www.forterrainc.com)

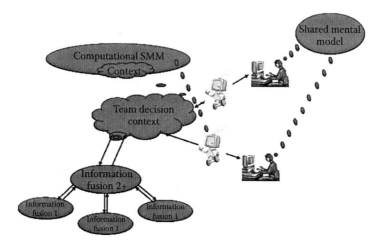

FIGURE 27.4 Concept of hybrid human team and multiagent team cognition.

used an approach of "watching the watchers" (viz., monitoring dynamic reports via Twitter, Facebook, and other web sites and seeking to verify or refute the emerging observations). This combined strategy enabled this student team to come in tenth. A description of this strategy, along with the winning strategy developed by a team from MIT, is provided in Ref. [55].

While the technology to support distributed team collaboration is rapidly advancing, advances are also being made in the study of team-based cognition and the dynamics of team mental models (see, e.g., Refs. [56–58]). Endsley et al. have conducted numerous studies involving how teams conduct situation awareness and decision making, particularly in military environments [58]. Warner et al. [57] and McNeese [59] have conducted experiments with human teams to understand the dynamics of group decision making and team cognition. Additional research will need to be conducted on the scalability of team cognition, especially using evolving social network tools.

Finally, it should be noted that research is being conducted on hybrid cognition involving human collaboration mixed with teams of intelligent agents. For example, Fan et al. [60] have explored the development of intelligent agents that emulate the behavior of human teams (viz., having different types of expertise, sharing information in a decision-making process, etc.) and how such teams of intelligent agents could interact with a team of human analysts/decision makers. This concept is illustrated in Figure 27.4. The intelligent agents use the recognition-primed decision (RPD) making paradigm for human cognition, developed by Klein et al. [61], which emulates how experts operate in real-world environments. John Yen and his associates have developed prototype systems of this type for military decision-making applications as well as emergency management situations.

27.5 Summary

Rapid changes in communications and mobile computing devices enable a new era of human observations. The combination of ubiquitous sensing and the proliferation of cell phone technology create a self-aware planet in which human and physical sensor observations can be combined for global situation awareness. In addition, advances in human–computer interaction technology and collaboration technologies enable humans to augment automated fusion systems by pattern recognition, semantic level contextual reasoning, and distributed ad hoc team collaboration. These lead to an emerging concept of human-centered information fusion in which humans are fully engaged in observations, reasoning and analysis, and collaboration. However, many challenges must be addressed to take advantage of these opportunities. It is anticipated that the data fusion and cognitive psychology communities will focus on these challenges for ultimate routine fusion of hard and soft data.

Acknowledgments

We gratefully acknowledge that this research activity has been supported in part by a Multidisciplinary University Research Initiative (MURI) grant (Number W911NF-09-1-0392) for "Unified Research on Network-Based Hard/Soft Information Fusion", issued by the U.S. Army Research Office (ARO) under the program management of Dr. John Lavery.

References

1. D. Hall and S. A. H. McMullen, *Mathematical Techniques in Multisensor Data Fusion*, 2 edn., Artech House, Boston, MA, 2004.
2. M. E. Liggins, D. L. Hall, and J. Llinas, *Handbook of Multisensor Data Fusion: Theory and Practice*, 2nd edn., CRC Press, Boca Raton, FL, 2008.
3. E. Cohen, C. Crane, J. Horvath, and J. Nagl, Principles, imperatives, and paradoxes of counterinsurgency. *Military Review,* March–April 2006: 49–53.
4. D. Hall and J. M. Jordan, *Human-Centered Information Fusion*, Artech House, Boston, MA, 2010.
5. D. Hall, M. McNeese, J. Llinas, and T. Mullen, A framework for hard/soft fusion, *Proceedings of the 11th International Conference on Information Fusion*, Cologne, Germany, pp. 1–8, June 30–July 03, 2008.
6. D. Hall, Challenges in hard and soft fusion: Worth the effort? *Proceedings of the SPIE Defense, Security and Sensing Symposium*, Orlando, FL, April 25–29, 2011.
7. J. Llinas, R. Nagi, D. Hall, and J. Lavery, A multidisciplinary university research initiative in hard and soft information fusion: Overview, research strategies and initial results, *Proceedings of the 13th International Conference on Information Fusion*, Edinburgh, U.K., July 2010.
8. *U.S. Army Counterinsurgency Handbook*. Skyhorse Publishing Inc., New York, pp. 3–4, 2007.
9. J. Burke, D. Estrin, M. Hansen, A. Parker, N. Ramanathan, S. Reddy, and M. B. Srivastava, Participatory sensing, *Proceedings of WSW'06 at SenSys'06*, Boulder, CO, October 31, 2006.
10. J. Palfrey and R. Gasser, *Born Digital: Understanding the First Generation of Digital Natives*, Basic Civitas Books, New York, 2008.
11. J. Howe, *Crowdsourcing: Why the Power of the Crowd Is Driving the Future of Business*, Crown Publishing Group, NY, 2008.
12. E. Waltz and J. Llinas, *Multisensor Data Fusion*, Artech House, Boston, MA, 1991.
13. D. Hall, J. Graham, L. More, and J. Rimland, Test and evaluation of soft/hard information fusion systems: A test environment, methodology and initial data sets, *Proceedings of the 13th International Conference on Information Fusion*, Edinburgh, U.K., 2010.
14. D. J. Saab, An ethno-relative framework for information systems design, *AMCIS 2008 Proceedings,* Toronto, Ontario, Canada, Paper 152, 2008.
15. D. McNeese, P. Bains, I. Brewer, C. E. Brown, E. S. Connors, T. Jefferson, Re. E. Jones, and I. S. Terrell, The Neocities simulation: Understanding the design and methodology used in a team emergency management simulation, *Proceedings of the Human Factors and Ergonomics Society 49th Annual Meeting*, Santa Monica, CA, pp. 591–594, 2005.
16. J. Meadows, Apollo's scientific legacy, *New Scientist*, 83(1165), 300, 1979.
17. R. Savill, Snails have homing instinct, amateur scientist discovers, *The Telegraph*, July 12, 2011.
18. A. L. Hughes and L. Palen, Twitter adoption and use in mass convergence and emergency events, *Proceedings of the 6th International ISCRAM Conference*, Gothenburg, Sweden, 2009.
19. NCZIF faculty *Participatory Sensing: A Review of the Literature and State of the Art Practices*, Technical report prepared by the staff of the Penn State Center for Network Centric Cognition and Information Fusion, November 20, 2009.
20. D. Hall, N. McNeese, and J. Llinas, H-space: Humans as observers. In *Human Centered Information Fusion*, D. Hall and J. Jordan (Eds.), Chapter 3, Artech House, Boston, MA, pp. 59–84, 2010.
21. M. Kaplan and E. Kaplan, *Bozo Sapiens: Why to Err Is Human*, Bloomsbury Press, New York, 2009.

22. D. J. Simons and C. F. Chabris, Gorillas in our midst: Sustained in-attentional blindness for dynamic events, *Perception*, 28, 1059–1074, 1999.

23. Crime reports, http://crimereports.com (2010 Public Engines, Inc., downloaded Apr. 26, 2012).

24. Crime mapping, http://www.crimemapping.com (The Omega Group, Apr. 26, 2012).

25. Spot crime reporting, http://www.spotcrime.com (2007–2012, Report see, Inc., Apr. 26, 2012).

26. J. D. Glater, Helping keep a city clean, and maybe safer, *New York Times*, January 18, 2009.

27. http://www.ushahidi.com/ (2008–2012 Ushahidi, Apr. 26, 2012).

28. E. Paulos, R. J. Honicky et al., Citizen science: Enabling participatory urbanism. In *Handbook of Research on Urban Informatics: The Practice and Promise of the Real-Time City*, M. Foth (Ed.), Information Science Reference, IGI Global, Hershey, PA, 2009.

29. C. Jones and S. Mitnick, Open source disaster recovery, *First Monday*, 2006, http://131.193.153.231/www/issues/issue11_5/jones/index.html

30. B. Schneiderman and J. Preece, 911.gov: Community response grids, *Science*, 315(5814), 944, 2007.

31. A. L. Hughes and L. Palen, Twitter adoption and use in mass convergence and emergency events, *International Journal of Emergency Management*, 6(3–4), 248–260, 2009.

32. K. Hamilton, *Literature Review of Issues in Participatory Sensing*, Penn State College of Information Sciences and Technology Technical Report, May 2011.

33. D. Hall, C. M. Hall, and S. A. H. McMullen, Perspectives on the human side of data fusion: Prospects for improved effectiveness using advanced human-computer interfaces. In *Handbook of Multisensor Data Fusion*, 2nd edn., M. Liggins, D. Hall, and J. Llinas (Eds.), Chapter 20, CRC Press, Boca Raton, FL, 2008.

34. C. Plaisant, The challenge of information visualization evaluation, *Proceedings of the Working Conference on Advanced Visualization Interfaces*, pp: 109–116, Gallipoli, Italy, 2004.

35. J. S. Yi, Y.-Ah. Kang, J. T. Stasko, and J. A. Jacko, Understanding and characterizing insights: How do people gain insights using visual information, *Proceedings of the 2008 Conference on Beyond Time and Errors: Novel Evaluation Methods for Information Visualization*, Article No. 4, Florence, Italy, ACM, New York, 2008.

36. C. Chen, *Information Visualization: Beyond the Horizon*, 2nd edn., Springer, New York, 2004.

37. E. Tufte, *The Visual Display of Quantitative Information*, Graphic Press, Levin, New Zealand, 2001.

38. E. Tufte, *Visual Explanation: Images and Quantities, Evidence and Narrative*, Graphics Press, Levin, New Zealand, 1997.

39. L. D. Rosenblum, *See What I'm Saying: The Extraordinary Powers of Our Five Senses*, W. W. Norton & Company, New York, 2010.

40. M. Ballora, B. Panulla, M. Gourley, and D. Hall, Sonification of web log data, *Proceedings of the 2010 International Computer Music Conference*, NYC/SUNY Stony Brook, New York, pp. 498–501, June 1–5, 2010.

41. M. Ballora and D. Hall, Do you see what I hear: Experiments in multi-channel sound and 3-D visualization for network monitoring? *Proceedings of the SPIE Conference on Defense, Security and Sensing*, Orlando, FL, April 5–9, 2010.

42. http://osl-www.colorado.edu/Research/haptic/hapticInterface.shtml (Univ. of Colorado, downloaded Apr. 26, 2012).

43. B. Menelas, M. Ammi, P. Bourdot, and S. Richir, Survey on haptic rendering of data sets: Exploration of scalar and vector fields, *Proceedings of the Virtual Reality International Conference (VRIC)*, Laval, France, 2008.

44. S. Pinker, *How the Mind Works*, W. W. Norton & Company, New York, 1997.

45. S. Pinker, *The Language Instinct,* Harper Perennial Modern Classics, New York, 2000.

46. S. Pinker, *The Stuff of Thought: Language as a Window into Human Nature*, Viking Adult, New York, 2007.

47. R. A. Burton, *On Being Certain: Believing You Are Right even When You're Not*, St. Martin's Press, NY, 2008.

48. http://www.wikipedia.org/ (Wikimedia Project, downloaded Apr. 26, 2012).

49. B. Pursel, Virtual world technologies, In *Human-Centered Information Fusion*, D. L. Hall and J. M. Jordan (Eds.), Chapter 10, Artech House, Boston, MA, 2010.
50. http://secondlife.com/ (Linden Lab, downloaded Apr. 26, 2012).
51. http://www.saic.com/products/simulation/olive/ (SAIC Corp. downloaded Apr. 26, 2012).
52. http://www.protonmedia.com/ (Proton Media Microsoft Partner, downloaded Apr. 26, 2012).
53. http://us.battle.net/wow/en/ (Blizzard Entertainment Inc., downloaded Apr. 26, 2012).
54. W. McGill, The gamification of risk management, internet blog at http://www.professormcgill.com/, downloaded on February 22, 2011.
55. https://networkchallenge.darpa.mil/rules.aspx (U.S. Govt. Defense Adv. Research Proj. Agency downloaded Apr. 26, 2012).
56. J. Tang, M. Cebrian, N. Giacobe, H.-W. Kim, T. Kim, and D. Wickert, Reflecting on the DARPA red balloon challenge, *Communications of the ACM*, 54(4), 78–85, April 2011.
57. N. Warner, M. Letsky, and M. Cowen, Cognitive model of team collaboration: Macro-cognitive focus, *Proceedings of the Human Factors and Ergonomics Society*, Cognitive Engineering and Decision-Making, New Orleans, LA, pp. 269–273, January 2005.
58. M. R. Endsley, W. B. Jones, K. Schneider, M. McNeese, and Mica Endsley, A model of inter- and intra-team situational awareness: Implications for design, training and measurement. In *New Trends in Cooperative Activities Understanding System Dynamics in Complex Environments*, M. McNeese, J. Salas, and M. Endsley (Eds.), Human Factors and Ergonomics Society Press, Santa Monica, CA, pp. 46–67, 2001.
59. M. McNeese, Metaphors and paradigms of team cognition: A twenty year perspective, *Proceedings of the Human Factors and Ergonomics Society Annual Meeting*, Denver, CO, pp. 518–522, January 2003.
60. X. Fan, B. Sun, S. Sun, M. McNeese, and J. Yen, RPD-enabled agents teaming with humans for multi-context decision making, *Proceedings of the fifth International Joint Conference on Autonomous Agents and Multi-Agent Systems*, Hakodate, Japan, ACM, New York, pp. 34–41, 2006.
61. G. Klein, J. Orasanu, R. Calderwood, and C. E. Zsambok, *Decision Making in Action: Models and Methods*, Ablex Publishing Co., Norwood, MA, 1993.

IV

Power Management

28 **Designing Energy-Aware Sensor Systems** *N. Vijaykrishnan, M.J. Irwin,*
   M. Kandemir, L. Li, G. Chen, and B. Kang 595
   Introduction • Sources of Power Consumption • Power
   Optimizations: Different Stages of System Design • Energy Reduction
   Techniques • Conclusions • Acknowledgments • References

29 **Operating System Power Management** *Vishnu Swaminathan*
   and Krishnendu Chakrabarty 609
   Introduction • Node-Level Processor-Oriented Energy Management • Node-Level I/O-
   Device-Oriented Energy Management • Conclusions • References

30 **An Energy-Aware Approach for Sensor Data Communication** *H. Saputra,*
   N. Vijaykrishnan, M. Kandemir, Richard R. Brooks, and M.J. Irwin 639
   Introduction • System Assumptions • Caching Based
   Communication • Experimental Results • Spatial Locality • Related
   Work • Conclusions • Acknowledgments • References

31 **Compiler-Directed Communication Energy Optimizations for Microsensor**
   Networks *I. Kadayif, M. Kandemir, A. Choudhary, M. Karakoy, N. Vijaykrishnan,*
   and M.J. Irwin 655
   Introduction and Motivation • High-Level Architecture • Communication
   Optimizations • Experimental Setup • Results • Our Compiler
   Algorithm • Conclusions and Future Work • References

32 **Jamming Games in Wireless and Sensor Networks** *Rajgopal Kannan, Costas*
   Busch, and Shuangqing Wei 679
   Introduction • Motivation • Research Issues • Multilayer Game Models • Novel
   Evaluation Metrics • Acknowledgments • References

33 **Reasoning about Sensor Networks** *Manuel Peralta, Supratik Mukhopadhyay,*
   and Ramesh Bharadwaj 693
   Introduction • Related Work • Perpetual Requirements Engineering for Sensor
   Networks • SOLj Example: Autoregulated Power Generation Network • Counterfactuals
   in Sensor Networks • Counterfactual Example: Autoregulated Power Generation
   Network • Concluding Remarks • Appendix • References

34 **Testing and Debugging Sensor Network Applications** *Sally K. Wahba, Jason O. Hallstrom, and Nigamanth Sridhar* 711
Introduction • Network Simulators • Network Testbeds • Debugging Systems • Conclusion • Glossary • References

Sensor networks have gained great importance in a wide spectrum of applications. Large variety of hostile environments require the deployment of large number of sensors for intelligent patient monitoring, object tracking, etc. Sensor nodes are typically battery operated and they possess a constraint on their energy, which is an important resource in sensor networks and other embedded systems. To maximize the sensor nodes' lifetime after their deployment, energy saving can be obtained by using static/dynamic power management techniques. In this part, we summarize the contributions by different authors on the different aspects of power management.

Vijaykrishnan et al. emphasize on designing energy aware sensor systems in Chapter 28. The chapter introduces the major sources of energy consumption, the levels at which energy optimization can be performed, and presents some representative optimizations that can be performed in the sensor network design space.

Swaminathan and Chakrabarty propose several node-level energy reduction techniques for the processor and I/O devices in real-time sensor nodes in Chapter 29. The problem of scheduling devices for minimum energy consumption is known to be *NP*-complete. They describe online algorithms to schedule the shutdown and wakeup for I/O devices in sensor nodes by making a few simplifying assumptions to develop near-optimal device schedules in reasonable periods of time.

Saputra et al. propose a new energy-efficient and secure-communication protocol that exploits the locality of data transfer in distributed sensor networks in Chapter 30. The focus of their work is on reducing the energy consumed due to the communication that happens between the nodes in a local cluster. Their approach also addresses another important issue of providing security for sensed data applications. Furthermore, they discuss other related works in energy consumption and improving security in sensor networks.

Kadayif et al. present a set of compiler-directed communication optimization techniques and evaluate from the energy perspective in Chapter 31. More importantly, the authors focus on a wireless sensor network environment that processes array-intensive codes and has the following salient features: present energy behavior of a set of array-intensive applications on a wireless microsensor network, explain how source-level communication optimizations can reduce the energy consumption during communication, present a strategy where the communication time is overlapped with computation time, and present a compiler algorithm that applies several communication optimizations in a unified framework for optimizing the energy spent in sensor network during execution.

Kannan and Iyengar have revised their chapter from the first edition. In Chapter 32, they structure the issue of reliable query routing (RQR) in a distributed sensor network by describing a model for reliable data-centric routing with data aggregation in sensor networks, where interest queries are disseminated through the network to assign sensing tasks to sensor nodes.

Peralta et al. provide a new chapter (Chapter 33) for the second edition of the book. They look at how to best adapt sensor networks to their dynamically changing environments. A secure operations language extension of Java is presented. This approach allows the system to reconfigure itself to unforeseen circumstances while maintaining, as much as possible, the functionality desired by the system designer. This adaptation is not unrelated to swarm logic approaches.

Finally, Wahba et al. finish this part with a discussion of sensor network debugging and test issues in Chapter 34. They survey the available sensor network simulation tools and testbeds. This is followed by a discussion of the debugging tools that are available.

In summary, this part has highlighted issues regarding maximization of sensor node's lifetime after deployment by the use of energy-saving power management techniques.

28

Designing Energy-Aware Sensor Systems

N. Vijaykrishnan
The Pennsylvania State University

M.J. Irwin
The Pennsylvania State University

M. Kandemir
The Pennsylvania State University

L. Li
The Pennsylvania State University

G. Chen
The Pennsylvania State University

B. Kang
The Pennsylvania State University

28.1 Introduction 595
28.2 Sources of Power Consumption 596
28.3 Power Optimizations: Different Stages of System Design..........597
28.4 Energy Reduction Techniques...598
      Supply Voltage, Frequency and Threshold Voltage Scaling • Shutting
      Down Idle Components • Computational Offloading • Energy
      Aware Routing
28.5 Conclusions.. 606
Acknowledgments...607
References...607

28.1 Introduction

Power and related metrics have become important design constraints in computing systems ranging from low-end embedded systems to high-end servers for various reasons. The limited improvements in the energy capacity of batteries and the rapid growth in the complexity of battery-operated systems have reduced the longevity of operation between battery recharges. Hence, optimizing energy consumption is of crucial importance in battery driven mobile and embedded devices. Energy optimization is specifically important in sensor nodes that may not be accessible after deployment for replacing the batteries. Hence, energy optimization is critical in determining the longevity of the sensor network. Due to constraints on physical size/weight, the energy capacity of the battery in sensor nodes is typically limited. Hence, energy optimization can either increase the lifetime of the node or enable more powerful operations at a sensor node. In order to tackle the limited battery capacities, many sensors scavenge energy from the external environment in addition to using batteries. However, the amount of power that can be harnessed this way is very limited (typically, in the range of $100\,\mu W$). Therefore, energy optimizations are important even in these environments. Energy consumption has a significant impact on the weight of the battery pack and consequently of the entire sensor node. For example, to limit the re-charge interval to 10 h, it requires a 0.5 lb battery to operate a system that consumes 1 W.*

Power consumption also influences the design of tethered high-end systems that may be used as base stations to process the data from individual sensor nodes. While energy consumption is not a major issue in these systems that do not rely on batteries for operation, power consumption influences the amount of current drawn from the power supply and influences the power supply grid design. It becomes more challenging and costly to design supply rails that supply larger currents in response to increased power consumption trends. Another concern arises due to the influence of power consumption on the thermal profile of the chip. A part of the power consumed from the supply rail is eventually dissipated in the form of heat. Thus power consumption influences the cost of packaging and cooling deployed in a system. For example, it may be possible to employ a cheaper plastic packaging for low power systems. Another consequence of the higher on-chip temperatures is the reduced reliability of the chip. Specifically, for every 10°C increase in junction temperature, the lifetime of a device is typically cut by half.

The objective of this chapter is to introduce the major sources of energy consumption, the levels at which energy optimization can be performed and present some representative optimizations that can be performed in the sensor network design space. The rest of this chapter is organized as follows. The next section introduces the various sources of power consumption. Section 28.3 provides the different levels at which energy optimization can be performed in the design and operation of a system. Representative examples of energy reduction techniques are provided in Section 28.4. Finally, conclusions are provided in Section 28.5.

28.2 Sources of Power Consumption

Power is consumed in the different portions of the sensor node such as the memory, communication blocks, mechanical parts, the compute processors and the sensing elements. Since a significant portion of sensor nodes are composed of electronic circuits, we primarily focus on power consumption of CMOS VLSI circuits in this chapter.

Power consumption can be classified into the three major components: switching power (P_{switch}), short circuit power (P_{sc}), and leakage power ($P_{leakage}$). The first two components are called as dynamic power as it is consumed only when there is activity in the circuits, i.e., when signals change from 0 to 1 and vice-versa. The power consumed over the entire execution time is defined as the energy consumption. The power (P) consumption of CMOS circuits can be expressed as shown below

$$P = P_{switch} + P_{sc} + P_{leakage}$$

$$= C_L V_{dd}^2 p_{0\rightarrow1} f_{clock} + t_{sc} V_{dd} I_{sc} p_{0\rightarrow1} f_{clock} + k_{design} V_{dd} I_{off} N_{transistor}$$

where
C_L is the capacitive load of the circuit
V_{dd} is the supply voltage
$p_{0\rightarrow1}$ is the switching frequency
t_{sc} is the short-circuit time
f_{clock} is the clock frequency
I_{sc} and I_{off} are the peak current during switching and the leakage current, respectively
k_{design} is a design specific parameter
$N_{transistor}$ is total number of transistors

Figure 28.1 illustrates the currents consumed due to the three components of power consumption in a simple CMOS circuit.

---

* This assumes the use of current Ni-Cd battery technology that offers a capacity of 20 Wh/lb.

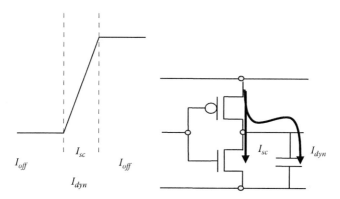

FIGURE 28.1 Components of power.

P_{switch} is consumed only when signals transition from 0 to 1, drawing current (I_{dyn}) from the power supply. However, switching power is dissipated as heat during both 0–1 and 1–0 transitions. Switching power can be reduced by decreasing the supply voltage, reducing the number of 0–1 transitions, reducing the capacitive load or by decreasing the clock frequency. Switching power is the dominant source of energy consumption in current 130 nm technology designs.

P_{sc} is consumed when both the pull down and pull up stacks of the CMOS circuits are conducting (see I_{sc} in Figure 28.1) during the period when the inputs are transitioning from 0 to 1, or from 1 to 0. The key to reducing short circuit power is slope engineering which involves reducing the time spent in signal transitions from 0 to 1 and vice versa. Fast rising and falling edges can help reduce the short-circuit power. Further, reducing the number of transitions also helps to reduce short-circuit power. Short circuit current forms a minor contribution in overall energy consumption in well designed circuits.

Unlike dynamic power consumption that has been the dominant form of energy consumption in CMOS-based circuits, leakage power, $P_{leakage}$, is consumed even in the absence of any switching activity (see I_{off} in Figure 28.1). With advancing technology, leakage current has been increasing exponentially [1]. Leakage current per device is a function of various parameters including temperature, supply voltage, threshold voltage, gate-oxide thickness and doping profile of the device. Specifically, leakage current increases with higher supply voltages, higher temperature, lower threshold voltage or thinner gate oxide. With continued progress in technology, devices are shrinking resulting in thinner gate-oxides and lower threshold voltages. Further, the increasing complexity of chips have resulted in a larger number of transistors on the chip and also higher on-chip temperatures due to larger power dissipation. Consequently, managing leakage energy has become more important. Leakage power can be reduced by shutting of supply voltage to idle circuits or by increasing the threshold voltage of devices.

28.3 Power Optimizations: Different Stages of System Design

The power consumption due to the different sources can be reduced through optimizations spanning different stages of a system design. At the lowest level of system design level, the fabrication process used for designing the system can be improved to reduce power consumption. While at the highest level, the software application can be tuned to reduce energy consumption. A brief description of optimization possible at different stages of the system design is given below:

- Process level: At the process level, factors such as the choice of the material for gate oxide (high-K vs. low-K dielectric), metal interconnect (e.g., copper vs. aluminum), doping concentration, device dimensions and device structure (single-gate vs. dual-gate) influence the power consumption characteristics. This level of optimization mainly involves process engineers and device engineers.

- Circuit/gate level: This optimization level deals with circuit design techniques. Supply voltage and frequency scaling techniques are the most widely used at this level. Voltage scaling yields considerable savings due to the quadratic dependence of power on supply voltage. However, the delay of the circuit increases with a reduction in the supply voltage due to a reduced driving strength (which is a function of the difference between supply and threshold voltage). Consequently, as supply voltage is reduced, the frequency of operation needs to be reduced. Since reducing supply voltage to the entire system can degrade performance, multiple-voltage techniques have been developed. In multiple voltage designs, timing-critical modules are powered at high voltage level, while the other modules are powered using a low voltage. Leakage energy problem is solved similarly with multiple-threshold voltage circuits. Transistors in the critical path use low threshold voltages for high performance, whereas those in the non-critical path use high threshold voltages to reduce leakage.
- Architectural level: In this level, larger blocks such as caches and functional units are the main subject. In complex digital circuits, not all of the blocks perform meaningful operations every clock. When a block is identified to be idle, it can be disabled to prevent useless but power consuming transitions. Circuit techniques such as clock gating provide ways to apply this technique. To tackle leakage power, idle blocks can be turned off by power-gating.
- Software level: An operating system can achieve significant power reduction by performing energy-aware task scheduling and resource management. Processors these days adopt multiple power modes which are initiated by operating systems. These are collectively called dynamic power management (DPM). Another important system component is the compiler. Compilers traditionally have been studied to generate efficient codes in terms of performance. Many of the performance optimization techniques also reduce power consumption. For example, spill code reduction result in both performance improvement and power reduction. There have been proposed power optimizing techniques which compromise performance, as well. Power-aware instruction scheduling technique can increase total number of cycles. However, the performance degradation has to be limited.

Efforts at optimizing the energy consumption in sensor networks span these different levels and range from designing energy-efficient circuits for sensors to designing energy-efficient algorithms to map onto the sensor networks [2–4]. These energy optimizations can also be classified as those that are performed at system design time and those that are employed at runtime. The optimization performed at process/circuit level are typically design time solutions, while those at architecture/software level can be either design time or run time optimizations.

28.4 Energy Reduction Techniques

Energy optimizations are typically targeted at the specific components of the sensor node that contribute a significant portion of overall energy consumption. The main components of interest in a sensor node are the processor, the memory, the communication component and the sensing elements. In the case of a mobile sensor, energy is also consumed by the components used to support locomotion.

In this section, we present representative energy optimizations targeted at the processor, memory and communication components of the sensor node. Due to the plethora of work on energy optimizations and due to space constraints, this section does not attempt to cover all efforts. Rather, our goal is to provide a sample of representative energy reduction techniques that are especially applicable to sensor systems.

28.4.1 Supply Voltage, Frequency and Threshold Voltage Scaling

Many techniques for controlling both active and standby power consumption have been developed for processor cores [5]. Table 28.1 shows the processor power design space for both active and standby power [6]. The second column lists techniques that are applied at design time and thus are part of the

TABLE 28.1 Processor Power Design Space

	Constant Throughput/Latency		Variable Throughput/Latency
	Design Time	Sleep Mode	Run Time
Active (Dynamic) [CV^2f]	Logic design Trans sizing Multiple V_{DD}'s	Clock gating	DFS DVS DTM
Standby (Leakage) [VI_{off}]	Stack effect Multiple V_T's Multiple t_{ox}'s	Sleep trans Multi-V_{DD} Variable V_T Input control	Variable V_T

circuit fabric. The last two columns list techniques that are applied at run time, the middle column for those cases when the component is idle and the last column when the component is in active use. (Run time techniques can additionally be partitioned into those that reduce leakage current while retaining state and those are state destroying.) The underlying circuit fabric must provide the "knobs" for controlling the run time mechanisms—by either the hardware (e.g., in the case of clock gating) or the system software (e.g., in the case of DVS). In the case of software control the cost of transitioning from one state to another (both in times of energy and time) and the relative energy savings needs to be provided to the software for decision making.

The best knob for controlling power is setting the supply voltage appropriately to meet the computational load requirements. While lowering the supply voltage has a quadratic impact on active energy, it decreases systems performance since it increases gate delay as shown in Figure 28.2. Choosing the appropriate supply voltage at design time for the entire component will minimize (or even eliminate) the overhead of level converters that are needed whenever a module at a lower supply drives a module at a higher supply.

The most popular of the techniques for reducing active power consumption at run time is dynamic voltage scaling (DVS) combined with dynamic frequency scaling (DFS). Most embedded and mobile processors contain this feature (triggered by thermal on-chip sensors when thermal limits are being approached or by the run-time system when the CPU load changes). Among the first approaches at exploiting DVS at run-time was proposed in Ref. [7]. Here, the authors propose a scheduling algorithm for minimizing energy consumption, where a process can be scheduled at different processor speeds, always keeping in mind the timely completion of such task. Many tasks in sensor environments may need to meet some hard real-time constraints. For example, if there is a radiation sensor in a nuclear plant, it needs to activate containment action with a given time constraint. In such environments, it is

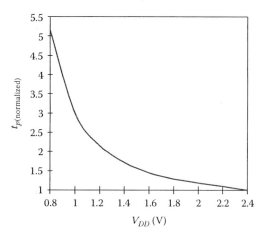

FIGURE 28.2 Delay as a function of V_{DD}.

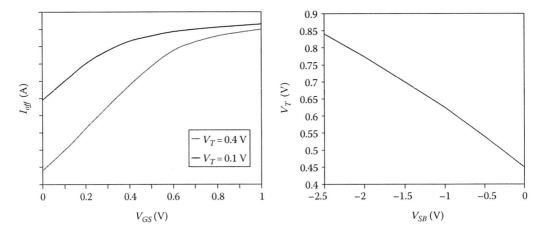

FIGURE 28.3 V_T effects.

important to ensure that the reduction in performance due to voltage scaling does not affect the ability to meet hard-real time constraints. An example of a DVS approach meeting hard real-time systems is presented in Ref. [8]. In their approach, the processor slack times are taken advantage of to power-down the processor, always paying attention to meet process deadlines.

DFS + DVS requires a power supply control loop containing a buck converter to adjust the supply voltage and a programmable PLL to adjust the clock [6]. As long as the supply voltage is increased before increasing the clock rate or decreased after decreasing the clock rate, the system only need stall when the PLL is relocking on the new clock rate (estimated to be around 20 ms).

Several techniques have evolved recently for controlling subthreshold current as shown in Table 28.1. Since increasing the threshold voltage, V_T, decreases subthreshold leakage current (exponentially), adjusting V_T is one such technique. As shown in Figure 28.3, a 90 mV reduction in V_T increases leakage by an order of magnitude. Unfortunately, increasing V_T also negatively impacts gate delay. As with multiple supply voltages, multiple threshold voltages can be employed at design time or run time. At run time, multiple levels of V_T can be provided using adaptive body-biasing where a negative bias on V_{SB} increases V_T as shown in Figure 28.3 [9]. Simultaneous DVS, DFS, and variable V_T has been shown to be an effective way to trade-off supply voltage and body-biasing to reduce total power—both active and standby—under variable processor loads [10].

Another technique that will apply to reducing standby power is the use of sleep transistors like those shown in Figure 28.4. Standby power can be greatly reduced by gating the supply rails for idle components. In normal mode (non-idle), the sleep transistors must present as small a resistance as possible (via sizing) so as to not negatively affect performance. In sleep mode (idle), the transistor stack effect [6] reduces leakage by orders of magnitude. Alternative, standby power can be completely eliminated by switching off the supply to idle components. A sensor node employing such a technique will require system software that can determine the optimal scheduling of tasks on cores and can direct idle cores to switch off their supplies while taking into account the cost (both in terms of energy and time) of transitioning from the on-to-off and off-to-on states.

28.4.2 Shutting Down Idle Components

An effective means for reducing energy consumption is in shutting down the components. In the case of mechanical components such as the disk, the spinning of the disks is stopped to reduce power consumption. In the case of processor datapath or memory elements, shutting down the component could mean gating the supply voltage from the circuit to reduce leakage energy or gating the clocking circuit to reduce dynamic energy. Use of idleness to transition an entity to a low energy mode is an issue

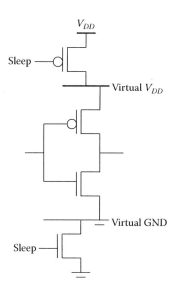

FIGURE 28.4 Gating supply rails.

that has been researched in the context of disks [11,12], network interfaces [13], and system events in general [14,15]. Many of these studies [12] have used past history (to predict future behavior) for effecting a transition. A similar strategy has been recently employed in the context of transitioning DRAM memory modules [16].

Shutting down idle components can be performed at different granularities. For example, supply gating can be performed for either the entire memory, a single bank or at an even finer granularity of a single memory block. The key issue is to determine when to shut down the component balancing the energy requirements with other tradeoffs such as performance or reliability degradation. For example, frequent spinning up and down of the disks can increase the wear and tear on the disk head and impact reliability. Reliability can also be a concern due to operation in the low energy mode. For example, memory circuits operating at a low-leakage energy mode are more susceptible to bit flips induced by radiation effects. There can be performance implications as well. When disks are shut down, data reads can be delayed for the disk to spin up to normal speed.

28.4.2.1 Leakage Control Techniques for Memories

The memory element plays a vital role in sensor nodes as it captures the state of the system. However, maintaining the supply voltage to the entire memory will result in a significant drain on the energy resources of the sensor node due to leakage power. Hence, techniques that only maintain power supply to portions of the memory storing data will be important. Hence, we focus on various approaches that exist to reduce leakage energy in cache memories in order to motivate similar techniques that could be applied in sensor nodes.

Approaches that target reducing instruction cache leakage energy consumption can be broadly categorized into three groups: (1) those that base their leakage management decisions on some form of performance feedback (e.g., cache miss rate) [17], (2) those that manage cache leakage in an application insensitive manner (e.g., periodically turning off cache lines) [18–20], and (3) those that use feedback from the program behavior [19,21,22].

The approach in category (1) is inherently coarse-grain in managing leakage as it turns off large portions of the cache depending on a performance feedback that does not specifically capture cache line usage patterns. For example, the approach in (1) may indicate that 25% of the cache can be turned off because of very good hit rate, but, it does not provide the guidance on which 75% of the cache lines are going to be used in the near future.

Approaches in category (2) turn off cache lines independent of the instruction access pattern. An example of such a scheme is the periodic cache line turn-off proposed in Ref. [18] that turns off all cache lines after a fixed period. The success of this strategy depends on how well the selected period reflects the rate at which the instruction working set changes. Specifically, the optimum period may change not only across applications but also within the different phases of the application itself. In such cases, we either keep cache lines in the active state longer than necessary, or we can turn off cache lines that hold the current instruction working set, thereby impacting performance and wasting energy. Note that trying to address the first problem by decreasing the period will exacerbate the second problem. On the plus side, this approach is simple and has very little implementation overhead.

Another example of a fixed scheme in category (2) is the technique proposed [20]. This technique adopts a bank based strategy. In this approach, when instruction fetches during execution moves from one cache memory bank to another, the hardware turns off the former and turns on the latter. Another technique in category (2) is the cache decay-based approach (its adaptive variant falls in category (3)) proposed by Kaxiras et al. [19]. In this technique, a small counter is attached to each cache line which tracks its access frequency. If a cache line is not accessed for a certain number of cycles, it is placed into the leakage saving mode. While this technique tries to capture the usage frequency of cache lines, it does not directly predict the cache line access pattern. Consequently, a cache line whose counter saturates is turned off even if it is going to be accessed in the next cycle. Since it is also a periodic approach, choosing a suitable decay interval is crucial if it is to be successful. In fact, the problems associated with selecting a good decay interval are similar to those associated with selecting a suitable turn-off period in Ref. [18]. Consequently, this scheme can also keep a cache line in the active state until the next decay interval arrives even if the cache line is not going to be used in the near future. Finally, since each cache line is tracked individually, this scheme has more overhead.

The approaches in category (3) attempt to manage cache lines in an application-sensitive manner. The adaptive version of the cache-decay scheme [19] tailors the decay interval for the cache lines based on cache line access patterns. They start out with the smallest decay interval for each cache line to aggressively turn off cache lines and increase the decay interval when they learn that the cache lines were turned off prematurely. These schemes learn about premature turn-off by leaving the tags on at all times. The approach in Ref. [22] also uses tag information to adapt leakage management.

In Ref. [21], an optimizing compiler is used to analyze the program to insert explicit cache line turn-off instructions. This scheme demands sophisticated program analysis and modification support, and needs modifications in the ISA to implement cache line turn-on/off instructions. In addition, this approach is only applicable when the source code of the application being optimized is available. In Ref. [21], instructions are inserted only at the end of loop constructs and, hence, this technique does not work well if a lot of time is spent within the same loop. In these cases, periodic schemes may be able to transition portions of the loop that are already executed into a drowsy mode. Further, when only select portions of a loop are used, the entire loop is kept in an active state. Finally, inserting the turn-off instructions after a fast executing loop placed inside an outer loop can cause performance and energy problems due to premature turn-offs.

Many of these leakage control approaches can be applied in the context of memory components in sensor nodes.

28.4.2.2 Multiple Low Power Modes

Instead of just having two modes of turn-on/turn-off as in the case of the leakage control mechanism explained above, the system could support multiple low power states. In this case, the task of choosing the appropriate power mode becomes more challenging. Typically, there are energy/performance tradeoffs that need to be considered in selecting the mode of operation. In order to illustrate these tradeoffs better and explain how multiple low power modes are supported, we will use the energy/performance tradeoffs that exist in operating the DRAM memory in different power modes as an example.

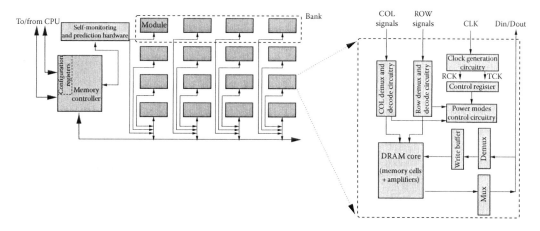

FIGURE 28.5 Memory system architecture.

Each energy mode is characterized by its *energy consumption* and the time that it takes to transition back to the active mode (*resynchronization time*). Typically, lower the energy consumption, higher the resynchronization time [23]. These modes are characterized by varying degrees of the module components being active. The major components of a DRAM module are the clock generation circuitry, ROW (row address/control) decode circuitry and COL (column address/control) decode circuitry, control registers and power mode control circuitry, together with the DRAM core consisting of the precharge logic, memory cells, and sense amplifiers (see Figure 28.5). The clock generation circuitry is used to generate two internal clock signals (TCK and RCK) that are synchronous with an external system clock (CLK) for transmitting read data and receiving write data/control signals. The packets received from the ROW and COL signals can also be used to switch the power mode of the DRAM. The details of the power modes are discussed below:

- *Active*: In this mode, the DRAM module is ready for receiving the ROW and COL packets and can transition immediately to read or write mode. In order to receive these packets, both the ROW and COL demux receivers have to be active. As the memory unit is ready to service any read or write request, the resynchronization time for this mode is the least (zero units), and the energy consumption is the highest.
- *Standby*: In this mode, the COL multiplexers are disabled resulting in significant reduction in energy consumption compared to the active mode. The resynchronization time for this mode is typically one or two memory cycles. Some state-of-the-art RDRAM memories already exploit this mode by automatically transitioning into the standby mode at the end of a memory transaction [23].
- *Napping*: The ROW demux circuitry is turned off in this mode, leading to further energy savings over the standby mode. When napping, the DRAM module energy consumption is mainly due to the refresh circuitry and clock synchronization that is initiated periodically to synchronize the internal clock signals with the system clock. This mode can typically consume two orders of magnitude less energy than the active mode, with the resynchronization time being higher by an order of magnitude than the standby mode.
- *Power-down*: This mode shuts off the periodic clock synchronization circuitry resulting in another order of magnitude saving in energy. The resynchronization time is also significantly higher (typically thousands of cycles).
- *Disabled*: If the content of a module is no longer needed, it is possible to completely disable it (saving even refresh energy). There is no energy consumption in this mode, but the data is lost. One could envision transitioning out of disabled mode by re-loading the data from an alternate location (perhaps another module or disk) and/or just performing write operations to such modules.

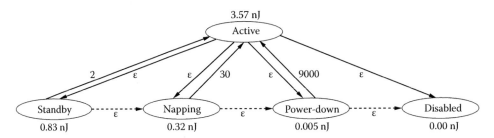

FIGURE 28.6   Power modes utilized.

When a module in standby, napping, or power-down mode is requested to perform a memory transaction, it first goes to the active mode and then performs the requested transaction. Figure 28.6 shows possible transitions between modes (the dynamic energy consumed in a cycle is given for each node) in our model. The resynchronization times in cycles (based on a cycle time of 2.5 ns) are shown along the arrows.

It is evident from the operation of the DRAM power modes that there are significant performance/power tradeoffs in selecting the power mode to operate. Schemes that have been employed to utilize these power modes typically predict the duration of the idleness of the memory module and then select the power mode such that the overheads in terms of energy/time in transitioning to and from a low-power mode amortizes the benefit gained in operating in the low power mode. Examples of such approaches are constant threshold and history based threshold predictors proposed in Ref. [16].

The rationale behind the constant threshold predictor (CTP) is that if a memory module has not been accessed in a while, then it is not likely to be needed in the near future (that is, inter-access times are predicted to be long). A threshold is used to determine the idleness of a module after which it is transitioned to a lower energy mode. Hence, using this scheme, a memory module is progressively transitioned from the active mode to the power down mode. The thresholds are typically chosen to be larger if the penalties for recovering to the active mode are large. There are two main problems with this CTP approach. First, we gradually decay from one mode to another (i.e., to get to power-down, we go through standby and napping), though one could have directly transitioned to the final mode if we had a good estimate. Second, we pay the cost of resynchronizing on a memory access if the module has been transitioned. In the history-based predictor (HBP), we estimate the inter-access time, directly transition to the best energy mode, and activate (resynchronize) the module so that it becomes ready by the time of the next estimated access.

Since sensor nodes are expected to have many components that will support multiple low power modes, we anticipate techniques such as CTP and HBP to be pertinent in this environment. An important consideration in devising such schemes is to estimate the effect of using these modes not on just individual components but on the entire sensor node. For example, performance degradations that result in order to save energy in the memory module should not be offset by the increased energy consumption in the rest of the components due to the prolonged operation.

28.4.2.3 Adaptive Communication Hardware

Similar to the memory elements, other components can also be partially shutdown to conserve power consumption. For example, portions of the pipeline in the processor can be shut down based on the data width of the computation or the number of bits used by the analog to digital converter can be varied based on desired accuracy of the data. We show how an analog-to-digital (ADC) converter, one of the components used by the communication system can be adapted to the required channel conditions.

The architecture of a traditional pipelined ADC can be easily modified for adapting its structure to the required resolution to conserve energy consumption. Each stage has the same function and its input

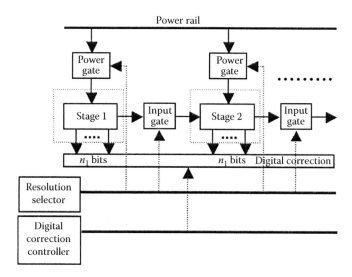

FIGURE 28.7 Adaptive bitwidth analog to digital converter.

is only correlated to the previous stage's output. The MSB (Most Significant Bit) is defined by left most stage of the ADC. Similarly, the LSB (Least Significant Bit) is defined by right most stage of the ADC. If unused stages could be shut down, significant power saving can be expected. Supply gating is used as the mechanism for reducing power consumption in the unused segments. There is no leakage current or static power consumption in the supply gated units. However, a recovery latency of 350 ns is incurred for activating a shutdown segment. Based on the required resolution, supply gating is applied to selected units in our adaptive system shown in Figure 28.7. Further, the input gates of the unused stages are disabled.

In order to determine the desired number of bits that need to be used by the ADC, the channel condition is determined using the receiving power. Next, the symbol rate is determined based on the channel condition to ensure the required bit error rate. This, in turn can be used to determine the number of stages of the adaptive pipelined ADC that can be shut down while still meeting the required resolution. Higher symbol rates are possible only when channel conditions are good and when the ADC supports higher resolution to detect the symbol correctly. Thus, when symbol rates are lower, it is possible to conserve power consumption by shutting down portions of the ADC.

Through the different examples that were presented in this section, we have reiterated the importance of shutting down idle components as an important scheme in conserving energy consumption.

28.4.3 Computational Offloading

In a wireless sensor nodes, it is important to optimize both computation and communication energy. There are different opportunities in such a system that allow tradeoffs between computation and communication energy costs. A lot of research has been carried out to exploit such tradeoffs to reduce the overall energy consumption (e.g., [24–26]). For example, Flinn and Satyanarayanan [24] have proposed an environment where applications dynamically modify their behaviors to conserve energy by lowering data fidelity and adjusting the partition of computation tasks between client and server.

A major goal of these approaches is to offload complex computational tasks from an energy-constrained node to a more powerful server. However, the cost of offloading the task to a remote server involves communication costs for transmitting the code and data as well as for receiving the results of the computation. Hence, the challenge in these schemes is deciding on whether the energy reduced by decreasing the number of computations is greater than the overhead incurred for communication.

Based on these tradeoffs, a decision should be made as where the computation will be performed—locally or remotely. While it is possible to make some of these decisions statically, the energy cost of computation and communication can vary based on the operating conditions and user supplied input for which accurate information can only be obtained at runtime. For example, the power amplifier of the transmitter in the mobile client should be tuned according to its distance from the server and the channel interference. This setting affects the energy cost of transmitting a bit. Similarly, the user input can also influence the energy costs of local computation and offloading. As an example, the complexity of a computation can depend on the magnitude of an input parameter (e.g., image size in an image processing application).

The energy consumption due to communication is dominated by either the receiving or transmitting energy based on the type of sensor network employed. When communicating over small distances (less than 5 m) in micro-sensor networks, the cost of transmission is relatively small in contrast to macro-sensor networks that need to transmit larger distance where transmission energy is significantly larger than receiving energy. Hence, when considering computational offloading algorithms in micro-networks, it is important to reduce the idle energy expended by the receivers when they are idling snooping for data. The need to offload computation will require the receivers to be active in order to receive the results back and may in itself cause a significant portion of overall energy. Hence, proposed approaches for computational offloading estimate the amount of time that the server will require for completing the offloaded computation before it returns the results and shuts down the receiver for that period.

28.4.4 Energy Aware Routing

Energy consumed due to communication between the different sensor nodes can consume a significant portion of the energy. Since, the energy consumption for transmitting is a function of the distance between the communicating nodes, there have been several efforts focusing on devising routing algorithms that minimize the distance of communication. A common approach to reducing energy consumption is the use of multi-hop routing. In multi-hop routing, nodes between the source and sink nodes of the communication are used as intermediary hops in the transmission. This approach is beneficial as the energy increases faster than linear with distance of transmission. Hence, having multiple transmissions of smaller distances is more energy-efficient than a single transmission of a longer distance.

Another approach to reduce the energy consumption is to partition sensors network into several clusters [3] and limit the communication outside the cluster to a selected node or a set of nodes in the cluster. This approach confines the communication of the other nodes to a small distance, and incorporates some intelligent processing within the local cluster to reduce the amount of communication outside the cluster.

There are different criteria for selecting the routing algorithms used in a sensor network. For example, a minimum energy path to reach the destination may not always be desirable as the optimal path to route a packet may require the use of an intermediary node that is running low on battery. In such a case, a higher energy path that avoids this node may be more desirable in order to prevent this node from becoming dead. In Ref. [27], a sub-optimal path chosen based on a probabilistic approach instead of always using the optimal path is shown to be preferable to prevent nodes from dying.

28.5 Conclusions

This chapter has emphasized the need for energy optimization and emphasizes the need for optimizations to span different design stages of the system. Optimizing the system energy requires a concerted effort spanning from the choice of the underlying process technology for fabrication of the system to the runtime software deployed on the sensor node. We have also introduced the major sources of energy consumption and explained various approaches at reducing the energy consumption due to these parts. Devising new energy optimizations continue to remain an important issue for sensor networks. Techniques aimed at limiting the energy consumption hold the key for increased deployment of sensor networks.

Acknowledgments

This work was supported in part by the NSF Career Awards CCR-0093082 and CCR-0093085 and NSF Grant 0082064 and a MARCO 98-DF-600 grant from GSRC. The authors also acknowledge the contributions of various graduate students from their lab who worked on several projects whose results are abstracted here.

References

1. J. A. Butts and G. Sohi. A static power model for architects. In *Proceedings of the 33th Annual International Symposium on Microarchitecture*, Monterey, CA, December 2000.
2. A. Chandrakasan et al. Design considerations for distributed microsensor systems. In *Proceedings of the IEEE 1999 Custom Integrated Circuits Conference*, San Diego, CA, May 1999, pp. 279–286.
3. D. Estrin. Wireless sensor networks: Application driver for low power distributed systems. In *Proceedings of the ISLPED*, Huntington Beach, CA, August 2001, p. 194.
4. W. Heinzelman, A. Sinha, A. Wang, and A. Chandrakasan. Energy-scalable algorithms and protocols for wireless microsensor networks. In *Proceedings of the International Conference on Acoustics, Speech, and Signal Processing*, Istanbul, Turkey, June 2000.
5. R. Brodersen, M. Horowitz, D. Markovic, B. Nikolic, and V. Stojanovic. Methods for true power minimization. In *International Conference on Computer Aided Design*, San Jose, CA, November 2002, pp. 35–42.
6. J. Rabaey. *Digital Integrated Circuits: A Design Perspective*. Prentice Hall, Upper Saddle River, NJ, 2003.
7. M. Weiser, B. Welch, A. Demers, and S. Shenker. Scheduling for reduced CPU energy. In *Proceedings of the 1st Symposium on Operating Systems Design and Implementation*, Monterey, CA, November 1994, pp. 13–23.
8. Y. Shin and K. Choi. Power conscious fixed priority scheduling for hard real-time systems. In *Proceedings of the 36 Design Automation Conference (DAC'99)*, New Orleans, LA, 1999.
9. D. Duarte, Y. F. Tsai, N. Vijaykrishnan, and M. J. Irwin. Evaluating run-time techniques for leakage power reduction techniques. In *Asia-Pacific Design Automation Conference*, Bangalore, India, January 2001.
10. S. M. Martin, K. Flautner, T. Mudge, and D. Blaauw. Combined dynamic voltage scaling under adaptive body biasing for lower power microprocessors under dynamic load. In *International Conference on Computer Aided Design*, November 2002, San Jose, CA, pp. 712–725.
11. F. Douglas, P. Krishnan, and B. Marsh. Thwarting the power-hungry disk. In *Proceedings of the Winter 1994 USENIX Technical Conference*, San Francisco, CA, 1994.
12. K. Li, R. Kumpf, P. Horton, and T. Anderson. A quantitative analysis of disk drive power management in portable computers. In *Proceedings of the Winter 1994 USENIX Technical Conference*, San Francisco, CA, 1994.
13. M. Stemm and R. H. Katz. Measuring and reducing energy consumption of network interfaces in hand-held devices. *IEICE Transactions on Communications*, Special Issue on Mobile Computing, 2000.
14. L. Benini, A. Bogliogo, S. Cavallucci, and B. Ricco. Monitoring system activity for OS-directed dynamic power management. In *Proceedings of the ACM ISLPED'98*, Monterey, CA, 1998.
15. L. Benini, R. Hodgson, and P. Siegel. System-level power estimation and optimization. In *Proceedings of the ACM ISLPED'98*, Monterey, CA, 1998.

16. V. Delaluz, M. Kandemir, N. Vijaykrishnan, A. Sivasubramaniam, and M. J. Irwin. DRAM energy management using hardware and software directed power mode control. In *Proceedings of the International Conference on High Performance Computer Architecture (HPCA)*, Nuevo Leone, Mexico, January 2001.

17. M. D. Powell, S. Yang, B. Falsafi, K. Roy, and T. N. Vijaykumar. Reducing leakage in a high-performance deep-submicron instruction cache. *IEEE Transactions on VLSI*, 9(1), 77–89, February 2001.

18. K. Flautner, N. Kim, S. Martin, D. Blaauw, and T. Mudge. Drowsy caches: Simple techniques for reducing leakage power. In *Proceedings of the 29th International Symposium on Computer Architecture*, Anchorage, AK, May 2002.

19. S. Kaxiras, Z. Hu, and M. Martonosi. Cache decay: Exploiting generational behavior to reduce cache leakage power. In *Proceedings of the 28th International Symposium on Computer Architecture*, Gothenburg, Sweden, June 2001.

20. N. Kim, K. Flautner, D. Blaauw, and T. Mudge. Drowsy instruction caches: Leakage power reduction using dynamic voltage scaling and cache sub-bank prediction. In *Proceedings of the 35th Annual International Symposium on Microarchitecture*, Istanbul, Turkey, November 2002.

21. W. Zhang et al. Compiler-directed instruction cache leakage optimization. In *Proceedings of the 35th Annual International Symposium on Microarchitecture*, Istanbul, Turkey, November 2002.

22. H. Zhou, M. C. Toburen, E. Rotenberg, and T. M. Conte. Adaptive mode control: A static power-efficient cache design. In *Proceedings of the 2001 International Conference on Parallel Architectures and Compilation Techniques*, Barcelona, Spain, September 2001.

23. 128/144-MBit Direct RDRAM Data Sheet, Rambus Inc., May 1999.

24. J. Flinn and M. Satyanarayanan. Energy-aware adaptation for mobile applications. In *The 17th ACM Symposium on Operating Systems Principles*, Kiawah Island, SC, December 1999.

25. Z. Li, C. Wang, and R. Xu. Computation offloading to save energy on handheld devices: A partition scheme. In *International Conference on Compilers, Architectures and Synthesis for Embedded Systems*, Atlanta, GA, November 2001, pp. 238–246.

26. U. Kremer, J. Hicks, and J. Rehg. A compilation framework for power and energy management on mobile computers. In *The 14th International Workshop on Parallel Computing*, Cumberland Falls, KY, August 2001.

27. R. C. Shah and J. Rabaey. Energy aware routing for low energy ad hoc sensor networks. In *IEEE Wireless Communications and Networking Conference (WCNC)*, Orlando, FL, March 17–21, 2002.

28. A. Sinha and A. Chandrakasan. Operating system and algorithmic techniques for energy scalable wireless sensor networks. In *Proceedings of the 2nd International Conference on Mobile Data Management*, Hong Kong, China, January 2001.

29. W. Ye, J. Heidemann, and D. Estrin. An energy-efficient MAC protocol for wireless sensor networks. In *Proceedings of the 21st International Annual Joint Conference of the IEEE Computer and Communications Societies*, New York, June 2002.

30. T. Kuroda. Optimization and control of v_{DD} and v_t for low power, high speed CMOS design. In *International Conference on Computer Aided Design*, November 2002, San Jose, CA, pp. 28–34.

29

Operating System Power Management

29.1 Introduction 609
    CPU-Centric DPM • I/O-Centric DPM
29.2 Node-Level Processor-Oriented Energy Management...............611
    The LEDF Algorithm • Implementation Testbed • Experimental
    Results
29.3 Node-Level I/O-Device-Oriented Energy Management............617
    Optimal Device Scheduling for Two-State I/O Devices • Online
    Device Scheduling • Low-Energy Device Scheduling of Multi-State
    I/O Devices • Experimental Results
29.4 Conclusions..636
References...637

Vishnu
Swaminathan
Duke University

Krishnendu
Chakrabarty
Duke University

29.1 Introduction

Energy consumption is an important design consideration for wireless sensor networks. These networks are useful for a number of applications such as environment monitoring, surveillance, and target detection and localization. The sensor nodes in such applications operate under limited battery power. Sensor nodes also tend to be situated at remote and/or inaccessible locations, and hence the cost of replacing battery packs is high when the batteries that power them fail.

One approach to reduce energy consumption is to employ low-power hardware design techniques [6,10,22]. These design approaches are static in that they can only be used during system design and synthesis. Hence, these optimization techniques do not fully exploit the potential for node-level power reduction under changing workload conditions, and their ability to trade-off performance with power reduction is thus inherently limited. An alternative and more effective approach to reducing energy in embedded systems and sensor networks is based on *dynamic power management* (DPM), in which the operating system (OS) is responsible for managing the power consumption of the system.

Many wireless sensor networks are also designed for *real-time* use. Real-time performance is defined in terms of the ability of the system to provide real-time temporal guarantees to application tasks that request such guarantees. These systems must therefore be designed to meet both functional and timing requirements [5]. Energy minimization adds a new dimension to these design criteria. Thus, while energy minimization for sensor networks is of great importance, energy reduction must be carefully balanced against the need for real-time responsiveness.

Recent studies have shown that the CPU and the I/O subsystem are major consumers of power in an embedded system; in some cases, hard disks and network transceivers consume as much as 20% of total system power in portable devices [14,19]. Consequently, CPU-centric and I/O-centric DPM techniques have emerged at the forefront of DPM research for wireless sensor networks.

29.1.1 CPU-Centric DPM

Designers of embedded processors that are used in sensor nodes now include variable-voltage power supplies in their processor designs, i.e., the supply voltages of these processors can be adjusted dynamically to trade-off performance with power consumption. *Dynamic voltage scaling* (DVS) refers to the method by which quadratic savings in energy is obtained through the run-time variation of the supply voltage to the processor.

It is well-known that the power consumption of a CMOS circuit exhibits a cubic dependence on the supply voltage V_{dd}. However, the execution time of an application task is proportional to the sum of the gate delays on the critical path in a CMOS processor. Since gate delay is inversely proportional to V_{dd}, the execution time of a task increases with decreasing supply voltage. The energy consumption of the CMOS circuit, which is the product of the power and the delay therefore exhibits a quadratic dependence on V_{dd}. In embedded sensor nodes, where peak processor performance is not always necessary, a drop in the operating speed (due to a reduction in operating voltage) can be tolerated in order to obtain quadratic reductions in energy consumption. This forms the basis for DVS; the quadratic dependence of energy on V_{dd} has made it one of the most commonly used power reduction techniques in sensor nodes and other embedded systems. When processor workload is low, the OS can reduce the supply voltage to the processor (with a tolerable drop in performance) and utilize the quadratic dependence of power on voltage to reduce energy consumption.

29.1.2 I/O-Centric DPM

Many peripheral devices possess multiple power states—usually one high-power working state and at least one low-power sleep state. Hardware-based timeout schemes for power reduction in such I/O devices have been incorporated into several device designs. These techniques shut down devices when they have been idle for a pre-specified period of time. A device that has been placed in the sleep state is powered up when a new request is generated.

With the introduction of the ACPI standard in 1997, the OS was provided with the ability to switch device power states dynamically during run-time, thus leading to the development of several new types of DPM techniques. Predictive schemes use various system parameters to estimate the lengths of idle periods for devices. Stochastic models with different probabilistic distributions have been used to estimate the times at which devices can be switched between power states. The goals of these methods, however, is to minimize the response times of devices. Indeed, many such probabilistic schemes see widespread use in portable and interactive systems such as laptop computers. However, their applicability in sensor systems, many of which require real-time guarantees, is limited due to a drawback inherent to probabilistic methods.

Switching between device power states incurs a time penalty, i.e., a device takes a certain amount of time to transition between its power states. In hard real-time systems where tasks have firm deadlines, device switching must be performed with caution to avoid the potentially disastrous consequences of missed deadlines. The uncertainty that is inherent in probabilistic estimation methods precludes their use as effective device switching algorithms in hard real-time systems whose behavior must be predictable with a high degree of confidence. Current-day practice consists of keeping devices in real-time systems powered up during the entirety of system operation; the critical nature of I/O devices operating in real-time prohibits the shutting down of devices during run-time.

In this chapter, we describe several node-level energy reduction methods for in wireless sensor networks. The first algorithm targets the processor in a sensor node. This algorithm is implemented on a laptop equipped with an AMD Athlon 4 processor and running the Real-time Linux (RT-Linux) OS.

Experimental power measurements validate and support our simulation results. A significant amount of energy is saved using the algorithm described here. We describe an optimal offline algorithm that generates device schedules for minimum energy consumption of I/O devices in hard real-time sensor nodes. The problem of scheduling devices for minimum energy consumption is known to be $\mathcal{NP}$-complete. However, by making a few simplifying assumptions, online algorithms can be developed to generate near-optimal device schedules in reasonable periods of time. Here, we describe two such online algorithms to schedule the shutdowns and wake-ups for I/O devices in sensor nodes that require hard real-time temporal guarantees.

29.2 Node-Level Processor-Oriented Energy Management

We are given a set $R = \{r_1, r_2, ..., r_n\}$ of n tasks. Associated with each task $r_i \in R$ are the following parameters: (1) an arrival time a_i, (2) a deadline d_i, and (3) a length l_i (represented as the number of instruction cycles). Each task is placed in the ready queue at time a_i and must complete its execution by its deadline d_i. The tasks are not preemptable. The CPU can operate at one of k voltages: $V_1, V_2, ..., V_k$. Depending on the voltage level, the CPU speed may take on k values: $s_1, s_2, ..., s_k$. The supply voltage to the CPU is controlled by the OS, which can dynamically switch the voltage during run-time. The energy E_i consumed by task r_i is proportional to $v_i^2 l_i$. The problem we address is defined as follows:

$\mathcal{P}_{cpu}$: Given a set R of n tasks, and for each task $r_i \in R$, (1) a release time a_i, (2) a deadline d_i, and (3) a length l_i, and a processor capable of operating at k different voltages $V_1, V_2, ..., V_k$ with corresponding speeds $S_1, S_2, ..., S_k$, determine a sequence of voltages $v_1, v_2, ..., v_n$ and corresponding speeds $s_1, s_2, ..., s_n$ for the task set R such that the total energy consumed $\sum_{i=1}^{n} v_i^2 l_i$ by the task set is minimized, while also attempting to meet as many task deadlines as possible.

29.2.1 The LEDF Algorithm

LEDF is an extension of the well-known earliest deadline first (EDF) algorithm [11]. The algorithm maintains a list of all released tasks called the ready list. These tasks have an absolute deadline associated with them that is recalculated at each release based on the absolute time of release and the relative deadline. When tasks are released, the task with the earliest deadline is selected for execution. A check is performed to see if the task deadline can be met by executing it at a lower voltage (speed). Each speed at which the processor can run is considered in order from the lowest to the highest. For a given speed, the worst-case execution time of the task is calculated based on the maximum instruction count. If this execution time is too high to meet the current absolute deadline for the task, the next higher speed is considered. Otherwise, a schedulability test is applied to verify that all ready tasks will be able to meet their deadlines when the current earliest-deadline task is run at a lower speed. The test consists of iterating down the ordered list of tasks and comparing the worst-case completion time for each task (at the highest speed) against its absolute deadline. If any task will miss its deadline, the selected speed is insufficient and the next higher speed for the current task is considered. If the deadlines of all tasks in the ready list can be met at the highest speed, LEDF assigns the lower voltage to the task and the task begins execution. When the task completes execution, LEDF again selects the task with the nearest deadline to be executed. As long as there are tasks waiting to be executed, LEDF schedules the one with the earliest absolute deadline for execution. Figure 29.1 describes the algorithm in pseudo-code form.

For a processor with two speeds, the LEDF algorithm has a computational complexity of $O(n \log n)$ where n is the total number of tasks. The worst-case scenario occurs when all n tasks are released at time $t = 0$. This involves sorting n tasks in the ready list and then selecting the task with the earliest deadline for execution. When more than two speeds are allowed, the complexity of LEDF becomes $O(n \log n + kn)$, where k is the number of speed settings that are allowed.

Procedure LEDF()

t_c: current time;

$S_h > S_{l1} > S_{l2} > \dots S_{lm}$: Available processor speeds

`schedulable = 1`

1. **if** ready_list $\neq$ NULL
2.   Sort task deadlines in ascending order;
3.   Select task τ_i with earliest deadline;
4.   **for** $S = S_{lm}$ to S_h
5.   **if** $t_c + \dfrac{l_i}{S} \leq d_i$ **then**
6. $t = t_c + \dfrac{l_i}{S}$
7.   **for** each task τ_u that has not completed execution
8.   **if** $t + \dfrac{l_u}{S_h} \leq d_u$ **then**
9. $t = t + \dfrac{l_u}{S_h}$
10.   **else**
11.   `schedulable = 0`
12.   **break**
13.   **endfor**
14.   **if** `schedulable = 1` **then**
15.   Schedule τ_i at S
16.   **break**
17.   **endif**
18.   **endfor**

FIGURE 29.1 The LEDF algorithm.

29.2.2 Implementation Testbed

29.2.2.1 Hardware Platform

The power measurement experiments were conducted on a laptop with an AMD Mobile Athlon 4 processor. AMD's PowerNow! technology offers greater flexibility in setting both frequencies and core voltages [2]. The 1.1 GHz Mobile Athlon 4 processor can be set at several core voltage levels ranging from 1.2 to 1.4 V in 0.05 V increments. For each core voltage there is a predetermined maximum clock frequency. The power states we chose to use in our scheduler and simulations are shown in Table 29.1. Although we use only three speeds in our experiments, an extension to using all five available speeds appears to be quite straightforward.

PowerNow! technology was developed primarily to extend battery life on mobile systems. We therefore conducted our experiments on a laptop system rather than a desktop PC. Instead of inserting a current probe into the laptop, we opted to simply measure system power during the experiments. The laptop's system power is drawn from the power converter at approximately 18.5 V DC. Instead of using an oscilloscope or digital ammeter to take exact CPU power measurements at very high frequencies, we chose the simpler approach of using a large capacitor to average out the DC current drawn by the entire laptop. This method works primarily due to the periodic nature of our tests. In a periodic real-time system, the power drawn

TABLE 29.1 Speed and Voltage Settings for the Athlon 4 Processor

Power State	Speed (MHz)	Voltage (V)
1	1100	1.4
2	900	1.35
3	700	1.25

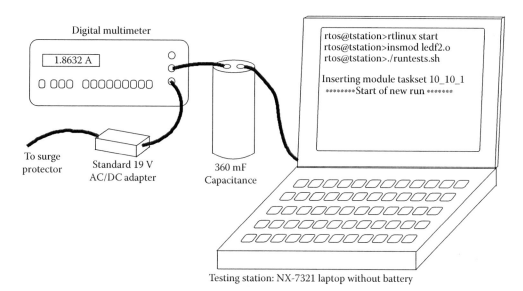

Digital multimeter

rtos@tstation>rtlinux start
rtos@tstation>insmod ledf2.o
rtos@tstation>./runtests.sh

Inserting module taskset 10_10_1
********Start of new run*******

To surge protector

Standard 19 V AC/DC adapter

360 mF Capacitance

Testing station: NX-7321 laptop without battery

FIGURE 29.2 Illustration of the experimental setup.

over one hyperperiod is roughly the same as the power drawn over the next hyperperiod as long as no tasks are added or removed from the task set. Since a fairly large amount of energy needs to be sourced and sunk by the capacitor at the different processor speeds and activity levels, we used a 30 V DC 360 mF capacitance (a 160 and a 200 mF capacitor in parallel). This capacitance proved capable of averaging current loads for power state periods ranging up to hundreds of milliseconds. When the processor power state switches at a lower rate than this, the current measurements taken between the AC/DC converter and the voltmeter readings fluctuate. Figure 29.2 illustrates our experimental hardware setup.

29.2.2.2 Software Architecture

We used RT-Linux [24] as the OS for our experiments. In addition to providing real-time guarantees for tasks and a periodic scheduling system, RT-Linux also provides a well-documented method of changing the scheduling policies. An elegant modular interface allows for easy adaptation of the scheduler module to use LEDF and then load and unload it as necessary. We used this feature of RT-Linux to swap LEDF for a regular EDF scheduler during power comparisons. Furthermore, RT-Linux uses Linux as its idle task, providing a very convenient method of control and evaluation for the execution of the real-time tasks.

LEDF sorts all tasks by their absolute deadlines and chooses the task with the EDF. If there are no real-time tasks pending, the Linux/Idle task is chosen and run at the lowest available speed. A timeout is then set to preempt the Idle task at the next known release time. Once a speed is identified for a task, the switching code is invoked if the processor is not already operating at that speed.

Switching the power state of a mobile Athlon 4 processor simply consists of writing to a model specific register (MSR). The core voltage and clock frequency at which the processor is to be set are encoded into a 32-bit word along with three control bits. Another 32-bit word contains the stop-grant timeout count (SGTC), which represents the number of 100 MHz system clocks during which the processor is stalled for the voltage and frequency changes. The maximum phase-locked loop (PLL) synchronization time is 50 μs and the maximum time for ramping the core voltage appears to be 100 μs. Calling the WRMSR macro then instruments the power state change. For debugging, the RDMSR macro was used with a status MSR to retrieve the processor's power state. Decoding the two 32-bit word values reveals the maximum, current, and default frequency and core voltage.

The RT-Linux high-resolution timer used for scheduling is based (in x86 systems) on the time-stamp counter (TSC). The TSC is a special counter introduced by Intel that simply counts clock periods in the CPU since it was started (boot-time). The *gethrtime*() RT-Linux method (and all methods derived from it) convert the TSC value into a time value using the recorded clock frequency. Thus, a simple calculation to determine time in nanoseconds from the TSC value would be the product of TSC and clock period. Since RT-Linux was initially developed without the need for dynamic frequency switching, the speed used for the calculation of time is set at boot time and never changed. Thus, when the processor is slowed to a low-power state with a lower clock frequency, the TSC counts at a lower rate. However, the *gethrtime*() method is oblivious to this and the measurement of time slows down proportionally. It is not clear what happens to the TSC, and thus how to measure time, during a speed switch. The TSC does appear to be incremented during some part of the speed switch, but the count is not a reliable means of measuring time. Re-calibrating the rate at which the TSC is incremented appears to be a non-trivial task which requires extensive re-writing of the RT-Linux timing code. Therefore, we chose to track time from within the LEDF module.

29.2.3 Experimental Results

We now present data from our power measurement experiments. In these experiments, we measure the total system power consumption of the laptop. Knowledge of CPU power savings, however, is useful in generalizing the results. CPU power savings can easily be derived from a set of experiments. In order to isolate the power used by the processor and system board, we can turn off all system components except the CPU and system board. We can then take a power reading when the CPU is halted. This power measurement represents the total system power excluding CPU power. We can then subtract this base power from all future power readings in order to obtain CPU power alone. However, halting a processor is far more complex than simply issuing a "HLT" instruction. Decoupling the clock from the CPU involves handshaking between the CPU and the Northbridge. We were unable to obtain sufficient documentation to implement this.

As an alternative method of estimating power drawn by the system board and components, the power consumption of the CPU with maximum load can be calculated from system measurements at two power states. This can be done by devising tests to isolate power drawn by the LCD screen, hard drive, and the portion of the system beyond our control. Once an estimate for system power is available, we can eliminate this from all our readings to get an approximation of the fraction of CPU power being saved.

Ratios for power consumption in different states can be calculated using the well known relationship for CMOS power consumption, i.e., $P = faCV_{dd}^2$, where P is the power, f is the frequency of operation, a is the average switching activity, C is the switching capacitance, and V_{dd} is the operating voltage. The switching capacitance and average switching activity is constant for the same processor and software, so we only consider the frequency and the square of the core voltage. It is also reasonable to assume that other components of the laptop (the screen and hard disk, for example) draw approximately the same current regardless of the CPU operating voltage. We therefore calculate that power state 2 uses approximately 76% as much power as power state 1 and power state 3 uses only 50.7% as much power as the maximum power state. The minimum power configuration for this processor is 300 MHz at 1.2 V, which consumes only 20% of the power consumed in the maximum power state.

In our case, we chose to compare a fully loaded processor operating at 700 MHz (with a core frequency of 1.25 V) and at 1100 MHz (with a core voltage of 1.4 V). The 700 MHz configuration uses $(700 \times 1.25^2)/(1100 \times 1.4^2)$, or 50.73% as much CPU power as the 1100 MHz configuration. For a given task running at 1100 MHz, the observed current consumption was 2.373 A. For the same task running at 700 MHz, we observed a current reading of 1.647 A. Assuming that the current consumption of the other components was approximately the same during both runs, the difference in CPU current consumption is 0.726 A. This means

$$I_{1100} - I_{700} = 0.726$$

$$\Rightarrow I_{1100} - 05073I_{1100} = 0.726$$

$$\Rightarrow I_{1100} = 1.474 \text{ A}$$

TABLE 29.2 Current Consumptions
of Various System Components

CPU (1100 MHz)	Screen	Disk	Current Drawn (A)
Idle	Off	STBY	1.5
Idle	Off	On	1.54
Idle	On	STBY	1.91
Idle	On	Sleep	1.9
Idle	On	On	1.97
Max load	Off	STBY	1.93
Max load	On	On	2.45

In other words, a measured difference of $(2.373 - 1.647) = 0.726$ A of current implies that the fully loaded CPU operating at 1100 MHz draws approximately 1.474 A. Knowing this, we deduce from the information in Table 29.2 that the system board and basic components draw approximately 0.456 A, and that under normal operation, the system (including the disk drive and display) draws about 0.976 A in addition to the load from the CPU. This estimation, although approximate, provides a useful method of isolating energy used by the CPU for various utilizations and scheduling algorithms.

We performed several experiments with three different versions of the scheduling algorithm and different task sets at various CPU utilization levels. We constructed a pseudo-random task generator to generate our test sets. Using the task generator, we created several random sets of tasks. The release times of the tasks are set to the beginning of a period and deadlines to the end of a period. Computation requirements for the tasks are chosen randomly and then scaled to meet the target utilization.

The tests programs consist of multiple threads that execute "for" loops for specified periods of time. The time for which these threads run can be determined by examining the assembly level code for each iteration of a loop. Each loop consists of five assembly language instructions which take 1 cycle each to execute. The random task set generator takes this into account when generating the task sets.

The simulator is a simple PERL program that reads in task data and generates the schedule which would be generated by the LEDF scheduler. It then takes user-supplied baseline power measurements and uses them to compute the power consumption of the task set. Summing up the fraction of the period spent in each state and multiplying it by the appropriate power consumption measurement produces the overall power consumption for the task set. As a reasonable representation of the load generated by the Linux/Idle task, the simulator assumes the Linux/Idle task to consume a certain amount of power whose value lies between the power consumptions of a fully-loaded and fully-idle system running at a given speed. This power value was determined by measuring the power consumption of the laptop with regular Linux running a subset of daemon processes in the background.

We used a single power-state version of LEDF (in effect, EDF) as a comparison point. These tests show the maximum power requirements for the amount of work (computation) to be done. We also used 2-and 3-speed versions of LEDF to observe the effect of adding additional power states. The 2-speed version used operating frequencies of 700 and 1100 MHz, and the 3-speed version incorporated an intermediate 900 MHz operating frequency. The CPU utilizations ranged from 10% to 80% in increments of 10%. The maximum utilization of 80% was necessary to guarantee that the Linux/Idle task had sufficient time available for control operations. Without forcing the scheduler to leave 20% of the period open for the Linux/Idle task, the shell became unresponsive, forcing a hard reboot of the machine between each test. We also implemented the cycle-conserving EDF (ccEDF) algorithm from [20] and compared our algorithm to it. This implementation of ccEDF uses a set of discrete speeds.

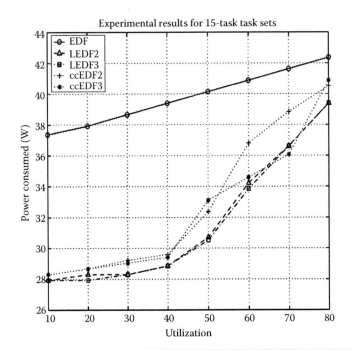

FIGURE 29.3 Heuristic comparison for 15-task task set.

The results are shown in Figure 29.3 for a 15-task task set. Each data point represents the average of three randomly generated task sets for a given utilization value and task set size. LEDF2 (LEDF3) and ccEDF2 (ccEDF3) refer to the use of two (three) processor speeds.

The power savings ranged from 9.4 W in a minimally utilized system to 2.6 W in a fully utilized system. The fully utilized system has lower power consumption under LEDF because LEDF schedules the non-real-time component at the lowest speed. Note, however, that up to the 50% mark the power savings remain over 9 W and remain in most cases over 7 W for 60% utilization. With a maximum utilization of 80%, the system can still save significant power with a reasonable task load.

A comparison between measured experimental results and simulation results is shown in Figure 29.4. In most cases, the simulated and measured values are the same or within 2% of each other. The simulation results thus provided a very close match to the experimental results, indicating that the simulation engine model accurately models the real hardware. Since the simulation engine does not take into account the scheduler's computation time, the fidelity of the results may degrade for very high task counts due to the extra cost of sorting the deadlines. In order to verify this, we evaluated LEDF with several randomly-generated task sets with different utilizations with the number of tasks ranging from 10 to 200 and measured the execution time of the scheduler for each task set. Our results show that the execution time of the scheduler was in the order of microseconds, while the task execution times were in the order of milliseconds. For increasing task set size, scheduler runtime increases at a very slow rate. Thus, scheduling overhead does not prove to be too costly for the power-aware version of EDF. For task sets with more than 240 tasks, the RT-Linux platform tended to become unresponsive. These results are shown in Table 29.3. The entries in the table correspond to task sets with 40% utilization, but with varying number of tasks. The other task sets we experimented with (task set utilizations of 50% and 80%) also exhibit the same trend in scheduler runtime and are not reproduced here. The scheduler overhead in Table 29.3 indicates the time taken by the scheduler to sort the task set and to identify the active task. Even though our implementation of LEDF is currently of $O(n^2)$ complexity and can be replaced by a faster $O(n \log n)$ implementation, it is obvious from Table 29.3 that scheduling overhead is negligible for over a hundred tasks for

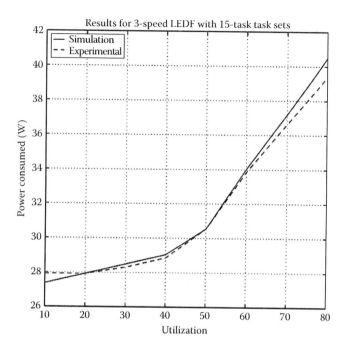

FIGURE 29.4 Comparison of experimental 3-state LEDF with expected results.

TABLE 29.3 Measured Scheduler Overhead for Varying Task Set Sizes

Number of Tasks	Measured Scheduler Overhead (ns)
10	1,739
20	1,824
30	1,924
60	3,817
120	6,621
180	10,916
200	12,243

utilizations ranging from 10% to 80%. For small task sets where the task set consists of a few hundred tasks, scheduling overhead is negligible compared to task execution times.

29.3 Node-Level I/O-Device-Oriented Energy Management

Prior work on DPM techniques for I/O devices has focused primarily on scheduling devices in non-real-time systems. The focus of these algorithms is minimizing user response times rather than meeting real-time task deadlines, and therefore these methods are not viable candidates for use in real-time systems. Due to their inherently probabilistic nature, the applicability of the above methods to real-time systems falls short in one important aspect—real-time temporal guarantees cannot be provided. Such methods perform efficiently in interactive systems, where user waiting time is an important design parameter. In real-time systems, minimizing response time of a task does not guarantee that its deadline will be met. It thus becomes apparent that new algorithms that operate in a deterministic manner are needed in order to ensure real-time behavior.

29.3.1 Optimal Device Scheduling for Two-State I/O Devices

In this section, we describe a non-preemptive optimal offline scheduling algorithm to minimize the energy consumption of the I/O devices in hard real-time systems. In safety-critical applications, offline scheduling is often preferred over priority-based run-time scheduling to achieve high predictability [25]. In such systems, the problem of scheduling tasks for minimum I/O energy can be readily addressed through the technique described here. This algorithm is referred to as the *Energy-Optimal Device Scheduler* (EDS). For a given job set, EDS determines the start time of each job such that the energy consumption of the I/O devices is minimized, while guaranteeing that no real-time constraint is violated. EDS uses a tree-based branch-and-bound approach to identify these start times. In addition, EDS provides a sequence of states for the I/O devices, referred to as the I/O device schedule, that is provably energy-optimal under hard real-time job deadlines. Temporal and energy-based pruning are used to reduce the search space significantly. Our experimental results show that EDS reduces the energy consumption of I/O devices significantly for hard real-time systems. We next define and describe some important terms and assumptions. First, we define the device scheduling problem $\mathcal{P}_{io}$ and describe the task and device models in greater detail.

We are given a task set $\mathcal{T} = \{\tau_1, \tau_2, \ldots, \tau_n\}$ of n tasks. Each task $\tau_i \in \mathcal{T}$ is defined by (1) an arrival time a_i, (2) a worst-case execution time c_i, (3) a period p_i, (4) a deadline d_i, and (5) a *device-usage list* L_i. The device-usage list L_i for a task τ_i is defined as the set of I/O devices that are used by τ_i. The hyperperiod H of the task set is defined as the least common multiple of the periods of all tasks. Without loss of generality, we assume that the deadline of a task is equal to its period, i.e., $p_i = d_i$.

A set $\mathcal{K} = \{k_1, k_2, \ldots, k_p\}$ of p I/O devices is used in the system. Each device k_i is characterized by

- Two power states—a low-power sleep state $ps_{l,i}$ and a high-power working state $ps_{h,i}$
- A wake-up time from $ps_{l,i}$ to $ps_{h,i}$ represented by $t_{wu,i}$
- A shutdown time from $ps_{h,i}$ to $ps_{l,i}$ represented by $t_{sd,i}$
- Power consumed during wake-up $P_{wu,i}$
- Power consumed during shutdown $P_{sd,i}$
- Power consumed in the working state $P_{w,i}$
- Power consumed in the sleep state $P_{s,i}$

Requests can be processed by the devices only in the working state. All I/O devices used by a task must be powered-up before the task starts execution.

In I/O devices, the power consumed by a device in the sleep state is less than the power consumed in the working state, i.e., $P_{s,i} < P_{w,i}$. Without loss of generality, we assume that for a given device k_i, $t_{wu,i} = t_{sd,i} = t_{0,i}$ and $P_{wu,i} = P_{sd,i} = P_{0,i}$. The energy consumed by device k_i is given by

$$E_i = P_{w,i} t_{w,i} + P_{s,i} t_{s,i} + M P_{0,i} t_{0,i}$$

where
 M is the total number of state transitions for k_i
 $t_{w,i}$ is the total time spent by device k_i in the working state
 $t_{s,i}$ is the total time spent in the sleep state

Incorrectly switching power states can cause increased, rather than decreased, energy consumption for an I/O device. Incorrect switching of I/O devices is eliminated using concept of *breakeven time* [10], which is defined as the time interval for which a device in the powered-up state consumes an energy exactly equal to the energy consumed in shutting a device down, leaving it in the sleep state and then waking it up (Figure 29.5). If any idle time interval for a device is greater than the breakeven time t_{be},

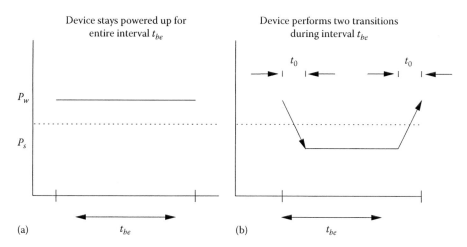

FIGURE 29.5 Illustration of breakeven time. The time interval for which the energy consumptions are the same in (a) and (b) is called the breakeven time.

energy is saved by shutting it down. For idle time periods that are less than the breakeven time, energy is saved by keeping it in the powered-up state.

Associated with each *task set* T is a *job set* $\mathcal{J} = \{j_1, j_2, \ldots, j_l\}$ consisting of all the instances of each task $\tau_i \in T$, where $l = \sum_{k=1}^{n} H/p_k$, where H is the hyperperiod and p_k is the period of task τ_k. Except for the period, a job inherits all properties of the task of which it is an instance. This transformation of a pure periodic task set into a job set does not introduce significant overhead because optimal I/O device schedules are generated offline, where scheduler efficiency is not a pressing issue.

For the sake of simplicity, we assume that the devices have only a single sleep state. An extension for devices with multiple low-power states is described in Section 29.3.1.2. In order to ensure that the states of the I/O devices are clearly defined at the completion of the jobs, we assume that the worst-case execution times of the tasks are greater than the transition time of the devices.

The offline device scheduling problem $\mathcal{P}_{io}$ is formally stated below:

- $\mathcal{P}_{io}$: Given a job set $\mathcal{J}$ that uses a set $\mathcal{K}$ of I/O devices, identify a set of start times $\mathcal{S} = \{s_1, s_2, \ldots, s_l\}$ for the jobs such that the total energy consumed $\sum_{i=1}^{p} E_i$ by the set $\mathcal{K}$ of I/O devices is minimized and all jobs meet their deadlines.

This set of start times, or schedule, provides a minimum-energy device schedule. Once a task schedule has been determined, a corresponding device schedule is generated by determining the state of each device at the start and completion of each job based on the its device-usage list.

There are no restrictions on the time instants at which device states can be switched. The I/O device schedule that is computed offline is loaded into memory and a timer controls the switching of the I/O devices at run-time. Such a scheme can be implemented in systems where tick-driven scheduling is used. We assume that all devices are powered up at time $t = 0$. Next, we describe the theory underlying the EDS scheduling algorithm.

29.3.1.1 Pruning Technique

We generate a schedule tree and iteratively prune branches when it can be guaranteed that the optimal solution does not lie along those branches. The schedule tree is pruned based on two factors—time and energy. Temporal pruning is performed when a certain partial schedule of jobs causes a missed deadline deeper in the tree. The second type of pruning—which we call *energy pruning*—is the central idea on

TABLE 29.4   Example Task Set $T1$

Task	Arrival Time	Completion Time	Period (Deadline)	Device-Usage List
τ_1	0	1	3	k_1
τ_2	0	2	4	k_2

TABLE 29.5   List of Jobs for Task Set $T1$ from Table 29.4

	j_1	j_2	j_3	j_4	j_5	j_6	j_7
a_i	0	0	3	4	6	8	9
c_i	1	2	1	2	1	2	1
d_i	3	4	6	8	9	12	12

which EDS is based. The remainder of this section explains the generation of the schedule tree and the pruning techniques that are employed. We illustrate these through the use of an example.

A vertex v of the tree is represented as a 3-tuple (i, t, e) where i is a job j_i, t is a valid start time for j_i, and e represents the energy consumed by the devices until time t. An edge z connects two vertices (i, t, e) and (k, l, m) if job j_k can be successfully scheduled at time l given that job j_i has been scheduled at time t. A path from the root vertex to any intermediate vertex v has an associated order of jobs that is termed a *partial schedule*. A path from the root vertex to a leaf vertex constitutes a *complete schedule*. A *feasible schedule* is a complete schedule in which no job misses its associated deadline. Every complete schedule is a feasible schedule (temporal pruning eliminates all *infeasible* partial schedules).

An example task set $T1$ consisting of two tasks is shown in Table 29.4. Each task has an arrival time, a worst-case execution time and a period. We assume that the deadline for each task is equal to its period. Task τ_1 uses device k_1 and task τ_2 uses device k_2. Table 29.5 lists the instances of the tasks, arranged in increasing order of arrival. In this example, we assume a working power of 6 units, a sleep power of 1 unit, a transition power of 3 units and a transition time of 1 unit.

We now explain the generation of the schedule tree for the job set shown in Table 29.5. The root vertex of the tree is a dummy vertex. It is represented by the 3-tuple $(0, 0, 0)$ that represents dummy job j_0 scheduled at time $t = 0$ with an energy consumption of 0 units. We next identify all jobs that are released at time $t = 0$. The jobs that are released at $t = 0$ for this example are j_1 and j_2. Job j_1 can be scheduled at times $t = 0$, $t = 1$, and $t = 2$ without missing its deadline. We also compute the energy consumed by all the devices up to times $t = 0$, $t = 1$, and $t = 2$. The energy values are 0, 8, and 10 units, respectively. (Figure 29.6 explains the energy calculation procedure.) We therefore draw edges from the dummy root vertex to vertices $(1,0,0)$, $(1,1,8)$, and $(1,2,10)$. Similarly, job j_2 can be scheduled at times $t = 0$, $t = 1$, and $t = 2$ and the energy values are 0, 8, and 10 units respectively. Thus, we draw three more edges from the dummy vertex to vertices $(2,0,0)$, $(2,1,8)$, and $(2,2,10)$. Note that job j_2 would miss its deadline if it were scheduled at time $t = 3$ (since it has an execution time of 2 units). Therefore, no edge exists from the dummy node to node $(2,3,e)$, where e is the energy consumption up to time $t = 3$. Figure 29.7 illustrates the tree after one job has been scheduled. Each level of depth in the tree represents one job being successfully scheduled.

We then proceed to the next level. We examine every vertex at the previous level and determine the jobs that can be scheduled next. By examining node $(1,0,0)$ at level 1, we see that job j_1 would complete its execution at time $t = 1$. The only other job that has been released at $t = 1$ is job j_2. Thus, j_2 can be scheduled at times $t = 1$ and $t = 2$ *after* job j_1 has been scheduled at $t = 0$. The energies for these nodes are computed and edges are drawn from $(1,0,0)$ to $(2,1,10)$ and $(2,2,14)$. Similarly, examining vertex $(1,1,8)$ results in vertex $(2,2,16)$ at level 2. The next vertex at level 1—vertex $(1,2,10)$—results in a missed deadline at level 2. If job j_1 were scheduled at $t = 2$, it would complete execution at time $t = 3$. The earliest time at which j_2 could be scheduled is $t = 3$; however, even if it were scheduled at $t = 3$, it would miss its deadline. Thus, scheduling j_1 at $t = 2$ does not result in a feasible schedule. This branch can hence be pruned.

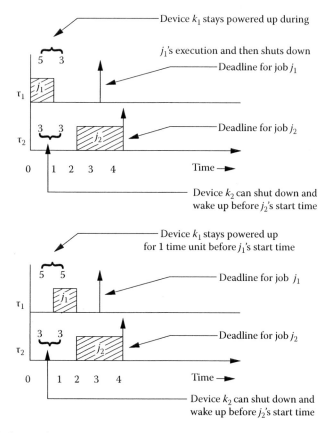

FIGURE 29.6 Calculation of energy consumption.

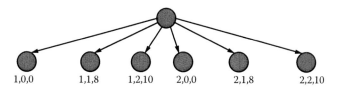

FIGURE 29.7 Partial schedules after 1 scheduled job.

Similarly, the other nodes at level 1 are examined and the unpruned partial schedules are extended. Figure 29.8 illustrates the schedule tree after two jobs have been scheduled. The edges that have been crossed out represent branches that are not considered due to temporal pruning.

At this point, we note that vertices (2,2,14) and (2,2,16) represent the same job (j_2) scheduled at the same time ($t = 2$). However, the energy consumptions for these two vertices are different. This observation leads to the following theorem:

Theorem 29.1

When two vertices at the same tree depth representing the same job being scheduled at the same time can be reached from the root vertex through two different paths, and the orders of the previously scheduled jobs along the two partial schedules are identical, then the partial schedule with higher energy consumption can be eliminated without losing optimality.

Proof

Let us call the two partial schedules at a given depth Schedule A and Schedule B, with Schedule A having lower energy consumption than Schedule B. We first note that Schedule B has higher energy consumption than Schedule A because one or more devices have been in the powered-up state for a longer period of time than necessary in Schedule B. Assume that i jobs have been scheduled, with job j_i being the last scheduled job. Since we assume that the execution times of all jobs are greater than the maximum transition time of the devices, it is easy to see that the state of the devices at the end of job j_i will be identical in both partial schedules. By performing a time translation (mapping the end of job j_i's execution to time $t = 0$), we observe that the resulting schedule trees are identical in both partial schedules. However, all schedules in Schedule B after time translation will have an energy consumption that is greater than their counterparts in Schedule A by an energy value E_δ, where E_δ is the energy difference between Schedules A and B. It is also easy to show that the energy consumed *during* job j_i's execution in Schedule A will always be less than or equal to j_i's execution in Schedule B. This completes the proof of the theorem.

The application of this theorem to the above example results in partial schedule B in Figure 29.8 being discarded. As one proceeds deeper down the schedule tree, there are more vertices such that the partial schedules corresponding to the paths to them from the root vertex are identical. It is this "redundancy" that allows for the application of Theorem 29.1, which consequently results in tremendous savings in memory while still ensuring that an energy-optimal schedule is generated. By iteratively performing this sequence of steps (vertex generation, energy calculation, vertex comparison and pruning), we generate the complete schedule tree for the job set. Figure 29.9 illustrates the partial schedules after three

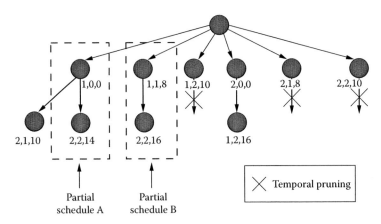

FIGURE 29.8 Partial schedules after two scheduled jobs.

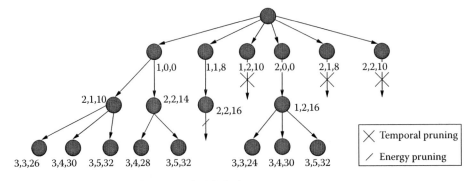

FIGURE 29.9 Partial schedules after three scheduled jobs.

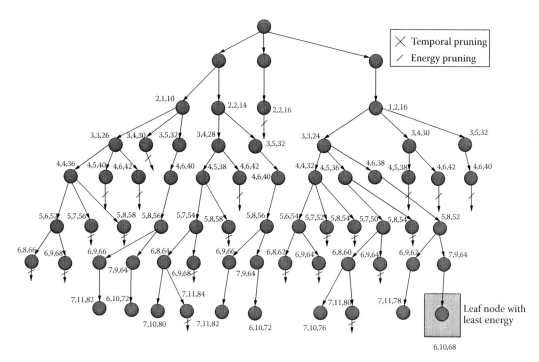

FIGURE 29.10 Complete schedule tree.

jobs have been scheduled for our example. The complete tree is shown in Figure 29.10. We have not shown paths that have been temporally pruned. The edges that have been crossed out with horizontal slashes represent energy-pruned branches. The energy-optimal device schedule can be identified by tracing the path from the highlighted node to the root vertex in Figure 29.10.

29.3.1.2 The EDS Algorithm

The pseudocode for EDS is shown in Figure 29.11. EDS takes as input a job set $\mathcal{J}$ and generates all possible non-preemptive minimum energy schedules for the given job set. The algorithm operates as follows. The time counter t is set to 0 and openList is initialized to contain only the root vertex (0,0,0) (Lines 1 and 2). In lines 3–10, every vertex in openList is examined and nodes are generated at the succeeding level. Next, the energy consumptions are computed for each of these newly generated vertices (Line 11). Lines 15–20 correspond to the pruning technique. For every pair of replicated vertices, the partial schedules are checked and the one with the higher energy consumption is discarded. Finally, the remaining vertices in currentList are appended to openList. currentList is then reset. This process is repeated until all the jobs have been scheduled, i.e., the depth of the tree equals the total number of jobs (Lines 25–28). Note that several schedules can exist with a given energy consumption for a given job set. EDS generates all possible unique schedules with a given energy for a given job set. One final comparison of all these unique schedules results in the set of schedules with the absolute minimum energy.

Devices with multiple low-power sleep states can be handled simply by iterating through the list of low-power sleep states and identifying the sleep state that results in the most energy-savings for a given idle interval. However, the number of allowed sleep states is limited by our assumption that the transition time from a given low-power sleep state is less than the worst-case completion time of the task.

29.3.1.3 Experimental Results

We evaluated EDS for several periodic task sets with varying hyperperiods and number of jobs. We compare the memory requirement of the tree with the pruning algorithm to the memory requirement

Procedure EDS($\mathcal{J}$, *l*)

$\mathcal{J}$: Job set.
l: Number of jobs.
openList: List of unexpanded vertices.
currentList: List of vertices at the current depth.
t: time counter.

```
1.  Set t = 0; Set d = 0;
2.  Add vertex (0,0) to openList;
3.  for each vertex v = (j_i, time) in openList {
4.      Set t = time + c_i;
5.      Find set of all jobs J' released up to time t;
6.      for each job j ∈ J' {
7.          if j has been previously scheduled
8.              continue;
9.          else {
10.             Find all possible scheduling instants for j;
11.             Compute energy for each generated vertex;
12.             Add generated vertices to currentList;
13.         }
14.     }
15.     for each pair of vertices v_1, v_2 in currentList {
16.         if v_1 = v_2 and
                partial schedule(v_1) = partial schedule(v_2) {
17.             if E_v1 > E_v2
18.                 Prune v_1;
19.             else
20.                 Prune v_2;
21.         }
22.     }
23.     Add unpruned vertices in currentList to openList;
24.     Clear currentList;
25.     Increment d;
26.     If d = l
27.         Terminate.
28. }
```

FIGURE 29.11   Pseudocode description of EDS.

TABLE 29.6   Experimental Task Set *T*1

Task	Execution Time	Period (Deadline)	Device List
τ_1	1	4	k_1, k_3
τ_2	3	5	k_2, k_3

of the tree without pruning. Memory requirement is measured in terms of the number of nodes at every level of the schedule tree.

The first experimental task set, shown in Table 29.6, consists of two tasks with a hyperperiod of 20. The device-usage lists for tasks were randomly generated. Expansion of the task set in Table 29.6 results in the job set shown in Table 29.7. Figure 29.12a and b shows the task and device schedules generated for the task set in Table 29.6 using the fixed-priority rate-monotonic scheduling algorithm [15]. Since device k_3 is used by both tasks, it stays powered up throughout the hyperperiod. The device schedule for k_3 is therefore not shown in Figure 29.12.

If all devices are powered up throughout the hyperperiod, the energy consumed by the I/O devices for any task schedule is 66 J. Figure 29.13 shows an optimal task schedule generated using EDS. The energy consumption of the optimal task (device) schedule is 44 J, resulting in a 33% reduction in energy consumption.

TABLE 29.7   Job Set Corresponding
to Experimental Task Set $T1$

	j_1	j_2	j_3	j_4	j_5	j_6	j_7	j_8	j_9
a_i	0	0	4	5	8	10	12	15	16
c_i	1	3	1	3	1	3	1	3	1
d_i	4	5	8	10	12	15	16	20	20

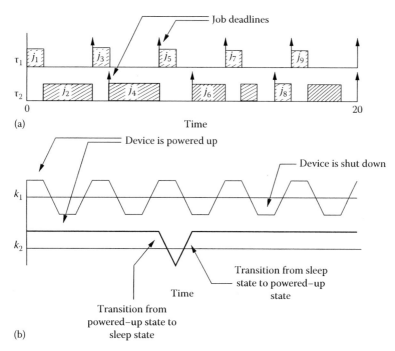

FIGURE 29.12   Task schedule for task set in Table 29.6 using RMA.

From Figure 29.12b, we see that device k_2 stays powered up for almost the entire hyperperiod and device k_1 performs 10 transitions over the hyperperiod. Moreover, device k_2 stays powered up even when it is not in use due to the fact that there is insufficient time for shutting down and powering the device back up. By examining Figure 29.13a and b, we deduce that minimum energy will be consumed if (1) the time for which the devices are powered up is minimized, (2) the time for which the devices are shutdown is maximized, and (3) the number of device transitions is minimized (however, if the transition power of a device k_i is less than its active [operating] power, then energy is minimized by forcing any idle interval for the device to be at least $2t_{0,i}$). In Figure 29.13b, no device is powered up when it is not in use. Furthermore, by scheduling jobs of the same task one after the other, the number of device transitions is minimized, resulting in the maximization of device sleep time. Our approach to reducing energy consumption is to find jobs with the maximum device-usage overlap and schedule them one after the other. Indeed, two jobs will have maximum overlap with each other if they are instances of the same task. This is the approach that EDS follows.

A side-effect of scheduling jobs of the same task one after the other is the maximization of task activation jitter (see Figure 29.13). In some real-time control systems, this is an undesirable feature, which reduces the applicability of EDS in such systems. However, it is clear that jobs of the same task must be scheduled one after the other in order to minimize device energy. It therefore appears that scheduling devices for minimum energy and minimizing activation jitter are not always compatible goals.

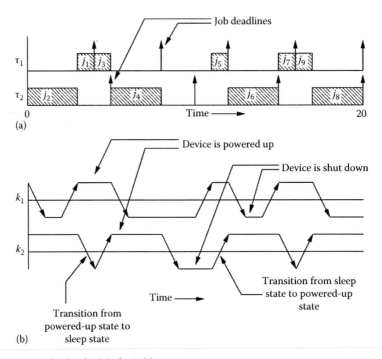

FIGURE 29.13 Optimal task schedule for Table 29.6.

TABLE 29.8 Percentage Memory Savings

Tree Depth i	No. of Vertices at Depth i		Memory Savings (%)
	EE	EDS	
1	7	7	0
2	4	4	0
3	20	14	30
4	18	12	61
5	76	24	68
6	156	26	83
7	270	18	93
8	648	24	96
9	312	8	97
Total	1512	158	90

In order to illustrate the effectiveness of the pruning technique, we compare EDS with an exhaustive enumeration method (EE) which generates all possible schedules for a given job set. The rapid growth in the state space with EE is evident from Table 29.8. We see that the number of vertices generated by EE is enormous, even for a relatively small task set as in Table 29.6. In contrast, EDS requires far less memory. The total number of vertices for EDS is 87% less than that of EE.

By changing the periods of the tasks in Table 29.6, we generated several job sets whose hyperperiods ranged from $H = 20$ to $H = 40$ with the number of jobs J ranging from 9 to 13. For job sets larger than this, EE failed due to lack of computer memory. EE also took prohibitively large amounts of time to run to completion. These experiments were performed on a 500 MHz Sun workstation with 512 MB of RAM and 2 GB of swap space. The results are shown in Table 29.9.

TABLE 29.9 Comparison of Memory Consumption
and Execution Time for EE and EDS

	No. of Vertices		Execution Time	
Job Set	EE	EDS	EE	EDS
$H = 20, J = 9$	1,512	158	<1 s	<1 s
$H = 30, J = 11$	252,931	1,913	2.3 s	<1 s
$H = 35, J = 12$	2,964,093	2,297	28.2 s	4.6 s
$H = 40, J = 13$	23,033,089	4,759	7 min 15 s	35.2 s
$H = 45, J = 14$	—	7,815	—	2 min 29.5 s
$H = 55, J = 16$	—	18,945	—	2 h 24 min 15 s
$H = 60, J = 17$	—	30,191	—	5 h 10 min 23.2 s

—, Failed due to insufficient memory.

For job sets with the number of jobs being greater than 17 jobs, the EDS algorithm failed due to insufficient memory. We circumvent this problem by breaking up the vertices generated at level 1 into several separate subproblems. Energy pruning is then performed within and across each subproblem. This is explained in greater detail in the next paragraph.

Let us consider our running example for pruning. Figure 29.7 illustrates the partial schedule tree after one job has been scheduled. The original EDS algorithm expands each of these nodes in a breadth-first fashion and then performs energy-based pruning across all nodes at the second level, as shown in Figure 29.8. At deeper levels, the number of nodes increases tremendously, thereby making excessive demands on memory. An enhancement to EDS that addresses the memory consumption issue is to expand only a single level-1 vertex at a time and perform temporal and energy pruning within this single subproblem. The memory requirement is therefore reduced significantly. The minimum-energy schedule derived from solving this single subproblem is then recorded. When the next subproblem is solved, energy pruning is performed both within the current subproblem and across all previously solved subproblems. The solution of a single subproblem results in a minimum-energy schedule with a given level-1 job. This energy value is used as an additional bound that is used for further pruning, even at intermediate depths, in succeeding subproblems. With this enhancement, we were able to solve job sets of up to 26 jobs. Even larger problem instances can be solved by breaking the vertices at lower levels into independent subproblems. Here, however, we restrict ourselves only to level-1 subproblems.

The results for the enhanced EDS algorithm are shown in Table 29.10. For this set of experiments, we used a PC running at 1.4 GHz with 512 MB of RAM.

The energy consumptions using EDS are also compared to the case where all devices are powered up. Each job in the job set uses one or more out of three I/O devices whose power values for each of are shown in Table 29.11. These values pertain to real devices that are currently deployed in embedded systems. The minimum-energy schedules generated by EDS result in energy savings of up to 45% for the larger job sets listed in the table. The growth of the search space (and corresponding increase in execution time) is also evident from the table. An important point to note here is that the use of the energy value of a complete schedule obtained from solving a single subproblem as a bound results in significant pruning at lower levels in the tree. Therefore, the time taken to search the final set of complete schedules for a minimum energy schedule is significantly reduced. This results in faster execution times for the enhanced EDS algorithm.

Finally, we discuss the impact of the assumption that $P_{sd,i} = P_{wu,i}$ and $t_{sd,i} = t_{wu,i}$ on energy consumption. If $P_{sd,i} < P_{wu,i}$ and $t_{sd,i} < t_{wu,i}$, we can expect to save more energy. If $P_{sd} < P_{wu}$, devices will not consume as much energy in transitioning between power states, and if $t_{sd} < t_{wu}$, devices can be powered-down sooner and can stay in the low-power sleep state for longer periods of time. Hence, without the assumption that $P_{sd,i} = P_{wu,i}$ and $t_{sd,i} = t_{wu,i}$, we can obtain greater savings in energy.

TABLE 29.10 Energy Consumption Using EDS

| Job Set | Energy Consumption (J) | | | Execution Time |
	Enhanced EDS	All Powered Up	$\%\Delta E_1 = \dfrac{E_{eds} - E_{apu}}{E_{apu}}$	Enhanced EDS
$H = 20, J = 9$	44.12	66.60	**−33.7%**	<1 s
$H = 30, J = 11$	60.92	96.9	**−37.1%**	<1 s
$H = 35, J = 12$	69.85	113.05	**−38.2%**	<1 s
$H = 40, J = 13$	78.17	129.20	**−39.4%**	<1 s
$H = 45, J = 14$	87.13	145.35	**−40.0%**	<1 s
$H = 55, J = 16$	104.33	177.65	**−41.2%**	<1 s
$H = 60, J = 17$	112.73	193.80	**−41.8%**	3.98 s
$H = 65, J = 18$	121.53	203.95	**−40.4%**	19.15 s
$H = 70, J = 19$	129.93	226.1	**−42.5%**	58.8 s
$H = 80, J = 21$	147.13	258.4	**−43.0%**	7 min 31 s
$H = 85, J = 22$	156.0	274.0	**−43.0%**	30 min 45 s
$H = 90, J = 23$	164.33	290.7	**−43.4%**	2 h 39 min 35 s
$H = 95, J = 24$	170.45	306.85	**−44.5%**	8 h 9 min 17.3 s
$H = 105, J = 26$	186.23	339.15	**−45.0%**	50 h 0 min 26.6 s

E_{eds}, energy consumption using EDS; E_{apu}, energy consumption with devices all powered up.

TABLE 29.11 Device Parameters Used in the Evaluation of LEDES and MUSCLES

Device k_i	Device Type	P_w (W)	$P_{sd}^{i,0} = P_{wu}^{i,1}$ (W)	$P_{sd}^{i,1} = P_{wu}^{i,2}$ (W)	$P_{sd}^{i,2} = P_{wu}^{i,3}$ (W)	t_0 (s)	$P_s^{i,1}$ (W)	$P_s^{i,2}$ (W)	$P_s^{i,3}$ (W)
k_1	HDD [8]	2.3	1.5	0.6	0.3	0.6	1.0	0.5	0.2
k_2	NIC [1]	0.3	0.2	0.05	—	0.5	0.1	3 m	—
k_3	DSP [3]	0.63	0.4	0.1	—	0.5	0.25	0.05	—

As we noted earlier, optimal device scheduling is an $\mathcal{NP}$-complete problem. However, by making a few simplifying assumptions, online polynomial-time algorithms that generate near-optimal solutions can be developed with relative ease. In the next section, we describe two such algorithms. These algorithms schedule the shutdowns and wake-ups of I/O devices such that energy is reduced, and also ensure that no real-time deadlines are missed.

29.3.2 Online Device Scheduling

In this section, we describe LEDES, a near-optimal, deterministic device scheduling algorithm for two-state I/O devices. We assume here that the start time for each job is fixed and known a priori. Under this assumption, the device scheduling problem $\mathcal{P}_{io}$ is redefined as follows:

- $\mathcal{P}_{io}$: Given the start times $\mathcal{S} = \{s_1, s_2 \ldots, s_n\}$ of the n tasks in a real-time task set T that uses a set $\mathcal{K}$ of I/O devices, determine a sequence of sleep/working states for each I/O device $k_i \in \mathcal{K}$ such that the total energy consumed $\sum_{i=1}^{p} E_i$ by $\mathcal{K}$ is minimized and all tasks meet their respective deadlines.

In the following sections, we describe the conditions under which device state transitions are allowed to minimize energy and ensure the timely completion of tasks. These conditions are different for different scenarios; the scenarios are dependent on the execution times of the tasks that comprise the task set and the number of sleep states present in a device. We begin by assuming that all task execution times are greater than the maximum transition time among all devices and all devices have only one sleep state. We then show that when devices have multiple power states, ensuring timeliness becomes more complex.

One notable advantage of online I/O device scheduling is that online DPM decision-making can exploit underlying hardware features such as buffered reads and writes. A device schedule constructed offline and stored as a table in memory precludes the use of such features due to its inherently deterministic approach. The flexibility of online scheduling enhances the effectiveness of device scheduling.

The need for deterministic I/O device scheduling policies is motivated in detail in Ref. [23]. It was shown in Ref. [23] that it is not possible to ensure timely completion of tasks without a priori knowledge of future device requests. A naive, probabilistic algorithm cannot be used for real-time task sets. We quantify the determinism required to make device scheduling decisions in hard real-time systems through the notion of *look-ahead*, which is a bound on the number of tasks whose device-usage lists must be examined before making a state transition decision, in order to guarantee that no task deadline is missed. Next, we present the *Low-Energy Device Scheduler* (LEDES) for online scheduling of I/O devices with two power states.

29.3.2.1 Online Scheduling of Two-State Devices: Algorithm LEDES

LEDES assumes that the execution times of all tasks are greater than the transition times of the devices they use. Under this assumption, the amount of look-ahead required before making wake-up decisions to ensure timeliness is easily bounded. We derive this result by presenting the following theorem from Ref. [23]:

Theorem 29.2

Given a task schedule for a set T of n tasks with completion times $c_1, c_2, \ldots c_n$, the device utilization for each task, and an I/O device k_1, it is necessary and sufficient to look ahead m tasks to guarantee timeliness, where m is the smallest integer such that $\sum_{i=1}^{m} c_i \geq t_{0,l}$.

In most practical cases, the completion times of tasks are greater than the transition times $t_{0,i}$ of device k_i. This leads to the following corollary to Theorem 29.2.

Corollary 29.1

Given a task schedule for a set T of tasks with completion times $c_1, c_2, \ldots c_n$, the device utilization for each task, and an I/O device k_j, it is necessary and sufficient to look ahead one task to ensure timeliness if the completion times of all tasks in T are greater than the transition time $t_{0,j}$ of device k_j.

The LEDES algorithm operates as follows (also see Figure 29.14). At the start of task τ_i (Line 1), devices not used by the next "immediate" tasks τ_i and τ_{i+1} are put in the sleep state (Lines 3 and 4). The time difference between the start of τ_{i+1} and the end of τ_i's execution is evaluated and compared with the transition time $t_{0,j}$ to determine whether k_j's wakeup can be guaranteed at τ_i's finish time. If k_j is powered down, then a wakeup decision must be made (Line 8). A device must be woken up at s_i if its wake-up cannot be deferred to τ_i's finish time. This is implemented in Line 12 and the device is woken up if needed.

If the scheduling instant at which LEDES is invoked is the completion time of τ_i (Line 11) and if k_j is powered up (Line 12), it can be shut down only if it can fully enter the powered down state before s_{i+1}, since there may be a need for it to be woken up again. If k_j is in the sleep state (Line 15) and is used by τ_{i+1}, it must be woken up to ensure the timely start of τ_{i+1}. These decisions are made for each device and the entire process repeats at each scheduling instant (although there is no mention of the break-even

Algorithm LEDES(k_j, τ_i, τ_{i+1})
curr: current scheduling instant;
1.   **if** curr $= s_i$
2.   **if** k_j is powered-up
3.   **if** $k_j \notin L_i \bigcup L_{i+1}$
4.   **shutdown** k_j
5.   **if** $k_j \in L_{i+1}$
6.   **if** $s_{i+1} - (s_i + c_i) \geq t_{0,j}$
7.   **shutdown** k_j
8.   **else**
9.   **if** $k_j \in L_{i+1}$ and $s_{i+1} - (s_i + c_i) < t_{0,j}$
10.   **wakeup** k_j
11. **if** curr $= s_i + c_i$
12.   **if** k_j is powered-up
13.   **if** $k_j \notin L_{i+1}$ and $s_{i+1} - $ curr $\geq t_{0,j}$
14.   **shutdown** k_j
15.   **else**
16.   **wakeup** k_j

FIGURE 29.14   The LEDES algorithm.

time in Figure 29.14, an implicit check is made to ensure that the idle period for a given device is always greater than the breakeven time).

A simple extension to LEDES can efficiently schedule devices that possess multiple sleep states with the ability to switch from any low-power state directly to the working state. Such a device can be viewed as a device with only two power states. Although the transition times from the sleep states to the powered-up state (and vice-versa) may be different, the correct sleep state to switch a device to is identified simply by performing a series of transition-time checks to verify that there is sufficient time to wake the device up if it is switched to the selected sleep state. However, LEDES cannot make full use of the available sleep states for devices which possess multiple sleep states, but do *not* possess the ability to jump to any sleep state from the powered-up state.

We next present a more general I/O-centric power management algorithm for hard real-time systems. This algorithm is called the *MU*lti-State Constrained *L*ow *E*nergy *S*cheduler (MUSCLES). MUSCLES can also schedule devices which do not have the ability to jump from the powered-up state to any sleep state. Therefore, we assume that at a device scheduling instant, a device may be switched from one power state to the next higher- or lower-power state, i.e., only a single transition is possible at any scheduling instant. In the next section, we describe the MUSCLES algorithm in greater detail.

29.3.3 Low-Energy Device Scheduling of Multi-State I/O Devices

In this section, we describe the MUSCLES scheduling algorithm. The properties of a real-time periodic task remain unchanged from Section 29.3.1. However, I/O device properties now include parameters to describe the different power states. These device properties are re-stated here for the sake of completeness. Each I/O device $k_i \in \mathcal{K}$ is now characterized by

- A set $\mathcal{PS}_i = \{ps_{i,1}, ps_{i,2}, \dots, ps_{i,m}\}$ of m sleep states
- A powered-up state $ps_{i,u}$
- Transition time from $ps_{i,j}$ to $ps_{i,j-1}$, denoted by $t_{wu}^{i,j}$
- Transition time from $ps_{i,j}$ to $ps_{i,j+1}$, denoted by $t_{sd}^{i,j}$
- Power consumed during switching up from state $ps_{i,j}$ to $ps_{i,j-1}$, denoted by $P_{wu}^{i,j}$
- Power consumed during switching down from state $ps_{i,j}$ to $ps_{i,j+1}$, denoted by $P_{sd}^{i,j}$
- Power consumed in the working state P_w^i
- Power consumed in sleep state $ps_{i,j}$, denoted by $P_s^{i,j}$

We assume, without loss of generality, that for each device $k_i \in \mathcal{K}$, $t_{wu}^{i,j+1} = t_{sd}^{i,j} = t_{0,i}$ and $P_{wu}^{i,j+1} = P_{sd}^{i,j} = P_{0,i}$. The total energy E_i consumed by device k_i over the entire hyperperiod is given by

$$E_i = P_w^i t_w^i + \sum_{j=1}^{m} P_s^{i,j} t_s^{i,j} + M P_{0,i} t_{0,i}$$

where

M is the number of state transitions
t_w^i is the total time spent by the device in the working state
$t_s^{i,j}$ is the total time spent by the device in sleep state $ps_{i,j}$

In order to provide conditions under which devices can be shut down and powered up, we first define a few important terms.

Inter-task time: The inter-task time IT_i for task τ_i is the time interval between the start of task τ_{i+1} and completion of task τ_i. Thus $IT_i = s_{i+1} - (s_i + c_i)$. There are two scheduling instants associated with a task τ_i. These correspond to the start and completion time of τ_i, respectively. For minimum-energy device scheduling under real-time constraints, it is not always possible to schedule devices at all scheduling instants. This is formalized using the notion of a valid scheduling instant.

Valid scheduling instant: The completion time of τ_i is defined to be a *valid scheduling instant* for device k_j if $s_{i+1} - (s_i + c_i) \ge t_{0,j}$. In other words, the completion time of τ_i is a valid scheduling instant if and only if $IT_i \ge t_{0,j}$. The start time of τ_i is always a valid scheduling instant. Thus, a task τ_i can have either one or two scheduling instants, depending on the magnitude of IT_i relative to the transition time $t_{0,j}$ of a device k_j. Valid scheduling instants are important for energy minimization. Wake-ups can be scheduled at these points to minimize energy and also ensure that real-time requirements are met. Consider the example shown in Figure 29.15. This figure shows two tasks τ_i and τ_{i+1} with the inter-task time $IT_i < t_{0,j}$. Assume that device k_1 (first used by task τ_{i+2}) is in state $ps_{1,1}$ at τ_i's completion time $(s_i + c_i)$. If a device were to be woken up at $s_i + c_i$, it would complete its transition to state $ps_{1,0}$ only in the middle of τ_{i+1}'s execution and would be in the higher-powered state for the rest of τ_{i+1}'s execution (i.e., until the next scheduling instant). If, on the other hand, the device were to be woken up at s_{i+1}, we can still ensure that the device is powered-up before task τ_{i+2} starts (with the assumption that $c_{i+1} > t_{0,1}$). However, the device stays in the lower-powered state until s_i, resulting in greater energy savings. Hence, we see that wake-ups at valid scheduling instants always result in lowered energy consumption. It is always preferable to wake a device up as late as possible in order to utilize the full potential of online device scheduling.

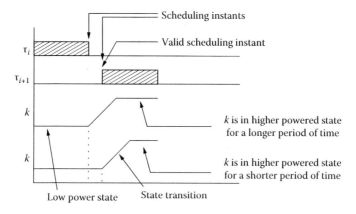

FIGURE 29.15   Illustration of an invalid scheduling instant.

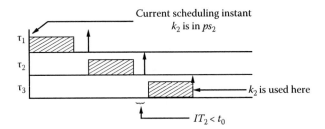

FIGURE 29.16   To show that look-ahead of one task is insufficient when devices have multiple sleep states.

In Section 29.3.1, we have shown that a look-ahead of one task is sufficient when devices have only one sleep state. However, a look-ahead of one task is not sufficient when where devices have multiple low-power sleep states. This is clarified through the example shown in Figure 29.16.

Figure 29.16 shows the execution of three tasks τ_1, τ_2, and τ_3. Assume that the start time of τ_1 is the current scheduling instant. Assume that tasks τ_1 and τ_2 do not use device k_2, which is in sleep state $ps_{2,2}$ at time s_1. An algorithm using a look-ahead of one task, i.e., looking ahead only to task τ_2, would erroneously decide that there is no need to wake k_2 up at time s_1. The same situation arises at scheduling instant $s_i + c_i$. At τ_2's start time (s_2), looking ahead to task τ_3, k_2 is switched to state $ps_{2,1}$. At τ_2's completion time, again looking ahead one task to τ_3, k_2 is switched-up to the powered-up state $ps_{2,u}$. However, if the intertask time IT_2 were less than $t_{0,2}$, k_2 would not have sufficient time to wake up, resulting in τ_3 missing its deadline.

From the above example, it is interesting to note that look-ahead represented as the number of *future tasks* is inadequate for devices with multiple low-power states. When devices have multiple states, look-ahead must be represented as the number of *valid scheduling instants* between tasks. In fact, the notion of look-ahead changes slightly when considering multiple-state I/O devices. Scheduling complexity thus increases with increasing look-ahead due to the additional computational burden of determining look-ahead. Hence, minimizing look-ahead makes the scheduler more efficient.

We now present an upper bound on the look-ahead necessary to ensure timeliness while making shut down decisions for a device [23].

Theorem 29.3

Consider an ordered set $T = \{\tau_1, \tau_2 ..., \tau_n\}$ of n tasks that have been scheduled a priori. Let $\mathcal{K} = \{k_1, k_2 ..., k_p\}$ be the set of p I/O devices used by the tasks in T. In order to decide whether to switch a device $k_i \in \mathcal{K}$ from state $ps_{i,j}$ to $ps_{i,j+1}$ at task τ_c's start or completion time, it is necessary and sufficient to look ahead L tasks, where L is the smallest integer such that the total number of valid scheduling instants associated with the sequence of tasks τ_c, τ_{c+1}, ... τ_{c+L-1} excluding the current scheduling instant is at least equal to $j + 1$. The device k_i can be switched down from $ps_{i,j}$ to $ps_{i,j+1}$ if no task τ_t, $c \leq t \leq c + L - 1$, uses device k_i.

If the intertask times of all tasks are less than the transition time $t_{0,j}$ for device k_j, Theorem 29.3 yields the following corollary.

Corollary 29.2

Suppose the intertask time IT_i is less than the transition time $t_{0,i}$ for every task $\tau_c \in T$. In order for a device $k_i \in \mathcal{K}$ to be switched down from state $ps_{i,j}$ to $ps_{i,j+1}$ at the start or completion time of task τ_c, it is necessary and sufficient to look ahead $j + 1$ tasks to ensure timeliness. Moreover, no task τ_t, $i \leq t \leq j$, must use device k_i.

On the other hand, if the intertask times for all tasks is greater than or equal to the transition time $t_{0,j}$, Theorem 29.3 leads to the following corollary.

Corollary 29.3

Suppose the intertask time IT_i is greater than or equal to the transition time $t_{0,j}$ for every task $\tau_c \in T$. In order for a device $k_j \in K$ to be switched down from state $ps_{i,j}$ to $ps_{i,j+1}$ at the start or completion time of task τ_c, it is necessary and sufficient to look ahead $[(j+1)/2]$ tasks to ensure timeliness. Moreover, device k_j must not be used by any task τ_t, $i \le t \le j$.

Look-ahead increases as the depth of the sleep-state increases. We next present an upper bound on look-ahead for making wake up decisions.

Theorem 29.4

Consider an ordered set $T = \{\tau_1, \tau_2, \ldots, \tau_n\}$ of n tasks and a set $K = \{k_1, k_2, \ldots, k_p\}$ of p devices used by the tasks in T. Suppose the first task after τ_c that uses device k_i is τ_{c+L}. The device $k_i \in K$ must be switched up from state $ps_{i,j+1}$ to $ps_{i,j}$ at the start or completion time of task τ_c if and only if the total number of valid scheduling instants including the current scheduling instant associated with the tasks $\tau_c, \tau_{c+1}, \ldots, \tau_{c+L-1}$ is exactly equal to $j+1$, where L is the look-ahead from the current scheduling instant.

Theorems 29.3 and 29.4 form the basis for the MUSCLES algorithm, which is described in the next section.

29.3.3.1 Online Scheduling for Multi-State Devices: Algorithm MUSCLES

For a precomputed task schedule, MUSCLES generates a sequence of power states for every device such that energy is minimized. It operates as follows (also see Figure 29.17): Let device k_i be in state $ps_{i,j}$ at scheduling instant s_m. MUSCLES finds the next task τ_L that uses k_i (Line 1). A check is then performed to test whether k_i can be switched down to a lower-powered state. This is done by ensuring that there

```
Algorithm MUSCLES(S, PS, kᵢ)
curr: current scheduling instant;
At sₘ:
1.    Find first task τ_L that uses device kᵢ;
2.    Compute number of valid scheduling instants X between sₘ and τ_L;
3.    if X ≥ j + 1
4.        switch down kᵢ from psᵢ,ⱼ to psᵢ,ⱼ₊₁;
5.    else if X = j
6.        wake up kᵢ from psᵢ,ⱼ to psᵢ,ⱼ₋₁;
At sₘ + cₘ:
7.    Find first task τ_L that uses device kᵢ;
8.    Compute number of valid scheduling instants X between sₘ and τ_L;
9.    if X ≥ j + 1
10.       switch down kᵢ from psᵢ,ⱼ to psᵢ,ⱼ₊₁;
11.   else if X = j and curr is a valid scheduling instant
12.       wake up kᵢ from psᵢ,ⱼ to psᵢ,ⱼ₋₁;
13.   else leave kᵢ in psᵢ,ⱼ.
```

FIGURE 29.17 The MUSCLES algorithm.

are at least $j + 1$ valid scheduling instants between the current scheduling instant and τ_L's start time. The presence of $j + 1$ valid scheduling instants implies that device k_i can be switched down from state $ps_{i,j}$ to $ps_{i,j+1}$ (Line 3). The absence of $j + 1$ valid scheduling instants precludes the shutting down of k_i to a lower-powered state; a check is then performed to test whether the device must be switched up. If exactly j instants are present, then the device must be switched up in order to ensure timeliness (Line 4). At the completion of a task τ_m, the same process is repeated. However, an additional check is performed to test if the current scheduling instant is a valid scheduling instant. This is done to minimize energy consumption. If the current scheduling instant is not a valid scheduling instant, the device is left in the same state until a valid scheduling instant (Line 10). MUSCLES guarantees that no task ever misses its deadline.

LEDES and MUSCLES are both polynomial-time algorithms. MUSCLES has a worst-case complexity of $O(pn^2)$, where p is the number of I/O devices used in the system and n is the number of tasks in the task set, and LEDES is $O(p)$. The complexity increases in MUSCLES because the amount of look-ahead, in terms of valid scheduling instants, for each device must be computed before any state-transition. Nevertheless, the relatively low complexity of MUSCLES makes online device scheduling for low energy and real-time execution feasible.

29.3.4 Experimental Results

We first evaluated LEDES and MUSCLES with several randomly-generated task sets with varying utilizations. The task sets consist of six tasks with varying hyperperiods and randomly-generated device-usage lists. These task sets are shown in Table 29.12. Since jobs may be preempted, we consider each preempted slice of a job as two jobs with identical device-usage lists. As a result, the number of jobs listed in each task set in Table 29.12 is an approximation. Each task set is scheduled using the rate-monotonic algorithm. The utilization of each task set is varied from 10% to 90% to observe the impact of slack on the energy consumption of the I/O devices.

While evaluating LEDES, we assumed that the single low-power sleep state for the devices corresponded to the highest-powered sleep state of the device. The energy consumptions at different utilizations for task set T_1 is shown in Figure 29.18. Figure 29.19 illustrates the percentage energy savings for each of the task sets obtained from the LEDES algorithm.

A study of Figure 29.18 reveals that the energy consumption using LEDES and MUSCLES increases with increasing utilization. This is because devices are kept powered up for longer periods of time within the hyperperiod. The resulting decrease in sleep time causes this increased energy consumption. However, we see that energy savings of over 40% can be obtained for task sets with low utilization and over 35% for task sets with high utilization. No task deadlines are missed at any utilization value.

One other important observation that can be made from the graphs is that the savings in energy obtained from MUSCLES *over* LEDES decreases with increasing utilization. This is because the number of valid scheduling instants decreases with increasing utilization. Thus, MUSCLES cannot place devices in deep sleep states as often in high-utilization task sets as it can in low-utilization task sets.

We also evaluated LEDES and MUSCLES with three real-life task sets. These task sets are used in an instrument navigation system (INS) [12], a computer numerical control (CNC) system [13], and an

TABLE 29.12 Evaluation Task Sets for LEDES and MUSCLES

Task Set	Approximate Number of Jobs	Hyperperiod
T_1	303	1,700
T_2	68,951	567,800
T_3	36,591	341,700

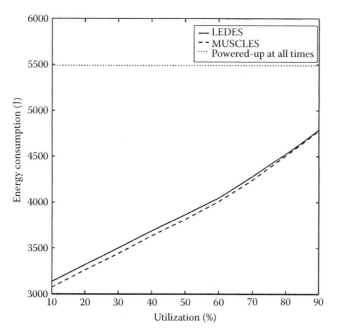

FIGURE 29.18 Comparison of LEDES and MUSCLES for task set T_1.

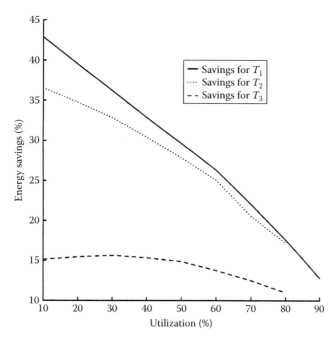

FIGURE 29.19 Energy savings using LEDES.

aviation platform (GAP) [17]. The assignment of devices to tasks in the task sets has been inferred from the functionality of the tasks. For example, task 2 in the GAP task set is a communication task that uses the NIC and task 7 is a status update task that performs occasional reads and writes, and therefore uses a hard disk.

Table 29.13 presents the energy consumptions for these task sets using LEDES and MUSCLES. The energy values here are expressed in units of joules, and they correspond to the energy consumption of

TABLE 29.13   Comparison of LEDES and MUSCLES Using
Real-Life Task Sets

Task Set	Energy (J)			% Savings	
	All Powered Up	LEDES	MUSCLES	LEDES	MUSCLES
CNC	403,104	197,140	117,604	51	70
INS	16.5×10^6	7.7×10^6	3×10^6	51	81
GAP	381×10^6	210×10^6	153×10^6	45	60

TABLE 29.14   Comparison of LEDES and EDS

Job Set	Energy Consumption (J)			$\Delta E_3 = \dfrac{E_{eds} - E_{ledes}}{E_{ledes}}$
	EDS	LEDES	Timeout	
$H = 20, J = 9$	44.12	59.69	60.21	−26.0%
$H = 30, J = 11$	60.92	75.29	85.23	−19.0%
$H = 35, J = 12$	69.85	88.4	100.87	−20.0%
$H = 40, J = 13$	78.17	102.65	108.76	−23.8%
$H = 45, J = 14$	87.13	116.9	130.43	−25.4%
$H = 55, J = 16$	104.33	145.4	155.5	−28.2%
$H = 60, J = 17$	112.73	159.65	170.43	−29.3%
$H = 65, J = 18$	121.53	173.9	192.76	−30.1%
$H = 70, J = 19$	129.93	188.15	216.8	−30.9%
$H = 80, J = 21$	147.13	216.65	240.98	−31.9%
$H = 85, J = 22$	156.0	230.9	252.43	−32.4%
$H = 90, J = 23$	164.33	245.15	270.32	−33.0%
$H = 95, J = 24$	170.45	259.4	282.53	−34.3%
$H = 105, J = 26$	186.23	287.9	315.76	−35.4%

E_{eds}, energy consumption using EDS; E_{ledes}, energy consumption using LEDES.

the I/O devices over the duration of a single hyperperiod. Using LEDES, we obtain an energy savings of 45% for the GAP task set. With MUSCLES, an energy savings of 80% is obtained for the INS task set. Owing to the low utilizations of real-life task sets, significant energy savings can be obtained by intelligently performing state transitions for I/O devices.

Finally, we compare LEDES with EDS and a simple timeout-based scheme. In the timeout-based scheme, a device is powered-down if it has not been used for a pre-specified interval of time (here, we assume that the timeout interval is 1 unit). However, a timeout-based scheme cannot be used in hard real-time systems since it cannot guarantee that jobs complete execution before their deadlines. Nevertheless, we compare our algorithms with the timeout method to highlight the effectiveness of our algorithms. These results are presented in Table 29.14. EDS performs better than LEDES and the timeout method for all experimental task sets. Moreover, the timeout method resulted in an average of 6.8 missed job deadlines over all our job sets.

29.4 Conclusions

Energy is an important resource in battery-operated sensor systems. For such systems that operate under real-time constraints, energy consumption must be carefully balanced with real-time responsiveness. In this chapter, we have described two approaches to energy minimization in sensor networks—node-level energy minimization and network-level energy minimization.

The node-level energy minimization techniques described here focus on minimizing the energy consumption of the processor and I/O devices in a sensor node, respectively. We described the implementation of a DPM scheme that uses an EDF-based scheduler to support real-time execution. The scheduler is efficient and can be easily integrated into the kernels of real-time OSs on sensor nodes. The LEDF algorithm provides significant energy savings in real-time systems.

In many embedded systems, the I/O subsystem is a viable candidate to target for energy reduction. Optimal device schedules for minimum energy consumption can be generated using the offline scheduling algorithm described here. However, the device scheduling problem is known to be $\mathcal{NP}$-complete. With the assumption that device scheduling decisions are made only at task starts and completions, online polynomial-time low-energy I/O device scheduling algorithms to generate near-optimal device schedules can be developed. The first online algorithm described here, called LEDES, efficiently schedules I/O devices that possess two power states—a high-powered working state and a low-powered sleep state. Even under this somewhat restrictive assumption, experimental results show that energy savings of over 40% can be obtained. A generalized version of LEDES, called MUSCLES, that schedules devices with more than two low-power sleep states has also been described here. Experimental case studies for real-life task sets show that energy savings of over 50% can be obtained by targeting the I/O subsystem for power reduction. The amount of energy that can be saved decreases with increasing task set utilization; nevertheless, energy savings of over 40% with these device scheduling algorithms in high-utilization task sets.

References

1. AMD Am79C874 NetPHY-1LP Low-Power 10/100 Tx/Rx Ethernet Transceiver Technical Datasheet.
2. AMD PowerNow! Technology, http://www.amd.com/us-en/Processors/ProductInformation/0,,30_118_756_8079^64,00.html
3. Analog Devices Multiport Internet Gateway Processor. http://www.analog.com
4. L. Benini, A. Bogliolo, G. A. Paleologo, and G. De Micheli, Policy optimization for dynamic power management, *IEEE Transactions on Computer-Aided Design*, 16(6), 813–833, June 1999.
5. G. C. Buttazzo, *Hard Real-Time Computing Systems: Predictable Scheduling Algorithms and Applications*, Kluwer Academic Publishers, Norwell, MA, 1997.
6. A. P. Chandrakasan and R. Broderson, *Low Power Digital CMOS Design*, Kluwer Academic Publishers, Norwell, MA, 1995.
7. E.-Y. Chung, L. Benini, and G. De Micheli, Dynamic power management using adaptive learning tree, *Proceedings of the International Conference on Computer-Aided Design*, San Jose, CA, pp. 274–279, 1999.
8. Fujitsu MHL2300AT Hard Disk Drive. http://www.fujitsu.jp/hypertext/hdd/drive/overseas/mhl2xxx/mhl2xxx.html
9. R. Golding, P. Bosh, C. Staelin, T. Sullivan, and J. Wilkes, Idleness is not sloth, *Proceedings of the USENIX Technical Conference on UNIX and Advanced Computing Systems*, New Orleans, LA, pp. 201–212, 1995.
10. C. Hwang and A. C. H. Wu, A predictive system shutdown method for energy saving of event-driven computation, *Proceedings of the International Conference on Computer-Aided Design*, San Jose, CA, pp. 28–32, 1997.
11. K. Jeffay, D. F. Stanat, and C. U. Martel, On non-preemptive scheduling of periodic and sporadic tasks with varying execution priority, *Proceedings of the Real-Time Systems Symposium*, San Antonio, TX, pp. 129–139, December 1991.
12. D. Katcher, H. Arakawa, and J. Strosnider, Engineering and analysis of fixed priority schedulers, *IEEE Transactions on Software Engineering*, 19, 920–934, September 1993.
13. N. Kim, M. Ryu, S. Hong, M. Saksena, C. Choi, and H. Shin, Visual assessment of a real-time system design: Case study on a CNC controller, *Proceedings of the Real-Time Systems Symposium*, Washington, DC, pp. 300–310, 1996.

14. K. Li, R. Kumpf, P. Horton, and T. Anderson, A quantitative analysis of disk drive power management in portable computers, *Proceedings of the USENIX Winter Conference*, San Francisco, CA, pp. 279–292, 1994.

15. C. L. Liu and J. Layland, Scheduling algorithms for multiprogramming in a hard real-time environment, *Journal of the ACM*, 20(1), 46–61, 1973.

16. J. W. S. Liu, *Real-Time Systems*, Prentice-Hall, Upper Saddle River, NJ, 2000.

17. D. C. Locke, D. Vogel, and T. Mesler, Building a predictable avionics platform in Ada: A case study, *Proceedings of the Real-Time Systems Symposium*, San Antonio, TX, pp. 181–189, 1991.

18. Y.-H. Lu, L. Benini, and G. De Micheli, Operating system directed power reduction, *Proceedings of the International Conference on Low-Power Electronics and Design*, Rapallo, Italy, pp. 37–42, 2000.

19. M. Newman and J. Hong, A look at power consumption and performance of the 3Com Palm Pilot, http://guir.cs.berkeley.edu/projects/p6/finalpaper.html

20. P. Pillai and K. G. Shin, Real-time dynamic voltage scaling for low-power embedded operating systems, *Proceedings of the Symposium on Operating Systems Principles*, Banff, Alberta, Canada, pp. 89–102, 2001.

21. T. Simunic, L. Benini, P. Glynn, and G. De Micheli, Event driven power management, *IEEE Transactions on Computer-Aided Design*, 20(7), 840–857, 2001.

22. M. B. Srivastava, A. P. Chandrakasan, and R. W. Broderson, Predictive system shutdown and other architectural techniques for energy efficient programmable computation, *IEEE Transactions on VLSI Systems*, 4, 42–55, 1996.

23. V. Swaminathan and K. Chakrabarty, Energy-conscious, deterministic I/O device scheduling in hard real-time systems, To appear in *IEEE Transactions on Computer-Aided Design of Integrated Circuits and Systems*, July 22, 2003.

24. The RT-Linux Operating System, http://www.fsmlabs.com/community/

25. J. Xu and D. L. Parnas, Priority scheduling vs. pre-run-time scheduling, *International Journal of Time-Critical Computing Systems*, 18, 7–23, 2000.

30

An Energy-Aware Approach for Sensor Data Communication

H. Saputra
*The Pennsylvania
State University*

N. Vijaykrishnan
*The Pennsylvania
State University*

M. Kandemir
*The Pennsylvania
State University*

Richard R. Brooks
Clemson University

M.J. Irwin
*The Pennsylvania
State University*

30.1 Introduction 639
30.2 System Assumptions 641
30.3 Caching Based Communication 642
30.4 Experimental Results 645
30.5 Spatial Locality... 648
30.6 Related Work...650
30.7 Conclusions...652
Acknowledgments...652
References...652

30.1 Introduction

Distributed sensor networks are envisioned to support various new applications that monitor and interact with the physical world [1–3]. These networks are made of many small interacting nodes that have computing, communication and sensing abilities. Many sensor network applications in eco-monitoring (forest fires, soil moisture) and disaster tracking (contaminant transport, volcanic plume flow tracking) require the sensors to be deployed in locations that are inaccessible (or expensive to access) requiring the sensor nodes to support wireless communication. Further, the sheer number of sensors deployed in distributed networks (in addition to inaccessibility to some nodes) makes it essential for the individual nodes to operate unattended for long duration. A major limiter to the lifetime of the sensor networks is the limited energy available in these nodes. The limited capacity of the battery pack on these nodes and the inability to replace the batteries in these unattended systems makes it important to conserve the energy consumption in the systems.

Communication is an important factor in determining the energy consumption in distributed sensor applications [4]. Energy consumed by transmitting data wirelessly over 10–100 m range has been found to consume as much energy as thousands of operations. Further, the overwhelming volume of sensed information and the need to aggregate data from the different sensed nodes for detecting events of interest leads to a large number of communications across the different nodes. In order to reduce the number of communications, many prior approaches limit the communication to within a set of local nodes called cluster. These clusters support some local processing of the sensed data and extract some useful information and limit the communication to the nodes outside the cluster to this extracted information. For example in a distributed sensor network that detects a vehicle and its possible route, a node that senses an event communicates with other neighboring nodes in its cluster. The information gathered locally in the cluster is used to calculate the position of an event and its possible route. Then this information is forwarded to the next cluster [5].

The focus of this work is on reducing the energy consumed due to the communication that happens between the nodes in a local cluster. Since local nodes frequently share data to identify events of interest, it is important to focus on this task. Due to their spatial proximity, the nodes in the same cluster often sense the same value because of spatial correlation. For example, the temperature values read by adjacent nodes within the same spatial region tend to be similar. Further, sensed data exhibits temporal locality. For example, the temperature values sensed by sensors deployed in a specific region modulate periodically based on the season and time of day. This work exploits this locality of values transmitted by the nodes within a cluster to reduce the energy consumption. Specifically, instead of transmitting the data values themselves, we maintain a cache of recently transmitted values within the cluster and transmit the index of the cached value whenever the same value needs to be transmitted again. Since transmitting the index of a small cache requires a smaller number of bits to be transmitted than actual values, our technique can provide significant energy savings.

In addition to reducing the energy consumption of communication, the proposed approach also addresses another important issue of providing security for sensed data transmissions. Due to the use of wireless transmission for communication, sensed data will be vulnerable to eavesdropping and tampering. While encryption of the communication data is one possible option for providing security, encryption/decryption is costly in terms of computational resources in terms of time and energy consumption [6]. In our approach, there is inherent security provided for some of the transmissions limiting the number of encryptions that are required for secure data transmissions. Specifically, a value that has locality is not transmitted as raw data but only as the index of the entry containing that value in the table, this provides security against eavesdropping. Hence, encryption can be limited to only transmitting actual values when the table is established. Note that in our system, the information of interest that we want to keep secure is the real sensed value. The remaining information such as the index value can still be observed. This means someone can still detect whether the node sensed the same data or not. This scenario also happens if we use a traditional cryptography, where the cipher text of the same plain text will be the same when using the same secret key.

We evaluate the proposed approach by modeling the energy consumption of a StrongArm processor based sensor node. Our experiments utilize synthetic data that exhibits varying degrees of locality in the values transmitted among the sensor nodes in order to model different types of application scenarios. Further, we vary the frequency of data transmission and the underlying encryption technique to quantify the energy savings that result from the proposed approach. The results from our evaluation show that the proposed technique reduces both the energy consumed by data transmissions and the energy consumed by encryption required for security.

The rest of this chapter is organized as follows. The next section explains our system and experimental methodology. Section 30.3 describes the proposed communication protocol employed for transmitting the sensed data among the nodes within a single cluster of a distributed sensor system. Experimental results showing the effectiveness of our approach is discussed in Section 30.4. Section 30.5 explains the enhanced communication protocol for spatial locality value. Section 30.6 discusses other related work in optimizing energy consumption and improving security in sensor networks. Finally, we provide conclusions in Section 30.7.

30.2 System Assumptions

Sensor network consist of hundreds of nodes that are deployed and connected to each other wirelessly. Each node typically consists of sensors, embedded processor and communication hardware.

Our sensor network consists of 32 clusters which each cluster contains 4 nodes. Each node in the same cluster is placed within the distance of d (500 m) as shown in the figure above. The characteristics of our sensor network are as follows:

- Sensors: Each node contains one or more sensors. In this work we assume two kinds of sensor networks. The first sensor network is a temperature sensor network which employs temperature sensors. We choose this environment for a network with high value locality in sensed data. The second sensor network is a vehicle tracking sensor network. This sensor network is based on the network build by Brooks [5]. Each node of this network contains of three different sensors: Acoustic, PIR, and Seismic. This application is representative of low value locality.
- Computational resources: The underlying system is based on a 59 MHz Strong Arm processor (SA-1100). Three major consumers of energy due to computations are the processor, memory accesses and memory leakage. The table used to cache the previous sensed values contains 16 entries with each entry holding 16 bits of data (i.e., we assume that the size of data generated by a sensor is: 16 bits).
- Communication: In this work, we use the following energy model adapted from Ref. [7]:

$$E_{Tx}(k,d) = E_{elec} \star k + \varepsilon_{amp} \star k \star d^2$$

$$E_{Rx} = P_{rec} \star T_{on}$$

where
E_{Tx} is the energy consumed when transmitting k bits of data within the distance of d
E_{Rx} is the energy consumed when receiving/listening for data during T_{on} seconds
E_{elec} is 50 nJ/bit
ε_{amp} is 100 pJ/bit/m^2
P_{rec} is 0.072 mW

As we can see, the energy consumption due to communication is proportional to the packet size and the square of the distance between two nodes. Our approach tries to reduce the packet size by exploiting the possible value locality that a sensor network might have. We cannot reduce the energy consumption by reducing the distance because we assume that our network is a non-mobile sensor network, meaning that nodes cannot move after they have been deployed. The details of our approach are explained in the next section. We assume that the receiver consumes the same amount of energy when it is either listening (and not receiving data) or receiving data. Further, we shutdown the receiver completely during idle periods determined by the message cycle time. Note that our proposed optimization only influences transmission energy and does not change receiver energy.

To protect the confidentiality of the data, sensor network needs to implement a cryptography algorithm. Different cryptography algorithms consume different amount of energy due to their different resource and computational requirements. Our approach tries to reduce the number of cryptography processes; both encryption and decryption, by using the index communication (see Section 30.3 for more details).

We use two different cryptography algorithms: one that uses more resources (Rijndael [8–10]) and the other that uses lesser resources (RC5 [11,12]). We simulate those two algorithms on JouleTrack energy simulator [13] in order to get their energy consumption when executing on a SA1100 processor. Rijndael encryption consumes on the average 156 μJ while RC5 encryption consumes 29 μJ.

30.3 Caching Based Communication

The basis of our communication protocol relies on each node in a cluster caching the *n* previous data values sensed in the cluster. Thus, all the sensor nodes in the cluster must have a coherent cache. If the new sensed value that needs to be transmitted matches with one of the cached values; only the index number of the matching cache entry is transmitted instead of the actual data value. When there is a cache miss, the node transmits the sensed data along with the index in the cache where this new data should be stored. Since we can have two different kinds of packets based on whether there is a hit or miss in the cache, the transmitted packet contains an additional bit to distinguish the packet type. The packet size when a cache miss occurs is given by

$$\text{Packet Size}_{\text{miss}} = 1 + k + \log_2 n$$

where
   k is the size of data generated by the sensors
   n is the number of entries in the cache

For instance: when we have 16 bit data values, and an 16-entry cache, a cache miss results in the transmission of 21 bits data (1 bit to indicate the packet type, 16 bits for the sensed value itself, and 4 bits for the cache index). The packet size transmitted in the case of a cache hit is given by

$$\text{Packet Size hit} = 1 + \log_2 n$$

So when there is a hit, we can reduce the packet size from 16 bits (without using caching) to 5 bits for the example considered above.

Figure 30.1 shows how our caching scheme works in the normal scenario, assuming that there is no communication lost. Nodes 1, 2, 3, and 4 are in the same cluster, so whenever one of these nodes senses an interesting event, it needs to send it to its neighboring nodes. Ptr 1, Ptr 2, Ptr 3, and Ptr 4 are the write pointer for the caches in nodes 1, 2, 3, and 4, respectively. After writing a data into a cache, this pointer is changed appropriately to point to the next available free entry. Also note that, we reclaim the oldest cache entries to provide the free entries when the cache becomes full. This will be explained further later.

Let us now consider an example. Initially, assume that there is data "A" in each node's cache. Then node 1 is ready to transmit a new sensed data "B." Since node 1 can not find "B" in its cache (cache miss),

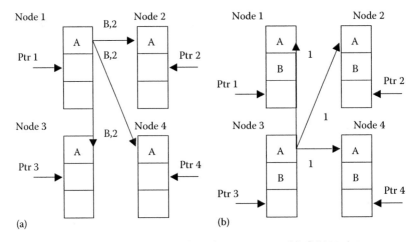

FIGURE 30.1 Normal communication scenario. (a) Node 1 wants to send "B." (b) Node 3 wants to send "A."

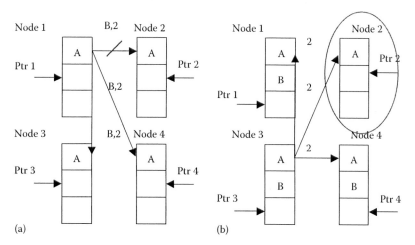

FIGURE 30.2 Lost communication Scenario A. (a) Node 1 wants to send "B." (b) Node 3 wants to send "B."

it needs to send the value "B" along with the value of Ptr 1 (which has a value of 2) as shown in Figure 30.1a. Then the other nodes in the cluster receive this data and update their caches. Now all the caches contain both "A" and "B," and the write pointers are changed to point the third location. Next, node 3 senses a new data value "A." Node 3 will generate a cache hit (because "A" is already in its cache), so it does not need to send "A" again instead it will send the pointer that points the location of "A" within its cache (Location: 1). So node 3 will send a packet that only contains the index, as shown in Figure 30.1b.

Since wireless networks can experience packet loss during transmissions, it is important to consider cases where packets are lost. Let us consider the scenario shown in Figure 30.2. Assume that node 1 wants to send "B" to nodes 2, 3, and 4 (Figure 30.2a). Node 3 and 4 receive this packet but node 2 does not receive it due to a communication loss. Write Pointers of node 1, 3, and 4 are changed because they have written a new, while write pointer of node 2 still points to the second location. Suppose node 3 wants to send "B" to the other nodes, it will send the index 2 (cache hit). Node 2 will detect that its cache is not coherent with the other nodes caches, because the location 2 of its cache is an invalid entry. Then node 2 updates the missing cache entries by making a request to node 3.

Another lost communication scenario is shown in Figure 30.3. The difference between this scenario and the previous scenario (Figure 30.2) is that node 3 wants to send a different data "C" that is not already in its cache. As shown in Figure 30.3a, node 3 will send a packet that contains both C and its Ptr 3 (C,3). Node 1 and 4 will receive this packet normally. On the other hand, node 2 will raise an exception because the index received from the packet does not match with its write pointer (Ptr 2). Then node 2 updates the missing cache entries by making a request to node 3.

Assuming we have a scenario represented in Figure 30.3 but now node 2 wants to send data "C" instead of node 3, see Figure 30.4. In this scenario, node 2 will send a packet that contains "C" and its Ptr 2, in this case the value of the packet will be (C,2). Then nodes 1, 3, and 4 will receive it and raise an error because the index contained in the packet is not the same as the write pointers of these nodes. In this case, all the cache values are flushed.

If we want to cache *n* previous data values, we should provide a cache that contains more than *n* entries, because if we have only n cache lines it can lead to a problem shown in Figure 30.5.

In Figure 30.5, initially all caches are full (so that all write pointers will point to the top location of caches) and node 1 wants to send data "D." Because that data is not in the cache, node 1 will create a packet that contains D and Ptr 1 (D,1) and send it to the other nodes. Assume that the packet does not reach node 2, then the write pointer of node 2 will not advance to the second location. Nodes 3 and 4 will receive it normally and update their write pointer. Next, if node 1 senses the value "D," it transmits the index 1 as shown in Figure 30.5b. Nodes 3 and 4 will look into their cache entry at location 1 and

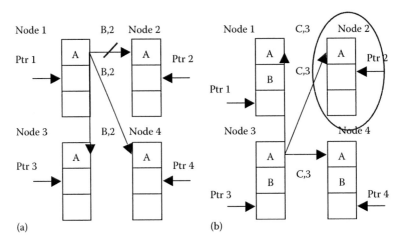

FIGURE 30.3 Lost communication Scenario B. (a) Node 1 wants to send "B." (b) Node 3 wants to send "C."

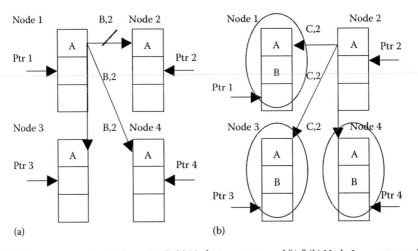

FIGURE 30.4 Lost communication Scenario C. (a) Node 1 wants to send "A." (b) Node 2 wants to send "C."

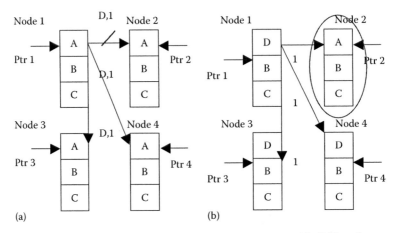

FIGURE 30.5 Incorrect index interpretation problem. (a) Node 1 wants to send "D." (b) Node 1 wants to send "D."

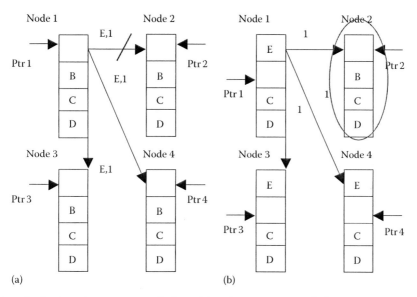

FIGURE 30.6 Packet lost when using n+ cache lines. (a) Node 1 wants to send "E." (b) Node 1 wants to send "E."

retrieve the correct value "D." However, node 2 will retrieve the wrong value "A." This happens because the value at location 1 of node 2's cache is valid.

To overcome this kind of problem, we use caches that have more number of cache lines than the actual number of unique values that we want to store (n+ caches). Specifically, if n is the number of unique values that we want to store and we provide m additional cache lines so that the caches will have at least m empty cache lines at any time. Whenever, there are n valid entries in the cache, the addition of a new entry results in the oldest entry in the cache being marked invalid. This provides m empty cache lines at all times and unless, m communications to a node are lost, the kind of problem illustrated in Figure 30.5 will not occur. Figure 30.6 shows the scenario when we have communication lost on the n+ caches. Node 2 in Figure 30.6 knows that its cache is not coherent with the other caches because location 1 in its cache has an invalid entry.

Since we send only the index value whenever we have cache hit, we do not need to apply an encryption process to that packet. We only apply an encryption whenever we have to send the original data value. This implies the number of cryptography processes needed is less than the original approach.

30.4 Experimental Results

Our experiments utilize the configuration and synthetic data generated as shown in Table 30.1. We generated different sets of data to vary the degree of locality in the values transmitted among the sensor nodes in order to model different types of application scenarios. While the sensor nodes that measure the temperature in the same cluster exhibit good value locality, other applications such as vehicle tracking exhibit poor value locality [5]. Further, we vary the gap between successive data transmission (called message cycle) between the different nodes. This again is a function of the application and can also vary during the course of an application. For example, the temperature measurements can be more frequent during the summer season when forest fires are more likely as opposed to the winter season. We also model different data loss probabilities to see how robust our energy savings are in the presence of varying degrees of data loss. The data loss parameter measures the percentage of transmissions where at least one of the cluster nodes does not receive the data. Finally, we used two different security algorithms: RC5 and Rijndael in our evaluation.

TABLE 30.1 Synthetic Data Configurations

Processor Type	Strong Arm SA-1100
Processor speed	59 MHz
Number of sensors	128 sensors
Number of clusters	32 clusters
Message size	16 bits
Table size	32 bytes
Lost rate	Vary
Value locality	Vary
Message cycle T	Vary
Distance	500 m

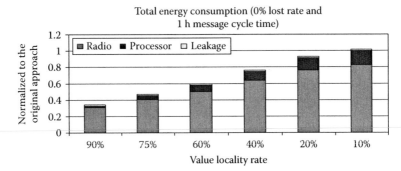

FIGURE 30.7 Energy consumption based on various value localities using RC5 cryptography. Original approach refers to transmitting data values always.

First, we evaluated the energy savings of our approach by varying the value locality assuming a perfect wireless channel as shown in Figure 30.7. Here a value locality of 10% means that a new sensed value has a 10% chance of hitting in the value cache. We can observe from Figure 30.7 that when we have high locality of 90%, we can reduce the energy consumption by more than 60%. When the locality is only 20%, we get less than 10% reduction in energy consumption. The energy savings accrue from the fewer bits communicated on cache hits and the fewer encryption/decryption employed in the nodes. We also account for the overhead spent in the additional bits for sending the index bits as well as the overhead for accessing the small tables in our measurements. The tables will also consume leakage energy immaterial of whether they are accessed or not, but this leakage is reduced to minimum by applying leakage control techniques during the idle time between messages [14]. Specifically, we observe that the communication energy alone reduces by 60% when the value locality is 90%. In contrast, the communication energy increases by 7% when value locality is only 10%. In order to illustrate the energy savings that accrue from the reduction in encryption process, Figure 30.8 shows how many encryption processes need to be executed for various value localities and data loss rates.

Figure 30.9 presents how the data loss rate affects the energy consumption. This shows that high data loss rates of more than 20% will make it difficult to maintain coherent value caches and hence may increase the energy consumption when value locality is poor. The results show that we can gain 4% energy reduction even with a 20% data loss rate if we can have 50% value locality. Obviously, our scheme works best when the value locality is high and data loss rate is low. Further, it is still a useful technique for reducing energy even with moderate value locality and less than 20% data loss rates.

Figure 30.10 shows how the message cycle time affects the total energy consumption of the sensor. Since the memory cells that store the cached value consume leakage energy, the energy expended in these cells increases proportional to the duration of their storage. Hence, the message cycle time is an

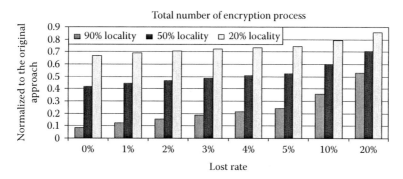

FIGURE 30.8 Number of encryption processes executed for various value localities and lost rates.

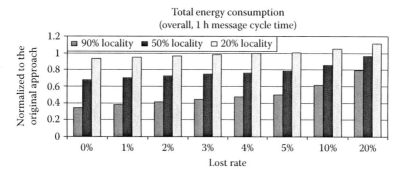

FIGURE 30.9 Energy consumption based on various lost rates and value locality using RC5 cryptography.

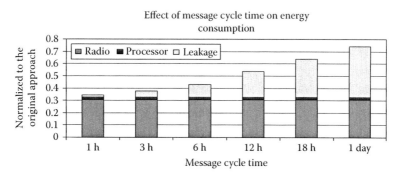

FIGURE 30.10 Energy consumption based on various message cycle time using 90% value locality and 0% lost rate.

important factor that determines the relative magnitude of energies consumed by the transmission of sensed values and that expended by leakage when caching the values in the sensor nodes. Note that reducing the leakage by employing leakage control techniques [14] does not completely eliminate leakage energy as data should be retained. From the results in Figure 30.10, we see that the energy reduction obtained from the communication energy reduction is greater than the overhead incurred by caching the data values for the different message cycle times ranging from 1 h to 1 day. We reduce about 25% energy consumption even with a 1 day message cycle time assuming that we have 90% value locality.

All of the results reported till now are based on the use of the RC5 cryptography algorithm for securing transmissions. The effectiveness of our scheme also depends on the complexity of the security algorithm. When a more complex algorithm such as Rijndael is employed, we observe more savings as

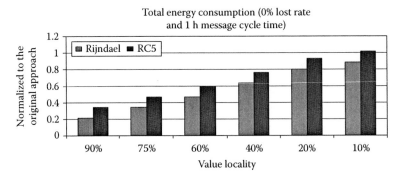

FIGURE 30.11   Energy consumption based on various value localities using Rijndael and RC5 algorithms.

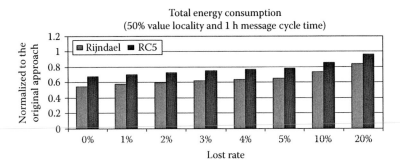

FIGURE 30.12   Energy consumption based on various lost rates using Rijndael and RC5 algorithms.

illustrated in Figure 30.11. Figure 30.12 shows the comparative energy savings for various data loss rates and 50% value locality when using Rijndael and RC5 algorithms. When using the Rijndael algorithm, we obtain around 20% energy savings even with 20% lost rate. On the other hand, RC5 algorithm only gives us 3% energy savings for the same configuration.

30.5 Spatial Locality

In previous sections, we use an index to send the same values that are already in the cache. In this section, we improved our scheme by introducing another packet format that is used to send a data value that is closely related to the latest one (spatial locality). If the absolute difference between the new sensed value and the previous value is less than some threshold, we send the difference instead of the new value itself. For instance, we use the threshold of 2^8 with the original data size of 16 bits. This means that if the new sensed value (V_{new}) is in the range of ($V_{prev} - 2^8 + 1$) and ($V_{prev} + 2^8 - 1$) where V_{prev} is the previous sensed value, we send the value difference ($V_{new} - V_{prev}$) instead of V_{new} in a smaller packet. Consequently, the overhead bits needed to identify what type of packet that is being sent requires 2 bits, for instance 00 for the original value packet, 01 for the index packet, and 10/11 for the value difference packet, one for a positive offset, and the other for a negative offset. The size of these packets is

$$\text{Packet Size}_{value} = 2 + k + \log_2 n$$

$$\text{Packet Size}_{index} = 2 + \log_2 n$$

$$\text{Packet Size}_{diff} = 2 + \text{Threshold bits} + \log_2 n$$

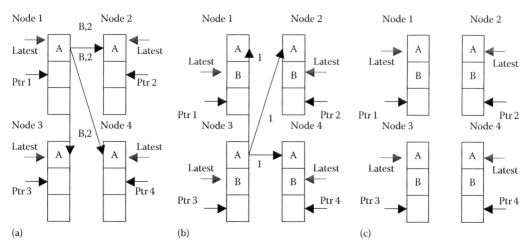

FIGURE 30.13 Spatial value locality using additional pointer. (a) Node 1 sends B. (b) Node 3 sends A. (c) Final cache entries.

As we can see from these equations, the additional packet also requires an index value ($\log_2 n$) since it is pointing the location where the new value will be stored. Note that this method is trying to exploit the possible spatial locality. Consequently if the spatial locality of an application is low, the energy consumption increases rather than decreases due to the additional bit in the value packet and the index packet; as compared to the temporal locality exploitation scheme. Another issue is the size of the threshold bits. It affects the potential for exploiting the spatial locality of an application. When we have a higher threshold, it means that more values will be considered as a value-different packet since the range is larger. However, a large threshold also requires a larger packet size for the value-difference packet.

The architecture of this approach is similar as the previous one. One cell of the cache is used to store the latest sensed value that is used for value-difference packet. We can eliminate this additional storage overhead by adding a new pointer in the cache to point the latest sensed or received value as shown in Figure 30.13.

The latest sensed value in Figure 30.13 is pointed by the Latest pointers. The Figure 30.13a shows the original cache entries with the latest value of A. Node 1 then sends a new sensed value, B. The latest pointer now points the value of B as shown in Figure 30.13b. Then node 3 sends an index value 1 (since it senses the already known value, that is A). After the other nodes receive this index value, their latest pointer will be updated so that they will point to the value of A as shown in Figure 30.13c.

Figure 30.14 shows the situation when node 3 from Figure 30.13 sends a new value using value-difference packet, that is node 3 sends a packet with the offset value of +C. After receiving this packet, the other nodes will store a new value D in the third location, where D is the sum of the value pointed by the Latest pointer, B and the offset (+C).

The situation when there is a packet loss is similar to the method we use in the temporal locality exploitation. The difference is that when the current sensed value is re-sent using the normal packet, the Latest pointers are updated to point to this index.

We use the same configuration as the previous scheme discussed earlier in this experiment. First, we evaluated the comparison of energy savings of our spatial locality approach and the previous approach by varying the value locality rates; assuming a perfect wireless channel and the threshold size is half of the data size as shown in Figure 30.15.

From Figure 30.15, we see that the value-difference packet is useful when we have low temporal locality rate. When we have only a 10% temporal locality rate and a 20% spatial locality rate, the proposed approach consumes the same amount of energy consumption as the original approach. However, if we have a 10% temporal locality rate and a 90% spatial locality rate, energy is reduced by 10% as compared to the original scheme. The original scheme is the method that does not apply any of our proposed schemes.

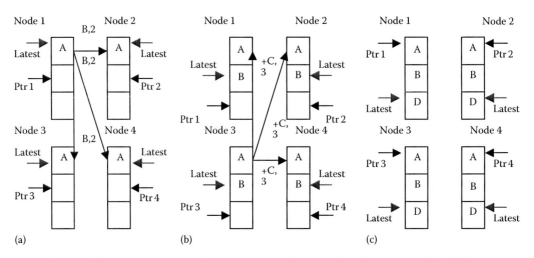

FIGURE 30.14 The situation when a node sends a value-difference packet. (a) Node 1 sends B. (b) Node 3 sends an offset +C. (c) Final cache entries (D = *Latest + C).

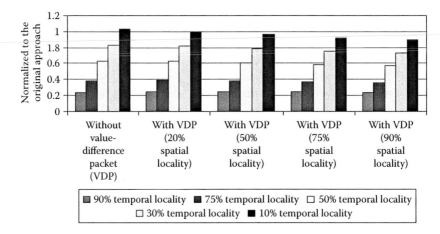

FIGURE 30.15 Energy consumption comparison between using value-difference packet and without using value-difference packet by varying locality rates. The original approach is the one that does not apply any of our proposed schemes.

Figure 30.16 shows the comparisons between these two approaches by varying the lost rate while Figure 30.17 shows the effect of varying the threshold size. Based on Figure 30.17, we have more reduction on the energy consumption when we reduce the size of the threshold assuming that the spatial locality rate remains the same. From these three figures (Figures 30.15 through 30.17), we can conclude that we can achieve more energy savings than without exploiting the possibility of spatial value locality.

30.6 Related Work

Several works have focused on optimizing routing protocol, optimizing energy consumption, and improving security protocol for distributed sensor network [6,14–18].

Shah and Rabaey [15] present energy aware routing for low energy sensor networks by considering network survivability as their primary metric. They show that the network lifetimes can be increase

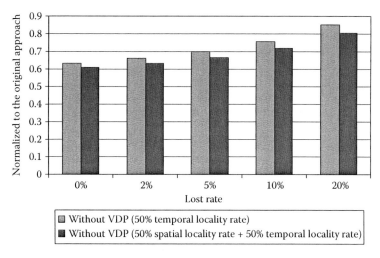

FIGURE 30.16 Energy consumption comparison between using value-difference packet (VDP) and without using value-difference packet by varying lost rates.

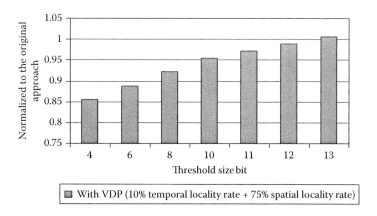

FIGURE 30.17 Energy consumption comparison between using value-difference packet (VDP) and without using value-difference packet by varying lost rates.

up to 40% over comparable schemes like directed diffusion routing. Nath and Niculescu [18] show that simple trajectories can be used in implementing important network protocols such as flooding, discovery and network management. This technique will reduce the number of transmissions needed for such network protocols. This reduction will also give lower energy consumption. The work of Ghiasi et al. [16] focuses on how we can cluster the sensor nodes such that the energy consumption can be optimized. Our work is different in that it specifically targets value locality due to spatial or temporal correlations for optimizing the communication across nodes in a cluster.

The study in Ref. [17] shows how we can reduce energy consumption of encryption algorithm used in sensor networks using dynamic voltage scaling. Because sensor networks usually have a lot of small sensors, its better to use very limited resources in each node in order to reduce the cost. However, when we use a system with limited capability [6], we have to consider using cryptography algorithms that require small resources, such as memory space. In Ref. [6], the authors show that using RSA algorithm on an Atmel processor (8-bit processor) is not possible due to the large resources requirements of the RSA algorithm. They present another security protocol that is suitable for sensor networks that use low capability processors, called SPINS. This protocol uses the modified RC5 cryptography algorithm. Our work is complementary to these efforts in that it attempts to eliminate some of the cryptographic operations by coding the transmitted data using the indexes in the cache.

30.7 Conclusions

In this paper, we presented a new communication protocol for sensor nodes to reduce the overall energy consumption. The proposed technique also provides an inherent security to the transmitted data thereby reducing the cost associated with cryptographic techniques. The energy savings of the proposed approach comes from exploiting the value locality that a sensor network exhibits when communicating data (transmitting) within its local cluster and from reducing the number of encryption processes required. The evaluation of our technique shows that it is effective in reducing energy under different application characteristics such as message intervals and value locality as well as wireless channel conditions.

Acknowledgments

This work was supported in part by NSF CAREER 0093085 and 0093082 and NSF Award 0082064, 0103583, and 0202007.

References

1. D. Ainsworth, 'Smart' sensor technologies promise big savings in state energy costs, *Berkeleyan*, May 2001.
2. P. Eng, Tiny fire marshals: Dust-sized sensors could provide early warnings of forest fires, *ABC News*.
3. S. Hollar, COTS dust, Master's thesis, University of California, Berkeley, CA, 2000.
4. D. Estrin, Comm 'n sense: Research challenges in embedded networked sensing, Presentation at *UCLA Computer Science Department Research Review*, Los Angeles, CA, April 27, 2001.
5. R. R. Brooks, Reactive sensor network, Applied Research Laboratory, Pennsylvania State University, University Park, PA.
6. A. Perrig et al., SPINS: Security protocols for sensor networks, *Proceedings of MOBICOM*, Rome, Italy, 2001.
7. W. R. Heinzelman et al., Energy-scalable algorithms and protocols for wireless microsensor networks, *Proceedings of the International Conference on Acoustics, Speech, and Signal Processing (ICASSP'00)*, Istanbul, Turkey, June 2000.
8. A. Hodjat and I. Verbauwhede, AES module C code written using the suggested NIST C, University of California, Los Angeles, CA.
9. J. Daemen and V. Rijmen, The block cipher Rijndael, *Smart Card Research and Applications, LNCS 1820*, J.-J. Quisquater and B. Schneier, Eds., Springer-Verlag, New York, 2000, pp. 288–296.
10. J. Daemen and V. Rijmen, Rijndael, the advanced encryption standard, *Dr. Dobb's Journal*, 26(3), 137–139, March 2001.
11. I. Kaplan, RC5 source code. http://www.bearcave.com/cae/chdl/rc5.html
12. B. Schneier, *Applied Cryptography* (2nd edn.). John Wiley & Sons, Chichester, U.K., 1996.
13. A. Sinha and A. P. Chandrakasan, JouleTrack—A web based tool for software energy profiling, *38th Design Automation Conference*, Las Vegas, NV, June 18–22, 2001.
14. V. Degalahal, N. Vijaykrishnan, and M. J. Irwin, Analyzing soft errors in leakage optimized SRAM designs, *Proceedings of the International Conference on VLSI Design*, New Delhi, India, January 2003.
15. R. C. Shah and J. Rabaey, Energy aware routing for low energy ad hoc sensor networks, *IEEE Wireless Communications and Networking Conference (WCNC)*, Orlando, FL, March 17–21, 2002.

16. S. Ghiasi et al., Optimal energy aware clustering in sensor network, *Sensor*, 2, 258–269, 2002.

17. L. Yuan and G. Qu, Design space exploration for energy-efficient secure sensor network, *Proceedings of the IEEE International Conference on Application-Specific Systems, Architectures, and Processors (ASAP'02)*, San Jose, CA, 2002.

18. B. Nath and D. Niculescu, Routing on a curve, *First Workshop on Hot Topics in Network (HotNets-I)*, Princeton, NJ, October 28–29, 2002.

19. V. Raghunathan et al., Energy aware wireless sensor networks, *IEEE Signal Processing Magazine*, 19(2), 40–50, March 2002.

20. R. Min et al., An architecture for a power-aware distributed microsensor node, *2000 IEEE Workshop on Signal Processing Systems (SiPS '00)*, Lafayette, LA, October 2000.

21. J.-P. Hubaux, L. Buttyán, and S. Čapkun, The quest for security in mobile ad hoc network, *ACM Symposium on Mobile Ad Hoc Networking and Computing*, Long Beach, CA, 2001.

22. H.S. Kim et al., Multiple access caches: Energy implications, *Proceedings of the IEE CS Annual Workshop on VLSI*, Orlando, FL, April 27–28, 2000, pp. 53–58.

23. A. Hodjat and I. Verbauwhede, Power measurements and energy efficient implementations of network security algorithms for wireless sensor networks, Annual Research Review 2001, Electrical Engineering Department, University of California, Los Angeles, CA, 2001.

24. J. Feng, F. Koushanfar, and M. Potkonjak, System architecture for sensor network issues, alternatives, and directions, *IEEE International Conference on Computer Design: VLSI in Computers and Processors, ICCD 2002*, Freiburg, Germany, 2002.

25. R. Min, T. Furrer, and A. Chandrakasan, Dynamic voltage scaling techniques for distributed microsensor networks, *Workshop on VLSI (WVLSI '00)*, Orlando, FL, April 2000.

26. A. Chandrakasan et al., Power aware wireless microsensor systems, *Keynote Paper ESSCIRC*, Florence, Italy, September 2002.

31

Compiler-Directed Communication Energy Optimizations for Microsensor Networks

I. Kadayif
The Pennsylvania State University

M. Kandemir
The Pennsylvania State University

A. Choudhary
Northwestern University

M. Karakoy
Imperial College London

N. Vijaykrishnan
The Pennsylvania State University

M.J. Irwin
The Pennsylvania State University

31.1 Introduction and Motivation..655
31.2 High-Level Architecture...657
      Language Support
31.3 Communication Optimizations...658
      Data Decomposition and Parallelization • Naive
      Communication • Message Vectorization • Message
      Coalescing • Message Aggregation • Inter-Nest Optimization
31.4 Experimental Setup..663
      Benchmark Codes • Modeling Energy Consumption
31.5 Results...665
      Energy Breakdown • Sensitivity Analysis • Impact of Inter-Nest
      Message Optimization • Impact of Overlapping Communication
      with Computation • Communication Error
31.6 Our Compiler Algorithm 675
31.7 Conclusions and Future Work...676
References..677

31.1 Introduction and Motivation

A networked system of inexpensive and plentiful microsensors, with multiple sensor types, low-power embedded processors, and wireless communication and positioning ability offers a promising solution for many military and civil applications. As noted in Ref. [1], technological progresses in integrated, low-power CMOS communication devices and sensors make a rich design space of networked sensors viable. Recent years have witnessed several efforts at the architectural and circuit level for designing and implementing microsensor based networks (e.g., see Refs. [2–6] and the references therein). While architectural/circuit-level techniques are extremely critical for the success of these networks, software optimizations are also expected to become instrumental in extracting the maximum benefits from the performance and energy behavior angles. In particular, large-scale data management techniques (at the computation and communication levels) are very important [7].

Optimizing energy consumption of a wireless microsensor network is important not only because it is not possible (in some environments) to re-charge batteries on nodes (due to environment-related issues), but also because software running on sensor nodes may consume a significant amount of energy. In broad terms, we can divide the energy expended during operation into two parts: computation energy and communication energy. To minimize the overall energy consumption, we need to minimize the energy spent in both computation and communication. While power-efficient customized wireless protocols (e.g., [8,9]) are part of the big picture as far as minimizing communication energy is concerned, we can also employ application level and compiler level optimizations that target at reducing communication energy.

In this chapter, we focus on a wireless microsensor network environment that process array-intensive codes. Note that array-intensive codes are very common in many embedded image and signal processing applications [10]. Focusing on a sensor network where the nodes form a two-dimensional mesh, we present a set of source code level communication optimization techniques and evaluate them from the energy perspective. Such networks can typically be found in several application domains such as vehicle tracking, CAD-based imaging, and building/road protection. While our communication optimization techniques can be applied by programmers, the existing optimizing compiler technology can also be used to automate them. Specifically, in this chapter, we make the following contributions:

- We present energy behavior of a set of array-intensive applications on a wireless microsensor network. Our results indicate that, for some applications, the computation energy dominates while some others communication energy dominates. More importantly, the communication energy profile is strongly dependent on how the arrays (datasets) are decomposed (distributed) across memories of sensor nodes.
- We explain how source-level communication optimizations can reduce the energy consumption during communication, and present experimental data. We show that, in some cases, optimizing communication energy aggressively can even shift the energy bottleneck from communication to computation.
- We report on results of our sensitivity analysis where we modify several parameters in our energy models. The objective of this analysis is to observe how these changes affect the energy benefits coming from the compiler optimizations.
- We present a strategy where we overlap the communication time with the computation time. Our experimental results indicate that such an overlap can reduce the leakage energy consumption of sensor nodes significantly. Note that leakage energy is expected to play a major role in upcoming process technologies [11].
- Based on the experimental data collected, we present a compiler algorithm that applies several communication optimizations in a unified framework for optimizing the energy spent in sensor network during execution.

This chapter is a first step in using high-level (source code level) compiler optimizations for reducing energy consumption in sensor networks. Our energy savings show that software level optimizations can be very useful in prolonging lifetime of these networks. Our communication optimizations are also, in a sense, complementary to the traditional signal mapping strategies in the literature (e.g., [12,13]).

The remainder of this chapter is organized as follows. Section 31.2 gives a high-level view of the architecture assumed in this study. Section 31.3 discusses communication optimizations considered in this work and explains how they help reduce energy consumption. Section 31.4 presents the computation and communication energy models used in our work. Section 31.5 gives experimental data that show the effectiveness of our approach. Section 31.6 puts our major observations into perspective and presents a unified compiler algorithm. Finally, Section 31.7 concludes the chapter with a summary of our major contributions and a brief discussion of future work on this topic.

31.2 High-Level Architecture

The upper part of Figure 31.1 shows the sensor network architecture assumed in this work. Basically, we assume that the sensor nodes are distributed over a two-dimensional space (area), and the distance between neighboring sensors are close to uniform throughout the network. In our study, each sensor node is assumed to have the capability of communicating with its four neighbors. Each node in this network can be identified using coordinates a (in the horizontal dimension) and b (in the vertical dimension), and can be denoted using P(a, b). Consequently, its four neighbors can be identified using P(a-1, b), P(a+1, b), P(a, b-1), and P(a, b+1). Note that the nearest-neighbor communication style matches very well with many real-world uses of microsensors, as the neighboring nodes are expected to share data to carry out a given task [14]. It should also be noted, however, that, although not discussed here, our software framework is able to handle general node-to-node or broadcast types of communications as well.

The lower part of Figure 31.1 illustrates the major components of a given sensor node. Each node in our network contains sensor(s), A/D converter, battery, processor core, instruction and data memories, radio, and peripheral circuitry. Since sensor networks can be deployed in time-critical applications, we do not consider a cache architecture. Instead, each node is equipped with fast (SRAM) instruction and data memories. The software support for such an architecture is critical. Our compiler takes an input code written in C and *parallelizes* it across the nodes in the network. After parallelization, each sensor node executes the same code (parameterized using parameters a and b) but works on a different portion of the dataset (e.g., a rectilinear segment of a multidimensional array of signals). In other words, the sensor nodes effectively exploit data parallelism. As explained below in detail, the compiler parallelizes

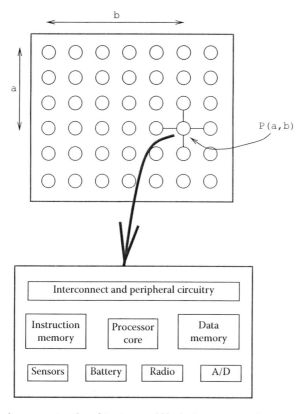

FIGURE 31.1   Assumed sensor network architecture and blocks in a sensor node.

each loop nest in the code using the data decomposition supplied by the programmer. For a given array, a data decomposition specifies how the elements of the array are decomposed (distributed) across the memories of sensor nodes. When an array element is mapped to the memory of a sensor node, that sensor node is said to *own* the array element.

This style of parallel sensor operation can be found in diverse application domains from vehicle tracking to earthquake studies. For example, in the vehicle tracking/detection domain, the sensors can be placed regularly to form a two-dimensional mesh structure in the area that needs to be protected [15]. A parallel application (vehicle detection software) continuously runs and checks to see whether there is a vehicle in the area, and if so the sensors collaboratively track it. During this tracking activity, sensors frequently engage in communication to share data for accurate tracking (and for vehicle identification). Obviously, minimizing inter-node communication can help reduce the overall communication energy dramatically.

31.2.1 Language Support

In order to express communication at the source language level, we assume that two communication primitives are available. The first primitive is of the form:

```
send DATA to P(c,d).
```

When executed by a node P(a,b), this primitive sends data (denoted by DATA) to processor P(c,d). The data communicated can be a single array element, an entire array, or (in many cases) an array region. We assume that this data is received by the node P(c,d) when it executes our other primitive:

```
receive DATA from P(a,b).
```

In order for a communication to occur, each send primitive should be matched with a corresponding receive primitive. Note also that all communication protocol-related activities are assumed to be captured within these primitives. If the size of the message indicated in these calls is larger than the packet size, the message is divided into several packets. This activity occurs within the send and receive routines. While the efficient implementation of these high-level primitives is an important topic in itself, it is beyond the scope of this chapter. This chapter rather deals with the problem of how these primitives can be used by an optimizing compiler. Our compiler framework takes an original code and after generating the node program (i.e., the program that will be run on sensor nodes) inserts these send/receive primitives automatically (without user involvement).

31.3 Communication Optimizations

31.3.1 Data Decomposition and Parallelization

We focus on applications where arrays of signals are processed by multiple sensor nodes in parallel. In such applications, given an array of signals, typically, each processor is responsible from processing a portion of the array. Note that this operation style matches directly to an environment where each sensor node is collecting some data from the portion of an area covered by it and processing the collected data.

Our parallelization strategy works on a single nest at a time; that is, each nest in the application code being optimized is parallelized independently of the other nests. In order to parallelize a given nest over sensor nodes, we need to perform two tasks: (1) decomposing arrays of signals across the memories of sensor nodes, and (2) distributing loop iterations across nodes. Note that array decomposition and loop iteration distribution, together, achieve parallelism across sensor nodes. In this chapter, a decomposition for an *m*-dimensional array of signals is specified by application programmer using a notation of the form [D1][D2]...[Dm] at the beginning of application, where each Di can be either an asterisk, meaning that the corresponding dimension is not decomposed or block, meaning that one block of

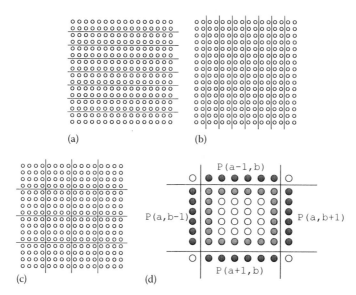

FIGURE 31.2 (a–c) Decomposition of a two-dimensional array of signals across sensor nodes. (d) Communication requirements of a sensor node P(a,b).

adjacent elements is assigned to each node memory. For example, Figure 31.2a shows how an array is decomposed across eight sensor nodes. Note that each sensor node takes a row-block of the array. This decomposition is expressed as [block][*] indicating that the first dimension is decomposed across the sensor nodes whereas the second one is not. Figure 31.2b, on the other hand, shows the [*][block] decomposition on eight sensor nodes. Finally, Figure 31.2c shows how an array is decomposed in both the dimensions (that is, a [block][block] decomposition) using 16 sensor nodes.

In this work, we adopt an array decomposition oriented parallelization strategy based on the owner-computes rule used by optimizing compilers [16]. In this strategy, an array element is updated (written) by only the node that owns it. Let us assume that $\{I\}$ is the set of iterations that will be executed by a given loop nest and that $\{D\}$ is the set of data elements (array elements) that will be used in the computation within the loop. Assume further that $dist(\{D\}) \rightarrow (p,\{D_p\})$ is a decomposition function that gives the set of data elements $\{D_p\}$ mapped to sensor node p. Let $r_0, r_1, r_2, \ldots, r_s$ be $(s + 1)$ references in an assignment statement in the loop where r_0 is the left-hand-side (LHS) reference and the remaining ones are right-hand-side (RHS) references. For a given reference r_k, we define a function $subs_{r_k}(.)$ as a mapping from $\{I\}$ to the set of array elements accessed by this reference, $\{D'_k\}$. That is, $\{D'_k\}$ is the set of array elements accessed through r_k when the iterations in $\{I\}$ are executed. Then, the local index set $\{I_p\}$ for node p can be defined as

$$\{I_p\} = \left\{ i \mid i \in subs_{r_k}^{-1}\left(\{D'_0\} \cap \{D_p\}\right) \right\}.$$

Note that $\{I_p\}$ represents the set of iterations that assign values to the elements of p accessed by the LHS reference r_0. That is, the iterations in $\{I_p\}$ are the ones that will be executed by sensor node p (according to the owner-computes rule) as these are the iterations that assign values to the array elements owned by p. It should also be noted that these iterations also access data elements using RHS references. We can express the set of these elements as

$$\{D_{RHS_p}\} = \bigcup_{k=1}^{s} subs_{r_k}(\{I_p\}).$$

Consequently, the elements in set $\{D_{RHS_p}\} - \{D_p\}$ are the elements that sensor node p needs to receive from other processors. This is called the receive-set. The send-set of a given sensor node can also be computed in similar manner. To summarize, our approach takes data (array) decompositions into account and, using the array references in the code and information about loop iterations, computes the elements that each processor needs to receive and send. Note that unless its $\{D_{RHS_p}\} - \{D_p\}$ set is empty a sensor node needs communication before completing its part of the workload [17].

31.3.2 Naive Communication

As explained above, during parallel execution, processors may need to engage in communication with each other. This is because in manipulating the array elements in its portion a node can require array elements that belong to the portions of arrays owned by some other nodes. Consider, for example, the following application code fragment which is to be parallelized across 256 sensor nodes (for illustrative purposes) that form a 16 × 16 grid:

```
for(i=2;i<=95;i++)
  for(j=2;j<=95;j++)
    u[i][j] = (v[i-1][j]+v[i+1][j]
             +v[i][j-1]+v[i][j+1])/4;
```

Assuming a [block][block] data decomposition for both 96 × 96 arrays u and v, a (non-boundary) node needs data from each of its four neighbors to compute new values for some of the elements in its portion of array u. This scenario is illustrated in Figure 31.2d for a node P(a,b). As will be demonstrated shortly, in general, there might be different ways of implementing the communications required by this fragment. In this section, we describe the most straightforward (naive) strategy; the following sections focus on communication optimizations with different levels of sophistication.

In the naive communication model, each sensor node executes a code fragment similar to the one shown below. Note that in addition to receiving elements from its neighbors, a node also sends elements to its neighbors. The following fragment is the code that will be executed by each node P(a,b). Figure 31.2d shows the set of elements that need to be received by node P(a,b) from its neighbors (shown as shaded dark) and the set of elements that will be sent by node P(a,b) to its neighbors (shown as lightly-shaded). In the code below, within the loops, node P(a,b) first sends the elements that will be needed by its neighbors, and then, receives the elements that it needs from its neighbors.

```
for(i=1;i<=6;i++)
  for(j=1;j<=6;j++)
    {
      if (i==1) send v[i][j] to P(a-1,b);
      if (i==6) send v[i][j] to P(a+1,b);
      if (j==1) send v[i][j] to P(a,b-1);
      if (j==6) send v[i][j] to P(a,b+1);
      if (i==1) receive v[i-1][j] from P(a-1,b);
      if (i==6) receive v[i+1][j] from P(a+1,b);
      if (j==1) receive v[i][j-1] from P(a,b-1);
      if (j==6) receive v[i][j+1] from P(a,b+1);
        u[i][j] = (v[i-1][j]+v[i+1][j]
                 +v[i][j-1]+v[i][j+1])/4;
    }
```

It should be noted that the original array sizes (for both u and v), 96 × 96, is decomposed across (16 × 16) sensor nodes and each node (after the decomposition) owns a 6 × 6 section of the arrays. Note also that, after applying the owners-compute rule, each node executes its local index set which consists of a total of 6 × 6 loop iterations. It should be stressed that in general it may not be possible to perform all communications (or to replicate some array elements) before the computation begins. This is because data dependences in the code can require that the new value of a data item needs to be computed first (by the owner node) before it can be communicated to a neighboring node.

31.3.3 Message Vectorization

Placing communication calls (the send/receive statements) within the innermost loop can increase the energy spent in communication dramatically. This is due to three main reasons. First, each such communication placement invokes a new communication for each iteration of the enclosing loops. Second, each such communication typically sends/receives a small number of array elements. For example, in the naive communication case above, each communication call sends (or receives) only a single array element. Assuming two bits for each array element (e.g., for an image that uses four colors), it is easy to see that naive communication causes a large communication overhead. Third, the if-statement used in the inner loop positions can degrade the performance of processor core in sensor node dramatically, thereby incurring an energy overhead. Consequently, an optimization that extracts communication from within loops and combines element messages per loop iteration in one vectorized message preceding the loop can be very useful. Such an optimization is termed as message vectorization [18,19] and can be applied based on the results of data dependence analysis [16]. More specifically, given a loop nest, this optimization is performed in two steps. In the first step, the nest is analyzed and the outermost loop level above which the communication can be performed is determined. Note that this is the loop level where element messages resulting from the same array reference may legally be combined (into a vectorized message). In the second step, the communication calls (send/receive statements) are inserted in the code. The following example shows the message vectorized version of our example above:

```
send v[1][1..6] to P(a-1,b);
send v[6][1..6] to P(a+1,b);
send v[1..6][1] to P(a,b-1);
send v[1..6][6] to P(a,b+1);
receive v[0][1..6] from from P(a-1,b);
receive v[7][1..6] from from P(a+1,b);
receive v[1..6][0] from from P(a,b-1);
receive v[1..6][7] from from P(a,b+1);
for(i=1;i<=6;i++)
  for(j=1;j<=6;j++)
    u[i][j] = (v[i-1][j]+v[i+1][j]
               +v[i][j-1]+v[i][j+1])/4;
```

Note that, in this message-vectorized fragment, the entire communication is hoisted above the nest. Note also that each send/receive call communicates six array elements.* While this optimization does not reduce the number of elements sent/received, it significantly reduces the number of times the communication will be performed. Since in sensor networks each new communication initiation has an energy cost, this optimization can reduce the communication energy costs significantly.

31.3.4 Message Coalescing

This is an optimization that targets at eliminating the communication of redundant data from one sensor node to another. If the sets of array elements that will be communicated due to two different references to the same array overlap (i.e., contain common elements), message coalescing transfers these

---

* A notation such as v[6][1..6] means all elements in the set {v[6][1],v[6][2],v[6][3],v[6][4],v[6][5],v[6][6]}.

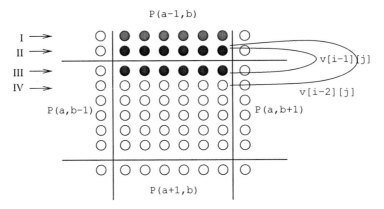

FIGURE 31.3 Overlapping receive sets.

elements only once [16]. It is typically applied after message vectorization. For example, consider the following program fragment:

```
for(i=2;i<=95;i++)
for(j=2;j<=95;j++)
  u[i][j] = (v[i-1][j]+v[i-2][j]
            +v[i][j-1]+v[i][j-2])/4;
```

Note that in this code fragment the communications due to v[i-1][j] and v[i-2][j] overlap. That is, the receive-sets due to these two references contain some common array elements. If each receive-set is vectorized independently, the same array element would be transferred twice. To illustrate this, let us consider Figure 31.3 which shows the communications due to references v[i-1][j] and v[i-2][j]. Note that to compute the new values of row III of its portion of array u, node P(a,b) needs the rows I and II (of array v) from its neighbor P(a-1,b). Similarly, to compute the new values of row IV (of array u), it needs the row II (of array v) from P(a-1,b). In other words, node P(a,b) requires the same row (that is, row II) from P(a-1,b) twice. Instead of performing a separate communication for each request, message coalescing combines these communications into a single (vectorized) message. The same scenario occurs with references v[i][j-1] and v[i][j-2] as well (when communicating with sensor node P(a,b-1)).

31.3.5 Message Aggregation

The two optimizations discussed so far try to reduce communication due to a single array of signals. That is, they are applied to each array independently. Message aggregation, in contrast, targets multiple arrays and tries to ensure that only one message is sent (from a given sensor node) to each sensor node. It is usually applied after message vectorization and message coalescing, and combines all data that will go to the same node into a single message. Note that to implement this optimization, an extra level of buffering might be required. More specifically, during code generation, the array elements to be aggregated are copied to a single buffer so that they can be sent as a single message. The receiving processor then copies the buffered data back to the appropriate locations in its memory. The following code fragment illustrates a case where message aggregation can be applied. In this fragment, there are communications due to two different arrays (v and w). Message aggregation combines these communications into one communication:

```
for(i=2;i<=95;i++)
  for(j=2;j<=95;j++)
    u[i][j] = (v[i-1][j]+w[i-1][j]
              +v[i][j+1]+w[i][j+1])/4;
```

It should be noted that these three communication optimizations, namely, message vectorization, coalescing, and aggregation, do not have too much impact on computation energy. Although these optimizations reduce the number of send/receive calls inserted in the code, their overall impact on computation energy is not expected to be significant. In fact, our experiments revealed that the maximum computation energy variance due to these optimizations was 1.2%.

31.3.6 Inter-Nest Optimization

A common characteristic of all the three communication optimization discussed in previous sections is that they work on a single nest at a time. While this might make the user's or compiler's job easier, in some cases, it may also lead to unnecessary communication (and extra energy consumption). One such scenario occurs, for example, when the sets of communications required by two successively executed nests overlap. To illustrate this, we consider the following code fragment:

```
for(i=2;i<=95;i++)
  for(j=2;j<=95;j++)
    u[i][j] = (v[i-1][j]+v[i+1][j]
              +v[i][j-1]+v[i][j+1])/4;
for(i=2;i<=95;i++)
  for(j=2;j<=95;j++)
    w[i][j] = (v[i-1][j]+v[i+1][j]
              +v[i][j-1]+v[i][j+1])/4;
```

In this code fragment, assuming the same data decomposition for all the arrays involved, the communication requirements for a given sensor node $P(a,b)$ is the same in both the nests. That is, in both the nests, the same processor needs to receive the same elements of array v. While message vectorization can optimize each nest individually, the communication due to the second nest would lead to a wasted energy consumption. An inter-nest optimization captures this redundant communication and optimizes it away. In this example, what this optimization does is just to eliminate the communication before the second nest. In general, however, applying this optimization can be more difficult. For example, if some elements of array v are updated between these two nests (by their owner nodes), these updated elements still need to be transferred (e.g., before the second nest above). In mathematical terms, if (for a given sensor node) D' is the receive set due to the first nest, D'' is the receive set due to the second nest, and Du is the set of updated elements (i.e., between these two nests), the elements in $D' \cup (D'' - Du)$ can be transferred before the first nest and the elements in Du can be transferred before the second one. Here, $\cup$ and $-$ refer to set union and set subtraction, respectively. While not all applications benefit from this optimization, in cases where it is applicable, the energy benefits might be very significant.

It should be mentioned, however, that unlike previous optimizations that target a single nest at a time, this optimization in general increases the computation energy consumed. This is because as explained above this optimization performs set arithmetic on array regions that come from different nests. These operations in general demand construction of new loops to collect the elements to be operated on, and are costly from both execution time and energy perspectives. In fact, as will be shown later in the chapter, in some cases the increase in the computation energy can offset potential energy gains coming from the optimized communication. Therefore, inter-nest optimizations should be applied with care.

31.4 Experimental Setup

31.4.1 Benchmark Codes

We use a set of array-intensive benchmark programs in our experiments. The salient characteristics of the codes in our experimental suite are summarized in Figure 31.4. The first, third, fourth, and sixth benchmarks are motion estimation codes. The second one is an alternate direction integral code. mxm

Benchmark	Input Size (kB)	Number of Arrays	Brief Description
3-step-log	295.08	3	Motion estimation
adi	271.09	6	Alternate direction integral
full-search	98.77	3	Motion estimation
hier	97.77	7	Motion estimation
mxm	464.84	3	Matrix multiply
parallel-hier	295.08	3	Motion estimation
tomcatv	174.22	9	Mesh generation
jacobi	312.00	2	Stencil-like computation
red-black SOR	156.00	1	Stencil-like computation

FIGURE 31.4 Benchmarks used in the experiments. The second, and third columns give, respectively, the total input size, and the number of arrays in the code.

and tomcatv are integer matrix multiplication code and a mesh generation code, respectively. The last two codes, Jacobi relaxation and red-black successive over-relaxation (SOR), contain stencil-like computations and reductions, the two techniques commonly used in image and video processing. Each array element is assumed to be 4 bit wide.

31.4.2 Modeling Energy Consumption

Dynamic energy consumption is due to switching of hardware components is dependent strongly on how different components of a sensor node are exercised by a given application [20]. We separate the system energy into two parts: computation energy and communication energy. Computation energy is the energy consumed in processor core (datapath), instruction memory, data memory, and clock network. In this work, we focus on a simple, single-issue, five-stage pipelined processor core which is suitable for employing in a sensor node. This core has instruction fetch (IF), instruction decode/operand fetch (ID), execution/address calculation (EXE), memory access (MEM), and write-back (WB) stages. We use SimplePower [21], a publicly-available, cycle-accurate energy simulator, to model the energy consumption in this processor core. The modeling approach used in SimplePower has been validated to be accurate (with an average error rate of 8.98%) using actual current measurements of a commercial DSP architecture [22].

We assume that each node has an instruction memory and a data memory (both are SRAM). The energy consumed in these memories is dependent primarily on the number of accesses and memory configuration (e.g., capacity, the number of read/write ports, and whether it is banked or not). We modified the Shade simulation environment [23] to capture the number of references to instruction and data memories and used the CACTI tool [24] to calculate the per access energy cost. The data collected from Shade and CACTI are then combined to compute the overall energy consumption due to memory accesses.

The clock generation circuit (PLL), the clock distribution buffers and wires, and the clock-load on the clock network presented by the clocked components are the main energy consumers for the clock network in our sensor node. We enhanced SimplePower to estimate the clock network energy consumption in each cycle by determining which parts of clock network are active and using the corresponding energy models for active components.

As our communication energy component, we consider the energy expended for sending/receiving data. The radio in the sensor nodes is capable of both sending data and, at the same time, sensing incoming data. We assume that if the radio is not sending any data, it does not spend any energy (omitting the energy expended due to sensing). After packing data, the processor sends the data to the other processor via radio. The radio needs a specific startup time to start sending/receiving message. We used the radio

Parameter	Value
$|P|$	160
Instruction memory	8 kB
Data memory	16 kB
P_{tx}	80 mW
P_{rx}	200 mW
T_{st}	450 μs
P_{out}	1 mW
L	250 bit
B	1 Mbps

FIGURE 31.5 Parameters used in our base configuration.

energy model presented by [7] to account for communication energy. In this model, the power equation of the radio is expressed as

$$P_{radio} = N_{tx} \left[P_{tx}(T_{on-tx} + T_{st}) + P_{out}T_{on-tx} \right] + N_{rx} \left[P_{rx}(T_{on-rx} + T_{st}) \right],$$

where

$N_{tx/rx}$ is the average number of times per second that the transmitter/receiver is used
$P_{tx/rx}$ is the power consumption of transmitter/receiver
P_{out} is the output transmit power that drives the antenna
$T_{on-tx/on-rx}$ is the time interval required to send/receive data
T_{st} is the startup time of the transceiver

Also, note that $T_{on-tx/on-rx} = L/B$, where L is packet size (message length in bits) and B is the data transmit/receive rate in bits per second.

Our *base configuration* uses the values given in Figure 31.5. The power values in this table are similar to those used in Refs. [5,7]. In all our experiments we maintain that P_{rx} is equal to $2.5P_{tx}$ (as the receiver has more circuitry than transmitter). That is, whenever P_{tx} is modified, P_{rx} is also modified accordingly. In Figure 31.5, $|P|$ denotes the total number of sensor nodes that participate in parallel execution of the application.

31.5 Results

Our presentation is in four parts. In the first part (Section 31.5.1), we give an energy breakdown (between computation and communication) when different data decompositions and communication optimizations are used. In the second part (Section 31.5.2), we present a sensitivity analysis where we modify several parameters used in our base configuration. In the third part (Section 31.5.3), we evaluate the effectiveness of inter-nest communication optimization. Finally, in the fourth part (Section 31.5.4), we quantify the potential energy benefits of overlapping computation and communication.

31.5.1 Energy Breakdown

Table 31.1 gives the energy breakdown for our benchmarks for three different data decompositions and communication optimizations. Since the computation energies for different decompositions and different communication optimizations are almost the same we report only one computation energy value for each benchmark. The column decomposition (in Table 31.1) refers to a decomposition

TABLE 31.1 Computation and Communication Energy Breakdown for Our Applications

Benchmark		Communication Energy			Computation Energy				
		Column	Row	Block	Instr. Mem.	Data Mem.	Clock	Datapath	Total
3-step-log	v	6,867	5,922	12,064	266,503	58,376	27,411	69,894	422,185
	v+c	5,449	5,449	5,449					
	v+c+a	3,559	3,559	3,559					
adi	v	173,196	173,196	82,886	38,204	7,734	8,037	19,541	73,516
	v+c	173,196	173,196	82,886					
	v+c+a	165,636	165,636	36,894					
full-search	v	127,615	123,835	189,985	343,946	84,058	31,132	84,654	543,788
	v+c	110,605	110,605	110,605					
	v+c+a	108,715	108,715	108,715					
hier	v	191,422	185,752	284,977	24,961	5,945	1,902	5,373	38,179
	v+c	165,907	165,907	165,907					
	v+c+a	164,017	164,017	164,017					
mxm	v	431,700	431,700	286,080	16,609	2,653	2,576	5,734	27,572
	v+c	367,440	367,440	146,976					
	v+c+a	367,440	367,440	146,976					
parallel-hier	v	63,807	61,917	94,992	168,742	36,775	15,326	40,626	261,468
	v+c	55,302	55,302	55,302					
	v+c+a	55,302	55,302	55,302					
tomcatv	v	153,780	61,512	58,934	18,877	3,788	3,195	7,844	33,706
	v+c	153,780	61,512	42,494					
	v+c+a	142,440	57,732	33,623					
jacobi	v	72,839	72,839	33,671	75,264	12,016	11,858	27,568	126,706
	v+c	72,839	72,839	33,671					
	v+c+a	72,839	72,839	33,671					
red-black-SOR	v	76,619	76,619	39,719	73,732	11,472	12,288	28,160	125,652
	v+c	76,619	76,619	39,719					
	v+c+a	76,619	76,619	39,719					

All energy values are in microjoules.

whereby each sensor node owns a column-block of each array; the other decompositions correspond to cases where the arrays involved in the computation are decomposed in row-block and block-block manner across sensor nodes. Also, we consider three different communication optimizations: message vectorization (denoted v), message vectorization + message coalescing (denoted v+c), and message vectorization + message coalescing + message aggregation (denoted v+c+a). Unless stated otherwise, all energy numbers given are for dynamic energy only (i.e., they do not include the leakage energy consumption).

We can make several observations from the numbers reported in Table 31.1. First, we see that in some benchmarks the computation energy dominates the communication energy, whereas in some benchmarks it is the opposite. This is largely a characteristic of the access pattern exhibited by the application under a given (data decomposition, communication optimization) pair. Based on these results, we can conclude that communication energy constitutes a significant portion of the overall energy budget. The second observation is that the communication optimizations save energy. For example, v+c improves the communication energy of the hier benchmark by 13.7% (over the only message-vectorized version) when column decomposition is used. As another example, the v+c+a version in tomcatv saves 7.2% communication energy over the v+c version. In fact, as can be seen from our results, in some cases, an

TABLE 31.2   Communication Energy with Naive Communication

Benchmark	Column	Row	Block
3-step-log	173,200,144	173,200,144	173,200,144
adi	2,429,856	2,429,856	971,942
full-search	360,833,632	360,833,632	360,833,632
hier	22,719,154	22,719,154	22,719,154
mxm	15,819,376	15,819,376	25,311,000
parallel-hier	28,866,690	28,866,690	28,866,690
tomcatv	2,024,880	809,952	570,206
jacobi	1,036,739	1,036,739	414,695
red black SOR	1,036,739	1,036,739	414,695

All energy values are in microjoules.

optimization can even shift the energy bottleneck from communication to computation. For example, in the adi benchmark, when block decomposition is used, the communication energy of the v+c version is larger than the computation energy. However, when we use the v+c+a version, the communication energy becomes less than the computation energy. Third, we can observe that data decomposition in some benchmarks makes difference in communication energy. For example, the block decomposition performs much better than column and row decompositions in adi. In fact, for this benchmark code, while communication energy dominates computation energy in column and row decompositions, communication energy is approximately half of the computation energy if the message optimizations are applied in conjunction with block decomposition.

It should be stressed that working with the naive communication strategy described earlier (that is, not using any communication optimization) may result in an intolerable communication energy. To illustrate this, Table 31.2 shows the communication energy consumption when naive communication is employed. Comparing these values with those in Table 31.1 emphasizes the difference between optimizing and not optimizing communication energy. These results clearly show that communication optimizations are vital to keep the energy consumption of a sensor network under control.

31.5.2 Sensitivity Analysis

To measure the sensitivity of communication optimizations to several parameters of the radio, we performed another set of experiments where only one parameter is modified at a time. The parameters modified are T_{st} (startup time), P_{tx} (transmitter power), and B (data rate). Note that some of these variations also help us (indirectly) evaluate the impact of different communication protocols. For example, increasing the number of error-control bits added by a protocol can be thought of as increasing P_{tx}. In this section, using the base configuration, we also experimented with different network sizes.

Figure 31.6 shows the effect of startup time on communication energy in the adi and jacobi benchmarks. From these graphs, we can make three observations. First, both the graphs show that the startup time has a larger impact with block decomposition as compared to row and column decomposition. This is due to the fact that, in block decomposition, the message lengths are in general smaller than those in column and row decompositions. Consequently, the startup time has a larger impact with block decomposition. The second observation is that a large startup time can bring the communication energy to the same order of magnitude as the computation energy. Our third observation emphasizes the importance of communication optimizations. One can see from the graph in Figure 31.6a that the v+c+a version when used in conjunction with the block decomposition prevent the communication energy from significantly increasing when T_{st} is increased (as compared to row and column decompositions). That is, message aggregation optimization can be vital in coping with the negative impact of large communication parameters. This is because what message aggregation does is to combine small

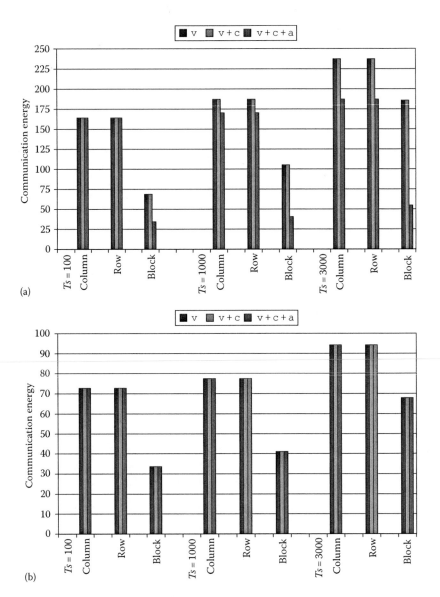

FIGURE 31.6 The effect of startup time on communication energy of adi (a) and jacobi (b).

messages from different arrays into a large message. In other words, it reduces the number of messages which, in turn, makes the communication behavior less sensitive to the startup time.

Figure 31.7 illustrates the effect of transceiver power on communication energy in the adi and full-search benchmarks. As can be seen clearly from these figures the communication energy increases almost linearly with the transceiver power (P_{tx}). This is because the transceiver power is very large as compared to the transmit power of the antenna (P_{out}), and is the main factor that determines the overall trend in communication energy.

The effects of data transmit/receive rate on communication energy of the adi benchmark and the jacobi benchmark are shown in Figure 31.8. Since an increase in transmit/receive rate reduces transmit/receive time, the radio will need to be active for a smaller period of time to send/receive the message, and consequently, communication energy is reduced. Also note that, for very high rates, the

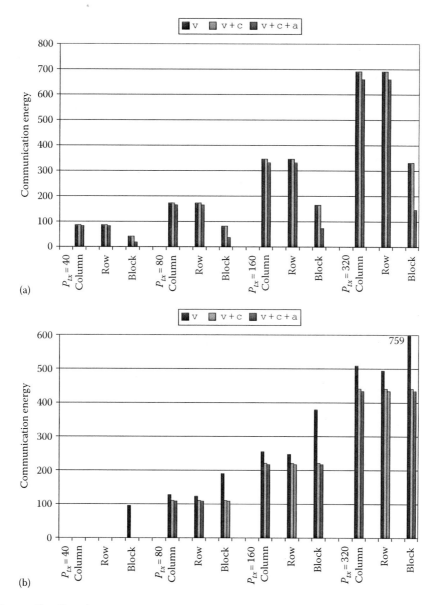

FIGURE 31.7 The effect of transceiver power on communication energy of adi (a) and full-search (b).

startup time balances or dominates the transmit/receive time, so energy overhead due to startup time plays a very critical role in total communication time. As a consequence, the number of messages (rather than total size of messages) determines the communication energy. While, in the jacobi benchmark, with a 1 MB/s transmit/receive rate, data communication energy for block decomposition is approximately half of the communication energy of column/row decomposition, at 10 MB/s rate (an extreme case), the communication energy is almost the same for column, row, and block decompositions.

   Recall that our base configuration consists of 160 sensor nodes. To see the impact of changing the network size on energy consumption, we performed another set of experiments. In these experiments, we changed the network size from 40 nodes to 640 nodes at regular intervals. The first graph

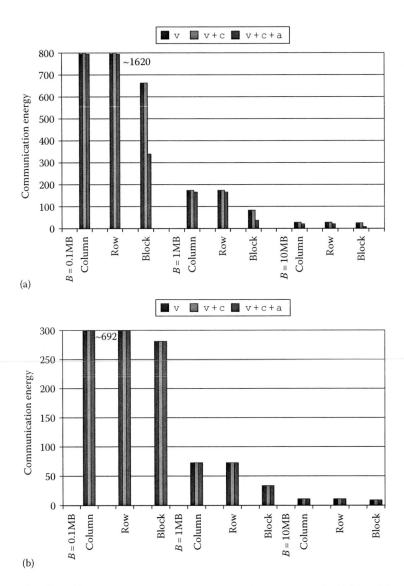

FIGURE 31.8 The effect of data transmit/receive rate on communication energy of adi (a) and jacobi (b).

in Figure 31.9a gives the communication energy consumption of adi when row (or column) decomposition is used. All other parameters are the same as in the base configuration (Figure 31.5). We see that the communication energy increases as the network size is increased. This is due to the increase in the number of messages. However, if the data is decomposed in block manner, we obtain the energy behavior plotted in the second graph in Figure 31.9b. We now observe that block decomposition in conjunction with message aggregation makes an important difference and limits the increase in communication energy. This is because, in this benchmark, the increase in number of messages is limited due to message aggregation. This example clearly demonstrates the importance of suitable combination of data decomposition and communication optimization. Therefore, we believe that future compilers/programming environments that target networks of microsensors should focus on reducing communication. Our experiments with other benchmark codes in our experimental suite also showed similar trends.

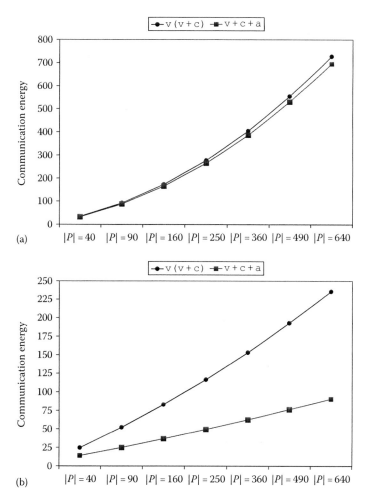

FIGURE 31.9 The effect of the number of sensor nodes on communication energy of adi. (a) Row/column decomposition. (b) Block decomposition.

31.5.3 Impact of Inter-Nest Message Optimization

As explained earlier, inter-nest optimizations try to eliminate the unnecessary communications by taking advantage of already communicated data. For example, if a set of array elements have already been communicated in a nest, they do not need to be communicated in the next nest unless they are modified between the two nests. The aggressiveness of such an optimization is measured in terms of the number of nests that can be considered at once. The least aggressive form considers only the neighboring nests, and does not try to take advantage redundant communication that happened, say, two nests earlier. Among our benchmark codes, tomcatv has an access pattern that can take advantage of this optimization. We experimented with five different versions of this benchmark. The first one (the aggressiveness level is 1) does not apply any inter-nest optimization, whereas the last one (the aggressiveness level is 5) can eliminate redundant data anywhere in the entire application code. As we move from levels 1 to 5, we use more and more inter-nest optimizations.

One important drawback of inter-nest optimizations is that the code after the transformations can become extremely complex. This is because applying inter-nest communication optimization requires hoisting a given communication in the code as much as possible so that it can be combined with communication coming from another nest (which will be executed earlier). As shown in the previous

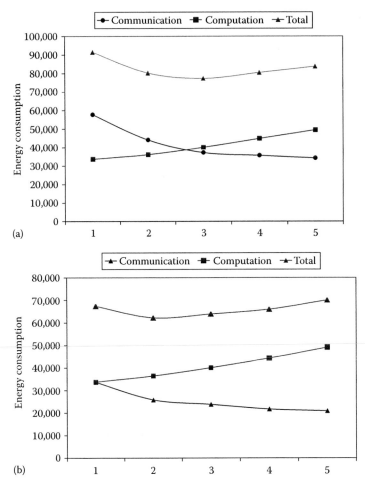

FIGURE 31.10 The effect of the inter-nest message optimizations on computation energy and communication energy of the v+c+a version of tomcatv. (a) Row decomposition. (b) Block decomposition.

research [25], this requires extensive inter-nest data dependence analysis (using a polyhedral tool such as Omega Library [26]) and code restructuring. This code re-structuring (basically, it is a process of combining different array segments that will be communicated in a single message) can render the resulting (communication-optimized) code very complex, thereby increasing the energy spent during computation in datapath and memory. Therefore, when inter-nest optimization is employed, there is a tradeoff between computation energy and communication energy.

To study this tradeoff, we give in Figure 31.10 the computation and communication energies of tom-catv optimized using inter-nest optimizations with different levels of aggressiveness. These graphs also show the overall (computation + communication) energy (denoted Total). We see from these results that there is an operating point (in terms of aggressiveness) that gives the best results from the total energy viewpoint. In this example code, this optimum point is 3 when row decomposition is used, and 2 when block decomposition is used. Going beyond this point (that is, trying to optimize communication energy even more aggressively) leads to so much increase in computation energy that cannot be offset by the decrease in communication energy. Therefore, these results illustrate the trad-eoff between communication and computation energy. Based on these results, we can conclude that an optimizing compiler should try to determine this optimal operating (optimization) point to achieve the best results.

31.5.4 Impact of Overlapping Communication with Computation

The results presented so far were obtained under the assumption that a processor sits idle during communication. In fact, we assumed that the processor, when waiting for the data to be sent, places itself (or a tiny OS does this for it) into a low-power mode where energy consumption is negligible. This is not very realistic as processor and memories typically consume some amount of leakage during these waiting periods. Obviously, the longer the waiting period, the higher the leakage energy consumption. Consequently, we can reduce the leakage energy consumption by reducing the time that the processor waits idle for the communication to be completed. This can be achieved by overlapping the computation with communication. Consider the following code fragment:

```
for(i=2;i<=N-1;i++)
  for(j=2;j<=N-1;j++)
    u[i][j] = (v[i-1][j]+v[i+1][j]
              +v[i][j-1]+v[i][j+1])/4;
```

If we assume a `[*][block]` data decomposition for u and v across M sensor nodes, each node (except the first and the Mth one) needs one column of array v from its left neighbor and one column of array v from its right neighbor. Note that each processor owns a $N \times N/M$ portion of array v (and u). Since there are no data dependences in this code, the processor can execute its loop iterations in any order. Consequently, it can divide its loop iterations into three disjoint groups: (i) the iterations that are needed to compute the values of the elements to be sent; (ii) iterations that work only on local data (i.e., do not need any communication); and finally (iii) the iterations that can be completed only after data have been obtained from its neighbors. Therefore, a given nest can be re-structured as

```
1. Execute iterations in (i);
2. Send data;
3. Execute iterations in (ii);
4. Receive data;
5. Execute iterations in (iii);
```

Note that this is very different from the straightforward execution, which can be summarized as

```
1. Send data;
2. Receive data;
3. Execute iterations in (i+ii+iii) in their original order;
```

Note also that in our current example above the first group (i) is empty set. However, there will still be gains in using the re-structuring above. As soon as a processor sends its data, instead of waiting for the data it needs (for executing iterations in (iii)), it continues with execution of iterations in (ii). Then, in step 4, it does not spend too much time. And finally, it executes the iterations in (iii). Note that by separating the send and receive and doing some useful work between them, the message waiting time is effectively reduced. To evaluate the energy savings due to this optimization, we applied it to three benchmarks from our experimental suite, and measured the percentage computation energy savings. The results shown in Figure 31.11 indicate that our approach can save around 20% of original computation energy of the v+c+a version. In these experiments, we have assumed that the leakage energy consumption per cycle of a component is 20% of the per access dynamic energy consumption of the same component. Since current trends indicate that leakage energy consumption will be more important in future, this optimization can be expected to be more useful with upcoming process technologies.

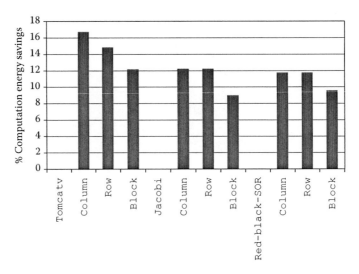

FIGURE 31.11 Percentage savings in computation energy due communication/computation overlapping.

31.5.5 Communication Error

Communication errors in a sensor network may cause a significant energy waste due to re-transmission. To see whether our optimizations better resist against this increase in energy consumption, we performed another set of experiments where we measured the energy consumption under a random error model. Since the results with most of the benchmarks are similar, here we focus only on `parallel-hier`. Figure 31.12 gives the communication energy consumption of the v+c+a version of this application assuming different error rates. For comparison purposes, we also give the communication energy consumption (in microjoules) of the v version without error. The results shown indicate that the v+c+a version starts to consume more energy than the error-free v version only beyond a 15% error rate. In other words, an application powered by our optimization can tolerate more errors (than the original application) under the same energy budget.

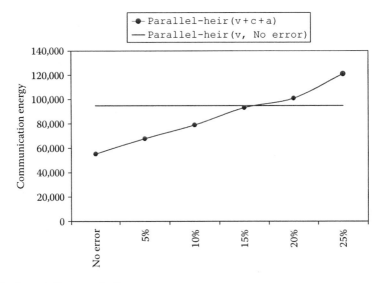

FIGURE 31.12 Impact of communication error.

31.6 Our Compiler Algorithm

In this section, we propose a compiler-based communication optimization algorithm for microsensor networks. Our algorithm combines the message vectorization, coalescing, aggregation, inter-nest optimization, and computation/communication overlapping in a unified framework. It takes as input a (sequential) C program annotated with data (array) decompositions. Its output is a parallelized node program that is to be executed in each sensor node. Each node program is parameterized using the (a,b) coordinates and contains optimized and automatically-inserted communication calls (send/ receive statements). Our algorithm is implemented using the SUIF compiler infrastructure [27] and performs the following steps:

- Using the owner-computes rule, each loop nest is parallelized. As explained earlier in the chapter, the owner-computes rule distributes loop iterations over sensor nodes in such a way that each sensor nodes executes iterations that assign values to the array elements it owns.
- Each loop nest is optimized using message vectorization, coalescing, and aggregation. Note that since these transformations do not affect computation energy significantly, we can apply them as aggressively as possible. In the ideal case, the compiler tries to obtain the v+c+a version of each nest in the code.
- Communication energy is optimized using inter-nest optimization. However, since using this optimization aggressively can cause a large increase in computation energy, our compiler takes a different approach. It first uses the optimization by considering only neighboring two nests. That is, it takes advantage of a previous communication if and only if the said communication occurs in the previous nest. The compiler then estimates the computation energy and communication energy. We estimate the computation energy using the publicly-available energy-aware compilation framework presented in Ref. [28]. This compiler estimates the energy consumption at the source-level for our single-issue, five-stage pipelined processor. The energy consumed in a datapath is dependent on the number, types, and sequence of instructions executed. Consequently, using the approach in Ref. [29], the compiler estimates this information and calculates the datapath energy consumption. The energy consumed in SRAM memories is dependent on the number of data and instruction accesses and the memory configuration (i.e., capacity, number of ports, etc.). The compiler estimates the number of instruction and data accesses (for our simple architecture) and then, using the configuration parameters, computes the energy that will be expended in memory. The energy consumed in a single cycle due to clock network depends on the parts of the clock network that are active. The PLL and the main clock distribution circuitry are normally active every clock cycle during execution. Therefore, the compiler captures the energy consumption due to those two components by estimating the number of cycles that the code would take. However, the participation of the clock-load varies based on the active components of the circuit as determined by the software executing on the system. For example, the clock to the SRAMs is gated (disabled) when they are not used. The compiler exploits the estimation techniques for the datapath and memories explained above to effectively account for this varying clock-load in a given cycle. This compiler-based energy estimation framework has been validated using the cycle-accurate architectural-level energy simulator (SimplePower), and found to be within 6% error margin while providing significant estimation speedup. Note that since our processor is very simple (which is suitable to be used in a sensor node), such a compiler-based, high-level energy estimation is possible.

Estimating communication energy within the compiler, on the other hand, involves determining the number of messages and the number of elements to be sent and received. The compiler extracts these data from the code (considering the optimization applied) and then using the analytical formulations from [7] along with energy parameters (as those in our base configuration) estimates communication energy. The overall energy estimation is, therefore, sum of the computation energy and communication energy estimations. The compiler then compares this estimate with the estimate obtained from the v+c+a versions of nests (i.e., without any inter-nest optimization), and checks whether applying inter-nest optimization reduces the *overall* (program-wide) energy consumption. If it does not, the this step in the algorithm stops, and the compiler proceeds with the next step. On the other hand, if it does, then the compiler increases the aggressiveness of inter-nest optimization (that is, it considers three neighboring nests at a time), and repeats the energy estimation. If this last energy estimation is smaller than the previous one (i.e., the one obtained considering only two nests at a time), the compiler continues to increase the level of aggressiveness, and so on; otherwise it terminates this step. In this way, the highest aggressiveness level with the minimum overall energy estimation is found. For example, this strategy successfully determines the optimum aggressiveness levels shown in Figure 31.10:

- Each loop nest is checked to see whether it can benefit from overlapping communication with computation. To do this, the compiler generates the sets (i), (ii), and (iii) as discussed in Section 31.5.4. If the set (ii) is not empty, this means that the nest in question can benefit from overlapping. If this is the case, the compiler restructures the nest accordingly.

Now, we would like to discuss three important points briefly. First, the algorithm discussed above works with a user-specified data decomposition. However, it is possible to add an outermost loop to this algorithm to enumerate some subset of possible data decompositions, and estimate energy consumption of optimizations under different decompositions. In this way, the compiler can determine the most suitable data decompositions from the energy perspective. Second, when we used the algorithm described above for optimizing communication energy of our benchmark codes, we found that, for a given (user-specified) data decomposition, this algorithm determines the optimized code with minimum overall energy consumption. Third, as mentioned earlier, to implement inter-nest optimization, our approach uses a polyhedral tool [26], which might take a significant amount of compilation time. To check this, we measured the time spent in compilation of each benchmark. We have found that when the optimization algorithm described above is used, the time spent in compilation varies between 3.4 and 5.1 s, averaging (across all benchmarks) in 4.4 s. Considering the large energy benefits at runtime, we believe that these compilation times are tolerable. Based on this discussion, we believe that this compiler algorithm can be part of a software framework for optimizing applications in sensor networks.

31.7 Conclusions and Future Work

Advances in CMOS technology and microsensors enabled construction of large, power-efficient networked sensors. The energy behavior of applications running on these networks is largely dictated by the software support employed such as operating systems and compilers. Our results presented in this chapter indicate that high-level communication optimizations can be vital if one wants to keep the communication energy consumption of sensor network under control. Our results also show that it is possible to adopt a unified compiler algorithm that combines message vectorization, coalescing, aggregation, inter-nest optimizations, and computation/communication overlapping.

While these optimizations are vital for reducing communication energy, there are also optimizations [5,10,20,30,31] that can target computation energy expended in sensor nodes. Our future work will study the interaction between communication and computation optimizations. Another promising research direction is implementation of efficient software-level communication primitives for sensor networks. Such primitives can improve the effectiveness of compiler, and at the same time, can lead to a better optimized code.

References

1. J. Hill, R. Szewczyk, A. Woo, S. Hollar, D. Culler, and K. Pister. System architecture directions for network sensors. In *Proceedings of the ASPLOS*, Cambridge, MA, 2000.
2. G. Asada, M. Dong, T. S. Lin, F. Newberg, G. Pottie, and W. J. Kaiser. Wireless integrated network sensors: Low power systems on a chip. In *Proceedings of the ESSCIRC'98*, Hague, the Netherlands, September 22–24, 1998.
3. B. Chen, K. Jamieson, R. Morris, and H. Balakrishnan. Span: An energy-efficient coordination algorithm for topology maintenance in ad hoc wireless networks. In *Proceedings of the ACM MOBICOM Conference*, Rome, Italy, July 2001.
4. A. Lim. Distributed services for information dissemination in self-organizing sensor networks. Special Issue on Distributed Sensor Networks for Real-Time Systems with Adaptive Reconfiguration, *Journal of Franklin Institute*, 338: 707–727, 2001.
5. R. Min, M. Bhardwaj, S.-H. Cho, A. Sinha, E. Shih, A. Wang, and A. Chandrakasan. Low-power wireless sensor networks. In *Proceedings of the VLSI Design'2001*, Bangalore, India, January 2001.
6. J. Rabaey et al. PicoRadio supports ad-hoc ultra-low power wireless networking. *IEEE Computer Magazine*, 42–48, July 2000.
7. E. Shih, S. H. Choo, N. Ickes, R. Min, A. Sinha, A. Wang, and A. Chandrakasan. Physical layer driven protocol and algorithm design for energy-efficient wireless sensor network. In *Proceedings of the 7th Annual International Conference on Mobile Computing and Networking*, Rome, Italy, July 16–22, 2001.
8. S.-H. Cho and A. Chandrakasan. Energy-efficient protocols for low duty cycle wireless microsensor networks. In *Proceedings of the ICASSP'2001*, Salt Lake City, UT, May 2001.
9. W. Ye, J. Heidemann, and D. Estrin. An energy-efficient MAC protocol for wireless sensor networks. In *Proceedings of the 21st International Annual Joint Conference of the IEEE Computer and Communications Societies*, New York, June 2002.
10. F. Catthoor, S. Wuytack, E. D. Greef, F. Balasa, L. Nachtergaele, and A. Vandecappelle. *Custom Memory Management Methodology—Exploration of Memory Organization for Embedded Multimedia System Design*. Kluwer Academic Publishers, Boston, MA, June 1998.
11. A. Chandrakasan, W. J. Bowhill, and F. Fox. *Design of High-Performance Microprocessor Circuits*. IEEE Press, Piscataway, NJ, 2001.
12. Y. Wang, A. R. Reibman, B. H. Juang, T. Chen, and S. Y. Kung. *Multimedia Signal Processing*. IEEE Press, Princeton, NJ, 1997.
13. D. J. Wang and Y. H. Hu. Multiprocessor implementation of real time DSP algorithms. *IEEE Transactions on VLSI Systems*, 3(3): 393–403, September 1995.
14. A. Sinha and A. Chandrakasan. Operating system and algorithmic techniques for energy scalable wireless sensor networks. In *Proceedings of the 2nd International Conference on Mobile Data Management*, Hong Kong, China, January 2001.
15. R. R. Brooks and S. S. Iyengar. *Multi-Sensor Fusion: Fundamentals and Applications with Software*. Prentice Hall, Upper Saddle River, NJ, 1998.
16. M. Wolfe. *High Performance Compilers for Parallel Computing*. Addison-Wesley Publishing Company, Menlo Park, CA, 1996.
17. I. Kadayif, M. Kandemir, A. Choudhary, and M. Karakoy. An energy-oriented evaluation of communication optimizations for microsensor networks. In *Proceedings of the International Conference on Parallel and Distributed Computing*, Klagenfurt, Austria, August 2003.
18. Z. Bozkus, A. Choudhary, G. Fox, T. Haupt, and S. Ranka. Compiling HPF for distributed memory MIMD computers. *The Interaction of Compilation Technology and Computer Architecture*. Eds., D. Lilja and P. Bird. Kluwer Academic Publishers, Norwell, MA, 1996.

19. Z. Bozkus, A. Choudhary, G. Fox, T. Haupt, and S. Ranka. Compiling distribution directives in a Fortran 90D compiler. In *Proceedings of the 5th IEEE Symposium on Parallel and Distributed Processing*, Dallas, TX, December 1993.

20. L. Benini and G. De Micheli. System-level power optimization: Techniques and tools. *ACM TODAES*, 5(2): 115–192 (2000).

21. N. Vijaykrishnan, M. Kandemir, M. J. Irwin, H. Y. Kim, and W. Ye. Energy-driven integrated hardware-software optimizations using SimplePower. In *Proceedings of the International Symposium on Computer Architecture*, Vancouver, British Columbia, Canada, June 2000.

22. R. Y. Chen, R. M. Owens, and M. J. Irwin. Validation of an architectural level power analysis technique. In *Proceedings of the 35th Design Automation Conference*, San Francisco, CA, June 1998.

23. B. Cmelik and D. Keppel. Shade: A fast instruction-set simulator for execution profiling. In *Proceedings of the 1994 ACM SIGMETRICS Conference on the Measurement and Modeling of Computer Systems*, Boulder, CO, May 1994, pp. 128–137.

24. S. Wilton and N. P. Jouppi. CACTI: An enhanced cycle access and cycle time model. *IEEE Journal of Solid-State Circuits*, 31(5): 677–687, 1996.

25. M. Kandemir, P. Banerjee, A. Choudhary, J. Ramanujam, and N. Shenoy. A global communication optimization technique based on data-flow analysis and linear algebra. *ACM Transactions on Programming Languages and Systems*, 21(6): 1251–1297, November 1999.

26. W. Kelly, V. Maslov, W. Pugh, E. Rosser, T. Shpeisman, and D. Wonnacott. The Omega library interface guide. Technical Report CS-TR-3445, CS Department, University of Maryland, College Park, MD, March 1995.

27. S. P. Amarasinghe, J. M. Anderson, M. S. Lam, and C. W. Tseng. The SUIF compiler for scalable parallel machines. In *Proceedings of the Seventh SIAM Conference on Parallel Processing for Scientific Computing*, San Francisco, CA, February 1995.

28. I. Kadayif, M. Kandemir, N. Vijaykrishnan, M. J. Irwin, and A. Sivasubramaniam. EAC: A compiler framework for high-level energy estimation and optimization. In *Proceedings of the 5th Design Automation and Test in Europe Conference*, Paris, France, March 2002.

29. M. Wolf, D. Maydan, and D. Chen. Combining loop transformations considering caches and scheduling. In *Proceedings of the International Symposium on Microarchitecture*, Paris, France, December 1996, pp. 274–286.

30. L. Benini, A. Macii, E. Macii, and M. Poncino. Synthesis of application-specific memories for power optimization in embedded systems. In *Proceedings of the DAC'00*, Los Angeles, CA, 2000, pp. 300–303.

31. T. Simunic, L. Benini, A. Acquaviva, P. W. Glynn, and G. De Micheli. Dynamic voltage scaling and power management for portable systems. In *Proceedings of theDAC'01*, Las Vegas, NV, pp. 524–529, 2001.

32. W.-T. Shiue and C. Chakrabarti. Memory exploration for low power, embedded systems. In *Proceedings of the Design Automation Conference*, New Orleans, LA, 1999.

32

Jamming Games in Wireless and Sensor Networks[*]

Rajgopal Kannan
Louisiana State University

Costas Busch
Louisiana State University

Shuangqing Wei
Louisiana State University

32.1 Introduction 679
32.2 Motivation..681
32.3 Research Issues...682
32.4 Multilayer Game Models..683
     Physical Layer Modeling • Network Layer Modeling
32.5 Novel Evaluation Metrics 689
     Price of Jamming • The Paradox of Jamming
     and Its Price • Convergence Time • Novel Analytical Methods
Acknowledgments...691
References..691

32.1 Introduction

Wireless technology has been making a profound and unprecedented impact on society and our daily lives. However, wireless transmissions are susceptible to interference and security breaches due to the open nature of wireless channels over which information is carried. In particular, there exists a class of threats launched by adversaries using radio interference, namely, jamming attacks [29]. Jamming attacks have already been identified as one of the prominent problems resulting in loss of network availability, increased latency, and reduced throughput due to significant deterioration of link quality.

In wireless sensor network applications, the nodes may have power and computing capability constraints. For example, a sensor node is typically operated with a battery that has limited energy capacity. To maximize the lifetime of the nodes, the time until nodes run out of power, it is important to minimize the utilization of individual nodes. Jammers can impact the energy utilization of sensor nodes, by either overworking the sensors or blocking specific paths in the sensor network. When nodes are not within mutual transmission range, information from the source nodes propagate via intermediate nodes to reach the destination. The problem of energy preservation relates directly to balancing the packet traffic for minimizing the node congestion. With load balancing, the lifetime of a battery-operated sensor network is prolonged, since the time that the first node runs out of energy is extended. The intermediate nodes aggregate data and relay aggregated data upstream. This in-network aggregation is done to reduce the number of packets transmitted and hence the energy consumption [19,21]. It is known (through radio technologies) that data transmission over wireless links is far more expensive than local computation on the same data. In this respect, in-network data aggregation is a very important concept for energy-constrained sensor nodes [11,16]. Jamming can

---

[*] This work was funded in part by grants from NSF under #CNS-1018273 and AFOSR #FA9550-10-1-0448.

impact the sensor network by either destabilizing the balanced traffic or inhibiting the data aggregation in intermediate nodes of the network.

In this chapter, we aim to investigate security aspects of networks that involve jamming. Traditionally, jamming has been viewed as a physical layer problem (jamming via radio interference). There is a large body of work analyzing classical physical layer countermeasures such as spread-spectrum and frequency hopping [23,29]. However, as we have recently shown in [31,32], the unique Nash equilibrium point in a zero-sum game with average total throughput as the payoff function is when both transmitter and jammer exploit full degrees of freedom all the time over multiple frequency bands. None of the two players are willing to take the frequency hopping option, which is surprisingly against traditional wisdom [29]. This is fundamentally due to the nature of the cost function in the underlying jamming game with finite power-constrained transmitter and jammer. This shows that considering new cost functions and jammer models can lead to surprising yet useful results.

In general, despite the extensive study of jamming problems in the physical layer using game theory [5,14,27,28], several important and critical issues arising due to the latest developments in wireless communications have not received deserved attention. These issues include lack of study of cost functions reflecting delay constraints, utilization of channel degrees of freedom in the presence of intelligent jammers, and information exchange between transceivers to be exploited by smart adversaries. There is certainly an urgent need to develop and analyze more sophisticated jamming models and techniques, as well as physical layer countermeasures.

More fundamentally, jamming at the physical layer has an obvious cross-layer impact in terms of costs and performance. "Local" jammers focused on disrupting communication at a single link can result in the disruption of existing routing paths, thereby impacting the network layer. As an extreme case, powerful jammers can completely jam specific links, thereby preventing nodes from forwarding packets on those links, reducing connectivity, and, perhaps, increasing network costs such as delay. However, as we describe later, such an outcome need not be the worst possible from the perspective of the jammer(s). Clearly in order to understand the behavior of jammers as well as their impact on network users, we need to accurately model the cost of jamming across layers as well as the objective functions (benefits) of jammers. In particular, we need to analyze the strategies of *smart* physical layer jammers whose objectives are to maximally deteriorate important higher level network performance metrics such as power consumption, delay, etc. Some of these strategies include subtly affecting network flows (by either introducing malicious flows or strategically creating "local" interference) resulting in increased congestion and lower "global" (network-wide) throughput. Thus, in order to accurately understand the global impact of jamming, we must develop more sophisticated cross-layer jamming models that help us analyze jamming over the strategy space of jammers as well as network users and thereby enable the development of effective countermeasures. The models should be flexible enough to account for different types of jammers, their resources, capabilities and constraints, the amount of information available, and whether they collaborate or not. They should capture realistic cross-layer cost functions to users along with jammer utility functions.

Once cost and utility functions have been developed, the impact of jamming on network performance can be studied using methodologies such as minimax optimization. However, these optimization techniques are not well suited for developing new perspectives on *strategic* jamming as well as countermeasures. The interplay between rational wireless transceivers and malicious jammers fits naturally within a game-theoretic framework with network outcomes representing natural equilibria between resource-constrained users and adversaries. While there is a large body of work on game-theoretic analysis of jamming for wireless networks, most of the work is focused on specific jamming games at various layers [9,30,33,34]. The existing literature does not accurately and comprehensively capture the cross-layer jammer–network user interaction via realistic cost and utility functions and jammer models that account for jammer resources, constraints, and collaborative capabilities as we propose to develop. One important metric that is not reflected in much of the jamming game literature is the correlated impact of jammer behavior at various layers. Game models should accurately capture the transition between cost and utility functions at different layers, how a specific action/strategy of a jammer at the physical layer is reflected in terms of costs and strategies of users at the network layer. In this chapter, we develop a model of jammer–network user interactions in terms of multiple layer games, where the output of a game at the lower layer is translated into a set of costs that determine the outcome of another game at the higher layer.

32.2 Motivation

Since the physical layer is an interface between the real transmission medium, that is, wireless channels, and upper layers in wireless networks, security issues that reside in the physical layer play a pivotal role in determining the overall security performance for any wireless network. But an obvious impact of jamming at the lower layers is to disrupt higher level network functions by rerouting traffic and increasing congestion on specific links. Thus, we need to develop new game models and utility functions that better reflect the network-wide interplay between jammers and users and capture the availability/unavailability of resources, availability and extent of common knowledge, and impact on higher layers. While there is a large amount of work on MAC layer jamming [9] describing different types of jammers (e.g., constant, random, reactive, and deceptive from [33,34]), we specifically focus on network layer issues here. In most cases, MAC layer jamming models do not incorporate intelligent jamming via waveform selection, but assume a much more simplified binary (success-failure) jamming process. However, we model jammers as intelligent adversaries at the physical layer using intelligent waveforms and assume the impact of jamming is not a binary process, but rather percolates up from the physical layer via modified traffic rates and flows. For example, in [12], we have developed online algorithms for intelligent antijamming over interference channels. In order to better understand the impact of jamming on networks, we assume the existence of a successful MAC protocol, that is, packet transmissions do not result in collisions, which implies that all lost packets are the result of jamming attacks, not the collision at the MAC layer. More specifically, the games we model should take into account the following issues.

Physical layer issues: (1) The utility/payoff functions should reflect the growing need for delivering delay-sensitive traffic over wireless networks, which cannot be easily captured in the traditional zero-sum game framework. (2) Jamming games should be modeled for a range of constraints on jammers including limited jamming power budget, as well as jammer knowledge about the feature or characteristics of communication signals between a sender and its receiver. Jammers may also possess side information about transmitters that can have a significant impact on jamming strategies. (3) The defensive strategy space of legitimate users could also be extended. The awareness of the presence of jammer, as well as the belief about what jammer possesses pertaining to the ongoing communication link, will consequently change the antijamming strategies and mechanisms taken by legitimate transceivers.

Network layer issues: While it is crucial for us to reexamine the jamming and antijamming strategies from a local point-to-point perspective in the physical layer, it is also of great importance to understand how the entire or regional network will respond to local jamming attacks. In particular, the disruption of a subset of communication links by local jamming attacks will have a far-reaching effect on network congestion in wireless networks. Figure 32.1 depicts a simple wireless network, where the underlying

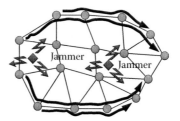

FIGURE 32.1 Jamming example.

wireless topology is abstracted as a communication graph using standard transmission and interference models with edges between wireless nodes in range. On the left part of Figure 32.1, we can see a network traffic flow in the network; suppose that each flow follows a different path in the network. On the right part of the figure, we can see the effects in the flow when jammers attack the network. In order to avoid the jammers, the flow is deflected to alternative routes where the congestion of the flows increases.

The antijamming strategy space is thus no longer restricted to the one including mechanisms implemented only in physical and link layers, such as more sophisticated coding/decoding schemes, power control, multiuser collaboration, ARQ and spreading in space, time, and frequency. Instead, the impact of jamming on the entire network performance in terms of metrics such as overall congestion can be alleviated via more intelligent routing. This requires us to formulate and investigate the interplay between a network of nodes and a set of jammers via routing and congestion games with more sophisticated utility functions that reflect link cost functions subject to physical layer constraints.

32.3 Research Issues

We want to study the nature, types, and ranges of equilibria of games formulated across the network spectrum in order to understand the fundamental limits of wireless network performance under coexisting users and jammers. The novelty of such research lies in the modeling of strategic rational jammers via games and in analyzing these games not in an isolated fashion but by explicitly accounting for the interaction between the physical and network layers via appropriate system parameters. Research issues that we examine in this context are as follows:

1. *Multilayer jamming models*: Development of novel jamming models that incorporate jammers' knowledge and resource capabilities and constraints; availability of local/global network information that affects how a jammer selects its objective function; and jammer's interference capabilities including smart interference over multiple edges and coordinated jamming actions that affect a network over longer time intervals via repeated games.

2. *Multilayer games*: At the physical layer, we need to develop new jamming game formulations via cost functions (e.g., based on delay and channel state information [CSI] constraints) that provide new insight into counterintuitive equilibria along with new *informative jamming games* under incomplete information. We need to formulate and study novel multilayer games where the impact of jammers and jamming games in the physical layer are translated upward via realistic higher level cost functions to network users and routing strategies, in particular, network routing and flow-control games: (a) in the presence of restricted-knowledge jammers themselves playing local games and (b) rational cooperating jammers with network-wide knowledge. We must consider both power-limited games, where an edge cost is proportional to the transmitter's power consumption, and throughput-limited games, where an edge cost is a function of the effective flow rate.

3. *Novel jamming metrics*:

   a. *Price of jamming*: Traditional stability and quality of equilibria metrics such as the price of stability/anarchy (*PoS/PoA*) are unlikely to fully capture the jammer–user interaction complexity of multilayer games. Thus, we propose novel equilibria metrics labeled *price of jamming* (*PoJ*) that will analytically measure the impact of a network of cooperative or uncooperative jammers. We use the *PoJ* to evaluate the range of jamming versus non-jamming equilibria along with network social costs. It is likely that some jamming games will be unstable (nonexistence of equilibria or cyclic under best-response dynamics). In such cases, we must develop and evaluate approximate games.

   b. *Paradox of jamming*: Remarkably, there are cases when jamming leads to better equilibria than otherwise. Motivated by this counterintuitive result, we will define and study a new metric labeled as *paradox of jamming* (*ProJ*), which measures the unintended *positive* consequences of jamming on the rest of the network. Are there networks that always have a positive

ProJ? Such networks are expected to be highly resilient to jamming. What are the underlying graph metrics that can best measure the *ProJ*, for example, can we derive upper and lower bounds on the *ProJ* based on the size of multicuts and multicommodity flows for a particular jamming strategy in the network?

4. *Innovative techniques*: To obtain tight results on the *PoJ*, we must develop innovative techniques for analyzing the range of equilibria. We have proposed some novel techniques based on the idea of expansion, support sets, and expansion chains in [6–8,13]. These methods are different from standard potential function methods and help us to correlate worst-case equilibrium costs with network size in a fundamental manner and obtain tight bounds on *PoA* and *PoJ* metrics. For analyzing physical layer jamming, we must use novel duality properties tailored to the underlying zero-sum outage probability game, as well as utilizing a zero-sum Bayesian game framework for solving *informative* jamming under incomplete information.

32.4 Multilayer Game Models

32.4.1 Physical Layer Modeling

32.4.1.1 Link-Wise Jamming Models (Single-Edge Jamming)

We first look at a system model for a single edge (i, k) between a transmitter (node i) and its receiver (node k) under a jamming attack, as depicted in Figure 32.2a and b. In this model, the source sends a signal vector $\mathbf{X}$ to its receiver that is attenuated by a multiplicative vector $\mathbf{h}_{i,k}$, where multiplication is operated over the components of associated vectors. Such a model has the flexibility of allowing us to study both narrow band and wideband communications, subject to how we define channel coefficient vectors. The attenuation factor captures both multipath fading effect and propagation loss. The mean squared value of the nth component of this attenuation vector $h_{i,k}(n)$ is determined by the distance $d_{i,k}$ between the transmitter and its receiver, for example, $E\left[\left|h_{i,k}(n)\right|^2\right] \propto 1/d_{i,k}^2$ [25]. The jammer could eavesdrop the

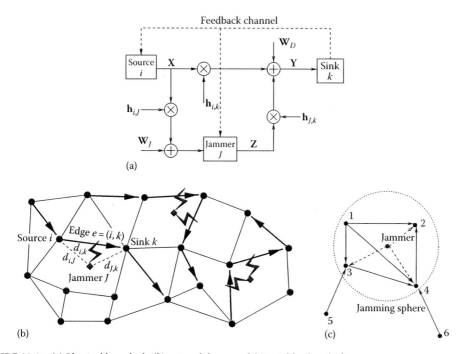

FIGURE 32.2 (a) Physical layer link, (b) network layer, and (c) neighborhood of jammer.

forward communication link through its own channel from the source node. As a result, a properly designed jamming signal $\mathbf{Z}$ is introduced as an interference to the receiver through the jamming link attenuated by $\mathbf{h}_{J,k}$ whose mean squared value is as follows: $E\left[\left|h_{J,k}(n)\right|^2\right] \propto 1/d_{J,k}^2$ variance.

In addition to the possibly eavesdropped signal from the forward link, the jammer could also get knowledge about other system parameters such as channel fading coefficients $\mathbf{h}_{i,k}$, measured and fed back by the sink node to its transmitter. These types of side information will allow the jammer to come up with more effective jamming signals interfering the link (i, k) to the maximum extent. As a local defensive measure, the source and sink nodes seek collaboration to adaptively change transmission parameters, such as transmission power or transmission rate, and corresponding receiver structure to optimize their performance under the jamming attack. Such an interaction between jammer and transceiver can be modeled as a non-cooperative game, for example, zero-sum game. For single-edge jamming, we propose two game models with the following payoff functions: (1) *Game-Payoff function 1*: Delay-constrained throughput game in block-fading channels where the payoff function is frame error rate of the link (i, k), which can be associated with the outage probability under a given transmission rate from source [1]; (2) *Game-Payoff function 2*: Non-delay constrained achievable rate game in ergodic fading channels, where the payoff function is the total achievable rate of the link (i, k) over either narrow or broadband channels [31].

Connection to network layer: The consequence of a jamming game played on a single edge $e = (i, k)$ will no longer be restricted only to this particular link in the context of the overall network topology. From the network layer perspective, edge e could be on multiple paths determined by an upper layer protocol carrying aggregate traffic with flow $F(i, k)$. Let $R_{i,k}^\star$ denote the effective (equilibrium) link capacity across edge e when the physical layer game reaches equilibrium (guaranteed under a mixed strategy since this is a zero-sum game). We will translate this to an edge cost function $C_{i,k}$ to be considered in the decision made by the network layer protocols with regard to path and flow allocation. We propose two novel network layer edge cost functions directly derived from the physical layer.

- *Proposed $C_{i,k}^P$ in terms of transmission power*: To let flow $F_{i,k}$ go through the edge (i, k) without a specific delay constraint, provided the incurred delay is not unbounded, the effective rate $R_{i,k}^\star$ satisfies $R_{i,k}^\star = \alpha F_{i,k}$, where $\alpha > 1$ is a constant [4]. Thus, for game-payoff function 1, we can derive the following lower bound for transmission power $P_{T,i}$ at node i:

$$P_{T,i} \geq C_{i,k}^P \triangleq \left[P_J^\star \left(\frac{d_{i,k}}{d_{J,k}} \right)^2 + \sigma_w^2 d_{i,k}^2 \right] \exp\{\alpha F_{i,k}\} \tag{32.1}$$

  where

  $d_{i,k}$ and $d_{J,k}$ are the distance from node i and the jammer to node k, respectively
  $P_J^\star$ is the average equilibrium jamming power
  σ_w^2 is the variance of the additive Gaussian noise

- *Proposed $C_{i,k}^T$ in terms of throughput*: Under Kleinrock independence approximation and Pollaczek–Khinchin (P-K) formula, we compute the average number of data units in service at edge (i, k) as

$$C_{i,k}^T = \frac{F_{i,k}}{R_{i,k} P_{succ}^\star \left(R_{i,k}^\star \right) - F_{i,k}} \left(1 - \frac{F_{i,k}}{2R_{i,k}^\star} \right) \tag{32.2}$$

where we assume fixed-length packet with Poisson arrival process of rate f_e and fixed-length channel codeword under regular Automatic Repeat-reQuest (ARQ) protocol [4]. In Equation 32.2, $1 - P_{succ}^\star \left(R_{i,k}^\star \right)$ denotes the equilibrium decoding error probability in game-payoff function 2 (based on outage probability). Under the edge cost model $C_{i,k}^T$ in Equation 32.2, we can see that a malicious jammer can affect the edge cost through decreasing $P_{succ}(R_{i,k})$ for a particular $R_{i,k}$, which consequently restricts how much total flow can go through this edge. The transmitter can defend against such an attack via selecting a proper $R_{i,k}$ and corresponding adaptive transmission strategies in physical layer to counter-control $P_{succ}(R_{i,k})$. If the total flow $F_{i,k}$ is prescribed, the selection of optimal $R_{i,k}^*$ will also be a function of flow.

32.4.1.2 Neighborhood Jamming Models (Multiple Edge Jamming)

We can consider jamming to be *informative* or *uninformative*. In our uninformative jamming model over multiple edges, the jammer does not know the codebook and the exact signal set adopted by the communicator due to unsuccessful eavesdropping, but still targets flows over multiple edges. We assume that flows are scheduled by the MAC layer over orthogonal channels in time or frequency so that there are no collisions from packet transmissions with interference caused only by jamming. For instance, as shown in Figure 32.2c, a jammer could jam flows on edges (1, 2) and (4, 2), which share a common receiver, independently, as they are on orthogonal channels or target flows over multiple edges, for example, (5, 3) or (6, 4), which are both within the jamming sphere. Here, the issue for the jammer is how to split its total jamming power over the multiple receivers/edges over orthogonal frequencies/slots, implying that network layer allocation of flows affects physical layer jammer's strategy and, consequently, affects transmitter's countermeasure schemes. Once an edge (i.e., a flow) is picked by a jammer, the single-edge-based physical layer jamming game will be played between the jammer and associated transceiver pair. Therefore, the interaction between the transmitter and a local jammer pervades through physical and network layers. For simplicity in this illustration, we do not consider the superposition of jamming interference, that is, there are a finite number of jammers positioned sparsely in a network area such that jamming spheres do not overlap. Therefore, when the jamming scope is extended from a single-edge to a multiple-edge neighborhood, a jammer must target its flow attacks appropriately for maximal impact. We need to study different jammer strategies and countermeasures of picking targeted edges/flows, as well as associated jamming powers and jamming waveforms based upon jammer knowledge of local network topology and corresponding flow distribution, under a total power constraint.

32.4.2 Network Layer Modeling

We abstract the network as a flow routing graph $G = (V, E)$, with $|V| = n$ nodes and $|E| = m$ edges (links). In order to isolate and better understand the impact of jamming, we assume that jamming at the network layer is felt through congestion on overlapping routing paths, while wireless interference between nonoverlapping paths is assumed to be handled by the MAC layer. Let J denote a set of jammers that are positioned around the network so that each jammer $j \in J$ affects a subset of edges $E_j \subseteq E$ and respective adjacent nodes $V_j \subseteq V$. Note that J is not a subset of V. We further assume *sparse jammers*: $V_j \cap V_j' = \emptyset$, and $E_j \cap E_j' = \emptyset$, where $j, j' \in J$; namely, a node or edge in G is affected by at most one jammer. We propose the following jammer models in the network layer:

- *Stand-alone jammers*: We define such jammers as those working in isolation to attack links and receiver nodes within their transmission range. Each jammer $j \in J$ has available power budget P_j and can decide how to use it in order to maximize its malicious goal of jamming the edges E_j. A stand-alone jammer still plays a game in the physical layer whose outcome affects the network layer links that further affect the whole network layer flows. There can be the following subcases of stand-alone jammers: (a) *single link jammer*, where each jammer j focuses on a specific link $e \in E_j$, and the jammer dedicates its power P_j to affect the flow on link e and (b) *neighborhood jammer*, where each jammer j distributes its power to affect multiple links in E_j.
- *Collaborative jammers*: Jammers collaborate with each other to maximize their effect in the network layer. The coordination between the jammers can be achieved either in a centralized or in a distributed manner. The benefit to the jammers is that they can conserve their power through appropriate

power scheduling and use it wisely to jam the network more effectively. Let $P_{j,t}$ denote the amount of power that jammer j uses at time t. Consider the following subcases: (a) *discretized power jammers*, in which the jammers are scheduled so that at any moment of time t, $P_{j,t} \in \left\{0, P_j^{\max}\right\}$, where $P_j^{\max}$ is the maximum power available to j, and (b) *continuous power jammers*, where the jammers are scheduled so that at any moment of time t, $P_{j,t} \in [0, P_{\max}]$. An important problem is the optimal scheduling and power allocation of jammers in the context of the repeated games described in the next subsection. This problem is NP-hard and we need to develop bicriteria approximation algorithms for scheduling and power allocation that maximizes the PoJ metric.

- *Malicious nodes*: A type of attack that appears only at the network layer is from malicious nodes that infuse artificial flow into the network with the main purpose to increase the congestion in the hope to decrease the network throughput and even bring down the network. The malicious nodes are not jammers in the strict sense of jamming the physical layer signal but are normal nodes in the network with a malicious purpose, namely, $J \subseteq V$. Network layer games with malicious nodes have been studied in [2] as variations of the classic congestion game model with the sum of edge congestions. They quantify the effect of the malicious players with respect to the *price of malice* that measures the rate of network deterioration. Originally, the price of malice was defined in [20], in the context of virus inoculation games.

32.4.2.1 Routing Games Background

A routing game in graph G is a tuple $(G, \mathcal{N}, \mathcal{P})$, where $\mathcal{N} = \{1, \ldots, N\}$ is a of N rational players, where each player sends flow (set or stream of packets) f_i from source node u_i to destination node v_i, and $\mathcal{P}$ are the strategies of the players. We have that $\mathcal{P} = \bigcup_{i \in \mathcal{N}} \mathcal{P}_i$, where $\mathcal{P}_i$ denotes the *strategy set* of player i, which is a collection of available paths in G for the player to use from u_i to v_i. We distinguish two types of flows: *atomic*, where the flow of a player uses one path, and *splittable*, where the flow may use multiple paths. For atomic flows, a *pure strategy* for player i is any path $p \in \mathcal{P}_i$ available to player i. A *pure strategy profile* in atomic flows is any *routing* (collection of paths) $\mathbf{p} = [p_1, p_2, \ldots, p_N]$, where $p_i \in \mathcal{P}_i$ is the path chosen by player i. A *mixed strategy* in atomic flows for player i is a probability distribution pr_i, such that $\sum_{p \in \mathcal{P}_i} pr_i(p) = 1$. A *mixed strategy profile* in atomic flows is a collection of probability distributions $[pr_1, pr_2, \ldots, pr_N]$. These definitions can be extended to splittable flows.

We proceed with definitions of cost functions. For simplicity of presentation, consider atomic flows with pure strategies; all the definitions can be extended to splittable flows and mixed strategies. For game R and routing $\mathbf{p}$, the *social cost* (or *global cost*) is a function of routing $\mathbf{p}$, and it is denoted $SC(\mathbf{p})$. The *player or local cost* is also a function on $\mathbf{p}$ denoted $pc_i(\mathbf{p})$. We use the standard notation $\mathbf{p}_{-i}$ to refer to the collection of paths $\{p_1, \ldots, p_{i-1}, p_{i+1}, \ldots, p_N\}$ and $(p_i; \mathbf{p}_{-i})$ as an alternative notation for $\mathbf{p}$ that emphasizes the dependence on p_i. Player i is *locally optimal* (or *stable*) in routing $\mathbf{p}$ if $pc_i(\mathbf{p}) \leq pc_i\left(p_i'; \mathbf{p}_{-i}\right)$ for all paths $p_i' \in \mathcal{P}_i$; that is, there is no better alternative path for the player. A greedy move by a player i is any change of its path from p_i to p_i' that improves the player's cost, that is, $pc_i(\mathbf{p}) > pc_i\left(p_i'; \mathbf{p}_{-i}\right)$. *Best response dynamics* are sequences of greedy moves by players. In the aforementioned notations, when the context is clear, we will drop the dependence on $\mathbf{p}$; for example, we will write pc_i instead of $pc_i(\mathbf{p})$.

A routing $\mathbf{p}$ is in a *Nash equilibrium* if every player is locally optimal. Nash-routings quantify the notion of a stable selfish outcome. In the games that we study, there could exist multiple Nash equilibria. A routing $\mathbf{p}^*$ is an optimal pure strategy profile if it has minimum attainable social cost: for any other pure strategy profile $\mathbf{p}$, $SC(\mathbf{p}^*) < SC(\mathbf{p})$. We quantify the quality of the Nash equilibria with the *PoA* (sometimes referred to as the coordination ratio) [15], and the *PoS* [22]. Let $\mathbf{P}$ denote the set of distinct Nash equilibria, and let SC^* denote the social cost of an optimal routing $\mathbf{p}^*$. Then,

$$\text{(Price of anarchy) } PoA = \sup_{\mathbf{p} \in \mathbf{P}} \frac{SC(\mathbf{p})}{SC^*}, \quad \text{(Price of stability) } PoS = \inf_{\mathbf{p} \in \mathbf{P}} \frac{SC(\mathbf{p})}{SC^*}.$$

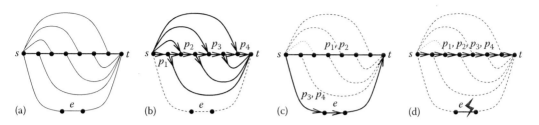

FIGURE 32.3 Network layer equilibrium examples. (a) Network configuration, (b) optimal solution, (c) nash equilibrium, and (d) jammer.

For an illustration of the aforementioned definitions, Figure 32.3 depicts various outcomes in an atomic *bottleneck* routing game with four players going from s to t with flow $f_i = 1$. In this game, $pc_i(\mathbf{p}) = \max_{e \in p_i} f_e$, also denoted as C_i, which is the flow on the worst congested edge in path p_i, where f_e denotes the total flow (congestion) on edge e. Figure 32.3b depicts the coordinated optimal solution with $SC^* = 1$. Figure 32.3c depicts a Nash equilibrium where two pairs of two players choose the same path. Each player has cost $pc_i = 2$ since each edge of their path is used by two paths. Since no edge is used by more than two paths, the social cost is $SC = 2$. Since this is the worst Nash equilibrium, the PoA for the case in Figure 32.3c is $PoA = 2/1 = 2$ (since the social cost of any Nash equilibrium does not exceed 2 and the optimal social cost is 1). The price of stability is $PoS = 1$, since there is a Nash equilibrium with optimum social cost 1 (the case of Figure 32.3b).

32.4.2.2 Routing Games with Jammers

Here, we describe new classes of routing games that incorporate jammers $\mathcal{J}$, and that we denote as $(G, \mathcal{N}, P, \mathcal{J})$.

32.4.2.2.1 Bottleneck Routing Games

Each player's utility cost is the worst congestion on its path edges [3,6–8]. In the atomic case, player i has utility cost function $pc_i = C_i$, namely, the worst congestion on any edge in the player's path. The social cost is $SC = \max_{e \in E} f_e$, which is also denoted as C, the maximum congestion on any edge in the network. It is well known that C directly relates to the optimal schedule of batch packet problems [10,17,18,24]. In [3], the authors observed that in wireless networks the maximum congested edge is related to the lifetime of the network since the nodes adjacent to high congestion edges transmit a large number of packets that result in higher energy utilization. High congestion edges also result in congestion hot spots that slow down the network and increase its vulnerability to malicious attacks. Thus, minimizing the maximum congested edges results in more secure networks. In [3], the authors consider splittable and atomic flow. They prove the existence and nonuniqueness of equilibria, that the PoA may be unbounded for specific *edge congestion functions*, which depend on the number of paths that use the edge. If the edge congestion function is polynomial with degree p, then they bound the *PoA* with $O(m^p)$, where m is the number of edges in the graph.

In our previous work [8], we studied atomic bottleneck games. We proved that the PoS is 1. We showed that the PoA is bounded by $O(L + \log n)$, where L is the maximum allowed path length in P, and n is the number of nodes in the network. We also prove that $\kappa \leq PoA \leq c(\kappa^2 + \log^2 n)$, where κ is the size of the largest edge-simple cycle in the graph and c is a constant. Some of the techniques that we propose (e.g., expansion) were used in an initial form in [8]. In an extension of this work, we consider in [6] wireless networks with routing classes $Q_1, \ldots, Q_\psi$, where each class j has a service cost S_j. Each path belongs to exactly one routing class. The player i cost function is $C_i + S_i$, where S_i is the cost of the service class of the path of i, and C_i is the congestion experienced by player i by considering only the paths in the same service class. The social cost function is the maximum player cost. We showed in [6] that such games stabilize with best response dynamics and the PoS is 1. The PoA is bounded by $O(\min(C^*, S^*) \cdot \psi \log n)$, where $C^* + S^*$ is the optimal social cost. Such games can be used to provide approximations to the social cost function $C + D$, where D is the maximum path length in the routing. An improved version of this work is [7].

Based on our previous study of bottleneck games [6–8], we study new types of bottleneck games that incorporate jammers. As we explain in the following (in the *PoJ* metric description), when a jammer attacks an edge, it may force the players to deviate in alternate routes with higher congestion. For example, Figure 32.3d depicts a situation where a jammer attacks edge e in an atomic bottleneck game. With the jammer, there is a worse Nash equilibrium with congestion 4, increasing by a factor of 2 with respect to the worst equilibrium congestion without the jammer. We need to formulate appropriate metrics and study systematically such effects of jammers. We must also consider splittable flow and mixed equilibria.

32.4.2.2.2 Power Routing Games

These types of games are derived directly from Equation 32.1 of our physical layer model. For the atomic games scenario, we consider a player cost function that is proportional to the total energy that the player spends on its chosen path:

$$pc_i = \sum_{e \in p_i} (\beta_e \cdot P_e^J + \gamma_e) e^{\alpha_e \cdot f_e}, \tag{32.3}$$

where
  P_e^J is the power devoted by a jammer toward edge e
  f_e is the total flow (congestion) on edge e from all paths that use e
  β_e depends on the position of the jammer with respect to the endpoints of e
  γ_e is the thermal noise of the wireless channel
  α_e is a constant that appears in Equation 32.1

Note that if e is not jammed, $P_e^J = 0$. Equation 32.3 is proportional to the power that the player i spends in the network in order to transmit flow f_i.

In [13], we examined a scenario without jammers and a simplified version of the player cost function: $pc_i = \sum_{e \in p_i} 2^{f_e}$. For social cost, we use the worst congested edge (bottleneck) metric, $SC = C$ and obtain near optimal PoA. We used the expansion technique to obtain this result (see the following). The expansion technique can be extended now to incorporate jammers using Equation 32.3.

32.4.2.2.3 Delay Routing Games

Here, we use Equation 32.2 from the physical layer to model the player costs. With this game, we want to capture more precisely the effects of the jammers in the delays of the flows. For every edge e, we associate the respective cost C_e^T, as defined in Equation 32.2. We divide the edges E into two classes: those affected by the jammers E^J and those not affected by the jammers $E - E^J$. Those edges that are affected by the jammers in E^J have associated a cost $\widehat{C}_e^T$, which takes into consideration the effect of the jammers in probability of success of the rate $P_{succ}^* \left(R_e^* \right)$. Hence, for player cost function, we use a bottleneck type of metric:

$$pc_i = \max \left(\max_{e \in p_i \cap E^J} \widehat{C}_e^T, \max_{e \in p_i \cap (E - E^J)} C_e^T \right)$$

For social cost function, we will consider the highest congested edge in network C as in the bottleneck games. Alternatively, we can consider as social cost the maximum player cost, which incorporates the effect of the jammers. We explore whether we can apply the expansion technique or potential function analysis to study such games.

32.4.2.2.4 Repeated Games

Here, we examine the interplay between network layer games and physical layer games. We consider games where players take turns in alternate phases in the physical and network layer. In the *physical layer phase*, the jammers interact locally and play the zero-sum games described in the physical layer modeling. In the *network layer phase*, we let the jammers and players interact in a routing game (G, $\mathcal{N}$, P, $\mathcal{J}$) in the network layer until a Nash equilibrium is reached. The two phases alternate until stable states (Nash equilibria) are reached in both layers. We need to explore the convergence properties of these games as well as the quality of the resulting equilibria.

32.5 Novel Evaluation Metrics

32.5.1 Price of Jamming

We propose the PoJ as a new metric that compares the worst equilibria with and without jammers. Figure 32.3d depicts a situation where a jammer attacks edge e in an atomic bottleneck game. For simplicity, we assume that e is completely blocked, that is, routes through e have infinite cost. In this situation, there is a new worst Nash equilibrium, depicted in the figure, with social cost $SC = 4$. Thus, with the jammer, we can have worse equilibria, which reflects worse network performance. At the same time, the optimal solution remains the same even with the jammer, namely, $SC^{\star} = 1$. Let PoA_J denote the new price of anarchy considering the jammer; clearly, $PoA_J = 4/1 = 1$. We propose the notion of PoJ to determine the effect of jamming. The PoJ is the ratio of the PoA with jammers to the PoA without jammers. Namely,

$$PoJ = \frac{PoA_J}{PoA}.$$

The PoJ for the example of Figure 32.3d is $PoJ = 4/2 = 2$, since $PoA_J = 4$ and $PoA = 2$. Here, the PoJ measures the extent to which a jammer can damage the Nash equilibria of the game. A $PoJ > 1$ obviously represents a favorable outcome for jammers. Do there exist classes of networks and utility functions with small $PoJs$? Note that this is just one possible definition for the PoJ; we can formulate alternative definitions (for instance, where the PoJ is the ratio of the PoA_J versus the price of stability without the jammer) and evaluate the types of realized networks.

32.5.2 The Paradox of Jamming and Its Price

There are several examples of unintended consequences from standard game theory literature, for example, Braess's Paradox [26], which illustrates how adding a low-cost link in some networks can actually lead to costlier equilibria. We would like to investigate whether similar phenomena exist in jamming games. For example, can we find networks where jamming a set of links actually leads to *better* equilibria for the rational players? Such networks are expected to be highly resilient to jamming. Second, given the capability, does it always make sense for a jammer to be maximally malicious, that is, jam as many edges as possible? Surprisingly, the answer is yes, for the former and no, for the latter. We label this phenomenon as the ProJ. We can also define a corresponding metric that is inversely proportional to the PoJ. ProJ larger than 1 implies the existence of the paradox. This is illustrated in the following example.

Consider Figure 32.4a that shows an atomic bottleneck game on a six-node cycle. Except for node 6, which has no packet to transmit, each node i has a packet to send to $i + 1$. For atomic bottleneck games, the best equilibrium is when each node sends its packet clockwise via edge $(i, i + 1)$. This is also the social optimum and the social cost is thus $SC^{\star} = 1$. The worst cost equilibrium arises when each node sends its packet counterclockwise along the longest path. The equilibrium cost is $SC = 5$. The PoA of this network (without jammer) is thus 5. Now consider a jammer jamming edge $e = (6, 1)$. If $C_e^J > 5$ (congestion on e with jammer), then the only equilibrium in this jammed network has cost 1 (the shortest paths). Thus, the PoA of the network with jammer is 1. Clearly, the PoJ of this network is less than one: $PoJ < 1$, which implies the occurrence of the ProJ. The presence of the jammer on edge e has positive consequences for the rest of the players. Now consider Figure 32.4b, which is the same six-node cycle as Figure 32.4a except that each node now has a packet to send to its diagonally opposite node, that is, 1–4, 2–5, etc. Suppose a jammer can jam any number of edges in this network with each jammed edge costing $C_e^J \geq 6$.

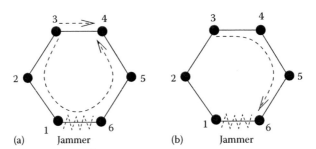

FIGURE 32.4 Jamming paradox.

Player cost functions are $\max\left(C_e^J, C_e^{NJ}\right)$ along its path, where C_e^{NJ} is just the edge congestion without jammer. Suppose the payoff for the jammer is the maximum edge congestion in the jammed network, that is, the cost of the worst equilibrium. It is easy to see that the jammer gains nothing by jamming multiple edges, since rational players do not have better alternatives and thus do not switch paths. The jammer can, in fact, obtain its maximum payoff of 6 by jamming a single edge.

These examples motivate us to formally study the PoJ and develop metrics to capture these paradoxical effects. In particular, we believe that investigating the following questions will enable us to design games and networks that are resilient to jamming: Is the PoJ universally applicable, that is, can we find jamming games on all networks (graphs) that exhibit this property? Conversely, are there classes of networks on which jamming never has paradoxical results regardless of the specific cost function? For any given network (graph), can we always find jamming and non-jamming utility functions that exhibit this property? What are the crucial network graph characteristics that give rise to these effects? The answer to those questions will enable us to design games and networks whose communication graphs contain useful antijamming substructures.

32.5.3 Convergence Time

A very fundamental question in game theory is whether games converge to Nash equilibria and how fast they can be reached. Here, we must examine the case where the network starts with an arbitrary state and players use best response dynamics and explore how the jammers affect the convergence to Nash equilibria. Several of these questions are associated with the PLS complexity class of local search problems [22]. In [3], it is shown that finding the best Nash equilibrium that minimizes the social cost is a NP-hard problem. We need to investigate the effects of the jammers also in the complexity of computing Nash equilibria.

32.5.4 Novel Analytical Methods

32.5.4.1 Edge Expansion Technique

A formal technique that we have used extensively in our previous work on bottleneck games [6–8] and power games [13] to bound the PoA is *expansion*. The basic idea is to use certain properties of the cost functions in order to construct a chain of edges $E_0, E_1, E_2, \ldots$, such that $|E_{i+1}| \geq r \cdot |E_k|$; in other words, the edges in the chain expand with rate r, and we call this an *expansion chain*. When k is sufficiently large, the length of the chain must be relatively small, since $|E| = m$ (e.g., constant k gives $\log m$ expansion steps). Each E_i maintains certain properties about the congestion of its edges in the form of a lower bound (e.g., we may have that the edges in E_i have congestion at least $C - i$, where C is the largest congestion in the Nash equilibrium). Each set E_i is *supported* by E_{i+1} in the sense that without E_{i+1} the congestion in E_i cannot be maintained. In other words, the edges in E_{i+1} are obtained from the alternative paths of the players that use the edges in E_i. The chain starts with some edge $\{e\} = E_0$ having congestion C. Since the chain has small length, the congestion C must be small too, since otherwise there would not be enough edges in the graph to support the expansion chain sets and also preserve the lower bounds on the congestion. We can use the expansion method in order to provide bounds on the PoJ.

32.5.4.2 Potential Function Analysis

A fundamental technique in game theory is potential function analysis. Some games are called *potential games* if they have associated with them a potential function $F(\mathbf{p})$, where $\mathbf{p}$ is some routing state. A potential function can help in different ways. First, it can help to prove the convergence to a Nash equilibrium using best response dynamics, using monotonicity properties of F (e.g., see our previous works [6–8]). Second, it can help to quantify the quality of the Nash equilibrium in terms of PoA by linking the potential function to the optimal social cost function SC^* [22].

Acknowledgments

We are grateful to NSF and AFOSR for the support provided for this research under grants #CNS-1018273 and AFOSR #FA9550-10-1-0448.

References

1. G. Amariucai, S. Wei, and R. Kannan. Gaussian jamming in block-fading channels under long term power constraints. In *IEEE International Symposium on Information Theory (ISIT)*, Niece, France, July 2007.
2. M. Babaioff, R. Kleinberg, and C.H. Papadimitriou. Congestion games with malicious players. In *Proceedings of the 8th ACM Conference on Electronic Commerce 2007 (EC'07)*, San Diego, CA, June 2007, pp. 103–112.
3. R. Banner and A. Orda. Bottleneck routing games in communication networks. *IEEE Journal on Selected Areas in Communications*, 25(6):1173–1179, 2007. [Also appears in INFOCOM'06.]
4. D.P. Bertsekas and R.G. Gallager. *Data Networks*. Prentice Hall, Englewood Cliffs, NJ, 1992.
5. M.H. Brady, M. Mohseni, and J.M. Cioffi. Spatially-correlated jamming in Gaussian multiple access and broadcast channels. In *Proceedings of the 40th Annual Conference on Information Sciences and Systems*, Princeton, NJ, March 2006.
6. C. Busch, R. Kannan, and A.V. Vasilakos. Quality of routing congestion games in wireless sensor networks. In *Proceedings of the 4th International Wireless Internet Conference (WICON)*, Maui, HI, November 2008.
7. C. Busch, R. Kannan, and A.V. Vasilakos. Approximating congestion + dilation in networks via quality of routing games. *IEEE Transactions on Computers*, 2011.
8. C. Busch and M. Magdon-Ismail. Atomic routing games on maximum congestion. *Theoretical Computer Science*, 410(36):3337–3347, August 2009.
9. Y. Chen, W. Xu, W. Trappe, and Y.Y. Zhang. *Securing Emerging Wireless Systems: Lower Layer Approaches*. Springer, New York, 2009.
10. R. Cypher, F. Meyer auf der Heide, C. Scheideler, and B. Vöcking. Universal algorithms for store-and-forward and wormhole routing. In *Proceedings of the 28th ACM Symposium on Theory of Computing*, Philadelphia, PA, 1996, pp. 356–365.
11. C. Intanagonwiwat, R. Govindan, and D. Estrin. Directed diffusion: A scalable and robust communication paradigm for sensor networks. In *MobiCom '00: Proceedings of the 6th Annual International Conference on Mobile Computing and Networking*, New York, 2000, pp. 56–67. ACM.
12. R. Kannan, S. Wei, C. Busch, and A. Vasilakos. Online algorithms for maximizing link transmission quality over a jammed wireless channel. *IEEE Percom-IQ2S Workshop on Quality of Service*, Galveston, TX, March 2009.
13. R. Kannan, C. Busch, and A.V. Vasilakos. Tight price of anarchy of (super) polynomial bottleneck congestion games. In *Proceedings of the 2nd International Conference on Game-Theory in Networking—GameNets*, Shanghai, China, April 2011.

14. A. Kashyap, T. Basar, and R. Srikant. Correlated jamming on MIMO Gaussian fading channels. *IEEE Transactions on Information Theory*, 50:2119–2123, September 2004.
15. E. Koutsoupias and C. Papadimitriou. Worst-case equilibria. In *Proceedings of the 16th Annual Symposium on Theoretical Aspects of Computer Science (STACS)*, Trier, Germany, March 1999, pp. 404–413. Volume 1563 of *LNCS*. Springer-Verlag.
16. B. Krishnamachari, D. Estrin, and S.B. Wicker. The impact of data aggregation in wireless sensor networks. In *ICDCSW '02: Proceedings of the 22nd International Conference on Distributed Computing Systems*, Washington, DC, 2002, pp. 575–578. IEEE Computer Society.
17. F.T. Leighton, B.M. Maggs, and S.B. Rao. Packet routing and job-scheduling in O(*congestion + dilation*) steps. *Combinatorica*, 14:167–186, 1994.
18. T. Leighton, B. Maggs, and A.W. Richa. Fast algorithms for finding O(congestion + dilation) packet routing schedules. *Combinatorica*, 19:375–401, 1999.
19. S. Madden, M.J. Franklin, J.M. Hellerstein, and W. Hong. Tag: A tiny aggregation service for ad-hoc sensor networks. *SIGOPS Operating Systems Review*, 36(SI):131–146, 2002.
20. T. Moscibroda, S. Schmid, and R. Wattenhofer. When selfish meets evil: Byzantine players in a virus inoculation game. In *Proceedings of the 25th Annual ACM SIGACT-SIGOPS Symposium on Principles of Distributed Computing (PODC 2006)*, Denver, CO, July 2006, pp. 35–44.
21. E.F. Nakamura, A.A.F. Loureiro, and A.C. Frery. Information fusion for wireless sensor networks: Methods, models, and classifications. *ACM Computing Surveys*, 39(3):9, 2007.
22. N. Nisan, T. Roughgarden, É. Tardos, and V.V. Vazirani. *Algorithmic Game Theory*. Cambridge University Press, Cambridge, U.K., 2007.
23. R. Poisel. *Modern Communications Jamming Principles and Techniques*. Artech House, Norwood, MA, 2003.
24. Y. Rabani and É. Tardos. Distributed packet switching in arbitrary networks. In *Proceedings of the Twenty-Eighth Annual ACM Symposium on the Theory of Computing*, Philadelphia, PA, May 22–24, 1996, pp. 366–375.
25. T.S. Rappaport. *Wireless Communications: Principle and Practice*, 2nd edn. Prentice Hall, Englewood Cliffs, NJ, 2002.
26. T. Roughgarden and É. Tardos. How bad is selfish routing. *Journal of the ACM*, 49(2):236–259, March 2002.
27. S. Shafiee and S. Ulukus. Capacity of multiple access channels with correlated jamming. In *Proceedings of the Military Communications Conference, MILCOM*, Atlantic City, NJ, October 2005, pp. 218–224.
28. S. Shafiee and S. Ulukus. Correlated jamming in multiple access channels. In *Proceedings of the Conference on Information Sciences and Systems*, Baltimore, MD, March 2005.
29. M.K. Simon, J.K. Omura, R.A. Scholtz, and B.K. Levitt. *Spread Spectrum Communications Handbook*, Revised Edition. McGraw-Hill, New York, 1994.
30. D. Slater, P. Tague, M. Li, and R. Poovendran. A game-theoretic framework for jamming attacks and mitigation in commercial aircraft wireless networks. In *AIAA Infotech and Aerospace Conference*, Seattle, WA, April 2009.
31. S. Wei and R. Kannan. Jamming and counter-measure strategies in parallel Gaussian fading channels with channel state information. In *IEEE MILCOM*, San Diego, CA, November 2008.
32. S. Wei, R. Kannan, V. Chakravarthy, and M. Rangaswamy. CSI usage over parallel fading channels under jamming attacks: A game theory study. *IEEE Transactions on Communications*, 60(4):1167–1175, 2012.
33. W. Xu, W. Trappe, Y. Zhang, and T. Wood. The feasibility of launching and detecting jamming attacks in wireless networks. In *ACM MobiHoc*, Urbana-Champaign, IL, 2005.
34. W. Xu, T. Wood, W. Trappe, and Y. Zhang. Channel surfing and spatial retreats: Defenses against wireless denial of service. In *ACM WiSe*, Philadelphia, PA, 2004.

33

Reasoning about Sensor Networks

33.1 Introduction 693

33.2 Related Work 695

33.3 Perpetual Requirements Engineering for Sensor Networks 695
SOLj: Secure Operations Language-JAVA • SOLj Events •
SOLj Definitions • SINS

33.4 SOLj Example: Autoregulated Power Generation Network.......698
General Aspects • System Description • System Operation

33.5 Counterfactuals in Sensor Networks..699
Language of Counterfactual Theory • Formal Description of SOLj
Agent Network • Program Transformers • Kripke Versioning
Model • Interpreting the Counterfactual Implication • Fragment
of Counterfactual Logic • Temporal Logic Fragment

33.6 Counterfactual Example: Autoregulated Power Generation
Network..705
Formal Characterization of a SOLj Network's Behavior • Source
Code Change • Proof

33.7 Concluding Remarks..707

33.A Appendix: Theoremhood of CNT within VC 707

References...709

Manuel Peralta
Louisiana State University

Supratik
Mukhopadhyay
Louisiana State University

Ramesh Bharadwaj
Naval Research Laboratory

33.1 Introduction

Sensor networks are embedded networked systems that receive percept streams from the environment and constantly react to them. From a software-based point of view, modifications done to any system Sensor Network Systems (SNS) should be performed under utmost caution as SNSs are often deployed in *mission-critical applications*. The latter means that any disruption that inhibits the system in satisfying its operational semantics will definitely yield catastrophic results. Also, sensor networks are required to react dynamically to ever-changing environmental conditions without human intervention. Therefore, one should strive to provide methods that ensure the correctness of SNSs under reconfiguration is preserved, that is, the SNS respects its operational requirements, while the system structure changes in response to an evolving environment.

For SNS applications, requirements dynamically change in a rapid, unpredictable, and continuous fashion. In applications such as those driving search and rescue missions, any operational expansion/contraction requires dynamic reorganization of the system. For these application scenarios, any downtime resulting from upgrade of the control system leads to unacceptable disruption of service. As a result, continued availability of such systems in a mission-critical setting, even under dynamically changing requirements, is of utmost importance. We need to develop techniques, tools, and methods that can build, manage, and maintain SNS systems whose requirements keep changing perpetually during their life cycle. Such systems should be able to autonomously reengineer themselves rapidly online under changing requirements, with minimal or no disruption in service, and yet meeting all constraints of timeliness, cost, and performance in a reasonable way.

Traditional software development methodologies assume that requirements are well understood and available in the form of a formal or rigorous specification of the required system behavior. However, this assumption fails to hold for software that is meant to control SNS applications deployed in rapidly evolving scenarios. While it is possible to develop logically precise requirements for software computing mathematical functions, the behavior of a software system depends on extraneous factors that are not usually foreseen during its development. These include factors such as platform of deployment (e.g., the word length of the machine on which the software is run), the communication protocols used, the amount of memory available, etc. In software solving real-world problems, such extraneous factors are compounded by those from the system's physical environment that expect the software to cope with dynamically changing business constraints. In some cases, given time and money, it is possible to get the original developers to update the software to meet the changed business requirements. However, in certain cases, such an update may not be possible even with adequate time and money (e.g., the original developers may have moved away from the technology and it may not be possible to acquire a suitable team to build on their work). In these situations, the system must be phased out with millions of dollars in software development cost wasted.

Conventional approaches used in industry are inadequate in rapidly evolving mission-critical scenarios due to (1) the unpredictable nature of the evolution of the system requirements and (2) dynamically changing situational environments driven by the dynamics of the market/mission partners, including rapid mergers, disinvestment, formation of coalitions, noisy and unpredictable communication channels, and cyber attacks aimed at disruption of the network. We use the phrase "perpetual requirements engineering" to denote an approach where we address dynamically changing requirements throughout the software life cycle, including autonomous rapid reconfiguration and persistent redeployment of distributed software systems without disruption to service commitments, in an expeditious manner. Our approach improves on the agile development paradigm. Traditionally, agile development has been used successfully for software projects with rapidly changing requirements. One example of the agile development approach is extreme programming. In agile development, activities generally alternate between modeling and coding, with major portions of the design being generated as implementation proceeds. Traditional agile development approaches, however, suffer from a lack of automated support. Manual effort is needed to incorporate any changes in the requirements into a software artifact. Extensive manual refactoring of code is often needed to ameliorate the effects of dynamically changing code requirements. System updates are developed manually, at a huge cost and in an untimely and unpredictable manner. This is one inhibiting factor on the scalability of agile development methods, thereby making them suitable only for development projects of a small or moderate size. This is also the reason why traditional agile development methods tend to succeed only when an experienced development team is available. Lack of dynamic adaptation means frequent shipping of new code in response to ever-changing requirements. This not only adds to the cost but also results in increased downtime of deployed software due to frequent updates to the running code. While the involvement of the customer is an essential component of the agile development method, in many situations, it becomes difficult for the development team to stay in touch with the customer, especially in the postdeployment stage (e.g., consider projects whose development is outsourced).

In large software projects with dynamically changing requirements as usually encountered in the development of SNS applications, instead of manually performing development iterations every time a requirement changes, we need to implement techniques and strategies for automatic incremental update of system or subsystem components comprising the deployed software, in response to changing requirements. We need a requirements engineering paradigm that can automatically reflect incremental changes in the requirements by a dynamic reconfiguration and persistent update of the running software.

33.2 Related Work

Significant research has been performed in massively distributed environment-aware computing (also known as "swarm computing" [4,12]), in particular for creating and reasoning about swarm programs. Most of these works have been focused on developing programming paradigms, tools, and languages for swarm computing. EnviroTrack [1], an object-based distributed middleware system, raises the level of programming abstraction for distributed s by providing a convenient and powerful interface to the application developer geared toward tracking the physical environment. Menezes and Tolksdorf [12] study different abstractions in the field of swarms. However, none of these works are concerned with the problem of developing formal methods for building sensor-based systems that provide provable guarantees of meeting their requirements.

Römer et al. [16] survey middleware challenges in the area of wireless s. According to Ref. [16], adaptability and data-centric communication should be important issues in coordinating services in networks that involve wireless sensors. We augment the desirable properties of coordination frameworks for wireless sensor networks stated in Ref. [16] with the capability of intelligent data/service fusion.

This chapter also deals with applying the theory of counterfactuals to s; this theory allows us to reason about hypothetical situations. It has been used for in Philosophy and Political Science [14] for decision making in a hypothetical environment. In Physics, it has been used for reasoning about measurements in quantum mechanics [13,18]. The main idea exposed by Fisler et al. [5] is to gain knowledge regarding the effects of changing access control policies before actually making such changes. The work of Fisler et al. is similar to the one presented in this chapter as it tries to find the effects of a change a priori. The work of Chockler et al. [3] employs counterfactual reasoning also in the context of model checking. In this instance, the authors emphasize their work in coverage issues. They use counterfactual reasoning to enhance the coverage information. This work differs from ours given that they use counterfactual logic to explore alternative scenarios, whereas, in our case, we explore a single alternative version given an initial version.

The approach given by Guo and Subramaniam [7] exhibits a method by which change impact analysis is modeled and verified in a distributed setting. This approach is loosely based on model checking. Their model is, in essence, a network of state machines that communicate either via shared variables or queues. Changes are modeled as adding and/or deleting transitions from the composite state machine that represents the distributed system. The work in Ref. [7] is similar to ours in the context of two aspects: (1) the authors are formally representing change in a system; however, their approach targets distributed computations and is based on model checking, whereas our approach targets sequential computation, and it is based on theorem proving and (2) this approach prunes the global state space by using *partial order reduction* in order to infer the valid transitions when a change occurs, whereas our approach deals with change at the source code level, and the validity of the change is inferred by our logical calculus. In Ref. [19], Subramaniam et al. enhance the approach shown in Ref. [7]. The changes are still represented by adding and/or deleting transitions of a composite state machine. However, this work addresses the issue of test suite coverage when changes occur. This approach detects the affected tests based on whether or not these include the affected transitions. Using formal verification techniques similar to the ones presented in Ref. [7], the authors are able to reduce the total regression test suite based on which tests are relevant after a given change. Our approach goes in a different direction by formally characterizing the source-code change and determining if the changes to the current source code version are logically consistent with its properties and the future desired properties.

33.3 Perpetual Requirements Engineering for Sensor Networks

In the context of perpetual requirements in SNSs, we have developed an approach that relies on two technological elements. The first is Secure Operations Language for Java (SOLj) [2], which is an event-based domain-oriented synchronous programming extension of Java used to write the specification for agents, which, in this context, are the software counterparts of one or more sensors. The second element is the Secure Infrastructure for Networked Systems (SINS). SINS a is a distributed run-time system in which the SOLj agents are deployed. The SINS runtime and framework provide a set of security policies that help avoid compromising the SOLj agents' behavior and interactions.

33.3.1 SOLj: Secure Operations Language-JAVA

Given that SOLj is an extension of Java, it is presented as modular extensions to its core language, that is, Java. A *module* comprises the specification unit in SOLj; it is composed of type definitions, variable declarations, service declarations, assumptions and guarantees, and definitions. In the future, we will use the word *agent* to denote a module instance. A SOLj module may include *attributes* as they are described in the following:

`deterministic` Declares a module that does not exhibit nondeterministic behavior. The compliance with this attribute is checked by the SOLj compiler.

`reactive` Declares a module that will cause a state change only when its (visible) environment fires an event via a state change or by invoking a method. Also, this attribute denotes that the module's response to an event will happen in the next immediate step.

The type definition section contains user-defined types as well as enumerated types. The Java comment `//@type definitions` precedes the "type definitions" section. It provides SOLj directive to the compiler indicating the start of the type definition section. The variable declaration section defines three types of variable, which are explained in the following:

> *monitored variables*: variables in the environment that influence the agent's behavior
> *controlled variables*: variables in the environment that are changed by the agent's behavior
> *internal variables*: variables that reflect the agent's internal state

The Java comment `//@Services` precedes the "service declarations" section, which declares the agent's external interface. It contains the methods that realize the services the module provides. For each method declaration within a service, the SOLj language provides the capability of declaring the corresponding preconditions and postconditions that denote the conditions under which each service should start and terminate. The preconditions and postconditions are encoded as arithmetic expressions and type declarations. A type declaration is denoted by a type judgment expression `T:x`, where x is a variable and T is a type. These constraints are enforced at runtime in a dynamic manner. Also, each service invocation must use a "continuation variable." This variable includes a boolean field called "done," which is assigned a value of `true` once the service invocation is ready to provide a return value.

The `//@Assumptions` comment denotes the starts of the *assumptions* section, which includes assumptions that determine the agent's correct way to operate. If any of these assumptions is violated by the environment, the agent's execution is aborted. The `guarantees` section contains the agent's required safety properties. The `definitions` section provides *update functions*, which denote variable definitions. These specify the corresponding values for internal and controlled variables. For the sake of future references, we will distinguish between *monitored variables*, whose values are given by the environment, and *dependent variables*, whose values are the result of SOLj agents' computations. These values are obtained using the values of monitored variables and (possibly) past-dependent variables.

33.3.2 SOLj Events

Based on the Software Cost Reduction (SCR) method's SCR Abstract Language (SAL), SOLj has been provided with the capability of defining events [10]. Intuitively, SCR events can be interpreted as state changes. Moreover, the occurrence of an event is triggered when a variable has its value changed. This is done by update functions that change the dependent variable's value:

$$@T(c) \quad =^{def} \quad \neg\ @PREV(c) \wedge c$$
$$@F(c) \quad =^{def} \quad @PREV(c) \wedge \neg c$$
$$@C(c) \quad =^{def} \quad @PREV(c) \neq c$$

An initialization method (init) assigns starting values to all dependent variables (controlled or internal). Each module contains an init method. Each dependent variable is updated by just one of the update functions. Moreover, a *dependency relation* is induced by the interplay between update functions and dependent variables. We denote this relation with D_m. Let a and b be two dependent variables, then we say that $(a,b) \in D_m$ if and only if a is updated by the function corresponding to b. The fact that a may depend on the previous values of other variables and itself does not influence the dependency relation. Also, a *dependency graph* may be derived from D_m by interpreting the set of dependent variables as the nodes and each $(a,b) \in D_m$ as the edges.* For each module, we need to consider its corresponding dependency graph as acyclic.

From a simplified point of view, a SOLj program executes in a sequence of *steps*, and each step is preceded by the triggering of an event. The module's dependency relation induces the order in which the variable updates and service invocations are carried out. Each computation step may be decomposed as follows:

1. The environment or the agent itself triggers an event in a nondeterministic manner.
2. Each agent *responds* to this event by modifying the values of its dependent variables.

From an external point of view, the updates and service invocations can be thought of happening in a synchronous manner (as it is dictated by the *synchrony hypothesis* and exemplified by languages such as Esterel and LUSTRE [8]). The latter implies that all dependent variable updates and service invocations that occur as a response to an event triggering happen before the next event is triggered.

33.3.3 SOLj Definitions

The main part of a SOLj module is the definitions section. This section declares and defines the update function that corresponds to each dependent variable. Update functions provide the value of the updated dependent variable. An update function's body is comprised of *return* statements which are constrained by *conditional expressions*. These are activated by event triggers initiated by the environment and/or the agent. Syntactically, these conditional expressions are denoted by Java conditional expressions with the difference that the guards are *SOLj events*.

SOLj expressions can be service invocations such as A:B(varList)^cont. The identifier A denotes the name/URL of the service, B denotes the name of the invoked method, varList denotes the set of formal parameters that are provided to the invoked method, and cont denotes the unique continuation variable associated with the invocation. When the "done" field in the continuation is assigned "true," the service invocation has terminated. Finally, a compiler derives Java code from the SOLj definitions. This derived code executes on the SINS.

33.3.4 SINS

SOLj module instances interact in a runtime environment called SINS. Generally, a SINS implementation consists of a set (one or more) of *SINS virtual machines* (SVM). Each of these virtual machines acts as a container for one or more agents in a given host node. Distributed SVMs communicate among themselves using the *Agent Control Protocol* (ACP) [20] with the purpose of exchanging agent and control information. A supplementary protocol, known as the Module Transfer Protocol (MTP), takes care of the code distribution, digital signatures, authentication, and code integrity. Complying with locally enforced security policies, SOLj agents are allowed to access local resources in a host. Compliance with these security policies is verified using an inductive theorem prover. Observer agents (termed "security agents") are in charge of enforcing other safety property and security requirements. These agents monitor the execution of application-specific agents and also engage in corrective actions once a violation is detected.

---

* The notion of a dependency relation is easily extended to the entire system.

33.4 SOLj Example: Autoregulated Power Generation Network

The following example will be employed as a didactic tool in order to illustrate how our SOLj–SINS-based approach looks like in a practical setting.

33.4.1 General Aspects

In the realm of power generation and considering a very simple perspective, we have the interplay of three interrelated variables: voltage, current, and resistance. In large-scale power distribution networks, such interplay of these variables can mean the difference between efficient or wasteful power distribution. It would be desirable, thus, to have a sensor-network-based system, which would adequately react to changes in this quantities in a dynamic manner.

33.4.2 System Description

Our simplified sensor network for power generation system is comprised mainly of two types of agents:

  Distribution line agent (DistLineAgent): This agent type is in charge of communicating the state of the distribution lines to the GenAgents, that is, it reports the conditions regarding voltage and resistance. Figure 33.1 shows the SOLj source code for this type of agent.
  Generation-engine agent (GenAgent): This agent type is responsible of regulating the electrical current that is input into the electrical distribution system based on the voltage and resistance on distribution lines. Figure 33.2 shows the SOLj source code for this type of agent.

33.4.3 System Operation

Let us assume that we are dealing with an electrical distribution network comprised of several GenAgents and multiple DistLineAgents. Furthermore, let us assume that the environment temperature variations (the material's own reaction to conducting current and environmental heat) dilate and contract the inner core of the conducting cables. Therefore, the conducting cables' resistance is also variable. From general physics, we know that these three quantities are interrelated by the following formula:

$$I = \frac{V}{R}$$

where
  I stands for the electrical current in the system (measured in Amperes)
  V stands for the system's voltage (measured in Volts)
  R stands for the conducting cables' resistance (measured in Ohms)

The DistLineAgents via sensors placed along the distribution lines have the knowledge of the actual current and resistance along the distribution line system, and they communicate this information back to the set of GenAgents, which, in turn, regulate the electrical generators that provide the input electrical current to the distribution network. When, for instance, the environment exhibits a high temperature (due to heat wave perhaps), the resistance in the distribution lines will increase. Provided that the voltage remains the same, the correct reaction of the system would be to increase power production. The GenAgents monitor the resistance and can thus react to this change in the distribution system by actuating the generation engines to produce more power.

```
//@Module Declaration
deterministic reactive module DistLineAgent

//@Type definitions
Voltage = floatStream;
Current = floatStream;
Resistance = floatStream;

//@Controlled variables
Voltage currVoltage;
Current currCurrent;
Resistance currRes;

//@Monitored variables
Voltage observedVoltage;
Current observedCurrent;
Resistance observedRes;

//@definitions
//@initialization
void init(){
      currVolatage = null;
      currCurrent = null;
      currRes = null;
}

//@update functions
Resistance outputResistance(){
      if(@C(observedRes) && @T(observedRes >= thresholdRes))
      return currRes;
}
Voltage outputVoltage(){
      if(@C(observedVoltage) && @T(observedVoltage >= thresholdVoltage))
          return currvoltage;
}

Current outputCurrent(){
      if(@C(observedCurrent) && @T(observedCurrent >= thresholdCurrent))
          return currCurrent;
}
```

FIGURE 33.1   SOLj code for the `DistLineAgent` module.

33.5 Counterfactuals in Sensor Networks

Counterfactual logic aids us in reasoning about statements that are not matter of fact. Lewis [11] provided a sound and complete inference system and also stated that this logic was a decidable one. We have created a logical calculus based on counterfactual logic. This calculus allows us to express properties that would take place in a future version of a given program and verify that indeed these properties would hold if the changes between the new and old versions were applied.

33.5.1 Language of Counterfactual Theory

As it is shown in Ref. [11], we will shortly define the language that comprises the logic of counterfactuals (p_i *denotes a propositional variable*):

$$\phi ::= p_i \mid \neg\phi \mid \phi \wedge \phi \mid \phi \vee \phi \mid \phi \rightarrow \phi \mid \phi \Box\!\!\rightarrow \phi$$

The counterfactual expression $\phi \Box\!\!\rightarrow \psi$ should be read as follows: *if it had been the case that* ϕ, *it would have been the case that* ψ. Therefore, if we had a statement that contained an anteced that expressed

10 Manuel Peralta[1] and Supratik Mukhopadhyay[1] and Ramesh Bharadwaj[2]

```
//@Module Declaration
deterministic reactive module DistLineAgent

//@Type definitions
Voltage = floatStream;
Current = floatStream;
Resistance = floatStream;
Revolutions = floatStream;

//@Controlled variables
Revolutions rpm;

//@Monitored variables
Voltage currVoltage;
Current currCurrent;
Resistance currRes;
Revolutions currRevs;

//@Services
//@ Revolutions EGenerator:RevUpEngine(revDelta)
//@ pre = (currCurrent >= thresholdCurrent) &&
//@        (currVolate <= thresholdVoltage)
//@
//@ post = currRevs > rpm

//@ Revolutions EGenerator:RevDownEngine(revDelta)
//@ pre = (currCurrent < thresholdCurrent) ||
//@        (currVolatage > thresholdVoltage)
//@
//@ post = currRevs < rpm

//@Update Functions
Revolutions UpdateRPMs(){
    if(@C(currVoltage) && @C(currCurrent))
        return EGenerator:RevUpEngine(revDelta)^revUpCont;
else
        return EGenerator:RevDownEngine(revDelta)^revDownCont;
```

FIGURE 33.2 SOLj source code for GenAgent.

the properties of a given program and the required changes to create the new version and the consequent expressed the properties that the new version would exhibit, then it seems plausible that we can encode such a statement using counterfactual logic.

33.5.2 Formal Description of SOLj Agent Network

Given what was presented in earlier sections, we can abstract a sensor network, by considering the set of SOLj agents that realize its behavior. Let M_i denote the syntactically correct code for a given agent, then we can denote a SOLj agent network by the following parallel composition:

$$SN_0 \triangleq M_1 \,||\, M_2 \,||\ldots||\, M_n$$

A new version of a given sensor network will be created when at least the code of one agent is changed (as indicated by our *program transformer approach*). Hence, a new version of a given sensor network is given by the following expression:

$$SN_1 \triangleq M_1 \,||\, M_2 \,||\ldots||\, M_i[s_j/s_j'] \ldots ||\, M_n$$

where

 M_i is the changed SOLj agent module

 s_j and s_j' are the changed (swapped) statements within M_i's specification code

In general, any number of modules within the parallel composition may be changed by an arbitrary number of statement swaps. In terms of logical descriptions, we can also denote the SOLj agent network by the following extended conjunction:

$$\Psi_{SN} \triangleq \overset{n}{\underset{i=1}{\wedge}} \wedge M_i$$

where each M_i is denoted by

$$\Psi_{M_i} \triangleq \overset{m}{\underset{j=1}{\wedge}} f(j) = Stat_j$$

where

 $f(j)$ is the (uninterpreted) function symbol that denotes line in the SOLj agent's code

 $Stat_j$ is a well-formed statement in the SOLj specification language

The two latter expressions yield that the logical expression corresponding to a SOLj network will be

$$\Psi_{SN} \triangleq \overset{n}{\underset{i=1}{\wedge}} \overset{m}{\underset{j=1}{\wedge}} f_i(j) = Stat_j$$

The latter expression does not imply that all SOLj agents have the same number (i.e., m) of statements. We can freely assume that m is just the number that denotes the longest SOLj module in the system.

33.5.3 Program Transformers

In Ref. [6], the author defines the notion of *predicate transformer*; it can be defined as a first-order logic formula that yields (via *existential quantifier elimination*) the weakest precondition for a given command (a statement in an imperative language) and its corresponding postcondition. Inspired by the latter notion, we have found a way to logically express program transformation using what we call *program transformers*.

 Let Ψ_{SN} denote the conjunction of formulas shown in Section 33.5.2. Furthermore, we choose to model possible program transformations using the following formula:

$$
\begin{aligned}
(\forall u \in U)(u \neq u_i) \wedge (u \neq u_j) \quad &\rightarrow \quad (f(u) = f'(u)) \\
&\wedge \quad (f(u_i) = f'(u_j)) \\
&\wedge \quad (f(u_j) = f'(u_i))
\end{aligned}
$$

The latter expression denotes the existence of a *new* SOLj module f' that happens to differ from our original SOLj module f only by swapping two statements (i.e., u_i and u_j). In the rest of the section, we shall use Ψ_S to refer to the implication mentioned earlier. With Ψ_{SN} defined earlier, we can formally express our *program transformer expression* as

$$\Psi_T \triangleq (\exists f)(\exists R)\Psi_{SN} \wedge \Psi_S$$

Notice that Ψ_T is a *second-order logic* formula since we are quantifying over one function symbol and one relation symbol. Also, let us assume that we have a fixed SOLj module f_0. By definition, f_0 is a finite list of statements in *Stat*. Thence, we can logically express $f_0: U \to Stat$ by a finite conjunction of equalities as it is expressed in the following:

$$\Phi_f \triangleq \overset{1}{\underset{i \in U}{\wedge}} \wedge f_0(i) = Stat_i$$

where
  Φ_f is the logical formula that represents f_0
  $Stat_i$ is a well-formed statement in SOLj

Therefore, we can "apply" Ψ_T to Φ_f by joining them by conjunction that produces the following formula:

$$\Psi_T \wedge \Phi_f \triangleq (\exists f)(\exists R)\Psi_P \wedge \Psi_S \wedge \Phi_f$$

Notice that we can only eliminate the quantifier f in the aforementioned formula. As a result, we obtain a logical expression that denotes f_0's new version (f'). Also, since we cannot eliminate the quantified variable R (an n-ary relational symbol) because elimination of relation symbols is still an open problem in second-order logic. Therefore, the resulting formula for f' is relatively more complex compared to the one for f_0.

Also, do notice that, we may only eliminate the quantified f in the latter formula. The resulting formula will be the logical expression that denotes our new version of f_0, namely, f', however, notice that, since we cannot eliminate the quantified variable R (as elimination of n-ary quantified relation symbols is an open problem), the resulting expression is not quite the corresponding formula for f' in the same manner Φ_f denoted f_0.

33.5.4 Kripke Versioning Model

Lewis [11] defines the semantics of propositional counterfactual logic using a multiple-world-based model. Therefore, we have opted to use a similar model to interpret our program transformations. In this instance, each version of the program will represent a possible world. Once we apply a *program transformer* (as it was defined in the last section), we produce the formula for a new version, that is, the transformation establishes the relationship between two possible worlds in our model (the two program versions). In what follows, we take the liberty of writing $s_i \leftarrow s_j$ to represent the fact that the statement s_i was swapped by statement s_j. The following definition provide a formal interpretation of our counterfactual logic.

Definition 33.1 (Kripke Version Model)

A Kripke Version Model $\mathcal{R}$ is a triple $\langle \mathcal{P}, \Rightarrow, P_0 \rangle$ where

  1. *$\mathcal{P} = \{P_k\}_{k \in \mathbb{N}}$ is the set of all n-line programs that are the different program versions.*
  2. *P_0 is the initial program.*

3. $\Rightarrow \subseteq \mathcal{P} \times \mathcal{P}$ *is a binary relation defined the set of all possible program versions, where $\Rightarrow$ is the smallest relation such that the following properties hold:*

   a. $s_i \leftarrow s_i \triangleq$ *statement s_i is left unchanged. This stands for the do-nothing transformation.*

   b. $s_i \leftarrow s_j \triangleq$ *statement s_j replaces statement s_i, where $s_j \in P_k$. We usually call this primitive transformation, a swap.*

   c. $s_i \leftarrow s_j \triangleq$ *statement s_j replaces statement s_i, where $s_j \notin P_k$. Thence, s_k is a new statement.*

   d. $(\forall i) s_i \in P_k$ *can be changed only once.*

Moreover, we have assumed that the relation $\Rightarrow$ is *reflexive*, *symmetric*, and *transitive*. We justify the latter statement in the following:

1. *Reflexivity*: For any program $P_i \in \mathcal{P}$, it is obvious that the *do-nothing transformation* will yield that any program can be transformed into itself. Therefore, $P_i \Rightarrow P_i$ given that for all $s_j \in P_i$, $P_i = P_i[s_j/s_j]$.

2. *Symmetry*: For any program $P_i, P_j \in \mathcal{P}$, any of the aforementioned transformations can be reversed, and thence, $P_i \Rightarrow P_j$ implies $P_j \Rightarrow P_i$.

3. *Transitivity*: For any program $P_i, P_j, P_k \in \mathcal{P}$, applying two or more transformations to a program will yield intermediate versions; this is equivalent to transforming the initial version by composing the transformations into one. Thus, $P_i \Rightarrow P_j$ and $P_j \Rightarrow P_k$ imply that $P_i \Rightarrow^+ P_k$, where $\Rightarrow^+ \triangleq \Rightarrow \circ \Rightarrow^{n-1}$ and $n > 1$.

33.5.5 Interpreting the Counterfactual Implication

Based on what was stated earlier, the main purpose of the model we have defined is to provide the interpretation for our counterfactual-based *program transformers*. To that end, we let P_0 denote the source code of a given program. Furthermore, we assume we have a counterfactual statement that expressed a change being done to the program as $\phi \square\mapsto \psi$, where

- ϕ stands for assertions regarding P_0 and some transformation $s_i \leftarrow s_j$ that implies that $P_1 = P_0[s_i/s_j]$.
- ψ stands for assertions regarding P_1.

Thus, following the model-theoretical interpretation proposed by Lewis in Ref. [11], our version of the counterfactual implication is interpreted as

$$\mathcal{R} \Vdash \phi \square\mapsto \psi \tag{33.1}$$

where $\mathcal{R}$ denotes our previously defined *Kripke Versioning Model*. Moreover, letting α_i, β_i denote propositional statements about the structure of P_i and P_j, respectively, then, we can state that

$$\phi \triangleq \left(\bigwedge_{i=1}^{n} \alpha_i \right) \wedge (P_i \Rightarrow^+ P_j)$$

$$\psi \triangleq \bigwedge_{j=1}^{m} \beta_j$$

The $\Rightarrow^+$ symbol denotes the positive/transitive closure regarding the relation $\Rightarrow$. Moreover, from an initial version, source-code version P_0, we generate future alternative versions via the application of one ore more transformations. The current version's properties and the changes made to this version imply the desired properties the new version would have. Therefore,

$$\mathcal{R} \Vdash \phi \square\mapsto \psi \triangleq (\exists_{\min} k \in \mathbb{N}) \left(\bigwedge_{i=1}^{n} \alpha_i \right)$$

$$\wedge (P_0 \Rightarrow^k P') \rightarrow (\wedge_{j=1}^{m} \beta_j)$$

The aforementioned expression means that there is a minimal number of transformation steps such that provided that the properties of our initial program $P_0(\dot{1}_{i=1}^n \wedge \alpha_i)$ and the transformation that leads from the initial version to the desired future version are true, then it follows that the desired properties (encoded in our logic) of the future version also hold.

33.5.6 Fragment of Counterfactual Logic

Given that we are striving to provide programmers with a mechanical procedure and a tool whose purpose is to aid in reasoning about the properties that would hold for future versions of a given program, we provide a logical calculus that will enable us to infer such properties in a algorithmic manner. Thus, we provide the set of inference rules taken from Ref. [11] and known as *VC* logic; these rules will help us formalize and realize our approach's proof theoretical fragment:

1. $\dfrac{\vdash \phi \rightarrow \psi, \phi}{\vdash \psi}$ Modus ponens

2. $\dfrac{\vdash \neg\psi \square\!\!\rightarrow \psi}{\vdash \phi \square\!\!\rightarrow \psi}$ Vacuity

3. $\phi \square\!\!\rightarrow \phi$ Reflexivity rule

4. $\dfrac{\phi \square\!\!\rightarrow \psi}{\phi \rightarrow \psi}$ Counterfactual-elimination rule

5. $\dfrac{\vdash (\dot{1}_{i=1}^n \lambda_i) \rightarrow \psi}{\vdash [\dot{1}_{i=1}^n (\phi \square\!\!\rightarrow \lambda_i)] \rightarrow (\phi \square\!\!\rightarrow \psi)}$ Conditional deduction rule

6. $\dfrac{\vdash \square(\phi \rightarrow \psi)}{\vdash \phi \square\!\!\rightarrow \psi}$ Counterfactual necessity theorem

The reader may wonder about the *counterfactual necessity theorem* (CNT). First of all, a simple intuitive argument can be made in order to show why this is a theorem of *VC*. If we recall Lewis's system of spheres [11] and taking into account that *VC* subsumes many of the standard proof-theoretical system of modal logic, then it is easy to see that the strict implication $\Upsilon(\phi \rightarrow \psi)$ will hold for all worlds in the system of spheres. Hence, the counterfactual implication $\phi \square\!\!\rightarrow \psi$ follows immediately by definition.

If the reader is still unconvinced about the theoremhood regarding the CNT, it is suggested he or she skips to the appendix and/or review [9] in which the authors give sufficient arguments about why CNT is a theorem of *VC*.

33.5.7 Temporal Logic Fragment

Since a SOLj network exhibits a concurrent form of computation, we have to employ a more expressive inference system than the one used for sequential computations (i.e., first-order logic). Thence, we need a fragment of temporal logic in order to encode properties of a SOLj network. Such fragment is given in the following*:

1. $\dfrac{\vdash \phi}{\vdash \square\phi}$

2. $\Upsilon\phi \leftrightarrow \phi \wedge \square\Upsilon\phi$

3. $\Upsilon(\phi \rightarrow \psi) \rightarrow (\Upsilon\phi \rightarrow \Upsilon\psi)$

4. $\square(\phi_1 \vee \phi_2) \leftrightarrow \square\phi_1 \vee \square\phi_2$

5. $\dfrac{\vdash \phi_1 \rightarrow \square(\phi_1 \vee \phi_2)}{\vdash \phi_1 \rightarrow \phi_1 \mathcal{U}\phi_2}$

---

* ϕ ranges over all the well-formed first-order logic expressions.

33.6 Counterfactual Example: Autoregulated Power Generation Network

Using the agent types defined in Section 33.3, we will briefly show how to apply our counterfactual-based approach to a practical situation.

33.6.1 Formal Characterization of a SOLj Network's Behavior

As it has been stated before, a SOLj network is a *reactive system* and thus it continually responds to different stimuli. In Ref. [15], we used counterfactual logic to assert properties of sequential programs. However, in this instance, we are dealing with a perpetually functioning reactive system, and, thus, we need to integrate *temporal logic* constructs into our logic.

In the most simple case, a run in our system is just an infinite sequence of the basic actions (function/service calls) performed by the network agents. Therefore, the following expression denotes a run of the system we specified in Section 33.4:

$$\rho_1 \triangleq [(oC \wedge oR \wedge oV) \wedge (rD \wedge \neg rU \vee rU \wedge \neg rD)]$$

where *oC*, *oR*, and *oV* denote the calls to the `outputCurrent`, `outputResistance`, and `outputVoltage` update functions, respectively. While *ru* and *rD* denote calls to the `revUpEngine` and `revDownEngine` services. Thus, in terms of temporal logic, the run-time behavior exhibited by the agents defined in Figures 33.1 and 33.2 is

$$\Box \rho_1$$

which is interpreted as an infinite sequence of calls to the services shown in ρ_1.

33.6.2 Source Code Change

Let us assume that there has been a change in the requirements of the system, which imply that the `genAgent` modules need only to augment the generators' revolution when the current voltage exceeds certain threshold and decrease the revolutions when the current goes below certain threshold. In terms of our logic, the actual state of the `genAgent` module is given by the following list:

- $f(1) =$ "`if (@C(currVoltage) && @C(currCurrent))`"
- $f(2) =$ "`return EGenerator:RevUpEngine(revDelta)^revUpCont`"
- $f(3) =$ "`else`"
- $f(4) =$ "`return EGenerator:RevDownEngine(revDelta)^revDownCont`"

Our intended change would alter the latter formulas into the ones shown in the following list:

- $f'(1) =$ "`if (@C(currVoltage))`"
- $f'(2) =$ "`return EGenerator:RevUpEngine(revDelta)^revUpCont`"
- $f'(3) =$ "`if (@C(currCurrent))`"
- $f'(4) =$ "`return EGenerator:RevDownEngine(revDelta)^revDownCont`"

The desired basic behavior for this new version of the SOLj agent network is given by the following expression:

$$\rho_2 \triangleq [(oC \wedge oR \wedge oV) \wedge (rD \vee rU)]$$

Let Ψ_{SN} represent the current network structure in the same manner as the formulas defined in Section 33.5.2. Also, let $\mathrm{swap}_{1,3}^{SN}$ denote the following formula:

$$
\begin{aligned}
(u \neq 2) \wedge (u \neq 4) \quad &\rightarrow \quad (f(u) = f'(u)) \\
&\wedge \quad (f(1) = f'(3)) \\
&\wedge \quad (f(3) = f'(1))
\end{aligned}
$$

33.6.3 Proof

In this section, we show the counterfactual implication statement that encodes the desired program transformation and its effects. Let Ψ_{SN}^1 and Ψ_{SN}^2, respectively, denote the logic expressions for the current and desired configurations of the SOLj network. Furthermore, let ρ_1 and ρ_2 denote the operation sequences that, respectively, correspond to Ψ_{SN}^1 and Ψ_{SN}^2. Moreover, let $\mathrm{swap}_{1,3}^{SN}$ be defined as before (i.e., the formula that encodes the change between Ψ_{SN}^1 and Ψ_{SN}^2). In an intuitive way, our claim may be formulated as follows:

Given the initial structure of the system $\left(\Psi_1^{SN}\right)$ and the desired change $\left(\mathrm{swap}_{1,3}^{SN}\right)$, then the system would eventually transition from the initial behavior (denoted by $\Upsilon\rho_1$) to the desired behavior (denoted by $\Upsilon\rho_2$.)

Hence, using the already known fragments of counterfactual and temporal logics, we can symbolize the latter assertion as follows:

$$
\Psi_{SN}^1 \wedge \mathrm{swap}_{1,3}^{SN} \Box \!\!\rightarrow (\Box\rho_1 \mathcal{U} \Box\rho_2)
$$

SOLj network axioms: We need to assume several facts in order to derive the desired counterfactual proof. First of all, we readily assume that the formulas that denote each version configuration (i.e., Ψ_{SN}^1 and Ψ_{SN}^2) materially imply their respective behaviors. Thence,

1. $\Psi_{SN}^1 \rightarrow \Box\rho_1$
2. $\Psi_{SN}^2 \rightarrow \Box\rho_2$

Also, by definition of Υ in the fragment of temporal logic we are using, we can assert

1. $\Psi_{SN}^1 \rightarrow \rho_1 \wedge \bigcirc\Box\rho_1$
2. $\Psi_{SN}^2 \rightarrow \rho_2 \wedge \bigcirc\Box\rho_2$

Using propositional logic, we can *weaken the consequent* on both implications, and, hence, we have

1. $\Psi_{SN}^1 \rightarrow \bigcirc\Box\rho_1$
2. $\Psi_{SN}^2 \rightarrow \bigcirc\Box\rho_2$

Temporal logic proof: The following step requires us to recall the definition of both ρ_1 and ρ_2:

$$
\rho_1 \triangleq [(oC \wedge oR \wedge oV) \wedge (rD \wedge \neg rU \vee rU \wedge \neg rD)]
$$

$$
\rho_2 \triangleq [(oC \wedge oR \wedge oV) \wedge (rD \vee rU)]
$$

By simple propositional logic, we can assert that $\rho_1 \rightarrow \rho_2$. Since temporal logic allows us to use the *necessitation rule*, then we know $\Upsilon(\rho_1 \rightarrow \rho_2)$, which, in turn, entails $\Upsilon\rho_1 \rightarrow \Upsilon\rho_2$. Since we already know that $\Psi_{SN}^1 \rightarrow \Box\rho_1$, therefore, by transitivity of material implication, we can state that $\Psi_{SN}^1 \rightarrow \Box\rho_2$. Again, taking into account the definition of Υ and by *weakening the consequent*, we can assert

1. $\Psi_{SN}^1 \rightarrow \bigcirc\Box\rho_1$
2. $\Psi_{SN}^1 \rightarrow \bigcirc\Box\rho_2$

Notice that it is the case that ρ_1 implies ρ_2; however, they are not equivalent. Hence, the system could exhibit either of the two behaviors. Thence, given the latter implications and by weakening and strengthening the consequent and antecedent, respectively, we can assert the following:

$$\Psi_{SN}^1 \wedge \text{swap}_{1,3}^{SN} \rightarrow \bigcirc \square \rho_1 \vee \bigcirc \square \rho_2$$

Strengthening the antecedent in the latter step may seem arbitrary, but regarding our *program transformers*, it means that we are applying the code transformation to the first version of the SOLj Network. The temporal logic fragment we are employing allows us to assert $\Psi_{SN}^1 \rightarrow \bigcirc(\square \rho_1 \vee \square \rho_2)$. Based on the last step and using temporal logic again, we can assert what follows:

$$\Psi_{SN}^1 \wedge \text{swap}_{1,3}^{SN} \rightarrow \square \rho_1 \mathcal{U} \square \rho_2$$

Counterfactual proof: Temporal logic allows us to use the *necessity rule* and thus assert the following:

$$\square[\Psi_{SN}^1 \wedge \text{swap}_{1,3}^{SN} \rightarrow (\square \rho_1 \mathcal{U} \square \rho_2)]$$

Lastly, we employ the CNT from our counterfactual logic fragment and conclude that

$$\Psi_{SN}^1 \wedge \text{swap}_{1,3}^{SN} \square \rightarrow (\square \rho_1 \mathcal{U} \square \rho_2)$$

33.7 Concluding Remarks

We have introduced a framework for guaranteeing safe source-code changes regarding reactive systems (i.e., sensor networks). The principles exhibited by this framework are realized by two pieces of technology. The first is SOLj, which is a domain-centric language based in the SCR project. This language lets us precisely specify the desired behavior the reactive agents will exhibit during execution. In a nutshell, the SOLj language allows us to provide the functional specification regarding a given reactive system. Moreover, we are able to readily compile the agents' SOLj code into standard JAVA source code. The next piece of technology that allows us to realize our approach is known as SINS. SINS provides a run-time environment that enables a set of SOLj agents to run in a manner that does not compromise system's integrity (i.e., SINS helps us realize nonfunctional requirements as general security, access policies, etc.).

Also, we have shown that our counterfactual verification approach (defined for sequential cases in Ref. [15]) can also be used in simple reactive systems. We defined the language for counterfactual logic, its syntax and semantics via *program transformers* formulas and *Kripke structures*, respectively. In this instance, we were required to augment our logical language with a fragment of temporal logic (as shown in Ref. [17]). The reason for the latter is that reactive systems exhibit properties that are expressed in terms of *safety* and *liveness*, which can be conveniently expressed using temporal logic. We first conjectured and then showed that the desired change and the current structure of the SOLj code counterfactually implied the desired change by encoding it as a sentence in our logical language and then proving it was a theorem in the logic.

Appendix 33.A: Theoremhood of CNT within VC

Earlier in Section 33.6.3, we established how our claim was a theorem of our proposed logic. In this section, we used a theorem (CNT) whose proof was not given immediately. The purpose of this section is to serve as a complement to the simple model-theoretical justification we used. In what follows, we provide a twofold argument of why CNT is a theorem. Section 33.A.1 redefines the notion of *necessity* (in our case, the global temporal operator ϓ) in terms of the counterfactual conditional. Section 33.A.2 establishes the formal proof of CNT's theoremhood.*

33.A.1 Necessity in Counterfactual Logic

In Ref. [9], the authors propose and prove a definition of necessity based on the counterfactual implication. We provide such a definition in the following:

$$\Box\alpha \leftrightarrow (\neg\alpha\Box\!\!\!\rightarrow \bot)$$

Although the justification of such equivalence falls outside of the scope of this section, we can intuitively and readily justify it based on Lewis's model of spheres. A statement is necessary in this model when it cannot counterfactually entail a contradiction. Moreover, if there are no α-worlds in which contradiction holds, then α must be a necessary statement. A more involved explanation and a proof-theoretical justification of this can be found in Ref. [9].

In order to show the theoremhood of CNT, we need to prove an additional auxiliary equivalence:

$$(\neg\alpha\Box\!\!\!\rightarrow \bot) \leftrightarrow (\neg\alpha\Box\!\!\!\rightarrow \alpha)$$

By proving that the latter equivalence holds in *VC*, we would be in position to assert $ϓ\alpha \leftrightarrow (\neg\alpha\Box\!\!\!\rightarrow \alpha)$. Let α and β range over well-formed expressions in temporal logic. Let DWC be an abbreviation of the *Deduction Within Conditional* rule of *VC*. Last but not least, we expand our language's vocabulary with $\bot$ to denote proof-theoretical contradiction. The required proof is given in the following:

1. $(\alpha \wedge \neg\alpha) \rightarrow \bot$ $\bot$ Definition
2. $[(\neg\alpha\Box\!\!\!\rightarrow \alpha) \wedge (\neg\alpha\Box\!\!\!\rightarrow \neg\alpha)] \rightarrow (\neg\alpha\Box\!\!\!\rightarrow \bot)$ DWC in 1
3. $\neg\alpha\Box\!\!\!\rightarrow \neg\alpha$ Reflexivity
4. $\neg\alpha\Box\!\!\!\rightarrow \alpha$ Assumption
5. $(\neg\alpha\Box\!\!\!\rightarrow \neg\alpha) \wedge (\neg\alpha\Box\!\!\!\rightarrow \alpha) \wedge$ Intro in 3 and 4
6. $\neg\alpha\Box\!\!\!\rightarrow \bot$ Modus ponens in 2 and 5
7. $(\neg\alpha\Box\!\!\!\rightarrow \alpha) \rightarrow (\neg\alpha\Box\!\!\!\rightarrow \bot) \rightarrow$ Intro in 4–6
8. $\bot \rightarrow \alpha$ $\bot$ Triviality
9. $(\neg\alpha\Box\!\!\!\rightarrow \bot) \rightarrow (\neg\alpha\Box\!\!\!\rightarrow \alpha)$ DWC in 8
10. $[(\neg\alpha\Box\!\!\!\rightarrow \bot) \rightarrow (\neg\alpha\Box\!\!\!\rightarrow \alpha)] \wedge [(\neg\alpha\Box\!\!\!\rightarrow \alpha) \rightarrow (\neg\alpha\Box\!\!\!\rightarrow \bot)] \wedge$ Intro in 7 and 9
11. $(\neg\alpha\Box\!\!\!\rightarrow \bot) \leftrightarrow (\neg\alpha\Box\!\!\!\rightarrow \alpha)$ $\leftrightarrow$ Intro in 10

33.A.2 Proof of CNT

Using the latter section's results (33.A.1), we give the derivation that shows that CNT is a theorem of *VC*:

1. $\Box(\alpha \rightarrow \beta) \rightarrow [\neg(\alpha \rightarrow \beta)\Box\!\!\!\rightarrow (\alpha \rightarrow \beta)\Box]$ Definition, Elimination
2. $\bot \rightarrow (\alpha \rightarrow \beta)$ $\bot$ Triviality
3. $[\neg(\alpha \rightarrow \beta)\Box\!\!\!\rightarrow \bot] \rightarrow [\neg(\alpha \rightarrow \beta)\Box\!\!\!\rightarrow (\alpha \rightarrow \beta)]$ DWC in 2
4. $[\neg(\alpha \rightarrow \beta)\Box\!\!\!\rightarrow (\alpha \rightarrow \beta)] \rightarrow [\alpha\Box\!\!\!\rightarrow (\alpha \rightarrow \beta)]$ Vacuity
5. $\Box(\alpha \rightarrow \beta) \rightarrow [\alpha\Box\!\!\!\rightarrow (\alpha \rightarrow \beta)]$ Transitivity in 1 and 4
6. $[\alpha \wedge (\alpha \rightarrow \beta)] \rightarrow \beta$ Instantiation of Modus ponens

---

* The proofs given in this section are adaptations of those shown in Ref. [9]. We saw the need of expanding the proofs as we found that they were too succinct.

7. $[(\alpha\Box\!\!\rightarrow\alpha)\wedge(\alpha\Box\!\!\rightarrow(\alpha\rightarrow\beta))]\rightarrow(\alpha\Box\!\!\rightarrow\beta)$ DWC in 6
8. $\Upsilon(\alpha\rightarrow\beta)$ Assumption
9. $[\alpha\Box\!\!\rightarrow(\alpha\rightarrow\beta)]$ Modus ponens in 8 and 5
10. $\alpha\Box\!\!\rightarrow\alpha$ Reflexivity
11. $(\alpha\Box\!\!\rightarrow\alpha)\wedge[\alpha\Box\!\!\rightarrow(\alpha\rightarrow\beta)]\wedge$ Intro in 9 and 10
12. $(\alpha\Box\!\!\rightarrow\beta)$ Modus ponens in 7 and 11
13. $\Box(\alpha\rightarrow\beta)\rightarrow(\alpha\Box\!\!\rightarrow\beta)\rightarrow$ Intro in 8–12

Since CNT is indeed a theorem in *VC*, we can freely use it in our main proof. Additionally, although Hale and Hoffmann [9] use Υ as *metaphysical necessity* or simply alethic necessity, they also assert that any logic that complies with the axioms of system *K* will be able to incorporate this definition of Υ. Since both *VC* and temporal logic subsume *K*, we can assert that the redefinition of this operator based on the counterfactual implication is not conflictive.

References

1. Abdelzaher, T., Blum, B., Cao, Q., Chen, Y., Evans, D., George, J., George, S. et al. EnviroTrack: Towards an environmental computing paradigm for distributed sensor networks. In: *Proceedings of the 24th International Conference on Distributed Computing Systems*, Tokyo, Japan, pp. 582–589, 2004, http://dx.doi.org/10.1109/ICDCS.2004.1281625 (accessed on 23-05-2011).
2. Bharadwaj, R., Mukhopadhyay, S. SOLj: A domain-specific language (DSL) for secure service-based systems, pp. 173–180, March 2007, http://dx.doi.org/10.1109/FTDCS.2007.32 (accessed on 24-05-2011).
3. Chockler, H., Halpern, J.Y., Kupferman, O. What causes a system to satisfy a specification? *ACM Transactions on Computational Logic* 9(3), 1–26, June 2008.
4. Evans, D. Programming the swarm. Technical report, Naval Research Laboratory, Washington, DC, 2000, http://citeseerx.ist.psu.edu/viewdoc/summary?doi=10.1.1.36.6556 (accessed on 23-05-2011).
5. Fisler, K., Krishnamurthi, S., Meyerovich, L.A., Tschantz, M.C. Verification and change-impact analysis of access-control policies. In: *Proceedings of the 27th International Conference on Software Engineering, ICSE'05*, St. Louis, MO, pp. 196–205, ACM, New York, 2005.
6. Gries, D. The predicate transformer wp, *The Science of Programming (Monographs in Computer Science)*, Chapter 7, pp. 108–113. Springer, New York, February 1987.
7. Guo, B., Subramaniam, M. Formal change impact analyses of extended finite state machines using a theorem prover. In: *International Conference on Software Engineering and Formal Methods*, Cape Town, South Africa, pp. 335–344, 2008.
8. Halbwachs, N. Delay analysis in synchronous programs. In: *Computer Aided Verification, Lecture Notes in Computer Science*, Courcoubetis, C. (ed.), Vol. 697, pp. 333–346, Springer, Berlin, Germany, 1993, http://dx.doi.org/10.1007/3-540-56922-7_28 (accessed on 24-05-2011).
9. Hale, B., Hoffmann, A. *Modality: Metaphysics, Logic, and Epistemology*, Chapter 4, pp. 81–96, Oxford University Press, New York, May 2010.
10. Heitmeyer, C.L., Jeffords, R.D., Labaw, B.G. Automated consistency checking of requirements specifications. *ACM Transactions on Software Engineering and Methodology*, 231–261, April 1996.
11. Lewis, D.K. Logics, *Counterfactuals*, 2nd edn., Chapter 6, pp. 118–143, Wiley-Blackwell, Malde, MA, January 2001.
12. Menezes, R., Tolksdorf, R. A new approach to scalable linda-systems based on swarms. In: *Proceedings of the 2003 ACM Symposium on Applied Computing, SAC'03*, pp. 375–379, ACM, New York, Melbourne, Florida, 2003, http://dx.doi.org/10.1145/952532.952607
13. Mitchison, G., Jozsa, R. Counterfactual computation. *Proceedings of the Royal Society of London. Series A: Mathematical, Physical and Engineering Sciences* 457(2009), 1175–1193, May 2001.

14. Pearl, J. *Causality: Models, Reasoning, and Inference.* Cambridge University Press, Cambridge, U.K., March 2000.
15. Peralta, M., Mukhopadhyay, S. Code-change impact analysis using counterfactuals, *Computer Software and Applications Conference COMPSAC 2011, IEEE, 35th Annual,* Munich, Germany, pp. 694–699, 18–22 July 2011. doi: 10.1109/COMPSA C.2011.96.
16. Römer, K., Kasten, O., Mattern, F. Middleware challenges for wireless sensor networks. *SIGMOBILE Mobile Computing and Communications Review* 6(4), 59–61, October 2002, http://dx.doi.org/10.1145/643550.643556
17. Schneider, F.B., Temporal Logics, *On Concurrent Programming (Texts in Computer Science),* 1st edn., Chapter 3, pp. 55–74, Springer, New York, May 1997.
18. Skyrms, B.: Counterfactual definiteness and local causation. *Philosophy of Science* 49(1), 43–50, 1982.
19. Subramaniam, M., Guo, B., Pap, Z. Using change impact analysis to select tests for extended finite state machines. In: *International Conference on Software Engineering and Formal Methods,* Hanoi, Vietnam, pp. 93–102, 2009.
20. Tressler, E. Inter-agent protocol for distributed sol processing. Technical report, Naval Research Laboratory, Washington, DC, 2002.

34

Testing and Debugging Sensor Network Applications

34.1 Introduction 711
34.2 Network Simulators..712
    EmStar • ATEMU • TOSSIM • Avrora •
    DiSenS • SenQ • COOJA
34.3 Network Testbeds 716
    MoteLab • Kansei • Mobile Emulab • NESTbed
34.4 Debugging Systems 719
    Interactive Debugging • Postmortem Debugging •
    Static Analysis Techniques
34.5 Conclusion 723
Glossary 724
References...724

Sally K. Wahba
Clemson University

Jason O. Hallstrom
Clemson University

Nigamanth Sridhar
Cleveland State University

34.1 Introduction

Wireless sensor network technology has fundamentally transformed our capacity to monitor the physical world. Perhaps the most significant result of this transformation has been the discovery of a new frontier of application possibilities. Sensor networks are already changing the way we manage our water, maintain our public infrastructure, and safeguard our borders. This is just the beginning. Sensor network technology has just begun to breach the public market. Exciting opportunities—and applications—lie ahead.

The flip side of this technology's tremendous potential is the extraordinary challenge that it presents for application developers. Wireless sensor network development combines some of the most significant challenges in distributed computing and embedded system design. The applications are inherently parallel and distributed, executing on hardware platforms that are severely resource-constrained in memory, bandwidth, and power. At the same time, the demand for instrumentation density has resulted in inexpensive platforms that are relatively fragile—increasing the likelihood of both transient and persistent network faults.

These development challenges have propelled research into new testing and debugging tools designed specifically for wireless sensor network systems. Indeed, the development of such tools has been a major research emphasis in the sensor network community for over a decade. In this chapter, we survey the state of the art, giving consideration to the three main tool categories that have emerged. In increasing order of fidelity, we consider simulation systems, testbed systems, and in situ debugging systems.

34.2 Network Simulators

Network simulation has been used as a testing tool in networked and distributed systems for many years. The use of network simulators in testing sensor network applications is only natural. The primary goal of simulation is to provide developers with an inexpensive means of rapidly testing their applications before deployment. Simulators provide a controlled environment that allows developers to repeatedly evaluate their applications. Simulators also scale; they are not limited by physical deployment constraints. Users can test their applications with many sensor nodes.

Simulators vary in fidelity from high-level network simulators to cycle-accurate simulators. Some may incorporate physical radios to test node communication. Here we discuss some of the most representative efforts in sensor network simulation.

34.2.1 EmStar

One of the earliest simulators for wireless sensor networks is *EmSim*, part of the *EmStar* suite [19]. EmStar is a software environment for developing wireless sensor networks that run on 32 bit microservers. At the heart of EmStar lies a microkernel extension to Linux. The remainder consists of supporting libraries, services, and tools. The libraries allow users to interact with the microkernel. The services provide time synchronization, networking, and sensing capabilities. The tools support emulation, simulation, and visualization. EmSim, EmCee, and EmView are the supporting simulator, emulator, and visualizer, respectively. Native code for the target processor is simulated on virtual nodes running in parallel on the same machine.

EmSim is based on a hybrid approach that uses both simulation and emulation, using EmCee. As input, EmSim takes a configuration file that includes node IDs, network positions, and other parameters. During simulation, EmSim allows users to change the node configurations, including device position to study the effect of these changes on the simulation. To provide more accurate simulation of the radio, EmCee can be used to substitute simulated radio models with real radio channels on two testbeds. The first testbed is composed of 55 Mica2 [32] motes attached to a ceiling in a 4 ft grid. The second is a portable testbed composed of 16 Mica2 motes. While EmCee yields more accurate experimentation over radio links, the experiments are not reproducible.

When first developed, EmStar supported only applications developed in its software environment. Accordingly, *EmTOS* [20] was developed to extend EmStar and include heterogeneous systems composed of both motes and microservers. EmTOS supports the simulation of both nesC [18] and EmStar applications. It provides a wrapper library that encapsulates nesC applications, so they can utilize services of the EmStar system. When using EmTOS, each node has a simulation configuration file, defining node location and orientation, the radio channels used, and the software to execute on the node. During execution, EmTOS logs messages from both motes and microservers for later visualization and analysis. Data collected during simulation can then be analyzed and visualized by EmView.

EmSim provides several benefits to its users. It allows users to simulate applications running on both motes and microservers. Further, it allows developers to use actual radio models for more accurate experimentation. Finally, application code does not need to be changed when moving from simulation to actual deployment.

34.2.2 ATEMU

ATEMU [39] is a cycle-accurate simulator for Mica2 nodes that bridges the gap between network deployment and simulation. Similar to EmSim, ATEMU is based on a hybrid approach. Low-level operations of the processor, timers, and radio are emulated, while the wireless medium is simulated. ATEMU uses an XML configuration file as input. This file includes specifications of each sensor node, including hardware configuration (e.g., memory size, IO devices, and CPU), executable binary image, ID, and any events that could occur during simulation (e.g., significantly increasing a light sensor's reading at a certain time).

ATEMU initializes the simulated nodes accordingly. During each simulation cycle, ATEMU advances the clock of each node by one cycle and executes an instruction or polls an interrupt. Simulation continues this way until the user interrupts it, or a debugging breakpoint or watchpoint is reached. Breakpoints and watchpoints are available through XATDB, a graphical front-end and debugger. XATDB allows users to step through assembly- or C-level instructions. It also displays the values of registers, stack pointers, and local and global variables.

Users may choose ATEMU for simulating their applications because of the debugging features provided by XATDB. Further, application code does not need to be modified from simulation to deployment. However, since the simulation of ATEMU is cycle accurate, it imposes significant processing overhead, which hinders its scalability.

34.2.3 TOSSIM

TOSSIM [30,49] is a simulator for TinyOS [18,22,48] applications that overcomes the scalability limitations of ATEMU. TOSSIM has four main features. First, it scales to thousands of nodes. Second, it simulates the reactive nature of TinyOS applications. Third, it is relatively accurate in simulating node communications. Finally, it bridges the gap between algorithm simulation and implementation by simulating actual nesC code that can be deployed without modification.

TOSSIM consists of five parts. The first is a modified nesC compiler that compiles TinyOS applications into executables that can run on a developer's operating system. The remaining components are used by the generated executable. The second component is an event queue that models hardware interrupts on each node. The third is a library of emulated TinyOS hardware components, such as EEPROM, ADC, and some components of the radio stack. The fourth component includes wireless signal and sensor models. The final component is composed of services that allow external programs to interact with TOSSIM.

TOSSIM provides several features. First, similar to ATEMU, TOSSIM provides debugging features, such as breakpoints and step-through functions. Second, TOSSIM provides radio models with varying accuracy and complexity to suit the user's simulation needs. Third, TOSSIM includes a visualization tool, TinyViz for visualizing, controlling, and analyzing simulations. This includes displaying data contained in radio packets and sensor values. Users can also extend TinyViz by implementing their own visualizations.

Although TOSSIM overcomes the scalability limitations of ATEMU, it does not simulate applications at the low level of ATEMU. Further, it does not accurately simulate power consumption.

PowerTOSSIM [43] extends TOSSIM to overcome the latter limitation. It was the first scalable simulator that provides accurate power consumption estimates. PowerTOSSIM provides a TinyOS module, *PowerState*, that calculates the power consumption of hardware peripherals, such as EEPROM, sensors, LEDs, and the radio. The hardware emulators of TOSSIM are modified to call commands on PowerState to log the power consumption of hardware peripherals. The commands in PowerState are implemented based on accurate measurements (using an oscilloscope) of the actual power consumption of the hardware peripherals.

PowerTOSSIM implements supporting components that measure the power consumption of (1) radio communication, such as sending and receiving messages, (2) EEPROM activities, such as read and write operations, and (3) CPU execution (per instruction). Accurate power consumption estimation of the CPU is the most challenging component of PowerTOSSIM. To calculate CPU power consumption, application binaries are instrumented, and basic blocks are extracted. The basic blocks are then translated to their corresponding assembly instructions. The number of CPU cycles in each block is calculated, and the total cycle count for each simulated mote is derived. PowerTOSSIM overestimates and underestimates certain CPU commands, which makes the accuracy of predicting power consumption vary between 87% and 99.55%.

34.2.4 Avrora

Avrora [47] is a scalable, cycle-accurate simulator for sensor networks. Avrora combines the benefits of ATEMU's accuracy and TOSSIM's scalability; it is 20 times faster than ATEMU and scales up to 10,000 nodes. Similar to ATEMU and TOSSIM, Avrora emulates application binaries. In Avrora, each node is modeled as a Java thread with an event queue; events represent hardware interrupts. After each simulated instruction, the queue is checked to determine if an event should be fired.

When simulating radio communication, Avrora faces two problems. The first is the *sampling problem*, in which a simulated node may sample RSSI before all transmissions influencing RSSI have been sent (via the simulator). The second is the *send–receive problem*, in which a simulated receiver node may execute beyond a received message, while the simulated sender node has not yet transmitted the message. The sampling problem is solved using a wait-for-neighbors approach, in which a global data structure maintains the progress of each node; a node waits to inspect RSSI until the global data structure indicates that all relevant transmissions have been sent. The send–receive problem is solved using a synchronization interval approach, in which each node blocks after a certain interval and waits for all other nodes to reach the same local simulation time. Specifically, an interval represents the number of cycles it takes to transmit 1 byte (i.e., 3072 cycles). This delay preserves the causality relationship between senders and receivers.

PolarLite [25] improves the speed of Avrora by exploiting radio and MAC characteristics to decrease the number of node synchronizations during simulation. The first speedup is achieved when the radio is off. When a node's radio is off, the earliest possible communication time is predicted. No synchronization is needed until the radio wakes up and can actually be used to send and receive messages. The time it takes for the radio to wake up is larger than the 3072-cycle synchronization interval used by Avrora. Exploiting radio off time results in a speedup of two to three times when simulating networks consisting of up to 256 nodes.

The second speedup is achieved by exploiting the MAC-level backoff time. No messages are sent during a MAC backoff period, hence no synchronization is needed. To determine the backoff time, a probing mechanism based on instruction-level pattern matching is used. This technique exposes the instructions related to MAC-level backoff during simulation. The backoff time is calculated accordingly. Exploiting the MAC backoff time results in a speedup of 1.1–1.3 times when simulating networks consisting of up to 64 nodes.

Despite the achieved speedup, the performance of Avrora can be improved further. *SnapSim* [24] is a distributed cycle-accurate simulator based on Avrora and does not adhere to a fixed synchronization interval. SnapSim surpasses the performance of Avrora by 2–10 times for typical wireless sensor network applications. Similar to PolarLite, the improvement in performance is the result of reduced synchronizations. Specifically, the number of synchronizations is reduced to the number of transmissions in the simulated run. To achieve this, SnapSim relies on optimistic simulation with backtracking and re-execution.

When a simulation begins, each node saves a snapshot (i.e., a clone of the simulation thread) and probes for the next *global nearest transmission time* (i.e., the earliest transmission in the network). Nodes that pass the global nearest transmission time backtrack to the latest snapshot and re-execute until they reach the new nearest transmission time. After reaching the global nearest transmission time, nodes update their snapshots and continue probing for the next global nearest transmission time.

This algorithm improves simulation speed without affecting accuracy. Specifically, SnapSim improves the performance of Avrora up to 4000 times for applications with low radio utilization and between 2 and 10 times for typical network-centric applications.

34.2.5 DiSenS

DiSenS [51] is another distributed, cycle-accurate simulator. It is more scalable than TOSSIM and faster than Avrora, capable of simulating hundreds of nodes in real time, and thousands of nodes a bit slower. The distributed nature of DiSenS is different from that of Avrora. While nodes in Avrora are simulated as Java threads, in DiSenS, simulation is distributed over multiple clusters, and each cluster is capable of simulating multiple nodes. Node communication is simulated by intercluster communication. DiSenS uses a cycle-accurate hardware emulator to emulate an ATmega128L microcontroller, onboard flash, serial ID chip, radio chips, LEDs, and sensor boards. DiSenS also provides various pluggable radio and power models. This allows users to experiment with various models to achieve the desired simulation speed, scalability, and fidelity.

Similar to any distributed simulator, node synchronization is necessary. To reduce synchronization overhead, DiSenS uses node partitioning for parallel execution. Nodes that are likely to communicate together are grouped in the same cluster to minimize intercluster communication.

34.2.6 SenQ

SenQ [50] extends the QualNet [44] simulator to support the simulation of sensor network applications. The extension is composed of two main modules, a discrete event network simulator (DES) and a sensor node (SN) emulator. The DES includes an event queue, an event scheduler, various event handlers, and a simulation clock. It follows the typical discrete event simulation architecture. The SN emulator includes models for the operating system running on a sensor node, hardware devices, and corresponding drivers (e.g., radio, hardware clock, and ADC). Users can add operating system models. As a result, users can compare application performance as a function of the operating system. To increase simulation fidelity, SenQ includes models of sensing phenomenon, battery power, and clock drift.

Compared to the preceding simulators, SenQ provides models for battery power and clock drift. Further, it allows users to compare application performance while taking into account the effects of the operating system. However, SenQ is limited to simulating a network of 1,000 nodes, unlike Avrora and TOSSIM, which can simulate close to 10,000 nodes.

34.2.7 COOJA

COOJA [37] is a discrete event simulator that simulates a sensor network across multiple levels—network, operating system, and machine code. This, in turn, enables users to tune their algorithms, enhance operating system modules, and improve their implementations.

High-level network simulations are run in Java. Users can add their own radio models for network-level simulations. After simulation is complete, users can port their code to the Contiki operating system [14] to support operating system-level simulation. To achieve this, users provide COOJA with a configuration file describing, for each node, the available memory and supported hardware peripherals. During simulation, COOJA uses plugins that allow direct access to Contiki. This allows users to dynamically interact with the simulation. Users can also influence simulation by altering the configuration file without changing the underlying source code. Machine code instructions are emulated using MSPSim [16], a Java-based emulator for the MSP 430 microcontroller. An emulated node is allowed to run for a certain duration. Emulation results are passed to the simulator to determine the next node to be emulated. COOJA supports the simulation of all three levels at the same time (i.e., different nodes can be simulated at different levels).

COOJA has also been extended to support accurate network-scale power profiling [17]. Contiki's power consumption model and MSPSim are modified to provide more accurate power consumption modeling with minimal overhead. Power consumption is modeled by measuring how long a hardware component spends in various operating modes. The results are multiplied by the premeasured power consumption of the corresponding components/nodes.

To combine the benefits of simulation speed and repeatability with testbed realism, COOJA is extended to support a checkpointing mechanism [38]. The checkpointing mechanism is used to switch application execution between simulation (using COOJA and Contiki) and testbeds (using Tmote Sky motes [34]). Each node's state is maintained in both a testbed and a simulation. After a network checkpoint, the network state can be transferred from simulation to a testbed or vice versa to resume application execution. For example, users may choose to use simulation for debugging and visualization and then switch execution to a testbed when real hardware is used for radio messages.

Checkpointing is accomplished in three steps. First, the network is frozen to ensure that the network state is not affected during checkpointing. Second, a snapshot of the network is recorded. Finally, the network resumes execution. Each node's state consists of the following: (1) the compiled program code, if altered, (2) external flash, if altered, (3) variables in memory, (4) hardware timers, (5) LED state, and (6) radio status. At each checkpoint, a node's state can be altered. For example, faults can be injected to determine how an application behaves in the presence of faults. Further, altering a node's state enables users to debug an application if an error occurs during execution in a testbed. When an error occurs, a checkpoint is applied, the network is switched to a simulator and rolled back, and the simulation may retrigger the error for debugging.

34.3 Network Testbeds

Network simulation has become one of the most important diagnostic and evaluation instruments in the sensor network developer's toolbox. State-of-the-art simulators offer many of the same benefits afforded by traditional desktop debuggers, including rapid deployment, stepwise execution, statement and variable breakpoints, and run-time inspection. Moreover, developers can evaluate their applications on networks containing tens of thousands of simulated nodes—far beyond the scale that is typically possible when using real hardware.

Simulation alone, however, is often insufficient as an experimentation platform. While the fidelity of sensor network simulators continues to improve, there are important physical factors that are yet to be simulated faithfully. In particular, existing sensor network simulation platforms are incapable of capturing the inherent complexity of low-power wireless links, environmental influences (e.g., physical obstructions, temperature-induced clock variability), and low-level hardware dynamics (e.g., some peripheral interactions, interrupt behaviors). Physical experimentation remains a necessity.

Yet, physical experimentation presents obstacles of its own. First, the cost of the hardware presents an economic barrier to large-scale experimentation. Second, application installation and network deployment are time-consuming manual processes that typically preclude rapid experimentation and debugging. Finally, the hardware provides virtually no visibility into the state and behavior of the network, presenting a significant diagnostic and evaluation challenge.

Network testbeds are designed to support physical experimentation while mitigating these three obstacles. In its most common usage, the term *network testbed* refers to a permanently deployed collection of hardware infrastructure designed to support network experimentation, along with software services for managing the use of this hardware and the experiments executed on it. A number of 802.11 testbed efforts have been reported in the literature (e.g., [5,27,35,42]), which is no surprise given the overlap in experimentation challenges these networks present. More recently, testbeds designed specifically to suit the requirements of wireless sensor network debugging and experimentation have been reported. Here, we consider some of the most representative efforts.

34.3.1 MoteLab

Harvard University's *MoteLab* testbed [21,52] was among the first wireless sensor network testbeds described in the literature and remains one of the most well-recognized community resources for evaluating wireless sensor network systems. The hardware infrastructure has evolved over time, beginning with 26 Mica2 sensor nodes distributed across three floors of a campus building. Today, the network consists of 184 TMote Sky nodes, each attached to a TMote Connect [33].* The TMote Connect

---

* The TMote Connect is a rebranded LinkSys NSLU2, originally intended as a network-attached storage hub. The TMote Connect replaces the default LinkSys firmware.

provides TCP/IP connectivity to two attached TMote Sky nodes. The sensor nodes are attached and powered via USB, and the Connect is attached to the local network via a wired Ethernet port. The firmware provides support for remote reprogramming of the connected sensor nodes, as well as bidirectional communication. In the MoteLab design, all of the Connect devices are coordinated through a centralized testbed server.

The MoteLab server provides a web interface for configuring testbed experiments and collecting experimental data. The testbed adopts a batch scheduling design, enabling end users to submit experiments for later execution. Configuring a new experiment involves uploading a set of TinyOS application binaries and mapping those binaries to individual nodes. End users may also upload Java class files responsible for parsing any serial data transmitted by the application(s).* These classes are used to support data logging during the experiment run.

When an experiment is ultimately executed, the server reprograms the network based on the application mapping provided by the user. During execution, all serial traffic is routed back to the testbed server and logged for later download. This is typically used to support the collection of diagnostic and evaluation data. In addition, one sensor node is connected to a network-enabled digital oscilloscope. If the oscilloscope is enabled through the web interface, it is used to collect and log power consumption data at a fixed rate during the run. The testbed also provides services for interacting with individual nodes during execution. Specifically, it provides an open TCP/IP port; when connected, end-user applications receive all transmitted serial data. Similarly, messages may be transmitted to individual nodes through this port.†

MoteLab provides two additional services that are worth noting. First, the testbed server implements a health monitoring service that tracks device programming failures and periodically tests communication with the TMote Connect devices. Sensor nodes that cannot be reached are automatically moved to an offline status. Second, the testbed periodically executes a testbed experiment that measures the wireless connectivity among nodes. These data are used to generate graphical connectivity maps that may be useful in interpreting the results of other network experiments.

As we will see, the MoteLab architecture has served as a model for many of the sensor network testbed systems that have followed.

34.3.2 Kansei

The Ohio State University's *Kansei* testbed was completed shortly after the initial release of MoteLab and designed specifically to support large-scale evaluation of heterogeneous wireless sensor networks. The hardware infrastructure is organized into stationary, portable, and mobile *arrays*. The stationary array consists of 210 Extreme Scale Motes (XSMs) [2], each connected via a custom interface to an independent Stargate computer [10]. In addition, 150 of the Stargate nodes include a USB-attached TMote Sky. This combination enables experimentation over three independent radio interfaces simultaneously (i.e., low-power 916 MHz, IEEE 802.15.4, 802.11(b)). In addition to its own 802.11(b) wireless interface, each Stargate is attached to a PC cluster via Ethernet to support multitier device reprogramming (i.e., TMote Sky, XSM, Stargate) and bidirectional communication. The devices are arranged in a regular grid within a warehouse near Ohio State's campus. The portable arrays are designed to support in situ field experimentation and integration with experiments executed on the stationary array. While the portable arrays are intended to be application specific, Kansei includes a default array of 50 Trio nodes [15], TMote-based devices with an extended sensing suite. Finally, the mobile array consists of five robotic platforms that navigate on a plexiglass surface mounted above the stationary array. Each robot is linked to a Stargate device with coupled XSM and TMote Sky nodes. High-resolution overhead cameras provide tracking and ground truth support.

---

* The class files are generated by the end user using the *message interface generator* utility included with TinyOS.
† These features are implemented using the *Serial Forwarder* application included with TinyOS.

Kansei's software system is architecturally similar to MoteLab, designed around a centralized master controller, back-end configuration and logging database, and web-based front-end for experimentation scheduling and monitoring. It is, however, more advanced in several important dimensions. Most importantly, Kansei supports a much richer experimentation interface. As a heterogeneous testbed, it enables end users to map their application images across the entire device hierarchy, from XSMs and Stargates to Trios and mobile robots. The master controller coordinates with array controllers executing on the Stargate nodes to achieve this. Further, Kansei supports multiphased experimentation. End users can specify a series of experimentation phases, each consisting of a unique set of applications and device mappings.

Another important distinction is that Kansei provides native support for hybrid simulation. First, the infrastructure provides services for injecting sensor data streams at individual nodes within the stationary array. These streams are introduced at the sensor interface, effectively replacing physical sensing elements with virtual sensing elements. These services may be used to inject real field data or synthetic model-based data. Kansei also provides support for *stream scaling*; input streams can be scaled up to fill a desired time or space through temporal and spatial shifts. Second, the testbed supports direct integration with TOSSIM. The infrastructure provides support for placing outcalls from within TOSSIM to the physical testbed (e.g., sampling, communication) and synchronizing the executed physical events with the simulation clock.

Finally, it is worth noting that Kansei improves upon the basic health monitoring features provided by MoteLab. Specifically, Kansei implements the *Chowkidar* health monitoring service [3]. The service operates in the presence of periodic faults; upon termination, it guarantees the identification of node and link failures, as well as node restarts. As testbed infrastructures continue to scale, this type of service will become increasingly important.

34.3.3 Mobile Emulab

Developed several years prior to MoteLab, the *Emulab* environment developed at the University of Utah remains one of the most comprehensive testbed systems for IP-based network experimentation [53]. The environment provides access to both local and distributed computing resources through a rich set of virtualization services. End users specify the desired network topology (e.g., local-area links, wide-area links), link dynamics (e.g., bandwidth, loss rate), operating system images, and other parameters via a dialect of the popular *ns* scripting language [31]. The Emulab infrastructure maps the desired topology to physical resources and virtual components, including emulated nodes that realize the desired link characteristics—by, for example, introducing packet delays and/or losses. The infrastructure handles termination detection, policy enforcement, and experiment swapping, including device image restoration.

Proximate to the development of Kansei, Emulab was extended to support networking experiments involving mobile sensor nodes [26]. The extended hardware set includes 25 Mica2 nodes mounted along the walls and ceiling of an indoor laboratory, each connected to a serial bridge to support device reprogramming and bidirectional communication. The mobility platform is similar to Kansei's mobile array; it includes six mobile robots with onboard Stargate computers, each attached to a Mica2 node. Six high-resolution cameras are mounted overhead to provide tracking support. In addition to the features provided by way of integration with Emulab (e.g., device reprogramming, automated logging, device communication), the mobility extension supports robust path programming. End users specify the requested device paths as part of their configuration script. The overhead cameras are used to support device localization. The resulting coordinates are provided as feedback to a path planning service executed on the mobile Stargate computers. The result is a robust and adaptive path planning service.

While Emulab remains one of the most widely used environments for IP-based experimentation, the current status of the sensing and mobility enhancements is unclear.

34.3.4 NESTbed

Clemson University's *Network Embedded Sensor Testbed* (NESTbed) [12,13], a relatively recent effort, is distinguished from the preceding systems by its source-centric design and emphasis on interactivity. The testbed abandons batch-style scheduling in favor of a single-user, exclusive access model. The architecture is service-oriented, providing a rich set of Java RMI services [36] for interacting with the fixed network infrastructure. The intent is to provide an abstraction that presents the sensor network as a single virtual device. The supporting hardware consists of 80 wall-mounted TMote Sky nodes arranged in a regular grid within a laboratory on Clemson's campus. Each device is connected via USB to a centralized server through a series of power-controlled hubs. As in the previous architectures, the USB links are used to provide power and enable device reprogramming and bidirectional communication.

The service-oriented design enables the NESTbed system to be accessed through multiple front-ends, including experimentation-specific interfaces. Two default interfaces are included as part of the standard software distribution. The first is a graphical user interface designed to provide "point-and-click" access to testbed services. For example, individual applications are mapped to target devices by "dragging" the corresponding application images onto the appropriate device icons. The second is a shell interface that provides Linux-like shell commands for interacting with the testbed; hardware and project settings are represented using a file system abstraction. Beyond the obvious interface differences between the two front-ends, the shell tool provides a Linux-like scripting interface for automating experimentation and data collection tasks.

The NESTbed system provides features analogous to those offered by other sensor network testbeds, including device reprogramming, data logging, and TCP/IP-based message injection and reception. A key difference, however, is that end users upload application *source* materials rather than binaries. While the difference may seem minor, the source-centric design enables a new set of testbed services tailored for interactive debugging and experimentation. When an application is uploaded by an end user, the source materials are parsed by the testbed server to identify the constituent program modules, variables, functions, and message structures. The resulting metadata is stored as part of the user's project to enable drill-down navigation through the source materials, as well as configuration of the testbed's data inspection services. The source materials are then instrumented to support interaction with the testbed server. Specifically, the NESTbed weaves in source materials to support device configuration (e.g., reset state, radio power settings) and run-time variable access. The latter functions enable end users to query the values of program variables at runtime or modify their values to inject state faults. Similar features are provided for the message structures identified during the application parse. In this way, the NESTbed provides deep source-level access to the devices in the test network at runtime, providing many of the same functions as modern desktop debugging environments.

The NESTbed visualization enhancement [11] is a more complex example of the functionality benefits afforded by the testbed's source-centric design. The enhancement enables end users to select a set of program modules and functions to be inspected by the testbed server. The selected set is then instrumented to capture a record of the application's run-time invocation sequence, as well as any messages transmitted among nodes. When the user terminates the experiment, the resulting trace data is collected by the testbed server and used to generate a series of UML sequence diagrams that capture the control flow within and among nodes. The extension provides low-level visibility into the run-time behavior of sensor network applications.

The NESTbed's source-centric, interactive design enables a host of debugging and experimentation services that naturally complement the services provided by traditional testbed architectures.

34.4 Debugging Systems

One of the most significant advantages of testing sensor network software on a testbed is that developers get to see how their software behaves on real hardware. The process of moving from network simulators to network testbeds for testing sensor network software is akin to peeling an onion—more details about the behavior of the system are revealed. As more details are revealed, more sources of faults are also revealed. For a developer, it is essential to identify these sources of faults at each level and find ways of addressing these sources in such a way that when the system is deployed in the field, the likelihood of the faults occurring is minimized.

Traditional debugging systems, for example, on a PC, are typically *source-level debuggers*. The developer writes the source code and steps through the execution of the code. These debugging sessions can be interactive, where the application reaches a certain point in its execution (breakpoint) and stops temporarily, giving the developer a chance to explore the state of the system at that time. This kind of debugging, while effective for single threads of execution, does not translate directly to debugging distributed systems. Over the years, a number of different approaches to debugging distributed sensor network systems have been proposed.

At a high level, debugging systems for distributed sensor network systems can be classified into three categories. The first category includes systems that aim to provide a debugging experience that is similar to the experience of debugging software on a PC. These *interactive debugging* systems typically have a way of introducing additional code into the application that can then facilitate interrogation of the application at runtime. The second category includes systems that support *postmortem debugging*. In these systems, execution traces and/or statistics are collected during an application's execution and are analyzed with the aim of identifying bad behavior. The final category includes systems that depend on *static analysis techniques* for source-code instrumentation. The instrumentation introduced into the application typically monitors execution and prevents the application from entering unsafe states. In the remainder of this section, we review salient points about each of these categories of debugging systems and discuss representative systems.

34.4.1 Interactive Debugging

A key function that any interactive debugging system must provide is the ability for a developer to query the state of a running application. The developer must be able to establish points during the program's execution and, when such a point is reached, be able to interrogate the state of the application. One of the earliest approaches that enabled such an ability is the *Marionette* tool suite [54]. Marionette is essentially an embedded remote procedure call (RPC) system that can interact with applications implemented in nesC running the TinyOS operating system. The system is based on a fat client/thin server architecture, where the sensor node acts as the server and the developer's PC acts as the client.

Marionette allows a developer to open an interactive session with one or more sensor nodes in the network. Using a Python application on the PC, the developer can interact with the nesC application running on the sensor node. The primary mode of interaction involves the operations *peek* and *poke*. The *peek* operation allows the developer to query the current value of a given variable in the program. The *poke* operation allows the developer to set the value of a given variable. These operations are useful for the developer to see the effect of state changes on the entire application. The *peek* and *poke* operations are implemented using embedded RPC. The developer also has the ability to send function invocations to the running application.

Marionette does not require the application code to be changed. The tool suite, during the compilation process, creates an XML file that contains the name, type, location information of each variable in the program, as well as information about each function. This XML file is used by the Python client on the PC (PyTOS) to pick out specific variables and functions to query.

While Marionette provides visibility into the application's execution, this visibility is limited to the execution of the program on one node at a time. Even if the developer is interacting with more than a single node in a single session, the nodes themselves act independently, and the interactions between them are not captured. As such, Marionette is not a complete solution to the distributed debugging problem in sensor networks.

Marionette served as a precursor for the *Clairvoyant* system [55]. Clairvoyant is a more complete debugging solution aimed at networks of sensor nodes. As a source-level debugger, Clairvoyant allows developers to set breakpoints in the code just like they would in a regular sequential program. The developer is presented with an interface similar to that of the GNU Debugger (*gdb*) [46]. However, it is different from *gdb* in that the developer has a view of the entire network and is able to issue commands to the network at large.

This system is different from Marionette in that the developer can issue commands to the network that also capture interactions among nodes. These global commands are sent to all nodes via the broadcast medium. For example, the *gstop* (for global stop) command can be issued to pause execution of all nodes in the network. At this point, all nodes in the network are paused, and the developer can interact with any or all of the nodes in the network to query application state.

Clairvoyant also supports hardware access commands that allow developers to debug the impact of interrupts on applications. This is of particular importance for sensor networks because errors and faults caused by interrupts are often nearly impossible to predict and plan for. In addition to being able to query current variable values, Clairvoyant provides access to arbitrary memory locations both in RAM and the external flash. (Since sensor nodes typically have limited amounts of RAM, it is not uncommon for applications to depend on external flash.)

Finally, in addition to breakpoints, Clairvoyant supports the concept of *logpoints*. Every time a variable is accessed, these accesses can be logged. These logpoints are useful in identifying memory corruption errors and stack overflow errors. The lack of a full memory management unit (MMU), which is common in PC systems, makes this logging service a significant addition.

34.4.2 Postmortem Debugging

Most problems in sensor networks have to do with interactions among nodes in the network. A number of systems have been proposed that trace these interactions and identify potential faults in the network. While these systems allow developers to monitor events during execution, much of their utility is in examining these event traces for ex-post facto analysis.

The *Sympathy* system [40,41] is one of the earliest. Sympathy is targeted almost exclusively at sink-based sensor networks in which nodes in the network acquire sensor data and send the data to a base station via multi-hop routing. In addition, these applications generally have a known period at which each node reports data to the network. Based on this knowledge, Sympathy analyzes the amount of data arriving at the sink from each portion of the network.

When Sympathy examines packet metrics and determines that one or more nodes in a section of the network is not reporting the quantity of data expected, a root cause analysis is initiated. Based on an analysis of the packet statistics, Sympathy identifies the root cause of the fault as being one of seven kinds: (1) node crash, (2) node reboot, (3) no neighbors, (4) no route, (5) bad path to node, (6) bad node transmit, and (7) bad path to sink. The root cause diagnosis and packet metrics are reported to the system administrator. The system itself does not take any corrective action.

Another system that works on collected event traces is *DustMiner* [28]. This system collects traces that are more detailed than the simple packet metrics that Sympathy collects. Specifically, DustMiner collects complete traces of all events that occur on each sensor node. The system is primarily designed as a debugging tool to verify application correctness in a testbed before deployment. Reporting these traces on the same wireless medium that is used by the application would severely influence application behavior; the amount of data collected is simply too large. The collected event traces are reported to the administration system via a wired backchannel.

DustMiner is a data mining solution for application debugging in sensor networks; the focus is largely on node interactions. The developer encodes the expected behavior of the application in the form of event trace segments. Possible bad behaviors are also encoded as event trace segments. These segments, along with the event traces collected from the network, are supplied as inputs to the a priori data mining algorithm [1]. This mining operation identifies sequences of events occurring in various sections of the network where a fault shows up.

This is a heavyweight solution. However, the solution does have a high likelihood of identifying bugs in network protocols that might prove elusive otherwise. In particular, bugs that may manifest themselves only in rare cases in a real deployment can be identified using this approach. As such, this system is well suited to verifying the correctness of protocols that require high fidelity.

Yet another system proposed as a postmortem debugging solution is the *Macrodebugger (MDB)* [45]. *Macroprogramming* refers to the technique of programming application logic for the network as a whole. This is in contrast to the approach where application behavior is represented as a collection of individual node-level programs. A macroprogram simply describes the high-level actions that the network as a whole performs. The macroprogram compiler is responsible for translating this into chunks of code that each node executes. *MacroLab* [23] is a macroprogramming language for sensor networks that is fashioned after the MATLAB® language. The Macrodebugger system provides debugging support for MacroLab.

MDB is essentially a "record and playback" system. During execution of the program, the system collects event logs. These event logs are then played back during the debugging session. As such, the debugger has access to the complete history, as well as the future of the entire application at any given point in the application's execution. Based on this knowledge, the designers of MDB report the system as supporting "time travel"—the programmer can step forward through the execution of the program, as well as step back to analyze why the application made a certain decision.

34.4.3 Static Analysis Techniques

In addition to the debugging techniques discussed so far, there are other techniques and tools that have been proposed that allow a developer to include checks to safeguard against certain kinds of failures. These techniques use some form of static program analysis in order to identify and/or safeguard portions of the source code that have the potential to fail.

Safe TinyOS [8] is a version of the TinyOS operating system that is aimed at building software that can safeguard against common errors, particularly memory-related errors. The microcontroller platforms that are commonly used in sensor network applications typically have small amounts of memory (e.g., MicaZ—4 kB, TelosB—10 kB) and do not have a MMU. Handling memory is left to the programmer. Safe TinyOS enables *safe execution* by automatically inserting run-time checks in the code that ensure that memory faults do not occur.

Safe TinyOS introduces a set of keywords that a programmer can use to annotate nesC code to enable memory safety. These annotations serve as inputs to *Deputy* [7], a source-to-source translator for C programs that introduces run-time checks in the code based on the annotations. For example, the *SAFE* annotation can be used to mark a pointer as one that should not be used in pointer arithmetic. A common error that shows up in sensor network software has to do with array accesses—the array index that is being accessed may not be valid. Almost always, however, such illegal accesses can be detected by a combination of compile-time and run-time checks. The *COUNT* annotation can be used to inform Deputy to introduce a check to make sure that the array access is using a location that is inside the array, thereby preventing an overflow.

Deputy operates on C code, not nesC code. Therefore, in order to use Safe TinyOS annotations, the programmer needs to use a modified version of the nesC compiler, one that is aware of the new key words that Safe TinyOS introduces. These annotations are preserved by the modified nesC compiler and are still present in the generated C program. The resulting C program is then provided to Deputy as input, which introduces the necessary run-time checks.

If one of the run-time checks does, in fact, identify a memory fault, it is typically not possible for the code to "correct itself." One available option is to reboot the node. However, rebooting a node in the network is an expensive operation. (When the node turns itself off, the rest of the network has to reconfigure itself to account for this departure, and when the node comes back online, a second reconfiguration is needed.) Moreover, if the failure is caused because of a code error, the failure will likely occur again after the reboot anyway. Safe TinyOS includes a mechanism to inform the programmer of the error. During compile time, the tool suite also keeps track of fault locations using *fault location identifiers (FLIDs)*. These FLIDs are kept in a lookup table on the development machine. When a node reports the FLID, the developer can use the *decode_flid* tool to locate the exact location in the source code that caused the memory fault.

The final step in the Safe TinyOS toolchain is whole-program optimization using *cXprop* [9], which reduces program size and run-time overhead. cXprop analyzes the use of pointers in the C program produced by Deputy and removes sections of code that it can show to be unreachable or useless.

Neutron [6] is an improvement to Safe TinyOS, also targeted at helping developers build memory-safe programs. The key improvement is in how the system responds when a memory fault is detected. While Deputy inserts checks in the code, and Safe TinyOS reports these errors using FLIDs, the developer must still get involved in almost every such incident. The only other alternative that Safe TinyOS provides is to reboot the node. In the Neutron system, on the other hand, the application is divided into several independent component pieces called *application recovery units (ARUs)*.

At compile time, the modified nesC compiler divides the application into several ARUs. This is based on an inference algorithm that identifies portions of the application that can be safely packaged into independent components. The delineation and isolation of the application into ARUs ensure that each such unit can be restarted without affecting the rest of the application. The kernel is run in its own ARU. Therefore, most application components can be restarted relatively inexpensively.

NodeMD [29] is a system that works on the Mantis operating system [4] for sensor networks. Mantis is a multithreaded OS for low-power sensor network architectures. NodeMD is a system that aims to "close the loop"—the system includes tools for detecting faults, notifying the developer about the faults, and offering the developer a chance to correct the fault source in an interactive debugging session.

The fault detection system in NodeMD is focused on two kinds of faults. The first kind includes those that result from stack overflow. In an embedded system with no MMU, any function call could potentially cause a stack overflow. The good news is that before a function is called, it is possible to check to see if that call will cause the stack to overflow. This is exactly the approach that NodeMD takes. At compile time, NodeMD calculates how much stack space each function in the application will utilize. Then, at runtime, before each function is called, a check is performed to see if that invocation is safe. The invocation proceeds only if it is safe. The second kind of fault that NodeMD can detect is related to deadlock and livelock. These situations are caused when different parts of a distributed application are waiting for each other. Most sensor network applications have periodic execution cycles. The canonical system design involves some form of a loop that executes periodically. The developer generally has a sense of how long each execution cycle is estimated to take. NodeMD uses these estimates to detect any deadlock or livelock situations. In practical terms, checkpoints are placed at specific locations in the program such that any violations of the execution time estimates can be detected.

When NodeMD detects a fault, a notification is sent to the developer. The notification includes a trace of how the system arrived at this point. Keeping track of *all* events on the node can be prohibitive in terms of memory utilization. Instead, the designers of NodeMD have identified a set of 15 major events that are recorded, including context switches, procedure calls/returns, hardware interrupts, timer events, and thread events.

Once the developer has been notified of a fault, NodeMD switches into an interactive debugging session. The developer can examine the event trace that led to the fault and perform diagnostic tasks. Once the developer has diagnosed the problem and found a way to fix the error, NodeMD allows for new code to be downloaded to the sensor node. Mantis OS allows for dynamic code update, and so the application can make use of this new code in many cases.

34.5 Conclusion

The challenges associated with sensor network development are so widely documented, the phrase "sensor networks are notoriously difficult to program" has nearly reached proverb status in the literature. These challenges have collectively served as one of the most significant research drivers in the community for over a decade. The result has been the development and adoption of innovative new tools tailored for testing and debugging sensor network systems. In this chapter, we surveyed some of the most representative efforts, spanning simulation systems, testbed systems, and in situ debugging systems. As sensor networks continue to make their way into the commercial market, this set can only be expected to grow.

Glossary

Network testbed: Hardware and software infrastructure designed to support network experimentation, including the services used to provision and monitor the hardware.
Network simulator: A software program that models the behavior of a network, usually in a controlled environment.
Network emulator: A software program that imitates the behavior of a network and is usually driven by actual hardware and radio models.

References

1. R. Agrawal and R. Srikant. Fast algorithms for mining association rules in large databases. In *Proceedings of the 20th International Conference on Very Large Data Bases*, Santiago, Chile, pp. 487–499, September 1994, Morgan Kaufmann Publishers Inc., San Francisco, CA.
2. A. Arora, R. Ramnath, E. Ertin, P. Sinha, S. Bapat, V. Naik, V. Kulathumani et al. ExScal: Elements of an extreme scale wireless sensor network. In *Proceedings of the 11th IEEE International Conference on Embedded and Real-Time Computing Systems and Applications*, Hong Kong, P.R. China, pp. 102–108, August 2005, IEEE Computer Society, Washington, DC.
3. S. Bapat, W.M. Leal, T. Kwon, P. Wei, and A. Arora. Chowkidar: Reliable and scalable health monitoring for wireless sensor network testbeds. *ACM Transactions on Autonomous Adaptive Systems*, 4:3:1–3:32, February 2009.
4. S. Bhatti, J. Carlson, H. Dai, J. Deng, J. Rose, A. Sheth, B. Shucker, C. Gruenwald, A. Torgerson, and R. Han. MANTIS OS: An embedded multithreaded operating system for wireless micro sensor platforms. *Mobile Networks and Applications*, 10:563–579, August 2005.
5. B.A. Chambers. The Grid Roofnet: A rooftop ad hoc wireless network. Master's thesis, Massachusetts Institute of Technology, Cambridge, MA, May 2002.
6. Y. Chen, O. Gnawali, M. Kazandjieva, P. Levis, and J. Regehr. Surviving sensor network software faults. In *Proceedings of the ACM SIGOPS 22nd Symposium on Operating Systems Principles*, Big Sky, MT, pp. 235–246, October 2009, ACM, New York.
7. J. Condit, M. Harren, Z. Anderson, D. Gay, and G.C. Necula. Dependent types for low-level programming. In *Proceedings of the 16th European Conference on Programming*, Braga, Portugal, pp. 520–535, March–April 2007, Springer-Verlag, Berlin, Germany.
8. N. Cooprider, W. Archer, E. Eide, D. Gay, and J. Regehr. Efficient memory safety for TinyOS. In *Proceedings of the 5th International Conference on Embedded Networked Sensor Systems*, Sydney, New South Wales, Australia, pp. 205–218, November 2007, ACM, New York.
9. N. Cooprider and J. Regehr. Pluggable abstract domains for analyzing embedded software. *SIGPLAN Notes*, 41:44–53, June 2006.
10. Crossbow Technology Incorporated. Stargate datasheet. www.willow.co.uk/Stargate_Datasheet.pdf, 2003.
11. A.R. Dalton, S. Dandamudi, J.O. Hallstrom, and S.K. Wahba. A testbed for visualizing sensornet behavior. In *Proceedings of the 17th International Conference on Computer Communications and Networks*, St. Thomas, U.S. Virgin Islands, 7pp, August 2008. IEEE Computer Society, Washington, DC.

12. A.R. Dalton and J.O. Hallstrom. A file system abstraction and shell interface for a wireless sensor network testbed. In *Proceedings of the 3rd International IEEE/Create-Net Conference on Testbeds and Research Infrastructures for the Development of Networks and Communities*, Orlando, FL, 10pp, CD–ROM, Washington, DC, May 2007, IEEE Computer Society.

13. A.R. Dalton and J.O. Hallstrom. An interactive, source-centric, open testbed for developing and profiling wireless sensor systems. *International Journal of Distributed Sensor Networks*, 5(2):105–138, 2009.

14. A. Dunkels, B. Grnvall, and T. Voigt. Contiki—A lightweight and flexible operating system for tiny networked sensors. In *Proceedings of the 1st IEEE Workshop on Embedded Networked Sensors*, Tampa, FL, 8pp, November 2004, IEEE, Washington, DC.

15. P. Dutta, J. Hui, J. Jeong, S. Kim, C. Sharp, J. Taneja, G. Tolle, K. Whitehouse, and D. Culler. Trio: Enabling sustainable and scalable outdoor wireless sensor network deployments. In *Proceedings of the 5th International Conference on Information Processing in Sensor Networks*, Nashville, TN, pp. 407–415, April 2006, ACM Press, Washington, DC.

16. J. Eriksson, A. Dunkels, N. Finne, F. Österlind, and T. Voigt. MSPSim—An extensible simulator for MSP430-equipped sensor boards. In *Proceedings of the 4th European Conference on Wireless Sensor Networks*, Delft, The Netherlands, 2pp, January 2007, Springer-Verlag, Berlin, Germany.

17. J. Eriksson, F. Österlind, N. Finne, A. Dunkels, N. Tsiftes, and T. Voigt. Accurate network-scale power profiling for sensor network simulators. In *Proceedings of the 6th European Conference on Wireless Sensor Networks*, Cork, Ireland, pp. 312–326, February 2009. Springer-Verlag, Berlin, Germany.

18. D. Gay, P. Levis, R. von Behren, M. Welsh, E. Brewer, and D.E. Culler. The nesC language: A holistic approach to networked embedded systems. In *Proceedings of the ACM SIGPLAN 2003 Conference on Programming Language Design and Implementation*, San Diego, CA, pp. 1–11, June 2003, ACM Press, New York.

19. L. Girod, N. Ramanathan, J. Elson, T. Stathopoulos, M. Lukac, and D. Estrin. EmStar: A software environment for developing and deploying heterogeneous sensor-actuator networks. *ACM Transactions on Sensor Networks*, 3(3):13, August 2007.

20. L. Girod, T. Stathopoulos, N. Ramanathan, J. Elson, D. Estrin, E. Osterweil, and T. Schoellhammer. A system for simulation, emulation, and deployment of heterogeneous sensor networks. In *Proceedings of the 2nd International Conference on Embedded Networked Sensor Systems*, Baltimore, MD, pp. 201–213, November 2004, ACM, New York.

21. Harvard University. MoteLab: Harvard network sensor testbed. motelab.eecs.harvard.edu (accessed: May 2011).

22. J. Hill, R. Szewczyk, A. Woo, S. Hollar, D.E. Culler, and K. Pister. System architecture directions for networked sensors. In *Proceedings of the 9th International Conference on Architectural Support for Programming Languages and Operating Systems*, Cambridge, MA, pp. 93–104, November 2000, ACM Press, New York.

23. T.W. Hnat, T. Sookoor, P. Hooimeijer, W. Weimer, and K. Whitehouse. MacroLab: A vector-based macroprogramming framework for cyber-physical systems. In *Proceedings of the 6th ACM Conference on Embedded Network Sensor Systems*, Raleigh, NC, pp. 225–238, November 2008, ACM, New York.

24. H. Jiang, J. Zhai, S.K. Wahba, B. Mazumder, and J.O. Hallstrom. Fast distributed simulation of sensor networks using optimistic synchronization. In *Proceedings of the 7th IEEE International Conference on Distributed Computing in Sensor Systems*, Barcelona, Spain, page to appear, pp. 1–8, June 2011, IEEE Computer Society, Washington, DC.

25. Z.Y. Jin and R. Gupta. Improving the speed and scalability of distributed simulations of sensor networks. In *Proceedings of the 8th International Conference on Information Processing in Sensor Networks*, San Francisco, CA, pp. 169–180, April 2009, ACM Press, Washington, DC.

26. D. Johnson, T. Stack, R. Fish, D.M. Flickinger, L. Stoller, R. Ricci, and J. Lepreau. Mobile Emulab: A robotic wireless and sensor network testbed. In *Proceedings of the 25th IEEE International Conference on Computer Communications*, Barcelona, Spain, pp. 1–12, April 2006, IEEE Computer Society, Washington, DC.

27. A. Karygiannis and E. Antonakakis. mLab: A mobile ad hoc network testbed. In *Proceedings of the 1st Workshop on Security, Privacy, and Trust in Pervasive and Ubiquitous Computing*, Santorini Island, Greece, pp. 88–97, July 2005, Diavlos S.A. Athens, Greece.

28. M.M. Khan, H.K. Le, H. Ahmadi, T. Abdelzaher, and J. Han. Dustminer: Troubleshooting interactive complexity bugs in sensor networks. In *Proceedings of the 6th ACM Conference on Embedded Network Sensor Systems*, Raleigh, NC, pp. 99–112, November 2008, ACM, New York.

29. V. Krunic, E. Trumpler, and R. Han. NodeMD: Diagnosing node-level faults in remote wireless sensor systems. In *Proceedings of the 5th International Conference on Mobile Systems, Applications and Services*, San Juan, Puerto Rico, pp. 43–56, June 2007, ACM, New York.

30. P. Levis, N. Lee, M. Welsh, and D.E. Culler. TOSSIM: Accurate and scalable simulation of entire TinyOS applications. In *Proceedings of the 1st ACM Conference on Embedded Networked Sensor Systems*, Los Angeles, CA, pp. 126–137, November 2003, ACM Press, New York.

31. S. McCanne and S. Floyd. Network simulator ns-2. www.isi.edu/ nsnam/ns, 1997.

32. Memsic Corporation. Mica2 datasheet. www.memsic.com/support/documentation/wireless-sensor-networks/category/7-d atasheets.html?download=147%3Amica2 (accessed: May 2011).

33. Moteiv Corporation. Tmote Connect datasheet. sentilla.com/files/pdf/eol/tmote-connect-datasheet.pdf, 2006.

34. Moteiv Corporation. Tmote Sky datasheet. sentilla.com/files/pdf/eol/tmote-sky-datasheet.pdf, 2006.

35. E. Nordström, P. Gunningberg, and H. Lundgren. A testbed and methodology for experimental evaluation of wireless mobile ad hoc networks. In *Proceedings of the 1st International Conference on Testbeds and Research Infrastructures for the Development of Networks and Communities*, Trento, Italy, pp. 100–109, February 2005, IEEE Computer Society, Washington, DC.

36. Oracle. JDK 5.0 Remote Method Invocation (RMI). download.oracle.com/javase/1.5.0/docs/guide/rmi, 2004.

37. F. Österlind, A. Dunkels, J. Eriksson, N. Finne, and T. Voigt. Cross-level sensor network simulation with COOJA. In *Proceedings of the 31st IEEE Conference on Local Computer Networks*, Tampa, FL, pp. 641–648, November 2006, IEEE, Washington, DC.

38. F. Österlind, A. Dunkels, T. Voigt, N. Tsiftes, J. Eriksson, and N. Finne. Sensornet checkpointing: Enabling repeatability in testbeds and realism in simulations. In *Proceedings of the 6th European Conference on Wireless Sensor Networks*, Cork, Ireland, pp. 343–357, February 2009, Springer-Verlag, Berlin, Germany.

39. J. Polley, D. Blazakis, J. McGee, D. Rusk, and J.S. Baras. ATEMU: A fine-grained sensor network simulator. In *Proceedings of the 1st Annual IEEE Communications Society Conference on Sensor and Ad Hoc Communications and Networks*, Santa Clara, CA, pp. 145–152, Washington, DC, October 2004. IEEE Computer Society.

40. N. Ramanathan, K. Chang, R. Kapur, L. Girod, E. Kohler, and D. Estrin. Sympathy for the sensor network debugger. In *Proceedings of the 3rd International Conference on Embedded Networked Sensor Systems*, San Diego, CA, pp. 255–267, November 2005, ACM, New York.

41. N. Ramanathan, E. Kohler, and D. Estrin. Towards a debugging system for sensor networks. *International Journal of Network Management*, 15:223–234, July 2005.

42. D. Raychaudhuri, I. Seskar, M. Ott, S. Ganu, K. Ramachandran, H. Kremo, R. Siracusa, H. Liu, and M. Singh. Overview of the ORBIT radio grid testbed for evaluation of next-generation wireless network protocols. In *Proceedings of the Wireless Communications and Networking Conference*, New Orleans, LA, pp. 1664–1669, March 2005, IEEE Computer Society, Washington, DC.

43. V. Shnayder, M. Hempstead, B. Chen, G.W. Allen, and M. Welsh. Simulating the power consumption of large-scale sensor network applications. In *Proceedings of the 2nd International Conference on Embedded Networked Sensor Systems*, Baltimore, MD, pp. 188–200, November 2004, ACM, New York.
44. Scalable Network Technologies. Qualnet, May 2011. http://www.scalable-networks.com/products/qualnet/, May 2011.
45. T. Sookoor, T. Hnat, P. Hooimeijer, W. Weimer, and K. Whitehouse. Macrodebugging: Global views of distributed program execution. In *Proceedings of the 7th ACM Conference on Embedded Networked Sensor Systems*, Berkeley, CA, pp. 141–154, November 2009, ACM, New York.
46. The GNU Foundation. GDB: The GNU project debugger. http://www.gnu.org/software/gdb/, 2011.
47. B.L. Titzer, D.K. Lee, and J. Palsberg. Avrora: Scalable sensor network simulation with precise timing. In *Proceedings of the 4th International Symposium on Information Processing in Sensor Networks*, Los Angeles, CA, April 2005, ACM Press, Washington, DC.
48. UC Berkeley. TinyOS home page. www.tinyos.net (accessed: May 2011).
49. UC Berkeley. TOSSIM—TinyOS documentation. docs.tinyos.net/index.php/TOSSIM (accessed: May 2011).
50. M. Varshney, D. Xu, M. Srivastava, and R. Bagrodia. SenQ: A scalable simulation and emulation environment for sensor networks. In *Proceedings of the 6th International Conference on Information Processing in Sensor Networks*, Cambridge, MA, pp. 196–205, April 2007, ACM Press, Washington, DC.
51. Y. Wen, R. Wolski, and G. Moore. DiSens: Scalable distributed sensor network simulation. In *Proceedings of the 12th ACM SIGPLAN Symposium on Principles and Practice of Parallel Programming*, San Jose, CA, pp. 24–34, March 2007, ACM, New York.
52. G. Werner-Allen, P. Swieskowski, and M. Welsh. MoteLab: A wireless sensor network testbed. In *Proceedings of the 4th International Conference on Information Processing in Sensor Networks*, Los Angeles, CA, pp. 483–488, April 2005, IEEE Computer Society, Washington, DC.
53. B. White, J. Lepreau, L. Stoller, R. Ricci, S. Guruprasad, M. Newbold, M. Hibler, C. Barb, and A. Joglekar. An integrated experimental environment for distributed systems and networks. In *Proceedings of the 5th Symposium on Operating Systems Design and Implementation*, Boston, MA, pp. 255–270, December 2002, USENIX Association, Berkeley, CA.
54. K. Whitehouse, G. Tolle, J. Taneja, C. Sharp, S. Kim, J. Jeong, J. Hui, P. Dutta, and D. Culler. Marionette: Using RPC for interactive development and debugging of wireless embedded networks. In *Proceedings of the 5th International Conference on Information Processing in Sensor Networks*, Nashville, TN, pp. 416–423, April 2006, ACM, New York.
55. J. Yang, M.L. Soffa, L. Selavo, and K. Whitehouse. Clairvoyant: A comprehensive source-level debugger for wireless sensor networks. In *Proceedings of the 5th International Conference on Embedded Networked Sensor Systems*, Sydney, New South Wales, Australia, pp. 189–203, November 2007, ACM, New York.

Index

A

Acoustic/seismic signals
    ML beamforming, 348–349
    parametric methods, 346–348
    time-delay estimation methods, 352–353
    time-delay-type methods (*see* Time-delay-type
        methods)
ACP, *see* Agent control protocol (ACP)
Ad hoc virtual enterprises, 34
Aeroacoustic sensor networks
    AOA estimation
        AMI algorithm, 265
        CSM-MUSIC approach, 264–265
        performance analysis, 264
        with scattering, 261–262
        wideband beamforming, 264
        without scattering, 262–264
    autoregressive processes, 247
    detection and classification
        harmonic amplitude, 280
        harmonic signature, 280–282
        Neyman–Pearson detection criterion, 283
        noncentral chi-squared distributions, 282, 283
        probability density function, 280, 281
        probability distribution, 280
        saturation variation, 280, 281
        signal saturation effects, 283
        vehicle types identification, 280
    differential Doppler estimation, 247
    distributed sensor arrays
        array of arrays model, 267–269
        Cramér-Rao Bounds, 270–273
        source localization, 266, 267
        subspace algorithms, 266
        threshold coherence, 266
    extinction coefficients
        effective integral length scale, 258
        scattered signals, coherence, 260
        second-moment extinction coefficient, 257, 258
        sensor spacing, 257
        structure-function parameters, 259
        turbulence-induced distortions, 257
        turbulence quantities and inverse extinction
          coefficient, 259
        wavefronts impinging, 257
        zeroth-order Bessel function, 258
    Gaussian random process model, 247
    geometry, 248, 249
    moving sources tracking, 279
    multiple frequencies and sources, 260–261
    narrowband model
        with scattering, 253–256
        without scattering, 250–253
    signal-coherence characteristics, 246
    source signal modification, 247
    TDE
        BAE systems, 277
        CSD matrix, 274
        differential time delays, 279
        partial signal coherence, 273
        PSD and coherence estimation, 277–278
        threshold coherence *vs.* bandwidth, 275–276
        time duration, 276
        Ziv-Zakai bounds, 274
    time-varying factor, 246
    transmission loss, 248
    turbulent scattering, 246
Agent control protocol (ACP), 697
Agile development, 694
AML, *see* Approximated maximum-likelihood (AML)
Analog-to-digital (A/D) converter, 41
Analytic signal space partitioning (ASSP), 308, 310
Angle of arrival (AOA) estimation
    AMI algorithm, 265
    CSM-MUSIC approach, 264–265
    performance analysis, 264
    with scattering, 261–262
    wideband beamforming, 264
    without scattering, 262–264
ANN, *see* Artificial neural networks (ANN)
Application recovery units (ARUs), 723
Approximated maximum-likelihood (AML), 348,
        367, 369

Area-based algorithms
  edge detection, 73–75
  high-pass filter, 73, 74
  low-pass filter, 73–74
  median filter, 73, 74
  morphological operators, 76
  3×3 neighborhood of pixel, 72, 73
  4-neighbors of pixel, 72, 73
Array manifold interpolation (AMI) algorithm, 265
Array system performance analysis
  acoustic sources, computer-simulated results
    AML algorithm, 353, 356
    convex hull of array, 354
    range estimation error, 354–355
    single traveling source scenario, circular array, 355
    two-dimensional polar grid-point system, 355
    wideband MUSIC method, 356
  CRB, source localization (*see* Cramér-Rao bound)
  robust array design
    linear matrix inequality, 364
    MATLAB programming code, 363
    nonrobust conventional array design, 364–365
    SDP, 363–364
    uniformly spaced circle array, 364–365
Artificial neural networks (ANN), 315, 316, 421–422
ARUs, *see* Application recovery units (ARUs)
ASSP, *see* Analytic signal space partitioning (ASSP)
ATEMU simulator, 712–713
Athlon 4 processor, 612
Atmospheric effects
  airlight phenomenon, 221, 222, 227, 230
  atmosphere extinction, 222
  atmosphere luminance, 223
  electromagnetic radiation, 219
  heuristics and non-physics-based solutions
    CIBR, 235
    cloud detection, 235
    dark object subtraction, 236
    digital elevation map, 236
    haze optimization transformation, 236, 237
    haze perfection, 236
    image denoising, 239
    image restoration, 239
    IR and MMW radar images, 238, 239
    local weighted averaging method, 239
    motion detection algorithm, 237
    MTI bands, 235
    multi-image fusion, 237
    satellite imagery, 235–236
    sensor system components, 238
    virtual cloud point removal, 236, 237
    wavelet transform, 237
    whiteness property, 235

  infrared sensors, 224–225
  LADAR sensors, 226
  light attenuation, 220–221
  Mie scattering theory, 220
  MMW radar sensors, 225–226
  multispectral sensors, 226
  physics-based solutions
    aerosols scattering, 228
    dichromatic atmospheric scattering model, 226
    digital elevation model, 231
    fog removal system, 227–229
    fuzzy logic based approach, 234
    geometric-photometric model, 232, 233
    image dehazing, 232
    inverse intensity chromaticity space, 230
    iterative contrast restoration scheme, 229
    raindrops detection, 232, 233
    satellite imagery, 231, 232
    single-image systems, 231
    vehicle detection algorithm, 230
    wavelength-based equation, 227
    YIQ transform, 230
  sonar sensors, 226
  underwater environments, 223
  visible spectrum camera, 219, 220, 224
Automation systems
  characteristics, 378–379
  data fusion
    advantages, 380–381
    definition, 377
    goal-seeking paradigm, 382–383
    state-transition paradigm, 382
    system structure representation, 381
  major aspects, 378–379
  operational problems, 379–380
  security management, discrete automation
    asymmetric cryptography, 383
    consequences, 384
    digital signatures, 383
    evaluation mapping, 384
    evaluation set, 384
    reflection, 384
    tolerance function, 384–385
    uncertainties, 384–385
Autonomous guided vehicles (AGVs), 421
Autoregulated power generation network
  counterfactual proof, 707
  secure operations language-java
    distribution line agent, 698
    formal characterization, 705
    generation-engine agent, 698
    network axioms, 706
    system operation, 698–699
    voltage, current, and resistance, 698
  source code change, 705–706
  temporal logic proof, 706
Avrora simulator, 714

B

Backpropagation algorithm, 393
Batch estimation
  convergence criteria test, 444–445
  ill-conditioned matrix, 445
  processing flow, 443–444
  weighted least squares solution
    linearized iterative solution, 442–443
    non-linear differential equation, 440
    optical tracker equation, 440–441
    state transition matrix, 442–443
    target position and velocity, 440
    Taylor series, 441
Baum–Welch algorithm, 569, 571
Bayesian scheme, 537
  decentralized data fusion
    Bayes theorem, 530–531
    classical estimation techniques, 531–532
    likelihood function, 530
    probability distribution functions, 530
  distributed multi-sensor systems
    entropy, 536–537
    independent opinion pool, 535
    likelihood opinion pool method, 535–536
    linear opinion pool, 534–535
    mutual information, 537–538
Bayesian source number estimation, 294–295
Bayes theorem, 298–299
Beamforming
  acoustic wideband beamformer, iPAQS
    implementation, 367–369
  array system performance analysis (*see* Array
    system performance analysis)
  DOA estimation and source localization
    (*see* Direction-of-arrival estimation)
  historical background, 335–336
  narrowband *vs.* wideband, 336–337
  narrowband waveforms
    angular transfer function, 339
    broadband white Gaussian interferer, 341
    DOA angle, 340–341
    Gaussian noise, 341
    MVDR, 340–342
    output noise variance, 338
    output SNR, 338–339
    polar plot, 340–341
    receiver SNR, 338
    single tone waveform, 337
    spatial Nyquist criterion, 340
    time delay expression, 339
  radar wideband beamformer, subband approach,
    366–367
  wideband waveforms
    complex-valued weights per sensor channel,
      342–343
    Mth-order tapped delay line, 342

received waveforms, 342
source waveform, 341
Biological sensors, 4
Blind source separation (BSS), 292–293
Block–circulant–circulant–block (BCCB), 548
Block–Toeplitz–Toeplitz–block (BTTB), 548
Boltzmann distribution, 416
BSS, *see* Blind source separation (BSS)

C

Camera calibration, 83–84
CAMSHIFT algorithm, *see* Continuously adaptive
  mean shift (CAMSHIFT) algorithm
Causal state splitting and reconstruction (CSSR)
  algorithm
  conditional probability distributions, 572
  description, 573–576
  MGRS, 576
  side channel analysis, 576–577
Cellular automata model
  complexity classes, 152
  crossing tracks, 155–157
  definition, 151
  FACS concepts, 153
  intersecting tracks, 159–161
  linear tracks, 154–155
  network pathology effects
    Bayesian net track formation, 166, 167, 169
    EKF track, 165, 166, 168
    false-positives, 166, 167
    pheromone approach formation, 165
    pheromone track, 163–166, 168
  network traffic
    data packet propagation, 161, 164
    plot packet density, 161, 163
    random detections, 161, 162
    traffic jam formation, 161, 162, 164, 165
  nonlinear crossing tracks, 158–159
  packet speed, 153
  particle-hopping models, 152, 153
  traffic system, 152
Charge coupled device (CCD), 541
Chebyshev's inequality based multisensor data fusion
  Bayes rule, 554
  decision-rule-based fusions, 566
  definition, 553–554
  local threshold adjustment method
    critical values and phase changes, 560–561
    detection matrix construction, 559
    false-alarm and hit rate recalculation, 560
  problem formulation
    covariance, binary sensor decisions, 557–559
    signal attenuation model, 555–557
  sensor node, 553–554
  simulation, 564–566
  threshold-OR fusion method, 554, 561–564

Civilian applications
  agricultural crop monitoring, 16
  aircraft control-surface embedded sensing, 16
  climate monitoring, 16
  home security illegal-entry sensing, 16
  industrial machinery wear sensing, 16
  long-term in situ medical monitoring of people and
      animals, 16
  personnel heartbeat detection, 17–18
  rare and endangered species monitoring, 16–17
  traffic control sensing, 16
Clairvoyant system, 720–721
Closest point of approach (CPA) method, 6
Cluster formation process, 112, 113
Collaborative jammers, 685–686
Collaborative signal and information processing
      (CSIP)
  automated software synthesis, 204
  belief state, 194
  combinatorial tracking approach
      contour tracking, 201–202
      definition, 199
      shadow edge tracking, 202–203
      target counting, 199–201
  distributed embedded sensing systems, 204
  fixed node, 194, 195
  IDSQ
      individual target tracking, 195–196
      information-based approaches, 196–199
  information-driven approach, 203
  leader node, 194, 195
  optimization, 192, 193
  target detection, 203
  tracking scenario, 193–194
Collaborative tracking network (ColTraNe)
  acoustic sensor, 168
  CPA events, 169
  data routing, 168
  maintenance techniques, 169
  multiple target-tracking, 175
      Bowtie simulation, 179–185
      clump range, 179, 186
      X path simulation, 179, 180
  passive infrared sensor, 168
  seismic sensor, 168
  Twentynine Palms
      data transmission requirements, 175, 176
      disadvantages, 170
      geometric centroid approach, 170, 175
      track length, 170–174
  wireless communication, 168
Commercial-scale fossil power plant
  drift error, 505–507
      calibrated sensor data, 503–504
      sensor data calibration correction, 504–505
      thermal-hydraulic turbulence, 504
      uncalibrated sensor data, 503–504

  filter matrices, 502–503
  filter parameters and functions, 501–502
  temperature sensors, 500–501
  thermal-hydraulic noise and process transient, 501
  throttle steam temperature, 501, 503
  zero-mean fluctuating error, 505–507
Constrained least-squares (CLS) method, 349
Continuous analog signal sampling, 50–52
Continuously adaptive mean shift (CAMSHIFT)
      algorithm
  advantages, 481
  fast alignment algorithm, 483
  histogram transformation geometry, 484
  MSE, 484
  object and trajectory points, feedback loop, 485
  omni and planar camera, 481
  planar-image projection, 482
  probability distribution, 479–481
  projection ratio, 482
  second moment distribution, 480
  sector area, 483–484
  single measurement vector, 479
COOJA simulator, 715–716
COUNT annotation, 722
Counterfactual necessity theorem (CNT), 704, 708–709
Cramér-Rao bound (CRB)
  lower bound, 356
  signal-to-noise ratio
      array matrix, 360
      bandwidth and center frequency, 363
      Fisher information matrix, 360–361
      Gaussian signal, 363
      penalty matrix, 360–361
      polar array matrix, 361–632
      polar penalty matrix, 362
      sensitivity analysis, 359
      signal characteristics and array geometry, 358
      traveling source scenario, 362
  time-delay error
      array matrix, 358
      DOA, 358
      Fisher information matrix, 357
      inverse Fisher information matrix, 357–358
      root-mean-square error, 358–359
      traveling source scenario, 358–359
CSIP, *see* Collaborative signal and information processing
CSSR algorithm, *see* Causal state splitting and
                  reconstruction (CSSR) algorithm
Cycle-conserving EDF (ccEDF), 615

D

Data compression, 85
Data fusion, 94
Data registration
  closely mounted sensors and temporal changes,
      487–488

coordinate transformations
    affine transformations, 456
    quaternions, 456–457
    transformation matrix, 458–459
    viewer-centered coordinate system, 458
2D sensor readings, 455
ensemble registration, 488–489
feature level fusion, 485
feature selection
    Canny Edge detector, 474
    computational complexity, 469
    correlation-type registration systems, 473
    Daubechies-4 wavelet transform, 470–471
    feature detection, 471
    feature extraction, 472
    hierarchical registration correction, 472–473
    SIFT, 473–474
    user interface, 470
    WaveReg system, 469–470
image registration methods (*see* Image registration)
mapping function, 455–456
meta-heuristic approaches
    fitness function, 468–469
    Gaussian noise model, 464
    gene pools, elite reproduction scheme, 467
    genetic algorithm, 465–466
    geometric relationship, sensor readings, 464
    noise variance, 465–466
    simulated annealing, 467–468
    tabu search algorithm, 465–466
    terrain model, 465
    TRUST, 468–469
multi-resolution image mosaic algorithm, 485–486
multi-sensor fusion, 485
non-rigid body registration, 489–490
objective functions
    Chi-square distribution, 463
    Gaussian distribution, 463
    gray scale values, 462
    for uniform noise, 464
person detection system, 487
registration, definition, 455
video streams
    CAMSHIFT algorithm (*see* continuously adaptive mean shift (CAMSHIFT) algorithm)
    catadioptric sensor geometry, 475
    four point method, 478–479
    mirror cross-section, 475
    mirror surface slope, 476
    model *vs.* flow velocities, 478
    omni-directional image, 477
    optical flow equation, 477
    projection geometry, 476
    projective transformation, 478
    quadratic first-order differential equation, 476
    virtual perspective image, 477

Daubechies-4 wavelet transform, 470–471
Debugging systems
    interactive system, 720–721
    postmortem debugging, 721–722
    source-level debuggers, 720
    static analysis techniques, 722–723
Decimation, 52–54
Decision fusion, 95
Defense Advanced Research Projects Agency (DARPA), 7
Deterministic maximum-likelihood (DML), 346
Digital filters
    baseline wander removal, 49–50
    frequency response, 47, 48
    structures, 47
Digital-to-analog (D/A) converter, 41
Direction-of-arrival (DOA) estimation
    estimation and source localization
        acoustic/seismic signals (*see* Acoustic/seismic signals)
        array signal processing, 343
        far-field geometry, 343
        iterative algorithms, 344
        RF signals, 344–346
    time-delay error, 358
Discrete cosine transform (DCT), 547
Discrete event network simulator (DES), 715
Discrete Fourier transform (DFT)
    algorithm, 46
    circular shift operation, 46
    DTFT, 45
    FFT, 47
    finite-length sequence, 45, 46
    frequency response, 45
    properties, 46
    z-transform, 44
Discrete-time Fourier transform (DTFT), 45
Discrete-time system theory
    LTI system, 43
    properties, 43
    sequences, 42
    SISO system, 43
DiSenS simulator, 714–715
Distributed computing fault tolerance algorithms, 139
Distributed decision making, 109
Distributed embedded sensing systems, 204
Distributed multi-sensor systems
    entropy, 536–537
    independent opinion pool, 534
    likelihood opinion pool method, 535–536
    linear opinion pool, 534–535
    mutual information, 537–538
Distributed multi-target detection
    BSS, 292–293
    source number estimation (*see* Source number estimation)

Distributed sensor actuator network (DSAN) system
  advantages, 22
  taxonomy
    algorithm execution, 26, 27
    communication, 28–30
    data management, 27
    deterministic system, 23
    distributed services, 28
    exception management, 26–27
    function and implementation, 23
    input/output aspects, 24–26
    node operating system, 28
    nondeterministic system, 23
    processing architecture, 27–28
    quasi-deterministic system, 23
    security, 30
    system attributes, 30
    system integration, 30
    system interfaces, 27
Distributed source number estimation
  evaluation metrics, 299–300
  experimental results
    computation time, 302–303
    detection probability, 302–303
    experimental vehicles, 300
    histogram, 300, 302
    kurtosis, 302–303
    log-likelihood function, 300, 302
    Sensoria WINS NG-2.0 sensor nodes,
      300–301
    sensor laydown, 300–301
  hierarchy, 296–297
  network traffic, 303
  posterior probability fusion, Bayes theorem,
    298–299
Down-sampling, 52–54
DSAN system, *see* Distributed sensor actuator network
  system
DustMiner system, 721
Dynamic frequency scaling (DFS), 599
Dynamic power management (DPM)
  CPU-centric, 610
  I/O-centric, 610
Dynamic space–time clustering method, 113–114
Dynamic voltage scaling (DVS), 599, 610

E

Electro-optic (EO) sensing system, 541
e-Machines
  algorithm, 515, 516
  determination, 515–516
  initial automaton, 515
  PFSA, 515
Embedded systems, 33
EmCee emulator, 712
EmSim simulator, 712

EmTOS, 712
EmView visualizer, 712
Energy-aware sensor systems
  architectural level, 598
  circuit/gate level, 598
  energy consumption, 595–596
  energy reduction techniques
    computational offloading, 605–606
    DFS, 599
    DVS, 599
    energy aware routing, 606
    gate delay, 599
    idle components shutting down (*see* Idle
      components shutting down)
    processor power design space, 598, 599
    scheduling algorithm, 599
    subthreshold current control, 600
    supply rails gating, 600, 601
    V_T effects, 600
  power consumption sources, 596–597
  process level, 597
  software level, 598
Energy-optimal device scheduler (EDS) algorithm,
  618, 623
Environmental effects
  atmospheric dynamics
    acoustic environmental effects, 209–210
    diurnal cycle, 208
    EM waves, 211
    katabatic winds, 209
    optical waves, 211–212
    seismic environmental effects, 210–211
    thermal plumes, 208–209
  bulk Richardson index, 212
  calibration algorithm, 207–208
  chem/bio plume detection, 212
  chemical and biological sensor, 212
  chemical and biological weapon, 212, 213
  error bars, 207
  Kolmogorov spectrum, 213, 214
  large-scale weather model, 213
  sound waves propagation, 214–216
  temperature and humidity, 207
  temperature sensor, 208
Error regression, 400–401, 403
Estimation techniques
  batch estimation, 439 (*see also* Batch estimation)
  estimation processing flow, 431–432
  least squares method, 430
  linear minimum mean square error method, 431
  maximum likelihood method, 431
  multi-sensor parametric data, 429
  optimization approach
    conjugate gradient method, 438
    derivative methods, 438
    direct methods, 437–438
    indirect methods, 437

multiple dimensional Newton–Raphson method, 438
one-dimensional state vector, 438–439
sequential processing, 439
  feedback-control system approach, 446
  filter divergence and process noise, 449–450
  linearized sequential technique, 445
  maneuvering targets, 450–451
  non-linear formulation, 450
  processing flow, 448–449
  sequential weighted least squares solution, 446–447
  software tools, 451
state vector estimation, 432
state vector value, 429
system models
  coordinate systems, 434
  non-positional estimation, 433
  positional estimation, 432–433
  truncated Taylor series expansion, 433
target-tracking problem, 430
t criteria
  Bayesian weighted least squares criterion, 435
  best fit, 434
  conditional probability, 436
  maximum likelihood criteria, 436
  mean square error formulation, 436
  predicted observation, 434
  symmetric matrix, 435
  weighted least squares expression, 434–435
Evaluation mapping, 383–384
Evaluation set, 383–384
Exhaustive enumeration (EE) method, 626
Expectation maximization (EM), 571
Extreme scale motes (XSMs), 717

F

Fast Fourier transform (FFT), 47
Fault detection and isolation (FDI) algorithm, 496
Fault location identifiers (FLIDs), 723
Feature extraction
  definition, 77
  edges, 78
  feature stacking, 80, 81
  Hough transform, 78–79
  multisensorial oceanographic imagery, 80
  ROI, 80
  segmentation, 79–80
  statistical and structural features, 77–78
  target tracking, 80
Feature/information processing, 60
Feature selection, 102
Feedforward sigmoidal networks, 392–393
Finite impulse response (FIR) filter, 342
Fitness function, 462–464

Formal language measurement, 513
  exclusive-OR operation, 516–517
  weighted counting measure, 517
Free agents in a cellular space (FACS) concepts, 153
Fuzzy k-means clustering algorithm, 88
Fuzzy logic and fuzzy sets, 422–423

G

Generalized cross-correlation (GCC) approach, 353
Genetic algorithms, 415–416
Geometric transformations, 83
Group delay, 45

H

Hidden Markov models (HMMs), 134
  Baum–Welch algorithm, 569, 571
  conditional probability, 570
  continuous action recognition, 574–575
  CSSR algorithm
    algorithm description, 573–576
    conditional probability distributions, 572
    MGRS, 576
    side channel analysis, 576–577
  random process, 570
  state structure and transition matrix, 570
High-definition television (HDTV), 547
HMMs, *see* Hidden Markov models (HMMs)
Human-centric information fusion, 581–582
  collaborative decision making, crowd sourcing
    gamification concept, 588
    information fusion system, 587
    Mechanical Turk service, 588
    multiagent team cognition, 589
    OLIVE tool, 588
    recognition-primed decision, 589
  computer vision systems, 582
  human observers, 583–584
  human observing/reporting types, 585
  hybrid human/computer analysis, 586–587
  measures of performance, 583
  social cognition team, 582–583
  social networking tools, 582
Human-computer-interaction (HCI) devices, 378, 380
Human tracking, 60

I

Idle components shutting down
  adaptive communication hardware, 604–605
  cache memory, leakage control techniques, 601–602
  mechanical components, 600
  multiple low power modes, 602–604
  supply gating, 601
IDSQ algorithm, *see* Information-driven sensor query algorithm

Igloo White sensing system, 5
Image creation
    analog-to-digital images, 65–67
    dimensionality, 64
    sensor components, 64–66
    spectrum, 62–64
Image domains, 67–69
Image registration, 81–82
    clustering technique, 461
    feature matching, 459
    features and detection methods, 459, 460
    multiresolution methods, 461
    mutual information method, 461
    optimization problem, 462
    principle components analysis, 461
    reference image, 459
    relaxation-based methods, 461
    resampling and transformation, 459
    transform model estimation, 459
Image resampling, 86, 459
Images and video and transmission, 85
Independent opinion pool method, 534
Infinite impulse response (IIR), 43
Information-driven sensor query (IDSQ)
                algorithm, 7
    individual target tracking, 195–196
    information-based approaches, 196–199
Information theory context
    Bayesian framework (*see* Bayesian scheme)
    data fusion
        sensor administration research, 526
        sensor fusion research, 525–526
    distributed networks, 523
        data fusion, 524
        information processing in, 524, 525
        resource administration, 524
        vs. single sensor platform, 524
    probabilistic framework, 526
        bayesian scheme, decentralized data fusion,
                530–533
        distributed detection theory, 532–533
        sensor data model, single sensor, 527–529
Infrared and ultraviolet sensors, 4
Interpolation, 52, 54
Isolation fusers, 399–400

J

Jamming games, 680
    convergence time, 690
    edge expansion technique, 690–691
    multilayer (*see* Multilayer jamming game models)
    network layer issues, 681–682
    paradox of, 682–683, 689–690
    physical layer issues, 681
    potential function analysis, 691
    price of, 682, 689–690

K

Kalman filter, 439; *see also* Sequential processing
Kansei testbeds, 717–718
Karhunen Loeve transform (KLT), 547
k-nearest neighbor (k-NN) algorithm, 315, 316,
                319, 320
Kripke versioning model, 702–703
Kurtosis analysis, 327–328

L

Laplace approximation method, 294–295
Large-scale data management techniques, 655
Laser-range finder, 60, 62
Likelihood opinion pool method, 535–536
Linear and shift (time) invariant (LTI) system, 43
Linear least square estimation, 497
Linear minimum mean squares error (LMMSE), 547
Linear opinion pool method, 534–535
Linear programming, 423–425
Link-wise jamming models, 683–685
Lipschitz property, 393
Local threshold adjustment method
    detection matrix, 559
    false-alarm and hit rate, 560
    phase changes, 560–561
Low-energy device scheduler (LEDES) algorithm,
                629–630

M

Macrodebugger (MDB) system, 722
Magnetic sensors, 4
Marine Corps Air–Ground Combat Center
                (MCAGGC), 11
Maximum a posteriori (MAP), 531, 546, 566
Maximum entropy partitioning (MEP), 312
Maximum Likelihood (ML) estimation, 531
Measurement-based statistical fusion methods
    Bayesian risk function, 388–389
    correlation coefficient method, 389
    distributed decision fusion methods, 387
    distributed detection, 397
    door detection, ultrasonic and infrared sensors,
                397–398
    empirical risk minimization methods
        empirical estimate, 392
        feedforward sigmoidal networks, 392–393
        vector space methods, 394
    expected error, 389–390
    expected square error, 399
    fuser outputs, 391
    fusion rule estimation, 388
    Haar kernel, 395
    isolation fusers, 399–400
    joint conditional distribution, 391–392

localization-based fusers
 background and source measurement, 404
 δ-robust localization method, 408
 false-alarm probability, 404
 Gaussian source detection problem, 404
 Lipschitz constant, 408–409
 Lipschitz smoothness conditions, 404
 localization algorithm, 406–407
 low-level radiation sources, 404
 missed detection probability, 404
 source parameter estimation, 406
 SPRT detection, 405–406
 state packing number, 407–408
measurement error, 389–390
metafusers, 409–410
Nadaraya–Watson estimator, 394–395
noisy function estimators, 396–397
pattern recognition, 388
probability densities, 391
projective fusers, 400–401, 403
smoothness modulus, 395–396
systematic error, 389–390
Meta-protocol, 34
Micro-air vehicles (MAV), 10
Micro electro mechanical systems (MEMS), 543
Microsensor networks
 array-intensive codes, 656
 benchmark codes, 663–664
 communication energy, 656, 664–665
 communication errors, 674
 communication optimization
  data decomposition and parallelization, 658–660
  inter-nest optimization, 663, 671–672
  message aggregation, 662–663
  message coalescing, 661–662
  message vectorization, 661
  naive communication, 660–661
 compiler algorithm, 675–676
 computation energy, 656, 664
 energy breakdown, 665–667
 energy consumption optimization, 656
 high-level architecture, 657–658
 overlapping communication, 673–674
 sensitivity analysis
  adi and full-search benchmarks, 668, 669
  adi and jacobi benchmarks, 370, 667, 668
  block decomposition, 670, 671
  energy overhead, 669
  row/column decomposition, 670, 671
Mie scattering theory, 220
Military applications
 artillery and gunfire localization, 7–8
 computation-and communications-intensive
  multiple-target tracking situations, 7–8
 CPA method, 6
 decision fusion, 11
 extended Kalman filter, 6

IDSQ algorithm, 7
Igloo White sensing system, 5
imaging-based classification and identification, 13–15
MAV, 10
network-centric warfare, 5
SID, 5
UAV, 8–9
vehicle classification, 11–13
Vietnam-era Igloo White program, 5, 6
weapons platforms, 4
Military Grid Reference System (MGRS), 576
Minimum Mean Square Error (MMSE), 531–532
Minimum variance desired response (MVDR) method, 340–342
Mobile Emulab testbeds, 718
Mobile robots, behavior recognition
 experimental procedure
  Brainstem™ microcontroller, 317
  distributed pressure sensors layout, 316–317
  experimental setup, 316–317
  Haar wavelet sensor readings and plot, 318
  motion types, parameters, 317–318
  piezoelectric pressure sensor, 318
  Pioneer 2AT and Segway RMP, 317–318
 experimental results
  computational complexity, 323
  feature extraction methods, 323
  polynomial kernel, 320
  robot type and robot motion classification, 320–322
 pattern analysis, 319–320
Model order estimation, 292; *see also* Source number estimation
Module transfer protocol (MTP), 697
Monte Carlo method, 554
MoteLab testbed, 716–717
MTP, *see* Module transfer protocol (MTP)
Multilayer jamming game models, 682
 network layer modeling (*see* Network layer modeling, jamming games)
 physical layer modeling
  link-wise jamming models, 683–685
  neighborhood jamming models, 685
Multi-object classification, 109
Multiple dimensional Newton–Raphson method, 438
Multiple edge jamming, 685
Multi-resolution image mosaic algorithm, 485–486
Multi-resolution outdoor dual camera system, 60, 63
Multispectral imagery (MSI), 549
Multi spectral scanners (MSS), 542
Multispectral sensing
 array technology, 543–544
 color images, 548–549
 data acquisition
  radiometric resolution, 543
  spatial resolution, 542–543
  spectral resolution, 543

disadvantages, 542
mathematical model, 544–548
   image reconstruction formulation, 546–547
   superresolution, 547–548
super array sensing, 543
Multi-state constrained low energy scheduler
   (MUSCLES) algorithm, 630, 633–634
MVDR, *see* Minimum variance desired response
   (MVDR) method

N

Nadaraya–Watson estimator
   noisy function estimators, 396–397
   smoothness modulus, 394–395
Nagel–Schreckenberg model, 152
Nash equilibrium, 680
Natural language processing (NLP), 513
Neighborhood jamming models, 685
Network embedded sensor testbed (NESTbed), 719
Network layer modeling, jamming games
   collaborative jammers, 685–686
   malicious nodes, 686
   routing games
      background, 686–687
      bottleneck routing games, 687–688
      delay routing games, 688
      physical/network layer phase, 688–689
      power routing games, 688
      repeated games, 688–689
   stand-alone jammers, 685
Network simulators
   ATEMU, 712–713
   Avrora, 714
   COOJA, 715–716
   DiSenS, 714–715
   EmStar, 712
   high-level network simulator, 712
   SenQ, 715
   TOSSIM, 713
Network testbeds
   Kansei, 717–718
   Mobile Emulab, 718
   MoteLab, 716–717
   NESTbed, 719
Neyman–Pearson detection criterion, 283
Night Vision and Electronic Sensor Directorate
   (NVESD), 542
Node-level energy reduction methods
Node-level I/O-device
   experimental results
      energy consumptions, 635–636
      energy savings, 634–635
      LEDES and MUSCLES, 634–635
      LEDES *vs.* EDS, 636
   low-energy device scheduling, 630–634
   online device scheduling, 628–630

optimal device scheduling
   breakeven time, 618–619
   device parameters, 627–628
   EDS algorithm, 623–624
   EE method, 626
   energy consumption, 618, 627–628
   experimental task set, 624–625
   I/O device schedule, 618
   memory consumption *vs.* execution time,
      626–627
   memory requirement, 623–624
   memory savings percentage, 626
   minimum energy, 625–626
   power consumption, 618
   pruning technique (*see* Pruning technique)
   task schedule, 624–625
Node-level processor
   experimental results
      ccEDF, 615
      CPU power savings, 614
      current consumption, 614–615
      expected results *vs.*3-state LEDF,
         616–617
      power consumption, 614–615
      pseudo-random task generator, 615
      scheduling overhead, 616–617
      15-task task set, 616
   implementation testbed
      hardware platform, 612–613
      software architecture, 613–614
   LEDF algorithm, 611–612
NodeMD system, 723
Noise-free communication link, 108
Noise removal, 76–77

O

Object classification
   data-averaging classifier, 106
   decision-fusion classifier, 106
   feature selection, 102
   hard decision fusion, 105–106
   power spectral density estimation, 106, 107
   soft decision fusion, 103–105
   sub-optimal decision-fusion classifier, 108
Object detection
   hard decision fusion, 100–102
   soft decision fusion, 98–100
Object tracking, 93–94
Operating system power management
   DPM (*see* Dynamic power management (DPM))
   energy consumption, 609
   energy management
      node-level I/O-device (*see* Node-level
         I/O-device)
      node-level processor (*see* Node-level processor)
   real-time performance, 609

Optimal device scheduling
    breakeven time, 618–619
    device parameters, 627–628
    EDS algorithm, 623–624
    EE method, 626
    energy consumption, 618, 627–628
    experimental task set, 624–625
    I/O device schedule, 618
    memory consumption *vs.* execution time, 626–627
    memory requirement, 623–624
    memory savings percentage, 626
    minimum energy, 625–626
    power consumption, 618
    pruning technique (*see* Pruning technique)
    task schedule, 624–625
Optimal linear combination fuser, 409
Optimal minimum-variance filter, 497
Outdoor surveillance, 60, 62

P

Parameter estimation
    moving target resolution, 116
    peak, sensor types
        AAV data, close-up, 126, 127
        CPA table, 124
        GPS table, 124
        infrared, seismic, and acoustic data, 126
        node table, 123, 125
        sensor peak and ground truth data, 126, 127
        sensor peak clusters *vs.* time window size, 126, 128
    self-organization of network, 112–113
    SIF technique
        classification algorithm, 119
        data-classification algorithm, 117
        Fourier analysis, 116
        offline stages, 116, 118
        online stages, 116–118
        PCA, 116
        target identification, 119
        time series window, 117–119
    space–time neighborhood concept, 112
    stationary target localization (*see* Stationary target localization)
    velocity and position
        computed angle *vs.* true angle, 114, 115
        computed speed *vs.* true speed, 114, 115
        dynamic space–time clustering method, 113–114
        quality of estimation, 114, 115
Particle-hopping models, 152
Pattern matching subimage filter, 87
Person detection system, 60, 61
Pioneer 2AT, 317–318
Point-based operations
    bitplane slicing, 72
    contrast stretching, 70, 71

    conversion, 70
    histogram equalization, 71
    image averaging, 72
    image subtraction, 72
    inversion, 70, 72
    level slicing, 72
    thresholding, 70
Postprocessing, 60
Power-aware computing, 33
PowerTOSSIM simulator, 713
Preprocessing, 60
Principal component analysis (PCA), 116
Probabilistic finite-state automata (PFSA), 514
    construction, 314–315
    symbol-state image conversion, 312–314
Probability density function (PDF), 554
Probability distribution functions, 530
Projective fusers, 400–401, 403
Pruning technique
    complete schedule tree, 623
    energy, 619–620
    energy consumption, 620–621
    partial schedule
        one schedule job, 620–621
        three scheduled jobs, 622–623
        two schedule jobs, 621–622
    task set T1, 620
    temporal, 619
    vertex, 620
Pseudocoloring technique, 86

Q

Quadratic kernel, 328

R

RC5 cryptography, 646–648, 652
Real-time Linux (RT-Linux), 613
Real-time monitoring and control
    adaptive filter features, 508
    applications, 495
    FDI algorithm, 496
    multiple hypotheses testing
        Markov states, 509
        posteriori probability, 508–510
        Radon–Nikodym derivative, 511
    redundant measurements, 496
    sensor calibration (*see* Commercial-scale fossil power plant)
    sensor redundancy, 495
    signal calibration and measurement estimation
        bias and scale factor errors, 497
        calibrated measurement, 498
        calibration filter modifications, 500
        covariance matrix, 498
        density function, 499

intra-sampling failures and false alarms, 500
parity vector, 497
positive-definite diagonal matrix, 499
posteriori probability, 498–499
recursive algorithm, 497
unbiased weighted least squares estimation, 498
uncalibrated measurements, 496
Receiver operative characteristic (ROC), 564
heterogeneous system, 566
homogeneous system, 565
Recursive filter process, 448
Region of interest (ROI), 80, 553
Regularized Structured Total Least Squares(RSTLS), 549
Remote procedure call (RPC) system, 720
Reversible-jump Markov chain Monte Carlo (RJ-MCMC) method, 295

S

Safe TinyOS, 722–723
Sampling rate conversion
down-sampling, 52–54
up-sampling, 52, 54
Scale invariant feature transform (SIFT), 473–474
SCR, *see* Software cost reduction (SCR)
SDF, *see* Symbolic dynamic filtering
SDP, *see* Semi-definite programming (SDP)
Second order cone problem (SOCP), 364, 366
Secure operations language-java (SOLj)
autoregulated power generation network
distribution line agent, 698
generation-engine agent, 698
system operation, 698–699
voltage, current, and resistance, 698
definition, 697
events, 696–697
secure infrastructure for networked systems, 697
Segway RMP, 317–318
Seismic intrusion detectors (SID), 5
Semantic information extraction
anomalies, 520
behavior recognition, 517
experimental verification, 517
behavior recognition software demonstration, 519–520
circling behavior, 519
pressure sensitive floor, 518
pressure sensor data, 518
formal language measurement, 513
exclusive-OR operation, 516–517
weighted counting measure, 517
natural language, 513
semi-empirical model, 513
symbolic dynamics
conversion, formal languages, 514
e-machines determination, 515–516

Semantic information fusion (SIF) technique
classification algorithm, 119
data-classification algorithm, 117
Fourier analysis, 116
offline stages, 116, 118
online stages, 116–118
PCA, 116
target identification, 119
time series window, 117–119
Semi-definite programming (SDP), 363–364
SenQ simulator, 715
Sensor data communication, 639
caching scheme, 642–645
characteristics, 641
cryptography algorithms, 641
dynamic voltage scaling, 651
encryption process, 646–647
energy consumption, 639
index communication, 641
message cycle time, 646–647
packet size, 648
RC5 cryptography, 646–648, 652
spatial value locality, 649
StrongArm processor, 640
synthetic data configurations, 645–646
VDP, 649–651
wireless channel, 646
Sensor data model, single sensor
measurement vector, 527–528
observation model, 528–529
state vector, 527
Sensor fusion system, 87
Sensoria WINS NG-2.0 sensor nodes, 300–301
Sensor Information Technology (SensIT) program, 7
Sensor measurements
coherence time, 95, 96
spatial coherence regions, 95
temporal point sources, 97–98
zero-mean Gaussian stationary process, 96
Sensor network system (SNS), 693
counterfactual theory
autoregulated power generation network (*see* Autoregulated power generation network)
counterfactual necessity theorem, 704
interpretation of, 703–704
Kripke versioning model, 702–703
language of, 699–700
necessity based on, 708
program transformers, 701–702
SOLj agent network, 700–701
temporal logic fragment, 704
secure operations language-java (*see* Secure operations language-java)
Sequential processing method, 439
feedback-control system approach, 446
filter divergence and process noise, 449–450
linearized sequential technique, 445

maneuvering targets, 450–451
non-linear formulation, 450
processing flow, 448–449
sequential weighted least squares solution, 446–447
software tools, 451
Service abstraction, 381
SFNNP, *see* Symbolic false nearest neighbor partitioning (SFNNP)
Shalizi's method, PFSA, 514
Shoe-box-sized Sensoria WINS NG sensors, 192
SIFT, *see* Scale invariant feature transform (SIFT)
Signal attenuation model
heterogeneous system, 557
sensor deployment, 555
sensor measurements, 558
Simulated annealing method, 416–417
Single-edge jamming, 683–685
Single-input–single-output (SISO) system, 43
Single target detection, 291
Smart camera network, 84–85
SnapSim simulator, 714
SNS, *see* Sensor network system (SNS)
SOCP, *see* Second order cone problem (SOCP)
Soft computing techniques
artificial neural networks, 421–422
combinatorial optimization problems, 415
fuzzy logic and fuzzy set, 422–423
genetic algorithms, 415–416
linear programming, 423–425
simulated annealing method, 416–417
Tabu search, 420–421
TRUST
cost function, 418
dynamic system, 418
global optimization method, 417
subenergy tunneling term, 418–420
terminal repeller term, 419
Software cost reduction (SCR), 696
Source number estimation
centralized source number estimation
Bayesian, 294–295
ICA-based algorithms, 293
posterior probability, 294–295
sample-based approach, 295
variational learning, 295–296
distributed source number estimation (*see* Distributed source number estimation)
Spatial filtering, *See* Area-based algorithms
Speed of propagation, 349–352
Stand-alone jammers, 685
Stationary target localization
signal strengths, 120
accurate and inaccurate estimates, 121, 122
acoustic energy field, 121, 123
acoustic sensor values, engine, 122, 125
theoretical energy field, 121, 124
time delays, 120–121

Statistical maximum-likelihood (SML), 346
Support vector machines (SVM), 319, 320, 328, 331
Swarm computing, 695
Symbolic dynamic filtering (SDF)
definition of symbolic dynamics, 309
distributed dynamical systems, 307
low-complexity algorithms, 308
mobile robots, behavior recognition (*see* Mobile robots, behavior recognition)
pattern classification, 315–316
PFSA, feature extraction (*see* Probabilistic finite-state automata)
real-time analysis, 307
target detection and classification
binary classification, 324–325
characteristic signatures, 323
feature extraction, 325–326
flow chart, 325–326
number of feature vectors, 325
PIR data, 329–331
seismic data, 326–329
signal preprocessing, 325
test scenarios, 324
tree-structure, 324
typical data collection scenario, 323
time series partitioning, 308
time series-wavelet domain transformation, 310–311
wavelet surface profiles, 311–312
Symbolic false nearest neighbor partitioning (SFNNP), 308, 310
Sympathy system, 721
System optimization factors, 34

T

Tabu search, 420–421
Target tracking
Bayesian belief net, 147–148, 151
cellular automata model (*see* Cellular automata model)
ColTraNe (*see* Collaborative tracking network)
computation environment, 134–135
dependability analysis, 176–177
extended kalman filter, 149–150
filter equations, 145
noise covariance matrix, 145, 147
position estimation, 144
target position bounding ellipse, 145, 146
target position uncertainty, 145, 146
velocity estimation algorithm, 144
fusion center switching strategy, 133
HMM, 134
intercluster tracking framework
candidate track information, 138
data association, 136
entity identification, 136
multiple sensor entity tracking, 136
network-embedded entity tracking, 137–139

object classification, 136
object detection, 136
track prediction, 136
JDL data fusion model, 132–133
local parameter estimation, 139–142
mass algorithm, 134
motion algorithm, 134
multisensor tracking, 133
optimal sensor selection, 133
pheromone routing, 142–143, 148–149
resource parsimony, 178–179
Time-delay estimation (TDE)
BAE systems, 277
CSD matrix, 274
differential time delays, 279
partial signal coherence, 273
PSD and coherence estimation, 277–278
threshold coherence *vs.* bandwidth, 275–276
time duration, 276
Ziv-Zakai bounds, 274
Time-delay-type methods
CLS method, 349
differential time delays, 349
pseudo-inverse of matrix, 350–351
third order constraint equation, 351–352
Time-stamp counter (TSC), 613–614
TMote Connect, 716–717
Tolerance function, 383–385
Tor network traffic, 576–577
TOSSIM simulators, 713
Total least squares (TLS), 547
Traditional pattern-matching techniques, 513
Transmission delays, 33

U

Uniform linear array, 339–341
Uniform partitioning (UP) methods, 312
United States Department of Defense (DoD), 4
Unmanned aerial vehicle (UAV), 8–9
Up-sampling, 52–54

V

Value-difference packet (VDP), 649–651
Vector space methods, 394
Very high-definition (VHD), 547
Vietnam-era Igloo White program, 10
Virtual Internet test bed (VINT), 134

W

Wavelet domain–based image fusion system, 89
Wavelet-transformed space partitioning (WTSP), 308, 310
Worst-case scenario, 611

X

XATDB debugger, 713

Z

Zero-knowledge HMM identification algorithm, *see* Causal state splitting and reconstruction algorithm
Zero-mean Gaussian stationary process, 96